Physiology of Nutrition and Environmental Stresses on Crop Productivity

Other Related Books

Advancements in Micronutrient Research	*Hemantaranjan, A.*
Advances in Plant Physiology Vol. 1-14	*Hemantaranjan, A.*
Advances in Stress Physiology of Plants	*Panda, S.K.*
Environmental Physiology	*Hemantaranjan, A.*
Microbial Physiology	*Reddy, S. Ram*
Physiology of Bamboo	*Shanmughavel, P.*
Physiology of Crop Plants	*Gardner, F.P.*
Physiology of Fungi	*Verma, R.N.*
Plant Physiology and Plant Molecular Biology in New Millennium	*Hemantaranjan, A.*
Plant Physiology: Research Methods	*Narwal, S.S.*
Research Methods in Plant Science : Allelopathy Vol. 5 : Plant Physiology	*Narwal, S.S.*
Physiology of Plants under Abiotic Stress and Climate Change	*Hemantaranjan, A.*

Physiology of Nutrition and Environmental Stresses on Crop Productivity

Editor

A. Hemantaranjan

Department of Plant Physiology
Institute of Agricultural Sciences
Banaras Hindu University,
Varanasi – 221 005 INDIA

Published by:

Scientific Publishers (India)
5 A, New Pali Road, P.O. Box 91
Jodhpur 342 001 (India)

E-mail: info@scientificpub.com
Website: www.scientificpub.com

Branch Office
Scientific Publishers (India)
4806/24, Ansari Road, Daryaganj
New Delhi - 110 002 (India)

Reprint from : Advances in Plant Physiology, Vol. 14 (2013)
ISSN: 0972-9917

ISBN: 978-81-7233-874-9

Typeset by Rajesh Ojha
Printed in India

Preface

Plants are one of the most captivating and important groups of organisms living on Earth. They serve as the conduit of energy into the biosphere, provide food, and shape our environment. If we want to make development in understanding how these essential organisms function and build the foundation for a more sustainable future, then we need to apply the most advanced technologies available to the study of plant life. The preparation of ***Physiology of Nutrition and Environmental Stresses on Crop Productivity*** has been done exclusively due to praiseworthy contributions from Scientists of Distinction in unambiguous fields. Undeniably, the mission of the *International Treatise Series* on Plant Physiology is to diagnose the shortcomings under varied environments and to scientifically adopt precise and meaningful research for increasing the productivity, profitability and sustainability of agriculture in stressful environments by developing new germplasm for both traditional and alternative crops and by improving plant responses to abiotic stresses and to changes in the global environment by determining and developing technology to ameliorate, metabolic limitations by biological processes most sensitive to abiotic stress factors. Indeed, our objective is to publish innovative science of value across the broad disciplinary range of the treatise. I reaffirm that this plan has been undertaken with a view to strengthen the identical efforts to recognize the outcome of meticulous research in some of the very sensible and stirring areas of ***Plant Physiology-Plant Molecular Physiology/Biology-Plant Biochemistry*** for holistic development of the science of agriculture and crop production under changing environment. I am fervent to keep on the rarity and the prologue of excellent new ideas ensuring that the treatise calls to the best science done across the full extent of modern plant biology, in general, and plant physiology, in particular. In this book with inventive applied research, attempts have been made to bring together much needed eighteen review articles by thirty four contributors mainly from premier institutions of India and abroad for this volume. This volume addresses the challenges of environmentally healthy production of bulk and quality food, fodder, fiber and energy which are of ongoing international concern as well as highlights the major environmental factors that affect plant growth-development and explores ecologically sound approaches to solving stress-related problems.

This book has been broadly subdivided into three relevant sections, which on the whole highlight the necessity to find out evidence from the fields of plant nutrio-physiology (physiology of plant mineral nutrients) and molecular interventions for crop improvement under changing climate. The *Section I* has been devoted for seven select reviews on *Nutrio-physiology and Crop Productivity*. The *Chapter 1* portrays the important aspect of micronutrient nutrition, the phytosiderophores, which are generated in response to iron and eventually to other metal deficiencies and the very interesting systems are the central subject of the review written by the noted scientist from Argentina. Different microorganisms as well as fungi and bacteria have solved this problem, secreting high-affinity iron-binding compounds called siderophores. The *Chapter II* discusses the recent developments in boron nutrition research particularly related to physiological responses and boron homeostasis by an experienced Indian scientist. In the past decade understanding molecular mechanisms of boron transport and identification of B transporters has suggested that plants sense and respond to the boron conditions and regulate transporters to maintain B homeostasis and alleviate B deficiency and toxicity problems.

In light of breakthroughs, the crop phenology and plant nutrition of *Gerbera* have been thoroughly reviewed in *Chapter 3* by the Nagaland University's expert scientist belongs to the School of Agricultural Sciences and Rural Development. *Gerbera* is the latest sensation to Indian Floriculture, commercially grown throughout the world in a wide range of climatic conditions. Growing media for Gerbera are claimed to be highly porous with abundance in organic matter content and well drained to provide aeration to the root system. The improvement in physiochemical properties also aides in promoting the root growth and hastening the nutrient absorption by crop plants. A sincere effort in this regard, has been made to put forth some leaf nutrient diagnostic values suiting to *Gerbera*. Correspondingly in the field of nutrio-physiology, biofertilizers are the promising alternative for the biological management of soil fertility and productivity. Among them mycorrhiza have a great potential in biomass production and soil enrichment on wastelands, which help in mobilizing nutrients for uptake by plants especially those less mobile in soil solutions like 'P'. They also help in uptake of water and protect roots from pathogens. The arbuscular mycorrhiza are by far the commonest in agricultural and horticultural plants. Scientists with expertise have adequately gathered meaningful pieces of information in *Chapter 4 on* mycorrhiza in view of its potential use for augmenting soil fertility and crop productivity. Subsequently, nutrient use efficiency for maximization of crop productivity has been thoroughly reviewed in *Chapter 5.* Further, in *Chapter 6,* the NBPGR scientists have highlighted the impact of mineral toxicity on crop productivity based on convincing evidences, which is in fact the most important and alarming aspect of mineral nutrition especially under changing environment. In addition, significance of nickel as an essential mineral element in plant metabolism has been reviewed in *Chapter 7* that accomplishes certain aforementioned vital issues of the *Section I* of this volume.

In the *Section II* there are over ten substantial review articles related to *Plant Responses to Changing Environment and Environmental Stresses.* The objectives of the first review of this section as *Chapter 8* is to understand the monitoring and role of important CO_2 fluxes - photosynthesis and respiration in relation to the carbon pools present in the vegetation and soil and how they affect global carbon cycle and finally, the complexities and uncertainties in monitoring, understanding the ecosystem responses due to multiple processes to climate change and their implications in terms of implementation of policy issues have been analyzed by the elite team of IARI, New Delhi. Next to this, in *Chapter 9,* devoted authors have given an interesting account of cross-talk signaling and survival strategies in plants under changing environment, which enriches knowledge of researchers and contributes significantly towards the theme of this section. In *Chapter 10,* the NBPGR scientist has satisfactorily described plants as indicators of air pollution. The acute response and chronic effects of pollutants have been explained in terms of agriculture where chronic effects can manifest in the form of foliar chlorosis, retarded growth, retarded flowering, foliar/ flower abscission, reduced biomass and economic yield. Besides these, reputed scientists from the stream of Biotechnology have wonderfully illustrated ROS modulated responses during pathogen infestation in plants in *Chapter 11* that highlights the generation and scavenging of ROS during pathogen infestations and ROS modulated responses during plants defence, as transgenic approaches indicate the involvement of several genes associated with ROS generating and scavenging systems, during plant-pathogen interactions. In addition, in *Chapter 12,* focus on calcium and how it regulates abiotic signaling pathways have been scientifically reviewed by veteran scientists of Indian Institute of Science, Bangalore with appropriate caption "Calcium Signaling During Abiotic Stress in Plants: Roles of Calcium Dependent Protein Kinases". When we look into the *Chapter 13,* a brilliant picture emerges out on the vital problem of water deficit stress and its management in groundnut compiled by specialists from the Directorate of Groundnut Research, Junagadh that provides essential component of whole crop simulation model and can be used to specify the most appropriate rate and time of specific developmental process to maximize yield. In this chapter an attempt has been made to combine all the knowledge of water deficit stress and its impact on groundnut agronomy and physiology and management practices to grow high yielding groundnut varieties with targeted yield and recommend the same to groundnut workers to increase the productivity. Another imperative article on mechanism of salinity stress tolerance in crop plants and recent developments has been appropriately compiled by illustrious workers of IARI, New Delhi and Directorate of Groundnut Research, Junagadh in *Chapter 14* where an attempt was made to combine the existing knowledge of salinity tolerance in plants and the future prospect of research in this area.

It is important to determine that how the rice yield are influenced by elevated atmospheric CO_2, as it would have crucial implications on food security. Scientist from Kerala has nicely presented the research paradigm and inference of studies on high temperature stress in rice in *Chapter 15* that definitely supplements this volume in learning more on future food security of that region. This chapter is followed by another illuminating *Chapter 16* on influence of heat stress on plant responses where authors focus on the stepwise consequences of heat stress on different metabolic pathways in plant system in respect to physiological, biochemical, biotechnological and molecular aspects. The final part of the *Section II* reviews the impact of air pollution on plant metabolism and discussed with precise evidences in different heads by scientists of NBPGR in *Chapter 17*. Last but not the least, the *Section III on Technological Innovations In Agriculture* carries single but a vital chapter on "Can Nanotechnology be Harbinger for Second Green Revolution?" in that the NBPGR scientist to his level best emphasized that the future of agriculture on a global scale remains a subject of intense debate in view of persistent food insecurity and related poverty which rural agricultural population is facing in the developing countries. Moreover, its trade is declining in comparison with other industries and its negative impacts on the environment are a subject of scrutiny and debate. Apparently, modern innovative technologies are essential in deal with these problems.

In this committed endeavour, I am delighted to express my genuine appreciation to the national and international Members of the Editorial Committee as well as to a number of distinguished authors along with young and talented contributors from well known institutions for bringing up this unrivalled, sensible, thoughtful and extensive treatise of international standard. In addition, I am extremely grateful to the *Fellow Members of the Indian Society for Plant Physiology, New Delhi* for their trustworthy moral support and valued suggestions from time to time. My esteem is very much due to the Hon'ble Vice-Chancellor Dr. Lalji Singh, Banaras Hindu University, the Director, Institute of Agricultural Sciences, the Dean, Faculty of Agriculture, and my admired colleagues of the Banaras Hindu University and other universities and institutes of the world for their consistent moral support in this massive task. Besides these, I have my sincere thanks to Mr. A. Ishan Rahul, Engineer, NTPC, Nagpur for his incredible perfectionism in the conception of the very sensible and exceptional cover page of this *Volume*.

Last but not the least, I am indebted to my affectionate mother and my all family members for their blessings and good wishes in this wide mission. Besides these, I am extremely thankful to the outstanding staff members of the Scientific Publishers (India), Jodhpur, India, for their genuine competence in the perfect printing of international standard and worldwide circulation. Above all, I supremely recall that this undertaking is in advance dedicated to the adoring memory of my revered father Late Dr. A. Chittaranjan Sahay.

Dr. A. Hemantaranjan
Professor of Plant Physiology
Fellow, I.S.P.P.

Department of Plant Physiology
Institute of Agricultural Sciences
Banaras Hindu University
Varanasi 221 005, INDIA
JUNE 30, 2013
E-mail : hemantaranjan@gmail.com.

Contributors

A. Hemantaranjan
Department of Plant Physiology, Institute of Agricultural Sciences,
Banaras Hindu University, Varanasi, India

A.K. Trivedi
National Bureau of Plant Genetic Resources, Regional Station
Bhowali (Niglat) – 263132, Nainital, Uttarakhand (INDIA)

A.L. Singh
Directorate of Groundnut Research, P.B. No. 5,
Junagadh – 362 001, Gujarat, India

Ajay Kumar Dixit
Department of Biochemistry, Indian Institute of Science,
Bangalore, Karnataka -560012, India

Anaytullah
Nimbkar Agricultural Research Institute, P.O. Box-44,
Lonand Road, Phaltan, Dist. Satara-415523, Maharashtra, India

Ashish Kumar Srivastava
Area Office, National Seeds Corporation Limited,
Court Compound, Kashipur, Samastipur, Bihar-848101, India

Ashutosh Sharma
Department of Biotechnology, Guru Nanak Dev University,
Amritsar, India

Bandana Bose
Department of Plant Physiology, Institute of Agricultural Sciences,
Banaras Hindu University, Varanasi-221005, UP, India

Beena R.
Department of Plant Breeding and Genetics,
Regional Agricultutal Research Station,
Kerala Agricultural University, Pattambi, Palakkad, Kerala, India

Chelliah Jayabaskaran
Department of Biochemistry, Indian Institute of Science,
Bangalore, Karnataka PIN-560012, India

D. Bhaduri
Directorate of Groundnut Research, P.B. No. 5,
Junagadh – 362 001, Gujarat, India

Dhiman Mukherjee
Regional Research Station, Uttar Banga Krishi Viswavidyalaya,
Darjeeling, West Bengal 734301, India

Enrique J. Baran
Centro de Química Inorgánica (CEQUINOR/CONICET,UNLP)
Facultad de Ciencias Exactas, Universidad Nacional de La Plata,
C. Correo 962, 1900-La Plata, ARGENTINA

J.D.S. Panwar
Ex. Head, Plant Physiology. IARI, New Delhi, India.

Jyoti Sati
National Bureau of Plant Genetic Resources,
Regional Station, Bhowali – 263132
Nainital, Uttarakhand (India)

K. Chakraborty
Directorate of Groundnut Research, P.B. No. 5,
Junagadh – 362 001, Gujarat, India

K.A. Kalariya
Directorate of Groundnut Research, P.B. No. 5,
Junagadh – 362 001, Gujarat, India

Kalikinkar Bandyopadhyay
Division of Agricultural Physics,
Indian Agricultural Research Institute (IARI),
New Delhi-110 012, India

Madan Pal Singh
Division of Plant Physiology,
Indian Agricultural Research Institute (IARI),
New Delhi-110 012, India

Man Singh
Directorate of Groundnut Research, P.B. 5,
Junagadh 362 001, Gujarat, India

Nalini Pandey
Plant Nutrition and Stress Physiology Laboratory,
Botany Department, University of Lucknow,
Lucknow 22607, India

Nisha Goswami
Directorate of Groundnut Research, P.B. No. 5,
Junagadh – 362 001, Gujarat, India

P. K. Nagar
Former Head, Biotechnology Division,
Institute of Himalayan Bioresource Technology,
Palampur, (H.P.), India

Pramod Kumar Singh
Plant Physiology Laboratory, Department of Botany,
Udai Pratap Autonomous College,
University of M. G. Kashi Vidyapeeth,
Varanasi-221002, UP, India

Pratap Kumar Pati
Department of Biotechnology, Guru Nanak Dev University,
Amritsar (Punjab), India

Preeti, Sudhir Kumar
Department of Botany J.V. College,
Baraut (Baghpat)

R.K. Sairam
Indian Agricultural Research Institute,
New Delhi – 110 012, India

R.N. Nakar
Directorate of Groundnut Research, P.B. No. 5,
Junagadh – 362 001, Gujarat, India

Rakesh Pandey
Division of Plant Physiology,
Indian Agricultural Research Institute (IARI),
New Delhi-110 012, India

Rokolhuii Keditsu
Department of Horticulture,
School of Agricultural Sciences and Rural Development,
Nagaland University,
Medziphema 797 106, Nagaland

Sananda Mondal
Department of Plant Physiology,
Institute of Agricultural Sciences,
Banaras Hindu University, Varanasi-221005, UP, India

Sananda Mondal
Department of Plant Physiology,
Institute of Agricultural Sciences,
B. H. U., Varanasi, (UP), India

Vijay Paul
Division of Plant Physiology,
Indian Agricultural Research Institute (IARI),
New Delhi-110 012, India

Vinay Sehgal
Division of Agricultural Physics,
Indian Agricultural Research Institute (IARI),
New Delhi-110 012, India.

Pratap Kumar Pati
Department of Biotechnology, Guru Nanak Dev University,
Amritsar (Punjab), India

Preeti, Sudhir Kumar
Department of Botany, J.V. College,
Baraut (Baghpat)

R.K. Sairam
Indian Agricultural Research Institute,
New Delhi – 110 012, India

R.N. Nakar
Directorate of Groundnut Research, P.B. No. 5,
Junagadh – 362 001, Gujarat, India

Rakesh Pandey
Division of Plant Physiology
Indian Agricultural Research Institute (IARI),
New Delhi-110 012, India

Rokolhuii Keditsu
Department of Horticulture,
School of Agricultural Sciences and Rural Development,
Nagaland University,
Medziphema 797 106, Nagaland

Sananda Mondal
Department of Plant Physiology,
Institute of Agricultural Sciences,
Banaras Hindu University, Varanasi 221005, UP, India

Sananda Mondal
Department of Plant Physiology
Institute of Agricultural Sciences,
B. H. U., Varanasi, (UP), India

Vijay Paul
Division of Plant Physiology
Indian Agricultural Research Institute (IARI),
New Delhi-110 012, India

Vinay Sehgal
Division of Agricultural Physics,
Indian Agricultural Research Institute (IARI),
New Delhi-110 012, India

CONTENTS

Section - III
Technological Innovations In Agriculture

Section I:

NUTRIOPHYSIOLOGY AND CROP PRODUCTIVITY

1

PHYTOSIDEROPHORES AND RELATED SYSTEMS: METAL UPTAKE BY PLANTS

Enrique J. Baran

Centro de Química Inorgánica (CEQUINOR/CONICET,UNLP)
Facultad de Ciencias Exactas, Universidad Nacional de La Plata,
C. Correo 962, 1900-La Plata, ARGENTINA
E-mail: baran@quimica.unlp.edu.ar

INTRODUCTION

Iron is the most widely diffused transition metal essential to all the forms of life, from bacteria to man but, iron uptake and transport in different organisms is a relatively complex task (Baran, 1995; Fraústo da Silva and Williams, 1991; Williams and Fraústo da Silva, 1996). Although one of the most abundant elements in the earth crustes (Holleman and Wiberg, 1958; Williams and Fraústo da Silva, 1996), iron is not readily available in the soil in which it is found mainly in the form of insoluble oxohydroxide polymers. These oxide systems are quite stable and not very soluble at neutral pH, thus free Fe(III) in an aerobic aqueous environment is limited to an equilibrium concentration of approximately 10^{-17} M (Guerinot and Yi, 1994).

Therefore, the access to iron for bacteria, fungi and other micro-organisms and even for plants is difficult. Different microorganisms as well as fungi and bacteria have solved this problem, secreting high-affinity iron-binding compounds called siderophores. Now-a-days, more than 200 naturally occurring siderophores have been isolated and characterized (Bertini *et al.*, 1994). Siderophores are generally classified by the ligands used to chelate ferric iron. The major groups of siderophores posses the catecholate or hydroxamate residues for binding (Fig. 1); members of the first group are generally found in some bacteria whereas those of the second group are generated by higher organisms such as fungi and yeasts (Neilands, 1975; Hughes, 1982; Baran, 1995). Both ligands are weak acids (pK ~9) and present only coordinating oxygen atoms, generating 3:1 siderophore/Fe(III) complexes with octahedral coordination (Neilands, 1975). These Fe(III) complexes of siderophores present very high formation constants (log K_f values ranging between 22.9 to 52) and, consequently, siderophores are some of the best ligands for the complexation and stabilization of ferric ions (Devez *et al.*, 2009). An additional advantage of these high affinity Fe(III) chelators is that, under aerobic conditions, they will chelate Fe(II) cations and rapidly auto- oxidize them to the corresponding iron(III) species (Liu and Hider, 2002).

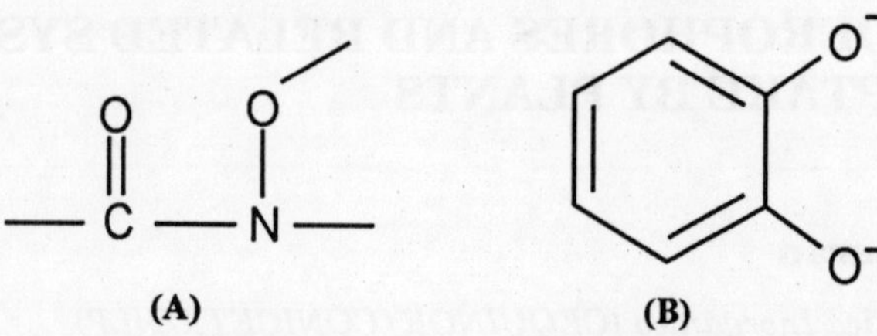

(A) (B)

Fig. 1. Types of siderophores: catecholate (**A**) and hydroxamate (**B**).

Apart from the catecholate and hydroxamate systems, in certain cases citric acid (Fig. 2) can also act as a siderophore (Gerinot, Meidl and Plessner, 1990; Palma Fett, LeVier and Guerinot, 1998). On the other hand, a variety of siderophores use this and other α-hydroxycarboxylic acids as the fundamental structural building blocks, generating different derivatives of citric acid (Okujo *et al.*, 1994; Vraspir and Butler, 2009; Budzikiewicz, 2010).

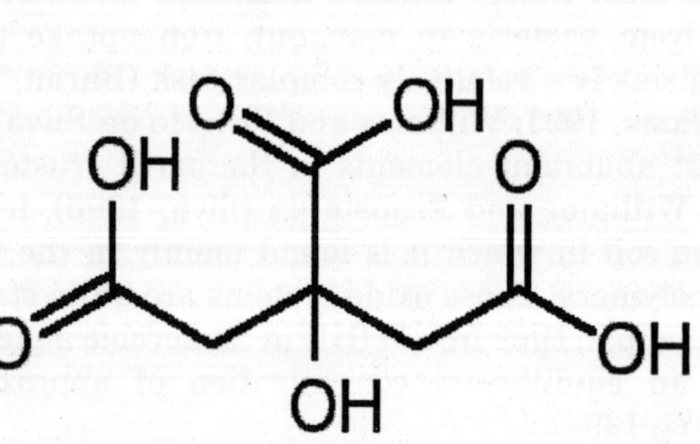

Fig. 2. Schematic structure of citric acid.

A very peculiar, interesting and important siderophore is desferrioxamine B, schematically depicted in Fig. 3, a fungal siderophore isolated from *Streptomyces pilosus*. It is a hydroxamate-based hexadentate trivalent metal chelator and is presently the therapeutic agent of choice for the clinical treatment of chronic iron overload. It possesses an extremely high affinity for Fe(III) and a much lower affinity for other metal cations present in biological fluids such as Ca(II), Mg(II) or Zn(II). The Fe(III) complex of desferrioxamine B is kinetically inert and presents a relatively low lipophilicity (Liu and Hider, 2002; Baran, 2010).

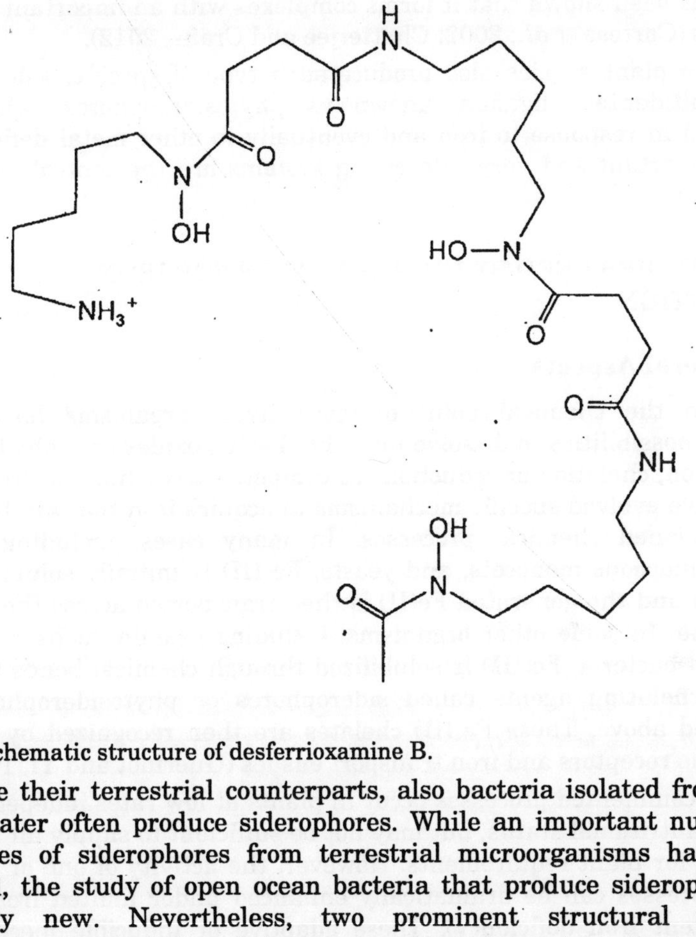

Fig. 3. Schematic structure of desferrioxamine B.

Like their terrestrial counterparts, also bacteria isolated from open ocean water often produce siderophores. While an important number of structures of siderophores from terrestrial microorganisms have been reported, the study of open ocean bacteria that produce siderophores is relatively new. Nevertheless, two prominent structural features characterize the majority of the marine siderophores investigated so far. One class of marine siderophores contains α-hydroxycarboxylic acid moieties, in the form of citric acid or β-hydroxyaspartic acid whereas the other class contain a unique peptidic head group, that coordinates Fe(III), appended by one of a series of fatty acids of different chain lengths (C12-C18) (Butler, 2005; Sandy and Butler, 2009; Vraspir and Butler, 2009). Interestingly, one of the distinguishing characteristics of the marine siderophores containing an α-hydroxycarboxylic acid moiety is the UV photoreactivity of their Fe(III) complexes, which leads to oxidation of the ligand and reduction of Fe(III) to Fe(II); the oxidized photoproduct can still coordinate Fe(III) (Butler, 2005; Vraspir and Butler, 2009).

In all the so far mentioned siderophores, oxygen is the binding atom to the metal center. For this reason it is interesting to mention that a very small sulfur containing siderophore, namely pyridine-2,6-bis (thiocarboxylic acid), is also known. It is produced and released by several strains of the genus *Pseudomonas* when grown in an iron deficient environment, and it has been shown that it forms complexes with an important number of metals (Cortese *et al.*, 2002; Chatterjee and Crans, 2012).

Some plant species also produce such type of specific, siderophore-like, multidentate ligands, known as phytosiderophores, which are generated in response to iron and eventually to other metal deficiencies. These important and very interesting systems are the central subject of this review.

2. STRATEGIES USED BY HIGHER PLANTS FOR IRON ACQUISITION

2.1. General Aspects

From the chemical point of view, living organisms have three different possibilities to dissolve insoluble Fe(III) oxides or oxohydroxides: protonation, chelation or reduction. To compete successfully for iron, organisms have evolved specific mechanisms to acquire iron that are based on the mentioned chemical processes. In many cases, including dicots, nongraminaceous monocots, and yeasts, Fe(III) is initially solubilized by reduction and the generated Fe(II) is then transported across the plasma membrane. In some other organisms, including graminaceous monocots, fungi and bacteria, Fe(III) is solubilized through chemical bonds to high-affinity chelating agents called siderophores or phytosiderophores, as mentioned above. These Fe(III) chelates are then recognized by specific membrane receptors and iron transport ensues (Guerinot and Yi, 1994).

The commented processes occur in plants at low rates independent of the iron-nutritional status, but may not be sufficient to supply all the iron adequate for plant requirements. However, the activity of one or more of these processes can be dramatically enhanced under limited iron conditions (latent iron deficiency). These adaptive or inducible mechanisms ("iron-deficiency stress responses") are suppressed when the iron requirement of the plant is satisfied, avoiding eventual iron toxicity in the plant caused by excessive iron uptake (Römheld, 1987).

Qualitative differences exist in the plant kingdom among genotypes in their iron deficiency-induced adaptation mechanisms and these differences can be classified according to two strategies (Römheld and Marschner, 1986; Römheld, 1987; Guerinot and Yi, 1994; Palma Fett, LeVier and Guerinot, 1998).

In **Strategy I**-type plants respond to iron deprivation with both morphological and physiological changes. Morphological modifications include enhanced development of lateral roots and differentiation of specialized transfer cells, increasing the surface area for reduction and transport of iron. Enhanced reduction of Fe(III) to Fe(II) under iron-deficiency stress seems to be the most typical feature of this strategy and is characterized by an increased acidification of the rhizosphere and the activation of plasma membrane-bound reductases.

Strategy II-type systems are characterized by an iron deficiency-induced release of specific Fe(III) chelating compounds (phytosiderophores) and a high affinity uptake system (transport protein) that transports the Fe(III) chelates as intact molecules.

The distribution of these two strategies in the plant kingdom shows a strong systematic order and it became evident that the chemotaxonomic boundary for the two strategies is not between dicotyledons and monocotyledons as suggested in the older literature, but between most higher plants (Strategy I) and grasses (Gramineae, Strategy II) (Römheld, 1987). Besides, it is considered that Strategy II is more efficient than Strategy I as, for example, grasses can grow on calcareous soils that will not support the growth of dicotyledons. This can probably be related to the fact that Strategy II is less pH-dependent than Strategy I and it has been observed that the release of phytosiderophores and Fe(III) uptake are only slightly depressed at high soil pH (Römheld and Marschner, 1986; Römheld, 1987; Guerinot and Yi, 1994).

2.2. Mechanistic Details of Strategies I and II

2.2.1. Strategy I

Under iron-deficient conditions, nongraminaceous plants basically release reductants or chelators and enhance proton excretion in the rhizosphere, and finally transport the generated Fe(II) through the plasma membrane (Guerinot and Yi, 1994; Mori, 1999; Grotz and Guerinot, 2006; Aguado-Santacruz *et al.*, 2012). This implies that, at least, three different and concurrent steps (pH-lowering at the rhizosphere, Fe(III) reduction, and Fe(II) transport) are involved in this mechanism.

In order to decrease the pH-value, increasing the amount of soluble Fe(III), plants extrude protons into the rhizosphere and it is usually admitted that these proton pumps are driven by plasma membrane ATPases (Römheld, Müller and Marschner, 1984; Gaxiola, Palmgren and Schumacher, 2007). For every one unit drop in pH, iron becomes a thousand fold more soluble. Although the protein responsible for releasing protons into the rhizosphere has not yet been identified, several proton-ATPases of the ***AHA*** (**A**rabidopsis ***H'***-**A**TPase) family are induced in the roots of iron-deficient plants (Grotz and Guerinot, 2006).

The requirement of a reduction step in cellular iron uptake is clearly established and the existence of plasma membrane ferric reductase activity has been described in both prokaryotic and eukaryotic cells (Dancis *et al.*, 1990). FRO2 is the enzyme responsible for the plasma membrane Fe(III) chelate reductase activity in plants that is induced under iron-deficiency in the roots of *Arabidopsis* (Robinson *et al.*, 1999). FRO2 belongs to a superfamily of flavocytochromes that transport electrons across membranes. Furthermore, overexpression of FRO2 has demonstrated that Fe(III) chelate reductase activity is the rate-limiting step in iron acquisition (Grotz and Guerinot, 2006).

Another interesting question is: what is the substrate for the Fe(III) chelate reductase? It is likely that various microbial siderophores can serve as substrates, as shall be discussed in Sect. 2.2.3.

Finally, once Fe(III) is reduced, the ferrous iron must be transported across the plasma membrane. The types and the overall characteristics of the involved transporters are so far not clearly known. One of such transporters is the so-called IRT1 ((Iron Regulated Transporter), Eide *et al.*, 1996), which is the major root iron transporter in *Arabidopsis* (Vert et al., 2002).

2.2.2. Strategy II

The Strategy II-plants use a chelation based strategy to acquire iron. In response to Fe-deficiency, Fe(III) binding non-protein amino acids termed phytosiderophores are synthesized and released. The characteristics of these molecules shall be discussed in detail in Sect. 3. of this review.

These plants also induce a high-affinity uptake system for the Fe(III)-phytosiderophore complexes that transport them into the root as intact complexes (Ma and Nomoto, 1996; Palma Fett, LeVier and Guerinot, 1998). One of the best characterized transporter families of this type is the YSL family. YSL proteins are believed to mediate the uptake of metals that are complexed with phytosiderophores or nicotianamine. From the biochemical point of view, the best studied member of this family is YS1 from maize. ZmYS1 protein accumulates in the roots and leaves of Fe-starved plants and functions as a proton-coupled symporter to transport Fe(III)/phytosiderophores. Different transport studies indicate that ZmYS1 might have other substrates in addition to iron, including zinc, copper and nickel (Roberts *et al.*, 2004; Schaaf *et al.*, 2004; Colangelo and Guerinot, 2006).

Interestingly, under certain circumstances plants may be use simultaneously both described Strategies to obtain iron from the soil. This is, for example, the case of rice which, in addition to having the ability to transport Fe/phytosiderophore complexes, is also able to transport Fe(II). This fact is not surprising since rice grows in water-logged soils that tend to be rich in Fe(II). Currently it is not known whether other Strategy II plants, growing in soils with limited Fe(II) concentrations are also capable to induce the Strategy I response (Grotz and Guerinot, 2006; Morrisey and Guerinot, 2009).

2.2.3. Utilization of microbial siderophores by plants

The relative importance of microbial siderophores in plant iron nutrition and the possible mechanisms by which plants make use of siderophores-bound iron have often been discussed but remain controversial (Palma Fett, LeVier and Guerinot, 1998). Notwithstanding, it has often been admitted that microbial siderophores can serve as substrates for Fe(III) chelate reductase (Guerinot and Yi, 1994).

A reasonable speculation in this context is that siderophore-complexes of low stability could act as a readily accessible source of Fe(III) for root exudates and enhance iron uptake. Besides, microbial siderophores may also serve indirectly after degradation by microbes and release of Fe(III).

On the other hand, it has been speculated that certain bacteria can utilize Fe-phytopsiderophore complexes as a source of carbon and iron (Guerinot and Yi, 1994).

Attempts to determine whether Strategy II plants that normally release their own siderophores can also utilize microbial siderophores have yielded conflicting data. Notwithstanding, in an interesting study designed to investigate the ability of Fe-rhizoferrin to serve as an iron source for barley (*Hordeum vulgare* L.) and corn (*Zea mays* L.) plants, it was demonstrated that iron from this source is taken up by these Strategy II plants. The mechanism involves ligand exchange between the ferrated microbial siderophores and phytosiderophores, which are then taken up by the plant (Yehuda *et al.*, 1996).

3. PHYTOSIDEROPHORES AND RELATED SYSTEMS

3.1. General Aspects

As discussed above, in the case of Strategy II-type mechanism, plants use a chelation strategy based on the generation and utilization of phytosiderophores. These chelating agents have been identified as hydroxyl- and amino-substituted iminocarboxylic acids, such as mugineic and avenic acid (Römheld, 1987; Devez, Achterberg and Gledhill, 2009) which are chemically related to nicotianamine, a non-proteinaceous amino acid with widespread occurrence in higher plants, and which is the main precursor of the biosynthesis of these phytosiderophores.

In recent years it has also been established that phytosiderophores can chelate other metals, apart from Fe(III), and, therefore, it has been suggested to use the more general term phytometallophores instead of phytosiderophores (Marschner, Römheld and Kissel, 1986; Grusak, Pearson and Marentes, 1999; Aguado-Santacruz *et al.*, 2012).

Another interesting general aspect of phytosiderophores activity is its apparent light and/or temperature dependence. The secretion of phytosiderophores from iron-deficient barley and wheat roots was found to occur within a specific period of time, from 4 to 6 hours, which usually begins about 2-3 hours after the onset of light (Marschner, Römheld and Kissel, 1986; Yehuda *et al.*, 1996). It was later suggested that temperature, rather than a light signal, is the trigger for the initiation of the phytosiderophores release (Yehuda *et al.*, 1996); however, it has not been clarified yet if other plant species share the same diurnal rhythm of phytosiderophores release. As phytosiderophores constitute a type of amino acid, they may be degraded by rhizosphere microorganisms and, consequently, their secretion during short periods may be considered as a strategy for protecting them from microbial degradation and/or to minimize loss by diffusion into the bulk soil (Ma and Nomoto, 1996). Additionally, the effects of light appear to have great importance in relation to various other aspects of iron transport in plants (Mori, 1998).

From the point of view of the biosynthesis of phytosiderophores, L-methionine is clearly established as the primary precursor (Mori and Nishizawa, 1987; Ma and Nomoto, 1993a) and nicotianamine as the key intermediate precursor molecule.

3.2. Nicotianamine

As stated above, nicotianamine is the key intermediate in the biosynthesis of phytosiderophores and has been found in all plants so far investigated (Rudolph *et al.*, 1985; Takahashi *et al.*, 2003; Callahan *et al.*, 2006). It has an important chelation activity for different metal cations including Fe(II) and Fe(II). Although its precise roles in higher plants remain unclear, as it is not secreted by roots it is thought to play a role in the internal transport of iron and other metals and may also be function as an iron-scavenger to protect cells from oxidative stress. It appears also involved in long-distance copper transport, plant reproduction and detoxification of heavy metals (von Wirén *et al.*, 1999; Takahashi *et al.*, 2003; Schuler and Bauer, 2011).

3.2.1. Biosynthesis of nicotianamine

S-adenosyl-methionine is generated by action of the enzyme SAM-synthase on L-methionine. The subsequent condensation of three amino-carboxypropyl groups of three S-adenosyl-methionine molecules, catalyzed by action of the enzyme nicotianamine synthase, generates the nicotianamine molecule, which structure is schematically depicted in Fig. 4. During the enzymatic reaction three covalent bonds are broken to release three aminocarboxypropyl groups from S-adenosyl-methionine and three new covalent bonds are formed, including an internal cyclization leading to the formation of an azetidine ring (Curie *et al.*, 2009; Dreyfus *et al.*, 2009).

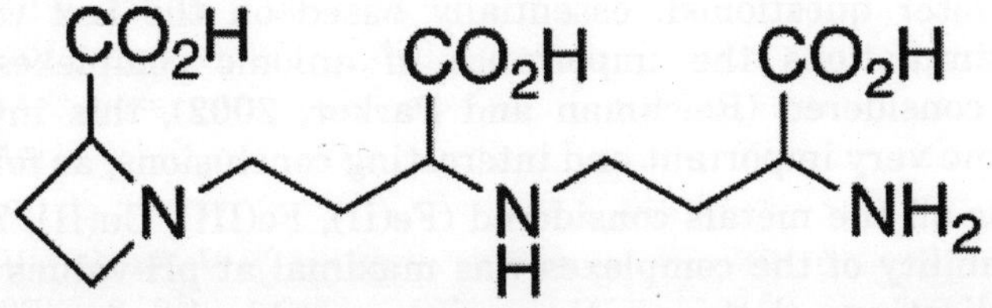

Fig. 4. Schematic structure of nicotianamine.

The total synthesis of nicotianamine has also been performed (Fushiya *et al.*, 1981; Matsuura, Hamada and Shioiri, 1994; Shioiri, Hamada and Matsuura, 1995)

3.2.2. Metal complexes of nicotianamine

The presence of three amino and three carboxyl groups in the nicotianamine molecule allows the formation of a hexadentate coordination that drives the formation of very stable octahedral chelates with different central metal cations.

The formation of metal complexes of nicotianamine has initially been investigated *in vitro* and it was shown that the divalent transition metal cations of the first series (3d series) form relatively stable complexes, as shown by the values of the respective stability constants presented in Table 1 (Beneš *et al.*, 1983; Schreiber, 1986).

Table 1. Stability constants (log K) of the M(II)-nicotianamine complexes of 3d-metals.

M(II)	Mn	Fe	Co	Ni	Zn	Cu
log K	8.8	12.1	14.8	16.1	14.7	18.6

Interestingly, the stability constants of these complexes follow the well-known Irving-Williams series (Fraústo da Silva and Williams, 1991; Porterfield, 1993; Baran, 2012a), i.e. the constants are in the order Mn(II) < Fe(II) < Co(II) < Ni(II) < Cu(II) > Zn(II). In addition, more recently it has been shown that nicotianamine can also complex Fe(III) *in vitro* with a high affinity, and in this case the log K = 20.6 (von Wirén *et al.*, 1999).

The ESR spectrum of the 1:1 Cu(II):nicotianamine complex suggests the presence of a distorted octahedral environment with axial symmetry, whereas the superhyperfine splitting points to the presence of three N-donor atoms in the coordination sphere (Mino *et al.*, 1981; Sugiura and Nomoto, 1984).

A very detailed and solid potententiometric and spectrophotometric study, complemented with computer simulations, were performed to characterize the interactions of nicotianamine with different metal cations, and specially to attain a deeper insight into its interaction with Fe(II) and Fe(III) and the pH-dependence of the stability of the different complexes (von Wirén *et al.*, 1999). Although some of the obtained results have been later questioned, essentially based on the fact that in the computer simulations the importance of anionic complexes was not adequately considered (Reichman and Parker, 2002), this investigation provided some very important and interesting conclusions, as follows:

- For all the metals considered (Fe(II), Fe(III), Cu(II), Zn(II)), the stability of the complexes was maximal at pH-values above 6.5, indicating that nicotianamine would be more likely a "symplastic" chelating agent for metals.
- The Cu(II) complex is very stable in mild acidic conditions, which strongly supports the possible Cu(II)/nicotianamine interactions in an "apoplastic" environment such as the xylem sap.
- *In vitro*, nicotianamine chelates Fe(III) with an affinity 10^8 times greater than that with which it binds Fe(II) although the kinetic stability of the Fe(III) complex is much lower than those of the corresponding Fe(II) complex. There is not evidence of autoxidation of the Fe(II) complex, even when O_2 was bubbled through the solution. This result is very interesting as for, example, the EDTA complex of Fe(II) is rapidly autoxidized in air to the corresponding Fe(III) species (Zang and van Eldik, 1990)
- If equimolecular amounts of Fe(II) and Fe(III) were added to nicotianamine at pH = 7.0 and the complexes separated by capillary electrophoresis, only the Fe(II) complex was detected, indicating that at this pH-value the chelator preferentially scavenge Fe(II) even in the presence of Fe(III), and the Fe(II)/nicotianamine complex will persist by virtue of its kinetic stability.
- Both, the Fe(II)/nicotianamine and the Fe(III)/nicotianamine complex, are not highly active Fenton reagents, as determined from their ability to mediate H_2O_2-dependent oxidation of deoxyribose.
- At pH-values around 5.5 carboxylic acids, like citrate, would be the main chelators of metal cations whereas at neutral pH-values nicotianamine would be almost the exclusive chelator. Besides, unlike Fe(II)/nicotianamine, the Fe(II)/citrate complex rapidly autoxidizes.

Most recently, metal complexes of nicotianamine have been studied *in vitro* with mass spectrometry techniques, namely electrospray ionization time-of-flight mass spectrometry, investigating the formation of different complexes by measuring free nicotianamine and metal/nicotianamine complexes at different pH-values (Rellán-Alvarez, Abadía and Alvarez-Fernández, 2008). This work confirmed the computer simulations mentioned above concerning the pH-dependent stability of several metal/nicotianamine complexes, the high stability of the Fe(II) complex, and the ligand exchange between citrate and nicotianamine at pH = 5.5.

The pH-dependence of the nicotianamine complexes of Mn(II), Fe(II), Cu(II) and Zn(II) was also investigated by Stephan *et al.* (1998). They found that the complexes remain undissociated in the alkaline pH range (pH > 7) and begin to dissociate with increasing proton concentration. The dissociation of the Mn(II) complex begins at about pH 6.5 and that of the Fe(II) complex at pH 5.41. These authors also measured the octanol/water partition coefficients of the free amine and of the four mentioned metal complexes in order to verify to which extent they could penetrate membranes. The partition coefficient for the free ligand is practically 1 and is drastically reduced in the metal complexes. This behavior is probably the result of a change of charge of the molecule and not of a change in molecular conformation. Under the used experimental conditions, the nicotianamine molecule is virtually neutral, and thus, as lipophilic as possible. In contrast, the metal complexes bear one negative charge which increases their hydrophilicity. These results suggest that free nicotianamine could permeate membranes faster than its metal complexes if membrane transport proteins are not involved in their movement.

All the commented studies clearly demonstrate the biochemical properties of nicotianamine and its importance in plant physiology. It is likely an important molecule for the phloem loading and/or unloading processes for Fe, Mn, Cu and Zn, although the technical difficulties to access to reasonable amounts of phloem sap, compatible with biochemical approaches, has not allowed further developments in this direction so far. Its importance for Fe metabolism in plants is undoubtedly, as suggested by the properties and characteristics of the Fe(II) and Fe(III) complexes of this molecule (Curie *et al.*, 2009).

It has been suggested that nicotianamine may have an important role in the chelation and transport of nickel in hyperaccumulator plants (Curie et al, 2009) as well as in the detoxification and/or tolerance of nickel (Kim *et al.*, 2005, Callahan *et al.*, 2006).

Recently, it has also been observed that under copper deficiency, but not under copper excess, certain plants induced the biosynthesis of nicotianamine, pointing again to the importance of this molecule in the accumulation and transport of this essential element (Irtelli, Petrucci and Navari-Izzo, 2009).

3.3. Mugineic Acid

Mugineic acid is so far probably the most investigated phytosiderophore and plays a central role in iron uptake by graminaceous plants (Strategy II plants). Since it was found that roots of water-cultured rice and oats contained an iron-chelating substance (Takagi, 1976) and mugineic acid was identified in barley (Takemoto *et al.*, 1978) some other analogs have been isolated and identified from various graminaceous species and cultivars and all these compounds are often generically termed as "mugineic acids" (Ma and Nomoto, 1996; Grusak, Pearson and Marentes, 1999) although they are, obviously, chemically different from them, but show very close structural relationships.

Fig. 5. Schematic structure of mugineic acid.

The schematic structure of mugineic acid is shown in Fig. 5 and the mentioned structural relationships with the other phytosiderophores are presented schematically in Fig. 6.

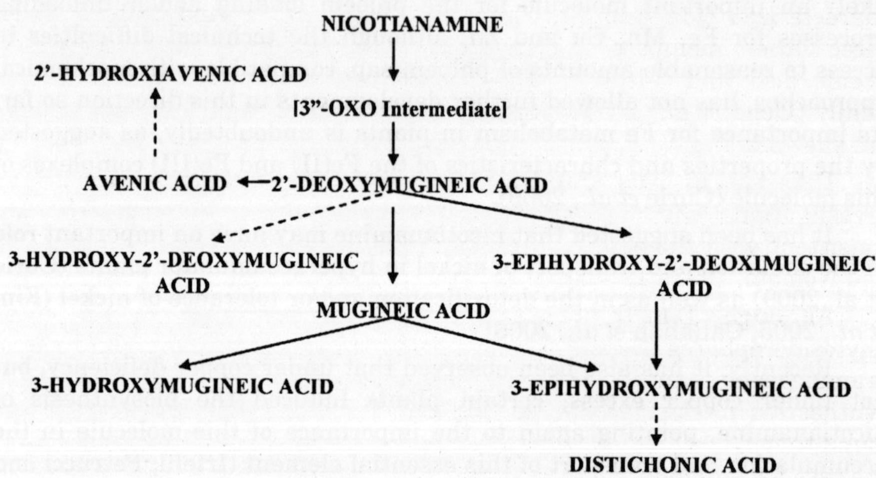

Fig. 6. Structural relationships between the different phytosiderophores (adapted from Ueno *et al.*, 2007).

The crystal structure of the acid was determined by single crystal X-ray diffractometry (Takemoto *et al.*, 1978), using a material obtained from root washings of water-cultured barley (*Hordeum vulgare* L.), using a chromatographic procedure (Sugiura and Nomoto, 1984). Crystals of the acids were obtained as small platelets by water recrystallization of the crude material. Its crystallographic analysis shows that mugineic acid belongs to the orthorhombic crystal system, space group $P2_12_12_1$ and Z = 4 molecules in the crystallographic unit cell, with the molecules in extended *trans-trans* form. The structural characteristics were additionally confirmed by ^{1}H and ^{13}C NMR-spectroscopy (Takemoto *et al.*, 1978). A detailed analysis of these NMR spectra, together with those of related compounds, was published later (Sugiura and Nomoto, 1984).

3.3.1. Biosynthesis of mugineic acid and related phytosiderophores

As stated in the previous Section, the precursor for the biosynthesis of siderophores is nicotianamine, which by action of the key enzyme NAAT (nicotianamine aminotransferase), which is a specific enzyme of graminaceous plants, transforms nicotianamine in an intermediate keto-acid, usually called 3"-oxo intermediate (containing an oxo-group instead of the amino group at the 3" C-atom). This intermediate is then transformed in 2'-deoxymugineic acid by action of the enzyme DMAS (deoxy-mugineic acid synthase). The 2'-deoxymugineic acid is the immediate precursor of the mugineic acid family of phytosiderophores, as hydroxylation of this molecule, at different positions, generates the different derivatives (Ma and Nomoto, 1996; Mori, 1999; Takahashi *et al.*, 2003). Initially, apart from the 2'-deoxymugineic acid precursor, other five compounds: mugineic acid, avenic acid, 3-hydroxymugineic acid, 3-epyhydroxymugineic acid and distichonic acid were isolated and characterized (Ma and Nomoto, 1996; Mori, 1998). And, subsequently, three other members (3-epihydroxy-2'-deoxymugineic acid, 2'-hydoxy-avenic acid and 3-hydroxy-2'-deoxymugineic acid) were added to this family (Ueno *et al.,* 2007). The schematic structures of these compounds are shown in Fig. 7.

Different *in vivo* and *in vitro* biosynthesis studies, supported by investigations with isotope labeled molecules (^{2}H, ^{13}C and ^{15}N) are consistent with the general synthetic schema briefly described above (Ma and Nomoto, 1993a,1996).

Although L-methionine is the initial precursor of all these phytosiderophores (Sect. 3.1.), its measured concentration in the roots of graminaceous plants is extremely low when compared to the amounts of synthesized phytosiderophores (Ma and Nomoto, 1996). However, it was found that L-methionine is recycled during these synthetic processes, as is the case for ethylene production (Ma *et al.*, 1995)

The total synthesis of mugineic acid and of some other of the related phytosiderophores has also been performed by different routes, as reviewed by Shioiri, Hamada and Matsuura (1995).

3.3.2. General characteristics of mugineic acid as a ligand

From the point of view of coordination chemistry mugineic acid and the related phytosiderophores are very interesting and powerful chelating agents. Mugineic acid possesses seven potential coordinating groups, i.e. three carboxylato groups, two hydroxyl groups, the azetidine N-atom and the secondary amine N-atom (cf. Fig. 5). The molecular conformation allows, obviously, the easy generation of multidentate binding and chelate generation with different cationic metal centers.

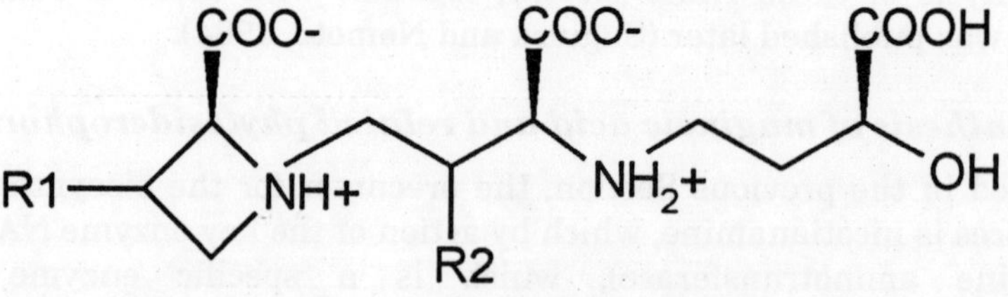

R1 = H; R2 = H — 2'-Deoxymugineic acid
R1 = H; R2 = OH — Mugineic acid
R1 = OH; R2 = H — 3-Hydroxy-2'-deoxymugineic acid (3-Epihydroxy-2'-deoxymugineic acid)
R1 = OH; R2 = OH — 3-Hydroxymugineic acid (3-Epihydroxymugineic acid)

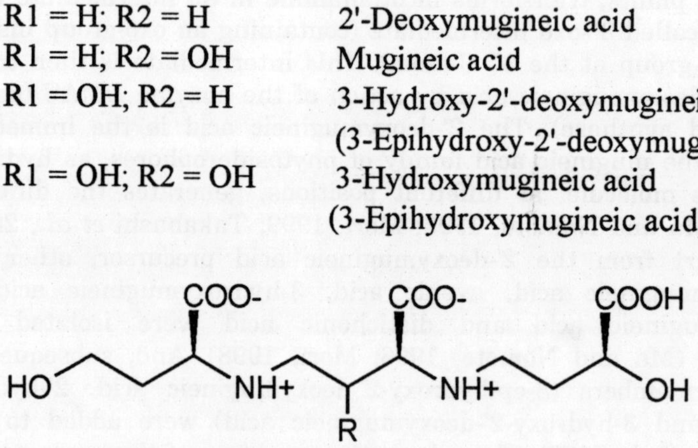

R = H — Avenic acid
R = OH — 2'-Hydroxyavenic acid

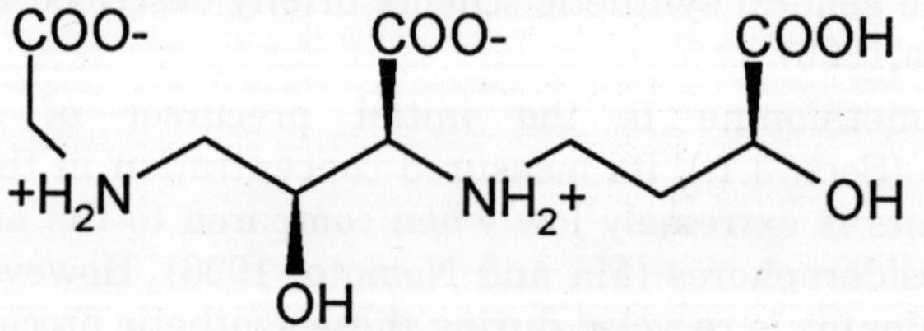

Distichonic acid

Fig. 7. Schematic structures of the phytosiderophores related to mugineic acid.

The protonation constants of the acid were determined by potentiometric titration at 20°C and working in 0.1 N KNO_3. Determined values were: pK_1(COOH) = 2.39, pK_2 (COOH) = 2.76, pK_3 (COOH) = 3.40, pK_4 (>NH^+) = 7.78 and pK_5 (NH_2^+) = 9.55 (Sugiura et al., 1981).

From the point of view from Pearson's theory of hard and soft acids and bases (HSAB) (Porterfield, 1993; Baran, 2012a, b) the available coordinating atoms are "hard" bases. Therefore, mugineic acid shows a clear preference for biologically relevant "hard" (Mg(II), Ca(II), Mn(II), Fe(III)) or borderline (Fe(II), Cu(II), Zn(II), Co(II)) cations. Besides, Cd(II), which is also often found in soils is a characteristic "soft" cation and, therefore, it presents a scarce affinity for ligands containing only "hard" coordination centers.

3.3.3. Metal complexes of mugineic acid

Although complexes of a number of metal cations with mugineic acid as ligand have been prepared and characterized by different physicochemical methods, it is not clear, in all cases, if such interactions occur also at the plants level. Therefore, in this Section we present, firstly, the so far known information about the naturally occurring metal uptake by phytosiderophores and, subsequently, a detailed analysis of the synthetic metal complexes of mugineic acid is given.

3.3.3.1. Metal uptake by mugineic acid and related phytosiderophores

As previously discussed, Strategy-II plants use a chelation based mechanism to acquire iron, inducing the generation of phytosiderophores in response to a Fe- deficiency complemented with a high-affinity uptake system for the transport of the Fe(III)/phytosiderophore complexes that transport them into the root as intact complexes.

The chemical nature of phytosiderophores can differ among plant species and even cultivars, but in general one distinct phytosiderophore dominates in each genotype (Kawai, Takagi and Sato, 1988; Römheld and Marschner, 1990; Römheld, 1991).

On the other hand, it has often speculated if the deficiency of other micronutrients also induces the release of phytosiderophores and during the years an important number of experimental studies were performed in order to answer this question.

The cases of Zn(II) and Cu(II) are probably the most investigated ones (Grotz and Guerinot, 2006). In the case of zinc it has been admitted for some years that the liberation of phytosiderophores in response to iron deficiency, may also favor the chelation of Zn(II) from the soil but it remains unclear if zinc deficiency induces phytosiderophores liberation (Ma and Nomoto, 1996), although this possibility has often been emphasized (Römheld, 1991; Zhang, Römheld and Marschner, 1991) and most recently it has been shown that greater amounts of phytosiderophores being exuded under Fe than under Zn deficiency, indicating that the triggering process is more direct in case of iron deficiency, while a more complicated and/or slower sequence of events may be necessary to trigger exudation of phytosiderophores under zinc deficiency (Rengel and Römheld, 2000).

On the other hand, different studies confirmed that phytosiderophores released by Fe-deficient graminaceous plant species efficiently chelate and mobilize not only Fe, but also Zn, Cu, Mn, Ni and, eventually,

Cd (Treeby, Marschner and Römheld, 1989; Römheld, 1991; Grant *et al.*, 1998; Grusak, Pearson and Marentes, 1999; Schaaf *et al.* 2004). On the other hand, the inhibitory effects of copper, zinc and cobalt upon iron-uptake were also investigated, showing that Cu(II) was the most effective inhibitor (Ma and Nomoto, 1993b). It has also been shown that Fe and Zn deficiencies can enhance Mn uptake in certain plant species (Pittman, 2005).

In a very detailed study of the absorption of zinc by maize (*Zea mays* L), von Wirén, Marschner and Römheld (1996) admit the induction of phytosiderophores by zinc deficiency. More interesting, they found that there are two independent mechanisms contributing to Zn-uptake in Fe-efficient maize, one of which is repressible by chelation and the other of which is dependent on a phytosiderophores transport system. It is assumed that in the absence of excess phytosiderophores the Zn/phytosiderophore complex is in equilibrium with the free Zn^{2+} species and its ligand, allowing the uptake of this free metal cation via a putative zinc transporter. With increasing concentration of free phytosiderophores, the chelated zinc species predominates and is transported across the root plasma membrane via the putative Fe/phytosiderophore transporter. Notwithstanding, it can not be excluded that Zn/phytosiderophores were recognized and transported by a separate transport protein (see also Grotz and Guerinot, 2006).

It has been suggested that other metals may also posses two or more relatively specific transport systems: high affinity systems that are active in metal limiting conditions and low affinity systems that functions when substrates are more abundant (Guerinot and Eide, 1999).

The interaction of root exudates from wheat and maize with Co(II), Zn(II) and Mn(II) was investigated using tracer techniques (^{57}Co, ^{65}Zn, ^{54}Mn). It was found that with increasing time a gradual transformation from ionic metal species to higher molecular forms occur (Merckx et al. 1986).

Although not very much detailed information about copper uptake in plants is so far available, in recent years important advances have been produced in relation to the transport of this essential micronutrient in the plant kingdom (Himelblau and Amasino, 2000; Grotz and Guerinot, 2006). The most interesting aspect of these transport systems is the involvement of specific metallochaperones.

Metallochaperones are soluble metal receptor proteins which garner metals from their very small homeostatic concentrations inter- or intracellularly and deliver those metals to the metalloenzyme that requires them for activity. The chaperone must be a thermodynamically stable complex as it approaches its target apoprotein, but the metal within the complex must be kinetically labile to allow fast metal ion exchange for its proper transport and distribution (Rosenzweig, 2000).

Although it has been shown that cadmium often interferes the iron uptake and transport (Morrissey and Guerinot, 2009) it has been speculated that phytosiderophores may be participate in Cd(II) uptake and accumulation by plants, similarly as they do with zinc and taking into account that both, zinc and cadmium, are members of group 12 of the periodic system. Notwithstanding, no definitive answers to these possibilities have been provided. In a detailed study of this problem Shenker *et al.* (2001) demonstrated that although phytosiderophores not only facilitates Fe uptake, but also led to coincidental uptake, translocation and accumulation of other metal cations in thc plant shoot, these effects were not observed for cadmium. These authors suggested that the presence of competing micronutrients such as Zn(II), Cu(II) or Mn(II) prevented Cd(II) chelation by phytosiderophores or, eventually, that Cd(II) do not form sufficiently stable complexes with these chelators. This last supposition is in line with coordination chemistry arguments provided by the HSAB-theory, as discussed in Sect. 3.3.2.

Due to the fact that aluminum is the most abundant metal in the earth's crust (Holleman and Wiberg, 1958; Williams and Fraústo da Silva, 1996) and because of the chemical similarities between Al(III) and Fe(III), it may be possible that Al(III) competes with Fe(III) for the phytosiderophores binding sites in the rhizosphere. Therefore, it seems very important to clarify this aspect of phytosiderophores mediated metal uptake in plants. A detailed investigation on this problem was recently performed by Yoshimura *et al.* (2011), by means potentiometric titration experiments. It was found that mugineic acid and epi-hydroxymugineic acid bind Al(III), in contrast to nicotianamine which does not interact with this cation under normal physiological conditions.

Furthermore, this investigation also showed that the equilibrium formation constants of the phytosiderophores/Al(III) complexes are appreciably smaller than those of the corresponding Fe(III) complexes, pointing to a higher selectivity of phytosiderophores for Fe(III) than for Al(III). These findings are consistent with the general observation that Al(III) complexes of O- and N- donor ligands posses weaker formation constants than the corresponding Fe(III) complexes. Thus, desferrioxamine B binds Fe(III) 10^6 times tighter than Al(III) (Ackrill and Day, 1985) and also EDTA binds Al(III) with a weaker formation constant than Fe(III) (Martell and Smith, 1974).

3.3.3.2. Synthetic metal complexes of mugineic acid

The investigation of coordination complexes of mugineic acid seems of central importance to attain a wider insight into the form in which this phytosiderophore interacts with the different metal cations relevant to plant physiology, and such knowledge is also important for a better comprehension of the uptake mechanisms involved in the activity of phytosiderophores. Approximately stability constants for the respective complexes of some of the most relevant cations (Sugiura *et al.*, 1981) are presented in Table 2 and compared again with the values of the corresponding nicotianamine complexes.

Table 2. Stability constants (log K) of M-mugineic acid complexes, compared with those of the corresponding nicotianamine complexes

Complex	Mn(II)	Fe(II)	Fe(III)	Co(II)	Ni(II)	Zn(II)	Cu(II)
Mugineic Acid		8.1	18.1			10.7	18.3
Nicotianamine	8.8	12.1	20.6	14.8	16.1	14.7	18.6

This comparison is very interesting, because apart of the comparable stability of the respective Cu(II) complexes, the constants for the Fe(II), Fe(III) and Zn(II) complexes are appreciably higher in the case of nicotianamine. It has been suggested that these differences arise by the presence of a terminal primary amino group in nicotianamine instead of the hydroxo group found at this same position in mugineic acid (Schreiber, 1966). This suggestion is in agreement with the relative positions of the mentioned donor groups ($-NH_2$ and OH^-) in the spectrochemical series of ligands (Porterfield, 1993) and in the case of the Cu(II) complexes this effect is probably smaller, because the donor groups are located in the axial positions, involving longer Cu-L distances. Moreover, the respective electronic spectra support all these comments (Mino *et al.*, 1981). Besides, the differences in coordination between the Fe(III) complexes of nicotianamine and mugineic acid are also reflected in the corresponding ESR spectra (Mino *et al.*, 1983).

From the structural point of view, only two mugineic acid complexes have so far been characterized by means of X-ray single crystal diffractometry, namely the Cu(II) and the Co(III) complex but interesting and valuable information has also been obtained for the Fe(III) complex.

Single crystals of the copper(II) complex were obtained mixing diluted equimolecular solutions of mugineic acid and $CuSO_4{\cdot}5H_2O$ and the obtained blue residue was recrystallized from ethanol-isopropyl alcohol (Nomoto *et al.*, 1981). The 1:1 complex crystallizes in the monoclinic crystal system, space group $P2_1$ and Z = 2 molecules in the crystallographic unit cell (Mino *et al.*, 1981; Nomoto *et al.*, 1981; Sugiura and Nomoto, 1984). The Cu(II) ion is in a distorted octahedral coordination, with mugineic acid acting as a hexadentate ligand. The basal plane is conformed by the azetidine N-atom, the secondary amine N-atom and two O-atoms, involving to the two terminal carboxylate groups; the axial positions are occupied by two O-atoms, one of them belonging to the intermediate carboxylate group and the other one, from the terminal hydroxyl group. This metal-to-ligand interaction is schematically depicted in Fig. 8. The equatorial bond lengths are 1.945 and 1.965 Å for the two Cu-O bonds and 2.033 and 2.018 Å for the Cu-N bonds, whereas the two axial bonds are somewhat longer (2.477 and 2.557 Å), a result which is in agreement with the well-known Jahn-Teller distortion for a $3d^9$ molecular system (Porterfield, 1993).

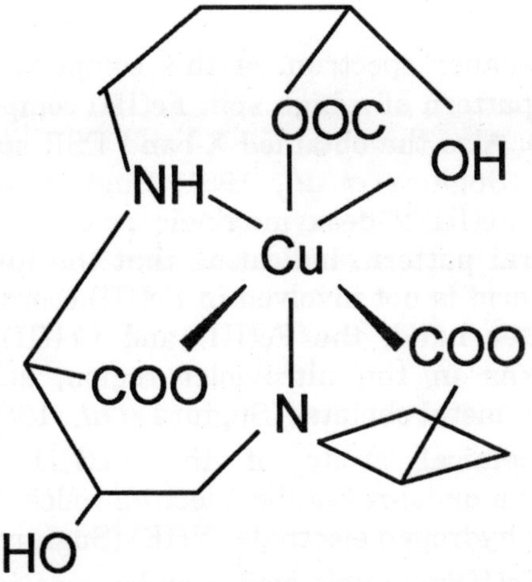

Fig. 8. Schematic structure of the Cu(II) complex of mugineic acid.

The ESR spectrum of the complex, measured at 77 K and pH = 7 is compatible with the proposed structural model and shows a typical Cu(II) hyperfine pattern with approximately axial symmetry and the superhyperfine splitting confirms the participation of two nitrogen donors in coordination. Also, the electronic and circular dichroism (CD) spectra are compatible with a tetragonally distorted octahedral Cu(II) complex (Mino *et al.*, 1981; Sugiura and Nomoto, 1984).

Although the crystal structure of the Fe(III)/mugineic acid complex has not as yet been established by X-ray diffractometry, relevant information on its coordination characteristic can be derived from the structure of the corresponding Co(III) complex. This complex was obtained by reaction of diluted equimolecular solutions of mugineic acid and $Na_3[Co(CO_3)_3]$ and purified by gel filtration. Single crystals were obtained by recrystallization from dioxane-water (Mino *et al.*, 1983).

The 1:1 Co(III):mugineic acid complex crystallizes in the monoclinic crystal system, space group $P2_1$ and Z = 4 molecules in the crystallographic unit cell, with two slightly different complex moieties (A and B). The coordination sphere of Co(III), CoO_4N_2, is nearly octahedral and similar to that described above for the Cu(II) complex but the axial bond lengths are appreciably shorter than in this last complex (Co-O distances between 1.90 and 2.05 Å); the other distances are comparable although somewhat shorter than in the Cu(II) complex. Another especially interesting aspect is that in the Co(III) an additional deprotonation of mugineic acid occurs, and the terminal hydroxyl group is also deprotonated.

An 1:1 Fe(III):mugineic acid complex can be prepared by mixing diluted aqueous solutions of the acid and a ferric salt at pH = 7 or by slow oxidation of the corresponding Fe(II) complex with oxygen (Mino *et al.*, 1983).

The ^{57}Fe-Mössbauer spectrum of this complex, obtained at 194 K, shows the typical pattern of a high spin Fe(III) complex (ΔE_Q = 0.24 and δ_{Fe} = 0.39 mm/sec). Also the obtained X-band ESR spectrum is typical of high spin Fe(III) (Sugiura *et al.*, 1981; Sugiura and Nomoto, 1984). Interestingly, the Fe(III)/2'-deoxymugineic acid complex shows a very similar ESR spectral pattern, indicating that the intermediate alcoholic group of mugineic acid is not involved in Fe(III) coordination (Mino *et al.*, 1983). On the other hand, the Fe(III) and Co(III) complexes showed similar CD patterns in the ultraviolet region, suggesting structural resemblance of both metal chelates (Sugiura *et al.*, 1981).

The electrochemical study of the Fe(III) complex by cyclic voltammetry shows a quasi-reversible electron redox wave with $E_{1/2}$ = 102 mV (vs. the normal hydrogen electrode, NHE) (Sugiura *et al.*, 1981).

Besides, the Fe(II)/mugineic acid complex was investigated by ^{57}Fe-Mössbauer spectroscopy, at 110 K. The observed hyperfine parameters (ΔE_Q = 2.88 and δ_{Fe} = 1.16 mm/sec) are characteristic for high-spin Fe(II) compounds (Sugiura *et al.*, 1981; Sugiura and Nomoto, 1984).

In order to investigate the behavior of mugineic acid and its complexes in solution, high resolution NMR studies were performed with the free acid and its diamagnetic Zn(II) and Co(III) complexes, as proton nuclear magnetic resonance has proven to be of limited value for the study of the Cu(II) and Fe(III) complexes, due of their paramagnetic effects. The conformational analysis, using chemical shifts and vicinal coupling constants, confirmed the hexadentate coordination of the investigated complexes and shows important conformational changes of the acid after coordination. Besides, the structural conformation of the Co(III) complex in aqueous solution corresponds well to that of its commented crystal structure, determined by X-ray diffractometry (Iwashita *et al.*, 1983).

Recently, a theoretical study of metal complexes of mugineic acid, using density functional theory methods combined with a polarizable continuum model was performed (Kato *et al.*, 2011). Calculations were performed for Ca(II), Mn(II), Fe(II), Fe(III), Zn(II) and Cu(II) complexes. The computational results show that the metal coordination structure of the respective Fe(II) and Fe(II) complexes is somewhat different, a conclusion which may be of interest in relation to the mechanism of iron release in cells. On the other hand, it has also been found that the explicit inclusion of mugineic acid/water interactions is a key factor for obtaining reliable computational results, suggesting the importance of hydrogen bonds in the stabilization and molecular conformation of the complexes.

3.4. Other Phytosiderophores

In comparison to mugineic acid the information so far accumulated for the other phytosiderophores remains relatively scarce. Some basic information concerning avenic acid, 2'-deoxymugineic acid 3-hydroxymugineic acid and distichonic acid, including general spectral characteristics, were summarized by Sugiura and Nomoto (1984).

Most recently, the potentiality of different chromatographic tools combined with other modern analytical techniques, for the separation, identification and characterization of phytosiderophore mixtures has been clearly demonstrated (Fan *et al.*, 1997; Xuan *et al.*, 2006; Ueno *et al.*, 2007).

3.4.1. Metal complexes of other phytosiderophores

During one of the above mentioned studies (Xuan *et al.*, 2006) Fe^{II}, Fe^{III}, Zn^{II}, Cu^{II} and Ni^{II} complexes of 2'-deoxymugineic acid were isolated from plant samples and the stability of these complexes was also tested.

In another study, the capacity to solubilize Fe^{III} by the novel phytosiderophores 3-hydroxy-2'-deoxymugineic acid and 2'-hydroxyavenic acid was assayed, showing a similar behavior as 2'-deoxymugineic acid and avenic acid, and suggesting the formation of 1:1 Fe : phytosiderophores complexes. Thus, it is clear that hydroxylation of phytosiderophores does not change the capacity to solubilize Fe^{III} from insoluble iron deposits (Ueno *et al.*, 2007).

In a previous similar study, it was shown that 2'-deoxymugineic acid and its hydroxylated derivatives, mugineic acid and 3-epi-2'-deoxymugineic acid also posses analogous Fe^{III}-solubilizing potential and that the introduction of OH-groups into the phytosiderophore skeleton increases Fe(III)/chelate stability in acid environments (von Wirén, Khodr and Hider, 2000).

Acknowledgment

The author is an Emeritus Professor from the Universidad Nacional de La Plata and the continuous support from this University is gratefully acknowledged.

REFERENCES

Ackrill P. and J.P. Day. 1985. Desferrioxamine in the treatment of aluminium overload, *Clin. Nephrol.* **24:** 594-597.

Aguado-Santacruz, G.A.; B. Moreno-Gómez, B. Jiménez-Francisco, E. García-Moya and R.E. Preciado-Ortiz. 2012. Impacto de los sideróforos microbianos y fitosideróforos en la asimilación de hierro por las plantas: Una síntesis, *Rev. Fitotec. Mex.* **35:** 9-21.

Baran, E.J. 1995. *Química Bioinorgánica*, McGraw-Hill Interamericana de España S.A., Madrid.

Baran, E.J. 2010. Chelation therapies: A chemical and biochemical perspective, *Curr. Med. Chem.* **17:** 3658-3672.

Baran, E.J. 2012a. Phytochelatins: Natural Chelating Agents Involved in Plant Protection. In: *Advances in Plant Physiology* (Ed. H. Hemantaranjan), Vol.13, pp. 389-414, Scientific Publishers, Jodhpur.

Baran, E.J. 2012b. Quelatoterapias: Avances Recientes y Perspectivas. In: *Aplicaciones de los Compuestos Metálicos en Medicina* (Eds. D. Gambino, V. Moreno and M. Navarro), pp. 383-445, EAE-LAP Lambert Academic Publishing GmbH & Co KG, Saarbrücken.

Beneš, I., K. Schreiber, H. Ripperger and A. Kircheiss. 1983. Metal complex formation by nicotianamine, a possible phytosiderophore, *Experientia* **39:** 261-262.

Bertini, I., H.B. Gray, S.J. Lippard and J.S. Valentine. 1994. *Bioinorganic Chemistry*, University Science Books, Mill Valley, CA.

Budzikiewicz, H. 2010. Microbial siderophores, *Progr. Chem. Org. Nat. Prod.* **92:** 1-75.

Butler, A. 2005. Marine siderophores and microbial iron mobilization, *BioMetals* **18:** 369-374.

Callahan, D.L., A.J.M. Baker, S.D. Kolev and A.G. Wedd 2006. Metal ion ligands in hyperaccumulating plants, *J. Biol. Inorg. Chem.* **11:** 2-12.

Chatterjee, P.B. and D.C. Crans 2012. Solid-to-solid oxidation of a vanadium(IV) to vanadium(V) compound: chemistry of a sulfur-containing siderophore, *Inorg. Chem.* **51:** 9144-9146.

Colangelo, E.P. and M.L. Guerinot 2006. Put the metal to the petal: metal uptake and transport throughout plants, *Curr. Op. Plant Biol.* **9:** 322-330.

Cortese, M.S., A. Paszczynski, T.A. Lewis, J.L. Sebat, V. Borek and R.L. Crawford. 2002. Metal chelating properties of pyridine-2,6-bis(thiocarboxylic acid) produced by *Pseudomonas* spp. and the biological activity of the formed complexes, *BioMetals* **15:** 103-120.

Curie, C., G. Cassini, D. Couch, F. Divol, K. Higuchi, M. Le Jean, J. Misson, A. Schikora, P. Czernic and S. Mari 2009. Metal movement within the plant: contribution of nicotianamine and yellow stripe 1-like transporters, *Ann. Bot.* **103:** 1-11.

Dancis, A., R.D. Klausner, A.G. Hinnebusch and J.G. Barriocanal 1990. Genetic evidence that ferric reductase is required for iron uptake in *Saccharomyces cerevisiae*, *Mol. Cell Biol.* **10:** 2294-2301.

Devez, A., E. Achterberg and M. Gledhill 2009. Metal ion-binding properties of phytochelatins and related ligands. In: *Metal Ions in Life Sciences* (eds. A. Sigel, H. Sigel and R.K.O. Sigel), Vol. 5, pp. 441-481, Royal Society of Chemistry, Cambridge.

Dreyfus, C., D. Lemaire, S. Mari, D. Pignol and P. Arnoux 2009. Crystallograhic snapshots of iterative substrate translocations during nicotianamine synthesis in archaea, *Proc. Natl. Acad. Sci. U.S.A.* **106:** 16180-16184.

Eide, D, M. Broderius, J. Fett and M.L. Guerinot 1996. A novel iron-regulated metal transporter from plants identified by functional expression in yeast, *Proc. Natl. Acad. Sci. U.S.A.* **93:** 5624-5628.

Fan, T.W.M., A.N. Lane, J. Pedler, D. Crowley and R. H. Higashi 1997. Comprehensive analysis of organic ligands in whole root exudates using nuclear magnetic resonance and gas chromatography-mass spectrometry, *Anal. Biochem.* **251:** 57-68.

Fraústo da Silva, J.J.R. and R.J.P. Williams 1991. *The biological chemistry of the Elements*, Clarendon Press, Oxford.

Fushiva, S., S. Nakatsuyama, Y. Sato and S. Nozoe 1981. Synthesis of nicotianamine and a related compound, derivatives of azetidine-2-carboxylic acid, *Heterocycles* **15:** 819-822.

Gaxiola, R.A., M.G. Palmgren and K. Schumacher 2007. Plant proton pumps, *FEBS Lett.* **581:** 2204-2214.

Grant, C.A., W.T. Buckley, L.D. Bailey and F. Selles 1998. Cadmium accumulation in crops, *Can. J. Plant Sci.* **78:** 1-17.

Grotz, N. and M.L. Guerinot 2006. Molecular aspects of Cu, Fe and Zn homeostasis in plants, *Biochim. Biophys. Acta* **1763:** 595-608.

Grusak, M.A., J.N. Pearson and E. Marentes. 1999. The physiology of micronutrient homeostasis in field crops, *Field Crops Res.* **60:** 41-56.

Guerinot, M.L. and D. Eide. 1999. Zeroing in on zinc uptake in yeast and plants, *Curr. Op. Plant Biol.* **2:** 244-249.

Guerinot, M.L., E.J. Meidl and O. Plessner 1990. Citrate as a siderophore in *Bradyrhizobium japonicum*, *J. Bacteriol.* **172:** 3298-3303.

Guerinot, M.L. and Y. Yi 1994. Iron: Nutritious, noxious, and not readily available, *Plant Physiol.* **104:** 815-820.

Himelblau, E. and R.M. Amasino 2000. Delivering copper within plant cells, *Curr. Op. Plant Biol.* **3:** 205-210.

Holleman, A.F. and E. Wiberg 1958. *Lehrbuch der Anorganischen Chemie*, Walter de Gruyter & Co., Berlin.

Hughes, M.N. 1981. The Inorganic Chemistry of Biological Processes, J. Wiley, Chichester.

Irtelli, B., W.A. Petrucci and F. Navari-Izzo 2009. Nicotianamine and histidine/proline are, respectively the most important copper chelators in xylem sap of *Brassica carinata* under conditions of copper deficiency and excess, *J. Exper. Botany* **60:** 269-277.

Iwashita T., Y. Mino, H. Naoki, Y. Sugiura and K. Nomoto. 1983. High-resolution proton nuclear magnetic resonance analysis of solution structures and conformational properties of mugineic acid and its metal complexes, *Biochemistry* **22:** 4842-4845.

Kato, M., S. Izuka, T. Fujihara, A. Nagasawa, S. Kawai, T. Tanaka and T. Takayanagi. 2011. Electronic structure calculation study of metal complexes with a phytosiderophore mugineic acid, *Inorg. Chim. Acta* **370:** 304-310.

Kawai, S., S. Takagi and Y. Sato. 1988. Mugineic acid-family phytosiderophores in root-secretions of barley, corn and sorghum varieties, *J. Plant Nutrit.* **11:** 633-642.

Kim, S., M. Takahashi, K. Higuchi, K. Tsunoda, H. Nakanishi, E. Yoshimura, S. Mori and N.K. Nishizawa. 2005. Increased nicotianamine biosynthesis confers enhanced tolerance of high levels of metals, in particular nickel, to plants, *Plant Cell Physiol.* **46:** 1809-1818.

Liu, Z.D. and R.C. Hider. 2002. Design of iron chelators with therapeutic applications, *Coord. Chem. Rev*. **232:** 151-171.

Ma, J.F. and K. Nomoto. 1993a. Two related biosynthetic pathways of mugineic acids in graminaceous plants, *Plant Physiol.* **102:** 373-378.

Ma, J.F. and K. Nomoto. 1993b, Inhibition of mugineic acid-ferric complex uptake in barley by copper, zinc and cobalt, *Phys. Plant.* **89:** 331-334.

Ma, J.F. and K. Nomoto. 1996. Effective regulation of iron acquisition in graminaceous plants. The role of mugineic acids as phytosiderophores, *Phys. Plant.* **97:** 609-617.

Ma, J.F., T. Shinada, T. Matsuda and K. Nomoto. 1995. Biosynthesis of phytosiderophores, mugineic acids, associated with methionine cycling, *J. Biol. Chem*. **270:** 16549-16554.

Marschner, H., V. Römheld and M. Kissel. 1986. Different strategies in higher plants in mobilization and uptake of iron, *J. Plant Nutrit*. **9:** 695-713.

Martell A.E. and R.M. Smith (1974). *Critical Stability Constants*, Plenum Press, New York.

Matsuura, M., Y. Hamada and T. Shioiri. 1994. Total synthesis of 2'-deoximugineic acid and nicotianamine, *Tetrahedron* **50:** 9467-9470.

Merckx; R., J.H. van Ginkel, J. Sinnnaeve and A. Cremers. 1986. Plant induced changes in the rhizosphere of maize and wheat. II. Complexation of cobalt, zinc and manganese in the rhizosphere of maize and wheat, *Plant Soil* **96:** 95-107.

Mino, Y., T. Ishida, N. Ota, M. Inoue, K. Nomoto, T. Takemoto, H. Tanaka and Y. Sugiura. 1983. Mugineic acid-iron(III) complex and its structurally analogous cobalt(III) complex: Characterization and implication for absorption and transport of iron in graminaceous plants, *J. Am. Chem. Soc.* **105:** 4671-4676.

Mino, Y., T. Ishida, N. Ota, M. Inoue, K. Nomoto, H. Yoshioka, T. Takemoto, Y. Sugiura and H. Tanaka. 1981. Molecular structure and spectroscopic properties of a copper(II) complex with mugineic acid, a novel amino acid from graminaceous plants, *Inorg. Chem.* **20:** 3440-3444.

Mori, S. 1998. Iron transport in graminaceous plants. In: *Metal Ions in Biological Systems* (eds. A. Sigel and H. Sigel), Vol. 35, pp. 215-238, Marcel Dekker, New York.

Mori, S. 1999. Iron acquisition by plants, *Curr. Op. Plant Biol*. **2:** 250-253.

Mori, S. and N. Nishizawa. 1987. Methionine as a dominant precursor of phytosiderophores in *Graminaceae* plants, *Plant Cell Physiol*. **28:** 1081-1092.

Morrisey, J. and M.L.Guerinot. 2009. Iron uptake and transport in plants: The good, the bad, and the ionome, *Chem. Rev.* **109:** 4553-4567.

Neilands, J.B. 1975. Microbial iron transport compounds (siderochromes). In: *Inorganic Biochemistry* (ed. G. L. Eichhorn), pp. 167-202, Elsevier, Amsterdam.

Nomoto, K., Y. Mino, T. Ishida, H. Yoshioka, N. Ota, M. Inouk, S. Takagi and T. Takemoto. 1981. X-ray crystal structure of the copper(II) complex of mugineic acid, a naturally occurring metal chelator of graminaceous plants, *J. Chem. Soc. Chem. Comm.* **1981:** 338-339.

Okujo, N., Y. Sakakibara, T. Yoshida and S. Yamamoto 1994. Structure of acinetoferrin, a new citrate-based dihydroxamate siderophore from *Acinetobacter haemolyticus*, *BioMetals* **7:** 170-176.

Palma Fett, J., K. LeVier and M.L. Guerinot 1998. Soil microorganisms and iron uptake by higher plants. In: *Metal Ions in Biological Systems* (eds. A. Sigel and H. Sigel), Vol. 35, pp. 187-214, Marcel Dekker, New York.

Pittman, J.K. 2005. Managing the manganese: molecular mechanisms of manganese transport and homeostasis, *New. Phytol.* **167:** 733-742.

Porterfield, W.W. 1993. *Inorganic Chemistry.A Unified Approach*, 2nd. Edit., Academic Press, San Diego.

Reichman S.M. and D.R. Parker. 2002. Revisiting the metal-binding chemistry of nicotianamine and 2'-deoxymugineic acid. Implications for iron nutrition in Strategy II plants, *Plant Physiol.* **129:** 1435-1438.

Rellán-Alvarez, R., J. Abadía and A. Alvarez-Fernández. 2008. Formation of metal-nicotianamine complexes as affected by pH, ligand exchange with citrate and metal exchange: a study by electrospray ionization time-of-flight mass spectrometry, *Rapid Commun. Mass Spectrometry* **22:** 1553-1562.

Rengel, Z. and V. Römheld 2000. Root exudation and Fe uptake and transport in wheat genotypes differing in tolerance to Zn deficiency, *Plant Soil* **222:** 25-34.

Roberts L.A., A.J. Pierson, Z. Panaviene and E.L. Walker 2004. Yellow stripe 1, expanded roles for the maize iron-phytosiderophores, *Plant Physiol.* **135:** 112-120.

Robinson, N.J., C.M. Procter, E.L. Connolly and M.L. Guerinot 1999. A ferric-chelate reductase for iron uptake from soils, *Nature* **397:** 694-697.

Römheld, V. 1987. Different strategies for iron acquisition in higher plants, *Physiol. Plant.* **70:** 231-234.

Römheld, V. 1991. The role pf phytosiderophores in acquisition of iron and other micronutrients in graminaceous species: An ecological approach, *Plant Soil* **30**: 127-134.

Römheld V. and H. Marschner 1986. Evidence for a specific uptake system for iron phytosiderophores in roots of grasses, *Plant Physiol.* **80:** 175-180.

Römheld, V. and H. Marschner 1990. Genotypical differences among graminaceous species in release of phytosiderophores, *Plant Soil* **123:** 147-153.

Römheld, V., C. Müller and H. Marschner 1984. Localization and capacity of proton pumps in roots of intact sunflower plants, *Plant Physiol.* **76:** 603-606.

Rosenzweig, AC. 2000. Metallochaperones: Bind and deliver, *Chem. Biol.* **9:** 673-677.

Rudolf, A., R. Becker, G. Scholz, Z. Procházka, J. Toman, T. Macek and V. Herout 1985. The occurrence of the amino acid nicotianamine in plants and microorganisms. A reinvestigation, *Biochem. Physiol. Pflanz.* **180:** 557-583.

Sandy, M. and A. Butler. 2009. Microbial iron acquisition: Marine and terrestrial siderophores, *Chem. Rev.* **109:** 4580-4595.

Schaaf, G., U. Ludewig, B.E. Erenoglu, S. Mori, T. Kitahara, N. von Wirén 2004. ZmYS1 functions as a proton-coupled symporter for phytosiderophore and nicotianamine-chelated metals, *J. Biol. Chem.* **279:** 9091-9096.

Schreiber, G. 1986. Identification and characterization of an endogenous cytometallophore of general distribution in plants, *Pure Appl. Chem.* **58:** 745-752.

Schuler, M. and P. Bauer 2011. Heavy metals need assistance: the contribution of nicotianamide to metal circulation throughout the plant and the *Arabidopsis* NAS gene family, *Front. Plant Sci.* **2:** article 69, 1-5.

Shenker, M., T.W.-M. Fan and D.E. Crowley 2001. Phytosiderophores influence on cadmium mobilization and uptake by wheat and barley plants, *J. Environ. Qual.* **30:** 2091-2098.

Shioiri, T., Y. Hamada and M. Matsuura 1995. Total synthesis of phytosiderophores, *Tetrahedron* **51:** 3939-3958.

Stephan, U.W., I. Schmidtke, V.W. Stephan and G. Scholz 1996. The nicotianamine molecule is made-to-measure for complexation of metal micronutrients in plants, *BioMetals* **9:** 84-90.

Sugiura, Y. and K. Nomoto 1984. Phytosiderophores. Structure and properties of mugineic acids and their metal complexes, *Struct. Bonding* **58:** 107-135.

Sugiura, Y., H. Tanaka, Y. Mino, H. Yoshioka, N. Ota, M. Inoue, K. Nomoto, H. Yoshioko and T. Takemoto 1981. Structure, properties and transport mechanism of iron(III) complex of mugineic acid, a possible phytosiderophore, *J. Am. Chem. Soc.* **103:** 6979-6982.

Takagi, S. 1976. Naturally occurring iron-chelating compounds in oat- and rice-root washings, *Soil Sci. Plant Nutr.* **12:** 423-433.

Takahashi, M., Y. Terada, I. Nakai, H. Nakanishi, E. Yoshimura, S. Mori and N.K. Nishizawa. 2003. Role of nicotianamine in the intracellular delivery of metals and plant reproductive development, *Plant Cell* **15:** 1263-1280.

Takemoto, T., K. Nomoto, S. Fushiya, R. Ouchi, G. Kusano, H. Ikino, S. Takagi, Y. Matsuura and M. Kakudo. 1978. Structure of mugeinic acid, a new amino acid possessing an iron chelating activity from roots washings of water-cultured *Hordeum vulgare* L., *Proc. Jpn. Acad.* **54B:** 469-473.

Treeby, M., H. Marschner and V. Römheld. 1989 Mobilization of iron and other micronutrient cations from a calcareous soil by plant-borne, microbial, and synthetic chelators, *Plant Soil*, **114:** 217-226.

Ueno, D., A.D. Rombola, T. Iwashita, K. Nomoto and J.F. Ma. 2007. Identification of two novel phytosiderophores secreted by perennial grasses, *New Phytol.* **174:** 304-310.

Vert, G. N. Grotz, F. Dedaldechamp, F. Gaymard, M.L. Guerinot, J.F. Briat and C. Curie. 2002. IRT1, an Arabidopsis transporter essential for iron uptake from the soil and plant growth, *Plant Cell* **14:** 1223-1233.

von Wirén, N., H. Khodr and R.C. Hider. 2000. Hydroxylated phytosiderophores species posses an enhanced chelate ability and affinity for iron(III), *Plant Physiol.* **124:** 1149-1158.

von Wirén, N., S. Klair, S. Bansal, J.F. Briat, H. Khodr, T. Shioiri, R.E. Leigh and R.C. Hider. 1999. Nicotianamine chelates both Fe(III) and Fe(II): Implications for metal transport in plants, *Plant Physiol.* **119:** 1107-1114.

von Wirén, N., H. Marschner and V. Römheld. 1996. Roots of iron-efficient maize also absorb phytosiderophore-chelated zinc, *Plant Physiol.* **111:** 1119-1125.

Vraspir, J.M. and A. Butler. 2009. Chemistry of marine ligands and siderophores, *Annu. Rev. Mar. Sci.* **1:** 43-63.

Williams, R.J.P. and J.J.R. Fraústo da Silva, J.J.R.1996. *The Natural Selection of the Chemical Elements*, Clarendon Press, Oxford.

Xuan, Y., E.B. Scheuermann, A.R. Meda, H. Hayen, N. Von Wirén and G. Weber. 2006. Separation and identification of phytosiderophores and their metal complexes in plants by zwitterionic hydrophylic interaction liquid chromatography coupled to electrospray ionization mass spectrometry, *J. Chromatogr. A* **1136:** 73-81.

Yehuda, Z., M. Shenker, V. Römheld, H. Marschner, Y. Hadar and Y. Chen. 1996. The role of ligand exchange in the uptake of iron from microbial siderophores by gramineous plants, *Plant Physiol.* **112:** 1273-1280.

Yoshimura, E., H. Kohdr, S. Mori and R.C. Hider. 2011. The binding of aluminum to mugineic acid and related compounds as studied by potentiometric titration, *Biometals* **24:** 723-727.

Zang, V. and R. van Eldik. 1990. Kinetics and mechanism of the autoxidation of iron(II) induced through chelation by ethylenediaminetetraacetate and related ligands; *Inorg. Chem.* **29:** 1705-1711.

Zhang, F.S., V. Römheld and H. Marschner. 1991. Diurnal rhythm of release of phytosiderophores and uptake rate of zinc in iron-deficient wheat, *Soil Sci. Plant Nutr.* **37**: 671-678.

2

UPDATE ON BORON - PHYSIOLOGICAL RESPONSES AND HOMEOSTASIS IN PLANTS

Nalini Pandey

Plant Nutrition and Stress Physiology Laboratory, Botany Department, University of Lucknow, Lucknow 220 607, India

INTRODUCTION

Boron belongs to Group III of the periodic table and is a metalloid with properties intermediate between metals and electronegative non-metals. Boron is an essential micronutrient for plant growth and metabolism. The essentiality of B for plants was first established by Agulhon (1910) and later by Warington (1923). The physiological and biochemical roles of boron in growth and development of plants have been extensively reviewed (Nable *et al.*, 1997; Bolanos *et al.*, 2004a; Camacho-Critobal *et al.*, 2008a; Herrera-Rodríguez, *et al.*, 2010). Boron is primarily essential in higher plants for maintaining the structural integrity of cell walls due to its role in cross-linking of cell wall rhamnogalacturonan II (RGII) and pectin (O'Neill *et al.*, 2004). This is because boric acid and borate are capable of forming complexes with compounds having two hydroxyl groups in *cis*-configuration and result in formation of borate esters with apiose residues of RG-II (Kobayashi *et al.*, 1996). Boron deficiency affects vegetative and reproductive growth of plants resulting in inhibition of cell expansion, death of meristem and reduced fertility. Boron deficiency harms numerous physiological processes, including sugar transport, cell wall synthesis, lignification, cell wall structure, carbohydrate metabolism, RNA metabolism, respiration, IAA metabolism, phenol metabolism and membrane transport (Parr and Loughman, 1983) and oxidative damage (Cakmak and Romheld, 1997) all leading to cell death. In recent years role of B in cell signaling has been proposed due to its interaction with transcription factors which could explain the different physiological processes affected by B deficiency (González-Fontes *et al.*, 2008).

The deficiency and toxicity of B is widespread in the agricultural soils worldwide and leads to heavy losses in crop yield. In soil solution B exists primarily as boric acid [$B(OH)_3$], which can be easily leached under high rainfall conditions (Shorrocks, 1997; Yan *et al.*, 2006) leading to deficiencies in plants. On the other hand under low rainfall B is not sufficiently leached and accumulates to levels which are toxic for plants as in arid and semiarid regions (Reid, 2007a). The problem is further compounded since the range of deficiency and toxicity levels are very narrow in plants and a concentration of B deficient for one plant may well be toxic for the other. Therefore controlling the B availability in soils for plant growth is important. To avoid deficiency and toxicity problems, it is important for plants to maintain their tissue B concentrations within an optimum range by regulating transport processes (Tanaka and Fujiwara, 2007). In the past decade understanding molecular mechanisms of B transport and identification of B transporters has suggested that plants sense and respond to the B conditions and regulate transporters to maintain B homeostasis and alleviate B deficiency and toxicity problems. The chapter discusses the recent developments in B research particularly related to physiological responses and B homeostasis.

2. BORON IN SOIL

Boron occurs naturally as calcium and sodium borates derived from minerals. Borates are found in the oceans, sedimentary rocks, coal, shale, and some soils. Borates are naturally released into the environment from the oceans, volcanic activity and other geothermal releases such as geothermal steam, and weathering of clay-rich sedimentary rocks. The average concentration of boron in rocks varies from 5 mg kg^{-1} in basalts to 100 mg kg^{-1} in shales. Sedimentary rocks generally have higher levels of B than igneous rocks. Low B content is found in soils derived from acid igneous soils, fresh water sedimentary deposits, and in coarse-textured soils low in organic matter. The average concentration of B in soil range from 2 to 10 mg Kg^{-1}. Boron availability in soil is influenced by soil texture, temperature and most important by soil pH. Boron occurs as undissociated boric acid ($B(OH)_3$) at pH below 7 and it dissociates to a borate anion ($B(OH)_4$) at alkaline pH. The undissociated boric acid is easily lost by leaching during high rainfall resulting in B deficiency in the plants growing in neutral or slightly acidic soils. In general, B becomes more available to plants with decreasing pH. In conditions of low rainfall B is not sufficiently leached from soil and may become toxic for plants (Reid 2007a). The other important sources for toxic B concentrations are irrigation water, waste from surface mining, fly ash and industrial chemicals (Nable *et al.*,1997; Kot, 2009).

Boron occurs in soil in four forms - water soluble, adsorbed, organically bound, and fixed in clay and mineral lattices. In coarse textured soils low in clay and low in organic matter B is highly mobile and subject to leaching losses. The water soluble B is available to the plants. The range between the deficient and toxic levels of B necessary for normal plant growth is very narrow and depends on the soil type, plant species, source of irrigation as well as environmental conditions. Shorrocks (1997) categorized water soluble B in soils to be deficient (<0.25 μg g^{-1}), inadequate (0.25 -0.5μg g^{-1}), sufficient (0.5 1-1 μg g^{-1}), high (1.1 -2 μg g^{-1}) and toxic (>2.0 μg g^{-1}).

3. BORON IN PLANTS

3.1 Boron Concentration in Plants

The range of deficient to toxic level of B is narrower than that for any other nutrient element and depend on different plant groups. Thus both deficient and excessive levels could be encountered in a crop during a single season. In many cases deficiency and sufficiency values overlap (Table 1). Plants differ in sensitivity to B deficiency. Most cruciferous, leguminous and solanaceous crops show higher susceptibility to B deficiency. Dicotyledonous plants were reported to have a higher B requirement and B contents than monocotyledons (Marshner, 1995). Tissue concentration more than 20 mg B Kg^{-1} dry weight in dicotyledons compared with 5-10 mg B Kg^{-1} in graminaceous monocotyledons growing on the same soil are often interpreted to indicate greater uptake ability in dicotyledons (Gupta, 1993). Boron levels in rice plant tissue are considered deficient if concentration is < 5 mg kg^{-1}, sufficient if concentration is about 6-15 mg kg^{-1} and toxic if concentration is > 30 mg kg^{-1}. Average B content in most plants is 20 mg kg^{-1} on dry weight basis. Boron is unevenly distributed within plants and highest levels are found in reproductive structures such as anthers, stigma and ovaries.

Table 1. Critical boron concentration (mg B kg^{-1}1 dry wt) or concentration ranges for diagnosis of B deficiency in plants

Species	Plant part	Critical B concentration range
Sunflower (*Helianthus annuus* L.)	Upper fully expanded leaves	10-13
Soybean (*Glycine max* L.)	Youngest open leaf	10-12
Oil seed rape (*Brassica napus* L.)	Youngest open leaf	10-14
Black gram (*Vigna mungo* L.)	Youngest open leaf	15-17
Green gram (*Vigna radiata* L.)	Youngest open leaf	16-18
Maize (*Zea may* L.)	Youngest emerging leaf	5-7

Based on their B requirement, plants could be divided into three groups: graminaceous plants, which have the lowest demand for B; the remaining monocots and most dicots with an intermediate requirement; and latex-forming plants, with the highest B requirement among plant species. Thus B requirements vary with plant type; in monocotyledons species, leaf B content ranges from 1 to 6 mg kg^{-1}; in most dicotyledons from 20 to 70 mg kg^{-1} and in dicotyledons with latex systems from 80 to 100 mg kg^{-1}. According to Bell (1997) some crop species such as sunflower, soybean, sugar beet, celery, apple and maize are considered to be relatively high boron demanding species (Benton, 2003). Toxicities have been reported in many plant species when the tissue concentration exceed 100 mg kg^{-1} B.

3.2 Boron Deficiency Symptoms

Generally, symptoms of B deficiency appear first in terminal shoots in the form of small, deformed leaves, borne on severely condensed branches. The growth of apical meristems is arrested and the shoot apex turns necrotic, giving rise to multiple axillary branches, which eventually turn necrotic giving the B deficient plants a bushy appearance. Stem of B deficient plants also turn brittle and show longitudinal splits of the cortex. The petiole and stem of B deficient plants show longitudinal and transverse splits or corky ridges (Bell, 1997). Some characteristic boron deficiency symptoms in plants are: 'Top sickness' of tobacco; 'Corky core' or 'internal cork' or 'drought spot' of apple; 'Water core' of turnip; 'Hard fruit' of Citrus; 'Yellows' of alfa alfa; 'Hen and chicken' of grapes; 'Heart rot' of sugarbeet; 'Stem crack' of celery; 'Hollow stem' of cauliflower and broccoli; 'Brown heart' of swade and 'Tipburn' of chinese cabbage (Gupta and Cutcliffe,1973).

In plants highly susceptible to B deficiency like the dicots sunflower, tomato, squash, alfalfa, inhibition of root growth and degeneration of meristematic regions appear quickly and simultaneously; in moderately susceptible dicots (pea, soybean, lupine), degeneration of growing points is delayed. However small grains and grasses (rye, oat, wheat, timothy) have normal vegetative growth and show boron deficiency symptoms only during formation of reproductive organs. Other than vascular plants, B is required by diatoms, some species of marine algal flagellates, and *Cyanobacteria* dependent on heterocysts for nitrogen fixation.

Sunflower is one of the most sensitive crops to low B supply, developing characteristic B deficiency symptoms on leaves, stems and reproductive parts (Asad *et al.,* 2003). These deficiency symptoms first become evident on the younger leaves, which develop a bronze colour and become hardened, malformed and necrotic, the stem becomes corky, the capitulum deformed and poor seed set results. Some of the effects of B deficiency in maize were appearance of white translucent streaks in the middle of the leaf lamina, reduced leaf area, failure of young emerging leaf to unroll, short and thick internodes and cessation of apical growth. The deficiency symptoms of B observed in periwinkle include reduction in growth, leaf area and chlorosis thickening and curling of leaves, terminal chlorosis, downward cupping of young emerging leaves, shortening of internodes, and death of growing tips of stem (Pandey and Archana, 2009).

4. BORON UPTAKE AND TRANSLOCATION

Heavy losses are observed in yield and quality of many crops that contain insufficient or toxic levels of B in soil. Thus the mechanisms that are involved in B uptake and distribution in plants can be critical to their productivity and important advances in the molecular aspects that control these processes have been reported in the recent years. It is accepted that plants take up B from soil in form of boric acid. The exact mechanism of B uptake by plant roots suggests that both active and passive uptake may occur. Dannel *et al.*(1997) and Pfeffer *et al.* (1999) presented evidence to suggest that B uptake involves both passive and active mechanisms, the predominant mechanism being determined by the availability of B. Dannel *et al.* (1997, 2000) proposed that B uptake is active in sunflower roots when the concentration in the external solution is low (1 μM), and that passive uptake dominates when the concentration is 100 μM. They observed a significantly higher concentration of B in the xylem exudate compared to the external solution when plants were provided with 1 μM B. This did not occur in plants grown with 100 μM B. Thus when plants are adequately supplied with B, its uptake is passive, but when B is in short supply, it is taken up by an active, metabolic process. It has been shown that B deficiency induces energy dependent, high affinity transport system which contributes to a B concentration mechanism in roots (Pfeffer *et al.,* 1999; Dannel *et al.*, 2000). Thus depending on B availability, boric acid uptake by roots is performed either by passive diffusion across lipid bilayer, facilitated transport by major intrinsic protein (MIP) channel, or actively by an energy dependent high-affinity transport system induced in response to low B supply, which is mediated via BOR transporters (Dannel *et al.,* 2000 and 2002; Nakagawa *et al.*, 2007).

4.1 Passive Uptake

Uptake of B by roots under conditions of adequate or excessive B availability, is mediated through a passive process that involves mostly B diffusion across lipid bilayer (Brown *et al.*, 2002; Tanaka and Fujiwara 2007). The evidence for passive transport came from studies which demonstrated that B uptake is not affected by metabolic inhibitors (DNP, KCN), and temperature (ranging from 2 to 47 °C) does not affect uptake in a manner consistent with carrier mediated uptake (Brown and Hu, 1994).The lipid permeability coefficient for boric acid supports the idea that B can cross membranes by a passive process to satisfy plant B requirements (Brown and Shelp, 1997; Dordas and Brown, 2000).

The uptake by a passive mechanism may involve both B diffusion across lipid bilayer and facilitated permeation of boric acid via the MIPs (major intrinsic proteins) of which aquaporins are a sub-group. Plants contain MIPs with four sub families small basic intrinsic proteins (SIPs), PIP (which are localized in the plasma membrane), TIP (which are found in the tonoplast) and the nodulin 26 (NOD26)–like intrinsic proteins (NIPs) (Tyerman *et al.*, 2002; Wallace *et al.,* 2006). The NIP subfamily is specific for plants and nine NIP genes have been identified in Arabidopsis. NIP5;1 a member of aquaporins in *Arabidopsis* is a plasma membrane boric acid transporter expressed in the cells of root epidermis, cortex, and endodoermis. In B deficient conditions expression of the NIP5;1 transcript has been shown to be transcriptionally upregulated. NIP5;1 acts as a major boric acid channel involved in B uptake from the root surface under conditions of B limitation (Takano *et al.*, 2006). In rice, OsNIP3;1 shows a close homolog sequence to AtNIP5;1 and it has also been identified as a boric acid channel required for efficient growth under B-deficient conditions (Hanaoka and Fujiwara, 2007). Tanaka *et al.*(2008a) showed that NIP6;1 the most similar gene to NIP5;1, encodes a B channel which functions in B transport in developing shoot tissues of Arabidopsis. NIP6;1 transcript accumulation is elevated in response to B deprivation in shoots but not in roots. They suggested that NIP6;1 functions in xylem–phloem transfer for preferential distribution of B into young growing tissues. Recently An *et al.*(2012) cloned an NIP5-like gene (named as CiNIP5) from trifoliate orange. Analysis of gene expression indicated that CiNIP5 expressed mainly in the roots of citrus seedlings, and it increased under B-deficiency but was suppressed under B toxicity. Additionally, the responses of CiNIP5 expression to B deficiency was increased more in B deficiency tolerant Carrizo citrange when compared to the less tolerant Fragrant citrus and a possible role of CiNIP5 was discussed in the mechanism of B-uptake efficiency of different rootstocks under B-deficiency.

4.2 Active Uptake

The active absorption of B is supported by the fact that B uptake was inhibited by both metabolic inhibitors and cold treatment in roots (Pfeffer *et al.*, 1999; Dannel *et al.*, 2000). The first B transporter involved in the process of xylem loading was identified as BOR1 in *A. thaliana*, which accumulated in the plasma membrane of pericycle cells under low B conditions (Takano *et al.*, 2001). Takano *et al.* (2001) demonstrated that the wild-type *A. thaliana* distributed B preferentially to young leaves, while such preferential transport was not evident in the *bor1-1* mutant. BOR1 is expressed in shoots, but it is not known how BOR1, an efflux transporter of B, regulates B distribution among leaves. In addition, *bor1-1* mutant plants showed a reduced transport of B to the shoot under B deficiency when compared to wild type *Arabidopsis* plants (Takano *et al.*, 2002). All these findings demonstrate that BOR1 is an efflux-type B transporter for xylem loading under low B limitation (Takano *et al.*, 2002). It is possible that BOR1 may be involved in directing B from the xylem to phloem for preferential delivery to young leaves. Uptake studies indicated that xylem loading is the key step for B accumulation in shoots with a low external B supply and that the *bor1-1* mutant is defective in this process (Takano *et al.,* 2002). BOR1 is an efflux-type boron transporter for xylem loading and is essential for protecting shoots from boron deficiency. The identification of BOR1 (boron exporters) confirmed that B transport was a process mediated not only by passive diffusion but also by transporters whose activity was regulated in response to B conditions.

Other BOR1-like genes have been identified in *Eucalyptus* (Domingues *et al.*, 2005) and rice (Nakagawa *et al.*, 2007). Nakagawa *et al.* (2007) described a B transporter, Os BOR1, in rice (*Oryza sativa* L.) which is an orthologue of *A. thaliana* BOR1 and plays a critical role in B acquisition by roots and translocation of B into shoots. Os BOR1 is a plasma membrane-localized efflux transporter of B and is required for normal growth of rice plants under conditions of limited B supply. Recently Pérez-Castro *et al.* (2012) characterized an efflux B transporter, VvBOR1, a homolog to AtBOR1. Transcriptional analyses during grapevine reproductive development show that the VvBOR1 gene is preferentially expressed in flowers at anthesis and a direct correlation between the expression pattern and B content in grapes was established, suggesting the involvement of this transporter in B accumulation in grapevine berries.

4.3 Boron Translocation

Boron translocation from roots to shoot, as also the intracellular compartmentalization of B is reported to be influenced by the level of B supply (Dannel *et al.*, 1998; Pfeffer *et al.*, 1997, 2001; Noguchi *et al.*, 2000, Li *et al.,* 2001). Retranslocation of B to sites that do not lose water rapidly (inflorescence, fruits, etc) involves phloem (Shelp *et al.*, 1995). Boron can also be transported via phloem to both reproductive and vegetative tissues (Shelp *et al.*, 1995; Matoh and Ochiai, 2005), although this capacity varies among species (Brown and Shelp, 1997). Huang *et al.*(2008) demonstrated that B is transported from mature leaves into actively growing reproductive organs via phloem. Evidence suggests that the principal factor that confers phloem B mobility to a plant species is the synthesis of sugar alcohols and the subsequent transport of the B-sugar alcohol complex in the phloem to sink tissues (Hu *et al.*, 1997). The capacity to use B present in mature tissues is now known to enhance species tolerance to transient B deficiency in the growth medium (Brown and Hu, 1996).

The mechanism that has been suggested to mediate phloem transport of B involves the formation of B-diol complexes as transport molecules (Brown and Hu, 1996; Hu *et al.*, 1997). B can readily bind to *cis*-hydroxyl groups of sugar alcohols (mannitol and sorbitol), which allow B to be transported through phloem. In addition, it has been observed that tobacco transgenic plants with enhanced sorbitol levels had higher capacity to transport B by phloem and increased tolerance to B deficiency (Bellaloui *et al.* 1999; Brown *et al.*, 1999). To demonstrate the critical role of sugar alcohols in B transport and to determine if the capacity for within-plant B mobility could be transferred to species in which B is normally immobile, Brown *et al.* (1999) engineered tobacco (*Nicotiana tabacum*) with the gene for sorbitol production, S6PDH, and determined that sorbitol production enhanced B mobility and tolerance of tobacco to B stress. However, B transport via phloem, especially to young tissues, also occurs in plants that are not able to produce these types of carbohydrates (Stangoulis *et al.*, 2001a; Takano *et al.*, 2001; Matoh and Ochiai, 2005).

5. BORON HOMEOSTASIS

Boron sufficient plants retain a larger proportion of B in roots than B deficient plants. Short term deprivation of B causes more marked decrease in B concentration in shoots than in roots (Dannel *et al.*, 1998; Li *et al.*, 2001). Pfeffer *et al.* (2001) attributed this to possible induction of high affinity B uptake mechanism involved in B transport from free space into the inside of the cell. While a relatively large proportion of the total cell B has been shown to be localized to the cell wall as water insoluble residue (Hu and Brown, 1994; Pfeffer *et al.*, 2001), its compartmentation is reported to be determined by the level of B supply, the plant organ and the genotype (Dannel *et al.*, 2002). Cultivars of the same species can differ in B uptake by 6-fold even when grown under identical environmental conditions. Probably the mechanism of differential uptake could be due to differential membrane permeability of boric acid between different cultivars and differences in the expression of channel proteins that contribute to cultivar differences in B uptake (Dordas and Brown, 2000).

Nakagawa *et al.* (2007) reported that disruption of Os BOR1 reduced B uptake and xylem loading of B. Os *BOR1* was detected in the roots of wild-type plants under low B conditions, but not under normal conditions, suggesting regulation of protein accumulation in response to B nutrition. They demonstrated that Os BOR1 is required both for efficient B uptake and for xylem loading of B. When *AT* BOR1 was over-expressed in *A. thaliana* plants, shoot growth and fertility were increased compared with wild-type plants when grown under low B conditions (Miwa *et al.*, 2006). Elevated levels of BOR1 presumably increased translocation of B from roots to shoots. This elevated B transport to shoots is likely to be beneficial for plants with low B supply. It has been observed that over-expression of BOR1 enhanced shoot growth but not root growth under low B supply. However enhanced expression of NIP5;1 improved root elongation in low B environments. Kato *et al.* (2009) demonstrated that the combination of enhanced expression of BOR1 and NIP5;1 enabled *A. thaliana* plants to become remarkably tolerant to B limitation in terms of the growth of both shoots and roots. From these experiments, it could be speculated that a limiting process for root and shoot growth under low B conditions is B uptake by NIP5;1 and xylem loading by BOR1 in *A. thaliana* wild-type plants. Under B-limited conditions, the combination of NIP5;1 for uptake and BOR1 for efflux into xylem is likely to be key to the efficient transcellular transport of B from the soil solution to the xylem (Fig. 1). BOR1 probably generates a concentration gradient between root cells and the medium which is essential for NIP5;1 to facilitate B uptake into root cells. It has also been demonstrated that homologues of *A. thaliana* BOR1 are involved in tolerance of high B in other plant species.

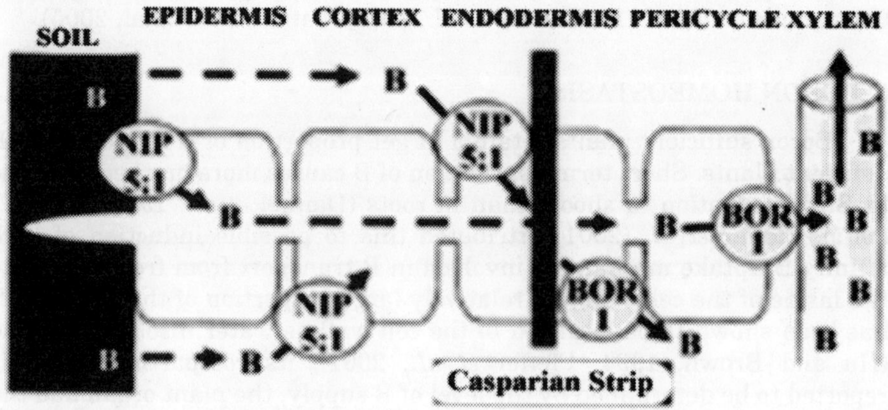

B uptake facilitated by NIP5;1, a boric acid channel; B concentration to xylem by BOR1, a B exporter

Figure. 1. A schematic model of B transport in *A. thaliana* roots under B limitation. Under B limitation, NIP5;1 increases the permeability of boric acid to cell membranes, and facilitates influx of B into root cells from the soil. BOR1exports B out of the cells toward the xylem against the concentration gradient. Under high levels of B supply, expression of both NIP5;1 and BOR1 is decreased by transcriptional and post-transcriptional regulation, respectively. The downregulation of NIP5;1 and BOR1 might be beneficial for avoidance of overloading of high concentrations of B to shoots. After Miwa and Fujiwara, (2010)

Hayes and Reid (2004) observed that the high B-tolerant barley (*Hordeum vulgare* L.) cultivar, Sahara, maintained low B concentrations in roots by active B efflux, resulting in decreased accumulation of B in shoots and elevated tolerance of high B through anion (borate) exchange or an anion channel. They showed that B-tolerant cultivar Sahara, was able to maintain half the concentrations of root B lower than the B-sensitive cultivar, Schooner. This resulted in xylem concentrations that were approximately 64% lower and leaf concentrations 73% lower in the tolerant cultivar. In Schooner, this concentration was similar to the external medium, whereas in Sahara, the root concentration was maintained at a lower concentration. For this to occur, B must be actively extruded from the root in Sahara, and this is presumed to be the basis for B tolerance in barley.

As discussed above, BOR1 is an efflux-type borate transporter required for the transport of B from roots to shoots under low B supply (Takano *et al.*, 2002). Accumulation of BOR1 protein decreases in response to a high B concentration in the medium. Upon exposure to high B, BOR1 is incorporated into endosomes and transported to the vacuole for degradation. This suggest that BOR1 is not involved in B tolerance in *Arabidopsis* (Takano *et al.*, 2005). Miwa *et al.* (2007) found that the expression of At*BOR4*, an *Arabidopsis* paralog of *BOR1*, generates plants that are tolerant to toxic B levels and improved plant growth. BOR4 is a polarly localized efflux type borate exporter that enhances B efflux from roots, is not degraded at post translational level and its over production was positively correlated with B tolerance.

Sutton *et al.* (2007) performed a quantitative trait locus (QTL) analysis of two barley cultivars differing in their tolerance of high B conditions, and found that the mapped region contained a BOR1 homologue. They named this gene *Bot1* and demonstrated that it conferred tolerance of high B in *S. cerevisiae,* as was reported for At-BOR1. Sutton *et al.* (2007) suggested that increased expression of BOT1 exporter in roots leads to reduction in B concentrations in roots. This cultivar contains 3.8 times more *Bot1*gene copies than the B-intolerant variety Clipper, and *Bot1* transcript levels in Sahara are about 160-fold and 18-fold higher in roots and leaf blades, respectively, when compared with Clipper (Sutton *et al.*, 2007). *Bot1* transcript levels are consistent with a lower net entry of B into the barley root as well in a higher B disposal from leaf guttation through hydathodes (Sutton *et al.*, 2007). Therefore, the ability of Sahara cultivar to maintain lower shoot B concentration is due to a mechanism of active B efflux from the root and *Bot1* encodes a functional B efflux transporter responsible for B tolerance.

Reid (2007b) cloned Hv-*BOR2* and Ta-*BOR2* from barley and wheat, respectively which were similar to At*BOR1* and provided tolerance of high B. Furthermore, Reid and Fitzpatrick(2009) noticed that necrosis in leaves was observed at higher leaf B concentrations in the tolerant cultivars in barley and wheat. It was shown that B concentrations in leaf protoplasts were lower in the tolerant cultivar, suggesting different partitioning of B within the shoot. It is likely that B transporters export toxic B out of cells into the apoplast and alleviate B toxicity symptoms in the cytoplasm of leaf cells. The tolerance of high B in plants is due to B exclusion and expression of some B transporters regulated responses. Boron exporters encoded by BOR1 homologues not only reduce B uptake by roots as a primary B tolerance mechanism, but also redistribute toxic B in leaves to confer further tolerance of high B.

Thus plants sense internal and external B conditions and regulate B transport by modulating the expression and/or accumulation of these transporters (Takano *et al.*, 2008; Miwa and Fujikura, 2010). The expression of genes encoding these transporters is finely regulated in response to B availability in the environment to ensure tissue B homeostasis. Furthermore, plants tolerant to stress produced by low B or high B in the environment can be generated through altered expression of these transporters. Borate exporter that enhances B efflux from roots would be important for the improvement of crop productivity in soils containing excess B, which are distributed in arid areas of the world.

6. PHYSIOLOGICAL ROLES OF BORON

Several physiological and biochemical roles of B in growth and development of plants have been characterized Dugger (1983), Parr and Loughman (1983), Römheld and Marschner (1991), Marschner (1995), Blevins and. Lukaszewski(1998), Camacho-Cristóbal *et al.* (2008) and Herrera-Rodríguez *et al.* (2010). Parr and Loughman (1983), reported varied physiological impairments as a result of B deficiency such as sugar transport; cell wall synthesis; lignification; cell wall structure; carbohydrate metabolism; RNA metabolism; respiration; indole acetic acid (IAA) metabolism; phenol metabolism, and membrane integrity. In addition, it has been reported that B deficiency may impair ascorbate metabolism (Lukaszewski and Blevins, 1996), induce oxygen activation (Marschner, 1995; Cakmak and Romheld, 1997) and affect nitrogen metabolism (Camacho-Cristóbal and González-Fontes 2007). The formation of complexes between B and compounds having *cis*-hydroxly groups (diols) such as pectic substances in cell walls, glycoproteins or glycolipids in membranes, and *o*-diphenols lead to most morphological and physiological changes caused by B deficiency. The well described roles of B are its involvement in (a) cell wall intactness and synthesis, possibly by formation of B-pectin complexes (Hu *et al.*, 1996; Loomis and Durst, 1991, 1992) and (b) plasma membrane integrity (Blevins and. Lukaszewski, 1998).

6.1 Cell Wall Stabilization

Plants contain B both in a water-soluble and insoluble forms of which the soluble form fluctuates with the quantity of B supplied, while insoluble B does not. The appearance of B deficiency symptoms coincides with the decrease of water-insoluble B which is the functional form while the soluble B represents the surplus. A close relationship between the primary cell walls and B levels revealed that 90% of the functional B was found to be bound to the cell wall fraction (Hu and Brown, 1994). Thus the initial symptoms of B deficiency in plants include abnormalities in cell wall and middle lamella organization and point to a relationship between B and primary wall pectic polysaccharides (Brown and Hu, 1997; Blevins and Lukaszewski, 1998; Brown *et al.*, 2002). It was observed that plants of poaceae (in which primary walls contain quantitatively lower amounts of pectin), have a much lower B requirement than the dicotyledons and non-graminaceous monocotyledons (Hu and Brown, 1994; Kobayashi *et al.*, 1996). Matoh *et al.* (1996) observed that plants' B requirements are correlated with the amounts of pectin present in the primary wall and that borate bridging could explain many of the characteristics of B deficient and B toxic plants. Formation of cis-diol complexes with hydroxyl groups of cell wall carbohydrates and/or glycoproteins to form borate esters was proposed as a mechanism for crosslinking cell wall polymers (Fig. 2). This type of bonding could account for brittle leaves of B deficient plants, while plants grown with high B levels produce leaves that are plastic or elastic in their response to bending (Hu and Brown, 1994).

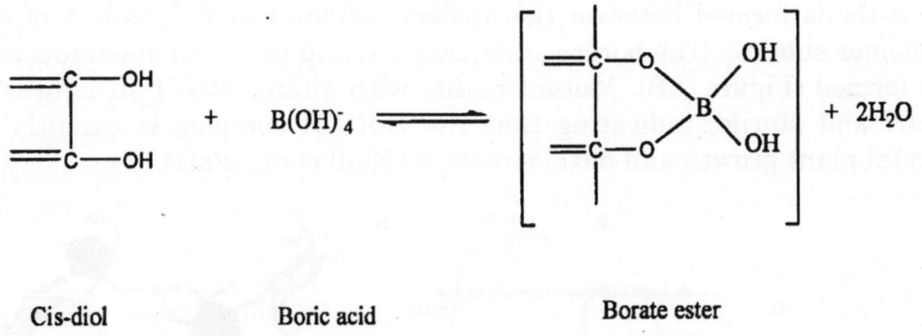

Cis-diol + $B(OH)_4^-$ ⇌ Borate ester + $2H_2O$

Fig. 2. Formation of *cis*-diol complex between boron and sugars.

Loomis and Durst (1992) observed that borate forms the most stable diesters with *cis*-diols on a furanoid ring and apiose, found in cell walls of a variety of plant species within both dicots and monocots can be the key sugar moiety for borate-cross linking cell wall polymers, while ribose, present in abundance in ribonucleotides, is likely involved in the chemistry of B toxicity. Another possible candidate for borate cell wall bridging is fucose. Diatoms and certain algae that require B contain fucose in walls (Matsunga *et al.*, 2004). Higher plants, other than the gramineae, have cell wall xyloglucans and rhamnogalacturonans with terminal fucose moieties (Kobayashi *et al.*, 1999). In addition, many of the glycoproteins secreted into the apoplast are also fucosylated.

A structural basis for the relationship between B and primary wall pectin was provided by the demonstration that most of the wall-bound B is present as a tetravalent 1:2 borate-diol ester that cross-links two molecules of rhamnogalacturonan II (RG-II) (Kobayashi *et al.*, 1996). Matoh *et al.*, (1996) found at least 98% of the insoluble B in tobacco cells bound to the cell wall as the borate diester with RG-II regions of pectic polysaccharides. Rhamnogalacturonan-II was first described in 1978 and has been found in the primary walls of dicotyledenous and monocotyledenous plants and gymnosperms (O'Neill *et al.*, 2004). RG-II is also present in the walls of some lower plants (ferns, horsetails, and lycopods). RG-II is a low molecular mass (5-10 Kd) pectic polysaccharide and contains eleven different glycosyl residues. The backbone of RG-II contains at least 8 1,4-linked α-D-Gal*p*A residues. Two structurally distinct disaccharides (C and D) are attached to C-3 of the backbone and two structurally distinct oligosaccharides (A and B) are attached to C-2 of the backbone (Fig. 3A). Using cultured sycamore cells, etiolated pea stems, and red wine as sources of RG-II-B complex, O'Neill *et al.* (1996) found that the borate ester was located on C-2 and C-3 of two of the four 30-linked apiosyl residues of dimeric RG-II and that borate cross-links two chains of RG-II to form a dimer in the primary walls of angiosperms and gymnosperms lycophytes, and pteridophytes. Ishii *et al.* (1999) also showed that the dimer contains a single borate diester cross link and that the ester is formed between the apiosyl residue in side chain A of each monomer subunit. The borate atom is chiral and thus two diasteroisomers are formed (Figure. 3B). Mutant plants with altered RG-II structures are dwarf and sterile, indicating that the B-RG-II complex is essential for normal plant growth and development (O'Neill *et al.,* 2001).

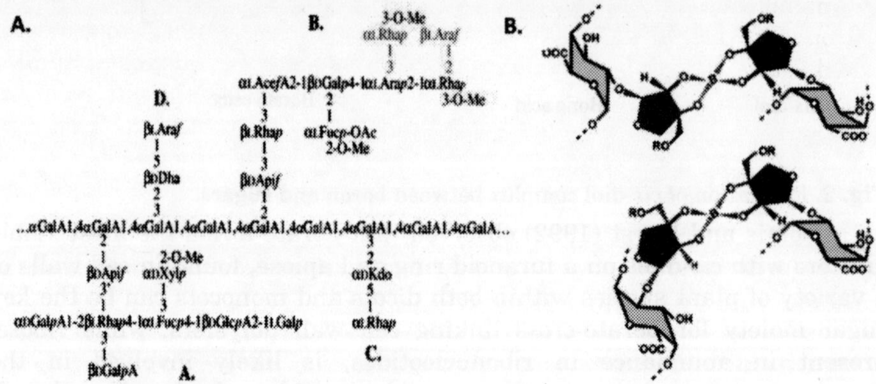

Fig. 3. The structure of rhamnogalacturonan (RG-II) with four side chain A, B, C and D found in the walls of plants (A). RG-II as a borate cross- linked dimer (B). Adapted from O'Neill *et al.* (2004).

6.2 Cell Membrane Permeability

Studies have demonstrated roles of B in membrane functions such as maintenance of membrane potentials and membrane permeability (Blevins and Lukaszewski, 1998; Wang *et al.*, 1999), functioning of enzymes and other proteins of the plasma membrane, transport processes across the membrane, and membrane integrity (Cakmak and Römheld, 1997; *Goldbach et al.*, 2001; *Brown et al.,* 2002; Goldbach and Wimmer, 2007). The role of B on membrane H^+ transport by generating a proton gradient across the plasmalemma was observed by Blaser-Grill *et al.* (1989), Ferrol and Donaire (1992) and Roldin *et al.* (1992). Decrease in proton extrusion by microsomal membranes isolated from the roots of B deficient plants was associated with decreased activity of H^+ ATPase (Ferrol *et al.*, 1993). A direct interaction between B and membranes was also indicated by the decrease in the activity of Fe-reductase (Goldbach *et al.*, 1990) and plasma membrane–bound oxidoreductase activity, observed within minutes of B deprivation (Barr *et al.*, 1993; Wimmer, 2000).

Pollard *et al.* (1977) observed restoration of ATPase activity and K uptake in B deficient roots after B resupply. Morré *et al.*(1986) identified the activity of auxin-sensitive plasma membrane NADH oxidase as ascorbate free radical (AFR) oxido-reductase and showed that by stimulating NADH oxidase, B could be involved in keeping ascorbate reduced at the cell wall/membrane interface. Boron supply to low B plants leads to hyper polarization of root membranes, and stimulation of ferricyanide-dependent HC release, ATPase activity, NADH oxidase activity, and ion transport (Goldbach *et al.*, 1990; Ferrol *et al.*, 1992, 1993; Barr *et al.*, 1993). Barr and Crane (1991) showed that addition of exogenous boric acid to low B cells caused an instant stimulation of the plasma membrane NADH oxidase. These changes associated with membrane function, may probably be the effect of B on physical properties of membrane proteins and could be due to B complexing with polyhydroxy groups of membrane components. Cakmak *et al.* (1995) also observed leakage of K, sucrose, phenolics, and amino acids in B deficient sunflower leaves and suggested a role for B in maintaining and stabilizing the integrity of plasma membranes by complexing membrane constituents by H-bonding or ester formation with glycolipids or glycoproteins which could keep enzymes or channels in an optimum conformation and anchored in the membrane.

A direct role of B in maintaining membrane structure is likely through *cis-diol* complexation with glycoproteins, which are structural constituents of the plasma membrane Effects of B deficiency, pointing to a structural role of B in membrane stabilization (Bolaños *et al.*, 2004a), are an altered permeability for potassium (K) and sugars (Parr and Loughman, 1983; Wang *et al.*, 1999), a damage of the peribacteroid membrane in nodules (*Bolanos* et al., 1994, 2004b) or a change in membrane bound Ca levels (Mühling *et al.*, 1998; Wimmer and Goldbach, 1999). Bolanos *et al.*(2004a) hypothesized that the primary role of B in biological systems is the stabilization of molecules with *cis*-diol groups.

Many workers showed that a direct role of B in maintaining membrane structure is likely through complexation with hydroxylated ligands such as glycoproteins and glycolipids which are structural constituents of the plasma membrane (Goldbach *et al.,* 2001; Brown *et al.*, 2002; Goldbach and Wimmer, 2007). Redondo-Nieto *et al.* (2007) showed that at least three potentially B-binding membrane glycoproteins were undetected in B-deficient pea nodules, which indicated that B and certain membrane glycoproteins are involved in membrane processes associated with general cell growth. Surface proteins attached to the membrane via a glycosyl-phosphatidylinositol anchor such as arabidogalactan proteins

(AGP) have also been suggested to be B-binding structures (Goldbach and Wimmer, 2007). A rapid decrease in the expression of several AGP genes in B deficient *Arabidopsis* roots was also observed by Camacho-Cristóbal *et al.* (2008b). Thus B might exert its action in membranes not only by stabilizing of membrane-molecules with *cis*-diol groups but also by regulating the expression of genes involved in membrane structure and function.

6.3 Oxidative Stress

As discussed above B is of crucial importance for the maintenance of structural integrity of plasma membranes. Membrane function could also be affected by the accumulation of oxidative free radicals (including ·OH) in cells, which is one of the indirect consequences of B deficiency in root and leaf cells (Cakmak and Römheld, 1997). There is possibility that the B may also protect plasma membranes against peroxidative damage by toxic O_2 species. Role of B in reactive oxygen species mediated cellular damage have been investigated in a number of studies (Garcia-Gonzales *et al.* 1988; Cakmak *et al.* 1995; Cakmak and Römheld, 1997; Pfeffer *et al.*, 1998; El- Shintinawy *et al.*, 1999; Liu and Yang, 2000). Oxidative stress in plants maybe imposed by redox imbalance resulting from B deficiency induced changes in cell wall structure. Enhanced generation of toxic O_2 species in B deficient tissues can be expected as a result of enhanced production of quinones. Ouinones produced by oxidation of phenolics can undergo one electron reduction forming a semiquinone radical. Subsequent reoxidation of the radical produces the superoxide radical (O_2·) and further toxic O_2 species such as hydrogen peroxide (H_2O_2) and hydroxyl radical (OH·) (Appel and Hirt, 2004. In B deficient tissues, generation of toxic O_2 species can also result directly from phenolics during their oxidation. For example, enzymatic oxidation of catechin by mushroom tyrosinase and potato phenolase produced both O_2^- and H_2O_2 (Jiang and Miles, 1993). Accumulation of phenolic compounds, particularly caffeic acid and quinones, which are highly reactive, leads to enhanced generation of the superoxide ions (O_2^-), which are known to cause peroxidative damage to cellular membranes.

The antioxidants ascorbic acid and SH-compounds are involved in detoxification of toxic O_2 species and dissipation of excessive light energy in chloroplasts (Foyer *et al.*, 1994). Decreased levels of ascorbic acid and -SH compounds enhance the generation of phenols/quinones in B deficient plants. Quinones formed by PPO can be reduced back to phenols by ascorbic acid and –SH compounds. The ascorbic acid and –SH compounds are consumed (oxidized) during the development of B deficiency and interfere with phenol oxidation and polymerization of quinones to brown pigments or with quinone-driven free radical generation. Lukaszewski and Blevins (1996), for the first time reported that B deficiency affects ascorbate metabolism. They observed that in B deficient squash roots ascorbate concentrations declined and that external supply of ascorbate overcome the root growth inhibitors induced by B deficiency. Cakmak and Römheld (1997) also observed in shoot tips and young leaves of sunflower that concentration of total ascorbic acid declined markedly with severity of B deficiency symptoms. Boron stimulates the activity of auxin sensitive NADH oxidase (semidehydroascorbate reductase), which catalyses the transfer of electrons to ascorbate free radical (Barr and Crane, 1991; Barr *et al.*, 1993). Inhibition of NADH oxidase in membranes of B deficient plants could explain the decrease in ascorbate concentration reported in B deficient plants (Lukaszewski and Blevins, 1996; Cakmak and Römheld, 1997). Decrease in ascorbate level in B deficient plant could also result from enhancement of ascorbate oxidase activity under conditions of B deficiency (Brown, 1979). This decrease in ascorbic acid was because the concentration of oxidized ascorbic acid increased, while reduced ascorbic acid decreased during B deficiency. This indicated that the biosynthesis of ascorbic acid was reduced and the consumption (oxidation) of ascorbic acid was stimulated during B deficiency. Therefore, decreases in concentration of non-protein SH- and reduced ascorbic acid by boron deficiency can also be a reflection of an induced oxidative stress caused by toxic O_2 radicals.

Cakmak and Römheld (1997) did not observe any significant change in total SOD activity in leaves of B deficient sunflower but observed a decrease in levels of glutathione reductase (GR) activity in response to B deficiency. As compared with B sufficient plants, the decreases in activity of glutathione reductase by B deficiency were about 3.5 fold, while activity of PPO was increased by a factor of 5.5. The decrease in glutathione reductase by B deficiency can be critical because it can sensitize B deficient cells to peroxidative damage triggered by enhanced production of quinones and toxic O_2 species. Glutathione reductase is essential for maintenance of high concentrations of glutathione and is directly involved in H_2O_2 detoxification (Foyer *et al.,* 1994). Possibly, the decrease in activity of glutathione reductase is related to the binding of quinones to NADPH binding sites of the enzyme (Bironaite *et al.,* 1989). It seems that the decreases in levels of ascorbic acid, SH-compounds and glutathione reductase during the course of B deficiency can enhance the development of B deficiency induced cell damage. A similar beneficial effect of ascorbic acid in the incubation solutions for B deficient leaves was however not observed by Pfeffer *et al.* (1998) and they concluded that the increased membrane permeability in B deficient leaves is not caused by low concentrations of ascorbate in the leaf apoplasm. In the same study it was suggested that B does not maintain plasma membrane integrity by complexing phenols and inhibiting PPO activity to prevent damage by oxygen free radicals or by regulating ascorbate metabolism to provide enough reduction equivalents in the apoplasm (Pfeffer *et al.*, 1998; Ruiz *et al.*, 1999; Cara *et al.*, 2002).

Studies have shown that the cellular antioxidant pool decreases significantly in sunflower or squash under B deficiency (Cakmak *et al.*, 1995; Lukaszewski and Blevins, 1996; Cakmak and Römheld, 1997) and that oxidative stress responsive genes are up-regulated in low B-acclimated tobacco cells (Kobayashi *et al.*, 2004). Koshiba *et al.* (2009) described the physiological responses of suspension-cultured tobacco BY-2 cells to B deficiency, and have shown for the first time that ROS and lipid peroxides are accumulated in these cells as compared to control cells, and showed a slight but significant decrease in the cellular ascorbate pool. They observed that supplementing the media with lipophilic antioxidants effectively suppressed the death of B-deprived cells, suggesting that the oxidative damage is the immediate and major cause of cell death under B deficiency. They demonstrated that B deprivation, and hence probably a defective cell wall structure, leads to oxidative damage.

Under B toxicity increased membrane permeability, MDA content and H_2O_2 accumulation and decrease in unsaturated fatty acid contents, indicating enhanced lipid peroxidation have been reported by Karabal *et al.* (2003) in barley and by Molassiotis *et al.* (2006) in shoot tips culture of the apple. A two fold increase in ascorbic acid and increased concentration of α-tocopherol in leaves of orange trees under B excess (irrigated with channel water containing high boron) was reported by Keles *et al.* (2004). Antioxidant enzyme catalase (CAT) activity was depressed in the leaves and stems of the explants exposed to high B in apple rootstock (Molassiotis *et al.*, 2006; Sotiropoulos *et al.*, 2006) while POD activity was enhanced. Karabal *et al.* (2003) investigated two barley cultivars- one sensitive to B toxicity (Hamidiye) while another tolerant (Anadolu). They found that in both the cultivars, remarkably higher constitutive CAT and ascorbate peroxidase (APX) activity was observed in leaf tissues. The tolerant cultivar Anadolu, demonstrated significantly higher CAT and APX activity in leaves and root tissues under B toxicity.

Molassiotis *et al.* (2006) reported increase in SOD activity at 3 and 6 mM B in apple *in vitro* and demonstrated the induction of a Cu/ Zn SOD and Mn-SOD isoform in leaves and stems of explants. Recently Han *et al.*(2009) reported the B deficiency and excess induced oxidative damage in leaves of citrus, as indicated by the increased content of TBA reactive compounds and concluded that the reduction of CO_2 assimilation by B excess is probably caused by a combination of factors such as oxidative damage, reduced photosynthetic enzyme activities and impaired electron transport capacity.

Ardic *et al.* (2009) observed significant differences in the antioxidant systems of the roots of two chickpea (*Cicer arietinum* L.) cultivars differing in tolerance to drought under toxic B conditions. These results suggest that roots of tolerant Gökce are better protected from B-stress induced oxidative stress due to enhanced SOD, CAT and POX activities under high B levels). Recently observed accumulation of H_2O_2 and TBARS content in both leaves and roots under deficient and excess B supply suggested oxidative damage due to excessive production of ROS in *Brassica* seedlings (Pandey and Archana, 2012). Increased activity of antioxidative enzymes-

SOD, CAT and POD along with PPO (polyphenol oxidase) was also observed under B stress (Pandey and Archana, 2011, 2012).

6.4 Nitrogen Metabolism

The essentiality of B for N_2 fixation was observed in the heterocyst of the cyanobacterium *Anabaena* PCC 7119 (Mateo *et al.*, 1986; García-González *et al.*, 1988, 1990) and in the vesicles of actinomycetes of the genus *Frankia* (Bolãnos *et al.* 2002). These microorganisms require B for the stability of their envelopes that protect nitrogenase from inactivation by oxygen when grown under N_2-fixing conditions. Loss of nitrogenase activity under B deficient conditions was explained by the destruction of nitrogenase by O_2. The altered O_2 status of the B deficient heterocysts was also apparent by the increased activity of SOD, CAT and POD (Garcia-Gonz´ales *et al.*, 1988). Heterocysts are capable of nitrogen fixation because they maintain a reducing (low O_2) environment which is possible because of a thick envelope comprised of glycolipids. Mutants of *Cyanobacteria* deficient in these envelope glycolipids fail to fix nitrogen. Within 6 h following B removal, glycolipid content dropped considerably indicating that B stabilizes the inner glycolipid layer of heterocyst envelopes and thus retards O_2 diffusion (Garcia-Gonzáles *et al.*, 1991).

Yamagishi and Yamamoto (1994) and Bolãnos *et al.(*1996) found the early stages of nodule development in soybean to be sensitive to B deficiency. Inadequate supply of B during early stages of nodule development limited both the development of the root nodules and nitrogen fixation. Bolãnos *et al.* (1996) found that during nodule development in B deficient pea plants, the number of *Rhizobia* infecting the host cells and the number of infection threads were reduced and the infection threads developed morphological aberrations. They suggested the involvement of B in *Rhizobium* legume cell surface interaction favoring nodule development. Boron is needed for the targeting of nodule specific plant derived glycoproteins which act as signals for bacteroid differentiation into a N_2- fixing form (Bolãnos *et al.*, 2004b). In addition, the cell walls of B deficient nodules have low levels of hydroxyproline/proline rich proteins such as ENOD2, which results in a higher oxygen diffusion into the nodules and the consequent inactivation of nitrogenase (Bonilla *et al.*, 1997).

The involvement of B in nitrogen assimilation such as reduced nitrate reductase (NR) activity and enhanced accumulation of nitrate have been described in B deficient plants (Kastori and Petrovic,1989; Ramón *et al.*, 1989; Shen *et al.,* 1993), and are due to the possible role of B in the *de novo* synthesis of the NR protein or facilitation of nitrate absorption (Ruiz *et al.,* 1998a). Tobacco plants subjected to a severe B deficiency had a significant decrease in leaf NR activity and nitrate concentrations in comparison with control plants (Camacho-Cristóbal and González-Fontes, 1999). More recently it has been shown that short term B deficiency led to a decline in root and, particularly leaf nitrate contents without affecting NR activity (Camacho- Cristóbal and González-Fontes, 2005; 2007). This decreased nitrate content was attributable to the lower net nitrate uptake rate found in B deficient plants, probably as a consequence of the drop in the levels of root plasma membrane H^+-ATPase (*PMA2)* transcript during the B deficient treatment (Camacho-Cristóbal and González-Fontes, 2007). In addition, B deficiency may also promote ammonium assimilation via asparagine synthetase in tobacco roots (Camacho-Cristóbal and González-Fontes, 2007). Mahboobi *et al.* (2002), reported the decreased (15-17%) activity of nitrate reductase in root and leaf tissues of barley and wheat cultivars compared to control subjected to boron toxicity. Eraslan *et al.* (2007) suggested the increased activity of nitrate reductase with increasing B concentration in tomato and pepper plants.

6.5 Secondary Metabolism

Boron has been shown to affect the concentration of phenolic compounds in plants. Accumulation of phenolics is a characteristic of B deficient tissues and occurs as a result of the enhanced synthesis and inhibition of phenols in cell wall synthesis (Loomis and Durst, 1991; Cakmak and Römheld, 1997). A comparative study of plants showing differential susceptibility to B deficiency showed that the concentration of total phenol is increased in plants sensitive to B deficiency (sunflower and corn) and in plants that had medium tolerance to the deficiency (peas) but B deficiency had little effect on the content of phenolic substances in plant that were tolerant to the B deficiency. Thus impairments in phenol metabolism and increases in levels of phenolics and polyphenol oxidase activity are typical indications of B deficiency in B sensitive plant species such as *Helianthus annuus* (Cakmak and Romheld, 1997). Accumulation of phenolics in B deficient tissues can activate enzymes that use phenolics as substrates. It has been shown that quinones are more toxic than the phenols from which they are derived (Mason and Wasserman, 1987; Pillinger *et al.* 1994). Cakmak *et al.* (1995) suggested a role of boron in preventing the oxidation of phenols by forming B- phenol complexes.

Formation of cis-diol complexes between B and sugars plays a decisive role in accumulation of phenolics in B deficient tissues. By forming complexes with 6-phosphogluconic acid, B inhibits activity of 6-phosphogluconate dehydrogenase (Dugger, 1983), and increases free (uncomplexed) 6- phosphogluconic acid in B deficient tissues. As a result under B deficient conditions, the substrate flux is shifted from glycolysis into the pentose phosphate pathway (PPP), which leads to enhanced phenol synthesis (Lee and Aronoff, 1967; Dugger, 1983). Under B deficient conditions the activity of glucose-6-phosphate dehydrogenase, the first enzyme of the PPP, is also enhanced much more than that of 6-phosphogluconate dehydrogenase (Gomes-Rodriguez *et al.*, 1987). Enhancement in concentration of phenolics in B-deficient tissues as a result of either stimulation of the pentose phosphate pathway Dugger (1983) or restriction in biosynthesis of phenolic alcohols (Pilbeam and Kirkby, 1983) is less pronounced in graminaceous monocots as compared with the dicotyledons.

The impairment in the cellular functions is closely associated with accumulation of phenolics in B-deficient tissues as most phenolics are known to be phytotoxic to roots and shoots (Vaughan and Ord, 1990). In B deficient plants, phenolic compounds such as caffeic acid, chlorogenic acid, ferulic acid, vanillic acid and coumarins accumulate (Camocho-Crsitóbal *et al.*, 2004). Increased activity of enzymes involved in phenylpropanoid biosynthesis, such as poyphenol oxidase (Pfeffer *et al.,* 1998; Camacho-Cristóbal *et al.*, 2002; Pandey and Archana, 2012), and phenylalanine ammonia lyase (Cakmak *et al.*, 1995; Ruiz *et al.*, 1998b; Camacho-Cristóbal *et al.*, 2002) also enhance accumulation of phenolic compounds. Enhanced accumulation of phenolics (Cakmak and Römheld, 1997) and increased activity of phenol–oxidizing enzymes such as polyphenol oxidase (Pfeffer *et al.*, 1998) in B deficient tissues may lead to increased production of quinones. Quinones are known to be more toxic than phenols and react with oxygen to generate reactive oxygen species and cause peroxidative damage to membrane constituents. The accumulation of substantial amounts of quinones in B deficiency-sensitive plant species has been considered as a major cause of cell damage and cessation of growth under B deficient conditions. The quinones produced by PPO are subsequently polymerised to brown pigments. Possibly, browning of B-deficient tissues is a consequence of enhanced phenol accumulation and PPO activity. Boron involvement in metabolism of phenolics and lignin biosynthesis contributes to strengthening of plant defense against pathogenic infections. Boron deficiency induced changes in phenol metabolism of plants produce a more favourable environment for the pathogens. Exposure of wheat plants to B deficiency predisposes them to infection by powdery mildew. Boron increases the disease resistance in plant such as potato scab disease, ergot on barley and damping off fungi on tomato and cabbage (Bergmann, 1984; Shorrocks, 1984).

Boron deprivation not only induces enhanced accumulation of the existing phenolics but also synthesis of new phenolic compounds (Camacho-Cristóbal *et al.*, 2002, 2004). Camacho-Cristóbal *et al.* (2005) reported increased accumulation of putrescine in tobacco plants deprived of B. Dixit *et al.* (2002) demonstrated increase in curcumin content of rhizomes of turmeric (*Curcuma longa*) grown under boron deficiency. Camacho-Cristòbal *et al.*(2002) showed an accumulation of caffeoyl putrescine in leaves from B-deficient tobacco plants which suggests a possible role of B in polyamine metabolism. Camacho-Cristòbal *et al.*(2005) observed that B deficiency increased the levels of conjugated polyamines in both leaves and roots in comparison to control and attributed this accumulation to structural alterations of the cell wall pectin and cytoskeleton found under B deprivation leading to altered binding between soluble polyamine and cell wall pectins

6.6 Water Relations

Boron regulates the intake of water into the cell and B deficient plants exhibit water stress. Radeva *et al.* (1980) reported that treating alfalfa plant with boric acid increased transpiration and water content of plants by decreasing the water deficit during dry periods and improving the water utilization by the plant. Sharma *et al.* (1984a) showed that B deprivation of cabbage plants led to decrease in stomatal opening associated with increase in diffusive resistance and decrease in transpiration rate. In cauliflower, B deficiency effect on leaf diffusive resistance and transpiration could, to a great extent, be reversed by resupply of B. Boron deficient plants shows decreased moisture percentage, less succulence, less metabolic activity, low water potential, poor stomatal pore opening, transpiration and a lower growth rate, in comparison to B sufficient plants (Sharma and Ramchandra, 1990). Boron stressed plants show stomatal closure and enhanced accumulation of proline which are typical features of water stressed plants (Pandey and Archana, 2009). Eraslan *et al.* (2007) also observed increased accumulation of proline content in tomato and pepper plants grown under B toxicity. The observed decrease in water potential and significantly high accumulation of proline in the leaves of plants receiving B deficient and excess supply, are suggestive of water stress. Sharma *et al.* (1984) and Sharma and Ramachandra (1990) observed that B deficient plants had high tissue water content in leaves probably due to high concentration of hydrophilic colloids or accumulation of their precursors such as soluble nitrogenous fractions, sugars and starch, which leads to an increase in bound water by lowering of matric potential in the leaves.

Pandey and Archana (2012) recently that observed decrease in total water content and WSD (water saturation deficit) along with increase in RWC (relative water content), SWC (specific water content), WUC (water use efficiency) and significantly high accumulation of proline in leaves and roots of B deficient and toxic *Brassica* plants are suggestive of impaired water metabolism. The increased SWC and RWC of tissues under B stress might be due to apoplastic bound water and high concentrations of hydrophilic colloids formed/synthesized under stress conditions and this might be the cause of less WSD and increased WUC of tissues. The higher WUC and low total water content in roots than leaves of B stressed seedlings suggested more water stress in the roots. Recently B has been suggested to play a role in water uptake by affecting the functionality of aquaporins (Martinez-Bastias *et al.*, 2004, 2010). Aquaporins are water channels that allow water to flow freely across cell membranes. Water channels in the plasma membrane of root cells play an important role in root hydraulic conductivity and plant– water relationships, particularly in response to changing environmental factors (e.g. drought, nutrient deficiency, low temperature) (Tyerman *et al.*, 2002). Expression of specific water channel proteins in plasma membranes can change the cellular osmotic water permeability coefficient up to 3-fold. Severe B deficiency causes a rapid decrease in the activity of the plasma membrane water channel proteins (ZmPIP1 aquaporins) in the root apex of maize and transgenic tobacco (Goldbach *et al.*, 2002; Martinez-Ballesta *et al.* (2008). This impact may lower the ability of roots to maintain hydraulic conductivity and result in water imbalance.

6.7 Reproduction

Boron plays an important role in reproductive development. The effect of B is more pronounced on the reproductive growth especially flowering, fruit and seed set and seed yield than vegetative growth even under mild deficiency (Sharma *et al.*, 1981; Sherrell, 1983; Loomis and Durst, 1992; Noppakoonwong *et al.,* 1997). There are several reports in a number of crops which demonstrated that B can be deficient and have a significant effect on yield even when there are no vegetative signs of deficiency and even when B concentration is at the adequate range (Perica *et al.*, 2001; Nyomora *et al.*, 1999; Hanson, 1991; Asad *et al.*, 2003). Boron deficiency symptoms during reproductive stage include shedding of buds, flowers and developing fruits and seeds as well poor fruit/seed quality and low viability of seeds. In plants subjected to B deficiency both number and size of flowers are severely restricted (Adams *et al.,* 1975; Zhang *et al.,* 1994). Decrease in the number of flowers in B deficient plants may be caused by the breakdown of the apical meristem restricting emergence of flower bearing branches and premature bud abscission. The reproductive parts of flowers such as anthers, ovary and stigma show aberrations even without prior expression of symptoms of B deficiency during vegetative growth. The floral parts possess a much higher concentration of B than vegetative parts when plants are not adequately supplied with B (Adams *et al.,* 1975; Xu *et al.,* 1993; O'Neill *et al.*, 1996; Dell and Huang, 1997), and indicate an involvement of B in plant reproductive development. Thus while wheat vegetative parts show no obvious leaf symptoms or growth reduction of B deficiency even when the young leaves have less than 2 mg B kg^{-1} dry matter, the B requirement in anthers for successful grain set is 10 mg B kg^{-1} dry matter (Rerkasem *et al.,* 1997). The boron concentration in *Brassica napus* organs have been reported to be 40, 38 and 17 mg g^{-1} of dry weight in pistils, stamens and leaves respectively (Dell and Huang, 1997).

Since B is known to be immobile poor supply of B to the reproductive organs may be responsible for limitation in reproductive development and seed yield of B deficient plants. Since flowers do not transpire as rapidly as leaves, the slow rate of transpiration because of the foliar coverings of the reproductive parts and the inflorescence architecture may result in passive delivery of B to the reproductive organs (Hansen and Breen, 1985). Environmental constraints such as temperature, light, humidity and availability of water could also restrict transport of B during the critical stages of reproductive development (Dell and Huang, 1997). Poor translocation of B from leaves and other mature tissues to the floral parts

(Brown and Shelp, 1997) and poor access of the pollen grains and the embryo sacs to the vascular supply are major factors for poor B delivery to the reproductive organs (Rawson and Nippakoonwonge, 1996; Dell and Huang, 1997; Brown *et al.,* 2002). As the vascular supply to the anther terminates at the tapetum and that to the ovules at the hypostate, the male and the female gametophytes are deprived of access to B supply directly through the vascular channels.

Boron requirement for flowering depends on male gametogenesis which has been reported to be particularly sensitive to low B supply especially in cereals/compact inflorescence species than those with axillary inflorescence (Rerkasem and Loneragan, 1994; Dell and Huang, 1997). Boron deficiency can result in plants being functionally male sterile (Rerkasem and Loneragan, 1994). However cases of female sterility have also been reported (Coetzer and Robbertse, 1987). Several studies have shown an involvement of B in microsporogenesis and male fertility (Zhang *et al.,* 1994; Rerkasem *et al.,* 1997; Huang *et al.,* 2000; Asad *et al.*, 2003; Pandey and Gupta, 2012). Small stamens with abnormally developed tapetum have been observed in B deficient oilseed rape (Zhang *et al.,* 1994). In B deficient maize plants emergence of tassels and anthers was delayed, the anthers lacked sporogenous tissue and many stamens turned into floral appendages or staminodes (Agarwala *et al.,* 1981; De Wet *et al.*, 1989). In wheat florets each sterile spikelet appears to be normal with fully developed lemma and palea but anthers were absent or poorly developed being small and mis -shapen with few malformed pollen grains (Rerkasem *et al.,* 1991; Cheng and Rerkasem, 1993). In severely deficient wheat, anthers developed as small arrow-shaped structures evoid of cells in the anther locules and exhibit inhibition of floret fertility (Huang *et al.,* 2000).

Boron deficient plants of oilseed rape show an abnormally developed tapetum (Zhang *et al.*, 1994) and arrest of microsporogenesis beyond the pollen mother cell stage (Xu *et al.,* 1993). Anatomical investigation of pollen development in a B-sensitive (SW41) and B-insensitive (Sonora) wheat failed to detect any abnormality in pollen until after the uninucleate vacuolate stage and at anthesis pollen grains were sterile (Rerkasem *et al.,* 1997). Visible impairment in pollen development in B deficient wheat did not appear until after the microspores were released from the tetrads. The effect of B deficiency is expressed structurally later in pollen development and at anthesis the pollen grains in well-formed anthers appear empty, misshapen, shrivelled, or may be normal in shape but lack reserves of storage materials such as starch (Dell and Huang, 1997). Rawson (1996) found that the critical period for pollen development extends from after emergence of the flag leaf to shortly after the flag leaf has become fully expanded and the flag leaf plays an important role in translocation of carbohydrate for pollen development. In wheat, low B did not alter the structural appearance of the mature pollen wall under the transmission electron microscope (Rerkasem *et al.,* 1997a). By contrast,

Smith *et al.* (1997) noted that the exine was less extensively developed in pollen from B-deficient avocado. Huang *et al.* (2000) showed that there may be two phases of pollen development during which B deficiency impairs pollen viability: (1) meiosis to young microspore; and (2) mitosis-I to pollen maturation. They showed that the period from premeiotic interphase through meiosis to late tetrad was the most sensitive stage of microsporogenesis in wheat while the period from mitosis-I to II during which starch accumulation occurred in pollen grains was less sensitive. During the first phase, impaired formation of pollen cell walls and cell expansion may be the primary consequence of B deficiency whereas during the second phase, carbohydrate delivery necessary for starch accumulation may result in pollen grains poor in starch content.

Several other studies substantiate a role of B in microsporogenesis and male fertility. Pollen germination and pollen tube growth are particularly sensitive to B and pollen viability is impaired by B-deficiency (Duggar, 1983; Bergmann, 1984; Cheng and Rerkasem, 1993). Robbertse *et al.* (1990) showed that petunia pollen tubes grew towards an increasing B concentration, exhibiting a "boronotropic" response. Boron probably promotes pollen germination by affecting H^+-ATPase activity, which initiates pollen germination and tube growth (Feijò *et al.*, 1995, Obermeyer and Blatt, 1995). *In vitro* germination tests also show that B was required for pollen germination and pollen tube growth in wheat (Cheng and Rerkasem, 1993; Rawson, 1996). Low *in vitro* germination of pollen grains is reflected in poor fertilization in B deficient plants (Huang *et al.*, 2000). Boron is known to bind with hydroxyl-rich organic molecules like sugar more rapidly and is thus involved in proper pollen tube formation and elongation. In absence of B pollen tubes may burst due to B function in cell wall structure of the pollen tube (Brown *et al.*, 2002). Boron deficiency may also alter the deposition of callose, cellulose and pectate in cell walls of anther and pollen cells leading to slowing or complete cessation of pollen tube elongation (Wang *et al.*, 2003). Pandey and Gupta (2012) observed that pollen grains of B deficient plan black gram plants were small in size, non-viable and failed to germinate as compared to those of B sufficient plants (Fig. 4 A-D).

Boron plays essential roles in the structure and function of cell walls and cellular membranes (Cakmak and Romheld, 1997; Matoh, *et al.*, 1997; Ishii *et al.*, 1999) and as such a role for B in pollen cell walls would be inevitable. Cytoskeletal changes induced by B deficiency observed in tips of meristematic cells (Baluska *et al.*, 2002; Yu *et al.*, 2002) suggest possible involvement of B in cytoskeletal changes preceding pollen germination and pollen tube growth. Matoh *et al.* (1998) detected borate-rhamnogalacturonan-II (RGII) complex in the cell walls of pollen tubes. It has been reported that the walls of growing tips of pollen tubes are composed predominantly of pectic polysaccharides including RG-II and that the normal growth of pollen tubes requires borate- and calcium-dependent cross-linking of pectin (O'Neill *et al.*, 2004). Recently, Iwai *et al.*

(2006) showed that *NpGUT1* gene encoding a glycosyltransferase thought to be involved in the borate cross-linking of pectin RG-II. is required for male and female reproductive tissues and fertilization and altering the composition of the main side chain A in RG-II in the vicinity of the apiosyl residue provokes important developmental and reproductive disorders. Delmas *et al.* (2008) have recently shown that the synthesis of the 3-deoxy-D-manno-2-octulosonic acid (Kdo) component of RG-II is required for pollen tube growth and elongation. Their observations using the Kdo-8-P synthase (AtkdsA2-V and AtkdsA2-S) gene expression are strikingly similar to the effects induced by *NpGUT1* misexpression and are a new example of pollen tube defect related to the RGII within the plant cell wall.

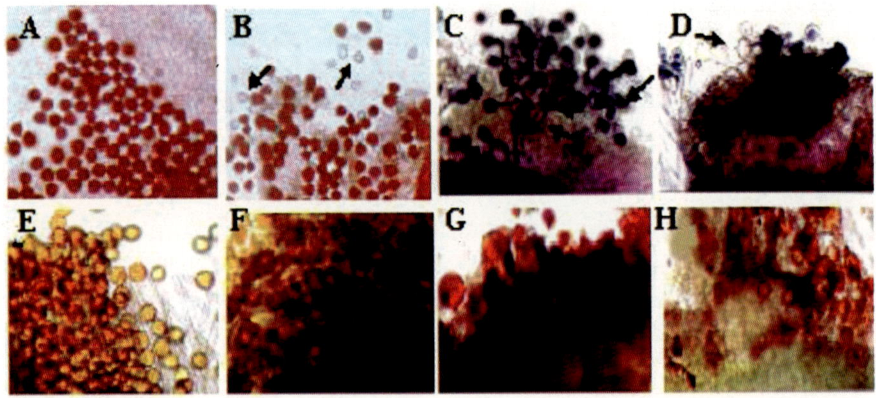

Fig. 4. A-H. Pollen grains of B sufficient plants (A) and B deficient plants (B) showing smaller and non-viable pollen grains (Arrows). Pollen grains germinating on stigma (Arrows) of control plants showing pollen tubes (C). Non-viable pollen grains of B deficient plants (D) which fail to germinate (Arrows). Histochemical localization of peroxidase (E, F) and esterase (G, H) in B sufficient (E, G) and B deficient (F, H) plants. Adapted from Pandey and Gupta (2012)

Besides microsporogenesis and pollen germination B also affects pollen receptivity of the stigma, pollen tube growth through the stylar tract and development of the megagametophyte. Non-receptiveness of silks in B deficient maize plants was found to result in poor setting of grains (Vaughanm, 1977). The stigmatic papilla of B deficient oilseed rape showed morphological aberrations (Xu *et al.,* 1993) and developed abnormal embryo sacs. Boron deficiency is reported to inhibit the development of ovules in cotton (Birnbaum *et al.,* 1974). High concentration of B in stigma and style is also reported to cause inactivation of callose by forming a B-callose complex at the interface of the pollen tubes and styles (Lewis, 1980). Enhanced accumulation of phenolic compounds on the stigmatic surface was found to inhibit the germination of pollen grains on the stigma of B deficient *Campsis grandiflora* (Dhakre *et al.,* 1994). In wheat, Rerkasem *et al.* (1993) reported that impaired development of anthers and poor pollen germination results in a failure of fertilization in B deficient wheat. Recently Pandey and Gupta (2012) observed that B deficiency affects stigma receptivity, pollen tube growth and development of megagametophyte. They reported the requirement of B for pollen-stigma interaction conducive to pollen germination and fertilization. They observed increase in peroxidase and decrease in esterase activity in stigmatic exudates in B deficient plants (Fig. 4 E-H) which probably resulted in poor pollen germination and fertilization leading to poor seed set. They also observed that additional supply of B increased esterase activity and stigmatic exudation which is required for proper pollen adhesion and germination. Esterase is known to be involved in rupture of stigmatic cuticle for proper pollen tube penetration through stylar tube for successful fertilization (Hiscock *et al.*, 2002).

Role of B in reproductive development of plants has a direct bearing on seed yield (Mozafarm 1993; Rerkasem *et al.,* 1993; Cheng and Rerkasem, 1993; Rawson, 1996). Under acute B deficiency, the crop fails to set seeds. Low grain yield of wheat is attributed to male sterility caused by reduced transport of B to the flowering parts, where it is critically required for microsporogenesis and pollen fertility (Rerkasem, 1996; Rawson, 1996; Subedi *et al.,* 1998). Boron is also important for post fertilization development and seed maturation. Seeds of low B plants showed poor germination and lack of vigor (Bell *et al.,* 1989). In sunflower Chatterjee and Nautiyal (2000) found that deprivation of B supply, even as late as the time of anthesis, produces morphological aberrations in seeds and reduces the seed content of non-reducing sugars, starch and oil. Sinha *et al.* (1999) observed enhanced accumulation of phenolic compounds and decrease in oil content in seeds of low B *Sesamum* plants. It is commonly accepted that floral and fruiting organs are especially sensitive to B deficiency (Brown *et al.*, 2002; Dell and Huang, 1997; Dell *et al.*, 2002). In many crops there is much higher demand for B during flowering and seed set even in crops where the B levels in leaves are in the adequate range. It was found that in B-deficient seeds during germination there is an increase in the abnormal seedlings and a decrease in germination percentage (Bell *et al.*, 1989, 1990; Rerkasem *et al.,* 1997b). This effect was reversed by supplemental B which enhanced seed germination and seed vigor (Bell *et al.*, 1990; Rerkasem *et al.*, 1997).

It was reported that there was an increase in fruit set and yield with B foliar applications (Nyomora *et al.,* 1999; Perica *et al.,* 2001). Foliar application of boron during the reproductive phase of sunflower leads to enhancement in seed yield because of B involvement in reproductive processes (Asad *et al.* 2003) and improves fruit yield and quality in strawberry (Singh *et al.* 2007). Dordas (2006) observed the effect of foliar B application on pod development, pod set, seed set, seed yield, yield components (pod number per inflorescence, seed number per pod, seed development, seed weight), and on seed quality (seed germination and seed vigor) and found that the seed yield was increased by an average of 37% and seed germination and seed vigor increased by 27% in alfalfa as compared with the untreated control. They also observed that the critical B levels for alfalfa used for forage production is below that for seed production and foliar B application can improve the seed yield and seed quality of alfalfa grown for seed production. The fact that B application increased seed vigor indicates that seeds with adequate B supply can germinate and produce seedlings with better ability to grow and withstand any adverse environmental conditions. This is in agreement with the suggestion that seeds that were developed in plants with adequate nutrient supply show high germination percentage and also have high seed vigor (Welch, 1999). Pandey and Gupta (2012) observed that foliar supply of B to B deficient black gram plants not only improved seed yield and seed vigor but also improved the seed quality in terms of storage seed proteins (albumin, globulin, 18 glutenin and prolamin) and carbohydrates (sugars and starch) in black gram.

7. BORON TOXICITY

Boron toxicity is a worldwide problem that significantly limits crop yield in alkaline and saline soils together with a low rainfall and very scarce leaching in agricultural areas of Australia, North Africa, and West Asia. In addition, B-rich soils also occur as a consequence of irrigation water containing high levels of B (Nable *et al.*, 1997) or fertilization of B to avoid deficiency (Takano *et al.*, 2008) since the concentration range between B deficiency and toxicity is very narrow.

A reduced growth of shoots and roots is typical of plants exposed to high B levels. The B toxicity symptoms are generally similar for most plants and appear in the leaves as marginal or tip chlorosis, which is followed by necrosis. Boron tends to accumulate in the leaf margins of the dicotyledons and leaf tips of monocotyledons and therefore most B toxicity effects are known to first appear in these regions and then progress along the leaf margins and into the lamina (Tanaka and Fujiwara, 2008). In pea B toxicity results in burning of the edges of the old leaves. In radish B excess is characterized by marginal yellowing and burning of lower leaves. In red clover (*Trifolium pratense* L.) toxicity symptoms of B include burning and yellowing of the older leaf edges (Gupta, 1983). Boron toxicity in snap bean (*Phaseolus vulgaris* L) results in the marginal chlorosis of the older trifoliate leaves while unifoliate leaves were chlorotic with intermittent marginal necrosis. In tomato plants, brown and necrotic lesions appeared in leaflet tips, which curled inwards (Eraslan *et al.*, 2007). Leaves show tip and marginal burning and yellowing between the veins in maize and apical and marginal chlorosis in older leaves in periwinkle plants toxic in B (Pandey and Archana, 2009). Recently, Óztúrk *et al.* (2010) reported the loss in seed yield in some canola cultivars under boron toxicity.

Boron toxicity exerts different effects on very diverse processes in vascular plants, such as altered metabolism, reduced root cell division (Liu *et al.*, 2000), lower leaf chlorophyll contents and photosynthetic rates, and decreased lignin and suberin levels (Nable *et al.*, 1997; Reid, 2007b). Although the physiological basis for B toxicity is not clear, three main causes have been proposed based on the the ability of B to bind compounds with two hydroxyl groups in the *cis*-configuration): (i) alteration of cell wall structure; (ii) metabolic disruption by binding to the ribose moieties of molecules such as adenosine triphosphate (ATP), nicotinamide adenine dinucleotide(reduced form) (NADH) or nicotinamide adenine dinucleotide phosphate(reduced form) (NADPH); and (iii) disruption of cell division and development by binding to ribose, either as the free sugar or within RNA (Reid *et al.*, 2004). The accumulation of high leaf B might lead to retardation of many cellular processes, and reduce the ability of cells to resist photo-oxidative stress and result in membrane leakiness, peroxidation of lipids and proline accumulation (Eraslan *et al.*, 2007; Pandey and Archana, 2012). A number of workers have reported oxidative damage and enhanced antioxidative enzyme activity and increased aascorbate and glutathione levels in B toxic plants (Gunes *et al.*, 2006; Molasiastos *et al.*, 2006 ; Cervilla *et al.*, 2007; Eraslan *et al.*, 2007; Pandey and Archana, 2011, 2012).

7.1 Boron Toxicity and Salinity

The concentration of B may be toxic for plants when they are irrigated with water containing high levels of B in saline soils showing poor drainage or plants are grown in soils with a natural presence of high concentrations of salts and B, usually in semiarid and arid regions characterized by low rainfall and poor drainage (Marschner, 1995; Nable *et al.*, 1997). Thus B toxicity and salinity may occur simultaneously (Nable *et al.*, 1997).

Some reports showed no additive effects of B and salinity on shoot weight of different cultivars (Bingham *et al.*, 1987; Grattan *et al.*, 1997; Greive and Poss, 2000). However, additive effects have been also proposed and the interaction of B and salinity declined the rate of germination and limited growth in maize and sorghum plants (Ismail, 2001). Salinity increased leaf injury due to B toxicity and membrane permeability increased with increase in B levels (Alpaslan and Gunes, 2001). However, it was also observed that the combined B toxicity and salinity caused less severe toxic effects on growth than what would be expected from the additive effects each, probably due to reduced uptake of B in the presence of chloride and reduced uptake of chloride in the presence of B (Yermiyahu *et al.*, 2008). The increased soil salinity may also reduce B movement to the plants and hence result in a reduction of B toxicity symptoms. Eraslan *et al.* (2008) also showed that B concentration in tissues of spinach plants was decreased by NaCl. Thus antagonism between salinity and B may be the result of decreased toxicity of B in the presence of NaCl, reduced toxicity of NaCl in the presence of B, or both together. Reduction of B accumulation in leaves in the presence of salinity has been reported (Alpaslan and Gunes, 2001; Apostol *et al.*, 2002; Yermiyahu *et al.*, 2007) and could be the result of the reduced rates of transpiration in plants where B is transported via xylem as consequence of the osmotic effect of the salt. The concentration of Na^+ in leaves decreased with increasing addition of B to the soil, probably due to the inhibition in root growth and reduction in root density caused by the B treatment.

Martinez-Ballesta *et al.* (2008a) proposed that under presence of both B and salt stress, boric acid could affect the activity of specific membrane components regulating the functions of certain aquaporin isoforms and ATPase as possible components of the salinity tolerance mechanism. Under saline conditions an optimum water balance is important to maintain plant homeostasis and for this the aquaporins maybe involved. At high external B levels, significant B transport is known to occur through the plasma membrane aquaporins (Dordas *et al.*, 2000; Dordas and Brown, 2001). Thus B uptake across the plasma membrane, by permeation through the lipid membrane and aquaporins, may be greatly influenced by the plant tolerance to salinity, through the associated changes in root hydraulic conductivity (Wimmer *et al.*, 2003). Bastias *et al.* (2004) reported an ameliorative effect of B on salinity due to increase in aquaporin functionality. They observed that aquaporin mediated uptake of water as well as B was reduced in NaCl-treated plants, but supplementing B increased their uptake. Martinez-Ballesta *et al.* (2008b) observed enhanced activity of specific membrane composition and enhanced water uptake due to abundance of ZmPIP1 and ZmPIP2 aquaporin contents when B was added to plants grown under saline conditions as possible components of the salinity tolerance mechanism producing a beneficial effect. It has also been reported that increased B and calcium supplies enhance crop salt tolerance and improve yield production in saline soils (Ballesta *et al.*, 2010), which could be useful for agriculture. The ameliorative effect of B on salinity and on the activity of specific membrane components regulating the functions of certain aquaporin isoforms and ATPase as possible components of the salinity tolerance mechanism is thus important

8. BORON AND CELLULAR SIGNALING

Boron has in recent years been assigned a role in cellular signaling. The established role of B as essential for cell wall structure imply that primary effect of B deficiency is to disturb structural organization of cell wall. Boron deficiency has been reported to affect the expression level of genes related to nitrogen metabolism (Redondo-Nieto *et al.,* 2007;

Camacho-Cristobal and González-Fontes, 2007), oxidative stress (Kobayashi *et al.*, 2004), B uptake (Takano *et al.*, 2006; Kasajima and Fujiwara, 2007), and cell wall (Camacho-Cristobal *et al.*, 2008b). However, there is no direct evidence to explain how the signal from B deficiency is transferred to nuclei. Using suspension cell cultures of tobacco BY-2 cells, Kobayashi *et al.* (2004) proposed that a rapid signal transfer from the cell wall to the cytoplasm could be involved in gene induction after the cellular redox imbalance imposed by B deficiency. Koshiba *et al.* (2009) showed that the death of cell after B deprivation directly involves oxidative damage and is not due to a typical programmed cell death. They hypothesized that ROS are originally produced as a signal for disturbance of the cell wall structure, and build up to toxic levels unless B is resupplied. This results in a change brought about by insufficient crosslinking of pectic polysaccharides at RG-II regions, as the B-RG-II complex significantly contributes to the wall tensile strength (Ryden *et al.*, 2003). With prolonged deficiency of B cell wall becomes weaker and less resistant to turgor, and may lead to opening of mechanosensitive channels (Nakagawa *et al.*, 2007) and generation of signals for the altered cell wall structure. Goldbach and Wimmer (2007) suggested that changes in B concentrations may lead to a mechanical cascade of signals extending into the cytoplasm via the cell wall-plasma membrane-cytoskeleton continuum, by an altered conformation of the membrane bound arabinogalactan proteins. This hypothesis is supported by the fact that B deprivation led to an altered polymerization pattern of cytoskeletal proteins (actin and tubulin) assemblies (Yu *et al.*, 2001, 2003) and to an inhibition of the endocytic pathway for internalization of B-cross-linked RG-II pectins in brefeldin A-induced compartments (Yu *et al.*, 2002). The primary action of B defivcincy would be on an apoplastic target, which signals deeper into the cell via endocytosis-mediated pectin signaling along putative cell wall-plasma membrane-cytoskeleton continuum (Yu *et al.*, 2002).

It has been also reported that B deficiency induces the expression of stress-responsive genes (Kobayashi *et al.*, 2004) and *NIP5;1* gene (Takano *et al.*, 2006; Kasajima and Fujiwara, 2007) that codes a protein essential for B uptake when this micronutrient is limiting. It has been proposed that a quick signal transfer from the cell wall to the cytoplasm could be involved for gene induction after B removal (Kobayashi *et al.*, 2004). Furthermore, there is also the possibility that B is affecting calcium-mediated signaling (Bolanos *et al.*, 2004a; Lecourieux *et al.*, 2006). In addition, it has been reported that high B induces the expression of several genes in both roots and rosette leaves of *Arabidopsis* (Kasajima and Fujiwara, 2007). One of these genes has been identified as a zinc finger family transcription factor (At1g03770), which could regulate the expression of downstream genes involved in the physiological response pathways to B toxicity in plants (Kasajima and Fujiwara, 2007). This result might support the putative role of B as a cellular signal capable of interacting with transcription factors (González-Fontes *et al.*, 2008), which could explain why the expression of several genes involved in different physiological processes are rapidly affected when plants are subjected to B deficiency. The B levels probably modulates the transcription factor interaction with the target gene sequence which could be either by the direct linkage of B to the transcription factors or through molecules capable of binding B and thereby linking with the transcription factors to regulate gene expression. The proteins encoded by these B-modulated genes, or eventually their absence, would act in diverse physiological processes at any cellular level. For instance, B might be essential in the cell wall not only for borate-ester bridges between RG-II monomers, but also as a signal to allow the expression of enzymes needed in the synthesis and assembly of cell wall, such as arabinogalactan proteins and expansins. Thus B may exert its primary function not only through stabilization of molecules containing cis-diol groups but also as a cellular signal capable of interacting with TF, which could explain why so many physiological processes are affected when vascular plants are subjected to B deficiency.

9. CONCLUSIONS

Boron is an essential element for the growth and development of plants, and adequate B nutrition is crucial for crop production. Therefore, it is important to understand the role of B in plants as well as the molecular mechanisms in response to B deficiency and toxicity in order to improve the tolerance of crops to B stress. The primary role of B is the crosslinking of cell wall RG-II and pectin assembly and stabilization of molecules with *cis*-diol groups. Boron deficiency and toxicity result in impaired growth and poor yield and plants maintain their tissue B concentrations within an optimum range by regulating transport processes to maintain B homeostasis. This is done by transporters responsible for efficient B uptake by roots, xylem loading and B distribution among leaves have been described and include the NIPs (nodulin-26-like intrinsic proteins), boric acid channels, and BORs, B exporters. These transporters are required under B deficiency for efficient acquisition and utilization of B. At the same time high B levels in the environment are tolerated by transporters which export B from roots back to the soil. Plants sense internal and external B conditions and regulate B transport by modulating the expression and/or accumulation of these transporters. It is important that the expression of genes encoding these transporters is finely regulated in response to B availability in the environment to ensure tissue B homeostasis. Plants tolerant to stress produced by low B or high B in the environment can be generated through altered expression of these transporters and may be useful in designing plants or crops tolerant to soils containing low or high B. Recently, the possible role of B as a cellular signal through the interaction with transcription factors has been proposed and is important to gain insight into the physiological functions of B.

REFERENCES

Adams, P; Graves, C J. and Winson, G W. 1975. Some effects of copper and boron deficiencies on the growth and flowering of *Chrysanthemum morifolium* cv. Hurricane *J. Sci. Food. Agric.*, 26: 899-901

Agarwala, SC; Sharma, PN; Chatterjee, C. and Sharma, CP. 1981. Development and enzymatic changes during pollen development in boron deficient maize plants. *J. Plant Nutr.*, 3: 329-336.

Agulhon, H. (1910). Role of Boron in Plants. Ph. D. Thesis. University of Paris.

Alpaslan, M. and Gunes A. 2001. Interactive effects of boron and salinity stress on the growth membrane permeability and mineral composition of tomato and cucumber plants. *Plant Soil,* 236: 123-128.

An, J-C; Liu Y-Z; Yang, C-G; Zhou, G-F; Wei, O-J and Peng, S-A. 2012. Isolation and expression analysis of CiNIP5, a citrus boron transport gene involved in tolerance to boron deficiency. *Sci. Hort.*, 142: 149–154

Appel, HM. 1993. Phenolics in ecological interactions. The importance of oxidation. *J. Chem. Ecol.*, 9: 1521-1552

Apel, K. and Hirt, H. 2004. reactive oxygen species: metabolism, oxidative stress and signal transduction. *Annu. Rev. Plant Biol.* 55: 373-399.

Apostol, KG; Zwiazek, JJ. And MacKinnon, MD. 2002. NaCl and Na_2SO_4 alter responses of jack pine (*Pinus banksiana*) seedlings to boron. *Plant Soil,* 240: 321-329.

Ardıc, M; Sekmen, AH; Turkan, I; Tokur, S. and Ozdemir, F. 2009. The effects of boron toxicity on root antioxidant systems of two chickpea (*Cicer arietinum* L.) cultivars. *Plant Soil,* 314:99-108

Asad, A; Blamey, FPC. and Edwards, DG. 2003. Effects of boron foliar applications on vegetative and reproductive growth of sunflower. Ann. Bot. 92: 565-570.

Baluska, F; Hlavacka, A; Šamaj, J; Palme, K; Robnison, DG; Matoh, T; McCurdy, DW; Menzel, D. and Volkman, D. 2002. F-actin dependent endocytosis of cell wall pectins in meristematic root cells. Insights from brefeldin A – induced compartments. *Plant Physiol.*, 130:422-431

Baluška, F; Šamaj J; Wojtaszek, P; Volkmann, D. and Menzel, D. 2003. Cytoskeleton-plasma membrane-cell wall continuum in plants. Emerging links revisited. *Plant Physiol.*, 133:482-91.

Barr, R; Böttger, M. and Crane, FL. 1993. The effect of boron on plasma membrane electron transport and associated proton secretion by cultured carrot cells. *Biochem. Mol. Biol. Int.*, 31:31–39.

Barr, R. and Crane, FL. 1991. Boron stimulates NADH oxidase activity of cultured carrot cells. In: *Current Topics in Plant Biochemistry and Physiology.* Vol. 10. D.D., Randall, D.G. Blevins and C.D. Miles (Eds). Univ. Mo Press, Columbia, 290.

Bastías, E;, Fernández-García, N. and Carvajal, M. 2004. Aquaporin functionality in roots of Zea mays in relation to the interactive effects of boron and salinity. *Plant Biol.*, 5: 415–421

Bastías, E; Alcaraz-Lòpez, C; Bonilla, I; Martínez-Ballesta, MC; Bolanõs, L. and Carvajal, M. 2010. Interactions between salinity and boron toxicity in tomato plants involve apoplastic calcium. *J Plant Physiol.*, 167: 54-60

Bell, RW. 1997. Diagnosis and prediction of boron deficiency for plant production *Plant Soil*, 193: 149-168.

Bell, RW; McLay, L; Plaskett, D; Dell, B. and Loneragan, JF. 1990. Internal boron requirements of green gram (*Vigna radiata*). *In* Plant Nutrition – Physiology and Application. Ed. M L van Beusichem. pp 275–280. Kluwer Academic Publishers, Dordrecht, The Netherlands.

Bell, RW; McLay, L. Plaskett, D; Dell, B. and Loneragan, JF. 1989. Germination and vigour of black gram (*Vigna mungo* (L.) Hepper) seed from plants grown with and without boron. *Aust. J. Agric. Res.*, 40: 273–279.

Bellaloui, N; Brown, PH. and Dandekar, AM. 1999. Manipulation of in vivo sorbitol production alters boron uptake and transport in tobacco. *Plant Physiol.* 119, 735-741.

Benton, JJ. 2003. Agronomic handbook; management of crops, soils and their fertility. CRC Press LLC. USA.

Bergmann, W. 1984. The significance of micronutrient boron in agriculture. Symposium held by the Borax Group in the International Trade Centre of the GDR, Berlin.

Blaser-Grillm, J; Knoppik, D; Amberger, A and Goldbach, H. 1989. Influence of boron on the membrane potential in *Elodea densa* and *Helianthus annuus* roots and H+extrusion of suspension cultured *Daucus carota* cells. *Plant Physiol,*. 90: 280-284.

Blevins, DG. and Lukaszewski, KM. 1998. Boron in plant structure and function. *Annu. Rev. Plant Physiol. Plant Mol. Biol.*, 49: 481-500.

Birnbaum, EH; Beasley, CA. and Dugger WM. 1974. Boron deficiency in under-fertilised cotton (*Gossypium hirsutum*) ovules grown *in vitro*. *Plant Physiol.*, 54: 931–935.

Birnoaite, AD; Ceans, NK. and Kulys, JJ. 1989. The inhibition of the glutathione reductase by quinones. *Eur. J. Biochem.*, 178: 693-703.

Bolaños, L; Brewin, NJ. and Bonilla, I. 1996. Effects of boron on rhizobium-legume cell surface interaction on nodule development. *Plant Physiol.*, 110: 1249-1256.

Bolaños, L; Esteban, E; Delorenzo, C; Fernandezpascual, M; Defelipe, MR; Garate, A; Bonilla, I. 1994. Essentiality of boron for symbiotic dinitrogen fixation in pea (*Pisum sativum*) rhizobium nodules. *Plant Physiol.*, 104, 85–90.

Bolaños, L; Mateo, P. and Bonilla, I. 1993. Calcium-mediated recovery of boron deficient *Anabaena* sp. PCC 7119 grown under nitrogen fixing conditions. *J. Plant Physiol.*, 142, 513–517.

Bolaños, L; Lukaszewski, K; Bonilla, I. and Blevins, D. 2004a. Why boron? *Plant Physiol. Biochem.*, 42: 907-912.

Bolaños, L; Redondo-Nieto, M; Bonilla, I. and Wall, LG. 2002. Boron requirement in the *Discaria trinervis* (Rhamnaceae) and *Frankia* symbiotic relationship. Its essentiality for *Frankia* BCU110501 growth and nitrogen fixation. *Physiol. Plant.*, 115, 563-570.

Bolaños, L; Redondo-Nieto, M; Rivilla, R; Brewin, NJ. and Bonilla, I. 2004b. Cell surface interactions of Rhizobium bacteroids and other bacterial strains with symbiosomal and peribacteroid membrane components from pea nodules *Mol. Plant-Microbe Interact.*, 17, 216-223.

Bonilla, I. Mergold-Villaseñor, C; Campos, MG; Sánchez, N; Pérez, H; López, L; Castrejón, L; Sánchez, F. and Cassab, GI. 1997. The aberrant cell walls of boron deficient bean root nodules have no covalently bound hydroxyproline-/ proline-rich proteins. *Plant Physiol.*, 115: 1329-1340.

Brown, JC. 1979. Effects of Boron Stress on copper enzyme activity in tomato. *J. Plant Nutr.*, 1: 39-53.

Brown, PH; Bellaloui, N; Hu, H and Dandekar, A. 1999. Transgenically enhanced sorbitol synthesis facilitates phloem boron transport and increases tolerance of tobacco to boron deficiency. *Plant Physiol.*, 119: 17-20.

Brown, PH; Bellaloui, N; Wimmer, MA; Bassil, ES; Ruiz, J; Hu, H; Pfeffer, H; Dannel, F. and Römheld, V. 2002. Boron in plant biology. *Plant Biol.*, 4: 205-223.

Brown, PH. and Hu H. 1994. Boron uptake by sunflower, squash and cultured tobacco cells. *Physiol. Plant.*, 91: 435–441.

Brown, PH. and Hu, H. 1996. Phloem mobility of boron is species dependent. Evidence for phloem mobility in sorbitol rich species. *Ann. Bot.,* 77: 497-505.

Brown, PH. and Hu, H. 1996. Phloem mobility of boron is species dependent: evidence for phloem mobility in sorbitol-rich species. *Ann Bot.,* 77: 497-505

Brown, PH. and Hu, H. 1997. Does boron play only a structural role in the growing tissues of higher plants? *Plant Soil,* 196: 211–215

Brown, PH. and Shelp, BJ. 1997. Boron mobility in plants. *Plant Soil,* 193: 85-101.

Cakmak, I; Kurz, H. and Marschner, H. 1995. Short-term effects of boron, germanium and high light intensity on membrane permeability in boron deficient leaves of sunflower. *Physiol. Plant,* 95:11–18

Cakmak, I. and Romheld, V. 1997. Boron deficiency-induced impairments of cellular functions in plants *Plant Soil,* 193: 71–83, 1997

Camacho-Cristöbal, JJ; Anzellotti, D. and González-Fontes, A. 2002. Changes in phenolic metabolism of tobacco plants during short-term boron deficiency. *Plant Physiol. Biochem.*, 40: 997–1002.

Camacho-Cristoòbal, JJ. and González-Fontes, A. 1999. Boron deficiency causes a drastic decrease in nitrate content and nitrate reductase activity, and increases the content of carbohydrates in leaves from tobacco plants. *Planta,* 209:528–36.

Camacho-Cristòbal, JJ. and Gonzaález-Fontes, A. 2002. Relationship between the organic-nitrogen status and the nitrate content in leaves of wild-type and mutant tobacco plants. *J. Plant Nutr. Soil Sci.*,165: 347–351.

Camacho-Cristóbal, JJ; Maldonadob, JM. and González-Fontes, A. 2005. A Boron deficiency increases putrescine levels in tobacco plants. *J Plant .Physiol.*, 162: 921-928

Camacho-Cristóbal, JJ. and González-Fontes, A. 2007. Boron deficiency decreases plasmalemma H+-ATPase expression & nitrate uptake, and promotes ammonium assimilation into asparagine in tobacco roots. *Planta* 226: 443-451.

Camacho-Cristóbal, JJ; Rexach, J. and González-Fontes, A. 2008a. Boron in plants: deficiency and toxicity. *J. Integr. Plant Biol.*, 50: 1247–1255.

Camacho-Cristóbal, JJ; Herrera-Rodríguez, MB; Beato, VM; Rexach, J; Navarro-Gochicoa, MT; Maldonado, JM. and González-Fontes, A. 2008b. The expression of several cell wall-related genes in Arabidopsis roots is down-regulated under boron deficiency. *Env. Exp. Bot.*, 63: 351-358.

Camacho-Cristobal, JJ; Lunar, L; Lafont, F; Baumert, A. and Gonzélez-Fontes, A. 2004. Boron deficiency causes accumulation of chlorogenic acid and caffeoyl polyamine conjugates in tobacco leaves. *J. Plant Physiol.*, 161: 879-881.

Cara, FA; Sánchez, E; Ruiz, JM. and Romero, L. 2002. Is phenol oxidation responsible for the short-term effects of boron deficiency on plasma-membrane permeability and function in squash roots? *Plant Physiol. Biochem.*, 240:853–858.

Cervilla, LM; Blasco, B; Rios, JJ; Romero, L. and Ruiz, JM. 2007. Oxidative stress and antioxidants in tomato (*Solanum lycopersicum*) plants subjected to boron toxicity. *Ann. Bot.*, 100: 747-756.

Chatterjee, C. and Nautiyal, N. 2000. Developmental aberrations in seeds of boron deficient sunflower and recovery. *J. Plant Nutri.*, 23: 835-841.

Cheng, C., and Rerkasem, B. 1993. Effect of boron on pollen viability in wheat. *Plant Soil,* 155/156: 313-315.

Dannel, F; Pfeffer, H. and Römheld, V. 1998. Compartmentation of boron in roots and leaves of sunflower as affected by boron supply. *J. Plant Physiol.*, 153: 615-622.

Dannel, F; Pfeffer, H. and Rohmeld, V. 1997. Effect of pH and Boron concentration in the nutrient solution on translocation of boron in the xylem of sunflower. *In* Boron in Soils and Plants. Bell RW and Rerkasem B. (Eds.). Kluwer Academic Publishers, Netherlands. pp. 183–186.

Dannel, F; Pfeffer, H. and Römheld, V. 2000. Characterization of root boron pools, boron uptake and boron translocation in sunflower using the stable isotopes 10B and 11B. *Aust. J. Plant Physiol.*, 27: 397–405.

Dannel, F; Pfeffer, H. and Römheld ,V. 2002. Update on boron in higher plant - Uptake, primary translocation and compartmentation. *Plant Biol.,* 4: 193-204.

Dell, BL; and Huang , L.. 1997. Physiological response of plants to low boron. *Plant Soil,* 193:103–120.

Dell, B ; Huang. L. and Bell, RW. 2002. Boron in plant reproduction. p. 103–117. In: *Boron in plant and animal nutrition.* Goldbach HE.(ED.). Kluwer Academic/Pledum Publ., New York.

De Wet, E., Robbertse, PJ. and Groeneveld, HT. 1989. The influence of temperature and boron on pollen germination in *Mangifera Indica* L. *S. Afr. Tydskr, Plant Grond.*, **6**: 228-234.

Delmas F, Seveno M, Northey JGB. 2008. The synthesis of the rhamnogalacturonan II component 3-deoxy-D-manno-2-octulosonic acid (Kdo)is required for pollen tube growth and elongation. *J. Exp.Bot.*, 59: 2639–2647.

Dhakre G, Singh J, Chauhan SVS. 1994. Analysis of phenolics and boron in the stigma of seedless *Campsis grandiflora* K. Schum. *Ind J Exp Biol.,* 32: 816-18.

Dixit, D., Srivastava, NK. and Sharma, S. 2002. Boron deficiency induced changes in translocation of $^{14}CO_2$ -photosynthate into primary metabolites in relation to essential oil and curcumin accumulation in turmeric (*Curcuma longa* L.) *Photosynthetica,* 40: 109-113.

Domingues, DS; Leite, SMM; Farro, APC; Coscrato, VE; Mori, ES; Furtado, EL; Wilcken, CF; Velini, ED; Guerrini, IA; Maia, IG. and Marino, CL. 2005. Boron transport in *Eucalyptus*. 2. Identification in silico of a putative boron transporter for xylem loading in eucalypt. *Genet. Mol. Biol.*, 28: 625-629.

Dordas, C. 2006. Foliar Boron Application Improves Seed Set, Seed Yield, and Seed Quality of Alfalfa. *Agron. J.*, 98:907–913.

Dordas, C. and Brown, PH. 2000. Permeability of boric acid across lipid bilayers and factors affecting it. *J. Membr. Biol.* 175, 95-105.

Dordas, C. and Brown, PH. 2001. Evidence for channel mediated transport of boric acid in squash (*Cucurbita pepo*). *Plant Soil,* 235: 95-103.

Dordas C, Chrispeels MJ, Brown PH (2000) Permeability and channel-mediated transport of boric acid across membrane vesicles isolated from squash roots. *Plant Physiol.*, 124: 1349-1361.

Dugger, WM. 1983. Boron in plant metabolism. In: *Encyclopedia of Plant. Physiology*, New Ser. 15 B Inorganic Plant Nutrition, Part B. pp. 626-650.

El-Shintinawy, F. 1999. Structural and functional damage caused by boron deficiency in sunflower leaves. *Photosynthetica.* 36: 565-572.

Eraslan, F; Inal, A; Gunes, A. and Alpaslan, M. 2007. Boron toxicity alters nitrate reductase activity, proline accumulation, membrane permeability and mineral constituents of tomato and pepper plants. *J. Plant Nutr.*, 30: 981-994.

Eraslan, F; Inal, A; Pilbeam, DJ. and Gunes, A. 2008. Interactive effects of salicylic acid and silicon on oxidative damage and antioxidant activity in spinach (*Spinacia oleracea* L. cv. Matador) grown under boron toxicity and salinity. *Plant Growth Regul.,* 55: 207-219.

Feijó, JA; Malhó, R. and Obermeyer, G. 1995. Ion dynamics and its possible role during in vitro pollen germination and tube growth. *Protoplasma,* 187: 155–167.

Ferrol, N; Belver, A; Roldán, M; Rodriguez-Rosales, MP. and Donaire, JP. 1993. Effects of boron on proton transport and membrane properties of sunflower (*Helianthus annuus* L.) cell microsomes. *Plant Physiol.,*103: 763–69.

Ferrol, N. and Donaire, JP. 1992. Effect of boron on plasma membrane proton extrusion and redox activity in sunflower cells. *Plant Sci.* 86: 41-47.

Foyer, CH; Lelandais, M. and Kunert, K.J. 1994. Photooxidative stress in plants. *Physiol. Plant.,* 92: 696-717.

Garcia-Gonz´ales, M; Mateo, P. and Bonilla, I. 1988. Boron protection for O_2 diffusion in heterocysts of *Anabaena* PCC7119. *Plant Physiol,.* 87:785–89.

García-González, M; Mateo, P. and Bonilla, I. 1990. Effect of boron deficiency on photosynthesis and reductant sources and their relationship with nitrogenase activity in *Anabaena* PCC 7119. *Plant Physiol.,* 93: 560-565.

Garcia-Gonzáles, M; Mateo, P. and Bonilla, I. 1991. Boron requirement for envelope structure and function in *Anabaena* PCC 7119 heterocysts. *J. Exp. Bot.*, 42:925–929.

Goldbach,, HE; Hartmann, D. and Rötzer, T. 1990. Boron is required for the ferricyanide-induced proton release by auxins in suspension-cultured cells of *Daucus carota* and *Lycopersicon esculentum*. *Physiol. Plant.*, 80:114–118.

Goldbach, HE. and Wimmer, M. 2007. Boron in plants and animals: Is there a role beyond cell-wall structure? *J. Plant Nutr. Soil Sci.*, 170, 39-48.

Goldbach, HE; Yu, Q; Wingender, R; Schulz, M; Wimmer, M; Findeklee, P. and Baluska, F. 2001. Rapid response reactions of roots to boron deprivation. *J. Plant Nutr. Soil Sci.*,164: 173-181.

Goldbach, HE; Wimmer, MA; Chaumont, F; Matoh, T; Volkmann, D. and Baluska, F. 2002. Rapid responses of plants to boron deprivation. In: *Boron in plant and animal nutrition*. Goldbach HE, Rerkasem B, Wimmer M, Brown PH, Thellier M, Bell RW.(EDs). New York: Kluwer Academic, pp.167–180.

Gomez-Rodriguez, MV; Castillo, JLD. and Alvarez-Tinaut, MC. 1987. The evolution of glucose-6-dehydrogenase and 6-P-gluconate dehydrogenase activities and the ortho-diphenol content of sunflower leaves cultivated under different boron treatments. *J. Plant Nutr.*, 10: 2211-2229.

González-Fontes, A; Rexach, J; Navarro-Gochicoa, MT; Herrera-Rodríguez, MB; Beato, VM; Maldonado, JM. and Camacho-Cristóbal, JJ. 2008. Is boron involved solely in structural roles in vascular plants? *Plant Signal. Behav*. 3: 24-26.

Gupta, U C. 1979. Boron nutrition of crops. *Adv. Agron*., 31: 273–307.

Gupta, UC. 1983. Boron deficiency and toxicity symptoms for several crops as related to tissue boron levels. *J. Plant Nutr.* 6: 387-396.

Gupta, UC. and Cutcliffe, JA.1973. Boron nutrition of broccoli, brussels sprouts and cauliflower grown on Prince Edward Island Soils. *Can. J. Soil Sci.*, 53: 275-279.

Gunes, A; Soylemezoglu, G; Inal, A; Bagei, EG; Coban, S. and Sahin, O. 2006. Antioxidant and stomatal responses of grapevine (*Vitis vinifera* L.) to boron toxicity. *Scient. Hort.*,110: 279-284.

Hanaoka, H. and Fujiwara, T. 2007. Channel-mediated boron transport in rice. *Suppl. Plant Cell Physiol.*, 48: 844.

Han, S; Tang, N; Jiang, H; Yang, L; Yan, L. and Chen, L. 2009. CO_2 assimilation photosystem II photochemistry, carbohydrate metabolism and antioxidant system of citrus leaves in response to boron stress. *Plant Sci.,* 176: 143-153.

Hanson, EJ. and Breen, PJ. 1985. Xylem differentiation and boron accumulation in 'italian' prune flower buds. *J. Am. Hortic Sci.*.110:389–92

Hayes, JE. and Reid RJ. 2004. Boron tolerance in barley is mediated by efflux of boron from the roots. *Plant Physiol*. 136: 3376–3382.

Herrera-Rodríguez, MB; González-Fontes, A; Rexach, J; Camacho-Cristöbal, JJ; Maldonado JM. and Navarro-Gochicoa, MT. 2010. Role of boron in vascular plants and response mechanisms to boron stresses. *Plant Stress*, 4: 115-122.

Hiscock, SJ; Bown, D; Gurr, SJ. Dickinson, HG. 2002. Serine 1 esterases are required for pollen tube 2 penetration of the stigma in Brassica. *Sex Plant Reprod.,* 15: 65-74.

Hu, H; Penn, SG; Lebrilla, CB. and Brown, PH. 1997. Isolation and characterisation of soluble B-complexes in higher plants. The mechanism of phloem mobility of boron. *Plant Physiol.,* 113: 649-655.

Hu, H. and Brown, P H. 1994. Localization of boron in cell walls of squash and tobacco and its association with pectin. *Plant Physiol.,* 105: 681-689.

Hu, H; Brown, PH. and Labavitch, JM. 1996. Species variability in boron requirement is correlated with cell wall pectin. *J. Exp. Bot.,* 47: 227-232.

Huang, L; Bell, RW. and Dell, B. 2008. Evidence of phloem boron transport in response to interrupted boron supply in white lupin (*Lupinus albus* L. cv. Kiev Mutant) at the reproductive stage. *J. Exp. Bot.,* 59: 575-583.

Huang, L; Pant, J; Bell, RW; Dell, B. and Deane, K. 1996. Effects of boron deficiency and low temperature on wheat sterility. In: *Sterility in wheat in subtropical Asia: extent, causes and solutions.* Rawson, HM. and Subedi, KD. (Eds). Proceedings of a workshop 18–21, September 1995, Lumle Agricultural Research Centre, Pokhara, Nepal. Canberra: Australian Centre for International Agricultural Research, pp 90–101.

Huang, L; Pant, J; Dell, B. and Bell, RW. 2000. Effect of boron deficiency on anther development and floral fertility in wheat (*Triticum aestivum* L. 'Wilgoyne') *Ann. Bot.,* 85: 493-500.

Ishi, T.and Matsunaga, T. 1996. Isolation and characterization of boron rhamnogalacturonan II complex from suger beet pulp. *Carbohy. Res.,* 284: 1-9.

Ishii, T; Matsunaga, T; Pellerin, P; O'Neill, MA; Darvill, and Albersheim, P. 1999. The plant cell wall polysaccharide rhamnogalacturonan II self-assembles into a covalently cross-linked dimer. *J. Biol.Chem.,* 274:13098–13104

Iwai, H; Hokura, A; Oishi, M; Chida, H; Ishii, T; Sakai, S. and Satoh, S. 2006. The gene responsible for borate cross linking of pectin Rhamnogalacturonan-II is required for plant reproductive tissue development and fertilization. *PNAS.,* 103: 16592-16597.

Jiang, Y. and Miles, PW. 1993. Generation of H_2O_2 during enzymic oxidation of catechin. *Phytochem.* 33: 29-34.

Karabal, E; Yucel, M. and Oktem, H. 2003. Antioxidant responses of tolerant and senitive barley cultivars to boron toxicity. *Plant Sci.,* 164: 925-933.

Kasajima, I. and Fujiwara, T. 2007. Identification of novel *Arabidopsis thaliana* genes which are induced by high levels of boron. *Plant Biotechnol.,* 24:355-60.

Kastori, R. and Petrovic, N. 1989. Effect of boron on nitrate reductase activity in young sunflower plants. *J. Plant Nutr.,* 12: 621-627.

Kato Y, Miwa K, Takano J, Wada M, Fujiwara T. 2009. Highly borondeficiency-tolerant plants generated by enhanced expression of NIP5;1,a boric acid channel. Plant and Cell Physiology 50: 58–66.

Keles, Y; Oncel, I. and Yenice, N. 2004. Relationship between boron content and antioxidant compounds in citrus leaves taken from fields with different water source. *Plant Soil,* **265**: 345-353.

Kobayashi, M; Matoh, T. and Azuma, J. 1996. Two chains of rhamnogalacturonan II are cross-linked by borate-diol ester bonds in higher plant cell walls. *Plant Physiol.*, 110: 1017-1020.

Kobayashi, M; Mutoh, T. and Matoh, T. 2004. Boron nutrition of cultured tobacco BY-2 cells. IV. Genes induced under low boron supply. *J. Exp. Bot.*, 55: 1441–1443.

Kobayashi, M; Nakagawa, H; Asaka, T. and Matoh, T. 1999. Borate-rhamnogalacturonan II bonding reinforced by Ca^{2+} retains pectic polysaccharides in higher plant cell walls. *Plant Physiol.* 119:199–203

Koshiba, T; Kobayashi, M. and Matoh, T. 2009a. Boron Nutrition of Tobacco BY-2 Cells. V. Oxidative Damage is the Major Cause of Cell Death Induced by Boron Deprivation. *Plant Cell Physiol..* 50: 26–36.

Koshiba, T; Kobayashi, M. and Matoh, T. 2009b. Boron deficiency. How does the defect in cell wall damage the cells? *Plant Signal. Beh.*, 46: 557-558.

Kot, FS. 2009. Boron sources, speciation and its potential impact on health. *Rev. Env. Sci. Biotech.*, 8: 3-28.

Lee, SG. and Aronoff, S. 1967. Boron in plants. A biochemical role. *Science.*, 158: 798-799.

Lecourieux, D; Ranjeva, R. and Pugin, A. 2006. Calcium in plant defence-signalling pathways. *New Phytol.*, 171: 249-69.

Lewis, DH. 1980. Boron, lignification, and origin of vascular plants - a unified hypothesis. *New Phytol.*, 84: 209-229.

Li, C; Dannel, F; Pfeffer, H; Römheld, V. and Bangerth, F. 2001. Effect of boron starvation on boron compartmentation, and possibly hormone-mediated elongation growth and apical dominance in pea (*Pisum sativum*) plants. *Physiol. Plant.*, 111: 212-219.

Liu, P. and Yang, YA. 2000. Effects of molybdenum and boron on membrane on lipid peroxidation and endogenous protective systems of soybean leaves. *Acta Bot. Sinica,* 42: 461-466.

Loomis, WD. and Durst, RW. 1991. Boron and cell walls. *Curr. Top. Plant Biochem. Physiol.*, 10: 149-178.

Loomis, WD. and Durst, RW. 1992. Chemistry and biology of boron. *Bio Factors,* 3: 229- 239.

Lukaszewski, KM and Blevins, DG. 1996. Root growth inhibition in boron-deficient or aluminum-stressed squash may be a result of impaired ascorbate metabolism. *Plant Physiol.,* 112:1135-1140.

Mahboobi, H; Yucel, M. and Oktem, HA. 2002. Nitrate reductase and glutamate delydrogenase activities of resistant and sensitive cultivars of wheat and barley under boron toxicity. *J. Plant Nutr.,* 25: 1829–1837.

Marschner H. 1995. *Mineral Nutrition of Higher Plants*. 2nd edn. Academic Press, San Diego, pp. 379-396.

Martinez-Ballesta, MC; Bastias, E. and Carvajal, M. 2008a. Combined effect of boron and salinity on water transport: The role of aquaporins. *Plant Signal. Behav.*, 3: 844–845.

Martínez-Ballesta, MC; Bastías, E; Zhu, C; Schäffner, A; Gozalez-Murua, B; Gonzalez-Moro, C. and Carvajal, M. 2008b. Boric acid and salinity effects on maize roots. Response of aquaporins ZmPIP1 and ZmPIP2 and plasma membrane H^+-ATPase in relation to water and nutrient uptake. *Physiol. Plant.*, 132: 479–490.

Mason, TL. and Wassermann, BP. 1987. Inactivation of red beet β-glucan synthetase by native and oxidized phenolic compounds. *Phytochemistry*. 26: 2197-2202.

Mateo, P; Bonilla, I; Fernández-Valiente, E. and Sánchez-Maeso, E. 1986. Essentiality of boron for dinitrogen fixation in *Anabaena* sp. PCC 7119. *Plant Physiol.*, 81: 430-433.

Matoh, T; Ishigaki, K; Mizutani, M; Matsunaga, W. and Takabe, K. 1992 . Boron nutrition of cultured tobacco BY-2 cells. I. Requirement for and intracellular localization of boron and selection of cells that tolerate low levels of boron. *Plant Cell Physiol.*, 33 : 1135 – 1141 .

Matoh, T. 1997. Boron in plant cell walls. *Plant Soil.* 193: 59-70.

Matoh T; Dell, B; Brown, PH. and Dell, RW. 1997. Boron in plant cell walls. *Plant Soil*, 193: 59-70

Matoh, T; Kawaguchi, S. and Kobayashi, M. 1996. Ubiquity of a borate-rhamnogalacturonan II complex in the cell walls of higher plants. *Plant Cell Physiol.*, 37: 636-640.

Matoh, T. and Kobayashi, M. 1998. Boron and calcium, essential inorganic constituents of pectic polysaccharides in higher plant cell walls. *J. Plant Res.* 111: 179-190.

Matoh, T. and Ochiai, K. 2005. Distribution and partitioning of newly taken-up boron in sunflower. *Plant Soil*, 278: 351-360.

Matsunaga, T; Ishii, T; Matsumoto, S; Higuchi, M; Darvill; A; *et al.* 2004. Occurrence of the primary wall polysaccharide rhamnogalacturonan II in pteridophytes, lycophytes, and bryophytes: Implications for the evolution of vascular plants. *Plant Physiol.*, 134: 339-351.

Molassiotis, A; Sotiropoulos, T; Tanou, G; Diamantidis, G. and Therios, I. 2006. Boron induced oxidative dammage and antioxidant and nucleolytic responses in shoot tips culture of apple rootstock EM9 (*Malus domestiac* Borkh). *Envir Exp. Bot.*, 56: 54-62.

Morré, DJ; Navas, P; Penel, C. and Castillo, FJ. 1986. Auxin-stimulated NADH oxidase (semidehydroascorbate reductase) of soybean plasma membrane: role in acidification of cytoplasm? *Protoplasma*, 133:195–97

Mozafar, A.1993. Role of boron in seed production. In: *Boron and its Role in Crop Production.* U. C. Gupta (Ed), CRC Press, Boca. Raton FL. pp. 186–206.

Miwa, K. and Fujiwara, T. 2010. Boron transport in plants: co-ordinated regulation of transporters. *Ann. Bot.*,105:1103-1108.

Miwa, K; Takano, J. and Fujiwara, T. 2006. Improvement of seed yields under boron limiting conditions through overexpression of BOR1, a boron transporter for xylem loading, in *Arabidopsis thaliana*. *Plant J.*, 46: 1084-1091.

Miwa, K; Tanaka, M; Kamiya, T. and Fujiwara, T. 2010. Molecular mechanisms of boron transport in plants: involvement of Arabidopsis NIP5;1 and NIP6;1. *Adv. Exp. Med. Biol.,* 679:83-96.

Miwa, K; Takano, J; Omori, H; Seki, M; Shinozaki, K. and Fujiwara, T. 2007. Plants tolerant of high boron levels. *Science.* 30:1417.

Miwa, K; Takehiro, K. and Fujiwara, T. 2009. Homeostasis of the structurally important micronutrients, B and Si. *Curr. Opin. Plant Biol.,* 12: 307-311.

Mühling, KH; Wimmer, M. and Goldbach, HE. 1998. Apoplastic and membrane-associated Ca^{2+} in leaves and roots as affected by boron deficiency. *Physiol. Plant.,* 102: 179–184.

Nable, RO., Cartwright, B. and Lance, RCM. 1990. Genotypic differences in boron accumulation in barley: relative susceptibilities to boron deficiency and toxicity. In: *Genetic Aspects of Plant Mineral Nutrition.* Bassam, NEI (ED). pp. 243–251. Kluwer Academic Publishers, Dordrecht, The Netherlands.

Nable, RO; Bañuelos, GS. and Paull, JG. 1997. Boron toxicity. *Plant Soil,* 193: 181-198.

Nakagawa, Y; Katagiri, T; Shinozaki, K; Qi, Z; Tatsumi, H; Furuichi, T; et.al.(2006) Arabidopsis plasma membrane protein crucial for Ca^{2+} influx and touch sensing in roots. *Proc. Natl. Acad. Sci. USA.,* 104:3639-3644.

Nakagawa, Y; Hanaoka, H; Kobayashi, M; Miyoshi, K; Miwa, K. and Fujiwara, T. 2007. Cell-type, specificity of the expression of Os BOR1, a rice efflux boron transporter gene, is regulated in response to boron availability for efficient boron uptake and xylem loading. *Plant Cell,* 19:2624-2635.

Noguchi, K; Dannel, F; Pfeffer, H; Römheld, V; Hayashi, H. and Fujiwara T. 2000. Defect in root-shoot translocation of boron in *Arabidopsis thaliana* mutant *bor 1-1. Plant Physiol.,* 156: 751-755.

Nyomora, AMS; Brown, PH. and Freeman, M. 1997. Fall foliar applied boron increases tissue boron concentration and nut set of almonds. *J. Am. Soc. Hortic. Sci.,* 122:405–409.

Nyomora, AMS; Brown, PH. and Krueger, B. 1999. Effects of rate and time of boron application on almond tissue B concentration and productivity. *HortScience,* 34:242–245.

Perica, S; Brown, PH; Connell, JH; Nyomora, AMS; Dordas, C; Hu, H. and Stangoulis, JC. 2001. Foliar boron application improves flower fertility and fruit set of olive. *HortScience,* 36:714–716.

Obermeyer, G. and Blatt, MR. 1995. Electrical properties of intact pollen grains of *Lilium longiflorum*: characteristics of the non-germination grain. *J. Exp. Bot.,* 46: 803-813.

O'Neill, MA; Eberhard, S; Albersheim, P. and Darvill, AG. 2001. Requirement of borate cross- linking of cell wall rhamnogalacturonan II for *Arabidopsis* growth. *Science,* 294: 846-849.

O'Neill, MA; Ishii, T; Albersheim, P. and Darvill, AG. 2004. Rhamnogalacturonan II: structure and function of a borate cross-linked cell wall pectic polysaccharide. *Ann. Rev. Plant Biol.* 55: 109–139.

O'Neill, MA, Warrenfeltz, D; Kates, K; Pellerin, P; Doco, T; Darvill, AG. and Albersheim, P. 1996. Rhamnogalacturonan-II, a pectic polysaccharide in the walls of growing plant cell, forms a dimer that is covalently cross-linked by a borate ester. *In vitro* conditions for the formation and hydrolysis of the dimer. *J. Biol. Chem.*, 271: 22923-22930.

Öztürk, Ö; Soylu, S; Ada, R; Gezgin, S. and Babaoglu, M. 2010. Studies on differential response of spring canola cultivars to boron toxicity. *J. Plant Nutr.*, 33: 1141-1154.

Pandey, N. and Archana. 2009. Boron stress induced changes in water stress and stomatal morphology in *Zea mays* L. and *Catharanthus roseus* L. *I. J. Plant Physiol.* 14: 310-314.

Pandey, N. and Archana 2011. Physiological and biochemical changes in *Catharanthus roseus* L. in response to boron nutrition. *J Plant Nutr.*, 34: 1797–1810.

Pandey, N. and Archana 2012. Membrane damage in an oxygen-free radical-dependant manner induced via boron deficiency and toxicity in maize. *Plant Stress (accepted).*

Pandey, N. and Gupta, B. 2012. The impact of foliar boron sprays on reproductive biology and seed quality of black. *J Trace Ele. Med., Biol.*, http://dx.doi.org/10.1016/j.jtemb.2012.07.003.

Pérez-Castro, R; Kasai, K; Gainza-Cortés, F; Ruiz-Lara, S; Casaretto, JA; Peña-Cortés, H; Tapia, J; Fujiwara, T. and González, E. 2012. VvBOR1, the grapevine ortholog of AtBOR1, encodes an efflux boron transporter that is differentially expressed throughout reproductive development of *Vitis vinifera* L. *Plant Cell Physiol.*, 53: 485-494.

Parr, AJ; and Loughman, BC. 1983. Boron and membrane function in plants . In: Metals and micronutrients. uptake and utilization by plants. Robb, DA. and Pierpoint, WS. (Eds). Academic Press , New York . pp. 87 – 107.

Pfeffer, H; Dannel, F. and Römheld, V. 1998. Are there connections between phenol metabolism, ascorbate metabolism and membrane integrity in leaves of boron deficient sunflower plants? *Physiol. Plant.*, 104: 479-485.

Pfeffer, H.; Dannel, F. and Römheld,V. 1997. Compartmentation of boron in roots and its translocation to the shoot of sunflower as affected by short term changes in boron supply, In: *Boron in Soils and Plants.* Bell, RW. and Rerkasem, B. (Eds). Kluwer Academic Publishers, Dordrecht. pp. 203-207.

Pfeffer, H; Dannel, F. and Römheld, V. 2001. Boron compartmentation in roots of sunflower plants of different boron status: A study using the stable isotopes ^{10}B and ^{11}B adopting two independent approaches. *Physiol. Plant.*, 113: 346-351.

Pfeffer, H; Dannel, F. and Römheld, V. 1999. Are there two mechanisms for boron uptake in sunflower? *J. Plant Physiol.*, 155: 34-40.

Pilbeam, D J. and Kirkby, EA. 1983. The physiological role of boron in plants. *J. Plant Nutr.*, 6: 563-582.

Pillinger, JM; Cooper, JA. and Ridge, I. 1994. Role of phenolic compounds in antialgal activity of barley straw. *J. Chem. Ecol.*, 20: 1557- 1569.

Pollard, AS; Parr, AJ. and Loughman, BC. 1977. Boron in relation to membrane function in higher plants. *J. Exp. Bot.*, 28: 831–841.

Radeva, V; Topchieva, A. and Simeonova, L. 1980. Physiological and biochemical changes in alfalfa under boric acid treatment. *Rast. Nauk,*. 17: 3-10.

Ramón, AM; Carpena, Ruiz, RO. and Gárate, A. 1989. *In vitro* stabilization and distribution of nitrate reductase in tomato plants: incidence of boron deficiency. *J. Plant Physiol.,* 135: 126-128.

Rawson, HM. 1996. The developmental stage during which boron limitation causes sterility in wheat genotypes and the recovery of fertility. *Aust. J. Plant Physiol.,* 23: 709-717.

Rawson, HM. and Noppakoonwong, RN. 1996. Effect of boron limitation in a combination with changes in temperature, light and humidity on floret fertility in wheat. In: *Sterility in Wheat in Subtropical Asia: Extent, Causes and Solutions.* Rawson, HM. and Subedi, KD (Eds). ACIAR, Canberra, Australia, pp. 85-89.

Redondo-Nieto, M., Pulido, L., Reguera, M., Bonilla, I. and Bolaños, L. 2007. Developmentally regulated membrane glycoproteins sharing antigenicity with rhamnogalacturonan II are not detected in nodulated boron deficient *Pisum sativum. Plant Cell Env.,* 30: 1436–1443

Redondo-Nieto M, Rivilla R, El-Hamdaoui A, Bonilla I, Bolaños L (2001) Boron deficiency affects early infection events in the pea-Rhizobium symbiotic interaction. *Aust. J. Plant Physiol.,* 28: 819-823.

Reid, R. 2007a. Update on boron toxicity and tolerance in plants. In: *Advances in Plant and Animal Boron Nutrition.* Xu, F; Goldbach, HE; Brown, PH; Bell, RW; Fujiwara, T; Hunt, CD; Goldberg, S. and Shi, L. (Eds). Springer, Dordrecht, The Netherlands, pp. 83-90.

Reid, R. 2007b. Identification of boron transporter genes likely to be responsible for tolerance to boron toxicity in wheat and barley. Plant Cell Physiol. 48: 1673–1678.

Reid, R. and Fitzpatrick, K. 2009. Influence of leaf tolerance mechanisms and rain on boron toxicity in barley and wheat. *Plant Physiol.,* 151: 413–420.

Reid, RJ; Hayes, JE; Post, A; Stangoulis, JCR. Graham, RD. 2004. A critical analysis of the causes of boron toxicity in plants. *Plant Cell Environ.,* 25: 1405-1414.

Rerkasem, B; Bell, RW; Lodkaew, S. and Loneragan, JW. 1997. Relationship of seed boron concentration to germination and growth of soybean (*Glycine max*). *Nutr. Cycling Agroecosyst.,* 48: 217–223.

Rerkasem, B; Lordkaew, S. and Jamjod, S. 1991. Assessment of grain set failure and diagnosis for boron deficiency in wheat. In: *Wheat for non- traditional warm areas.* Saunders, DA. (ED) Mexico , CIMMYT, pp. 500–504.

Rerkasem, B; Netsangotop, R; Lordkaew, S. and Cheng, C. 1993. Grain set failure in boron deficient wheat. *Plant Soil,* 155/156: 309–312.

Rerkasem, B. and Loneragan. JF. 1994. Boron deficiency in two wheat genotypes in a warm, subtropical region. Agron. J., 86: 887–890.

Robbertse, PJ; Lock, JJ; Stoffbey, E. and Coetzer, LA. 1990. Effect of boron on directionality of pollen tube growth in *Petunia* and *Agapanthus. S. Afr. J. Bot.,* 56: 487-492.

Roldin, M; Belver, A; Rodriguez-Rosales, MP; Ferrol, N. and Donaire, JP. 1992. *In vivo* and *in vitro* effects of boron on the plasma membrane proton pump of sunflower roots. *Physiol. Plant.*, 84: 49-54.

Römheld, V. and Marschner, H. 1991. Functions of micronutrients in plants. In: Micronutrients in Agriculture, 2nd ed. Mordvedt, JJ; Cox,FR; Shuman, LM. and Welch RM. (EDs). SSSA Book Series, No. 4, Madison, WI, USA. Pp. 297–328.

Ruiz, JM; Baghour, M; Bretones, G; Belakbir, A. and Romero, L. 1998a. Nitrogen metabolism in tobacco plants (*Nicotiana tabacum* L.): role of boron as a possible regulatory factor. *Int. J. Plant Sci..* 159: 121-126.

Ruiz, JM; Bretones, G; Baghour, M; Ragala, L; Belakbir, A. and Romero, L. 1998b. Relationship between boron and phenolic metabolism in tobacco leaves. *Phytochemistry*, 48: 269–272.

Ruiz, JM., García, PC., Rivero, RM. and Romero, L. 1999. Response of phenolic metabolism to the application of carbendazin plus boron in tobacco. *Physiol. Plant.*,106: 151-157.

Ryden, P; Sugimoto-Shirasu, K; Smith, AC; Findlay, K; Reiter, WD. and McCann, MC. 2003. Tensile properties of Arabidopsis cell walls depend on both a xyloglucan cross-linked microfibrillar network and rhamnogalacturonan II-borate complexes. *Plant Physiol.*, 132:1033-40.

Sotiropoulos, TE; Molassiotis, A; Almaliotis, D; Mouhtaridou, G; Dimassi, K; Therios, I. and Diamantidis, G. 2006. Antioxidant responses of the apple rootstock mm 111 shoots cultured under high boron concentrations in vitro. *J Plant Nutr.*, 29: 575-583.

Sharma, CP; Sharma, PN. and Bisht, SS. 1984. Induction of water stress by boron deficiency. *Proc. VIth International Colloquium for the Optimization of Plant Nutrition* Vol. 4 (Martin-Prevel, Eds.), IAOPN/GERDT Montpellier, pp. 1303-1307.

Sharma, PN. and Ramachandra,, T. 1990. Water relations and photosynthesis in mustard plants subjected to boron deficiency. *Ind. J. Plant Physiol.*, 33: 150-154.

Shelp, BJ; Marentes ,E; Kitheka, AM. and Vivekanandan, P. 1995. Boron mobility in plants. *Physiol. Plant.*, 94: 356-361.

Shen, ZG; Liang, YC. and Shen, K. 1993. Effect of boron on the nitrate reductase activity in oilseed rape plants. *J. Plant Nutr.*, 16: 1229-1239.

Sherrell, CG. 1983. Effect of boron application on seed production of New Zealand herbage legumes. *New Zealand. J. Exp. Agric.*, 11: 113-117

Shorrocks, VM. 1984. Boron deficiency its prevention and cure. Borax holdings Limited, Borax House, Carlisle Place, London SWIP IHT, England.

Shorrocks, V. 1997. The occurrence and correction of boron deficiency. *Plant Soil*, 193: 21-148.

Sinha, P; Sharma, CP. and Chatterjee, C. 1999. Seed quality of sesame (*Sesamum indicum*) as influenced by boron. *Ind. J. Agric. Sci.* 69: 14-16.

Singh, R; Sharma, RR. and Tyagi, SK. 2007. Pre-harvest foliar application of calcium and boron influences physiological disorders, fruit yield and quality of strawberry (*Fragaria ananassa* Duch.). *Scient. Hort.* 112: 215-220.

Smith, TE; Hetherington, SE; Ascher, CJ. and Stephenson, RA. 1997. Boron deficiency of avocado 1. Effect on pollen viability and fruit set. In: *Proc. Intl. Symp. on Boron in soils and Plants.* RW, Bell. and Rerkasem, B. (EDs). Kluwer Academic, Dordrecht.

Stangoulis, JCR; Brown, PH; Bellaloui, N; Reid, RJ. and Graham, RD. 2001a. The efficiency of boron utilisation in canola. *Aust. J. Plant Physiol.*, 28: 1109-1114.

Subedi, KD; Gregory, PJ; Summerfield, RJ. and Gooding, MJ. 1998. Cold temperatures and boron deficiency caused grain set failure in spring wheat *(Triticum aestivum* L.) *Field Crops Res.*, 57: 277-288.

Sutton,T; Baumann, U; Hayes, J; Collins, NC; Shi, BJ. and Schnurbusch, T. 2007. Boron-toxicity tolerance in barley arising from efflux transporter amplification. *Science*, 318: 1446–1449.

Takano, J; Miwa, K; Yuan, LX; von Wiren, N. and Fujiwara, T. 2005. Endocytosis and degradation of BOR1, a boron transporter of *Arabidopsis thaliana*, regulated by boron availability. *Proc. Natl. Acad. Sci. USA.*, 102: 12276-12281.

Takano, J; Miwa, K. and Fujiwara, T. 2008. Boron transport mechanisms: Collaboration of channels and transporters. *Trends Plant Sci.*, 13: 451–457.

Takano, J; Noguchi, K; Yasumori, M; Kobayashi, M; Gajdos, Z; Miwa, K; Hayashi, H; Yoneyama, T. and Fujiwara, T. 2002. Arabidopsis boron transporter for xylem loading. *Nature,* 420: 337–340.

Takano, J; Wada, M; Ludewig, U; Schaaf, G; von Wiren, N. and Fujiwara, T. 2006. The Arabidopsis major intrinsic protein NIP5;1 is essential for efficient boron uptake and plant development under boron limitation. *Plant Cell,* 18: 1498–1509.

Takano, J; Yamagami, M; Noguchi, K; Hayashi, H. and Fujiwara, T. 2001. Preferential translocation of boron to young rosette leaves in *Arabidopsis thaliana* regulated by the BOR1 gene. *Soil Sci. Plant Nutr.*, 47: 345–357.

Tanaka, M. and Fujiwara, T. 2007. Physiological roles and transport mechanisms of boron: perspectives from plants. *Eur. J. Physiol.*, 456: 671–677.

Tanaka, M,; Wallace, IS; Takano, J; Roberts, DM. and Fujiwara, T. 2008. NIP6;1 Is a Boric Acid Channel for Preferential Transport of Boron to Growing Shoot Tissues in Arabidopsis. *Plant Cell,* 20.

Tyerman, SD; Niemietz, CM. and Bramley, H. 2002. Plant aquaporins: Multifunctional water and solute channels with expanding roles. *Plant Cell Environ.*, 25: 173–194.

Vaughan, AKF. 1977. The relation between the concentration of boron in the reproductive and vegetative organs of maize plants and their development. *Rhod. J. Agric. Res.*, 15: 163–70.

Vaughan, D. and Ord, BG. 1990. Influence of phenolic acids on morphological changes in roots of *Pisum sativum*. *J. Sci. Food Agric.*, 52:289–299.

Wallace, IS; Choi, WG; and Roberts, DM. 2006. The structure, function and regulation of the nodulin 26-like intrinsic protein family of plant aquaglyceroporins. *Biochim. Biophys. Acta.*, 1758: 1165–1175.

Wang, ZY; Tang, YL; Zhang, FS. and Wang, H. 1999. Effect of boron and low temperature on membrane integrity of cucumber leaves. *J. Plant Nutr.*, 22: 543–550.

Warington, K. 1923. The effect of boric acid and borax on the broad bean and certain other plants. *Ann. Bot.*, 37: 629-672.

Welch, RM. 1999. Importance of seed mineral nutrient reserves in crop growth and development. pp. 205–226. In: *Mineral nutrition of crops fundamental mechanisms and implications.* Rengel, Z. (ED). Food Product Press, New York.

Wimmer, MA. and Goldbach, HE. 1999. Influence of Ca^{2+} and pH on the stability of different boron fractions in intact roots of *Vicia faba* L. *Plant Biol.*, 1: 632–637.

Wimmer MA, Muhling KH, Lauchlii A, Brown PH, Goldbach HE (2003). The interaction between salinity and boron toxicity affects the subcellular distribution of ions and proteins in wheat leaves. Plant Cell Environ.: 26: 1267-1274.

Xu, H; Huang, Q; Shen, K. and Sten, Z. 1993. Anatomical studies on the effects of boron on the development of stamen and pistil of rape (*Brassica napus* L.). *Zhiwa Xuebao,* 35: 453-457.

Yamagishi M, Yamamoto Y (1994) Effects of boron on nodule development and symbiotic nitrogen fixation in soybean plants. *Soil Sci. Plant Nutr.*, 40: 265-274.

Yan, X; Wu, P; Ling, H; Xu, G; Xu, F. and Zhang, Q. 2006. Plant nutriomics in China: An overview. *Ann. Bot.*, 98: 473-482.

Yermiyahu, U; Ben-Gal, A; Kerren, R. and Reid, RJ. 2008. Combined effect of salinity and excess boron on plant growth and yield. *Plant Soil*, 304: 73-87.

Yermiyahu, U; Sarig, P. and Ben Gal, A. 2007. Boron toxicity in conjunction with salinity and rootstock in grapevine. *J. Hort. Sci. Biotech.*. 82: 547–554.

Yu, Q; Baluska, F; Jasper, F; Menzel, D. and Goldbach, HE. 2003. Short-term boron deprivation enhances levels of cytoskeletal proteins in maize, but not zucchini, root apices. *Physiol. Plant.*, 117:270-278.

Yu, Q; Hlavacka,, A; Matoh, T; Volkmann, D; Menzel, D. and Goldbach, HE. 2002. Short-term boron deprivation inhibits endocytosis of cell wall pectins in meristematic cells of maize and wheat root apices. *Plant Physiol.*, 130: 415–421.

Yu, Q; Wingender. R; Schulz, M; Baluska, F. and Goldbach HE. 2001 Short-term boron deprivation induces increased levels of cytoskeletal proteins in Arabidopsis roots. *Plant. Biol.*, 3: 335-340.

Zhang, X; Shen, Z. and Shen, K. 1994. Effect of boron on floral organ development and seed setting of rape seed (*Brassica napus* L.). *Turang Xuebas,* 31: 146-152.

3

GERBERA : CROP PHENOLOGY AND PLANT NUTRITION

Rokolhuii Keditsu

Department of Horticulture, School of Agricultural Sciences and Rural Development, Nagaland University, Medziphema 797 106, Nagaland

E-mail : rokokdts@yahoo.com

INTRODUCTION

Floriculture is increasingly regarded as one of the viable options towards diversification from the traditional field crops due to higher returns per unit area, and fast developing habit of "saying it with flowers". From a symbol of love, flowers have transformed into an industry, which is characterized by constant change, activities and excellence, unique in its character and function. Although the floriculture industry is growing at a faster rate compared to other horticultural based industries. Still there is vast scope to bridge the gap of demand and supply in domestic as well as world market in the field of floriculture (Kundu *et al.*, 1997). Floriculture industry of India (Kumar *et al.*, 2010) amounts to 167 thousand ha and 9.88 thousand tons (loose flowers) in terms of area and production respectively while the cut flowers figure to 47942 lakhs (as number) with West Bengal (44.3%), Karnataka (12.2%), Maharashtra (11.9%), Gujarat (10.6%) etc as major partners in cut flowers production, trade and economics.

Gerbera is the latest sensation to Indian Floriculture, commercially grown throughout the world in a wide range of climatic conditions. According to the global trends in floriculture, gerbera occupies the 4th place among cut flowers (Sujatha *et al.,* 2002). At present Gerbera comprises of 45 species, native to tropical Asia and Africa. Out of which, seven species were recorded in India distributed in temperate Himalayas from Kashmir to Nepal at an altitude of 1,300 to 3,200 meters. Gerbera species of Indian origin are: *Gerbera andria, G.kunzeana, G. languinosa, G. macrophylla, G. nivea, G. ovalifolia* and *G.poiloselloides*. The cultivated species in this genus are *Gerbera asplenifolia, G. aurantica, G. kunzeana* and *G. viridifolia*.

Botanically known as *Gerbera jamesonii* Bolus ex Hook, it belongs to the family Asteraceae, and is commonly called as Transvaal daisy, Barbeton daisy or African daisy. The genus *Gerbera* is named in honour of German naturalist Traugott Gerber, while the species *jamesonii* was named for Dr. L.S. Jameson, a British colonial statesman. Gerbera is a herbaceous perennial, native to South Africa only (Brickell and Zuk, 1997). Tourjee *et al*. (1994) revealed that the early development of *Gerbera jamesonii* as a floricultural crop was traced from its collection as a novelty in South Africa to its establishment as a commercial crop in the 1930's. Loser (1986) grouped Gerbera into single, semi-double and double cultivars based on flower heads. The single types have one or two rows of ray florets on the periphery of the disc and the rest are disc florets while in the double types, more than two rows of ray florets are present in the flowers which enhances its attractiveness. They are further divided into 'Standard', `Spider' and `Mini' depending on the shape and size of the florets.

In order to get maximum production of quality flowers, crop phenology vis-à-vis plant nutrition holds a paramount importance. While doing so plants need to be exposed to optimum climatic conditions and nourished with appropriate nutrients during the entire growing period. However, the prime constraint in Gerbera cultivation is that the flowers are available in the market for a limited period and so far no genuine effort has been made to extend its availability all throughout the year. Therefore, study on planting time to extend the number of harvesting is one such very promising option. Differences in planting dates often bring about variation in growth, flowering, yield and quality of gerbera. In a 3-year- trial on Gerbera conducted by Parthasarathy and Nagaraju (2003), it was observed and opined that the growth, development, flower bud initiation and flowering were faster during warmer period (April – May and June – July) whereas the shelf life of flowers was more during October – November. Similarly, studies conducted at Dharwad to evaluate the best planting time for Gerbera cv. Sath Bata showed a profound influence of staggered planting on vegetative as well as reproductive attributes, with July planting resulted in maximum flower size (Singh, 2001).

Growing media for Gerbera are claimed to be highly porous with abundance in organic matter content and well drained to provide aeration to the root system. The richness of soil in terms of organic matter helps in alteration of physico- chemical (Aishwath *et al.*, 2003) and biological properties (Peterson and Larton, 1982) to suit uninterrupted growth of Gerbera. The improvement in physiochemical properties also aides in promoting the root growth and hastening the nutrient absorption by crop plants (Chandra and Singh, 2010). Greater scope for integration of chemical fertilizers with organic supplements obtained through biological sources is reported by Raniperumal *et al*, 1988; Gupta *et al.,* 1996). The recent crisis and hike in the prices of inorganic fertilizers further warranted towards the use of organic manures in flower crops (Waheeduzzama *et al.*, 2007). Information is highly lacking with regards to changes taking place within the rhizosphere of the soil as a result of either manuring or inorganic fertilization alone, or even when combined application of both organic manures and inorganic fertilizers are pract-ced. The present studies have precisely taken note of this issue since soil health as a safeguard mechanism against any potential nutrient mining has proclaimed an apex agenda in any crop yield maximisation programme.

In the light of this breakthrough, precise diagnosis of nutrient constraints vis-à-vis plant nutrition in Gerbera is highly under doldrum, primarily on account of no plausible crop specific leaf nutrient standards, with very limited efforts made under soilless culture of Gerbera. This has further compounded the entire process of flower yield maximisation. A sincere effort in this regard, has been made to put forth some leaf nutrient diagnostic values suiting to Gerbera. The growth and development of gerbera depend on a number of factors, out of which planting time and the use of nutrients plays an important role in the growth, flowering and yield. Although, gerbera cultivation has bound to be a profitable enterprise, experimental data in their planting time and the effect of organic fertilizer along with inorganic fertilizers on the flower crops under Nagaland conditions are extremely lacking. Though some work has been done on these aspects, but results are highly location specific. In light of these breakthroughs, the crop phenology and plant nutrition have been thoroughly reviewed.

2. PHENOLOGY OF FLOWERING

A comprehensive description of flower development in Asteraceae can be found in Harris (1995). In the Gerbera capitulum, three types of flowers (or florets) can be distinguished: central disc, trans and marginal ray florets. Floret initiation in gerbera, as in most species of Asteraceae tribe Mutisieae (Harris, 1995), takes place in an acropetal or centripetal manner on the capitulum disk. Peripheral ray floret primordia are the first to appear, and their organogenesis precedes that of trans and disc floret primordial. At early stages, all flower primordial is morphologically indistinguishable; the differentiation of ray, trans and disc florets occurs at mid-developmental stages. The basic organ composition of the three types of florets is similar. Positionally orthologous to sepals, whorl 1 is occupied by bristly pappus structures, whorl 2 is composed of a bilabiate corolla with elongated, fused petals and two opposite petal rudiments, whorl 3 is composed of five fused stamens or five separate staminodes, and whorl 4 comprises of two fused carpels with one ovary and a single ovule. During flower development, anthers arrest at intermediate stages of development, resulting in the formation of non-functional staminodes in the highly zygomorphic ray and intermediate trans florets. The central disc florets are sexually perfect and less zygomorphic. In other Asteraceae species, disc florets may be completely actinomorphic while others may be homomorphic, having only ray or disc florets (Bremer, 1994).

During flowering in Gerbera, specialized features are found at the level of inflorescence development, indicated by the presence of different flower types, and at the level of organ differentiation, manifested by the special anatomy of pappus, bristals (sepals) corolla and the fused anthers. At early stages of flower primordial development, all flower types are indistinguishable, and the differentiation between marginal ray/trans flowers and central disc florets takes place later. There are three types of flowers that exist in gerbera composite blooms. The dark central cushion contains disc florets. Around this disc is a ring containing intermediate trans florets. Finally, the outer petals constitute a final ring of ray florets. Thus, Gerbera is today a powerful system for functional analysis of the general regulation of various aspects of flower development, organ differentiation and metabolism.

The phenology of flowering behaviour is characterized on the basis of:

i. time taken in terms of days consumed to bud emergence.
ii. time taken to reach bud burst stage from planting time,
iii. time taken from bud emergence stage to bud burst, and
iv. time taken for full bloom from planting time.

These four parameters were most favourable with June date of planting under humid tropical climate of North East India because of the most suitable agropedological settings achieved during this month of planting. These could be characterized as :

i. better soil moisture nearer to field capacity soil moisture content,
ii. nearby 12-15% of total rainfall taking place and ensuring the necessary moisture level within soil,
iii. triggered nutrient flow, and
iv. sunshine hours of 6-8 hours/day plus minimum variation in air temperature as diurnal variation (only 4.8°C) with relative humidity of 88.6%.

These were the optimum conditions for achieving good floral response in Gerbera if a sound foundation for better flower yield coupled with flower quality is to be laid exploiting the other co-factors very effectively and efficiently. These growing conditions could be easily transformed into agropedological analogues for exploiting upon the growing conditions with much costlier inputs (Keditsu, 2012).

It is equally interesting to pinpoint the other prevailing conditions as unsuitable for favourable flowering behaviour which was observed October and March dates of planting. In October date of planting, the monsoon rain continues to recede from September onwards, and reaches to almost a complete halt, and soil highly porous in nature, allowing leaching of otherwise stored water causing restricted movement of water and nutrients under continuously growing diurnal variation in temperature (7.9ºC). On the other hand, the soil moisture has depleted to maximum without much appreciable water and nutrient movement under highly limited sap flow movement coupled highly varying diurnal temperature (11.7ºC) incase of March date of planting. This is the maximum variation in diurnal temperature, and Gerbera plants had to exert lot more energy for both nutrient and water acquisition despite an extended day length (sunshine 8.5 hours/day). It is suggested that two different pathways for water uptake: a direct one through xylem vessels and an indirect one through the cavity of stem. However, any direct method of water uptake contributes to the growth of the plant (Meeteren, 1978). According to Yu *et al.* (1990), flowering time was negatively correlated with cut flower yield, thereby based on this they constructed a practical model to assess the efficiency of indirect selection for cut flower yield using flowering time as a market trait.

3. PLANTING TIME AND CROP PERFORMANCE

Planting time in cut flowers is considered to be the most important factor in the pretext of phenology. In a given soil and climatic conditions, the optimum planting time needs to be standardized so that the quality production could be improved besides maintaining the soil fertility. Performance of each Gerbera variety varied with respect to number of flowers per plant, stalk length and flower diameter due to several factors like genomic constitution, environmental factors, growing conditions, soil and inputs (Horn *et al.*, 1974). The influence of planting time on the performance of Gerbera vis-à-vis crop phenology should be seen from the angles of the changes that are brought into the physico-chemical properties (pH of the growing medium, oxygen supply and flooding, microbial conditions, nutrient supply level and the soil temperature) of the growing medium, the meteorological conditions, physiological conditions and other interrelated factors. All of them either individually or collectively influence the performance of Gerbera versus crop phenology in relation to planting time. These factors have been thoroughly reviewed

3.1 Effect of Substrate Properties

The performance of Gerbera is largely dependent upon substrate properties. Gerbera cultivars, Amber and Queen Rebecca were grown on rockwool slabs in hydroponic system with nutrient solution pH set at 5.0, 6.0 or 7.0. Total flower production/m^2 was greatest at pH 6.0 in cv Amber during 1990-91, but in 1991-1992 no significant effect was observed in cv. Amber or in Queen Rebecca in either year. Nutrient solution pH had no significant effect on stem length or flower weight (Guba, 1994). The Gerberas grown in polystyrene showed high rates of gas exchange and plant growth at the beginning of the cultivation period, but low flower production and quality (Papadopoulou *et al.,*1996). In another study, application of *B.subtilis* (0.5 – 1.0 x 10^6 cfu/ml) alone reduced flower yield/ plant in cv.I. by 9%, compared with controls, but the other treatments increased flower yield, with VAM + *B. subtilis* (post planting) giving the highest yield (15% more than controls). For cv.II, all treatments increased yield, with VAM + *B. subtilis* (post planting) being most effective (29% increase). It was concluded that application of *B. subtilis* and VAM can be beneficial for plant health and yield, but incase of Gerbera the effects are cultivar dependent (Jacob, 1997). Plantlets of *Gerbera jamesonii* cultivars Ensophy, Estel, Suset, Rita, Tamara and Beauty produced in vitro were transferred to a greenhouse and grown for one year. The plants were grown with two different culture methods: substrate culture (closed system) and soil culture. In substrate culture, the plants were transferred to the culture beds filled with four kinds of growing medium: rockwool, cocopeat, perlite, and vermiculite. Number of flowers/plant, flower height, flower weight, and flower diameter were far better in substrate culture than yin soil culture in both cultivars. Rockwool gave better results in number of flowers/plant, plant height, plant weight, and flower diameter in 'Ensophy', while no significant differences were observed among growing media in 'Estel'. Relatively minimum number of flowers/plant was produced in cocopeat in both 'Ensophy' and 'Estel'. The cv. Ensophy showed black spots on the petals under high air temperatures in summer time, while other cultivars showed no such symptoms (Hahn *et al,* 2001). River waste could be a potential substitute for sphagnum peat (Benedetto *et al.*, 2003). Effects of rhizosphere heating at 7-11, 15 and 20°C on flower yields and quality of 7 Gerbera cultivars (Terra Cerris, Terra Queen, Clementine, Terra Sun, Terra Mor, Terra Visa and Terra Fame) grown in soilless cultures were investigated during the winter season. Flower yield and quality significantly increased due to rhizosphere heating. The increments of flower yield and stem length were 71.34 and 19.14%, respectively, after rhizosphere heating at 20°C. The highest increments of flower yield and stem length were 90.98 and 28.32%, respectively (Wang *et al.*, 2001).

3.2 Effect of temperature

The effects of regulating growing medium temperature on the growth and flowering of *Gerbera hybrida* cv. Nini plants grown at 19-23°C in rockwool were investigated in relation to leaf water potential, photosynthesis and dry matter distribution. The number of inflorescences per crown was 1.8-2.0 and was almost constant in all temperature regimes. Water stress during the day was reduced by cooling the medium in summer, the consequent increase in stomatal conductance resulting in increased photosynthesis and dry matter accumulation in the crown, while heating the medium reduced dry matter accumulation in roots. The combination of cooling the medium in summer and heating it in winter did not have any additive effect on cut flower yield (Katsukawa *et al.,* 2000).

In another study, the periods from October to November and February to April appeared to be highly favourable for the production of flowers in the regions where minimum and maximum temperatures were around 10ºC and 26ºC, respectively (Kiranjeet *et al.,* 1997). Gerbera cultivars Lily, Blorosa, Rosabella, Goldspot and Lilabella were studied for flower production from April 1997 to May 1998 under open conditions at Ludhiana. The best cultivar was Goldspot, producing a monthly mean of 23.56 flowers per unit area (3.78 flowers per plant), and the best season was April 1998, when the highest number of flowers were obtained in all cultivars (Kumar and Kumar, 2000).

Table 1 Response of different planting time on the crop phenology of open field Gerbera (Pooled data)

	Vegetative growth			Floral characteristics			Flower yield	
	Number of leaves/ plant	Leaf area (cm^2)	Plant height (cm)	Full bloom from planting time (No. of days)	Flower size (cm)	Stalk length (cm)	Fresh weight of flower (g)	Flower yield (kg/m^2)
M_1 (March)	9.17	89.31	19.88	128.20	8.25	26.46	8.74	1.59
M_2 (April)	10.99	111.97	23.10	130.44	8.09	26.35	9.60	1.83
M_3 (May)	10.78	103.87	25.09	121.63	9.05	31.99	11.41	2.25
M_4 (June)	15.96	138.78	27.09	116.56	9.12	35.77	13.42	2.95
M_5 (July)	11.76	120.61	24.08	120.10	8.75	36.17	11.08	2.14
M_6 (Aug.)	10.08	66.59	20.86	129.96	8.61	29.89	9.07	1.70
M_7 (Sept.)	9.80	87.71	19.56	130.53	8.15	27.72	8.11	1.45
M_8 (Oct.)	9.94	74.97	19.71	133.56	7.82	27.65	8.08	1.46
CD (p=0.05)	3.36	11.54	1.67	2.20	0.30	2.42	0.38	0.20

Source : Keditsu (2012)

3.3 Effect of Plant Population

Optimum plant population and crop performance are inter-related. In a 2-year greenhouse trials with the cultivars Brigitte and Sylvia, the in vitro-raised plants were either transplanted to benches on 2nd, 4th and 12th June at 6 x 6 or 8 x 8 cm, or they were potted individually in 6- or 8-cm diameter pots filled with a 1:1 high peat [sphagnum] : sand mixture supplemented with a compound fertilizer, Azofoska, at 2 g/litre. The potted plants were then transferred to 22-cm tall pots filled with a 3:1 high peat:bark mixture and grown for a further 6 months. Generally the best results were obtained in 8-cm pots or by bench planting at 8 x 8 cm

(Pytlewski, 1989). Later, Shamasundaran *et al.* (2002) showed that a plot size of 1.08 square meters corresponding to a plant stand of 16, arranged in a rectangular or square shape, was ideal for cv. Diablo under greenhouse conditions. The spacing of 30x30 cm significantly produced the most number of flowers/plant/year (25.50), number of flowers/m²/season (102.40) and number 'A' grade flowers/m²/season (83.31). The spacing of 30 x 40 cm, however, produced flowers with longer stalk (52.50 cm) and larger diameter (11.60 cm). The plant spacing of 30 x 30 cm, therefore, can be recommended as the most spacing for quality production in gerbera (Gurav *et al.*, 2005). The four-row planting technique and the 30x30 cm spacing recorded the highest number of days to flowering (46.50 and 46.33 days, respectively), while the rest of the treatments recorded 45.00 days to flowering. The four-row planting technique and 30 x 20 spacing resulted in the greatest stem length (53.35 and 52.30 cm, respectively), while the three-row planting technique and 30 x 20 cm spacing recorded the maximum flower diameter of 10.05 and 9.73 cm, respectively (Gurav *et al.*, 2002a).

3.4 Effect of Photoperiod

Photoperiod and crop phenology operate in tandem for better crop performance. Three (3) Gerbera cultivars were grown at day/night temperatures of 21/14, 18/18 and 14/21°C under greenhouse. The average number of damaged flowers/m² in cultivars Fleur, Terrafame and Joyce were 14.6, 12.4 and 15.6 at 21/14°C and 12.6, 9.6 and 12.8 at 14/21°C; similarly flower weights were 16.8, 22.8 and 18.6, and 17.0, 24.5 and 19.5 under the 2 treatments. Under cold day treatments (14/21°) the flowers were slightly smaller than under the other treatments but their colour was more intense. There were few differences in stem length between treatments (Os *et al.*, 1989). The total porosity of the growing media decreased during the initial growth period of the plants and increased at the later stages of plant growth. The highest yield was obtained from plants grown under 20 h lighting period as a total of the whole experimental period. In the first part of the experimental period, when the natural light conditions were poor, plants grown under 20 h lighting period produced the highest yield. However, when the natural light conditions improved in the spring, the plants grown under 10 h lighting period, partly or continuously, produced the highest yield (Pettersen and Gislerod, 2003).

Studies were carried out on Gerbera cv. Happipot during autumn/ winter and on cv. Tempo during winter/spring in the year 1987. Both cultivars were grown under 30, 50 or 90 micro mol/m²s with a 16-h photoperiod or 60 micro mol/m²s with a 20-h photoperiod. Light treatments were provided by 400-W HPS lamps with control plants kept under ambient light conditions. The growth, development and flowering of Gerbera cv. Happipot were significantly increased under all light treatments compared with the control. Highest plant width, height, shoot dry weight and number of buds and flowers and lowest number of days to flowering were obtained under the 90 micro mol/m^{2}s for 16 h treatment. The 60 micro mol/m^{2}s for 16 h light treatment resulted in significantly higher flower number in cv. Tempo than in the control. The 90 micro mol/m^{2}s for 16 h light treatment reduced the number of days to flowering (production time) by 23 d and 11 d for cv. Happipot and Tempo, respectively. The various light treatments had more effect on plant growth, development and flowering in the autumn-winter study than in the winter-spring study and this was probably due to increased ambient light conditions during the winter-spring study (Gagnon and Dansereau, 1990). Farina *et al.* (1993) later observed that the split night temperature regime increased flower yields between 1 Jan. and 28 Feb. compared with control. This regime did not increase flower yields in the other 2 seasons. The split night temperature regime reduced heating time by 30-40% compared with the control. The pulsing night temperature regime reduced heating time by 30% compared with the control and increased flower yield by 16%.

Martinez *et al.* (1993) reported that night and day short-term (5h) substrate flooding decreased carnation assimilation rates in cultivars Orion and Diamonetta. The decrease in photosynthetic rates was greater in plants in which half the root system was flooded than in those in which the full root system was flooded. This decline was not linked to a decrease in stomatal conductance indicating that it was due to photosynthetic capacity. Flooding slightly lowered shoot water potential; this decrease was higher in treatments affecting full root systems. This suggests that water potential is not a factor causing the observed decline in photosynthesis. Noordwijk and Brouwer (1993) observed that *Gerbera jamesonii* cv Fleur roots had a higher gas-filled root porosity (E_p (% v/v)) when grown permanently in a non-aerated than aerated solution. The E_p of roots increased during 2 weeks when half the root system of a young plant was transferred to a non-aerated solution; in older plants this response was not seen. These observations suggested that gerbera is a sensitive cut flower crop to low oxygen level in the growing medium. In another study by Shin *et al.* (1994) the number of stems > 40 cm was highest in cv Gomarchi (94%) and lowest in Appelbloesm (20%). Abnormal flowers occurred under the high-temperature and short-day conditions at the end of the year, the incidence varying between cultivars; Queen Rebecca had the most abnormalities (20.6%) while Terravisa had none.

The highest number of flowers (32/plant) was produced by the plants maintained under complete cover but the difference in flower yield and flower quality was only numerically significant (Kaur *et al.*,1996). Another study suggested that Gerberas can be planted in unheated tunnels after 15 May, when the average 24-h temperature is maintained at 15ºC. The choice of cultivars is also very important (Lisiecka, 1996). Syros *et al.* (2000) observed that differences in photosynthetic parameters within the irrigation treatments and between cultivars were related to the increase of specific peroxidase activity and the appearance of new cationic peroxidase isoforms. In all experiments, the 12-h photoperiod produced the highest number of inflorescences if the same daily supplementary light integral was used. Doubling the daily light integral by extending the photoperiod to 24 h did not increase the flower yield except in the case of Estelle in one experiment. On the other hand, an increase in the daily light integral by increased supplementary light intensity strongly affected the number of inflorescences in the 12-h photoperiod. As the quantity of Gerbera flower yield was strongly affected by supplementary lighting regimes, the grower must be aware of different distributions of the same light energy over one day, which may lead to a change of about 45% in number of inflorescences. If only the number of inflorescences is considered, a combination of short photoperiod and high light intensity can be recommended (Autio, 2000).

Studies carried out on the pigmentation regulated by light in ray floret of Gerbera hybrids demonstrated that light is an important factor for pigmentation of ray floret petal in Gerbera and the petal itself acts as a light sensor site to perceive the light signal. From the different light qualities evaluated, blue light promoted gene expression of chalcone synthetase and dihydroflavonol-4-reductase, and red light enhanced the gene expression of chalcone synthetase, indicating the photoreceptors responding to blue and red light involved in the photoregulation of flower pigmentation in Gerbera (Meng *et al.,* 2004).

The simultaneous, long-term effects of three substrates (zeolite, perlite and their 1:1 mixture) and of two heating systems (lower and upper) upon the diurnal changes in photosynthesis of Gerbera (*Gerbera jamesonii*) cultivars Cyprus and Heart Breaker were investigated. The results revealed the higher net CO_2 assimilation rate and stomatal conductance than Heart Breaker, while transpiration rate was almost the same. Based on both stomatal conductance and transpiration rate, the zeolite-perlite mixture substrate exerted a more beneficial effect on the soil-plant water system. No change in chlorophyll levels, either timewise or between gerbera cultivars, was observed in response to different substrates and heating systems. Diurnal net CO_2 assimilation rate, followed only for Cyprus, was reduced significantly from morning to late afternoon, possibly as a result of stomatal conductance reduction (Issa, 2001).

The performance of Gerbera in terms of flower yield (an index of productivity) which take into account all the growing condition at their optimum use. The flower yield related parameters such as fresh weight of flowers, number of flowers/m^2 and flower yield/m^2, all three were registered to have maximum response with the planting done in the month of June on account of suitable soil moisture (201.0 g/kg) conditioned by good rainfall (223.5 mm), air temperature range of 26.9–31.7°C (4.8°C in diurnal variation), relative humidity of 88.6% and a maximum nutrient flow (152.6 mg/kg $KMnO_4$-N, 7.2 mg/kg Bray-P and 114.3 mg/kg NH_4OAc-K). These conditions collectively created the optimum desired requisites for the plants to remain metabolically active for an extended period within the growth period. With other dates of planting, these conditions were far from optimum. For example, all the three indices of flower yield viz., fresh weight of flower, number of flowers and the flower yield were minimum due to minimum soil moisture (184.1 g/kg), lowest rainfall (96.2 mm), air temperature with maximum variation (23.52–31.4°C) and lowest nutrient regime (131.6 mg/kg $KMnO_4$-N, 6.1 mg/kg Bray-P and 104.3 mg/kg NH_4OAc-K) despite comparatively longer day length (sunshine of 5.8 hours/day) and relative humidity of 87.5%. These observations lend a strong support in favour of necessity to have different soil nutrient and climatic analogues in relation to optimum flower yield of Gerbera, if the optimum growing conditions via suitable planting time are to be harvested maximally (Table 2).

Table 2 Soil and climatic analogue for optimum production of open field Gerbera

	Analogues for Soil			
Optimum available nutrients (mg/kg)	152.6 N – 7.2 P_2O_5 – 114.3 K_2O			
Suboptimum available nutrients	< 122.1 N - 6.1 P_2O_5 – 97.6 K_2O			
	Analogues for climatic conditions			
	Soil moisture (g/kg)	Rainfall (mm)	Sunshine (hrs)	Diurnal air temperature (°C)
Optimum	201.0	223.5	6.8	4.81 (26.9 – 31.7°C)
Suboptimum	< 182.3	< 96.2	< 5.8	7.9 (23.5 – 31.4°C)

Note : Relative humidity showed no distinct difference since it varied from 82.1 to 89.9% considering March to October dates of planting

Source : Keditsu (2012)

4. SOIL FERTILITY AND PLANT NUTRITION

Demographic pressure of a burgeoning population has kept researchers on their toes to find out possible alternatives of raising productivity of horticultural crops including flower crop like Gerbera on per unit land area and time basis. Nutrient management, especially on INM basis is an approach that seeks to increase both quality of production and protect the environment for posterity. It relies on nutrient application and conservation, new technologies to increase nutrient availability to plants, and the dissemination of knowledge between farmers and researchers (Palm *et al.*, 2001). In the past, nutrient management has been driven by the need for maximizing production. But now, nutrient management that is sustainable involves maximizing production, preventing on-site soil degradation, and minimizing off-site involvement of applied nutrients (Tagaliavini and Marangoni, 2002).

Organic amendments such as manure, compost biosolids, and humic substances provide a direct source of C for soil organisms as well as an indirect source via increased plant growth and plant residue return (Bunemann *et al.*, 2006). Organic and integrated production systems offer alternatives to conventional production systems (Curl *et al.*, 2002; Peck *et al.*, 2005) irrespective of soil and crop type. However, integrated production methods have yet to attain the same widespread farmers' acceptance as organic production methods. On the other hand, when the factors relating agro-pedological conditions, indirectly taken as planting time (which exploits all the growing conditions) is standardized, both INM and planting time could well translate into an improved flower yield coupled with ray floret and disc floret quality. In this review, efforts have been made to highlight the results accrued chronologically on the above two important issues relating to Gerbera quality production.

Variation in soil fertility and plant nutrition is the single most important factor that dictates both flower yield and quality in addition to influencing soil health. Baser and Soyergin(1998) studied the fertility of greenhouse soils growing cut flowers was studied using 36 soil samples were taken from rose, carnation, freesia and gerbera greenhouses. The soils were loam-textured, non-saline, slightly and strongly alkaline, of low to medium lime content, and with medium and high levels of organic matter. Phosphorus and K contents were adequate. Bacteria present in the rhizosphere of *Gerbera jamesonii*, *Dianthus caryophyllus* and *D. barbatus* were characterized in a compacted red ferralitic soil at San Jose de las Lajas, Cuba. The rhizobacteria studied (*Pseudomonas [Burkholderia] cepacia, P. fluorescens* and *Azospirillum brasilense*) exhibited chemoattraction to root exudates. Pseudomonas was dominant over *Azospirillum*, *Bacillus* and *Streptomyces*. *P. cepacia* exhibited the maximum chemotactic responses to root exudates of both genera after 14 days and 40 minutes of exposure (Hernandez *et al.*, 2000).

Study to estimate the population of *Azotobacter* in the rhizospheres of flower crops (Gladiolus, Chrysanthemum, Aster, Tuberose, Rose and Gerbera) showed sufficient population of *Azotobacter* was present in the rhizospheres of all the flower crops grown in medium black soil (pH 7.6) of Western Maharashtra Plain Zone, India. However, the highest population (18,589) of *Azotobacter* per gram of soil was observed in the rhizosphere of Gladiolus, whereas, the least (12,820) number was seen in the rhizosphere of rose. It is suggested that *Azotobacter* prefer slightly alkaline soil for their multiplication and growth (Debnath and Wange, 2003). Studies by Soroa *et al.* (2003) showed that treatment with *G. fasciculatum* (arbuscular mycorrhizal fungi were applied at 35 spores/g of soil and the rhizobacteria at 108 cfu/g of soil.) was the most effective, with significant increases in flower diameter (27.9%), yield and flower initiation (50 days earlier) compared with the untreated control. The greatest number of flowers was produced by the untreated control, but this was not significantly different from the numbers produced with *G. fasciculatum* plus *A. brasilense* and *B. cepacia* alone.

To improve the nutrient acquisition property, efforts were made to develop an efficient growing medium specifically adapted for the mineral content of the species. The adaptation was based on the elemental analysis of the adult plant. The growth of Gerbera in adapted media was compared with the growth on Murashige and Skoog's (normally used) and Driver and Kuniyuki Walnut media. The propagation of Gerbera was best in the adapted medium; plantlets were larger and easier to separate, resulting in a higher multiplication rate. The comparison of the mineral content of plants showed that the higher uptake of Ca, P and Fe could be responsible for the improved growth. There was no exhaustion of any mineral in the media used, so differences in relative concentrations should be responsible for better growth in adapted media. (Bouman and Tiekstra,2001). The effect of various nutrients (including N, P, K, Ca, Mg, Mo and biofertilizers) on the performance of *Gerbera jamesonii* was found conspicuously different (Soroa, 2005).

A study was conducted to investigate if Gerbera plants show significant differences in growth and flower production when grown in two different soilless growing systems, with (closed system) and without nutrient solution recycling (open system). Growth indices showed no significant differences between the two systems. Nutrient concentration inside the substrate was lower in the closed system, possibly due to recycling of nutrient solution. Analyzing mineral contents in dry matter, luxury consumption of nitrogen, phosphorus and potassium was found in the open growing system, but it did not have an effect on production and quality of flowers. Total flower production was 12.5 per plant, while diameter and stem length were 10-12 and 40 cm, respectively, in both systems. Moreover, the amount of nutrient lost per year was calculated measuring the nutrient concentrations in the drainage solution. Results showed that about 93, 6.8, 165, and 107 kg/ha/year of N, P, K, and Ca, respectively, were released into the environment (Ferrante *et al.,* 2000). The other results indicated that the K/Ca, Mg/Ca and P/ (SO_4^{2-} + NO_3^-) ratio in the reference nutrient solution should be higher than the values suggested for the supply solution (Savvas,2001).

4.1 Leaf Nutrient Norms

Precision based fertilization is the call of the day for any nutrient management programme irrespective of the crop. Leaf nutrient norm facilitates a plant diagnostician to identify precisely the hidden nutrient hunger in the plant at different growth stages, without expressing the nutrient deficiency symptoms. In Gerbera, Tjia and Joiner (1989) first described deficiency symptoms along with leaf nutrient concentration. Kacperska (1985) suggested optimum leaf nutrient concentration as: 2.38-2.80% N, 0.37-0.425 P and 3.43-4.63% K using Peter variety of Gerbera grown on acid soils of Poland.

The leaf samples collected from the youngest fully open leaf (first leaf) to the oldest healthy leaf (10 the leaf) before the start of flowering of carnation were analyzed for different nutrient concentration. Nitrogen concentration in leaves varied from 3.11 (10 th leaf) to 4.47% (1st leaf) and was 3.92% in the 3rd and the 4th leaves. Phosphorus concentration ranged from 0.13 (10th leaf) to 0.24% (1st leaf) with a concentration of 0.19% in the 3rd and the 4th leaves. Potassium concentration varied from 2.92 (10th leaf) to 4.72% (1st leaf) with a concentration of 3.89% in the 4th leaf. Calcium ranged from 0.93 (1st leaf) to 2.43% (10th leaf). Magnesium ranged from 0.39 (1st leaf) to 0.82 (9th leaf), while Sulphur varied from 0.26 (10th leaf) to 0.34% (4th leaf). Iron ranged from 89 (1st leaf) to 199 ppm (10th leaf), Manganese from 230 (1st leaf) to 399 ppm (7th leaf), Zinc from 36 (2nd – 4th leaf) to 54 ppm (10th leaf) and Copper from 10(10th leaf) to 26 ppm (1st leaf) (Bhargava and Singh, 2002).

Anjaneyulu (2008) applied DRIS on Gerbera, and suggested. The optimum leaf N ranges from 2.54 to 2.87%, P from 0.13 to 0.16%, K from 3.37 to 4.37%, Ca from .145 to 1.99% Mg from 0.56 to 0.74% and S from 0.17 to 0.22%. Jeong *et al.* (2009) suggested deficient level of N as 1.60%, P as 0.07% and K as 0.40% while optimum concentration was advocated as : 5.7% N, 0.55% P and 4.0% K in Gerbera. Recent studies by Keditsu (2012) suggested leaf nutrients norms as: 2.22% to 2.56% N, 0.18% to 0.25% P_2O_5 and 1.80% to 2.20% K_2O in relation to fruit yield of 2.50 to 3.75 flowers/m^2 using DRIS (Diagnosis and Recommendation Integrated System)-based analysis for open field Gerbera on Alfisol. These leaf nutrient standards can be used for efficient fertilizer programming and to correct the nutrients in question for obtaining good flower yield and quality of flowers under protected conditions

4.2 Organic Manures as Growing Medium

Flower crop like Gerbera, being basically a cut flower crop recognizes very responsively under both soilless and open field growing conditions. Gerbera cv Dutch Lion grown on rockwool slabs, peat in plastic bags or peat, a 5:7:8 compost:rice husk:peat mixture or a 1:1 compost:peat mixture in jute sacks was highest on the 1:1 compost:peat mixture in jute sacks from August to October (6-8 months) (Pinamonti *et al.*, 1996). Pinnamonti *et al.* (1997) compared the rockwool, white peat, and 8:7:5 mix of white peat, rice chaff and compost (v/v) and a 1:1 mix of white peat and compost in soilless cultivation of Gerbera daisy (*Gerbera jamesonii*). Compost amended growing media offered a clear advantages in terms of improved plant nutrition, better vegetative growth and increased in both quantitative and qualitative characteristics of production. The physical and physico-chemical properties of compost combined with white peat offered higher production potential than with pure rockwool or peat. Furthermore, the use of compost did not lead to dangerous increases in the content of heavy metals (Zn, Cu, Ni, Pb, Ca and Cr).

Marfa *et al.* (1998) used linear parametric programming (LPP) was used to establish the most suitable substrates mixtures as growing media for cut Gerberas and carnations, according to crop requirements (aeration, available water and salinity). Experimental values of easily available water and bulk density were quite closed to predicted values of the mixtures. The results showed that LPP can be used as an efficient tool to formulate mixtures in which cattle manure compost takes part. Researches by Anuje *et al.* (2004) evaluating different media for Gerbera cultivation under polyhouse conditions demonstrated that the medium consisting of red soil + farmyard manure in a 1:1 ratio recorded the maximum values for plant height and flower stalk length. In their research, the medium consisting of cocopeat and farmyard manure in a 1:1 ratio observed the maximum values for leaf number, flowers/plant, flower diameter and vase life as well as the time to earliest flowering. The thickest flower stalk was observed in the medium containing cocopeat + garden soil + farmyard manure in 1:1:1 ratio. Similar work in a glass house, showed that the best media was a mix of saw dust: ash bran: organic manure (v/v) in 7:7:6 ratio (Wang *et al.*, 2005).

Certain microorganisms present in compost and compost extracts such as *Trichoderma*, *Rhizobacteria*, and fluorescent *Pseudomonas* are known to stimulate plant growth (Sylvia, 2005). These microbes benefit plants through different mechanisms of action including the production of secondary metabolites such as antibiotics and hormone-like substances, the production of siderophores, antagonistic to soil-borne root pathogens, phosphate solubilization, and nitrogen fixation (Dubeikovsky *et al.*,1993). Manures applied to soil improve its quality by altering the chemical and physical properties, increase organic manure content, water holding capacity, overall diversity of microbes, provide macro- and micronutrients essential for plant growth and suppress diseases with indirectly contribute to plant growth enhancement (Scheurell and Mahaffe, 2004; Heather *et al.*, 2006). Such composts having microbes of twin utility hold more promise in INM package for cut flower like Gerbera.

The trials in Argentina, micropropagated gerbera plantlets (cultivars Pacific and Donga) were grown hydroponically in the greenhouse in an organic substrate (a mixture of manures, pine needles and soil) in black plastic pots in an inert substrate (perlite) in black plastic bags. The number of flowers produced per plant and per m^2 was higher in perlite than in the organic substrates and was higher in Pacific than in Donga. Plants of cv Pacific growing in perlite produced 140 flowers/m^2 between November and February. The plants grown in perlite had longer stems and larger and more uniform flowers than plants grown in organic substrates (Mascarini, 1998). Boztok (1995) observed that stem length and number of stems/m^2 were greater for Jaguar than Barones in 1:1 mixture of peat and sand. In both cultivars, these two parameters were greatest in the pumice stone medium and least in the peat/sand mixture. Capitulum diameter of the cut flowers was greater in Barones than in Jaguar. Overall quality was best on the pumice stone medium. The yield and quality of the gerbera cultivars viz., Fame, Party, Regina and Ximena were evaluated in a heated plastic greenhouse using perlite, zeolite, sand or rockwool as substrates. Plants grown on the perlite medium had the highest total yield for all varieties, with a peak of 75-80 flowers. The lowest yield performance was found in zeolite, while rockwool and sand were intermediate. In all cases, the cultivar Party showed the highest yield, with small differences compared to the other varieties. Quality was a function of substrate, variety and production year. For most cultivars, perlite produced better quality flowers during the first period (7-8 months) of production, but no quality differences were observed among the all substrates during the second period of production (Maloupa and Gerasopoulas, 1999).

Singh and Singh (2003) carried out an experiment to partially substitute the chemical fertilizers by organic manures and bio–fertilizers in Rose cv. Gruss – an – Teplitz, and recorded maximum plant height in plants treated with FYM + dose of NPK by chemical fertilizers + *Azotobacter*. In a study on the cut flower yield and quality of *G. jamesonii* by Thane *et al.* (2007) under greenhouse conditions, 100% of the recommended fertilizer rates (RFR; 150:50:150 kg NPK/m^2), 40 or 70% RFR + 0.5 g *Azotobacter*/m^2, 40 or 70% RFR + 0.5 g PSB [phosphate solubilizing bacteria]/m^2, 40 or 70% RFR + *Azotobacter* + 0.5 g PSB/m^2, and *Azotobacter* + PSB. The application of 70% RFR + *Azotobacter* + PSB resulted in the greatest flower stalk length (52.96 cm), flower stalk diameter (0.70 cm), flower diameter (9.20 cm), number of flowers/plant (7.22) and /m^2 (70.03), and vase life (8.95 days). Of late, due to environmental concerns and diminishing supplies of peat soils for horticultural substrates, cocopeat is being considered as a renewable substitute for use as a promising substitute in growing medium in horticulture (Aswath and Padmanabhai, 2004).

The results of an experiment on the effect of crop establishment and substrates on the performance of *Gerbera jamesonii* showed that high-ridge culture enhanced survival, number of tillers, leaf growth increment and cut flower quality compared with seedbed culture. The evaluated substrates also improved leaf vigour and cut flower quality, with the combination slag + dregs of mushroom + organic fertilizer at 3:1:1 ratio, being the most effective. The air permeability of the substrate was one of the major factors that affected plant growth (Song *et al.*, 2004). The results by Sindhu *et al.* (2010) revealed that the medium amended with soil + FYM + vermicompost + samridhi + sawdust took maximum number of days (9.81 days) for the appearance of first new leaf. Leaf length (35.45 cm) and leaf width (17.24 cm) were found to be significantly higher in medium amended with samridhi. Maximum plant height (29.89 cm), number of leaves/plant (31.10 leaves), number of flowers/plant (10.03 flowers), flower head diameter (10.82 cm), flower stalk length (59.20 cm), and vase life (13.17 days) were found highest in medium amended with samridhi. Response of different INM modules using organic manures and inorganic fertilizers has produced significant increase in the performance of open field Gerbera (Table 3).

Table 3 Response of different INM-module of substrate on performance of open field Gerbera (Pooled data)

Treatments	Vegetative growth			Floral characteristics			Flower Yyeld	
	Number of leaves/ plant	Leaf area (cm^2)	Plant height (cm)	Full bloom from planting time (No. of days)	Flower size (cm)	Stalk length (cm)	Fresh weight of flower (g)	Flower yield (kg/ m^2)
T_0	10.85	44.74	21.88	135.16	7.51	25.59	9.07	1.95
T_1	12.61	60.66	26.07	133.89	7.95	29.80	10.77	2.16
T_2	17.07	80.46	29.07	136.60	8.07	26.79	11.02	2.63
T_3	12.54	85.03	26.59	133.70	8.10	26.99	11.05	2.49
T_4	13.21	99.73	29.13	129.38	9.06	26.29	12.95	3.19
T_5	15.79	86.28	29.55	130.51	7.85	26.27	9.20	1.84
T_6	11.78	97.48	25.12	131.30	7.72	25.67	9.03	1.76
T_7	16.61	90.34	26.30	131.92	7.85	25.56	9.55	2.08
CD (p=0.05)	4.63	6.12	3.12	1.98	0.62	NS	0.98	0.13

T_0 – 100% RDF (60 kg N : 40 kg P_2O_5 : 60 kg K_2O/ha), T_1 – 50% RDF + 50% Coco pith (N-equivalent basis), T_2 – 50% RDF + 50% Pig manure (N-equivalent basis), T_3 – 50% RDF + 50% FYM (N-equivalent basis), T_4 – 50% RDF + 25% Pig manure + 25% FYM, T_5 – 50% RDF + 25% Cocopith + 25% Pig manure, T_6 – 50% RDF + 25% FYM + 25% Cocopith, T_7 – 50% cocopith + 25% Pig manure + 25% FYM

Source : Keditsu (2012)

4.3 Response of Vermicompost

Vermicomposting is the bio-oxidation and stabilization of organic matter involving the joint action of earthworms (Oligochaete annelids) and microorganisms (Aira *et al.*, 2007), thereby, turning wastes into a valuable soil amendment called vermicompost. The compost is rich in both macro- and micronutrients, N fixers, and humus-forming microorganisms (Bano *et al.*, 1987), besides acting as a bioconcentrator of heavy metals and toxic substances (Edwards and Thompson, 1973). Vermicomposting could be developed and applied as a useful tool for profitable utilization of organic wastes for:

i. organic pollution retardant by rapid reduction of bulk and elimination of offensive odour,

ii production of vermi-fertilizer for application in both annual as well as perennial crops for providing efficient nutrition, and

iii production of earthworm tissue systems for large-scale proliferation.

Native earthworm species which are surface feeders, easy to breed, and responsive to improved cultural techniques, are ideal for vermicomposting (Bugg, 1994; Wang *et al.*, 2007).

Chemical, microbial, and growth regulator analysis of earthworm casts by Grappelli *et al.* (1985) showed that these casts have nearly neutral soil pH (6.5) with 51.60% water content, 16.78% total C,1.37% inorganic C, 1.63% total N, 0.40% NO_3-N, 0.92% total PO_4-P, 0.14% available PO_4-P, 1.61% K_2O, 8.60% total Ca, 0.14% available Ca, 2.51% total Mg, 0.45% available Mg, 910.10 mg/kg Fe, 218.40 mg/kg Mn, 7.20 mg/kg Cu, 0.35 mg/kg B, 68.30 mg/kg Zn, 1.8×10^8 cells/g bacteria, 2.8×10^6 cells/g actinomycetes, 2.0×10^5 cells/g fungi, 2.75 μ m/g GA_3, 1.50 mm/g cytokinin (Indole pyruvic acid), and 3.80 μ m/g IAA on a dry weight basis. Studies later suggested an enrichment technique of vermicompost with *Azotobacter chroococcum*, *Azospirillum lipoferum*, and *Pseudomonas striata* for improved solubilization of rock phosphate (Kumar and Singh, 2001; Sailaja Kumari and Usha Kumari, 2002). Biochemical properties of vermicompost showed that activity of cellulolytic enzymes was higher in casts than in soil, while activities of urease, protease, and phosphatase were lower in worm casts than undigested soil (Zhang *et al.*, 2000). These observations suggested that earthworms use microorganisms as a secondary food resource (Zachariah and Chhonkar, 2004; Zhang *et al.*, 2004).

Bugg (1994) reported about the involvement of earthworms in the process of nitrification. Influence of two species of earthworm viz., EF and EE was studied by Talashilkar *et al.* (1999) on the changes in chemical parameters governing the compost maturity of local grass, mango leaves, and farm wastes. A decrease in the C: N ratio and an increase in humic acid, cation exchange capacity, and water soluble carbohydrate were observed up to 150 days of composting. In another study, Rao *et al.* (1997) observed considerable increase in available K extracted from the worm casts over non-ingested soil, due to partial conversion of non-exchangeable K with exchangeable form, as a result of shift in soil K equilibrium.

The benefits of earthworms on soil physical conditions are well documented (Atiyeh *et al.*, 2001; Mota *et al.*, 2007; Munnoli and Bhosle, 2008). These benefits include: mixing of organic manure from the surface into lower soil horizons, improvement of aggregate stability through castings, and hydraulic properties of the soil through the creation of permanent burrows (Lee and Pankhurst, 1992; Chan and Heenan, 1992; Friend and Chan, 1995; Trojan and Linden, 1998). Munnoli and Bhosle (2008) observed that microbes in 1 g of press mud derived vermicompost with 100 x 109 cfu/g held 200 g of soil in position. The actual beneficial effect varies with different species of earthworms (Lee, 1985). The exact reason is not clear, but this could be due to the higher organic carbon levels and better drainage found in the burrowed clays (Chan *et al.*, 1988). The higher organic C levels provide a larger food reserve, making it possible to support a larger number of earthworms and better drainage of water within the soil profile. Cast production can be as high as 50

Mg/ha/year (Lal and De Vleeschauwer, 1982). In both temperate and tropical regions, the worm casts have more favourable soil conditions for plant growth (Lal and Akinremli, 1983) while, there have been evidences of favourable effect of earthworm activity on the availability of N and P in the soil, relatively less is known about its effect on the behaviour of K. Zaller (2006) suggested vermicompost extract very effective as foliar spray in field grown tomato.

4.4 Response of Pig Manure

Vermicompost-based pig manure has been promoted as a viable alternative container media component for the horticulture industry. The incorporation of vermicompost of pig manure origin into germination media upto 20% v/v enhanced root and shoot weight, leaf area, and root : shoot ratios of both, tomato and French marigold seedlings; however, amendment with vermicompost had little influence on pepper and cornflower seedlings growth. Moreover, there was no affect on the germination of seeds of any species. When the seedlings of tomato, French marigold and cornflower were transplanted into 6 cell packs there was greater plant growth in the media amended with vermicompost compared to the controlled media, and the greatest growth when vermicompost was amended into both the germination and transplant media. This effect was increased when seedlings in the transplant media were irrigated with water containing fertilizer (Bachman and Metzger, 2008). Among the five growing media evaluated, growth and flower quality of Gerbera were better in soil with vermicompost followed by soil with FYM. Dalma was the prolific bearer with 70.24 flowers/plant/year followed by Savannah and Rosalin. Growing media had no significant effect on yield of flowers per plant probably due to uniform fertigation given to all media (Thangal *et al.*, 2009).

Giardini *et al.* (1992) reported that Poultry manure treatment significantly decreased bulk density and increased total micro porosity, infiltration capacity, and available water capacity. Overall efficiency of organic manures with respect to native plant nutrient availability in an acid Alfisol showed the following trend: Poultry manure > Pig manure > Farmyard manure. Other soil properties such as pH, humic, and fulvic carbon contents showed differential patterns with nutrient availability at different intervals. Suwaree *et al.* (2009) observed that the treatments of Pig manure or cow manure with chemical fertilizer showed the best performance with maximum values in leaf and flower number. The medium treated with pig manure extract also performed well. These results revealed a potential for growing pot gerbera using organic fertilizers alone like Pig manure.

4.5 Response of Cocopeat

Cocopeat is an organic amendment, and a multi-purpose growing medium, that provides new opportunities for potting mixture suppliers, seedling, nurseries, hydroponic growers and a dynamic source for substrate development for a variety of horticultural crops. Gerbera growers worldover have been using cocopeat very successfully as growing medium, especially in form of cocopeat disks. Cocopeat has the same role to play with respect to changes in physical and chemical properties of growing medium as it does with any other organic manure. The pH of cocopeat is 5.8-6.4 an ideal pH for any crop to grow including Gerbera. According to Pivot (1988), physical properties were improved with different rates of peat or absorbent rockwool mixed with the 2 raw materials. All mixtures were fertilized before cultivation of Gerbera, following usual recommendations. Cultivation in the 2 raw materials gave poor results compared with mixtures of those containing 40% (by volume) peat or 20% rockwool, or with the control (80% peat + 20% perlite).

Gerbera growers prefer cocopeat as growing medium on account of:

i. amendment behaviour,
ii. high level of water absorbing capacity,
iii. material free of weeds and pathogens,
iv. acts as soil conditioner,
v. seasonal product but available round the year, and
vi. Slow decomposing with good stability (Abad *et al.*, 2002).

Paradiso and Pascale (2008) observed that addition of coco fibre to the perlite raised the rate of net assimilation (13.7 versus 12.1 m^2/s) of Gerbera plants compared to perlite alone, and higher differences were observed during the warmer months. This was accompanied by higher leaf area coupled with number of stems (54 versus 49 stems/plant) and higher concentration of leaf nutrients. Aswath and Padmanabha (2004) reported that a medium containing 100% cocopeat supplied with nutrient solution having EC of 1.8 dS/m produced good quality Gerbera flowers in addition to higher concentration of N, P, K and Cu in index leaves analysed after 9 months of transplantation. Gupta *et al.* (2004) in another study found the best response of cocopeat + saw dust + sand (1:1:1 v/v) with respect to number of days required for bud initiation (75.0), flowers diameter (9.75 cm), flowers stalk length (35.31 cm), number of flowers/plant (29.55), number of flowers/m^2 (210.0) and vase life (13.35 days) with a net profit of Rs. 18,794/100 m^2.

Khalaj and Amiri (2012) observed that the medium containing perlite + peat + expanded clay (25% + 70% + 5%) had significant difference with other media in number of flowers, flower disk diameter, shoot diameter, shoot neck diameter, flower height and vase life. In this media, number of flowers, flower disk diameter, shoot diameter, shoot neck diameter, flower height and vase life were 207 per m^2/year, 12.4 cm, 0.8 cm, 0.58 cm, 54.5 cm, and 11.6 days, respectively. With respect to the quantitative and qualitative traits of gerbera flower, among the studied substrates in this experiment, mixture of perlite + peat + expanded clay (25% + 70% + 5%) was the best medium suitable for Gerbera.

Gerbera could be forced in various kinds of soils, however growing medium should be well aerated, with good water holding capacity, good drainage and good physical structure. In heavy soils without enough drainage, the development of root system is suppressed and plants are more susceptible to soil borne diseases (Beattie and White, 1992). Due to high lily requirements for growing medium, growers very often use boxes instead of planting bulbs directly into the soil on greenhouse beds. Sphagnum peat with different additives, such as soil, sand or bark is usually used in different proportions. Since the last few years cocopeat, also known as coir dust or coconut mesocarp, has been considered as a renewable sphagnum peat substitute for the use in horticulture (Pickering, 1997; Yau and Murphy, 2000). Cocopeat has good physical properties, high total pore space, high water content, low shrinkage, low bulk density and slow biodegradation (Evans *et al.*, 1996; Prasad, 1997). The results of many experiments revealed that cocopeat used alone, or as a component of soil medium, is suitable for roses (Blom, 1999), gerbera (Labeke and Dambre, 1998), many potted plants (Kreij and Leeuven, 2001; Meerow, 1995; Noguera *et al.*, 1997; Pickering, 1997; Treder and Nowak, 2002) and also for Gerberas. Due to usually high initial level of potassium and sodium in cocopeat, the fertilization program should be adjusted carefully to plant requirements.

The number of flowers produced per month was highest in the peat + perlite medium and lowest on pumice. Irrigation frequency had little effect on any of the parameters measured. Photosynthetic rate was higher in plants grown on soil than in those grown on the other media. Evapotranspiration was highest in plants grown on peat + perlite (Maloupa *et al.*, 1996). Studies with 3 Gerbera cultivars showed that cut flower yields and quality are higher on mineral wool than on peat (Chmeil and Wrega, 1996). In another study, Lenzi *et al.* (1998) assessed the effects of substrate (large 3-5 mm perlite particles, fine 0.5-2 mm perlite particles, or 2:1 ratios of large:fine perlite or large perlite:peat) and drainage (25% and 40% of total water added) on growth and performance of *Gerbera jamesonii*. Different drainage levels were achieved by varying the frequency of fertigation. Total flower production was similar in all substrates (averaging 435 flowers/m^2 over 2 years) and there was no significant difference between drainage treatments. Gerbera cv. Bora Bora was grown in pots of perlite or a 3:1 mixture of peat:perlite (v/v). Plants grown in the mixture had a 20% greater leaf mineral content and 80% higher cut flower yield than those grown in perlite alone (Rea *et al.*, 1999). Studies (Georget, 2000) were later carried out to compare the growth of *Hydrangea macrophylla*, *Ceanothus repens* and *Cupressocyparis leylandii* in pots filled with 3 substrates. Substrate 1 was composed of 40% peat + 60% pine bark, substrate 2 of 40% peat + 50% bark + 10% coir, and substrate 3 of 40% peat + 40% bark + 20% coir. Gerbera growth was best in substrate 1 (*C. repens*), substrate 2 (*C. leylandii*), and substrate 3 (*H. macrophylla*).

The effects of CO_2 concentration (350 and 1200 micro mol/mol), growing media (peat+perlite, 3:1; peat+coco dust, 1:1; and coco dust alone) and concentration of nutrient solution (EC 2.1 and 2.8 mS/cm) on the growth of gerbera (*Gerbera jamesonii*) cv. Queen Rebecca microcuttings, leaf nutrient content and the chemical properties of the growing media were investigated. Plants grown in peat + coco dust had the highest dry weights of leaves and roots. CO_2 enrichment increased all growth variables, except leaf number. Leaf nutrient content of gerbera cuttings and chemical properties of growing media after eight weeks of growth, with reference to all treatments, are discussed and presented as tables (Nowak and Gabryszewska, 2001).

Gerbera cv. Mammut was grown on coir pith media (coir pith + garden soil + farmyard manure (FYM), sawdust media (sawdust + garden soil + FYM), commercial mixture (sand + red soil + FYM), sand media (sand + FYM) and red soil media (red soil + FYM), and sprayed with gibberellic acid (GA) at 100, 150 and 200 ppm, and water (control). GA at 200 ppm produced the highest number of flowers per plant (7 flowers/ plant). The treatment combination consisting of GA at 200 ppm and coir pith media resulted in high yield and good quality flowers. Flowers produced from plants in the coir pith treatment had the longest vase life (Sekar and Sujata, 2001). Barreto and Jagtap (2002) assessed the substrate media namely, coco-peat, peat, soilrite, perlite, vermicompost, compost and garden soil in various proportions were assessed under pot culture of gerbera cv. Sangria under naturally ventilated polyhouse. Vegetative growth was best in cocopeat alone. The flower quality with respect to head and disc diameter, number of ray florets and stalk length was superior in the pots having the combination of peat and vermicompost (1:1, v/v). The highest flower yield with remarkable vase life was produced by cocopeat combined with either compost or vermicompost (1:1, v/v).

The effects of growing media (sphagnum peat, 90% white peat + 10% perlite, brown peat, 60% brown peat + 40% calcined clay, and coir dust) on the yield and quality of gerbera (*Gerbera jamesonii*) showed intensive growth in white peat + perlite and coconut peat during the first three months of cultivation. The flowering date was not significantly affected by the growing media. The highest flower yields (per plant) were obtained with coconut peat (41.9), brown peat + calcined clay (36.1) and white peat + perlite (34.9). The lowest flower yield was obtained with sphagnum peat, on which flowering was pronounced during the early period of cultivation but declined substantially after 6 months (Abou and Strojny, 2003).

Gerbera was grown through 24 months in five growing media: sphagnum peat; white peat 90% + perlite 10% (v.v.); brown peat; brown peat 60% + calcined clay 40% (v.v.), and coir dust. Cut flower production of gerbera was the highest in media: white peat 90% + perlite 10% (v.v.), brown peat 60% + calcined clay 40% (v.v.), and coir dust. These media were also favourable for flower quality. The lowest cut flower production was observed in white peat alone. However, flowering intensity was good at the beginning but in this medium decreased markedly after 6 months of cultivation, compared to the other media, when its physical properties changed significantly due to fast decomposition (Nowak and Strojny, 2003). The effects of growing media (sphagnum peat + perlite, brown peat and coco fibre) on the flower yield, stem length, inflorescence diameter and fresh weight of cut flowers of gerbera were determined. The total porosity, bulk density, shrinkage water capacity and air capacity of the growing substrates had significant effects on the number and weight of fresh flowers, with total porosity and bulk density having the most significant effects. A high correlation among the physical properties of the growing media was recorded (Nowak and Strojny, 2004).

Two organic culture media, i.e. perlite:peat:organic manure (v/v) at 9:7:4 and saw dust:ash bran:organic manure (v/v) at 7:7:6, was compared with a conventional soil culture method of cut flower of *Gerbera. jamesonii* in a greenhouse. The biomass, flower number and quality, and the concentrations of N, P, and K in leaves grown in various culture media were significantly different. The best result obtained from the recycling medium mixed with saw dust, ash bran and organic manure, and the second favourite medium was the new mixture of perlite, peat and organic manure. The resistance to cold of the plants was closely related to the concentrations of N and P in the leaves. The concentrations of N and P in the leaves were negatively related to the ratios of the plants injured by the coldness. The effect of P in improving resistance to cold was greater than that of N (Wang *et al.,* 2005). A study was conducted on *Gerbera jamesonii* grown on mixtures of cinder, mushroom residue, peat and saw dust with organic fertilizer to determine which among these mixtures of culture media is best suited for organic ecotype soilless culture of the species. Results showed that the growth of *Gerbera jamesonii* was most significant on the mixture of 3 parts cinder, one part mushroom residue, and 5 (kg) muck and a kg of compound fertilizer per m^3 medium, followed by mixture of 2 parts peat, 4 parts saw dust, 2 parts cinder, 5 parts mushroom residue and 5 parts muck and 1 part compound fertilizer per m^3 medium (Fu *et al.,* 2001).

Caballero *et al.* (2007) suggested that some peat-substitute substrates can provide Fe-chlorosis in sensitive plants, which has traditionally been ascribed to the elevated pH of the growing media. Sarkka *et al.* (2008) observed that the total yield of Gerbera under greenhouse conditions was highest (approx. 430 flowers/m^2/year) in DPG30(dark peat granules 30% + coarse sphagnum peat 70%), perlite and PG40 (Peat granules 40% coarse sphagnum peat 60%) and lowest in AApeat (extra coarse sphagnum peat) and PG10+E(peat granules 10% + Eriophorum peat 90%). The yield in AApeat decreased after the first year of cultivation and remained at that low level until the end of the experiment. In contrast, the yield in PG40 started to increase after one year. The prolonged cultivation time reduced the yield quality in all growing media. Physical parameters changed only little during the experiment. Irrigation, soil moisture, pH, EC and NO_3-N were recorded regularly. Peat based substrate stabilizers held their structure throughout the 2.5 years of cultivation (Sarkka, *et al.*, 2008). In another study, two horticultural by-products viz., coconut coir and spent mushroom compost were tested by Ahmad *et al.* (2012) for their suitability as growing media for cut flower production. No major difference in flower yield was observed between conventional media and both the horticultural by-products. However, the flower quality was improved when both of the substrate was used in combination with conventional media. Moreover incorporation of coconut coir and mushroom compost lowered pH and increased available organic matter in addition to improved plant nutrition.

4.6 Inorganic Fertilizer Requirement

Soil fertility problems associated with human-induced nutrient depletion are widespread worldwide (Tan *et al.*, 2005). The use of man-made inorganic fertilizers is a fundamental component of INM, yet it is often either under-used or over-used in the absence of information on soil test-based fertilization assessment. There is a multiplicity of methods and techniques currently available for determining nutrient requirement, emphasizing the importance attributed to an awareness of fertilizer requirements. The fertilizer requirement of annual or perennial crops depends upon the objectives of fertilization, whether the purpose is to grow the crop or feed the crop. Fertilizer experiments have not generally provided calibrated soil or leaf test data because of their short-term nature, the biennial or variable production of many tree crops, their narrow focus and the difficulty in demonstrating yield responses because tree crops have relatively low rates of nutrient removal over long periods of time. Nutrient balance is basically a sound approach in the development of fertilizer recommendation, and can be easily estimated from crop nutrient removal data.

Gerbera plants (cv. Joyce) supplied with a nutrient solution with a low K:Ca ratio (0.8-0.9:3.0-4.0 mM/litre) showed symptoms of K deficiency (shortened stems and leaf damage), whereas plants supplied with a high ratio solution (7.0:0.8 mM/litre) showed no symptoms of K deficiency. At both ratios, plants supplied with 0.6 mM/litre were of better quality than those supplied with 1.7 mM/litre (Kreij, 1989). Volpi and Farina (1987) recommended optimum fertilizer requirement as: 60 kg N, 48 kg P_2O_5 and 96 kg K_2O/1000 m^2 annually. Dufault *et al.* (1990) observed that the number of marketable flowers of Gerbera seedlings (cv Florist Strain Yellow) increased as both N and K rates increased up to 110 kg/ha, but as the N rate was increased to 220 kg/ha cull flower production increased. In the 2nd year, marketable and cull yields increased as N rate increased but increasing K rate had no effect on yields. Marketable and cull yields also increased as plant density increased from 24,000 to 72,000 plants/ha in both years. Flower size and quality were unaffected by planting density. N and K rates had no effect on flower size, quality or vase life in either year.

Aswath *et al.* (1997) found that very few characters were significantly influenced by different N application rates. Variety x N interaction were noticed for two characters. Split application of 100 kg N/ha in the first year was found to be the best for over all performance. Experiments on Gerbera cv. Maria cut flowers showed both vase life and flower diameter were influenced by the soil N and K content. The longest shelf life of cut flowers were observed from a crop grown on soil containing 33 mg N and 40 mg K/100 g soil (Alexe and Amariutei, 1998). In another study Hunmili and Paswan (2003) observed that NPK at 30:10:20 g/m^2 had the highest plant height (45.69 cm), leaf area per plant (3435.67 cm), days to first flower bud (84.60), number of flowers per plant (67.01), flower size (12.15 cm), stalk length (42.79 cm), vase life (26.35), and suckers per plant (30.40). The number of days to full bloom was highest (7.80) with no NPK fertilizers. Gurav *et al.* (2002b) observed that treatment with the highest level of N (20 g/m^2) resulted in maximum plant height (41.52 cm), stem length (50.35 cm), flower stem diameter (0.67 cm) and flower diameter (8.85 cm). Likewise, the highest levels of P (20 g/m^2) and K (20 g/m^2) also resulted in maximum plant height (40.93 and 40.02 cm), flowers per plant (30.62 and 29.00), stem length (48.63 and 45.53 cm), flower stem diameter (0.63 and 0.67 cm) and flower diameter (8.97 and 8.70 cm, respectively). Another studies by Sujatha *et al.* (2002) observed that fertigation with 80% straight fertilizers (urea, super phosphate and muriate of potash) or 80% water soluble fertilizers (with an NPK grade of 16:8:24) was the most suitable for growth. Flower stalk length was highest with 80% water soluble fertilizers (51.74 cm). Fertigation with 80% water soluble fertilizers recorded the highest flower diameter (10.20 cm). Ibiza recorded maximum flower diameter (10.29 cm). Fertigation with 80% straight fertilizers recorded the maximum length of ray florets (4.30 cm). Maximum length of ray floret was recorded in Ibiza (4.49 cm). The cumulative flower yield was highest with 80% water soluble fertilizers (89.4). The highest yields were obtained with 80% water soluble fertilizers in Ibiza (116.00) and minimum with 60% water soluble fertilizers in Rendezvous (44.00). Interaction effects due to fertigation and cultivars had a significant effect on cumulative yield.

A study was conducted by Mohariya *et al.* (2004) to investigate the effects of phosphorus (0, 7.5, 10 and 12.5 q/m^2) and potash (0, 10, 12.5 and 15 g/m^2) on the qualitative parameters and vase life of gerbera (*Gerbera jamesonii*) grown under polyhouse conditions. The highest values for flower stalk length, flower diameter, number of grade I flowers per plant and vase life were obtained with 12.5 g P_2O_5/m^2 and 15 g K_2O/m^2. Nayak *et al.* (2005) observed that a mixture of NPK at 4, 2 and 2 g/pot, respectively, was the best in respect of number of plantlets per pot (3.70), days to flower bud formation (137.20 days), number of flowers per plant (15.60) and diameter of flower (8.92 cm). With the increase of the amount of K_2O, the number of flowers per plant and diameter of flower decreased. The diameter of the flower stem was maximum (0.55 cm) under 2:2:2 g

NPK/pot. The flower stem length was the highest (36.60 cm) with NPK at 6, 1 and 2 g/pot, respectively. The parameter such as plant height, number of leaves per plant, flower yield, flower stalk length, flower diameter, flower stalk thickness and vase life of Gerbera increased, whereas the number of days before bud appearance and flowering decreased with increasing rates of phosphorus (Dalal *et al.*, 2005). The results of other studies by Deshpande *et al.* (2005) indicated that plant height, number of leaves per plant and number of flowers per plant were increased with increasing P and potash rates. The days required for flower bud appearance and flowering were minimum in higher P_2O and K_2O rates. All the observations were found to be superior in the plants treated with 12.5 g P_2O_5/m^2 and 15 g K_2O/m^2.

Higher concentrations of nitrogen increased the production of Gerbera flowers but the mean quality index and vase life did not show significant differences. The uniformness of the production was improved by a higher availability of nitrogen. Lower level of nitrogen tended to reduce the vase life (Mascarini *et al.*, 2005). In another study, flower yield, flower stalk, thickness and number of grade I and II flowers were highest with the application of 10 g N and 15g of P/m^2. The number of grade III flowers was highest (1.91) with the application of 5 g N/m^2. Application of 5 g N and 15 g P/m^2 resulted in the longest vase life of flowers (11.83 days) (Pimple *et al.*, 2006). A higher number of marketable flowers were harvested following 6 or 12 irrigations with 0.5 or 1-strength nutrient solutions compared with 2-strength solutions. Maximum flowers were harvested in May than in March and April (Hwang *et al.*, 1997).

5. CONCLUSION AND PERSPECTIVES

Gerbera cultivation gradually is experiencing paradigm changes, especially with respect to greater intervention of production driven technologies. This has induced a distinct change in per capita consumption of flowers. Thumping success of Gerbera under protected conditions has extensively discouraged growers to raise this crop under open field conditions, with the result, most of the studies have accrued out under protected conditions. The cut flowers like Gerbera is intensively cultivated, and substantial amounts of nutrients are anticipated to be removed with every harvest. The definitive role of soil-climate specific agro-techniques is being increasingly explored to sustain the flower production coupled with quality as an important flower trade attribute. Improvisation in of flower yield alongwith quality makes this maiden but important task, all the more challenging, especially in the context of vagaries associated with open field conditions. Exploiting the prevailing growing conditions (e.g. planting time) and maintaining water-nutrient synergy through an integrated nutrient supply system across crop phenophases are considered as the two most important aspects of Gerbera cultivation, and if addressed to their potential, are supposed to bring some discernible changes in the overall scenario of Gerbera cultivation in northeast India, especially in Nagaland. Unfortunately, soil as a finite source of nutrients has seldom been attempted to engineer through exogenous application of both organic manures (also carries beneficial microbes, and can function as effective bio innoculant) and inorganic fertilizers in such a way that it sustains the crop nutrient demand so effectively, keeping any potential possibility of nutrient mining at no risk. This is where, the concept like integrated nutrient management finds its utmost intervention. The benefits of such technique can only be harvested when all other growing conditions are maintained within optimum requirements.

The role of plant nutrition in improved crop performance is well established. But, its impact is to be carried forward to evaluate the possibility of extending the vase life further through elevated activity of antioxidant profile of crop. In this context, expanding the use efficiency of applied nutrients could play a vital role. Application of automated fertigation including the bubbler/pulse fertigation could find greater perspective in years to come both under greenhouse as well as open field conditions. Nutrient diagnostics, preferably cultivar-based will further pave the way towards precision oriented fertilization programme so that sustained improvement in productivity of Gerbera alongwith both flower quality and vase life turn more of a reality. Such collective efforts will surely lead towards developing export oriented production technology of Gerbera as a major cut flower domestically as well as globally.

REFERENCES

Abad, M.P., P. Noguera, R. Puchades, A. Maquieira and V. Noguera. 2002. Physico-chemical properties of some coconut dusts for use as a peat substitute for containerised ornamental plants. *Biores. Tech.*, **82**:241-245.

Abou, Hadid A.F. and Z. Strojny. 2003. Effect of different container media on the growth of gerbera. *Acta Hort.*, **608**:59-63.

Ahmad, I., T. Ahmad, A. Gulfam and M. Saleem. 2012. Growth and flowering of gerbera as influenced by various horticultural substrates. *Pak. J. Bot.*, **44**:219-299.

Aiswath O.P., S. Dravid and P.S. Deshmukh. 2003. Nitrate reductase activity and chlorophyll content in wheat as influenced by chemical fertilizers, biofertilizers and farm yard manure. *Annals Agric. Res.* India **24**(3):466-473.

Aira, M., F. Monroy and J. Dominguez. 2007. Microbial biomass governs enzyme activity decay during aging of worm-worked substrates through vermicomposting. *J. Environ. Qual.*, **36**:488-452.

Alexe, C. and A. Amariutei. 1998. Research on the influence of nutrient regime on keeping quality of Gerbera flowers. *Anal. Res. Inst. Veg. Flori.*, **15**:339-345.

Anjaneyulu, K. 2008. Diagnostic leaf nutrient norms and identification of yield limiting nutrients in gerbera grown under protected conditions using DRIS. *Indian J. Hort.*, **65**(2):176-179.

Anuje, A.A., S.R. Dalal, V.S.Gonge and R.M. Deshpande. 2004. Effect of growing media on growth, flowering and yield of gerbera under polyhouse conditions. *Orissa J. Hort.*, **32**(2):106-108.

Aswath, C. and P.Padmanabha. 2004. Effect of cocopeat medium and electrical conductivity on production of Gerbera. *J. Orn. Hort.*, **7**(11):15-22.

Aswath, C., V.A. Parthasarathy and G. Bhowmik. 1997. Response of gerbera to nitrogen. *J. Orn Hort.*, **3**(1/2):59-64.

Atiyeh, R.M., C.A. Edwards, S.O. Subler and J.D. Metzger. 2001. Pig manure vermicompost as a component of a horticultural bedding plant medium: effects on physico-chemical properties and plant growth. *Biores. Tech.*, **78**:11-20

Autio, J. 2000. Supplementary lighting regimes strongly affect the quantity of gerbera flower yield. *Acta Hort.*, **515**:91-98.

Bachman, G.R. and J.D. Metzger. 2008. Growth of bedding plants in commercial plotting substrate amended with vermicompost. *Biores. Tech.*. **99**:3155-3161.

Bano, K., R.D. Kale, and G.N. Gajanan. 1987. Culturing of earthworms. *Eudrilus eugineae* for cast production as biofertilizer. *J. Soil Biol. Ecol.*, **7**: 98-105.

Barreto, M.S. and K.B. Jagtap. 2002. Studies on polyhouse gerbera substrate. In "Proceedings of the National Symposium on Indian Floriculture in the New Millennium", pp. 173-176. Floriculture research trend in India Lal Bagh, Bangalore, February 25-27.

Baser, H. and S. Soyergin. 1998. A research on some fertility properties of the soils growing cut flowers in the greenhouses at Yalova province. *Hort. J.*, **11**(2): 79-85.

Beattie, D.J. and J.W. White 1992. Lilium–hybrids and species. In "The Physiology of Flower Bulbs". De Hertogh A.A. and Le Nard M. (Eds.), Elsevier, Amsterdam.

Benedetto, A.di., C. Boschi, R. Klasman and J. Molinari. 2003. Ornamental bedding plant growth in different growing media. *J. Orn. Hort.*, **9**(2): 171-177.

Bhargava, B.S. and K.P. Singh. 2002. Leaf sampling technique in carnation. In:"Proceedings of the National Symposium on Indian Floriculture in the New Millinium", pp. 298-299. Floriculture research trend in India Lal Bagh, Bangalore, Indian Society of Ornamental Horticulture February 25-27.

Bijay Kumar, Mistry, N.C., Brajendra Singh and Chander P. Gandhi. 2010. Indian Horticulture Database 2009, pp. 14-15.

Blom, T.J. 1999. Coco coir versus granulated roockwool and 'arching' versus traditional harvesting of roses in a recirculating system. *Acta Hort.*, **481**: 503-507.

Bouman, H. and A. Tiekstra. 2001. Mineral nutrition in tissue culture: influence on propagation and quality of the plantlets. In "Plant Nutrition Food Security and Sustainability of Agro Ecosystems through Basic and Applied Research", pp.316-317. Fourteenth International Plant Nutrition Colloquium, Hannover, Germany.

Boztok, S. 1995. Effect of different substrates on quality and production of gerbera (*Gerbera jamesonii*). *Ege. Univ. Fac. Agri.*, **32**(3): 57-64.

Bremer, K. 1994. Asteraceae: Cladistics and Classification. Portland: Timber Press.

Brickell, C. and J.D. Zuk. 1997. A-Z Encyclopedia of Garden Plants. The American Horticultural Society. D.K. Publishing Inc. New York.

Bugg, R.L. 1994. Earthworm update. *Sustain. Agric. News,* **7**: 1-4

Bunemann, E.K., G.D. Schwenka and L. Van Zwiten. 2006. Impact of agricultural inputs on soil organisms – a review. *Australian J. Soil Res.,* **44**:379-408.

Caballero, R., J. Ordovás, P. Pajuelo, E. Carmona and A. Delgado. 2007. Iron-chlorosis in Gerbera as related to properties of various types of compost used as growing media. *Commun. Soil Sci. Pl. Anal.,* **38**:2357-2369.

Chan, K.Y., W.D. Bellotti and W.P. Roberts. 1988. Changes in surface soil properties of Vertisols under dryland cropping in a semi-arid environment. *Austral. J. Soil Res.,* **26**: 509-518.

Chan, K.Y. and D.P. Heenan. 1992. Surface hydraulic properties of a red earth under continuous cropping with different management practices. *Austral. J. Soil Res.* **31**:13-24.

Chandra, Rabosh Shukla and Singh Keshav. 2010. Effect of vermicompost of different organic wastes on the germination, growth and productivity of certain crops. *J. Solid Waste Tech. & Manag,* **36**

Chmeil, H. and M. Wrega. 1996. Comparing gerbera (*Gerbera jamesonii* H. Bolus ex Hook) cultivation in a peat substrate and on mineral wool. *Adv. Agri. Sci.,* **429**:69-76.

Curl, C.L., R.A. Fenske, J.C. Kissel, J.H. Shirai, T.F. Moate, W. Griffith, G. Coronado and B. Thompson. 2002. Evaluation of take-home organophosphorus pesticide exposure among agricultural workers and their children. *Environ. Hlth. Persp.,* **110**:A787-792.

Dalal, S.R., V.S. Gonge, A.G. Pimple, Rajeshwari Deshmukh and Utgikar Swarupa. 2005. Response of phosphorus levels to growth, flowering, yield and flower quality of gerbera under polyhouse. *Ann Pl. Physiol.,* **19**(1): 120-121.

Debnath, S. and S.S. Wange. 2003. Azotobacter population in the rhizosphere of flower crops. *J. Soils Crops,* **13**(1): 189-190.

Deepak, Kumar and Ramesh Kumar. 2000. Seasonal response on gerbera cultivars. *J. Orn. Hort.,* New Series **3**(2):103-106.

Deshpande, R.M., S.R. Dalal, V.S. Gonge, A.D. Mohariya and A.A. Anuje. 2005. Effect of phosphorus and potash on growth, flowering and yield of gerbera under polyhouse conditions. *Crop Res.,* Hisar **29**(2): 268-271.

Dubeikovsky, A.N., E.A. Murdukhova, V.V. Kochetkov, F.Y. Polikarpova and A.M. Boronin. 1993. Growth promotion of black current softwood cuttings by recombinant strain *Pseudomonas florescense* BSP53a synthesizing an increased amount of indole-3-acetic acid. *Soil Biol. Biochem.,* **25**:1277-1281.

Dufault, R.J., T.L. Phillips and J.W. Kelly. 1990. Nitrogen and potassium fertility and plant populations influence field production of gerbera. *HortSci.,* **25**(12):1599-1602.

Edwards, C.A. and A.R. Thompson. 1973. Pesticides and soil fauna. *Pest. Rev.,* **45**: 1-79.

Evans, M.R., S. Konduru and R.H. Stamps. 1996. Source variation in physical and chemical properties of coconut coirdust. *HortSci.,* **31**: 965-967.

Farina, E., T. Paterniani, C. Cervelli, M. Palagi and L. Volpi. 1993. Gerbera cultivation using different night temperature regimes. *Prot. Crops.,* **22**(7): 69-74.

Ferrante, A., F. Malorgio, A. Pardossi, G. Serra and F. Tognoni. 2000. Growth, flower production and mineral nutrition in gerbera (*Gerbera jamesonii* H. Bolus) plants grown in substrate culture with and without nutrient solution recycling. *Adv. Hort. Sci.,* **14**(3): 99-106.

Friend, J.J. and K.Y. Chan. 1995. Influence of cropping on the population of a native earthworm and consequent effects on hydraulic properties of Vertisols. *Austral. J. Soil Res.,* **33**:995-1006

Fu, SongLing, Fu,YuLan and ZhengHui Gao. 2001. Selection of organic ecotype soilless culture media for *Gerbera jamesonii. Acta Hort.,* Sinica **28**(6): 538-543.

Gagnon, S. and B. Dansereau. 1990. Influence of light and photoperiod on growth and development of gerbera. *Acta Hort.,* **272**:145-151.

Georget, M. 2000. Research for new substrates [for use] in nurseries and cut flowers. *PHM Hort. J.,* **417**: 20-22, 24.

Giardini, L., F. Pimpini, M. Borin and G. Gianquinto. 1992. Effect of poultry manure on soil practices and crop quality. *J. Agri. Sci.,* Cambridge **118**:207-213.

Grappelli, A., U. Tomati, E. Gall and B. Vergari. 1985. Earthworm casting in plant propagation. *HortSci.,* **20**: 874-876.

Guba, W. 1994. The effect of pH in the root medium on production and quality of gerbera grown on rockwool. *J. Fruit Orn. Pl. Res.,* **2**(3):123-134.

Gupta, A.P., Anil, R.S. and R.P. Narwal 1996. Effect of various organic manure and fertilizer nitrogen on the performance of wheat. *Anal. Biol.* **12**(2):188-194.

Gupta, Y.C., E.E Quec Dein, S.R. Dhima and Jain Ritu. 2004. Standardization of growing media under protected environment for gerbera in mid hill of Himachal Pradesh. *J. Orn. Hort.,* **7**(1):99-102.

Gurav, S.B., S.S.D. Patil, M.T. Patil, S.M. Katwate and B.R. Singh. 2002a. Studies on planting technique and spacing for gerbera under low cost polytunnel. In "Proceedings of the National Symposium on Indian Floriculture in the New Millennium", pp. 317-318. Floriculture research trend in India, Lal Bagh, Bangalore, 25-27 February.

Gurav, S.B., B.R. Singh, S.M. Katwate and E.D. Yadav. 2002b. Standardization of fertilizer dose for flower production of gerbera under protected conditions. In "Proceedings of the National Symposium on Indian Floriculture in the New Millennium", 313-314. Floriculture research trend in India, Lal Bagh, Bangalore, 25-27-February.

Gurav, S.B., B.R. Singh, U.T. Desai, S.M. Katwate, D.S. Kakade and A.V. Dhane. 2005. Effect of spacing on yield and quality of gerbera (*Gerbera jamesonii* Bolus ex hoof. f) under polyhouse. *J. Orn. Hort.,* New-Series **8**(1): 62-64.

Hahn, EunJoo, Jeon, MinWha and Paek KeeYoeup. 2001. Culture method and growing medium affect growth and flower quality of several Gerbera cultivars. *Acta Hort.*, **548**: 385-391.

Harris, E.M. 1995. Inflorescence and floral ontogeny in Asteraceae: a synthesis of historical and current concepts. *Bot. Rev.*, **61**:193-278.

Halvey, A.H. and S. Mayak. 1981. Sensence and post harvest physiology of cut flowers Part II. *Hort. Rev.*, **3**:59-143.

Heather, M.D., G.S. Alexandra, and P.D. Richard. 2006. Compost and manure mediated impacts on soil borne pathogens and soil quality. *Soil Sci. Soc. Am. J.* **70**:347-358.

Hernandez, A., D.Garcia, M.R. Soroa and A.Z. Hernandez. 2000. Study of several genera of bacteria associated with the rhizosphere of crops of Gerbera (*Gerbera jamesonii*) and pink carnation (*Dianthus barbatus, Dianthus caryophyllus*). *Trop. Climate,* **21**(3):15-18.

Hoogerwerf, A. and W.G. Van Doorn. 1992. Number of bacteria in aqueous solutions used for post harvest handling of cut flowers. *Post Harvest Bio. Tech.* **1**(4): 295-304.

Horn, W., G. Wricke and W.E. Weber. 1974. Genotypic and environmental effects on characters expression in *Gerbera jamesonii. Gareten Sci.*, **39**(3): 289-300.

Hunmili, Terangpi and L. Paswan. 2003. Effect of NPK on growth and flowering of gerbera. *J. Orn. Hort.*, New Series **6**(1): 71-72.

Hwang, Intaek, Kim Joung Keun, Kim SeonKook, Kim WolSoo and Kim Kwangsoo. 1997. Effects of nutrient solution strength and irrigation frequency on rice-hulled media in gerbera. *RDA J. Hort. Sci.*, **39**(2):86-90.

Issa, M., G. Ouzounidou, H. Maloupa and H.I.A. Constantinidou. 2001. Seasonal and diurnal photosynthetic responses of two gerbera cultivars to different substrates and heating systems. *Sci. Hort.*, **88**(3): 215-234.

Jacob, M. 1997. Beneficial organisms increase yield and resistance of gerbera. *TASPO Hort. Mag.*, **6**(12): 56-57.

Jeong, K.Y., B. Whipker, J. McCall and C. Gunter. 2009. Characterization of nutrient disorders of Gerbera hybrid, 'Festival'. *Acta Hort.*, **843**:177-182.

Jong, J. De. 1985. Gerbera can be a long lasting cut flower. *J. Flori.*, **40**(43):111.

Kacperska, I. 1985. The effect of quantity of NPK fertilizer doses on the yield of Gerbera cv. Peter. *Orn. Pl.*, Series B **10**:105-114.

Katsukawa, K., M. Aoki, K. Inamoto, M. Doi, K. Matsuyama and H. Imanishi. 2000. Growth and flowering of gerbera plants grown on temperature-regulated rockwool medium in relation to photosynthesis and water potentials. *J. Soc. High Tech. Agri.*, **12**(4): 248-253.

Kaur, K., A.P.S. Gill and Ramesh Kumar. 1996. Effect of modified environments on plant growth and flowering production of gerbera. *Madras Agric. J.*, **83**(11): 681-683.

Keditsu Rokolhuii (2012) Effect of planting time and integrated nutrient management on the growth, flowering and yield of *Gerbera jamesonii* cv. Red Gem". Ph.D. Thesis. Nagaland University, Nagaland, India p. 1-210.

Khalaj, M.A. and M. Amiri. 2012. Effect of different substrates on quantitative and qualitative traits of gerbera (*Gerbera jamesonii*) in a nonrecyclying soilless culture system. *J. Sci. Tech. Greenhouse Cult.,* **2**(8): 47-55.

Kiranjeet, Kaur, A.P.S. Gill and Ramesh Kumar. 1997. Growth and flowering behaviour of gerbera (*Gerbera jamesonii*) in sub-tropical region of north-west plains of India. *J. Orn. Hort.,* **3**(1/2): 26-29.

Kreij, De C. 1989. Low potassium level causes serious deficiency in Gerbera. *J. Flori.,* **44**(9): 46-47.

Kreij, De C. and G.J.L. Leeuven. 2001.Growth of pot plants in treated coir dust as compared to peat. *Commun. Soil Sci. Pl. Anal.,* **32**: 2255-2265.

Kumar, V. and V.P. Singh. 2001. Enriching vermicompost by nitrogen fixing and phosphate solubilising bacteria. *Biores. Tech.,* **76**:173-175.

Kundu, K.K., V.K. Jaisingh and K.S. Suhag. 1997. India's floriculture exports – growth status, constraints and export strategies : An analysis. *Indian J. Agri. Marketing* **11**(1&2):37-40

Labeke, Van M.C. and P. Dambre Van. 1998. Gerbera cultivation on coir with recirculating of the nutrient solution: a comparison with rockwool culture. *Acta Hort.,* **458**: 357-362.

Lal, R. and O.O. Akinremli. 1983. Physical properties of earthworm casts and surface soil as influenced by management. *Soil Sci.,* **135**: 114.

Lal, R. and D. De Vleeschauwer. 1982. Influence of tillage methods and fertilizer application of chemical properties of worm castings in a tropical soil. *Soil Till. Res.,* **2**:37-49.

Lee, K.E. 1985. Earthworms: Their Ecology and Relationship with Soils and Land Use. Academic Press, Sydney, Australia, pp. 200-211.

Lee, K.E. and C.E. Pankhurst. 1992. Soil organisms and sustainable productivity. *Austral. J. Soil Res.,* **30**: 855-892.

Lenzi, A., M. Rinaldi, M. Biocca, M. Maletta and G. Pergola. 1998. Substrates and volumes of fertigation in soilless crops. *Prot. Crops.,* **27**(Supplement 12): 34-38.

Lisiecka, A. 1996. The yield of gerberas grown in cold tunnels in relation to the date of planting. *Agric. Sci. & Forest.,* **81**: 283-294.

Loser, H. 1986. New gerbera cultivars at Heidellorgg. *Deutscher Gartenbau,* **40**(32):1461-1464.

Maloupa, E., F.N. Fakhri, K. Chartzoulakis and D. Gerasopoulos. 1996. Effects of substrate and irrigation frequency on growth, gas exchange and yield of gerbera cv. Fame. *Adv. Hort. Sci.,* **10**(4): 195-198.

Maloupa, E. and D. Gerasopoulos. 1999. Quality production of four cut gerberas in a hydroponic system of four substrates. *Acta Hort.,* **491**: 433-438.

Marfà. O., J.M. Tort, C. Olivella, R. Caceres and F.X. Martinez. 1998. Cattle manure compost as substarate. II – conditioning and formulation of growing media for cut flower cultures. *Acta Hort.,* **469**:305-312.

Martíez, F.X., X. Fabregas, L Tapia and A. Caballero. 1993. Effects of short term substrates flooding on gaseous exchange of carnation. *Acta Hort.,* **342**:77-84.

Mascarini L. 1998. Gerbera cultivation in growing media. *Int. Hort.,* **6**(19):86-88.

Mascarini, L., G. Lorenzo and F. Vilella. 2005. Nitrogen concentration in nutrient solution, post harvest life and flowers commercial quality in hydroponic gerbera. *Acta Hort.*, **697**: 371-375.

Meerow, A.W. 1995. Growth of two tropical foliage plants using coir dust as a container medium amendment. *HortTech.*, **5**:237-239.

Meeteren, U. Van 1978. Water relations and keeping quality of cut Gerbera flowers. I. The cause of stem break. *Sci. Hort.*, **8**(1):68-74.

Meng, Xiang Chun., Xing Tim and XiaoJin Wang. 2004. The role of light in the regulation of anthocyanin accumulation in Gerbera hybrida. *Pl. Growth Regulat.*, **44**(3):243-250.

Mohariya, A.D., S.R. Dalal, V.S. Gonge and A.A. Anuje. 2004. Effect of phosphorous and potash on flower quality and vase life of gerbera grown under polyhouse conditions. *Orissa J. Hort.*, **32**(2):19-21.

Mota, L.C., U.V. Meeteren and C. Block. 2007. Comparison of the physical properties of vermicompost from paper mill sludge and green compost as substitutes for peat-based potting media. *Acta Hort.*, **819**: 78-84.

Munnoli, P.M. and S. Bhosle. 2008 Soil aggregation by vermicompost of pressmud and earthworms (*Oligochaete annelids*). *Curr. Sci.*, **95**:1532-1535.

Nayak, D., T. Mandal and N. Roychowdhury. 2005. Effect of NPK nutrition on growth and flowering of *Gerbera jamesonii* L. cv. Constance. *Orissa J. Hort.*, **33**(2): 11-15.

Noguera, P., M. Abad, R. Puchades, V. Noguera and J. Martinez. 1997. Physical and chemical properties of coir waste and their relation to plant growth. *Acta Hort.*, **450**: 365-374.

Noordwijk, M. van and G. Brouwer. 1993. Gas-filled root porosity in response to temporary low oxygen supply in different growth stages. *Pl. Soil*, **152**(2):187-199.

Nowak, J. and E. Gabryszewska. 2001. Mineral nutrient requirements and effects of CO_2 enrichment on Gerbera microcuttings. *J. Hort. Sci. Biotech.*, **76**(6): 670-673.

Nowak, J.S. and Z. Strojny. 2003. Effect of different container media on the growth of gerbera. *Acta Hort.*, **608**:59-62.

Nowak, J.S. and Z. Strojny. 2004. The effect of physical properties of organic growing media on cut flower yield of gerbera. Folia Universitatis Agriculturae Stetinensis, Agricultura, **94**: 133-138.

Olivella,C. C. Biel, M. Vendrell and R. Save. 2000. Hormonal and physiological responses of *Gerbera jamesonii* to flooding stress. *HortSci.*, **35**(2): 222-225.

Os, P.C.van., Koster, R. de and A.A.M. vander Wurff. 1989. Day/night temperatures affect production and quality of Gerbera. *J. Flori.*, **44**(7): 56-57.

Palm, C.A., C.N. Gachengo, R.J. Delve, G. Cadisch and K.E. Giller. 2001. Organic inputs for soil fertility management in tropical agroecosystems: application of an organic resource database. *Agri., Ecosyst., Environ.*, **83**: 27-42.

Papadopoulou, E., D. Gerasopoulos and E. Maloupa. 1996. Effect of the substrate and frequency of irrigation on growth, yield and quality of *Gerbera jamesonii Bolus* cultivated in pots. *Int. J. Agric. Sci.*, **126**(3): 297-302.

Paradiso, R. and S. Pascale de. 2008. Effort of coco fibre addition to perlite on growth and yield of cut gerber. *Acta Hort.,* **779**:529-534.

Parthasarthy, V.A. and V.Nagaraju. 2003. Evaluation of gerbera under mid-hills of Meghalaya. *J. Orn. Hort.* **6**(4):376-380.

Peck, G.M., P.K. Andrews, C. Ritcher and J.P. Reganold. 2005. Internationalization of the organic fruit market: The case of Washington States' organic apple export to be the European union. *Renew. Agri. Food Sys.,* **20**: 101-112

Peterson, H. and M. Larton. 1982. A comparative analysis of soil fauna population and their role in decomposition processes. *Oikos* **39**(3):287-288.

Pettersen, R.I. and H.R. Gislerod. 2003. Effects of lighting period and temperature on growth, yield and keeping quality of *Gerbera jamesonii Bolus. Europ. J. Hort. Sci.,* **68**(1):32-37.

Pickering J.S. 1997. An alternative to peat. *The Garden* **122**: 428-429.

Pimple, A.G., S.R. Dalal, D.R. Nandre, S.M. Ghawade, Utgikar Swarupa 2006. Yield and quality of gerbera influenced by nitrogen and phosphorus levels under polyhouse conditions. *Int. J. Agri. Sci.,* **2**(2):320-321.

Pinamonti, F., G. Stringari and G. Zorzi 1997. Use of composts in soilless cultivation. *Compost Sci. Utiliz.,* **5**(2): 38-46.

Pinamonti, F. T. Zanella and G. Zorzi. 1996. Composts and jute sacks for soilless cultivation. *Agrarian Informer,* **52**(40):47-52.

Pivot, D. 1988. Slugde and bark compost mixed or not with peat or rockwool for the cultivation of potted gerbera. *Acta.Hort.,* **221**:75-84.

Prasad, M. 1997. Physical, chemical and biological properties of coir dust. *Acta Hort.,* **450**: 21-29.

Pytlewski, C.A. 1989. Trials on the growing method and fertilization of young gerbera plants obtained from in vitro culture. III. Effect of growing method on plant growth and cropping. *Orn. Plants,* Series B. **14**: 49-54.

Raniperumal Baskaran, S., P. Duraiswamy and S. Chellanmutthu 1988. Soil test crop response studies with organic and inorganic nutrients on finger millet. *Madras Agric. J.,* **75**(5-6): 180-184.

Rao, S.C.H., S.A. Rao and P.N. Takkar. 1997. Potassium availability and release behaviour in earthworm casts and non-ingested soils. *J. Indian Soc. Soil Sci.* **45**: 310-314.

Rea, E., F.Pierandrei, P. Cantone. and M. Maletta. 1999. Mineral composition of soilless-cultured gerberas. *Prot. Crops,* **28**(6):71-75.

Rogers, M.N. 1973. A historical and critical review of post harvest physiology research on cut flowers. *Hort Sci.,* **8**:189-194.

Sailaja Kumari M.S. and K. Usha Kumari. 2002. Effect of vermicompost enriched with rock phosphate on the yield and uptake of nutrients in cowpea (*Vigna unguiculata* L. Walp.). *J. Trop. Agri.,* **40**:27-30.

Särkkä, L.E., P. Tuomola, O. Reinkainen and M. Heeranen. 2008. Long term cultivation of cut gerbera in peat-based growing media. *Acta Hort.,* **779**:423-430.

Savvas, D. 2001. Nutritional management of gerbera (*Gerbera jamesonii*) grown in a closed soilless culture system. *Acta Hort.,* **554**:175-182.

Scheurell, S.J. and W.F. Mahaffe. 2004. Compost tea as container medium drench for suppressing seedling damping-off caused by *P. ultimatum*. *Phytopathol.*, **94**: 1156-1163.

Sekar, K. and A. Sujata. 2001. Effect of growing media and GA3 on growth and flowering of gerbera (*Gerbera jamesonii* H. Bolus.) under naturally ventilated greenhouse. *South Indian Hort.*, **49**(Special): 338-341.

Shamasundaran, K.S., K.P. Singh and S.C. Mandhar. 2002. Standardization of optimum size and shape of plots for gerbera under naturally ventilated greenhouse condition. *Indian J. Hort.* **59**(4): 431-434.

Shin, H.K., Kim, J.Y., Shin, R.W. and J.Y. Ko. 1994. Seasonal fluctuation of yield and quality of cut flowers in gerbera hybrids. *J. Agri. Sci., Hort.*, **36**(1):444-452.

Sindhu S.S., D.B. Gholap, M.C. Singh and M.R. Dhiman. 2010. Effect of medium amendments on growth and flowering in gerbera. *Indian J. Hort.*, **67**(Special Issue): 391-394.

Singh, K.P. 2001. Agro-techniques for gerbera cultivation in India. *Floricult. Today* **5**:9-13.

Singh, K.P. and Sangama. 2002. Influence of cultivars and flower stalk length on vase life of gerbera. In Proceedings of the National Symposium on Indian Floriculture in the New Millennium", 324-325. Floriculture Research Trend in India Lal Bagh, Bangalore, 25,27 February.

Singh, Anil K. and Yesh Pal Singh 2003. Leaf nutrient status, growth and flower yield in rose as influenced by organic and inorganic sources. *J. Orn. Hort.*, **6**(3): 229-233.

Song, JunYang, Chang, ZongTang, Pu, YaFeng, Chen, HongWu, Qi, XiTing and MeiMei, Zhao. 2004. Effects of substrate and cultural methods on the growth of *Gerbera jamesonii*. *Acta Agric.*, Shanghai **20**(3):65-67.

Soroa, M.R.2005. *Gerbera jamesonii* L. Bolus. *Trop. Climate,* **26**(4):65-75.

Soroa, M.R., Cortes, S.L. and A.Hernandez. 2003. Study of the effect of application of biofertilizers on some variables of growth and yield in *Gerbera jamesonii* cv. Bolus. *Trop. Climate,* **24**(2):15-17.

Sujatha, K., Gowda, J.V.N. and M.M. Khan. 2002. Effects of different fertigation levels on gerbera under low cost greenhouse. *J. Orn. Hort.,* New Series **5**(1): 54-59.

Suwaree, Saijeen, Orapin Kaewman and Mookda Suksawat. 2009. Evaluation of media, organic and chemical fertilizer applications on growth of pot gerbera (*Gerbera jamesonii*) *As. J. Food Ag. Ind.*, Special Issue S51-S56.

Sylvia, E.W. 2005. The effect of compost extract on the yield of strawberries ad severity of *Botrytis cinerea*. *J. Sustain. Agri.*, **5**:57-68

Syros, T., T.Yupsanis, M. Omirou and A. Economou. 2000. Photosynthetic response and peroxidases in relation to water and nutrient deficiency in gerbera. *Environ. Exp. Bot.*, **52**(1): 23-31.

Tagaliavini, M. and B. Marangoni. 2002. Major and nutritional issues in deciduous fruit orchards of north Italy. *Hort. Tech.*, **12**: 26-31

Talashilkar, S.C., P.P. Bhangarath and V.B. Mahta. 1999. Changes in chemical properties during composting of organic residues as influenced by earthworms acidity. *J. Indian Soc. Soil Sci.,* **47**:50-53.

Tan, Z.X., R Lal and K.D. Wiebe. 2005. Global soil nutrient depletion and yield reduction. *J. Sustain. Agri.,* **26**:123-146.

Thane, S.R., R.Z. Shembalkar, S.A. Bhongle and S.A. Badge 2007. Effect of integrated nutrient management on flower quality, yield and vase life of gerbera (*Gerbera jamesonii* H. Bolus) grown under shade net conditions. *Plant Arch.,* **7**(2): 679-680.

Thangal, M., M.S. Ladaniya and V.S. Korikanthimath. 2009. Performance of gerbera varieties in different growing conditions of Goa. *Indian J. Hort.,* **66**(1):79-82.

Tjia, B. and J.N. Joiner. 1989. Growing gerberas. *Greenhouse Mang.,* **3**(6):67-70.

Tourjee, K.R., James Harding and Thomas G. Byrne 1994. Early development of Gerbera as a floriculture crop *HortTech,*. **4**:134-140.

Treder, J. and J. Nowak. 2002. U*se subfloor coconut cultivation discount plants. Adv. Agri. Sci. Problem Iss.* **485**: 335-358(in Polish with English abstract).

Trojan, M.D. and D.R. Linden. 1998. Macroporosity and hydraulic properties of earthworm affected as influenced by tillage and residue management. *Soil Sci. Soc. Am. J.,* **62**:1687-1692.

Volpi, I. and E. Farina. 1987. The nutrition of *Gerbera jamesonii* hybrid : Interaction relationships of the major plant nutrients. *Prot. Crops,* **16**(3):100-102.

Waheeduzzama, M., M. Jawaharlal, R. Arulmohiyar and K. Indhumathi. 2007. Integrated nutrient management practices to improve flower yield in anthurium (*Anthurium andrearum* Lind.). *J. Orn. Hort.,* **10**(1): 42-45.

Wang, GouLiang, Wu, ZhuHua., Tang, GengGuo, Wang, YouLiang and Li XuePing. 2001. The effects of rhizosphere heating on flower yields and quality of soilless growing Gerbera during winter. *Acta. Hort.,* Sinica **28**(2):144-148.

Wang, MinYan, Wu, LiangHuan, Zhu, WeiGen and Zhang YongYong. 2005. Study on good quality and high effective eco-organic culture medium of *Gerbera jamesonii. Acta Agric.,* Zhejiangensis **17**(5): 326-329.

Wang, L.K. Y.T. Hung and K.H. Li. 2007. Vermicomposting Process. In "Handbook of Environmental Engineering. Biosolids Treatment Processes", Vol, pp 689-704. Humana Press Inc, Taiwan,

Yau, P.Y. and R.J. Murphy. 2000. Biodegraded cocopeat as a horticultural substrate. *Acta Hort.,* **517**: 275-278.

Yu, Y., J. Harding. and T. Byrne 1990. Inheritance of flowering time and its application for indirect selection in the Davis population of Gerbera. *HortSci.* **25**(11):1357-1358.

Zachariah, A.S. and P.K. Chhonkar. 2004. Biochemical properties of compost as influenced by earthworms and feed material. *J. Indian Soc. Soil Sci.,* **52**: 155-159.

Zaller, J.G. 2006. Foliar spraying of vermicompost extract : effects on fruit quality and indicators of late blight suppression on field grown tomatoes. *Biol. Agri. Hort.,* **24**: 165-180

Zhang, W.J., J.X. Feng, J. Wu and K. Parker. 2004. Differences in soil microbial biomass and activity of six agrosystems with a management disturbance gradient. *Pedosp.*, **14**: 441-447

Zhang, B.G., T.T. Li, T.S. Shen, J.K. Wang and Z. Sun. 2000. Changes in microbial biomass C, N, and P and enzyme activities in soil incubated with the earthworms *Metaphire guillelmi* or *Eisenia foetida*. *Soil Biol. Biochem.*, **32**: 2055-2062.

4

MYCORRHIZA- ITS POTENTIAL USE FOR AUGMENTING SOIL FERTILITY AND CROP PRODUCTIVITY

Preeti, Sudhir Kumar and J.D.S. Panwar*

Department of Botany, J.V. College, Baraut (Baghpat)
** Ex. Head, Plant Physiology. IARI, New Delhi.*

INTRODUCTION

Our Country is a fast developing nation for the overall development and has proved that the country which was known for a begging bowl for food grains has turned up towards self-sufficiency and also has excess production to feed the present level of population. From about 50 million tonnes of food grain production in 1950-51, harvest in 2010-11 reached over 241 million tonnes, highest ever along with high production of fruits, vegetables, milk, egg, fish. This increased food production in the country has mainly been achieved through productivity gains, as the net sown area has remained static for the last four decades. Our burgeoning problems of fast population growth has been a serious concern and by 2020, India would require more than 280 million tonnes of food grains, 66 million tonnes of oil seeds, 95 million tonnes of fruits,150 million tonnes of milk and 11.5 million tonnes of fish. The foremost challenges of elevating malnutrition, enhancing productivity and crop diversification can be met through better resource management by developing more productive, more nutritious, but less input demanding crops (Ayyappan, 2011).

Recent developments in the field of globalizing market price and enhanced cost of production has drastically affected the crop production through costly resources with the decreasing available land for farming due to industrialization and infrastructure. The poor and marginal farmers are the real culprit with the high escalating cost of fertilizers particularly for phosphatic fertilizers with the withdrawal of subsidy on the fertilizers and has compelled the researcher to work on the less input demanding crops with high fertilizer use efficiency.

After the evolution of green revolution in late 1960 onwards, the farmers use the fertilizers as the developed varieties were responsive to fertilizers and other agriculture inputs and it was noticed that the phosphatic fertilizers which were applied to the soil were not fully utilized in the same crop year and remained adhere to the soil particles which were not readily available to the crops immediately. Such unavailable forms of phosphorus can be made available to the clients through the process of solubilizing and or mobilizing of phosphorus using the economical and chief source such as fungi (AM) and bacteria (Phosphorus solubilizing bacteria).

2. ROLE OF BIOFERTILIZERS IN PHOSPHATE SOLUBILIZATION AND MOBILIZATION

The 'P' requirement of crop is often met by addition of phosphatic fertilizers of which only 15-20% of added 'P' fertilizer is available to the plants as rest gets fixed in the soil as metal phosphates. There are certain microorganisms, which play a major role in the solubilization and uptake of native and applied phosphorus. These are otherwise referred to as phosphate solubilizing microorganisms (PSMs). PSMs include several bacteria, fungi, *Actinomycetes* & *Cyanobacteria*, dominantly species of *Bacillus, Pseudomonas, Aspergillus, Penicillium, Anabaena* etc. which bring about solubilization of bound phosphates in soil. These efficient cultures have been multiplied on mass scale and developed as inoculants. Their inoculation results in solubilizing the insoluble inorganic phosphates and can also mineralize organic phosphatic compounds present in soil. PSMs have also been reported to produce growth promoting substances.

Besides phosphate solubilization, there is another group of microorganisms which help in mobilizing nutrients for uptake by plants especially those less mobile in soil solutions like 'P'. They also help in uptake of water and protect roots from pathogens. The microorganisms responsible for this are certain fungi called AM fungi. These fungi form symbiotic association with the roots of plant important in agriculture, horticulture and tropical forestry. Of all the AM fungi, endomycorrhizae are more important as bio- fertilizers.

Biofertilizers are the promising alternative for the biological management of soil fertility and productivity. Among them mycorrhiza have a great potential in biomass production and soil enrichment on wastelands (Kashyap *et al.,* 2005). Mycorrhiza word, first used by a German Researcher Frank in 1885, originates from the Greek word myco, meaning (fungus) and rhiz, (root) (Verma *et al.,* 2008). Mycorrhizae have traditionally been classified into two types:- 'ecotrophic' (ectomycorrhiza) and 'endotrophic' (endomycorrhiza) on the basis of the fungal vegetative filaments (hyphae) in relation to the host root tissues.

VAM are the most important fungi from the agricultural point of view. The VA-mycorrhizae association has been reported in all plant families; although it may be rare or absent in families such as *cruciferae, chenopodiaceae, caryophyllaceae* and *cyperaceae* (Hirrel *et al.,* 1978, Ocampo *et al.,* 1979). The nomenclature refers to the formation of vesicles and arbuscules, typical morphological structures formed by these fungi. These fungi belong to the Endogonaceae, a family in the order *Endogonies.* The revised classification of the Endogonaceae given by Gerdemann and Trappe, (1974) include seven genera of which four are known to form VA-mycorrhizal association with plants-*Glomus, Gigaspora, Acaulospora* and *Sclerocystis.* They are distinguished by the morphology of their resting spores. Arbuscular mycorrhiza (AM) replaced the earlier term "vesicular-arbuscular mycorrhiza" (VAM) because not all endomycorrhizas of this type develop vesicles, but all form arbuscules. AM fungi form symbiotic associations with members of Bryophytes, Pteridophytes, Gymnosperms and Angiosperms throughout the world. The arbuscular mycorrhiza are by far the commonest in agricultural and horticultural plants.

Macro and micro taxonomic characteristics used for identification of AM fungi

Macro-characteristics	
Sporocarp morphology	Size, Shape, Peridium
Spore	Colour, Shape, Size
Subtending hypha	Shape, Width, Pore occlusion
Auxillary cells	Size
Mycorrhizal anatomy	Hyphal characters, Intraradical spores
Micro-characteristics	
Spore wall structure	Colour, Dimension, Number, Type
Spore germination	Ornamentation Direct, Indirect

Source – Khade 2008

2.1 Mass Culturing and Inoculum Production Of Mycorrhiza

One of the major bottlenecks for large-scale application of mycorrhiza is the unavailability of adequate inoculum. Since AM fungus is an obligate symbiont and it requires a host to complete its life cycle. There are various methods currently being used for mass-production of these fungi.

2.1.1 Conventional method or Soil-based inoculum production

Culturing AM fungi on living plants growing in disinfected soil has been the most frequently used technique for increasing the number of propagules. The conventional methods involve the use of starter inoculum/

material and grow on soil in pots using suitable trap plants. Further to bulk up on mass scale after selecting potential AM fungi the same can be propagated in pots followed by in raised beds (on-farm) under field conditions (Douds *et al.*, 2006).

2.1.2 Soilless substrate based system

AMF can be cultured on plants in suitable soilless media with appropriate nutrients. Hoagland's nutrient solution is currently being used routine to grow AMF with inert materials in pots. Bark, calcined clay, expanded clay, perlite, soilrite and vermiculite have been used as inert substrates. Pre-colonized roots, spores and hyphae of AMF could be produced in hydroponics or aeroponic systems. Pre-colonized plants on sterile substrate that are needed for these systems. However, plants can be infected directly in the aeroponic system. In hydroponics inoculum can be propagated using the nutrient film technique (NFT) by growing pre-colonized plants in a defined nutrient solution, which flows over the host roots.

Nutrient film technique (NFT) was adapted for the AMF inoculum production by Mosse and Thompson (1984). Culture host plants are placed on an inclined tray over which flows a layer of nutrient solution. In an aeroponic system, (a fine mist of defined nutrient solutions) pre-colonized roots are suspended in air and are bathed in (Hung and Sylvia, 1998). Lee and George (2005) proposed a modified nutrient film technique for large scale production of AM fungal biomass with the help of improved aeration by intermittent nutrient supply, optimum P supply, and use of glass beads as support materials. This method of AM multiplication is free from any substrate and can be sheared which resulted in to very high propagules number.

2.1.3. Root-organ culture or axenic culture of AM fungi

In vitro culture of AMF was achieved for the first time in early 1960s (Mosse, 1962). Since then various pioneering steps were aimed at axenic culturing of AM fungi. Becard and Fortin (1988) and Fortin *et al.* (2002) presented a detail evaluation of the root organ culture (ROC) technique and reported the basic improvements necessary for AM fungus colonization of roots. The root organ culture is a most attractive mass multiplication method for providing a pure, viable, rapid and contamination free inoculum using less space and has an advantage over the pot culture multiplication/conventional system (Fortin *et al.*, 2002); Cranenbrouck *et al.*, 2005; Tiwari *et al.*, 2003). The mono-specific strains available can be used directly as starting material for large scale inoculum production, a sole Petri dish culture being enough to generate several thousand spores and meters of hyphae within four months (Dalpe, 2004). The hairy roots formation occurs due to transfer of root inducing Ri-plasmid from the bacterium, *Agrobacterium rhizogenes*. This system offers a rapid, clean and more efficient method and opens up the avenues for AM fungi commercialization and field application.

Mycorrhizal inoculum can be applied in various forms. There are various formulations available and the active ingredients are typically spores, mycelium/hyphae and colonized root pieces. The inert materials (inactive carriers), can be used as pot mix to grow and bulking up the AM inoculum. The important thing for growers is that AM formulations should be easy in application. Some of the inocula are available in active form (on-going symbiosis) where one has to take caution and strictly follow the instructions given on the packet. The inoculum should be applied in the root zone through layering technique, not on the soil surface. The inoculum applied or left on the surface will quickly die in absence of growing roots/host.

3. MECHANISM OF ARBUSCULAR MYCORRHIZAL SYMBIOSIS AND ACTION

It is recognized that mycorrhizal association is a natural strategy that most plants have developed in their development process (Paul and Clark, 1996). Arbuscular mycorrhizal fungi are obligate biotrophs; their spores are the only plant-independent phase and they have an ability to germinate and arrest growth many times if plant derived signals are missing (Requena *et al.,* 2007). The process of the development of AM symbiosis varies with the plant and fungal species involved (Smith and Smith, 1990). AM spores germinate under appropriate water and temperature conditions, hyphae grow for 2-3 weeks and they respond to root present in their vicinity (Requena *et al.,* 2007). Also, root exudates from host root elicit hyphal growth due to presence of plant phenolic substances, in particular flavonoids (Nair *et al.,* 1991), some lypophylic molecules (Buee *et al.,* 2000) or strigolactones (Akiyama *et al.,* 2005). On contact with host root cell the fungal hyphal cells form appressoria (Paszkowski, 2006) that develop in to haustoria like structures (arbuscules) to penetrate the root cortical cell walls and form interface with host cytoplasm (Smith and Read, 1997). Here it undergoes extensive dichotomous branching in to a tree like fungal structure that may entirely fill the cortical region and provide an increased surface area for metabolic exchanges between the plant and the fungus. These arbuscules collapses after some days, leaving an intact cortical cell which is then able to host another arbuscules. The signals that induce intracellular fungal branching or arbuscules decay have not been identified (Paszkowaski, 2006).

AM fungi also produce vesicles which are structures that function as storage (lipids) and reproductive organs. They also interface directly with the soil by producing extra radical hyphae that may extend several centimeters out into the soil, which increases the potential of nutrient uptake, and possibly also water uptake (Auge, 2001). Except for species from the genera *Scutellospora* and *Gigaspora*, all AM fungi form intra-or intercellular storage organs, lipid-rich vesicles, to varying degrees in late phases of the symbiosis (Smith and Read, 1997). The arbuscules are the key features of AM and are responsible for nutrient exchange. They represent a dead end in the growth of AM fungi (Bonfante and Perotto, 1995), because they finally senesce and collapse after 4-10 days of symbiosis (Sanders *et al.,* 1977). The fungal structures are then degraded completely by the plant cell and the plant cell recovers its original morphology (Jacquelinet-Jeanmougin *et al.,* 1987). This way, cortical cells are able to allow a second fungal penetration and arbuscules formation. During colonization, the fungal arbuscules occupies a major portion of the plant cortex cell, but is separated from the cell protoplast by a part of the host plasma membrane, the periarbuscular membrane. This membrane completely surrounds the arbuscules, leading to up to a four fold increase of the surface of the plasma membrane. Although it originates from the plant plasma membrane, the periarbuscular membrane exhibits different properties. In particular, phosphate transporters were shown to be located specifically in the periarbuscular membrance (Harrison *et al.,* 2002). Moreover, a high amount of H+- AT Pase activity (Gianinazzi-Pearson *et al.,* 1991) accompanied by the highly acidic nature of the space separating plant and fungal plasma membranes has been found (Guttenberger, 2000). The space separating plant and fungal plasma membranes corresponds to a new apoplastic compartment and represents the symbiotic interface. It is continuous with the peripheral plant cell wall, but its structure differs from it (Peterson and Bonfante, 1994). The mixture of primary plant cell wall components helps arbusculated cells to maintain their ability to synthesize and secrete cell wall material. This material, however, does not assemble further to build up a secondary wall (Peretto *et al.,* 1995). It has been accepted that the establishment of the symbiosis must be the result of a continuous molecular 'dialogue' between plant and fungus, as exerted through the exchange of both recognition and acceptance signals (Azcon-Aguilar and Barea, 1997). The result of this dialogue finally depends on the genome expression of both partners (Gianinazzi *et al.,* 1994). In exchange of the nutrients the fungus is supplied with photosynthates (about 20% of the plant assimilation capacity) and these assimilates are either respired or released in to the soil via transfer of sugar from the root to their endosymbiotic fungi, which later translocated this carbon in form of lipids and sugar into the external mycelium spreading in the soil (Bago *et al.*, 2003). As a result of higher carbon drain, microbial rhizosphere composition is altered.

4. PHOTOSYNTHESIS AND ROOT PHOSPHATASE ACTIVITY

4.1 Photosynthesis

The total photosynthates availability in the plant depends upon photosynthetic surface and the rate per unit surface area, minus respiration and photo respiration (Sinha *et al.*, 1988). Shanta Kumari and Sinha (1972) observed that photosynthetic rate in mung bean, cowpea and chickpea attained a maximum value at flowering, it decline in post-flowering period when pod development commenced. Utilization of photosynthates and their translocation to different organs have been reviewed by many workers.

Minchin and Patel (1973) showed that only 26% of the carbon-assimilated in the process of photosynthesis remained in the over ground part of the plant, 42% was transported to the nodule bearing roots and 32% entered the nodules directly. An adequate supply of carbohydrate to nodules is a major constraints in biological N-fixation in legumes (Allison, 1965).In legume nodules, maximum rate of N-fixation are under conditions favorable to photosynthesis (Sheehy *et al.*, 1980, William *et al.*, 1982 and Rainbird *et al.*,1983, Vijaylakshami, 1988; Panwar *et al.*, 1986,1990).

Short term $^{14}CO_2$ labeling experiment indicate that carbon requirements of VAM and *Rhizobium* are supplied by the photosynthesis of host plant (Ho and Trappe, 1973; Cooper, 1984). The fungi convert host metabolites into specific fungal compounds, lipids and particularly abundant in mature arbuscules, vesicles and hyphae (Kinden and Brown, 1975; Bonfante-Fasolo, 1984). Glycogen granules are usually associated with these lipid containing structures and also found in young vacuolated hyphae and the finest arbuscules branches. The demand of host photosynthates depends upon the amount of fungus associated with the root and its metabolic rate. Mycorrhizal roots have higher respiratory activity than uninfected roots and infected root cells have higher cytoplasmic volume, number of mitochondria, soluble protein content and several enzyme activities (Oliver *et al.*, 1983; Cox and Sanders, 1974)

Mycorrhizal plants with low available P export up to 10% more photosynthates to their root than P sufficient non-mycorrhizal plants (Kucey and Paul, 1982; Koch and Johnson, 1984). This additional drain could result in reduction in yield, unless compensation mechanism operates. When P limits the rate of photosynthesis, the investment of carbon in fungal symbiosis leading to increased rate of P acquisition could in turn lead to considerably increased rate of photosynthesis. This would offset the additional carbon utilization in the tissue below ground. According to Pang and Paul (1980) the non mycorrhizal plants transferred 39% of the fixed ^{14}c beneath ground. The $^{14}Co_2$ produced by root-microbial system of the mycorrhizal *Vicia faba* was twice as great as that of the non-mycorrhizal plants. Kucey and Paul (1982) found that N-fixing nodules of non-mycorrhizal *Vicia faba* utilized 6% of the total photosynthates and those of mycorrhizal plants 12%. Nodulated plants with or without mycorrhizal infection fixed about 47% higher $^{14}Co_2$ than plants without microbial symbionts (Harris *et al.*, 1985). At early stages of mycorrhizal and *Rhizobium* infection the carbon drain, not yet compensated for, may induce a transitory negative growth response.

Bethlenfalvay *et al.* (1982) observed that during the first ontogenic stages of the association, VA- mycorrhizal fungus was parasitic as a result of carbohydrate demand on the host by the endophyte, whereas at later stages mycorrhizal fungus and soybean expressed associated mutualistically. Negative correlation of root hexose sugar concentration with VAM colonization, nodule activity and root P-content were observed, when soybean plants inoculated with one of the endophyte after the other has colonized the host root (Bethlenfalvay *et al.*, 1985).

Pacovsky *et al.* (1986) showed that rapid colonization by a VAM fungus would result in an enhanced P-status, but would lower the level of the carbohydrate in roots. *Bradyrhizobium* strains unable to store P but capable of storing as poly-ß- hydroxyl butyrate could have a competitive advantage under these conditions .The elevated Co_2 has been reported to alter the partitioning of carbon between the above and below ground components of the forest ecosystem. Rygiewicz *et al.* (1994) reported that the elevated Co_2 (525 and 700µ moles mol^{-1}) concentration increased the rate of colonization, number of mycorrhizae and the presence of hyphae in the soil. The higher degree of mycorrhizal infection is likely to be associated with higher carbohydrate level with fine roots of the higher Co_2 treatments.

4.2 Root Phosphatase Activity

Helal (1990) studied the varietal difference in root phosphatase activity as related to the utilization of the organic phosphate. Bean varieties showed significant difference in their uptake of phosphorus from inocitol hexaphosphate with a clear dependency on the phosphatase activity of their roots at pH-5. The result suggests that root phosphatase activity is a significant factor of nutritional efficiency under limited mineral phosphorus supply. Liu *et al.* (2004) studied the *Rhizosphere* effect and root growth of two maize genotypes with contrasting P efficiency at low P availability. The result showed that P treatment increased root biomass, root to shoot ratio, lateral root length and acid phosphatase activity in roots and on the root surface but reduced root exudation of organic acids and pH in *Rhizosphere* for both genotypes. It was concluded that efficient use of P in calcareous soil is related to large root system, greater ability to acidify the *Rhizosphere* and positive response of acid phosphatase production and excretion to low P conditions.

Mc Lachlan and De Marco (1982) studied acid phosphatase activity of intact roots and phosphorous nutrition in plants and its relation to phosphorous garnering by wheat and a comparison with leaf activity as a measure of phosphorous status. In actively growing plants leaf phosphatase activity was better related to total phosphorous uptake than was root phosphatase activity. At the second stage, where yields continued to increase but there was little change in the phosphorous content of the plant other than it's redistribution between component parts, root phosphatase was better related to total phosphorous uptake.

Grierson and Comerford (2000) studied the non-destructive measurement of acid phosphatase activity in the *Rhizosphere* using nitro cellulose membranes and image analysis. Over 95% of the acid phosphatase activity of the root system of *D. sessilis* was associated with cluster roots and between 20 to 32% of the active root surface. About 26% of the acid phosphatase activity of the root system of *P. taeda* was associated with mycorrhizal roots and unsupervised white root tips and less than 10% of the root surface was active, irrespective of root type.

Mc Lachlan (1980) studied the acid phosphatase activity of intact roots and phosphorous nutrition in plants. The phosphatase activity of wild and cultivated progenitors of modern wheat's ranged in optical density from 15.5 to 45.4 per g fresh weight. All the cultivated species had low phosphatase activities. Ohtsuka and Saka (1988) studied the undestructive measurement of acid phosphatase activity in the root of rice seedlings. The enzyme activity of the whole roots including seminal and crown ones at the 4th leaf stage plant was higher than at younger or older seedlings stage.

Tarafdar and Marschner (1994) studied the phosphatase activity in the *Rhizosphere* and *hyposphere* of VAM with wheat supplied with inorganic and organic phosphorous. In the root compartment acid phosphatase activity was much higher than alkaline phosphatase activity and both were slightly enhanced by mycorrhizal infection. Throughout the hyphae compartment phosphatase activities were distinctly higher in the presence of mycorrhizal plants particular with a supply of organic P which also increased the percentage of infected root length. Phosphatase activity was strongly correlated with hyphae length which was highest with inoculation site from the root surface. Saker *et al.* (1999) studied the relationship between phosphorous fractions phosphatase activity and fertility in three tropical rain forest soils. Root phosphatase activity was greater for plants grown in soil low in available phosphorous. Fernandez and Ascencio (1994) studied the acid phosphatase activity in bean and cow pea plants grown under phosphorous stress. Acid phosphatase activity was above the values reported for these species under P-stress, however, APA in these legumes appears not to be inducible by the low P concentration level used in the study. A higher APA was found in young leaves, and the expression of APA also showed intra specific variation.

Root phosphatase activity was enhanced with the application of AM fungi at 60 DAS in urd bean (Preeti, 2011). Johnson *et al.* (2005) studied the liming and nitrogen fertilization effects, phosphatase activities, microbial biomass and mycorrhizal colonization in upland grassland. A positive linear relationship was observed between root phosphatase activity and P concentration of plant shoots. Rao *et al.* (1999) studied the influence of organic amendments on the development of ectomycorrhizae and their efficiency in P uptake and seedlings growth of *Pinus kesiya*. Root phosphatase activity was more with mycobionts inoculated activity of root was observed in seedlings inoculated with *S. leteus* and grown in grass litter amended soil.

Chen *et al.* (2003) observed the effect of plant species on phosphorous availability in a range of grassland soils. There were no significant differences on the level of water soluble P between soils under ryegrass and radiate pine, but the level of Cp were generally lowered compared with original soils due to plant uptake. Kumar and Chandra (2008) observed the influence of PGPR and PSB on *Rhizobium leguminoserum* Bv.*viciae* strain competition and symbiotic performance of lentil. Combined inoculation treatment of *Rhizobium* sp. +PSB+LK-884 was significantly superior to *Rhizobium sp.* alone in plant dry weight at different intervals and P uptake by grain and straw. Singh and Shrivastava (1985) studied the nodular physiology of urd bean as affected by urd bean mosaic virus and effect on some enzymatic activity. The enzymatic activities were found higher in the nodules of soil grown plants than in the nodules of plants grown in sand.

Tarafdar and Marschner (1994) studied phosphatase activity of VAM fungi in the *Rhizosphere* (root compartment) and hyphosphere (hyphal compartment) in mycorrhizal and non-mycorrhizal wheat (*Triticum aestivum*) supplied with phosphorus (200 mg P Kg^{-1} soil), either in organic (Na-phytate) or inorganic [Ca $(H_2PO_4)_2$] form. They found that in root compartment, acid phosphatase activity was much higher than alkaline phosphatase activity, and both were slightly enhanced by mycorrhizal infection. Throughout the hyphae compartment, phosphatase activities were distinctly higher in presence of mycorrhizal plants particularly with a supply of organic-P which also increased the percentage of infected root length. Mycorrhizal infection increased plant's dry weight, P-content and total P uptake mycorrhizal contribution accounted for 24-33% with P supplied in inorganic form, and 48-59% with P-supplied in organic form.

The high affectivity for P acquisition of the external hyphae is related to (a) formation of polyphosphatase in the hyphae, thus maintaining low internal P concentrations, (b) the small hyphae diameter leading to a relatively larger soil volume delivering P per unit surface area, compared to the root surface area (Jung and Classen, 1989) and a correspondingly 2-6 times higher P-influx rate per unit length of hyphae (Jakobsen *et al.*, 1992), and (c) production of extra-cellular acid phosphatase which catalyzes the release of P from organic complexes in the soil. Rates of translocation of phosphorus by *Glomus mosseae* in association with various host plants are in the range of 0.1- 2.0×10^{-9} mol cm^2s^{-1} (Cooper and Tinker, 1981) and are a temperature sensitive process. This is stopped by cytochalasin-B (inhibit cytoplasmic streaming), and reduced when transpiration of the associated plants is prevented. Translocation of P occurs down a concentration gradient between P source in the external hyphae and a sink in the roots. Loading and unloading of P from hyphae vacuoles, coupled with cytoplasmic streaming, are important in maintaining the high rates of translocation.

4.2.1 Phosphatase activity

Acid phosphatase enzyme secreted mainly by the plant roots, causes hydrolysis of organic phosphates present in the soil as phosphate esters resulting in the release of inorganic P for plant uptake. Exudation of phosphatases increases when plants are grown under phosphorus stress (Radersma and Grierson, 2004). It was reported that when *Triticum aestivum* grown in an acidic P-deficient soil amended with Fe- P, the P-efficient genotypes had a greater acid phosphatase activity in the rhizosphere than the inefficient genotype, with phosphatase activity correlating positively with growth and phosphorus uptake (Marschner *et al.*, 2006). In an another study involving wheat, rye and triticale under phosphorus stress, showed that wheat plants exuded more acid phosphatase into the rhizosphere followed by rye and triticale. A significant positive correlation was also found between acid phosphatase activity (both extra – and intra – cellular) and root, shoot and total plant dry weight (Pandey, 2006).

4.2.2 Stimulation of assimilative activities essential for plant growth

Nitrate reductase (NR) is the first enzyme in the NO_3-assmilation pathway and probably represents the rate-limiting step in this process (Campbell, 1988). However, NR levels are drastically decreased by adverse environmental conditions such as drought stress (Sanchez and Aguirreolea, 1993). AM fungi have the gene set for assimilatory NO_3 reduction (Kaldorf *et al.*, 1994), and NR activity has been located in arbuscules-containing cells (Kaldorf *et al.*, 1998). In addition, levels of NR activity are higher in AM plants subjected to drought stress than in the non-mycorrhizal ones.

4.2.3 Mycorrhizal effect on plant roots

Unlike Arabidopsis, more than 80% of higher plants associate with mycorrhizal fungi, which elicit profound changes in the root morphology of host plants (Hetrick, 1991). In particular, ectomycorrhizae suppress root elongation and induce dichotomous branching of short lateral roots, culminating in the formation of coralloid structures resulting from higher – order dichotomous branching. All of these anatomical structures are variable depending on the plant and fungal species.

Once the fungus is established, root branching is suppressed, which makes the plant more dependent on the nutrients provided by the fungus (Hetrick, 1991; Price *et al.*, 1989). Whether this modification of root system architecture (RSA) is a direct consequence of symbiosis or an indirect effect of improved nutrient status of the plant is not clear.

However, it appears that symbionts can triggers RSA changes by promoting lateral root initiation very early in the interaction (Harrison, 2005). Moreover, the maize mutant *lrt-1* normally lacks lateral roots, but displays extensive lateral root development following inoculation with the mycorrhizae *Glomus mosseae* (Paszkowski *et al.*, 2002). Notably, many micro-organisms that interact with plants can produce plant hormone analogs. Thus, symbiotic association might employ hormone signaling pathways to regulate RSA.

5. VA-MYCORRHIZAE AND PLANT GROWTH

As VAM are ubiquitous in nature: they are commonly associated with most plants important in agriculture, horticulture and forestry. In agriculture, potential hosts occur in the economically important families; *viz, Gramineae* and *Leguminoseae*. It is also found in many tropical plantation crops, trees, walnuts and all temperate fruit trees. Although VAM being omnipresent, the establishment of successful entry points on host trees, establishment, development and spreading of internal and external hyphae and its response to various kinds of hosts is dependent on prevailing fungi, plant and environmental factors interaction.

Mycorrhizal development is greatly affected by the light received by the shoot of the host plant. Low irradiation and short day lengths can reduce the development of VAM (Daft and El-Giahmi, 1978; Hayman, 1974), but which also markedly reduce plant growth in uninfected plants. Johnson, (1976) noted that shading reduces development of infection in shade tolerant species but not in light demanding species. Defoliation greatly changes the soluble carbohydrate content of grass roots and result in relation of mycorrhizal development (Daft and El-Giahmi, 1978). The effect of temperature (Hetric *et al.,* 1984; Sieverding, 1983; Smith and Brown, 1979), light (Koide, 1985; Smith, 1980), pH (Abbott and Robson, 1985; Hayman and Taveres, 1985; Hepper, 1984 and Read *et al.,* 1976) and moisture (Ponder, 1983; Readhead, 1975) have been well documented. In addition to above factors, metals, salinity, organic matter distribution in depth and aeration also play key role in establishment and development of VA-mycorrhizal fungi.

In a symbiotic relationship between host and VA-mycorrhizal fungus, the former is autotrophic for its carbon requirement and the latter is heterotrophic. Being an obligate endosymbiont they have very little capacity of independent growth. Harley and Smith (1983) pointed out that growth responses to mycorrhizal inoculation will be determined by movement (translocation) of soil derived nutrients to the plant through the fungus and the consumption of carbon compounds derived from host by the fungus.

The main agronomic significance of VAM lies in their ability to supply the plants with extra nutrients, particularly phosphorus. Crush (1974) reported that endophyte host relationship varies between parasitism and mutualism depending on the available phosphorus levels. Maximal mycorrhizal effectiveness occurs when nutritional conditions for the host are sub-optimal but these become severely deficient and the fungus becomes ineffective.

The VA-mycorrhizal system involves three components, the plant, the soil and the fungus. Interactions between the three determine the mycorrhizal effect on plant growth. The establishment, development and functioning is greatly affected in highly fertile or heavily fertilized soil. Under such conditions the infection more usually dies out because of high internal nutrient levels of both the nutrients (nitrogen and phosphorus) can make the plant immune to infection (Mosse, 1973b; Sanders, 1975). Studies on VA-mycorrhizal fungi have shown that different mycorrhizal fungi are not affected to the same extent by an improved external nutrient supply to the host. A differential tolerance to phosphate supply is observed not only in relation to development of infection (Jasper *et al.,* 1979) but also in the ability of fungal species or isolate to improve plant growth at different levels of fertilization (Carling and Brown, 1980; Gianinazzi - Pearson, 1982). Mycorrhizal endophyte responded significantly to the host plants grown in soils, containing 3.5 ppm available phosphorus (Jalali, 1988). Later, he confirmed, increased phosphate transport by mycorrhizal roots, by incorporating labelled ^{32}P. Powell (1977) reported that endophyte species differ in their effect on plant growth and performance depends on interactions between fungus and soil and it not always related to infection levels or to the number of connecting links between soil, root and mycelium.

Phosphorus plays an important role in the control of VA-mycorrhizal infection of root systems (Menge *et al.,* 1977; Mosse, 1972, 1973; Sanders, 1975). Sanders and Sheikh (1983) established a correlation between the final percentage of infection and the quantity of phosphorus added to the soil. Studies with onion plants which received foliar application of P, showed that the P concentration within the plants plays a more important role than soil P levels (Sanders, 1975). Plenchette (1982) supported this finding by using a split root technique in which the two halves of individual sundan grass root systems were grown in soils with light and low P in the presence and absence of the mycorrhizal fungus, *Glomus fasciculatum.* Ratnayaka *et al.* (1978) found that under low levels of P, root exudation measured in terms of the net leakage of soluble amino acids and reducing sugars were significantly higher than under the high P levels. They also proposed the hypothesis saying that phosphorus inhibition of mycorrhizal symbiosis is associated with a membrane mediated decrease in root exudation. Hetrick *et al.,* 1984) reported that mycorrhizal corn plants were significantly larger at lower concentration of 'P' and higher levels of 'P' were detrimental to the mycorrhizae. Negative correlation of differences in colonization and mycorrhizal response in differently fertilized growth media with the logarithm of the equilibrium solution P-concentrations was reported by Bierman and Linderman

(1983). At extreme levels of available phosphate in the rooting medium the growth of non-mycorrhizal plants can exceed that of VA-mycorrhizal ones (Bethlenfalvay *et al.*, 1983; Kierman *et al.*, 1983; Smith, 1982) and for the same concentration of P in the tissues. VA-mycorrhizal plants can have lower yields than non-mycorrhizal plants (Pairunan *et al.*, 1980). Increased levels of added soluble phosphorus reduced the mycorrhizal infection and spore production. However, root, shoot and total plant dry weight were significantly greater in mycorrhizal plants, at all levels of added soluble P, than non-mycorrhizal ones (Krishna and Bagyaraj, 1982). The increase in the addition of soluble P above certain level, resulted in decreased length of infected root. This also resulted in decreased length of external hyphae per cm of infected root and the formation of vesicles (Abbott and Robson, 1984).

Same *et al.* (1983) reported a decrease in percentage of mycorrhizal infected root length with increasing P-supply. It was also accompanied by a reduction in soluble carbohydrates. Some mycorrhizal fungi tolerate higher levels of soluble phosphate than others (Pond *et al.,* 1984). In general, large P additions to the soil decrease mycorrhizal development and functioning. Adding soluble phosphate improves growth of mycorrhizal plants less than that of non-mycorrhizal ones (Daft and Nicolson, 1969; Holevas, 1966 and Khan, 1972).

5.1 Plant Growth and Phosphorus Nutrition

In soil low in available P mycorrhizal plants have higher rates of growth than non-mycorrhizal plants, root shoot ratio is often lower and shoot fresh weight, dry weight ratio is higher following infections and their beneficial effect has been reported by many workers in India and abroad (Mosse, 1973; Bagyaraj, 1990; Tilak and Diwedi, 1990 and Panwar, 1991). VAM has more extension of the root absorbing surface beyond the zone explored by the root hairs allowing more water and nutrients uptake under irrigated and dry land conditions (Allen *et al.,* 1981; Snellgrove *et al,* 1986; Panwar, 1991 and 1992). The VAM symbiosis can affect the stomatal behaviour of the host plant (Auge and Duan, 1991).

The effects are not usually apparent for several weeks after germination. The rate of plant growth is determined by interactions between mycorrhizal infection and number of nutritional and non-nutritional aspects of the symbionts physiology. The importance of VAM in the 'P' nutrition is evidenced that the inflow of 'P' from soil in to the roots of mycorrhizal plants is faster than in to non-mycorrhizal plants (Hall and Sanders, 1982; Sanders, 1975; Sander *et al.*, 1977; Smith, 1982; Smith *et al.*, 1986 and Tester and Smith, 1985). On soils low in available 'P' the amount of 'P' in the orthophosphate (Pi) fraction in both shoot and root is increased in mycorrhizal plants (Gianinazzi-Pearson, 1984 and Gianinazzi- Pearson and Gianinazzi, 1986).

One mechanism underlying the increased rate of uptake of 'P' is the efficiency with mycorrhizal roots exploits the soil profile with hyphae extending beyond the depletion zone surrounding the absorbing root and its root hairs (Clarkson, 1985; Nye and Tinker, 1977 and Tinker, 1975). This fits with what is known of the fraction that determines rates of nutrient supply to and uptake by roots. For non mobile nutrients such as 'P' (and to a lesser extent 'K' and NH^{+}_4) root growth, root radius, development of root hairs and initial concentrations in the soil solution are more important determinants of the rate of uptake than are the kinetic properties of the uptake system of the roots (Clarkson, 1985; Nye and Tinker, 1977) mycorrhizal modifications of the nutrients uptake properties of roots depends upon (a) development of extra material hyphae in soil (b) hyphae absorption of phosphate (c) translocation of 'P' through hyphae over considerable distance and (d) transfer of 'P' from the fungus to the root cells.

There is clear evidence that all these processes take place. So there is immense need to compare different fungus-host combinations grown under different environmental condition.

Not only 'P' but other nutrients like P, Zn, S, Ca and N can be translocated from the place exceeding the radius of any depletion zone likely to develop around and actively absorbing roots (Rhodes and Gerdemann, 1975, 1978 a,b).

Although growth responses to infection frequently do not occur when soil 'P' is high, it is clear that inflow via hyphae can continue and give rise to high 'P' concentration in shoot as well as in roots, in the absence of a growth response being due to operation of a limiting factor other than 'p' availability (Smith and Son, 1987; Smith, 1982; Smith *et al.,* 1986). Luxury 'P' accumulation could provide a storage pool used in later stage of plant growth.

Improved 'P' nutrition as well as direct fungal effects may be implicated in the enhanced uptake by plants of other macro nutrients such as K and S (Powell, 1975) and of micronutrients such as Cu and Zn (Cooper and Tinker, 1978) nutritional interactions can be complex. For example, K requirement is strongly influenced by the source of N (NO_3 or NH^{+}_4) and by the concentration of Na. The anion and cation ratio can be altered by mycorrhizal infection. Mycorrhizal plants accumulate lower amount of organic anions in their vacuole and may therefore regulate cytoplasmic pH by disposing of a higher proportion of OH^- generated during nitrate reduction in the soil (Raven and Smith, 1976). They might, therefore, be expected to have higher rhizosphere pH than non-mycorrhizal plant assimilating this N source.

Singh and Gupta (2006) studied the effect of bio-fertilizers at different levels of phosphorus on nodulation, yield and protein content in black gram and found that dual inoculation of *Rhizobium* and PSB was superior to individual inoculation of these bio-fertilizers. The highest values for number of root nodules per plant, grain and straw yields, and seed protein content were obtained when dual inoculation was combined with optimum P dose of 20 ppm. Shankar Narayanan and Rajeshwari (2005) studied the influence of *Rhizobium* and Phosphobacteria on the interaction of VAM, *Glomus mosseae* and *Meloidogyne incognita* on black gram and find that among the treatments, combined inoculation of *Rhizobium*, Phosphobacteria on VAM and nematode, increased all the growth parameters, phosphorus content, spore populations, mycorrhizae colonization and reduce the gall index and nematode population significantly.

Deka and Azad (2006) studied the cultural and biochemical characteristics and found considerable variation of culture in respect of configuration, margin, elevation and color. Dubey (2006) studied the interaction of integrating bio-agents, oil cake and fungicide in various modes of application for the better management of web blight of urd bean. Among 54 integrations evaluated soil application of *P. glabra* cake, seed treatment with integration of *T. viride*, Carboxin and *Rhizobium* were found superior as they increased seed germination by 31.6%, yield by 36% and reduced disease intensity by 94.8% and also significantly enhanced plant height, root length and number of root nodules in urd bean. Manivannan *et al.* (2003) studied the growth and growth analysis of rice - fallow-black gram as influenced by foliar application of nutrients with and without *Rhizobium* seed inoculation. They found that *Rhizobium* seed treatment and foliar application of Micro sol B recorded markably higher leaf area index, dry matter production, crop yield and relative growth rate.

Prasad *et al.* (2002) studied synergism among phosphate solubilizing bacteria, *Rhizobacteria* and *Rhizobium* in urd bean. *Rhizobium* +PSB-1 +PGPR were superior in terms of plant dry weight, grain yield and N and P uptake. Mishra and Tiwari (2002) studied the response of urd bean to applied phosphorus, sulphur and *Rhizobium* inoculation on udicustochrepts. The highest leghaemoglobin content was obtained with P. The greatest P uptake was obtained with P followed by S and *Rhizobium.* S uptake was greatest with S followed by P and *Rhizobium*. Chandra and Pareek (2002) studied the effects of *Rhizobacteria* in urd bean and lentil. *Azotobacter chroococcum* applied singly and in combination with *Rhizobium* enhanced nodulation in urd bean. Both treatments increased plant's dry weight and grain yield. Tanwar *et al.* (2003) studied the effect of phosphorus and bio-fertilizers on yield and nutrient content and uptake by black gram. The crop yield, N and P contents and N and P uptake increased with increasing P rate. Inoculation with the combination of the bio-fertilizers resulted in higher yield, N and P content and N and P uptake of the grain. Tanwar *et al.* (2003) studied economics of black gram cultivations as influenced by phosphorus levels and bio-fertilizers. The seed yield increased with increase in P application.

Tanwar *et al.* (2002) studied the effects of phosphorus and bio-fertilizers on growth and productivity of black gram. The interaction between P rate and bio-fertilizers was significant with regard to the number of nodules and seed yield. Kulshreshtha and Khan (2002) studied the effect of ozone on root colonization by VAM fungi and root nodulation on black gram. The suppression of growth and other parameters were less in symbiotic plants than in non-symbiotic plants. Higher values for growth and other parameters were recorded for dual inoculated plants than for single inoculated plants. Sinha and Sharma (2001) studied the response of black gram to phosphorus fertilization and *Rhizobium* inoculation in the hill soils of Assam. *Rhizobium* inoculation significantly increased the number and mass of the root nodules per plant compared to the control.

6. VA-MYCORRHIZAE AND PLANT NUTRITION

6.1 Phosphorus

In phosphate deficient soils plants infected with VAM usually take up more phosphorus and grow better than non-mycorrhizal plants (Sanders and Tinker, 1971, 1973). The amount of phosphorus required for maximum plant yield differs greatly for different soils and for different plant species grown in the same soil (Barrow, 1975). Rorison (1968) observed that *Deschampia flexuosa* grew continuously in concentrations as low as 10^{-7} M where as *Scabiosa columbaria* responded favourably at a concentration of 10^{-3} M where there was significant increase of phosphorus in the roots. Maximum translocation rate of phosphorus for onion was found to be 6.4×10^{-16} moles s^{-1} and the P flux in the hyphae about 2×10^{-10} moles cm^{-2} S^{-1} (Cooper and Tinker, 1978). Moreover, the phosphate in the root environment also influences the mycorrhizal establishment and effect of mycorrhizae on plant growth. Allen *et al.,* 1981) reported that in *Bouteloua gracilis,* the maximum infection was in the presence of phytate (75%) whereas lower root infection levels were noted in the presence of sodium phosphate (19%). They also found higher alkaline phosphatase activity and significantly increased chlorophyll concentrations with mycorrhizal infection with no change in a/b ratio.

Different species of *Endogonaceae* differ in their ability to form VA-mycorrhizal associations. There are abundant evidences of some species being more efficient than others (Abbott and Robson, 1978, 1981; Daft and Nicolson, 1969; Hall 1976; Kuo and Huang, 1982; Powell and Daniel, 1978; Rangeley *et al.,* 1982) *Glomus tenuis* was more efficient than the indigenous fungi at a particular location (Mosse, 1977). Sanders *et al.,* (1977) suggested that different species of VA-endophyte may have different growth promoting abilities.

Mycorrhizal and non-mycorrhizal roots appear to absorb phosphorus from the same labile pool of P. The specific activity of P in plants growing in ^{32}P labeled soils was the same for mycorrhizal and non-mycorrhizal plants (Hayman and Mosse, 1972; Powell, 1975; Sanders and Tinker, 1971). Pairuman *et al.* (1980) reported that the effect of mycorrhizae in stimulating P uptake was the same for poorly soluble sources of P.

Different plant species and varieties differ in their ability to extract phosphorus from the same soil (Ballard and Dean, 1941; Hall, 1978; Powell, 1982). Cultivars of wheat and pearlmillet are known to differ in dry matter production upon mycorrhizal inoculation (Azcon and Ocampo, 1981; Krishna and Bagyaraj, 1984).

There are several reports indicating that mycorrhizal inoculation increased not only the utilization of soluble phosphates like mono calcium phosphate and superphosphate (Mosse, 1973a; Powell and Daniel, 1978) but also less soluble forms like rock phosphate (Mosse, 1977) bone meal (Daft and Nicolson, 1966) and even hydroxyapetite and aluminium and iron phosphates (Nyabyenda, 1977; Fitter, 1977) observed growth of mycorrhizal plants in the presence of insoluble tri-calcium phosphate.

Bieleski (1973) reported that P-uptake may be influenced by the steepness of the phosphate gradient from root to leaves. In mycorrhizal plants the increased conversion of phosphate from inorganic to organic forms in the leaves could increase the inorganic phosphate gradient from root to leaves. Possibly increasing the phosphate sink and thus enhances P-uptake (Allen *et al.*, 1981).

6.2 Nitrogen

Nitrogen is an inevitable pre-requisite for increased growth following relief of phosphate stress. Mycorrhizal fungi have a potential role in acquisition of N by plants (Raven *et al.*, 1978; Smith, 1980). Mycorrhizal celery plants derived significantly more ^{15}N than control plants. Positive corrections have been established between ^{15}N in mycorrhizal plants with percent mycorrhizal fungal colonization, number of hyphal crossings through the mesh into the area of ^{15}N placement and total length of hyphae g^{-1} of soil. The mean flux of N through the hyphae of *G. mosseae* was 7.42×10^{-8} mol. N $cm^{-2}s^{-1}$ for the inorganic N-treatment over 30 days of period and 1.74×10^{-8} mol. N $cm^{-2}s^{-1}$ for the organic N treatment over on 88 days period (Ames *et al.*, 1983).

Rhodes and Gerdemann, (1978b) pointed out that N content of mycorrhizal legumes has been attributed mainly to *Rhizobium* activity or in non-legumes, to the effects of increased P-content of the mycorrhizal plants. VAM prefers NH_4^{+}- N over NO_3-N (Bowen and Smith, 1981, Brown *et al.*, 1981). Increases (Wallace *et al.*, 1982), decreases (Heys *et al.*, 1982, Nemec and Meredith, 1981, Pang and Paul, 1980) or no differences (Rose and Youngberg, 1981; Schenck and Hielnson, 1973) in the N-content of mycorrhizal plants relative to control have been observed. Ames *et al.*, (1984) reported that mycorrhizal and non-mycorrhizal sorghum did not difer significantly in total plant N or percent utilization of the applied N at any fertilizer dose. The percent N derived from native soil sources and the 'A' values were significantly higher for mycorrhizal plants compared to non-mycorrhizal plants at intermediate fertilizer rates indicating that the mycorrhizal sorghum plants derived N from a source that was less available to non mycorrhizal plants.

6.3 Other Nutrients

Mycorrhizal infection can also increase the uptake of any soil nutrients in addition to phosphorus and nitrogen that move to plant roots, primarily by diffusion (Gildon and Tinker, 1983; Timmer and Leyden, 1980) zinc (Gilmore, 1971; Swaminathan and Verma, 1979) and other heavy metals (Pacovsky *et al.*, 1986).

Increased absorption of potassium, sulphate, zinc and strontium 90 (calcium tracer) by mycorrhizal plants has been reported (Gray and Gerdemann, 1973; Powell, 1975). Mycorrhizal roots can absorb zinc (Pacovsky, 1986: Thakur and Panwar, 1995) and copper (Pacovsky 1986: Thakur, 1994) from solutions at faster rate than can non-mycorrhizal P fertilized roots. Buwalda *et al.* (1983) reported that mycorrhizal infection in wheat and barley increased the concentration of anions viz chloride, sulphate and nitrate while addition of phosphorus fertilizer did not have similar effect.

Tinker and Gildon (1983) concluded that copper and zinc supplied to the plants can be markedly improved by mycorrhizal formation. Changes in the levels of phytohormones in VA-mycorrhizal plants have been reported (Allen *et al.,* 1982). Although VAM have been shown to synthesize phytohormones (Barea and Azcon-Aguilar, 1982) their precise role in plant growth and metabolism has not been studied so far in detail.

Verma *et al.* (2004) stated that generally growth in terms of plant height, number of leaves/plants; number of primary branches/plants were recorded higher in both boron and zinc treated seeds over control, foliar spray of 0.5% $ZnSO_4$ solution alone, followed by seed soaking plus foliar spray of $ZnSO_4$ proved effective and recorded significantly higher plant height, number of branches and number of leaves than control during all the years of experimentation. Similar results were obtained by Hemanatrajan, *et al.* (2003) and Verma and Pandey (2003).

7. VAM-WATER STRESS-HOST RELATIONSHIP

Water is pre-requisite for the growth of any plant. Plants growing in a natural environment are rarely free from water stress.Lack of water exerts a controlling influence on crop distribution and productivity, as well (Fischer and Turner, 1978). Bartels and Caesar (1987) found that two weeks of drought during seed filling reduced the yield of soybean by 25%. Examination of VA-mycorrhizal symbiosis has indicated that VAM may improve host resistance to drought stress (Johnson and Hummel, 1985; Nelson and Safir, 1982a).

The first systematic examination of mycorrhizal influence on plant-water relations was conducted by Safir *et al.* (1971, 72) on soybean plants.

The external hyphae of the mycobiont greatly enlarge the potential surface area of water and nutrient scavenging systems of plants. Under low moisture levels, water absorption capacity of roots might be enhanced by mycorrhizae. Nye and Tinker (1977) observed that hyphae ramifying into the soil are likely to increase the absorbing area for water uptake even further and may also be able to by-pass the dry zones that often surround slow growing roots during periods of drought. Other less obvious influences of endogonaceous fungi on plant-water relations are fungal derived hormonal effects on stomatal physiology, increased vasculature for water conduction and lowered radial resistance to water flow in colonized root cortical tissues (Allen, 1982; Daft and Okusanya, 1973; Safir and Nelson, 1985).

7.1 Enhanced Phosphorus Nutrition

Resistance to water movement in mycorrhizal soybean roots was diminished 40% when compared with P-deficient non-mycorrhizal controls (Safir *et al.,* 1971, 1972). Nelson and Safir (1982a, b) reported that the hyphae do not make significant direct contributions to water uptake. Several studies have indicated that VAM-hyphae may not be able to transport the quantities of water necessary to affect shoot water economy (Cooper and Tinker, 1981; Sanders and Tinker, 1973).

Mycorrhizal effect on hydraulic conductivity and transpiration of various citrus root stocks could be replaced by P-fertilization. Above studies supported the nutritional hypothesis offered by Safir *et al.* (1971, 1972) there were no significant differences between mycorrhizal plants and non-mycorrhizal plants with adequate P-contents. Koide (1985) who arrived at the same conclusion with mycorrhizal sunflower, when plants with equivalent root lengths were compared, no effect of VAM-colonization on the intrinsic hydraulic properties of the soil/plant system over a wide range of transpiration rate was shown. Elevated transpiration rates exhibited by mycorrhizal sunflower were attributed to improved P-status.

Radin and Eidenbach (1984) suggested that low levels of foliar P limit leaf expansion in cotton by decreasing the hydraulic conductivity of the root system. They also observed that P-deficient plants displayed lower leaf water potential. Low level of soil moisture reduces the P-diffusion rate considerably so that mycorrhizal infection could improve the P-nutrition of plants even where the soil P-content is quite high. Radin (1984) reported that P-nutrition had only a very small effect or no effect on cotton leaf water relations when soil moisture was adequate. Transpirational flux, closely linked to stomatal diffusive conductance, has been reported to both decrease and increase in response to low P levels. Claims for direct hyphal participation in water uptake were made (Allen, 1982; Allen *et al.,* 1981; Hardie and Leyton, 1981).

Mycorrhizal colonization frequently alters root-to-shoot ratio and Bosse and Ellis (1985) reported that root-to-shoot ratio could conceivably affect foliar water relations. It is likely that potted plants with different under ground biomass, will respond differently under conditions of drought, as proportionately different amounts of water would be available to transpiring plant tissues. Additionally, type, size and branching patterns of root system can affect exploitation of a given soil mass by plant through changes in hyraulic conductivity and water flow-rates (Nye and Tinker, 1977).

VAM might alter the hormonal balance of the host (Brown *et al.*, 1981). Various workers have offered this possibility as an explanation for changed stomatal behaviour and water flow rates in mycorrhizal plants (Allen, 1982; Levy and Krikun, 1980). Cooper (1984) stated that mycorrhizal infection may possibly alter membrane permeability through increased production of photosterios, thereby influencing symplastic water transport.

VAM can tolerate a wide spectrum of soil moisture, as evidenced by their presence in arid deserts and aquatic habitats (Cooper, 1984). Many workers reported that P-uptake in non-mycorrhizal plants may be diminished in soils of low water content (Nelson and Safir, 1982b). The diffusion coefficient of phosphate in soil is linearly related to soil moisture content and therefore, mycorrhizal P-supplies are likely to be much more beneficial to hosts growing in dry than in moist soils (Fitter, 1985).

Colonization by *G. fasciculatum* could decrease resistance to water transport in *Bouteloua* up to 90% under drought conditions (Allen *et al.*, 1981). Leaf water potentials of mycorrhizal plants dropped more quickly than those of non-mycorrhizal plants as moisture stress increased and stomatal resistance remained lower in colonized plants. Cooper (1984) suggested that this influence of *G. fasciculatum* might have resulted from improved water uptake, increased photosynthesis or elevated cytokinin levels.

Acclimatization of plants to drought frequently allows the maintenance of stomatal conductance at lower leaf and soil water potentials. Allen and Boosalis (1983) reported that the aforesaid effect has also been produced by mycorrhizae; when mycorrhizal and non-mycorrhizal wheat plants of similar size were subjected to gradually drying soils. Stomata in non-mycorrhizal plants began closing at leaf water potentials of -1.8 MPa and were closed after four days, but in mycorrhizal plants they did not begin to close until water potential of -2.2 to -2.7 MPa were reached and were transpiring even after six to seven days. Hardie and Leyton (1981) noted wilting to non-mycorrhizal clover plants at higher soil water potentials (-0.8 to -1.2 MPa) and that of plants colonized by *G.mosseae* at -1.8 to -2.4 MPa. Increased transpirational flux and stomatal conductance, both during stress and recovery, and photosynthetic rate during recovery was recorded in mycorrhizal *Citrus jambhiri* (Levy and Krikun, 1980). They also noted decreased root growth ansd transpiration rate and reduced leaf water potential in mycorrhizal *G.jambhiri*. *Glomus deserticola* and *G.fasciculatum* increased size and yield of plants subjected to number of drought events. Similar observations were noticed with soybean and *G. fasciculatum* (Busse and Ellis, 1985). Bildusas *et al.* (1968) reported that colonized plants, preconditioned at a soil-water potential of -0.8 Mpa and sampled at pre and peak drought, displayed lower soil plant liquid flow resistance in comparision to non-mycorrhizal plants. They also found 64% greater stomatal conductance in mycorrhizal than in non-mycorrhizal plants after rewatering.

Mycorrhizal root colonization of plants subjected to drought is reported to be significantly greater than well watered plants at the lowest P level (Yocom *et al.,* 1987). The high P levels which reduce benefit of mycorrhizal colonization under well watered conditions may still be inadequate under severe drought conditions. Different species of VA-mycorrhizal fungi response differently to water stress (Yocom *et al.,* 1987). Krishna and Bagyaraj (1984) suggested that recovery of mycorrhizal plants from water stress is rapid. They also noted higher relative water content, lower amount of proline and lower stomatal resistance in mycorrhizal plants, 24 hours after rewatering. Bethlenfalvay *et al.* (1987) reported that under stress conditions nodule development and activity, transpiration, leaf conductance, leaf and root dry weights were greater in soybean plants infected with VAM fungi when compared to P-fed non-VAM plants.

7.2 VAM-Legume-*Rhizobium* Interaction and Nitrogen Fixation

It is well known fact that successful production of legumes depends on effective nodulation and nitrogen fixation. For maximizing crop yields, we have to select efficient strain(s) of specific *Rhizobia* and investigate the compatibility of host cultivars with the *Rhizobial* strain. Bacteriazation is particularly important in case of soybean which was recently introduced in India. Inoculation of seed with *Bradyrhizobium japonicum* at the time of sowing is imperative, since *Rhizobia* specific to soybean are lacking in soils beneficial responses of legumes to inoculation with *Rhizobial* cultures have been reviewed by several workers (Brockwell, 1977; Khalifa, 1987; Nutman, 1977; Rewari and Tilak, 1988; Subba Rao, 1976; Panwar and Thakur, 1995; Thakur and Panwar, 1995; Thakur, 1994; Thakur and Panwar, 1997).

Interaction between mycorrhiza, nodulation and nitrogen fixation have been studied in *Phaseolus* sp. (Daft and El-Giahmi, 1974; Hayman and Davy, 1977); *Medicago sativa* (Smith and Daft, 1977); Clovers (Powell, 1977); peanut (Bagyaraj *et al.,* 1979); White clover, chickpea (Subba Rao *et al.,* 1985) and mung bean (Thakur and Panwar, 1995, 1997).

The coexistence of a bacterium and fungus as root endophytes of legumes establishing a tripartite symbiotic association was first described by Janse (1986). Jones (1924) examined 18 species of nodulated legumes and found that 15 were colonized by VA-mycorrhizal fungi. Asai (1944) reported that mycorrhizae were a necessary pre-condition for nodulation in many legumes.

As a consequence of a simultaneous infection with *Rhizobium sp.* and mycorrhizal fungi, legumes are benefitted because of improved N and P supplied and also from the N-P interactions (Munns and Mosse, 1980). The double symbiotic association of *Rhizobium* and VA-mycorrhizal fungi in legumes not only reduces the inputs of synthetic fertilizers, thereby saving energy, but also appear to reduce the cost of the system itself in terms of photosynthates drain, *Rhizobium* and VAM are known to derive carbon (C) compounds from their host plants. Additional allocation of C to mycorrhizal roots of *Vicia faba* is reported to be equivalent to about 10% of total photosynthates (Pang and Paul, 1980). Kucey and Paul (1982) found that N_2–fixing nodules of non-mycorrhizal *V. faba* utlized 6% of total photosynthates and those of mycorrhizal plants 12%. The mycorrhizae respired or incorporated about 4% of total photosynthates in both nodulated or singly infected plants. Nodulated plants with or without mycorrhizal infection fixed about 47% higher $^{14}CO_2$ than plants without microbial symbionts (Harris *et al.,* 1985).

At early stages of mycorrhizal and *Rhizobial* infection the carbon drain, not yet compensated for, may induce a transitory negative growth response. Bethlenfalvay *et al.* (1982b) observed that during the first ontogenic stages of the association, VA-mycorrhizal fungus was parasitic as a result of carbohydrate demand on the host by the endophyte whereas at later stages mycorrhizal fungus and soybean associated mutualistically. Negative correlations of roots hexose sugar concentration with VAM colonization, nodule activity and root P content were observed, when soybean plants inoculated with one of the endophyte after the other has colonized the host root (Bethlenfalvay *et al.,* 1985). They also reported that non-symbiotic plants or dipartite association produced significantly higher dry weights than those of tripartite plants inoculated at planting.

The rates of CO_2 exchange, leaf area, P content were associated with P nutrition or the presence of VAM fungus; leaf fresh weight and root dry weight with N nutrition or the presence of *Rhizobium* and leaf N and starch content by both factors. The development of microsymbiont structure and nodule activity was significantly low in tripartite association than in plants colonized by one endophyte only. Mycorrhizal and nodulated plants usually have lower root/shoot ratio than plants inoculated with either symbionts alone (Redente and Reeves, 1981; Smith, 1980).

The role of phosphorus in nitrogen fixation by soybean has been described by Bethlenfalvay and Yoder (1981). At low levels of P, no nodulation was recorded whereas at higher P regime shoot dry weight was six times higher than those grown in lowest P regime. Increase in acetylene reduction activity with increasing amount of P has been observed (Bethlenfalvay and Yoder, 1981). Bethlenfalvay *et al.* (1983) reported that at lower levels of P mycorrhizal plants produced significantly higher rates of N_2-fixation, plant and nodule mass and P content, with increased plant growth and greater ratio of extra-to-intraradical mycelium (index for VAM fungal effectiveness).Mycorrhizal infection in soybean roots was found reduced from 25.4 to 5.2 vesicles/4 cm root length at high levels of P, but there was not effect on the number of nodules plant^{-1} and yield.Under N-deficient condition, even crops well infected with VAM may not benefit from the association increased growth of nodulating/non-nodulating isolines was observed due to mycorrhiza in the presence of N source; however, nitrate inhibited the effectiveness of mycorrhiza (Karunarathne *et al.,* 1987). Addition of combined nitrogen decreased VAM infection and nodulation in *Pisum sativum* (Lanowska, 1966) and *Trifolium subterraneum.*

Bayne and Bethlenfalvay (1987) observed that different endophyte isolates have differential effect on the development and function of their co-endophytes as well as host plant. Colonization of soybean roots by VA-mycorrhizal fungi varies significantly with cultivars. The colonization of non-nodulating isoline of Clark *rjl* was the highest (80%) whereas that of Williams 82' was lowest (57%). The studies conducted by Crush (1974) reaffirmed the effect of mycorrhizae in stimulating nodulation. Increased plant dry weight (Bethlenfalvay *et al.,* 1987); Daft and El-Giahmi (1974) weight of nodules, amount of nodular tissue, rate of acetylene reduction, N and P contents and concentration of leghaemoglobin.

By using labelled nitrogen sources ($^{15}N_2$), it was observed that nodulated root systems of mycorrhizal beans fixed more nitrogen than nodulated root system with non-mycorrhizal plants (Kucey and Paul, 1982). In alluvial sandy-loam soil of Delhi, it was observed that inoculation of chickpea *(Cicer arietinum)* with *Rhizobium sp.* or *Glomus fasciculatum* individually or in combination resulted in significantly higher amounts of nitrogen fixation. This increase of nitrogen fixation in plants was pronounced at 50 Kg P_2O_5/ha application (Subba Rao *et al.,* 1985). Ho and Trappe (1975) suggested that VAM fungi (*Glomus mosseae)* and *G. macrocarpum* are capable of reducing nitrate to nitrite and presumably possess the nitrate reductase system. With the capability for reducing nitrate these fungi assume increased symbiotic effectiveness in terms of nitrogen assimilation and translocation to the host. Plants did not nodulate unless their P-concentrations were at least 0.15% and also observed that mycorrhizal infection helped the plant to reach this required level.

Smith and Brown (1979) noted mycorrhizal clover roots have higher P concentration than did the non-mycorrhizal control plants. They also suggested that, the enhancement of nodule efficiency could be explained by the increase in root P, which accumulated mainly as polyphosphate. The higher rates of photosynthesis were found in symbiotic plants than non-symbiotic plants. Increased growth and yield in nodulated soybeans inoculated with the VA-mycorrhizal fungus were accompanied by improved P-uptake lower root to shoot ratios, better nodulation and nitrogenase activity and modification in the nodulation pattern during growth.

Kumar and Elamathi (2007) studied the effect of nitrogen levels and *Rhizobium* application methods on yield attributes, yield and economics of black gram. The maximum pod number, plant height, number of leaves, dry weight of plant, nodule number and pod number were obtained under the application of *Rhizobium.* Kalita *et al.* (2006) studied the evaluation of native *Rhizobium* from acid soils of Assam. *Rhizobium* isolate AR1 was the most resistant against 9 antibiotics. *Rhizobium* isolate AR1 was found to be more effective in terms of number of nodules (75 and 100), nodule dry weight (0.41 g and 0.38 g) and percentage of nitrogen content both in black gram and green gram. Performance of green gram with all the isolates was better with respect to the parameters than black gram. *Rhizobium* isolate AR1 also showed the highest nitrogenase activity in the root nodules of black gram and green gram (4.02 and 4.25 mole/g/h). Deka and Azad (2006) did the screening for effective strain of *Bradyrhizobium.* Among the 15 strains, Gn3 was the most effective strain giving the highest acetylene reduction assay (479.01 m mole/g/h) and leghaemoglobin content (0.125 mm). On the other hand, strain C10 showed the highest antibiotic efficiency in antibiotic resistance against nine different antibiotics and wide host range infectively nodulating cow pea, urd bean, mung bean and pigeon pea.The 48% of *Rhizobial* isolates grew at optimum temperature (37°C) and 1% grew at low temperatures while the remaining 51% of the isolates showed the optimum growth at 28°C. The two isolates of *Rhizobia* from cowpea, two from mung bean and one from chickpea were found acid-tolerant (pH of 4 and 5) (Jadhav *et al.*, 2006). A significant variation in *Rhizobial* sensitivity was observed in different isolates from different mutant plants in contrast to native strain. Significant variations were also observed in the physiological characteristics of *Rhizobium* after gamma irradiation. The variations might have appeared due to *Rhizobial* mutation inside the host plant and a direct interaction between the *Rhizobia* and altered host physiology/phenotype occurred (Raisinghani, 2004). The nitrogen content of the fresh seeds was 5.78% in the *Rhizobium* inoculated plants, while that in uninoculated plants was only 2.72%. The plants inoculated with the AH isolate showed better nodulation and nitrogen content compared to the plants inoculated with VM isolate. However, the SG isolate completely failed to produce nodules on the black gram (Reddy and Malliah, 2002). Saxena *et al.* (2002) studied the toxin production and competitiveness of *Bradyrhizobium* S.P (*Vigna*) strain S-24 when inoculated with a sensitive strain Mo1 in a sterile soil @ the rate of 1:1 and 1:5, occupied 92% and 88% nodules, respectively. The sensitive strains Mo1, UASB741 and CT2014 failed to form nodules on *Vigna radiata* cv. Pusa-105 in the presence of cell free supernatant of S24 broth indicating the involvement of Vigna toxin in providing competitive edge to S-24.

Sriramachandrasekharan and Vaiyapuri (2003) studied the influence of granular insecticide carbofuran in association with *Rhizobium* on nodulation, growth and yield of black gram. Irrespective of the levels of carbofuran, *Rhizobium*-inoculated black gram showed better growth and higher pod yield (50.3 g) and straw yield per pot (81.1 g) than the uninoculated crop. Balyan *et al.* (2002) studied the enhancing nodulation in *Vigna mungo* by applying higher quantity of *Rhizobium* in planting furrows and PSB. The inoculation of PSB along with *Rhizobium* gave higher nodulation rate, plant dry matter, yields and N-uptake than PSB inoculation alone. The result suggested that *Rhizobium* inoculated in furrows at higher rate gave better results and the *Rhizobium* enhanced the beneficial effects of PSB. Nagrajan and Balchandar (2001) studied the influence of *Rhizobium* and application of organic amendments on nodulation and grain yield of black gram and green gram in acid soil. Seed inoculation of *Rhizobium* and application of organic amendments enhanced biomass, root nodulation and grain yield. Bio digested slurry at 5 t/ha+ *Rhizobium* gave the greatest plant height (44.2 cm and 53.7 cm, for black gram and green gram, respectively), nodule no. (23.3 and 24.0), nodule weight (45.3 and 42.3 mg) and grain yield (758.3 and 732.0 kg/ha).

Shanmugam *et al.* (2001) studied the interaction of *Pseudomonas fluoresecens* with *Rhizobium* in the management of black gram root rot incited by *Macrophomina phaseolina*. Glass house and field studies showed that seed treatment and soil application of Pf-1 was the most effective treatment to reduce root incidence to increase yield.

8. POTENTIAL APPLICATIONS OF AM FUNGI

The AM fungi had a very wide application and can be used for improving the nutrient status and moisture availability from the deeper zone of the soil beyond the root extension and very well be used under moisture stress conditions, salt stress and drought, heavy metal bioremediation and also in increasing the capacity against pest and disease management, waste land reclamation and its potential can be applied in tissue culture too.

The major role of AM fungi is to absorb nutrients (macro and micronutrients) and water from the soil and to transfer them into their hosts (Sharma *et al.,* 1997; Mishra, 2008b). This mechanism relies on the existence of an acidic compartment in the periphery of arbuscules presumably identical with the peri arbuscular space separating the plant and fungal plasma membranes (Guttenberger, 2000). This acidification of the peri-arbuscular space corresponds to proton gradients across the fungal arbuscular and the plant peri-arbuscular membrane powering transport processes across these membranes. Enzymes responsible for the generation of such an acidification –H+-AT Pases-have been studied for a long time (Murphy *et al.,* 1997).

Besides phosphate, a number of mineral nutrients and most notably water are transported by AM fungi to their host plants. In the case of nitrate, AM fungi play an even more active role when compared to phosphate, because they are able to liberate nitrate from complex organic material within the soil (Hodge, 2000). Few proteins involved in nitrate transport in mycorrhizal roots have been found so far; amongst them a tomato nitrate transporter with increased transcript levels in AM colonized roots (Hildebrandt *et al.,* 2002). In addition, a fungal *(G. intraradices)* nitrate reductase has been cloned, which is expressed in AM roots and therefore might be involved in the transfer of nitrogen as well.

8.1 AM Fungi and Environmental Stress Management

Arbuscular mycorrhizal fungi (AMF) have been shown to decrease plant yield losses in saline and alkaline soil conditions (Kashyap *et al.,* 2004; Sharma *et al.,* 2005a; Murkute *et al.,* 2006). Shoot and root dry matter yields and leaf area were higher in mycorrhizal than in non-mycorrhizal plants. Total accumulation of P, Zn, Cu, and Fe was maximum along with chlorophyll content and relative water content in mycorrhizal than in non mycorrhizal plants under both control and medium salt stress conditions (Colla *et al.,* 2007). AM fungi could also influence plant hormones (Barea and Azcon-Aguilar, 1981). Improved salt tolerance of mycorrhizal plants can be mainly related the changes in physiological processes such as increased carbon dioxide exchangeable rate, transpiration, stomatal conductance and water efficiency. This may be due to increased uptake of nutrients with low mobility, such as P, Zn and Cu (Ruiz-Lozano *et al.,* 1996; Al-Karaki and Al-Raddad, 1997; Al-Karaki and Clark, 1998), and to improved water relations (Ruiz-Lozano and Azcon, 1995; Al-Karaki and Clark, 1998). Enhanced salt tolerance in AM symbiosis was mainly related with the elevated superoxide-dismutase (SOD), catalase (CAT), ascorbate peroxidase (APX), peroxidase (POD) activity by AMF which degrade more reactive oxygen species and so alleviated the cell membrane damages under salt stress.

The contribution of the AM symbiosis to plant drought tolerance results from a combination of physical, nutritional, physiological and cellular effects. Following are the mechanisms by which AM symbiosis can alleviate drought stress in host plants (Auge, 2001). Hyphae with a diameter of 2-5mm can penetrate soil pores inaccessible to root hairs (10-20 mm diameter) and so absorb water that is not available to non-mycorrhizal plants through root extension via mycelium/hyphae association.

An increase in proline accumulation has been observed in mycorrhizal plants subjected to drought (Azcon *et al.,* 1996; Goioechea *et al.,* 1998). It has also been shown that mycorrhizal colonization and drought interact in modifying free amino acid and sugar pools in roots (Auge *et al.,* 1992b). Finally, a greater osmotic adjustment has been reported in leaves of mycorrhizal basil plants than in non mycorrhizal ones during a period of lethal drought (Kubikova *et al.,* 2001). During soil drying, mycorrhizal plants often maintain higher gas exchange rates than non-mycorrhizal plants of similar size and nutrient status (Goicoechea *et al.,* 1998). Duan *et al.* (1996) have suggested that AM fungi probably increase the ability of the root system to scavenge water in dried soil, resulting in less strain on foliage, and hence higher stomatal conductance and shoot water content at particular soil water potential. On the other hand, evidence has come from other studies that arbuscular mycorrhiza may influence leaf transpiration, even after leaves have been detached (Green *et al.,* 1998).

Some studies have demonstrated that AM symbiosis development induces the expression of genes encoding aquaporins that facilitate the passive movement of water molecules down a water potential gradient. (Krajinski *et al.,*2000).

8.2 Mycorrhiza and Soil Heavy Metal Bioremediation

Arbuscular mycorrhizal (AM) fungi provide an attractive system to advance plant-based environmental clean-up. During symbiotic interaction the hyphal network functionally extends the root system of their hosts. Thus, plants in symbiosis with AM fungi have the potential to take up heavy metals (Pb, Cd, As) from an enlarged soil volume. At high concentrations heavy metals interfere with essential enzymatic activities by modifying protein structure or by replacing a vital element resulting in deficiency symptoms. Similarly, heavy metals are taken up via the fungal hyphae and can be transported to the plant. Thus, in some cases mycorrhizal plants can show enhanced heavy metal uptake and root-to-shoot transport (phytoextraction) while in other cases AM fungi contribute to heavy metal immobilization within the soil (phytostablilization). The result of mycorrhizal colonization on clean-up of contaminated soils depends on the plant-fungus-heavy metal combination and is influenced by soil conditions. Different mechanisms involved in heavy metal detoxification have been proposed, as chelating agents are secreted that bind metals in the soil, e.g. histidine and organic acids from the plant, glomalin from the fungus. Binding of heavy metal to cell wall components in plants and fungi, the plasma membrane was found as a living, selective barrier in plants and fungi. Specific and nonspecific metal transporters and pores in the plasma membrane of plants and fungi (active and passive import). Chelates in the cytosol, e.g. metallothioneins (plants and fungi), organic acids, amino acids, and metal-specific chaperons (shown for plants, assumed for AM fungi). Export via specific or nonspecific active or passive transport from plants or fungal cells, sequestration of heavy metal in the vacuole of plant and fungal cells and transport of heavy metal in the hyphae of the fungus. In arbuscules, metal export from the fungus and import into plant cells via active or passive transport and mycorrhizae remediate the heavy metal contaminated soils by mainly two processes namely phytostablilization and phytoextraction.

8.3 AM Fungi and Pest Management

Number of studies have demonstrated the beneficial role of AM fungi in plant growth and health (Bethlenfalvay and Linderman, 1992; Bagyaraj, 1992) and it appears that they are essential for the survival of plant species in many ecosystems (Allen, 1991). Root pathogens are a major limiting factor for plant production. Fungi belonging to the genera *Phytophthora, Fusarium, Phythium, Rhizoctonia*, some bacteria, nematodes are the most widespread among the various fruits and vegetable and pulse crops.

They kill the roots or reduce their ability to absorb water and nutrients by penetrating root tissues and producing toxins. Many reports indicate an interaction between AM fungi and plant pathogenic organisms (Mukerji, 1999; Singh *et al.*, 2000; Abdel-Fattah and Shabana, 2002). A host plant previously inoculated with AM fungi exhibits increased resistance to several root diseases (Linderman, 2000). Reduction of disease symptoms and severity has been reported in different fruits and vegetables with mycorrhiza-pathogen interaction for various fungal pathogens like *Phytophthora, Fusarium, Pythium, Rhizoctonia, Verticillium, Thielaviopsis, Aphanomyces, Pyrenochaeta,* bacterial pathogens like *Pseudomonas, Erwinia,* viruses and nematodes like *Radopholus, Heterodera, Rotylenchus, Pratylenchus, Tylenchulus, Meliodogyne* (Bharadwaj and Sharma, 2006 a;b). The influence of AM fungi on root pathogens vary greatly with the type of pathogen.

8.4 Mechanism of Interaction Between AM Fungi and Plant Pathogens

Several hypothesis have been put forth to explain the mechanism involved in reduction of disease severity from pathogenic infections by mycorrhizal fungi (Linderman, 2000). Depending on the disease and the environmental conditions, any or all mechanism may be involved.

The AM inoculated roots may control the infestation of disease and pest as a result of changes in the root morphology along with physiological and biochemical changes, enhanced nutrient status and root exudates inhibiting pathogens activity in the zone of root elongation enzyme production and biological interactions of mycorrhizosphere competition of carbon source and developing a plant defense response in the host plant roots.

Plants colonized by AM fungi develop extensive root system with higher number of lateral roots as compared to non-mycorrhizal plants (Berta *et al.*, 1993). Mycorrhizal infection enhances lignifications in root tissues, particularly in stellar region which may be responsible for restricting soil borne pathogens that invade the host roots (Dehne and Schonbeck, 1978). This strong vascular system which increases nutrient flow, provides greater mechanical strength and reduces the detrimental effect of pathogens (Hussey and Roncadori, 1982).

There is increase uptake of water and nutrients by mycorrhizal fungi which affects pathogenesis. Through increased P nutrition, AM fungi enhance root growth, increase the capacity for absorption and affect cellular process in roots (Smith and Gianinazzi-Pearson, 1988). Such changes may explain increased tolerance of mycorrhizal plants towards pathogens. Mycorrhizal roots show increased respiration than non AM roots (Dehne, 1982). Increased respiration rate of AM roots indicates higher metabolic activity which might enable mycorrhizal plants to be more resistant against root pathogens. Increased ethylene and DNA methylation have also been reported in AM roots (Dugassa *et al.,* 1996). Similarly the arginine level in mycorrhizal roots was reported to be six fold more due to an increased specific activity of enzymes in ornithine cycle and blocking of cycle by arginase enzyme system in such roots (Dehne *et al.,* 1978). Mycorrhizal development enhances the production of isoflavonoids which play major role in natural defense of plants (Morandi, 1996).

Root exudation pattern also changes in quality as well as quantity with mycorrhizal colonization which might play an important role in pathogenesis (Graham *et al.,* 1981). Root exudates directly affect the pathogen by inducing their germination or by inhibiting saprophytic and pathogenic activity. Root exudation is maximum in zone of elongation, the portion of root growth that is most susceptible to AM fungal colonization (Buwalda *et al.,* 1984).

Greater tolerance of mycorrhizal plants towards the pathogens is attributed to increased phosphate status and more vigorous growth of plants (Azcon-Aguilar and Barea, 1996). In addition to phosphates, AM fungi increase uptake of micronutrients (Fe, Zn, Mg, Mn etc).

Mycorrhizal fungi, soil fungal pathogens and plant parasitic nematodes, all occupy the same root tissue, resulting in direct competition for space which might explain the disease resistance in mycorrhizal plants (Azcon-Aguilar and Barea, 1996). Both mycorrhizal fungus and pathogen are dependent on host photosynthates for their growth. The major substrate for microbial activity in the rhizosphere or on *Rhizoplane* is organic C released by host roots (Azaizeh *et al.,* 1995). When mycorrhizal fungus has primary access to photosynthates, the higher C demand may inhibit pathogen growth (Azcon-Aguilar and Barea, 1996).

The concept of mycorrhiza sphere is based on the fact that mycorrhizae exert the strong influence on the micro flora in the rhizosphere (Bansal *et al.,* 2000). This results in microbial shift in the mycorrhiza sphere which could influence the health of plants (Whipps, 2001, Barea *et al.,* 2002). The number and proportion of bacterial antagonists against several root pathogens increase when AM association is established. More actinomycetes antagonistic to *Fusarium solani* and *Pseudomonas solanacearum* were isolated from the rhizosphere pot cultures of *G. fasciculatum* colonized plants than from non-mycorrhizal control (Secilia and Bagyaraj, 1987).

The symbiotic host-fungus relationship is characterized by the formation of haustoria like intracellular arbuscules which are successively degraded. This process represents digestion of the fungus by the host. For this degradation to occur, the fungal cell wall of the endophyte has to be attacked by the host. Therefore, the roots colonized by mycorrhizal fungus show high chitinolytic activity. These enzymes can be effective against other fungal pathogens as well. Increased activity of peroxides associated with epidermal and hypodermal cells has also been reported in mycorrhizal roots (Gianinazzi and Gianinazzi-Pearson, 1992).

The response of plants to AM fungi involves temporal and spatial activation of different defense mechanisms (Garcia-Garrido and Ocampo, 2002), however, how these defenses affect the functioning and development of AM remains unclear. A number of regulatory mechanisms of plant defense response have been described during establishment of AM symbiosis. These include (i). Elicitor degradation (ii) Modulation of second messenger concentration (iii) Nutritional and hormonal plant defense regulation (iv) Activation of regulatory symbiotic gene expression.

Colonization of roots by AM fungi brings about biochemical changes within host tissues. These biochemical changes include (i) Stimulation of the phenylpropanoid pathway (Harison and Dixon, 1993), (ii) Changes in levels of aliphatic polyamines (El Ghachtouli *et al.,*1995), (iii) Activation of defense related genes (Harison and Dixon, 1993, Franken and Gnadinger, 1994, Dumas-Gaudot *et al.,* 2000), (iv) Enhancement of certain hydrolytic enzyme activities (Dumas-Gaudot *et al.,* 1996), (v) synthesis of protein of unknown function (Dumas-Gaudot *et al.,* 2000), (vi) Elicitation of plant chitinase and B-1, 3 glucanase (Blee and Anderson, 1996) which are antifungal against pathogenic soil and root born fungi, (vii) Induction of chitinase isoforms by mycorrhiza. These chitinase release oligosaccharide elicitors from the chitinase AM fungal cell walls which in turn stimulates the general defense responses of plants (Cordier *et al.,* 1996), (viii) Increase ethylene production and DNA methylation by AM roots can be related to gene expression for higher resistance of plants against pathogens (Dugassa *et al.,* 1996), (ix) The level of phytoalexin medicarpin, coumestrol, daidzein, medicarpin-malonyl glucoside, formononetin, showed transient increase in roots of *Medicago truncatula* days after inoculation with *Glomus* species, thereby showing that phytoalexin and their precursors are activated by AM inoculation (Harison and Dixon, 1993) and (x) Phenolics may also play a role in protecting AM roots against pathogenic fungi (Grandmaison *et al.,* 1993). A continuous increase of total soluble phenols has been found in mycorrhizal roots of peanuts (Krishna and Bagyaraj, 1984).

8.5 AM fungi and Wasteland Reclamation Through Nutrient Mobilization and Improving Soil Physical Properties

Abbott and Robson (1991) stated that the relationship between levels of mycorrhizal colonization and soil physical and chemical properties are markedly variable. Mycorrhizal fungi, specifically AM fungi have been adapted to wide range of environmental conditions (Sharma *et al.,* 1996). AM fungi are aerobic and soil aeration has considerable impact on their distribution. The AM symbiosis produces extra radial hyphae that may extend several centimeter out in to the soil and secretes organic material that are substrate for other microbes (Singh and Tilak, 1995). These hyphae associated microbes frequently produce sticky material that causes soil material to stick together and improve soil structure. Further by nutrient uptake, they improve plant cover and root proliferation. Mycorrhiza brings about reclamation of wasteland by improving nutrient status of soils and by improving soil physical properties (Kashyap *et al.,* 2005).

The cycling of macro and micronutrients in an ecosystem is influenced by multiple interactions involving soil microbial populations. Arbuscular mycorrhizal (AM) associations play important roles in this nutrient cycling through their microbial activity and their involvement in plant nutrient acquisition. The extra radical mycelium connects plant roots to the surrounding soil microhabitat increasing the soil volume exploited by host plants. Mycorrhizal hyphae transport mineral nutrients over greater distances from depleted zones than do roots. Thus, under low nutrient conditions AM-colonized roots may have an enhanced uptake of relatively immobile macro and micronutrients (Subramanyan and Charest, 1998). In addition, mycorrhizal roots often have not only increased length but also modified root architecture (Berta *et al.,* 1995). Root/AM colonization often results in an enhanced uptake of relatively immobile metal-micronutrients such as Cu, Zn and Fe (Mishra, 2008b).

Soil structure is influenced by soil water content and its variation with time and plant growth can strongly influence the magnitude and frequency of wetting and drying cycles. Decreased water content typically increases contact points between primary particles and organic matter, resulting in increased soil cohesion and strength (Horn and Dexter, 1989; Horn *et al.,* 1994). Localized drying of soil, in close proximity to roots, promotes binding between root exudates and clay particles (Reid and Goss, 1982), directly facilitating micro aggregate formation. More efficient exploration of water by mycorrhizal fungi may lead to more extreme wet and dry cycles, which could have very strong consequences for soil aggregation. Additionally, because the symbiosis can allow leaves to fix more carbon during water stress (Duan *et al.,* 1996), carbon inputs into the soil would be expected to be increased, which might be especially important in more arid environments. The release of compounds from living roots can be strongly influenced by mycorrhizal fungi (Jones *et al.,* 2004), because these fungi can effect plant carbon metabolism (Douds *et al.,* 2000).

During development of AM, the fungal symbionts grow out from the mycorrhizal root to develop a complex, ramifying network into the surrounding soil which can reach upto 30m of fungal hyphae per gram of soil (cavagnaro *et al.,* 2005; Wilson *et al.,* 2009). The secretion by AM fungi of hydrophobic, 'sticky' proteinaceous substances, referred to as glomalin (Rillig *et al.,* 2002), also contributes to soil stability and water retention (Bedini *et al.*, 2009). The combination of an extensive hyphal network and the secretion of glomalin is considered to be an important element in helping to stabilize soil aggregates, thereby leading to increased soil structural stability and quality (Bedini *et al;* 2009; Caravaca, 2006).

8.6 AM Fungi and Rooting Behavior in Tissue Culture

The symbiotic association of AM fungi with plants often produces plant growth regulating hormones, mainly auxins, affecting the plant root physiology and subsequently plant growth. The phytohormones released by this mycorrhizal association, if applied in *vitro* in form of root extracts influences the growth of culture and rooting in shoot cultures. The root extract of AM *Morus alba* was found to be equivalent to IBA in rhizogenesis while the extracts of *Trigonella foenum graecum, Ricinus communis* and *Solanum melongena* were not very effective in rooting (Sharma *et al.,* 2005b)

9. PLANT GROWTH AND YIELD

It is well known fact that successful cultivation of the legumes depends, in part, on an effective symbiosis with *Rhizobium.* Significant seed-yield increases following *Rhizobium* inoculation in cowpea (Bajpai *et al.,* 1974; Bagyaraj and Hedge, 1978) mungbean (Ramaswamy and Nair, 1965; Oblisami *et al.,* 1976) and blackgram (Sahu and Behera, 1972; Lehri *et al.,* 1974). Inoculants prepared with more than on *Rhizobium* strains were found to be superior over single strain inoculants in some (Oblisami *et al.,* 1976), but not all studies (Bagyaraj and Hedge, 1978; Ramachandran *et al.,* 1980).

The high yields report in northern Nigeria suggest that groundnut nodules estimates of as much as 240 kg N/ha. or 80% of the plant's N-uptake (Dart and Krantz, 1977). Seth (1979) studied the field performance of five selected *Rhizobium* isolates of groundnut at three different locations in Gujarat. A yield increase was obtained with one isolate at one location and was equivalent to the yield obtained with applications of 25-50 kg N/ha to un inoculated plots.Gupta *et al.* (1982) showed a significant increase in grain yield (16.8%) and nitrogenase activity (44.4%) in an interaction study between four cultivars of chickpea and four strains of *Rhizobium* over uninoculated control.

Mycorrhizal development is greatly affected by light received by the shoot of the host plant. Low irradiation and short day lengths can reduce the development of VAM (Daft and El-Giahmi, 1978), which also reduce plant growth in uninfected plants.As VAM is an obligate endosymbiont, they have very little capacity of independent growth. Harley and Smith (1983) pointed out that growth responses to mycorrhizal inoculation was determined by translocation of soil derived nutrients to the plant through the fungus and the consumption of carbon compounds derived from host by the fungus. Maximal mycorrhizal effectiveness occurs when nutritional conditions for the host are suboptimal, but when these become severely deficient the fungus becomes ineffective.

Ratnayake *et al.* (1978) found that under low level of P, root exudation measured in terms of the net leakage of soluble amino-acids and reducing sugars was significantly higher than under the high P-levels. Increased levels of added soluble P reduced the mycorrhizal infection and spore production, however, root, shoot and total dry weight were significantly greater in mycorrhizal plants, at all levels of added soluble P, than non-mycorrhizal ones (Krishna and Bagyaraj, 1982). The increase in the addition of soluble P above certain level resulted in decreased length of infected root and external hyphae per cm. of infected root in addition of reduced formation of vesicles (Abbott and Robson, 1984).The effect of temperature on root growth and colonization were studied by different workers (Smith and Brown, 1979, Hetric *et al.*, 1984). Vogelzang *et al.* (1993) found that shoot dry weight and root fresh weight were highest at 30°C temperature in mung bean both in mycorrhizal and non-mycorrhizal plants. While at 30°C root length were highest in mycorrhizal but in non-mycorrhizal plants root length were more at higher temperature (38°C).

Azacon-Agauilar *et al.* (1979) recorded that *Glomus* inoculation improved the yield of alfalfa, whereas *Rhizobium* inoculation alone did not. In dual inoculation, shoot dry weight increased by 211% compared with control. So they concluded that mycorrhizal fungi *G. mosseae* is an effective **"biological fertilizer"** for *Medicago sativa.* The sweet vetch plants inoculated with either a combination of mycorrhizal fungi and rhizobia or mycorrhizal fungi alone grew taller, more leaf number and produced more above ground plant material than plants infected with rhizobia alone or control plants.

Bala and Singh (1983) reported that under Punjab soil conditions the magnitude of mycorrhizal association in legume and pulses depends on the ecological factors, host specificity and mycorrhizal types and further Singh *et al.* (1984) reported the distinct responses of lentil to dual symbiosis on a P-deficient soil is dependent on *Rhizobium* and *Glomus* sp. for N-fixation, N and P nutrition and growth.

Inoculation with *G. fasciculatum* significantly increased the root dry weight in both cowpea and pigeon pea. Application of P significantly increased the dry weights of shoot and root. Plants inoculated with both *Rhizobium* + *Glomus* and supplimented with P recorded highest dry weight of root and shoot (Manjunath and Bagyaraj, 1984). Mathew and Johari (1989) found that in mungbean when a tripartite inoculation of indigenous VAM (*G. caledonicum)* + *R. phaseoli* was used, root and shoot ratio significantly increased. Application of indigenous VAM alone did not result in any significant increase in total biomass, whereas a 25% increase occurred from a tripartite inoculum. The number of pods and leaves per plant was also greater for the mixed treatment. A significant increase in leaf length and width was also observed for all the treatments. Rawat *et al.* (1991) reported that the shoot dry weight significantly increased in mycorrhizal plants than non-mycorrhizal. *Rhizobium* also affects shoot dry weight over control, but mycorrhizal and *Rhizobial* strains independently and their combinations failed to show significant differences with respect to grain yield.

Cowpea inoculated with the VAM fungus and *Rhizobium* strain pairing produced significantly more pods and higher N and P-content than either single inoculations or uninoculated control (Thiagarajan *et al.*, 1992). Mahdi and Atabani (1992) also reported that inoculation of soybean and lablab bean with *Glomus mosseae* increased nodulation, dry matter yield and tissue N and P-contents more than triple super phosphate fertilizer, but greater responses were obtained from *G. Mosseae* combined with fertilizer. Interaction between VAM fungi and *Bradyrhizobium* affect shoot dry weight and P-content in *Colopogonium caeruleum* (Ikram *et al.*, 1993). The VAM inoculation has also been reported to have the carry over effect on the growth of lucern crop. However, maize inoculation fail to carry over the mungbean crop planted immediately after maize harvest and thus did not increase the root colonization and grain yield of the succeeding crop. Fresh inoculation of the mungbean with *G. mosseae* increased the grain yield over the uninoculated control (Khadge *et al.*, 1992).

There are controversial reports on the carry over effect of VAM inoculation on the succeeding crops. Under these circumstances the persistence or non-persistence of VAM fungi probably played an important role under different sets of environmental conditions. In presence of phosphorus in the soil up to certain level increased the VAM development where as a higher dose of phosphorus resulted in the poor performance of VAM. The presence of P up to certain level increased the N and P uptake in the plants when legumes were inoculated with *Rhizobium* and/or VAM. The synergistic effect was noticed when both biofertilizers were combined in legume crops.

Combined inoculation by *Glomus mossae* and *Azotobacter* improved plant growth and yield of pomegranate in field conditions (Aseri *et al.*, 2008). *Glomus clarum* and *G. etunicatum* with *Rhizobium* improved seedling growth of *Schizolobium amazonicum* in nursery conditions (Siviero *et al.*, 2008). Improvement in fodder production and quality in silvopastoral system by AM fungi consortia and native *Rhizobium* (Mishra, 2008b). *Glomus intraradiaces* and *Azospirillum brasilense* enhanced fodder production and quality of *Panicum maximum* in different cuttings with N fixer and AM consortia (Mishra *et al.*, 2008a). Dual inoculation by AM fungi (consortia) and *Azotabacter* (exotic) improved the acclimatization of tissue cultured saplings of *Morus alba* (Kashyap and Sharma, 2006). Combined inoculation by AM fungi consortia and native

Azotobacter was better in terms of biomass yield of multipurpose tree sp. and soil microbial biomass (Sharma *et al.,* 2005a). Native strains of bio inoculants (AM fungi consortia and *Azotobacter)* together were more adaptive to alkaline field conditions and also enhanced fruit yield and nutrient uptake in vegetable species (Kashyap *et al.,* 2005). *Glomus mossae* with bacterial sp. (*Azotobacter* + *Bacillus megateranium* (P solubilizers) + *Bacillus mucilaginous* (K solubilizers) improved growth and biomass of *Zea mays* (Wu *et al.,* 2005). With 15 ppm of IBA combined inoculation with AM fungi and native *Azotobacter* enhanced survival of *Morus alba* saplings under in-vivo saline conditions (Kashyap *et al.,* 2004). In *Casurina equisetifolia,* biomass yield and nutrient uptake was maximum in combined inoculation of *Glomus fasciculatum* + *Azospirillum* + Phosphobacterium + *Frankia* (Rajendran and Devraj, 2004). *G. intraradiaces* and *Rhizobium tropici* together improved plant growth and nutrient uptake in two native forage legumes (Shockley *et al.,* 2004). *Glomus fasciculatum* + *Rhizobium* + *Pseudomonas striata* improved plant yield and nutrient uptake but with *Penicillium* variable negative effect observed on all parameters (Zaidi *et al.,* 2003). In *Calliandra calothyrsus,* improvement in fodder production by *Glomus etunicatum* + *Rhizobium* (Lesueur *et al.,* 2001). AM fungi with *Azospirillum* had negative overall effect while with *Rhizobium* improved nodule no. rhizosphere microflora and nutrient uptake in alfalfa (Biro *et al.,* 2000). Decrease in spore germination, hyphal length and root infection of AM fungi observed with *Rhizobium.*

Combined inoculation by AM fungi and *Rhizobium* improved plant growth and yield of urd bean under rain fed field conditions (Preeti, 2011). Kumar and Sharma (2006) studied the effect of *Rhizobium* and co-inoculants VAM and PSB on yield and nutrient uptake in black gram. Results showed that inoculation of the bio fertilizers significantly increased grain and straw yield compared with uninoculated control. The highest grains and straw yields (223.5 and 229.8 g/micro plot, respectively) were obtained with the mixed inoculation of all 3 bio-fertilizers under 120 g/hectare. Hakeem *et al*., (2004) observed the effect of *Rhizobium* in association with different sources of sulphur on growth of black gram and chemical properties of soil. The application of gypsum at 60 kg/hectare + *Rhizobium* spp inoculation (C4I1) was superior over all treatments for growth and yield of black gram. The highest yield of 126.33 g/plot was obtained in treatment with C4I1. Gunasekaran *et al*., (2004) studied the synergism between *Rhizobium*, plant growth promoting rhizobacteria (PGPR) and phosphate solubilizing bacteria (PSB) in black gram. Inoculation of *Rhizobium* alone had given a grain yield of 655 kg/hectare where as inoculation control had recorded 472 kg/hectare only. Combination of all 3 organisms (*Rhizobium* + PGPR + PSB) had recorded the maximum nodules; plant biomass and grain yield (760 kg/hectare) which was 61% higher than individual inoculation of *Rhizobium* alone.

Sheikh *et al.* (2003) observed the effect of *Rhizobium* and N, P, K on growth, physio-morphological traits and yield of waste water grown black gram. Waste water application increased the leaf nitrate reductase activity, carbonic anhydrate activity and total chlorophyll contents of *Rhizobium* inoculated plants. The optimum fertilizer treatment was N:P:K at 10:30:20 kg/hectare and N,P,K treated plant showed better performance under inoculated conditions than uninoculated. Tomar *et al.* (2003) studied the residual effect of black gram inoculated *Rhizobium*, VAM and PSB on succeeding wheat crop. *Rhizobium* combined with VAM or PSB recorded a 10% increase in plant dry weight than *Rhizobium* alone. The highest residual effect on all the parameters including the wheat grain yield was recorded with combined inoculation of *Rhizobium* + PSB + VAM.

Kumari and Nair (2003) observed the use of acid tolerant strain of *Bradyrhizobium* spp for black gram and green gram. The results indicated that for acidic soils, the mere development of efficient native strain of *Rhizobium* or *Bradyrhizobium* alone was not sufficient but it should be along with a package of practices recommendation consisting of application of organic manure and liming to neutralize the soil pH. Dubey (2003) observed the integrated management of web-blight of urd bean and mung bean. The result showed reduced disease intensity and considerably enhanced plant vigour and grain yield of 'urd' and 'mung' bean.

Sampath Kumar and Ganesh Kumar (2003) studied the effect of VAM fungi and *Rhizobium* on growth and nutrition of black gram and cow- pea. It was concluded that inoculation with mycorrhizal fungi and *Rhizobium spp* is more effective in increasing legume growth and biomass production. Jain and Gupta (2002) observed the effect of *Rhizosphere* fungi on nodule number, shoot and root length of *Vigna mungo*. All fungi did not significantly affect root and shoot length.

Jayashree *et al.* (2000) studied the evolution of *Pseudomonas florescence* (P 1) against black gram and sesame root knot disease. Result showed higher plant growth, better native *Rhizobium* nodulation and grain yield. The *Rhizosphere* soil recorded a high P-1 population. Sharma *et al.* (2000) studied the effect of *Rhizobium* inoculation and nitrogen on growth, dry matter accumulation and yield of black gram. Growth, yield and dry matter accumulation increased N application and *Rhizobium* inoculation with the local strains giving the best results.

10. FUTURE INTERVENTION

AM undoubtedly plays a very important role in enhancing the nutrient and water uptake and has a wider applicability under various agro-climatic condition subjected to climatic change enhancing the tolerance in plants against various pathogens and insects, improving the soil physical conditions and wastelands managements. However, there is a need for exploitation of AM fungi with enhanced potential use such as-

1. **Screening of existing indigenous spp and exotic species for better plant- soil- fungus interaction.**
2. **Understanding the molecular characterization of plant-fungus relationship to better AM symbiosis.**
3. **Studies related to genetic diversity, molecular mechanism and beneficial characteristic to mass culturing under laboratory conditions.**
4. **Role of AM fungi under changing C-N ratio in soil health with CO_2 sequestration and climatic change and defense mechanism of resistance against pathogens.**
5. **Field testing of different AM strains for crop plants, horticultural and vegetable crops, afforestation and timber plantation even at nursery level also.**
6. **Improvement for mass culturing and inocula production techniques of efficient type of strains needs attention.**
7. **Educating the farmers and gardeners for the potential use of different spp/strains of AM fungi.**
8. **Detailed studies on physiological and biochemical changes brought about by the host-AM-interaction for better symbiosis needs immediate attention.**

REFERENCES

Abbott, L.K. and Robson, A.D. (1978). Growth of substerranean clover in relation to the formation of endomycorrhizas by introduction and indigenous fungi in a field soil. *New Physiol.* **8:** 575-587.

Abbott, L.K. and Robson, A.D. (1981). Infectivity and effectiveness of five endomycorrhizal fungi: competition with indigenous fungi in field soil. *Aust. J. Agric. Res.* **32:** 621-630.

Abbott, L.K. and Robson, A.D.(1984). The effect of mycorrhizas on plant growth. [In] VA Mycorrhizae, ed. C.L. Powell, D.J. Bagyaraj, pp. 113-130. Boca Raton, Fla: CRC Press 234 pp.

Abbott, L.K. and Robson, A.D.(1985). The effect of soil pH on the formation of VA-mycorrhizas by two species on *Glomus. Aust.J.Soil Res., 23:253-261.*

Abbot, L.K. and Robson, A.D. (1991). Factors influencing the occurrence of vesicular mycorrhizas. *Agric. Ecocyst.Environ.* **35:**121-150.

Abdel-Fattah, G.M. and Shabana, Y.M. (2002). Efficacy of the arbuscular mycorrhizal fungus *Glomus clarumin* protection of cow peaplants againtsroot rot pathogen *Rhizoctonia solani. J. Plant. Diseases and Protection,* 109 (**2**): 207-215.

Akiyama, K., Matsuzaki, K., Hayashi, H. (2005). Plant sesquiterpines induce hyphal branching in arbuscular mycorrhizal fungi. *Nature. 435:824-827.*

Al-Karaki, G.N., Al-Raddad, A. (1997). Effects of arbuscular mycorrhizal fungi and drought stres on growth and nutrient uptake of two wheat genotypes differing in drought resistance. *Mycorrhiza.* **7**:83-88.

Al-Karaki, G.N., Clark, R.B. (1998). Growth, mineral acquisition and water use by mycorrhizal wheat grown under water stress. *J. Plant. Nutr.* **21**:263-276.

Allen, M.F. (1982). Influence of vesicular-arbuscular mycorrhiza on water movement through *Bouteloua gracilis* (H.B.K.) Lag ex strend. *New Phytol, 91: 191-196.*

Allen, M.F. (1991). The ecology of mycorrhizae. *Cambridge University Press, Cambridge,* 184p.

Allen, M.F. and Boosalis, M.G.(1983). Effect of two species of VA-mycorrhizal fungi on drought tolerance of winter wheat. *New Phytol, 93: 67-76.*

Allen, M.F., Moore, T.S. (Jr.) and Christensen, M. (1982). Phytohormone changes in *Bouteloua gracilis* infected by vesicular-arbuscular mycorrhiza II. Altered levels of gibberellin-like substances and abscissic acid in the host plant. *Can. J. Bot.* **60:**468.

Allen, M.F., Smith, W.K., Moore, T.S. and Christensen, M. (1981). Comparative water relations and photosynthesis of mycorrhizal and non-mycorrhizal *Bouteloue gracilis* (H.B.K.) Lag ex Strend. *The New Phytologist.* **88:** 683-693.

Allison, L.E.(1965). Organic Carbon. In Methods of Soil-Analysis (Eds C.A. Black, D.D. Evans, J.L. White, L.E. Enminger and F.E. Clark) *Madison: Wisconsin: American Society of Agronomy.*

Ames, R.N., Porter, L. St. John, T.V. and Reid, C.P.P. (1984). Nitrogen sources and "A' values for vesicular-arbuscular and non-mycorrhizal sorghum grown at three rates of ^{15}N ammonium sulphate. *The New Phytologist,* **97:** 269-276.

Ames, R.N., Reid, C.P.P., Porter, L. and Cambardell, C. (1983). Hyphal uptake and transport of nitrogen from two ^{15}N labelled sources by *Glomus mosseae,* a vesicular-arbuscular mycorrhizal fungus. *The New Phytologist,* **95:** 381-396.

*Asai, T. (1944). Uber die mycorrhizenbildung des legumeson. *Pflanza. J. ph. Bot.* **13:** 463-485.

Aseri, G.K., Jain, N., Panwar. J., Rao, A.V., Meghwal, P.R. (2008). Biofertilizers improve plant growth, fruit yield, nutrition, metabolism and rhizosphere enzyme activities of pomegranate *(Punica gronatum* L.) in Indian Thar Desert. *Scientia Horticulturae* **117:** 130-135.

Auge, R.M. and Duan, X. (1991).Mycorrhizal fungi and non-hydraulic root signals of soil drying. *Plant Physiol.97:821-824.*

Auge, R.M., Stodola, A.J.W., Brown, M.S., Bethlenfalvay, G.J. (1992 b). Stomatal responses of mycorrhizal Cowpea and Soybean to short-term osmotic stress. *New Phytol.* **120**:117-125.

Auge, R.M. (2001). Water relations, drought and vesicular-arbuscular mycorrhizal symbiosis. *Mycorrhiza.* **11**: 3-42.

Ayyappan.S (2011). Way forward –ICAR News. Vol. **17(3)**: 20. July-September 2011.

Azaizeh, A., Marschner, H., Romheld, V. and Wittenmayer, L. (1995). Effect of a vesicular arbuscular mycorrhizal fungus and other soil microorganisms on growth, mineral nutrient acquisition and root exudation of soil-grown maize plants. *Mycorrhiza,* **5:** 321-327.

Azacon, C., Aguilar, G.D., Azcon, R and Barea, J.M.(1979). Endomycorrhizal fungi and *Rhizobium* as biological fertilizers for *Medicago sativa* in normal cultivation . *Nature.***279***: 325-327.*

Azcon-Aguilar, C. and Barea, J.M. (1996). Arbuscular mycorrhizas and biological control of soilborne plant pathogens-an overview of mechanisms involved. *Mycorrhiza,* **6:** 457-464.

Azcon-Aguilar, C. and Barea, J.M. (1997). Applying mycorrhiza biotechnology to horticulture: significance and potentials. *Scientia Horticulture.* **68**:1-24.

Azcon, R, Gomez, M. and Tobar, R.M. (1996). Physiological and nutritional responses by *Lactuca sativa* to nitrogen sources and mycorrhizal fungi under drought. *Biol. Fertil. Soils.* **22**:156-161.

Azcon, R. and Ocampo, J.A. (1981). Factors affecting the VAM infection and mycorrhizal dependency of thirteen wheat cultivars. *New Phytol.* **87:** 677-685.

Bago, B., Pfeffer, P.E., Abubaker, J., Jun, J., Allen, J.W., Brovillette, J., Douds, D.D., ammers, P.J., Shachar-Hill, Y. (2003). Carbon export from arbuscular mycorrhizal roots involves the translocation of carbohydrate as well as lipid. *Plant physiol.* **131**: 1496-1507.

Bagyaraj, D.J. (1990). Biological interactions between a mycorrhizal fungi and other beneficial soil organism. Proceedings of National Conference on Current trends in Mycorrhizal Research held at HAU, Hissar, Feb. 14-16, pp. 76-77.

Bagyaraj, D.J. (1992).Vesicular-arbuscular mycorrhiza: application in agriculture. pp. 359-374. In: *Methods in Microbiology*.Vol. **25** (Eds. Jr Morris, DJ Read and AK Varma). *Academic Press, London.*

Bagyaraj, D.J. and Hedge, S.V. (1978). Response of cowpea (*Vigna unguiculata* L. Walp) to *Rhizobium* seed inoculation. *Curr. Sci.* **47**: 543-549.

Bagyaraj, D.J., Manjunath, A. and Patil, R.B. (1979). Interaction between a VAM and *Rhizobium* and their effect on soybean in the field. *New Phytol* **82:** 141-145.

Bajpai, P.D., Lehri, L.K. and Pathak, A.N. (1974). Effect of seed inoculation with *Rhizobium* strains on the yield of leguminous crops. *Proc. Indian Nat. Sci. Acad.* **40**(b):571-573.

Bala, S. and Singh, O.S. (1983). Endomycorrhizal associations of legumes and pulses in Punjab.*Indian J. Ecol.* **10(2)**:242-247.

Ballard, S.S. and Dean, L.A. (1941). Soil studies with radioactive phosphorus: significance of biological measurements of the retention of applied phosphorus by soils. *Soil Science*. **52**:162-173.

Balyan, S.K., Chandra, R. and Pareek, R.P. (2002). Enhancing nodulation in *Vigna mungo* by applying higher quantity of planting ferrows and PSB. *Legume Research. Agricultural Research Communication Centre, Karnal, India: 25(3): 160-164.*

Bansal, M., Chamola, B.P., Sarwar, N. and Mukerji, K.G. (2000). Mycorrhizosphere: Interactions between rhizosphere microflora and VAM fungi. In: "Mycorrhizal Biology" (eds. Mukerji, K.G., Chamola, B.P. and Jagjit Singh), *Kluwer Academic Publishers,* NY, pp, 143-152.

Barea, J.M. and Azcon-Aguilar, C. (1981). Production of plant growth regulating substances by the vesicular-arbuscular mycorrhizal fungus *Glomus mosseae. Appl. Environ. Microbiol.* Vol. **43:** 810-813.

Barea, J.M., Azcon, R., and Azcon-Aguilar, C. (2002). Mycorrhizosphere interactions to improve plant fitness and soil quality. *Antonie van Leeuwenhoek International Journal of General and Molecular Microbiology* **81:** 343-351.

Barrow, N.J. (1975). The response of phosphate to two annual pasture species I. Effect of the soil's ability to absorb phosphate on comparative phosphate requirement, *Aust. J. Agric. Res.* **26:**137-143.

Bartels, M. and Caesar, K. (1987). The influence of drought stress in various growth phases on growth, flower drop and yield of soybeans *(Glycine max* L. Merr.). *J. Agron. Crop. Sci.* **158:**346-352.

Bayne, H.G. and Bethlenfalvay, G.J. (1987). The *Glycine-Glomus-Rhizobium* symbiosis: Interactions between the mycorrhizal and nitrogen-fixing endophytes. *Plant Cell Environment.* **10:**607-612.

Becard, G. and Fortin, J.A. (1988). Early events of vesicular-arbuscular mycorrhiza formation on RiT-DNA transformed roots. *New Phytol.* **108**:211-218.

Bedini, S., Pellegrino, E., Avio, L., Pellegrini, S., Bazzoffi, P., Argese, E. and Giovannetti, M. (2009). Changes in soil aggregation and glomalin-related soil protein content as affected by the arbuscular mycorrhizal fungal species *Glomus mosseae* and *Glomus intraradiaces. Soil Biol Biochem* **41**:1491-1496.

Berta, G., Fusconi, A. and Trotta, A. (1993). VA mycorrhizal infection and the morphology and function of root systems. *Environ. Expt.* Bot. **33:** 159-173.

Berta, G., Fusconi, A., Trotta, A, and Scannerini, S. (1995). Morphogenetic modifications induced by the mycorrhizal fungus Glomus strain E3 in the root system of *Allium porrum* L. *New Phytol.* **114:**207.

Bethlenfalvay, G.J. and Yoder, J.F. (1981). The *Glycine-Glomus Rhizobium* symbiosis I. Phosphorus effect on nitrogen-fixation and mycorrhizal infection. *Physiol. Plant.* **52:**141-145.

Bethlenfalvay, G.J., Bayne, H.G. and Pacovsky, R.S. (1983), Parasitic and mutualistic association between a mycorrhizal fungus and soybean: The effect of phosphorus on host-endophyte interactions. *Physiol. Planta.* **57:**543-548

Bethlenfalvay, G.J., Brown, M.S. and Stafford, A.E. (1985). *The Glycine-Glomus-Rhizobium* symbiosis. 11. Antagonistic effects between mycorrhizal colonization and nodulation. *Plant Physiol.* **79**:1054-1059.

Bethlenfalvay, G.J., Pacovsky, R.S. and Brown, M.S. (1982 b). Parasitic and mutualistic associations between a mycorrhizal fungus and soyabean: development of the endophyte. *Phytopathol.* **72**:894-897

Bethlenfalvay, G.J., Brown, M.S., Mishra, M.L. and Stafford, A.E. (1987). *Glycine-Glomus-Rhizobium* symbiosis. V effect of mycorrhiza on nodule activity and transpiration in soya bean under drought stress. *Plant Physiol.* **85**:115-119

Bethlenfalvay, G.J. and Linderman, R.G. 1992. Mycorrhizae in Sustainable Agriculture. *ASA Special Publication* No. 54, Madison, Wis. Bagyaraj, DJ. 1992. Vesicular-arbuscular mycorrhiza: Application in Agriculture. *Methods in Microbiol.* **24:** 360-373.

Bhardwaj, A. and Sharma, S. (2006a). Biological control of nematodes *Melidogyne incognita* in *Lycoperscion esculenta* with AM fungi and oil cakes. *Pathology Journal* **5(2)**: 166-172.

Bhardwaj, A, and Sharma, S. (2006b). Reducing Prequirement using AM fungi in Mulberry grown under alkaline conditions. *Journal of Agronomy.* **5(3)**:471-474.

Bieleski, R.L, (1973). Phosphate pools, phosphate transport and phosphate availability. *Ann. Rev. Plant. Physiol.* **24**:225-252.

Bierman, N.B. and Linderman, R.G. (1983). Effect of container plant growth medium and fertilizer phosphorus on establishment and host growth response to vesicular arbuscular mycorrhizae, *J. Amer. Soc. Hort. Sci.* **108**:962-971.

Bilduasas, I.J., Dixon, R.K., Fleger, F.L., and Stewart, E.L. (1968). Growth nutrition and gas exchange of *Bromus inermis* inoculated with *Glomus fasciculatum. New Phytol. 102:303-311.*

Biro, B., Pechy, K.K., Voros, I., Takacs, T., Eggenberger, P. and Strasser, R.J. (2000). Interrelations between *Azospirillum* and *Rhizobium* nitrogen fixers and arbuscular mycorrhizal fungi in the rhizosphere of alfalfa in sterile, AMF free or normal soil conditions. *Applied Soil Ecology* **15**: 159-168.

Blee, K. A., and Anderson, A. J. 1996. Defense related transcript accumulation in *Phaseolus vulgaris* L. colonized by the arbuscular mycorrhizal fungus *Glomus intraradices* Schenck & Smith. *Plant Physiol. 110:675-688*

Bonfante-Fasolo, P.(1984). Anatomy and morphology of VA-mycorrhizae. *New Phytol. 99: 5-32.*

Bonfante, P. and Perotto, S. (1995). Strategies of arbuscular mycorrhizal fungi when infecting host plants. *New Phytol.* **130**:3-21.

Bowen, G.D. and Smith, S.E. (1981). The effects of mycorrhizas on nitrogen uptake of plants. In: Clark, F.E., Rosswall, T. (eds.). Terrestrial Nitrogen Cycle. Process, Ecosystem, Strategies and Management Impacts. Ecological Bulletin No. 33, Swedish Natural Science Res. Council, Stockholm. pp. 237-247.

Brockwell, J. (1977). Application of legume seed inoculents. [In] Hardy, R.W.; Gibon, A.H. (ear), A treatise on dinitrogen fixation sec. IV. Agronomy and Ecology. Cambridge Univ. Press, Cambridge. U.K.

Brown, R.W., Brown, C.L. and Kormanil, P.P. (1981). Effects of VA-mycorrhizae and soil phosphorus on cytokinins and nutrient levels in *Plantanus occidentalis. [In] Absts. of 5th North Amer. Conf. on Mycorrhizae, Quebec. pp. 12.*

Buee, M., Rossingnol, M., Jauneau, A., Ranjeva, R. and BeCard, G. (2000). The presymbiotic growth of arbuscular mycorrhizal fungi is induced by a branching factor partially purified from plant root exudates. *Mol. Plant. Microbe Interact. 13: 693-698.*

Busse, M.D. and Ellis, J.R. (1985). Vesicular arbuscular mycorrhizal (*Glomus fasciculatum*) influence on soya bean drought tolerance in high phosphorus soil. *Can. J. Bot. 63: 2290-2294.*

Buwalda, J.G., Stribley, D.P. and Tinker, P.B. (1983). Increased uptake of anions by plants with vesicular-arbuscular mycorrhizas. *Plant Soil.* **71**:463-467.

Buwalda, J.G., Stribley, D.P. and Tinker, P.B. (1984). The development of endoymycorrhizal root systems.V. The detailed pattern of development of infection and the control of infection level by host in young leek plants. *NewPhytol.* **96**:411.

Campbell, W.J. (1988). Nitrate reductase and its role in nitrate assimilation in plants. *Physiol Plant* **74**:214-219.

Caravaca, F., Alguacil, M.M., Azcon, R. and Roldan, A. (2006).Formation of stable aggregates in rhizosphere soil of *Juniperus oxycedrus*: effect of AM fungi and organic amendments. *Appl. Soil Ecol.* **33**:30-38.

Cavagnaro, T.R., Smith, F.A., Smith, S.E. and Jakobsen, I. (2005).Functional diversity in arbuscular mycorrhizas: Exploitation of soil patches with different phosphate enrichment differs among fungal species. *Plant Cell Environ.* **28**:642-650.

Carling, D.E. and Brown, M.F. (1980). Relative effect of vesicular-arbuscular mycorrhizal fungi on the growth and yield of soybeans. *Soil Sci. Soc. Am. J.* **44**:218-232.

Clarkson, D.T. (1985). Factors affecting mineral nutrient acquisition by plants. *Ann. Rev. Plant Physiol. 36: 77-115.*

Chen, C.R., Condron L.M., Sinaj S., Davis M.R., Sherlock R.R. and Frossard E. (2003). Effects of plant species on phosphorus availability in a range of grassland soils. *Plant Soil. 256: 115-130.*

Colla, G., Rouphael, Y., Cardarelli, M., Tullio, M., Rivera, C. and Rea, E. (2007). Alleviation of salt stress by arbuscular mycorrhizal in Zucchini plants grown at low and high phosphorus concentration.In: *Biology and Fertility of Soils.* Vol.**44**. No.3, 501-509.

Cooper, K.M. (1984). Physiology of VA-mycorrhizal association pp 155-203. In VA mycorrhiza. Ed. C.L. Powell, D.J., Bagyaraj; pp. 113-130. Boca Roton, Fla: CRC Press,London.

Cooper, K.M. and Tinker, P.B. (1978). Translocation and transfer of nutrients in vesicular-arbuscular mycorrhiza II. Uptake and translocation of phosphorus, zinc and sulphur. *New Phytol.* **81**:43-52.

Cooper, K.M. and Tinker, P.B. (1981). Translocation and transfer of nutrients in vesicular-arbuscular mycorrhizas. IV. Effect of environment variables on movement of phosphorus. *New Phytol. 88: 327-339.*

Cordier, C., Gianinazzi, S. and Gianinazzi-Pearson, V. (1996). Colonization patterns ofroot tissues by *Phytophthora nicotianae* var. *parasitica* related to reduced disease in mycorrhizal tomato. *Plant Soil* **185**: 223-232.

Cox, G. and Sanders, F.E. (1974). Ultrastructure of host fungus interface in a vesicular-arbuscular mycorrhizae. *New Phytol. 73: 901-912.*

Cranen Brouck, S., Voets, L. and Bivort, C. (2005). Methodologies for in vitro cultivation of AM fungi with root organs. [In] Vitro Culture of Mycorrhizas, Declerck S, Strullu DG and Fortin A (Eds.) Springer Verlag Berlin Heidelberg pp. 342-375.

Crush, J.R. (1974). Plant growth responses to vesicular-arbuscular mycorrhiza VII. Growth and nodulation of some herbage legumes. *New Phytol. 73: 743-749.*

Daft, M.J. and El-Giahmi, A.A. (1974). Effect of endogone mycorrhiza on plant growth. VII. Influence of infection on the growth and nodulation in French bean *(Phaseolus vulgaris). New Phytol.* **73.**

Daft, M.J. and El-Giahmi, A.A. (1978). Effect of vesicular-arbuscular mycorrhiza on plant growth VIII. Effects of defoliation and light on selected hosts. *New Phytol.***72**:1333-1339.

Daft,M.J. and Nicolson,T.H (1966).Effect of endogone mycorrhiza on plant growth.*New Phytol.65:343-350.*

Daft, M.J. and Nicolson, T.H. (1969). Effect of *endogone* mycorrhiza on plant growth III. Influence of inoculum concentration on growth and infection in tomato. *New Phytol.* **68:** 953-961.

Daft, M.J. and Okusanya, B.O. (1973). Effect of *endogone* on plant growth IV. Influence of infection on the anatomy and reproductive development in fourhosts. *New Phytol.* **72:** 1333.

Dart, P.J. and Krantz, B.A. (1977). Legumes in the semi-arid tropics. In: Exploiting the legume *Rhizobium* symbiosis in tropical agricultural J.M. Vincent, A.S. Whitney and J. Bose (Eds). Univ. Hawaii Coll. Trop. *Agric. Misc. Publ.* **145**: 119-154.

Dehne, H.W. and Schonbeck, F. (1978). The influence of endotrophic mycorrhiza on plant diseases, 3, Chitinase-activity and ornithine-cycle. *J. Plant Dis. Protec.* **85:** 666-678.

Dehne, H.W., Schonbeck, F. and Baltruschat, H. (1978). The influence of endotrophic mycorrhiza on plant diseases. 3, Chitinase-activity and the ornithine-cycle, *Z.Pflanzenkrankh. pflanzenschutz* **85:**666-678.

Dehne, H.W.(1982). Interaction between vesicular arbuscular mycorrhizal fungi and plant pathogens. *Phytopath.***72**:1115-1119.

Deka, A.K. and Azad, P.(2006).Screening of efficient strains of *Bradyrhizobium. Indian J. Pulses Res .* **19**:79-82

Douds, D.D., Pfeffer, P.E. and Shachar-Hill, Y. (2000). Carbon partitioning, cost and metabolism of arbuscular mycorrhizae. [In] Douds DD, Kapulnik Y,eds. Arbuscular mycorrhizas physiology and function. Dordrecht, the Netherlands : Kluwer Academic Publishers, 107-130.

Douds, Jr., D.D., Nagahashi, G., Pfeffer, P.E., Reider, C. and Kayser, W.M. (2006). On-farm production of AM fungus inoculum in mixtures of compost and vermiculite. *Biores. Technol.* **97**:809-818.

Duan, X., Neuman, D.S., Reiber, J.M., Green, C.D., Saxton, A.M. and Auge R.M. (1996). Mycorrhizal influence on hydraulic and hormonal factors implicated in the control of stomatal conductance during drought. *Journal of Experimental Botany* **47:** 1541-1550.

Dubey, S.C. (2003). Integrated management of web blight of urd and mung bean. Indian Phytopath. Indian Phytopathological Society, New Delhi, India. **56**(4):413-417.

Dubey, S.C.(2006). Integrating bio agents with plant extract, oil cake and fungicide in various modes of application for the better management of web blight of urd bean. *Archives. Phytopath. Pl. Protec. Taylor & Francis, Abingdon, UK:* **39**(5): 341-351.

Dugassa, G.D., Von Alten, H. and Schonbeck, F. (1996). Effect of arbuscular mycorrhiza (AM) on health of *Linum usitatissimum* L. infected by fungal pathogens. *Plant Soil* **185:** 173-182.

Dumas-Gaudot, E., Gollotte, A., Cordier, C., Gianinazzi, S. and Gianinazzi-Pearson, V. (2000). Modulation of host defense systems. In: *Arbuscular Mycorrhizas: Physiology and Function.* Eds Kapulnik Y , DD Douds Jr Y (Kluver Academic Press, Dordrecht, The Netherlands),pp. 173-200.

Dumas-Gaudot, E., Slezack, S., Dasi. B., Pozo, M.J., Gianinazzi-Pearson, V. and Gianinazzi, S. (1996). Plant hydrolytic enzymes (chitinases and a-1,3-glucanases) in root reactions to pathogenic and symbiotic micro-organisms, *Plant and Soil* **185**:211-221.

El Ghachtouli, N., Paynot, M., Morandi, D., Martin-Tanguy, J. and Gianinazzi, S. (1995). The effect of polyamines on endomycorrhizal infection of wild type *Pisum sativum* cv. Frisson (nod+myc+) and two mutants (nod- myc+ and nod-myc-). *Mycorrhiza.* **5:** 189-192.

Fernandez, D.S. and Ascencio J. (1994). Acid phosphatase activity in bean and cowpea plants grown under phosphorous stress . *Journal. Plant Nutrit* .**17** (2 &3): 229-241.

Fischer, R.A. and Turner, N.C. (1978). Plant productivity in the arid and semi-arid zones. *Ann. Rev. Plant Physiol.* **29:** 277-317.

Fitter, A.H. (1977). Influence of mycorrhizal infection on competition for phosphorus and potassium by two grasses. *New Phytol* . **79**:119-125.

Fitter, A.H. (1985). Functioning of vesicular - arbucular mycorrhizas under field conditions. *New Phytol.***99**:257-265.

*Frank, A.B. (1985). Uber die auf Wurzei symbiose beruhende Evanahurg Gewisser baume durch unterirdische. *Plize. Ber. Dtsch. Bot. Ges.* **3**: 128-145.

Franken, P. and Gnadinger, F. (1994). Analysis of parsley arbuscular endomycorrhiza : infection development and mRNA levels of defense- related genes. *Mol. Plant. Microbe Interact.***7**: 612-620.

Fortin, J.A., Becard G., Declerck, S., Dalpe, Y., St. Arnaud, M., Coughlan, A.P. and Piche, Y. (2002). Arbuscular mycorrhiza on root-organ cultures. *Can. J. Bot.* **80**:1-20.

Garcia-Garrido, J.M. and Ocampo, J.A. (2002). Regulation of the plant defence response inarbuscular mycorrhizal symbiosis. *Journal of experimental Botany* **53:** 1377-1386.

Gerdemann, J.W. and Trappe, J. M. (1974). The Endogonaceae in the Pecific North West. *Mycol. Mem.* 5: 1-76.

Gininazzi, S. and Gianinazzi-Pearson, V. (1992). Cytology, histochemistry and immunocytochemistry as tools for studying structure and function of endomycorrhiza. *Methods. Microbiol.* **24**:109-139.

*Gianinazzi-Pearson., V. (1982). Importance des mycorrhizes dans la nutrition et la physiologie de plantes. [In] Gianinazzi, S., Gianinazzi-Pearson V. and Trouvelot, A. (eds). Les Mycorrhizes Partic Integrante de la Plate: Biologie et Perspective d' Utilisation Cles cologyes de 1' INRA Paris.**13**:51-59.

Gianinazzi-Pearson, V. (1984). Mycorrhizal effectiveness in phosphate nutrition: How, when and where? [In] *Proc. 6th North Amer. Conf. on Mycorrhizae Bend. Oregon pp.* 150-154.

Gianizazzi-Pearson, V. and Gianizazzi, S. (1986). The physiology of improved phosphorus nutrition in mycorrhizal plants. [In] *physiological and genetical aspects of mycorrhizae, pp. 101-109. Paris, INFRA.*

Gianinazzi-Pearson, V., Smith, S.E., Gianinazzi, S. and Smith, F.A. (1991). Enzymatic studies on the metabolism of vesicular-arbuscular mycorrhizasV. Is H+- AT Pase a component of ATP-hydrolysing enzyme activities in plant-fungus interfaces. *New Phytol.* **117**:61-76.

Gianinazzi-Pearson, V., Gollotte, A., Dumas-Gaudot, E., Franken, P. and Gianinazzi, S. (1994). Gene expression and molecular modifications associated with plant responses to infection by arbuscular mycorrhizal fungi. *[In] M.J. Daniels et al. (ed.), Advances in Molecular Genetics of Plant-Microbe Interactions. Vol 3., Kluwer, Dordrecht, pp. 179-186.*

Gildon, A. and Tinker, P.B. (1983). Interaction of vesicular-arbuscular mycorrhizal infection and heavy metals in plants II. The effects of infection on the uptake of copper. *New Phytol.* **95**:363-368.

Gilmore, A.E. (1971). The influence of endotrophic mycorrhizae on the growth of peach seedlings. *J. Amer. Soc. Hortic. Sci.* **96**:35-38.

Goicoechea, N., Szalai, G., Antol, M.C., Sachez-Daz, M. and Paldi, E. (1998). Influence of arbuscular mycorrhizae and *Rhizobium* onfree polyamines and proline levels in water-stressed alfalfa.*J.Plant Physiol.* **153**:706-711.

Grandmaison, J., Gyorgy, M., Olah, R.C. and Valentin, F. (1993). Characterization and localization of plant phenolics likely involved in the pathogen resistance expressed by endomycorrhizal roots. In: *Mycorrhiza.* Vol. **3**. No. 4. pp.155-165.

Gray, L.E. and Gerdemann, J.W. (1973). Uptake of ^{35}sulphur by vesicular-arbuscular mycorrhizae. *Plant Soil.* **39**:687-689.

Graham, J.H., Leonard, R.T. and Menge, J.A. (1981). Membrane-mediated decrease in root exudation responsible for phosphorus inhibition of vesicular-arbuscular mycorrhizae formation. *Plant Physiol.* **68:** 548-552.

Green, C.D., Stodola, A. and Auge, R.M. (1998). Transpiration of detached leaves from mycorrhizal and nonmycorrhizal cowpea and rose plants given varying abscisic acid, pH, calcium and phosphorus. *Mycorrhiza* **8**:93-99.

Grierson, P.F. and Comerford, N.B. (2000). Non-destructive measurement of acid phosphatase activity in the rhizosphere using nitrocellulose membranes and image analysis. *Plants Soil.***218**:49-57.

Gunasekaran, S., Balchardar, D. and Sudaram, K.M. (2004). Studies on Synergism between *Rhizobium*, plant growth promoting rhizobacteria and Phosphate solubilising bacteria in blackgram. *Biofertilizers Technology. Scientific Publishers (India), Jodhpur, India: 269-272.*

Gupta, R.P., Kalra, M.S. and Bhandari, S.C. (1982). Studies on chickpea-Rhizobium interaction . Biological Nitrogen Fixation. Proc. Nat. Symposium IARI, New Delhi, pp 544-555.

Guttenberger, M. (2000). Arbuscules of Vesicular-arbuscular mycorrhizal fungi inhabit an acidic compartment within plant roots. *Planta.* **211**:299-304.

Hakeem, S.A., Thomas, T. and Un-Nissa, S.S. (2004). Effect of *Rhizobium* in association with different sources of sulphur on growth of black gram and chemical properties of soil. *Bioved.:***15** (1/2): 111-114.

Hall, K.A. and Sanders, F.E. (1982). Effect of Benomyl on vesicular-arbuscular mycorrhizal infection of red clover (*Trifolium-pretense* L.) and consequences for phosphorus inflow. *J. Plant Nutri.***5:**1355-1367.

Hall, I.R. (1976). Response of *Coprosma robustat* to different forms of endomycorrhizal inoculum. *Trans, Br. Mycol. Soc.* **97:**409-411.

Hall, I.R. (1978). Effects of endomycorrhizas on the competitive ability of white clover. *Nz. J. Agric. Res.***21:**509-515.

Hardie, K. and Leyton, L. (1981). The influence of VA mycorrhiza on growth and water relation of red clover. *I. In phosphate deficient soil. New Phytol.* **89**:599-608.

Harison, M.J. and Dixon, R.A. (1993). Isoflavonoid accumulation and expresion of defence gene transcripts during the establishment of vesicular- arbuscular-mycorrhizal associations in roots of *Medicago trunculata. Mole. Plant Microbe Interac.* **6:**643-654.

Harley, J.L. and Smith, S.S. (1983). Mycorrhizal symbiosis. *Academic Press, New York and London.*

Harris,D., Pacovsky, R.S. and Paul, E.A. (1985). Carbon economy of soyabean – *Rhizobium-Glomus* association. *New Phytol.***101**:427-440.

Harrison, M.J., Dewbre, G.R. and Liu, J. (2002). A.phosphate transporter from *Medicago truncatula* involved in the acquisition of phosphate released by arbuscular mycorrhizal fungi. *Plant cell.* **14**:2413-2429.

Harrison, M.J. (2005). Signaling in the arbuscular mycorrhizal symbiosis. *Ann. Rev. Microbiol.* **59**:19-42.

Hayman, D.A. (1974), Plant growth response to VAM III. Effect of light and temperature. *New Phytol.***73:**71-80.

Hayman, D.S. and Davy, J.M. (1977). Rothamsted Annu. Report. Rothamsted Exp. Station, Rothamsted, England. pp. 238.

Hayman, D.S. and Mosse, B. (1972). Plant growth responses to vesicular-arbuscular mycorrhiza III. Increased uptake of labile P from soil. *New Phytol.* **71**:41-47.

Hayman, D.S. and Taveres, M. (1985). Plant growth responses to VAM XV. Influence of soil pH on the symbiotic efficiency of different endophytes. *New Physiol.* **100**:367.

Helal, H.M. (1990). Varietal differences in root phosphatase activity as related to the utilization of organic phosphates. *Plant. Soil. **123:*** 161-163.

Hemanatrajan, A., Tridevi, A.K., Kumar, S. and Kumar, A. (2003). Influence of B and Zn on active iron, enzyme activity, protein and oil content of soya bean. *National Seminar on Physiological Interventions for Improved Crop Productivity and Quality. Dec.12-14, 2003 at Tirupati, pp 154.*

Hepper, C.M. (1984). Regulation of spore germination of the VAM fungus *Acaulospora laevis* by soil pH. *Trans. Br. Mycol. Soc.* **83:** 154-156.

Hetrick, B.A.D., Hetrick, J.A. and Bloom, J. (1984). Interaction of mycorrhizal infection, phosphorus level and moisture stress in growth of field corn. *Can. J. Bot.* **62:** 2267-2271.

Hetrick, B.A.D.(1991). Mycorrhizas and root architecture. *Experientia. 47: 355-62.*

Heys, R., Reid, C.P.P., St. Johan, T.V. and Coleman, (1982). Effects of nitrogen and phosphorus on blue grama growth and mycorrhizal infection. *Oecologia.* **54:** 260-265.

Hildebrandt, U., Schemelzer, E. and Bothe, H. (2002). Expression of nitrate transporter genes in tomato colonized by an arbuscular mycorrhizal fungus. *Physiol. Plant. **115**:125-136*

Hirrel, M.C., Meharavaran, H. and Gerdemann, J.W. (1978). Vesicular - arbuscular mycorrhiza in the chenopodiaceae and cruciferae: do they occur? *Can. J. Bot.* **56**: 2813.

Ho, L. and Trappe, J.M. (1973). Nitrate reducing capacity of two VAM (Vesicular-Arbuscular-mycorrhizal). Fungi Micology **67:** 886-888.

Hodge, A. (2000). Microbial ecology of the arbuscular mycorrhiza. *FEMS Microbiology Ecology.* **32:** 91-96.

Hodge, A., Campbell, C.D. and Fitter, A.H. (2001). An arbuscular mycorrhizal fungus accelerates decomposition and acquires nitrogen directly from organic material. *Nature.* **413**:297-299.

Holevas, C.D. (1966). The effect of a vesicular-arbuscular mycorrhiza on the uptake of soil phosphorus by strawberry (*Fragaria* sp. var. cambridge Favourite). *Hort. Sci.* **41:**57-64.

Horn, R. and Dexter, A.R. (1989). Dynamics of soil aggregation in a desert loss. *Soil and Tillage Research.* **13:** 253-266.

Horn. R., Taubner, H., Wuttke, M. and Baumgartl, T. (1994). Soil Physical properties related to soil structure. *Soil and Tillage Research.* **30:** 187-216.

Hung, L.L. and Sylvia, D.M. (1998). Inoculum production of vesicular arbuscular mycorrhizal fungi in aeroponic culture. *Appl. Environ. Microbiol.***54**: 353-357.

Hussey, R.S. and Roncadori, R.W. (1982). Vesicular-arbuscular mycorrhiza may limit nematode activity and improve plant growth. *Plant Dis.* **66:** 9-14.

Ikaram, A., Mahmud, A.W., Chong, K. and Faizah, A.W. (1993). Growth responses of *Calopogonium caeruleum* to dual inoculation with vesicular arbuscular mycorrhizal fungi and *Bradyrhizobia*. *Field Crops Research. 31: 131-144.*

Jacquelinet-Jeanmougin, J., Gianinazzi-Pearson, V. and Gianinazzi, S. (1987). Endomycorrhizas in the Gentianaceae. II. Ultrastructural aspects of symbiont relationships in *Gentiana lutea* L. *Symbiosis.* **3**: 269-286.

Jadhav, S.D., Kadam, S.B., Deshmukh, R.K. and Manorama (2006). *In vitro* assessment of native *Rhizobia* from various legume crops at different pH and temperature levels. *Journal Soils Crops. **16**:1, 88-94.*

Jain, V. and Gupta, V.K. (2002). Effect of rhizosphere fungi on nodule number, shoot and root length of *Vigna mungo*. *Ind. Phytopath. **55**(3): 323-324.*

Jakobsen, I., Abbott, L.K. and Robson, A.D. (1992). External hyphae of Vesicular-Arbuscular Mycorrhizal fungi associated with *Trifolium subterraneum*. I. spread of hyphae and phosphorus inflow into. *New Phytol, **120**:371-380.*

Jalali, B.L. (1988). Host growth responses to vesicular-arbuscular mycorrhizal system in differential soil phosphate levels. In: *First Asian Conf. Mycorrhizae Centre for Advance Study Bot. Madras. Abst:*27.

*Janse, J.M. (1986). Les endophytes radicaux des quelques plants javanaises. *An Jard, Bot. Buitenz.* **14:** 53.

Jasper, D.A., Robson, A.D. and Abbott, L.K. (1979). Phosphorus and the formation of vesicular-arbuscular mycorrhizas. *Soil Biol. Biochem.* **11:** 501-505.

Jayashree, K., Shanmugam, V., Raguchander T., Ramanathan A. and Samiyappan, R. (2000). Evaluation of *Pseudomonas fluorescens* against black gram and sesame root-rot disease. *Journal. Biol Control.:* **14**(2): 55-61.

Johnson, C.R. and Hummel, R.L. (1985). Influence of mycorrhizae and drought stress on growth on *Poncirus citrus* seedlings. *Hort. Sci.* **20:** 754-755.

Johnson D., Leake J.R. and Read D.J. (2005). Liming and nitrogen fertilization effects on phosphatase activities, microbial biomass and mycorrhizal colonisation in upland grassland. *Plant. Soil.* **271**: 157-164.

Johnson, P.N. (1976). Effects of soil phosphate level and shade on plant growth and mycorrhizas. *Nz. J. Bot.* **14:** 333-340.

Jones, I.R. (1924). A mycorrhizal fungus in the roots of legumes and some other plants. *J. Agric. Res.* **29:** 449-470.

Jones, D.L., Hodge, A. and Kuzyakov, Y. (2004). Plant and mycorrhizal regulation of rhizodeposition. *New Phytologist.* **163:** 459-480.

Jung , K.A. and Claassen, N. (1989). Availability in soil and acquisition by plants as the basis for phosphorus and potassium supply to plants. *Zeitschrift fiir pflanzener nahrung and Bodenkunde.* ***152:*** *151-157.*

Kaldorf, M., Schemelzer, E. and Bothe, H. (1998). Expression of maize and fungal nitrate reductase in arbuscular mycorrhiza. *Mol. Plant-Microbe Interact.* **11:** 439-448.

Kaldorf, M., Zimmer, W. and Bothe, H. (1994). Genetic evidence for the occurence of assimilatory nitrate reductase in arbuscular mycorrhizal fungi. *Mycorrhiza.* **5:**23-28.

Kalita, R., Deka, A.K. and Azad, P. (2006). Evaluation of native *Rhizobium* from acid soils of Assam. *Legume Research:* **29**(3):157-162.

Karunaratne, P.S., Baker, J.H. and Barker, A.V. (1987). Benefits of mycorrhizae to soyabeans grown on various regimes of nitrogen nutrition. *J. Plant. Nutrition.* **10:** 871-886.

Kashyap, S. and Sharma, S. (2006). In Vitro selection of salt tolerant *Morus alba* and its field performance with bio inoculants. *Horticulture Science (Prague).* **33:**77-86.

Kashyap, S., Sharma, S. and Vasudevan, P. (2005). Effect of native bio inoculants on vegetable crops grown on alkaline soil under field conditions. *Journal of New Seeds.* **Vol. 7(3)**: 75-90.

Kashyap, S., Sharma, S. and Vasudevan, P. (2004). Role of bio inoculants in development of salt-resistant saplings of *Morus alba* (var. sujanpuri) in vivo. *Scientia Horticulturae.* **100:** 291-307.

Khadge, B.R., Ilag, L.L. and New, T.W. (1992). Effect of VAM inoculum carryover on the successive cropping of maize and mung bean. *Plant Soil.* **140**: 303-309.

Khalifa, F.M. (1987) Effect of nitrogen on nodulation and yield of soya bean under two system of production in Sudan. *Agric. Sci.* **108**:259-265.

Khan, A.G. (1972). The effect of vesicular-arbuscular mycorrhizal associations on growth of cereals I. Effects on maize growth. *New Phytol.* **71**: 613-619.

Khade, S.W. (2008). Taxonomy of arbuscular mycorrhizal fungi. *Mycorrhiza News.* **20(1)**:2-7.

Kierman, J.M. Mendrix, J.W. and Mavonek, D.M. (1983). Fertilizer induced pathogenicity of mycorrhizal fungi to sweet gum seedlings. *Soil Biol. Biochem.* **15:** 257-262.

Kinden, D.A. and Brown, M.R. (1975). Electron microscopy of Vesicular Arbuscular Mycorrhizae of yellow popular II. Intracellular and vesicles. *Can. J. Microbiol. 21:1768-1780.*

Koch, E. and Johnson, C.R .(1984). Photosynthate partitioning in split-root citrus seedling with mycorrhizal and non-micorrhizal root systems. *Plant Physiol. 75: 26-30.*

Koide, R. (1985). The nature of growth depressions in sunflower caused by VAM infection. *New Phytol.* **99:** 449-462.

Krajinski, F., Biela, A., Schubert, D., Gianinazzi-Pearson, V., Kaldenhoff, R. and Franken, P. (2000). Arbuscular mycorrhiza development regulates the mRNA abudance of Mtaqp1 encoding a mercury-insensitive aquaporin of *Medicago truncatula*. *Planta.* **211:** 85-90.

Krishna, K.R. and Bagyaraj, D.J. (1982). Influence of vesicular-arbuscular mycorrhiza on growth and nutrition of *Arachis hypogaea. Legume Res.* **5**:18-22.

Krishna, K.R. and Bagyaraj, D.J. (1984). Phenols in mycorrhizal roots of *Arachis hypogaea. Experientia.* **40**:85-86.

Krishna, K.R. and Bagyaraj, D.J.(1984). Phosphatase in the rhizosphere of mycorrhizal and non micorrhizal peanut. *In: Proc. 6th North Amer. Conf. on Micorrhizal. Blend. Oregon. pp. 372.*

Kubikova, E., Moore, J.L., Ownlew, B.H., Mullen, M.D. and Auge, R.M. (2001). Mycorrhizal impact on osmotic adjustment in *Ocimum basilicum* during a lethal drying episode. *J. Plant physiol.* **158**:1227-1230.

Kucey, R.M.N. and Paul, E.A. (1982). Carbon flow. Photosynthesis and Nitrogen fixation in micorrhizal and nodulated fababeans (*Vicia faba* L.) *Soil Biol. Biochem. 14:407-412.*

Kulshreshtha, M. and Khan, M.W. (2002). Effect of Ozone on root colonization by VAM fungi and root nodulation on black gram. *Indian Phytopath.* **55** *(3): 264-268.*

Kumar, R. and Chandra, R. (2008). Influence of PGPR and PSB on *Rhizobium leguminoserum* BV. viciae strain competition and symbiotic performance in Lentil. *World J. Agric Sciences. 4(3):297-301.*

Kumar, A. and Elamathi, S. (2007). Effect of nitrogen levels and *Rhizobium* application methods on yield attributes, yield and economics of black gram. *Intern J. Agric Sci.* **3***(1): 179-180.*

Kumar, Rajesh and Sharma, C.P. (2006). Effect of *Rhizobium* and co-inoculants VAM and PSB on yield and nutrient uptake in black gram. *Farm Science J.* ***15(1)**: 93-94.*

Kumari, K.S.M. and Nair, S.K. (2003). Use of acid tolerant strains of *Bradyrhizobium sp.* for black gram and green gram. *Legume Research.* ***26(1)**: 60-62.*

Kuo, C.G. and Huang, R.S. (1982). Effect of vesicular-arbuscular mycorrhiza on the growth and yield of rice stubble cultured soybean. *Plant Soil.* **64**:325-330.

Lanowska, J. (1966). Influence of different sources of nitrogen on the development of mycorrhizae in *Pisum sativam. Pamiet Pulawski.* **21**:365.

Lee, Y.J. and George, E. (2005). Development of a nutrient film technique culture system for arbuscular mycorrhizal plants. *Hort. Sci.* **40**:378-380.

Lehri, L.K., Gangwar, B.R. and Mehrotra, C.L. (1974). Bacteriazation experiements with *Rhizobium. J. Indian Soil Sci.* **22**:66-69.

Lesueur, D., Ingleby, K., Odee, D., Chamberlain, Joanne, Wilson, J., Manga, T.T., Sarrailh, J.M. and Pottinger, A. (2001). Improvement of forage production in *Colliandra calothyrsus:* Methodology for the identification of an effective inoculum containing *Rhizobium* strains and arbuscular mycorrhizal isolates. *Journal of Biotechnology.* **91**:269-282.

Levy, I. and Krikun, J.(1980). Effect of vesicular-arbuscular mycorrhiza in citrus jambhiri water relations. *New Phytol.* ***85**:25-32.*

Lewis, D. H. (1973). Concepts in fungal nutrition and the origin of biotrophy. *Biol. Rev.* **48**: 261-278.

Linderman, R.G. (2000). Effect of mycorrhizae on plant tolerance to disease. [In] Arbuscular mycorrhizas: physiology and function". (eds. Kapulnick, Y. and Douds Jr., D.D.), Kluwer Academic Press, pp. 345-366.

Liu, Y., Mi G., Chen F., Zhang, J. and Zhang, F. (2004). rhizosphere effect and root growth of two maize genotypes with contrasting P efficiency at low P availability. *Plant Science.* ***167**: 217-223.*

Mahadi, A. and Atabani, I.M.A (1992). Response of *Bradyrhizobium* inoculated soya bean and lablab bean to inoculation with Vesicular Arbuscular Mycorrhiza. *Expl. Agric.**28**:399-407.*

Manivannan, V., Thanunathan, K., Imayavaramban,V. and Ramanathan, N. (2003). Growth and growth analysis of rice fellow urd bean as influenced by foliar application of nutrients with and without *Rhizobium* seed inoculation . *Legume Research.* ***26(4)**: 296-299.*

Manjunath, A. and Bagyaraj, D.J. (1984).Response of pigeon pea and cowpea to phosphate and dual inoculation with VAM and *Rhizobium. Trop. Agric.* ***61(1)**:48-52.*

Marschner, P., Solaiman, Z. and Rengel, Z. (2006). Rhizosphere properties of Poaceae genotypes under P-limiting condition. *Plant Soil.**283**: 11-24.*

Mathew, J. and Johari, B.N. (1989). Effect of indigenous and introduced VAM fungi on growth of mung bean. *Mycol. Res.**92(4)**: 491-493.*

McLachlan, K.D. (1980). Acid phosphorous activity of intact roots and phosphorous nutrition in plants II. Variations among wheat roots. *Aust. J. Agric. Res.* ***31**: 441-448.*

McLachlan, K.D. and De Marco, D.G. (1982). Acid phosphatase activity of intact roots and phosphorous nutrition in plants. III. Its relation to phosphorus garnering by wheat and a comparison with leaf activity as a measure of phosphorus status. *Aust. J. Agric. Res. 33: 1-11.*

Menge, J.A., Lembright, H. and Johnson, E.L.V. (1977). Utilization on mycorrhizal fungi in citrus nurseries. *Proc. Int. Soc. Citric.* **1**:129-132.

Minchin, F.R and Patel, J.S. (1973).The carbon balance of a legume and the functional economy of its roots nodules. *J. Exp. Bot. 24:259.*

Mishra, S. (2008b). Handbook on Biofertilizers Based Fodder Production Technology. Sharma, V. (ed.) Pub. By-Director, Science and Society Division, Department of Science and Technology, New Delhi, pp. 91.

Mishra, S.K. and Tiwari, V.N. (2002). Response of urd bean to applied phosphorous, sulphur and *Rhizobium* inoculation on Udic Ustochrepts. *Indian J. Pulses Res. 15(2): 136-138.*

Morandi, D. (1996). Occurence of phytoalexins and phenolics compounds in endo mycorrhizal interaction and their potential role in biological control. *Plant and Soil.* **185**:241-251.

Mosse, B. (1962). The establishment of vesicular- arbuscular mycorrhiza under aseptic conditions. *J.Gen.Microbiol*. **27**: 509-520.

Mosse, B. (1972). Effect of different *Endogone* strains on the growth of *Plaspalum notatum. Nature* (London), **239**:221-223.

Mosse, B. (1973a). Advances in the study of vesicular-arbuscular mycorrhizas. *Ann. Rev.Phytopath.* **11**:171-176.

Mosse, B. (1973b). Plant growth responses to vesicular-arbuscular mycorrhiza IV. In soil given addition phosphate. *New Phytol.* **72:** 127-136.

Mosse, B. (1977). The role of mycorrhiza in legume nutrition in marginal soils. In: J.M. Vincent, A.S Whiteny and J. Bose (Editors), Exploiting the legume Bradyrhizobium symbiosis in Trropical Agriculture,University of Hawaii, Kabuli, Maui, pp. 275-292.

Mosse, B. and Thompson, J.P. (1984). Vesicular arbuscular endomycorrhizal inoculum production. I. Exploratory experiments with bean (*Phaseolus vulgaris)* in nutrient flow culture. *Canadian. J. Botany*. **62**: 1523-1530.

Mukerji, K.G. (1999). Mycorrhiza in control of plant pathogens: Molecular approaches. [In] "Biotechnological approaches in biocontrol of plant pathogens". (eds. Mukerji, K.G., Chamola, B.P. and Upadhyay, R.K.), Kluwer Academic and Plenum Publishers, New York, pp. 135-155.

Munns, D.N. and Mosse, B. (1980). Mineral nutrition of legume crops. [In] Advances in Legume science (Ed. by R.J. Summerfield and A.H. Bunting), pp. 115-125. Proceedings of the International Legume Conference, Kew.

Murphy, P.J., Langridge, P. and Smith, S.E. (1997). Cloning plant genes differentially expressed during colonisation of *Hordeum vulgare* L. by the vesicular-arbuscular mycorrhizal fungus *Glomus intraradices*. *New Phytol.* **135**:291-301.

Murkute, A.A., Sharma, S., Singh, S.K. (2006). Studies on salt stress tolerance of Citrus rootstock genotypes with arbuscular mycorrhizal fungi. *Horticulture Science.* **33(2)**:22-26.

Nagrajan, P. and Balchander, D. (2001). Influence of *Rhizobium* and organic amendments on nodulation and grain yield of black gram and green gram in acid soil. *Madras Agric.J.* **88**(10/12): 703-705.

Nair, M.G., Safir, G.R. and Siqueira, J.O. (1991). Isolation and identification of vesicular-arbuscular mycorrhiza-stimulatory compounds from clover (*Trifalium repens*) roots. *Applied Environ. Microbiol.* ***57**:434-439*

Nelson, C.E. and Safir, G.R. (1982 a). The water relations of well-watered, mycorrhizal and non-mycorrhizal onion plants. *J. Amer. Soc. Hortic. Sci.* **107**:271-274.

Nelson, C.E. and Safir, G.R. (1982 b). Increased drought tolerance of mycorrhizal onion plants caused by improved phosphorus nutrition. *Planta,* **154**:407-413.

Nemec, S. and Meredith, F.I. (1981). Amino acid content of leaves in mycorrhizal and non-mycorrhizal citrus root stocks. *Aust. Bot.* **47**:351-358.

Nutman, P.S. (1977). Recent Development in Nitrogen Fixation (Eds). Newton. W., Postgate. J.R and Rodriguez Barrue Co. Academic Press New York and London. pp 443-447.

*Nyabyenda, P. (1977). *Einfluss der Bodentemeratur und organischer staffe auf die wirkung der vesikular arburskularen mykorrhiza*. Ph.D. Dissertation, Univ. Gottingen,Gottingen, (West Germany).

Nye, P.H. and Tinker, P.B. (1977). *Solute movement in the soil-root systems.* Black well Scientific Publications, London.

Oblisami, G., Balalraman, K. and Natarajan, T. (1976). Effect of composite cultures of *Rhizobium* on two pulse crops. *Madras Agric.* J.**63**.587-589.

Ocampo, J. A., Martin, J. and Hayman, D.S. (1979). Influence of plant interactions on vesicular arbuscular mycorrhizal infections I. Host and non - host plants grown together. *New Phytol.* **85**: 27-35.

Ohtsuka, T. and Saka H.(1988). Undestrictive measurement of acid phosphatase activity in the root of rice seedlings. *Japan Jour. Crop Sci.* ***57** (4):722-727.*

Oliver, A.J., Smith, S.E., Nicholas D.J. D., Wallace, W. and Smith F.A. (1983). Activity of nitrate reductase on *Trifolium subterraneum* :effects of mycorrhizal infection and phosphate nutrition. *New Phytol.**94**:63-79.*

Pacovsky, R.S. (1986). Micro-nutrient uptake and distribution in mycorrhizal and phosphorus fertilized soybeans. *Plant Soil.* **95**:379-388.

Pacovsky, R.S., Paul, E.A. and Bethenfalvay, G.J.(1986). Response of mycorrhizal and P-fertilized soyabeans to nodulation by *Bradyrhizobium* and ammonium nitrate. *Crop. Sci.* ***26**:145-150.*

Pairunan, A.K., Rabson, A.D., Abbott, L.K. (1980). The effectiveness of vesicular-arbuscular mycorrhizas in increasing growth and phosphorus uptake of subterranean clover from phosphorus sources of different solubilities. *New Phytol.* **84**:327-338.

Pandey, R. (2006). Root-exuded acid phosphatase and 32Pi-uptake kinetics of wheat, rye and triticale under phosphorus starvation. *Journal Nuclear Agric. Biol.* ***35** (3-4): 168-179.*

Pang, P.C. and Paul, E.A. (1980). Effects of Vesicular Mycorrhiza on ^{14}C and ^{15}N distribution in nodulated faba beans. *Can. J. Soil Sci.* ***60**:241-250.*

Panwar, J.D.S. (1991). Effect of VAM and *Azospirillum brasilense* on photosynthesis, nitrogen metabolism and grain yield in wheat. *Indian J. Plant Physiol. 34: 357-361.*

Panwar, J.D.S. (1992). Effect of VAM and *Azospirillum brasilense* on metabolic changes and grain yield of wheat under moisture stress condition. *Indian J. Plant Physiol.35:163-167.*

Panwar, J.D.S., Abbas , S. and Sita Ram, (1990). Effect of Benzyl amino purine and girdling site on photosynthesis, translocation and nitrogen fixation in Y-shaped mung bean. *Indian J. Plant Physiol. 33(1):16-20.*

Panwar, J.D.S., Shukla, D.S. and Sirohi, G, S, (1986). Growth and developmental aspects in relation to yield of mung bean. *Indian J. Plant Physiol. 29(4): 312-317.*

Panwar, J.D.S. and Thakur, A.K. (1995). Physiological and Biochemical studies in Vesicular Arbuscular Mycorrhizae-*Rhizobium* inoculated mung bean plants under field condition. *Proc Mycorrhizae biofertilizers for the future.Tata Energy Res. Institute. pp 471-477.*

Paszkowski, U., Kroken, S., Roux, C. and Briggs, S.P. (2002). Rice Phosphate transporters include an evolutionarily divergent gene specifically activated in arbuscular mycorrhizal symbiosis. *Proc. Natl. Acad. Sci. U.S.A. 99:13324-29*

Paszkowski, V. (2006). A journey through signaling in arbuscular mycorrhizal symbioses 2006. *New Phytologist.* **172**:35-46.

Paul, E.A. and Clark, F.E. (1996). Soil microbiology and biochemistry. *San Diego, CA: Academic Press.* 340p.

Peterson, R.L. and Bonfante, P. (1994). Comparative structure of vesicular-arbuscular mycorrhizas and ectomycorrhizas. *Plant Soil.* **159**:79-88.

Peretto, R., Bettini, V., Favaron, F., Alghisi, P. and Bonfante, P. (1995). Polygalacturonase activity and location in arbuscular mycorrhizal roots of *Allium porrum* L. *Mycorrhiza*. **5**:157-163.

*Plenchette, C. (1982). Recherches surles endomycorrhizes a vesicules et arbuscules. Influence de la plante-hole du champign on et du phosphore sur 1. expression de la symbiose endomycorrizienne. *Ph.D. Thesis, p. 170. of Level Canada.*

Pond, E.C., Menge, J.A. and Jarrel, W.M. (1984). Improved growth of tomato in salinized soil by VAM fungi collected from saline soils. Mycologia, **76**:74-84.

Ponder, F. (1983). Soil moisture levels and mycorrhizal infection in black walnut seedlings. *Commun. Soil. Sci. Plant Anal.* **14:** 507-511.

Powell, C.L. (1975). Plant growth responses to vesicular-arbuscular mycorrhiza VIII. Uptake of P by onion and clover infected with different *Endogone* spore types in ^{32}P labelled soils. *New Phytol.* **75:**563-566.

Powell, C.L. (1977). Mycorrhizas in hill country soils II. Effect of several mycorrhizal fungi on clover growth in sterilized soils. *N.D.J. Agric. Res.* **20**:59-62.

Powell, C.L. (1982). Effects of kole and mustard crops on response of white clover to VA-mycorrhizal inoculation in pot trials. *Nz. J. Agric. Res.* **25**:461-464.

Powell, C.L. and Daniel, J. (1978). Mycorrhizal fungi stimulate uptake of soluble and insoluble phosphate fertilizer from a phosphate deficient soil. *New Phytol. 80:351-358.*

Prasad, H., Chandra, R., Pareek, R.P. and Kumar, N. (2002). Synergism among phosphate-solubilising bacteria, rhizobacteria and *Rhizobium* in urd bean. *Indian J. Pulses Res. 15(2):*131-135

Price, N.S., Roncadori, R.W. and Hussey, R.S.(1989).Cotton root growth as influenced by Phosphorus –nutrition and vesicular arbuscular mycorrhizas. *New Phytol. 111:61-66..*

Preeti, (2011). Effect of *Rhizobium* and VAM on growth and yield of urd bean (*Vigna mungo* (L.) Hepper) under rain fed conditions. *Ph.D Thesis in Botany, C.C.S. Univ. Meerut.*

Radersma, S. and Grierson, P.F.(2004). Phosphorus mobilization in agroforestry: organic anions, phosphatase activity and phosphorus fractions in the rhizosphere. *Plant Soil.* **259**: 209-219.

Radin, J.W. (1984). Stomatal responses to water stress and abscissic acid in phosphorus deficient cotton plants. *Plant Physiol. **76**:392-394.*

Radin, J.W. and Eidenbach, M.P. (1984). Hydraulic conductance as a factor limiting leaf expansion of phosphorus deficient cotton plants. *Plant Physiol. **75**:372-377.*

Rainbird, R.M., Atkins, C.A. and Pate, J.S. (1983). Diurnal variation in the functioning of Cowpea nodules. *Plant Physiol. **72**:308-312.*

Raisinghani, G. (2004). Antibiotic sensitivity of *Rhizobium* isolated from gamma irradiated mutant plants of *Vigna mungo* L. *Journal Phytol Res. **17**:1,109-110.*

Rajendran, K. and Devraj, P. (2004). Biomass and nutrient distribution and their return of *Casurina equisetifolia* inoculated with biofertilizers in farm land. *Biomass and Bioenergy* **26:**235-249.

Ramaswamy, P.P. and Nair, K.S. (1965). A study on the response to greengram to bacterial seed inoculation. *Madras Agric. J.* **52:**241-242.

Ramchandaran, K., Menon, M.R. and Aiyer, R.S. (1980). Effect of composite rhizobial culture inoculation on cowpea *(Vigna unguiculata* L.Walp.). *Indian J. Microbiol.* **20:** 220-224.

Rangeley, A., Draft, M.I. and Newbould, P. (1982). The inoculation of white clover with mycorrhizal fungi in unsterile soils. *New Phytol.* **92:**89-102.

Rao, C.S., Sharma, G.D. and Shukla, A.K.(1999). Influence of organic amendments on the development of ectomycorrhizae and their efficiency in P uptake and seedlings growth of *Pinus kesiya. Indian J. Forest. **22**(1):7-13.*

Ratanayaka, M., Leonard, R.T. and Menge, J.A. (1978). Root exudation in relation to supply to phosphorus and its possible relevance to mycorrhizal formation. *New Phytol.* **81:**543-552.

Raven, J.A. and Smith, F.A. (1976). Nitrogen assimilation and transportation to vesicular plants in relation to intracellular pH regulation. *New Phytol.* **76:**415-431.

Raven, J.A., Smith, S.E. and Smith, F.A. (1978). Ammonium assimilation and the role of mycorrhizas. In climax communities in Scotland. *Trans Bot. Soc.* (Ediburgh). **43:**27-35.

Rawat, A.K., Khara, A.K. and Vaishya, U.K. (1991). Dual effect of *Rhizobium* and mycorrhiza on nodulation and yield on chickpea. *Legume Research.* **14(3)**:150-152.

Read, D.J., Koucheki, H.K. and Hodgson, J. (1976). Vesicular-arbuscular mycorrhiza in natural vegetation system I. The occurence of infection. *New Phytol.* **77**:641-653.

Readhead, J. (1975). Endotrophic mycorrhizas in Nigeria-Some aspects of the ecology of the endotrophic mycorrhizal association of *Khaya grandifoliola* C.D.C. *[In] Sanders, F.E., Mosse, B. and Tinker, P.B. (eds.). Endomycorrhizas, Academic Press, New York and London, pp. 469-484.*

Reddy, A.S. and Mallaiah, K.V. (2002). Studies on the effect of Rhizobium inoculation in black gram. Bionoculants for Sustainable agriculture and forestry: Proceedings of National Symposium held on February 16-18, 2001. Scientific Publishers (India), Jodhpur, India: 93-98.

Redente, E.F. and Reeves, F.B. (1981). Interactions between vesicular-asbuscular mycorrhizae and *Rhizobium* and their effect on sweetvetch growth. *Soil Sci.* ***132(6)**:410-415.*

Reid. J.B. and Goss, M.J. (1982). Interactions between soil drying due to plant water use and decrease in aggregate stability caused by maize roots. *Journal of Soil Science* **33**:47-53.

Requena, N., Serrano, E., Ocon, A. and Brequeninger, M. (2007). Plant signals and fungal perception during arbuscular mycorrhizal establishment. *Phytochem.* ***68**:33-40.*

Rewari, R.B. and Tilak, K.V.B.R. (1988). Microbiology of pulses.[In] Baldev, B., Ramajunam, S.P. and Jain, H.K. (Eds). Pulses Crop. Oxford and IBN Publishing Company Pvt. Ltd. Delhi, pp 373-411.

Rhodes, L.H. and Gerdemann, J.W. (1975). Phosphate uptake zones of mycorrhizal and non-mycorrhizal onions. *New Phytol.* **75:** 555-561.

Rhodes, L.H. and Gerdemann, J.W. (1978a). Translocation of calcium and phosphate by external hyphae of vesicular-arbuscular mycorrhizae. *Soil Science.* **126:** 125-126.

Rhodes, L.H. and Gerdemann, J.W. (1978b). Hyphal translocation and uptake of sulphur by vesicular-arbuscular mycorrhizae of onions. *Soil Biol. Biochem.* **10**:355-360.

Rillig, M.C., Wright, S.F., Nicholas, K.A., Schmid, W.F., Torn, M.S. (2002). The role of arbuscular mycorrhizal fungi and glomalin in soil aggregation: Comparing Effects of five plant species. *Plant Soil.* **238**:325-333.

Rorison, I.H. (1968). The response to phosphorus of some ecologically distinct plant species I. Growth rates and phosphorus absorpiton. *New Phytol.* **67**:913-923.

Rose, S.L. and Youngberg, C.T. (1981). Tripartite associations in snowbrush *(Ceanothus veluntinus):* Effect of vesicular-arbuscular mycorrhizae on growth, nodulation and nitrogen fixation. *Can.J. Bot.* **59**:34-39.

Ruiz-Lozano, J.M., Azcon, R. and Gomez, M. (1995). Effects of arbuscular mycorrhizal *Glomus species* on drought tolerance:physiological and nutritional plant responses. *Appl. Environ. Microbiol.***61**:456-460.

Ruiz-Lozano, J.M., Azcon, R. and Gomez, M. (1996). Alleviation of salt stress by arbuscular mycorrhizal *Glomus species* in *Lactuca sativa* plants. *Physiol. Plant.* **98**:767-772.

Rygiewicz, P.T., Johnson, M.G., Storm, M.J. and Tingey, D.T. (1994). Life history and carbon partitioning of ectomycorrhizae on the roots under elevated Co_2 *Abst. First GCTE Science Conference held at Wood Hole, Massachusctts, U.S.A. 23-27, May 1994, pp105.*

Safir, G.R., Bover, J.S. and Gerdemann, J.W. (1971). Mycorrhizal enhancement of water transport in soybean. *Science,* **172**:581-583.

Safir, G.R., Bover, J.S. and Gerdemann, J.W. (1972). Nutrient status and mycorrhizal enhancement of water transport in soybean. *Plant Physiol.***49**:700-703.

Safir, G.R. and Nelson, C.E. (1985). VA-mycorrhizae: plant and fungal water relations. [In] Monila, R. (ed), Proc. of 6th North Amer. Conf. on Mycorrhizae. Forest Res. Laboratory, Corvallis, Oregon, pp. 161-164.

Sahu, S. and Behera, B. (1972). Note on effect on *Rhizobium japonicum* inoculation on cowpea, groundnut and greengram. *Indian J. Agron.* **17**:359-360.

Saker, M. L., Congdon, R. A. and Mayeok, C. R. (1999). The relationship between phosphorous fractions, phosphatase activity and fertility in three tropical rain forest soils. *Tropical Ecology. 40(2): 261-267.*

Same, B.I., Robson, A.D. and Abbott, L.K. (1983). Phosphorus, soluble carbohydrates and endomycorrhizal infection. *Soil Biol. Biochem.* **15**:593-597.

Sampathkumar,G. and Ganesh kumar, A.(2003). Effect of VAM fungi and *Rhizobium* on growth and nutrition of *Vigna mungo* L. and *Vigna unguiculata* L. *Mycorrhiza News. 14(4):15-18.*

*Sanchez-Diaz, M.and Aguirreolea, J. (1993). Efectos fisiologicos que causa la falta persistente de agua en los cultivos. *Phytoma.***51**:26-36.

Sanders, F.E. (1975). The effect of foliar applied phosphate on the mycorrhizal infection of onion roots. *[In] Sanders, Academic Press. London and New York. pp. 261-276.*

Sanders, F.E., Tinker, B.P., Black, R.L.B. and Palmerly, S.M. (1977). The development of endomycorrhizal root systems. I. Speed of infection and growth-promoting effects with four species of vesicular arbuscular endophyte. *New Phytol*. **78**:257-268.

Sanders, F.E. and Sheikh, N.A. (1983). The development of vesicular-arbuscular mycorrhizal infection in plant root systems. *Plant Soil.* **71**:223-246.

Sanders, F.E. and Tinker, P.B. (1971). Mechanism of absorption of phosphate from soil by *Endogone* mycorrhizas. *Nature.* **233**: 278-279.

Sanders, F.E. and Tinker, P.B. (1973). Phosphate flow into mycorrhizal roots. *Pest. Sci.* **4**:385-395.

Sanders, F.E., Mosse, B, and Tinker, P.B. eds. (1975). Endomycorrhizas. *London Academic. pp. 626.*

Sanders, F.E., Tinker, P.B., Black, R.L.B. and Palmerley, S.M. (1977). The development of endomycorrhizal root systems I. Spread of infection and growth promoting effects with four species of vesicular-arbuscular endophytes. *New Phytol.* **78**:257-268.

Sankaranarayanan, C. and Sundarbabu, Rajeshwari. (2005). Influence of *Rhizobium* and phosphobacteria on the interaction of VAM, *Glomus mosseae* and *Meloidogyne incognita* on black gram. *International J. Trop. Plant Diseases.* **19(1/2)** : 133-279.

Saxena, A.K., Shende, Rasika, Tyagi, S.P. and Tilak, K.V.B.R. (2002). Toxin production and competitiveness of *Bradyrhizobium* sp. Strain 524. *Indian J. Microbiol. **42**(4): 289-293.*

Schenck, N.C. and Hielnson, K. (1973). Response of nodulating and non-nodulating soybeans to a species of *Endogone* mycorrhiza. *Agron. J.* **65**:849-850.

Secilia, J. and Bagyaraj, D.J. (1987). Bacteria and actinomycetes associated with pot cultures of vesicular-arbuscular mycorrhizas. *Canadian J. Microbiology* **33:** 1069-1073.

Seth, R.D. (1979). Studies on the development of rhizobial inoculation for groundnut. *Ph. D Thesis, University of Agric Sci., Bangalore.*

Shanmugam, V., Renukadevi, P., Senthil, N., Raguchander, T. and Samiyappan, R.(2001). Integration of *Psedomonas fluorescens* with *Rhizobium* in the management of black gram root rot incited by *Macrophomina phaseolina. Plant Disease Research. **16**(1): 23-29.*

Shanta Kumari, P. and Sinha,S.K. (1972). Variation in chlorophylls and photosynthetic rates in cultivars of bengal gram. *Phosynthetica. **6**:189-194.*

Sheehy, J.E., Fishbeck, K.A., Dejong, T.M., Williams, L.E. and Phillips, D.A. (1980). Carbon exchange rates of shoots required to utilize available acetylene reduction capacity in soya-bean and nodules. *Plant Physiol. **66**:101-104.*

Sheikh Javid., Akhter, A., Khan, N.A. and Inam, A. (2003). Effect of *Rhizobium* and NPK on growth, physiomorphological traits and yield of waste-water grown black-gram. *Indian J.Pulses Res.**16**(1): 25-30.*

*Sieverding, E. (1983). Influence of soil water regime on VA-mycorrhiza II. Effect of soil temperature and water regime on growth, nutrient uptake and water utilization of *Eupatorium ordoratum.* Z. Acker, Pflanzenbau. **15**:56-67.

Singh, A., Ahlawat, I.P.S. and Saraf, C.S. (1984). Response of spring/summer green gram (*Vigna radiata* (L.) Wilczek) to irrigation and antitranspirants with and without mulching. *Indian J. Agron. **28**: 260-262.*

Singh, A.K. and Shrivastava, S.K. (1985). Nodular physiology of urd bean as affected by urd bean mosaic virus effect on some enzymatic activity. *Phyton (Austria). Vol. 25. Fasc **2**. 213-217.*

Singh, B. and Gupta, B.R. (2006). Effect of biofertilizers at different levels of phosphorus on nodulation, yield and protein content in black gram. *Farm Science Journal. **15** (1): 78-80.*

Singh, G. and Tilak, K.B.V.R. (1995). Biofertilizers in wastelands improvement. [In] Chandra, S., Khhana, Kheri (ed). Microbes and man. pp. 46-46. Dehradun, India.

Singh, R., Adholeya, A. and Mukerji, K.G. (2000). Mycorrhiza in control of soil borne pathogens. [In] "Mycorrhizal Biology" (eds. Mukerji, K.G., Chamola, B.P. and Jagjit Singh). Kluwer Academic Publishers, NY, pp. 171-196.

Sinha, D.D. and Sharma, A. (2001). Response of black-gram to phosphorous fertilization and *Rhizobium* inoculation in the hill soils of Assam. *Annals Agric. Res.22(1): 151-153.*

Sinha, S. K., Bhargava, S. C. and Baldev, B. (1988). Physiological aspects of pulse crops. [In]. Pulse crops. Eds. Baldev, B., Ramanujam, S. P. and Jain, H. K. Pub. Oxford and IBH, New Delhi.

Sharma, M.P., Gaur, A., Bhatia, N.P. & Adholya, A. (1996). Growth responses & dependence of *Acacia nilotica* var cupriformis on the indigenous arbuscular mycorrhizal consortium of a marginal wasteland soil. *Mycorrhiza.* **6**:441-446.

Sharma, S., Kashyap, S. and Vasudevan, P. (2005a). Effect of bioinoculants on biomass productivity under agroforestry systems. *Indian Journal of Biotechnology.* **(4)**: 156-160.

Sharma, S., Kashyap. S., Vasudevan, P. (2005b). In vitro rhizogenesis of *Morus alba* by mycorrhizal extracts under saline stress. *European Journal of Horticulture Scienc.e* **70(2)**: 68-74.

Sharma, S., Madan, M. and Vasudevan, P. (1995). Ectomycorrhizas, soil fertility and biomass productivity. *Proceedings of National Academy of Sciences*, vol. **LXV**:345-356.

Sharma, Shivesh., Upadhyay, R.G. and Sharma, C.R. (2000). Effect of *Rhizobium* inoculation and nitrogen on growth, dry matter accumulation and yield of black gram. *Legume Research.* **23***(1): 64-66.*

Shockley, F.W., MacGraw, R.L. and Garrett, H.E. (2004). Growth and nutrient concentration of two native forage legumes inoculated with *Rhizobium* and Mycorrhiza in Missouri, USA. *Agroforestry Systems* **60:** 137-142.

Siviero, M.A., Motta. A.M., Lima D.S., Birolli, R.R., Huh, S.Y., Santinoni, I.A., Miyauchi, M.Y.H., Zangaro, W., Nogueria, M.A. and Andrade, G. (2008). Interaction among N fixing bacteria and AM fungi in Amazonian legume tree *(Schizolobium amazonicum)* in field conditions. *Applied Soil Ecology* **39:** 144-152.

Smith, S.E. (1980). Mycorrhizas of autotrophic higher plants. *Biol. Rev.* **55**:475-510.

Smith, S.E. (1982). Inflow of phosphate into mycorrhizal and non-mycorrhizal *Triticum-subterrance* at different. *New Phytol.* **90**:293-303.

Smith, S.E. and Brown, G.D. (1979). Soil temperature, mycorrhizal infection and nodulation of *Medicago trunculata* and *Trifolium subtarraneaum. Soil Biol. Biochem.* **11**:469-473.

Smith, S.E. and Daft, M.J. (1977). Interactions of growth, phosphate content and N_2 fixation in mycorrhizal *Madicago sativa, Aust. J. Plant. Physiol.* **4**(3).

Smith, S.E. and Read, D.J. (1997). Mycorrhizal symbiosis, *second ed. Academic Press, London. U.K*

Smith, S.E. and Son, C.L. (1987). Inflow of phosphate into onion roots: interaction between photo irradiance phosphorus supply and mycorrhizal infection. [In] *Proc. 7th North American Conference on Mycorrhizae. pp. 265.*

Smith, S.E. and Gianinazzi-Pearson, V. (1988). Physiological interactions between symbionts in vesicular-arbuscular mycorrhizal plants. Annual Review of Plant *Physiology and Molecular Biology* **39:** 221-244.

Smith, S.E., St. John, B.J., Smith F.A. and Bromely, J.L. (1986). Effect of mycorrhizal infection on plant growth, nitrogen and phosphorus nutrition on glass house-grown *Allium cepa* L. *New Phytol.* ***103****: 359-373.*

Smith, S.E. and Smith, F.A. (1990). Structure and function of the interfaces in biotropics symbiosis as they relate to nutrient transport. *New.Phytol.* **114**: 1-38.

Snellgrove,R.C., Stribley, D.P.,Tinker, P.B. and Lawlor, D.W. (1986). The effect of vesicular arbuscular mycorrhizal infection on photosynthesis and carbon distribution in leek plants. *[In] physiological and genetical aspects of mycorrhizae. (Eds.) Gianinazzi-Pearson, V. and Gianinazzi, S. pp.421-424.*

Sriramachandra sekharan, M. V. and Vaiyapuri, V. (2003). Influence of granular insecticide-carbofuran in association with *Rhizobium* on nodulation, growth and yield of black-gram.*Journal. Ecobiol.* ***15****(2):107-111.*

Subba Rao, N. S. (1976). Field response of legumes in India to inoculation and fertilizer application In: Nutman, P.S. (Eds). Symbiotic Nitrogen Fixation in Plants. *International Biol. Programme, Cambridge University Press, London. 7:225-268.*

Subba Rao, N.S., Tilak, K.V.B.R. and Singh, C.S. (1985). Effects of combined inoculation of *Azospirillum brasilense* and vesicular arbuscular mycorrhiza on Pearl millet. *Plant Soil.* **84**:283.

Subramanian, K.S. and Charest, C. (1998). Arbuscular mycorrhizae and nitrogen assimilation in maize after drought and recovery. *Physiol Plant.* **102:** 285-296.

Swaminathan, K. and Verma, B.C. (1979). Response of three crop species to vesicular-arbuscular mycorrhizal infection on zinc deficient Indian soil. *New Phytol.* **84**:481-487.

Tanwar, S.P.S., Sharma, G. L. and Chahar, M. S. (2002). Effects of biofertilizers on the growth and productivity of black gram. *Annals. Agric Res.* ***23****(3):491-493*

Tanwar, S.P.S., Sharma, G. L. and Chahar, M. S. (2003). Effect of phosphorus and biofertilizers on yield nutrient content and uptake by black gram. *Legume Research.* ***26****(1):39-41.*

Tanwar, S.P.S., Sharma, G. L. and Chahar, M. S. (2003). Economics of black gram cultivation as influenced by phosphorous levels and biofertilizers. *Legume Research.* ***26****(2): 149-150.*

Tarafdar, J. C. and Marshner, H. (1994). Phosphatase activity in the rhizosphere and hydrosphere of VAM wheat supplied with organic and inorganic phosphorous. *Soil Biol. Biochem.* ***26****(2) pp:387-395.*

Tester, M. and Smith, S.E. (1985). Phosphate in flow into *(Trifolium subterraneum* L.). Effect of photo-irradiance and mycorrhizal infection. *Soil. Biol. Biochem.* **17**:807-810.

Thakur, A.K. (1994). Physiological studies on *Rhizobium*- VAM symbionts in mung bean. *M.Sc. Thesis P.G. School, IARI, New Delhi.*

Thakur, A.K. and Panwar, J.D.S. (1995). Effect of *Rhizobium*-VAM interaction growth and yield in Mungbean *(Vigna radiata* L. Wilczek) under field conditions. *Indian J. Plant Physiol.* **38**:62-65.

Thakur, A.K. and Panwar, J.D.S. (1997). Response of *Rhizobium* – vesicular arbuscular mycorrhizal symbionts on photosynthesis, nitrogen metabolism and sucrose translocation in green gram (*Phaseolus radiatus*).*Indian J.Agric Science.***67***(6):245-248.June, 1997.*

Thiagarajan, T.R., Ames, R.N. and Ahmad, M.H. (1992). Response of cowpea (*Vigna unguiculata*) to inoculation with co-selected vesicular-arbuscular mycorrhizal fungi and *Rhizobium* strains in field trials. *Can. J. Microbiol.***38***:573-576.*

Tilak, K.V.B.R. and Dwivedi, (1990). Nitrate reductase activity of vesicular arbuscular mycorrhizal fungi. *Proceeding of national seminar on mycorrhizas held at HAU, Hisar, 14-16, Feb. 1990, pp. 59.*

Timmer, L.W. and Leyden, R.G. (1980). The relationship of mycorrhizal infection to phosphorus-induced copper deficiency in sour orange seedlings. *New Phytol.* **85**:15-23.

Tinker, P.B. (1975). Soil chemistry of phosphorus and mycorrhizal effects on plant growth. [In] Sanders, F.E., Mosse, B., Tinker, P.B. (eds), Endomycorrhizas, Academic Press, London and New York, pp. 353-371.

Tinker, P.B. and Gildon, A. (1983). Mycorrhizal fungi. [In] Robb, D.A., Pierpoint, W.S. (eds), Metals and micronutrients: Uptake and Utilization by plants. Academic Press, New York. pp. 21-37.

Tiwari, P., Adholya, A. and Prakash, A. (2003). Commercialization of arbuscular mycorrhizal biofertilizer. [In] Fungal Biotechnology in Agricultural, Food and Environmental Application, D.K. Arora (ed.), Marcel Dekker Inc., NY, USA, 195-201.

Tomar ,A., Kumar, N., Pareek, R. P. and Chandra, R. (2003). Residual effect of black gram inoculated *Rhizobium*, VAM and PSB on succeeding wheat crop. *Indian J. Pulses Res.* **16***(2):141-143.*

Verma, C. B, Lallu and Yadav, R. S. (2004). Effect of boron and zinc application on growth and yield of pigean pea. *Indian J. Pulses Res.* **17***(2): 149-151.*

Verma, H. R. and Pandey , A. K. (2003). Effect of foliar spray of boron and iron on growth and yield of paddy. National Seminar on Physiological Intervention for Improved Crop Productivity and Quality. December, 12-14, 2003 at Tirupati, pp.159.

Verma, N., Chaturvedi, S. and Sharma, A.K. (2008). Mass application of AMF using soil less substrate. *Mycorrhiza News.* **20(1)**:11-15.

Vijaylakshmi. (1988). Influence of benzyladenine on photosynthesis and assimilate partitioning in mungbean under different source-sink relationship. *M.Sc. Thesis, Div. Plant Physiology P.G. School. I.A.R.I. New Delhi.*

Vogelzang, B., Parsons, H. and Smith, S. (1993). Separate effects of high temperature on root grow of *Vigna radiata* L. and colonization by the vesicular arbuscular mycorrhizal fungus *Glomus versiforme. Soil Biol. Biochem.* **25**(8):1127-1129.

Wallace, L.L., Mc Naughton, S.J. and Coughenour, M.B. (1982). The effects of clipping and fertilization on nitrogen nutrition and allocation by mycorrhizal and non-mycorrhizal *Panicum oloratum* L. a C_4 grass. *Oecologia.* **54**:68-71.

Whipps, J.M. (2001). Microbial interactions and biocontrol in the rhizosphere. *Journal of Experimental Botany* **52:** 487-511.

William, L. E., Dejong, T. M. and Phillips, D. A. (1982). Effect of changes in shoot carbon exchange rate on soya bean root nodule activity. *Plant Physiol. 69:432-436.*

Wilson, G.W.T., Rice, C.W., Rillig, M.C., Springer, A. and Hartnett, D.C. (2009). Soil aggregation and carbon sequestration are tightly correlated with the abundance of arbuscular mycorrhizal fungi: results from long term field experiments. *Ecol Lett.* 12:452-461.

Wu, S.C., Cao, Z.H., Li, Z.G., Cheung, K.C. and Wong, M.H. (2005). Effects of biofertilizer containing N-fixer, P and K solubilizers and AM fungi on maize growth; a greenhouse trial. *Geoderma* **125:** 155-166.

Yolande, D. and Marcia, M. (2004). Arbuscular mycorrhizal inoculum to support sustainable cropping systems. *Plant Management Network. Online. Crop Management doi* : 10.1094 / C M – 2004 – 0301 -09 - R.V.

Yocom, D. H., Boosalis, M. G. and Flessner, T. R.(1987). The effect of prolonged drought on the yield of mycorrhizal and mycorrhizal grain crops. *[In]Sylvia, D. M., Hung, L. L. and Graham, J. H. Mycorrhiza in Next Decade. Practical Applications and Research Priorities. North Amer. Conf. Mycorrhiza, Gainesville, Florida.*

Zaidi, A., Khan, Sagir, Md., and Md. Amil, (2003). Interactive effect of rhizotropic microorganisms on yield and nutrient uptake of chickpea *(Cicer arietinum* L.) *European Jour. of Agron.* **19:**15-21.

5

NUTRIENT USE EFFICIENCY FOR MAXIMIZATION OF CROP PRODUCTIVITY

Dhiman Mukherjee

Regional Research Station, Uttar Banga Krishi Viswavidyalaya, Darjeeling, West Bengal 734 301

INTRODUCTION

Efforts to increase efficient nutrient use by crops are of growing importance as the global demand for food, fibre and fuel increases and competition for resources intensifies. Invariably, many agricultural soils of the world are deficient in one or more of the essential nutrients needed to support healthy plants (Evans, 2009). Acidity, alkalinity, salinity, anthropogenic processes, nature of farming, and erosion can lead to soil degradation. Plant nutrients, which come primarily from chemical fertilizers, manure, and in some cases sewage sludge, are essential for crop production (Khouri *et al.*, 2011). When applied in proper quantities and at appropriate times, nutrients (especially nitrogen, phosphorus, and potassium) help achieve optimum crop yields. However, improper application of nutrients can cause water quality problems both locally and downstream. Further proper and balance nutrient is must for optimum economic yield of crop. Nutrient management is the practice of using nutrients wisely for optimum economic benefit, while minimizing impact on the environment. Efficient management of nutrient through high NUE is important aspect of plant nutrient management, because nutrient has direct effect on crop physiology and ultimately good yield (Mukherjee, 2002). Nutrient use efficiency improvements must always be evaluated in

the context of maintaining the effectiveness of nutrient inputs in supporting the efficiency of the cropping system.

Increased NUE in plants is vital to enhance the yield and quality of crops, reduce nutrient input cost and improve soil, water and air quality. There are very few "ideal" soils in the world- that is, soils that contain all of the essential nutrients in the proper balance required by crops. While animal manures are excellent at providing many of the essential nutrients for crops, their composition is rarely in balance with what the soil requires to adequately supply the plant's needs. Similarly, legume cover crops are good as a N source for subsequent crops, but provide no other additional nutrients that were not already in the soil. It is in everyone's best interest to utilize nutrients as efficiently as possible (Baligar *et al.*, 2001). However, accomplishing this goal- or even defining it- is difficult to achieve. In general, getting as much of a nutrient as possible into the harvested portion of a crop is the concept of efficient nutrient use. Tracking the recovery of applied nutrients is a key component to measuring nutrient efficiency. The definition of NUE in plants need to be clearly defined and carefully selected to reflect the end use. Much can be achieved by selecting nutrient efficient genotypes and to incorporate these in breeding programs. However, the poorly developed state of nutritional genetics of plants and its response to environmental variables and management practices and the difficulty of identifying nutrient efficiency traits by rapid and reliable techniques have contributed to a lack of progress and success in breeding plant cultivars with high NUE.

Plant species and cultivars within species differ in absorption and utilization of nutrients and such differences are attributed to morphological, physiological and biochemical processes in plants and their interaction with climatic, soil, fertilizer, biological and management practices. An improved NUE in plants can be achieved by careful manipulation of plant, soil, fertilizer, biological, environmental factors and best management practices. There is great need for a well coordinated, multi-disciplinary, team effort of plant geneticists and breeders, physiologists, biologists, agronomists, soil scientists, and chemists among other disciplines, to formulate an effective system to overcome the internal and external constraints that are contributing to lower nutrient use efficiencies and to make increased NUE in plants a reality. Additions of fertilizers and/or amendments are essential for a proper nutrient supply and maximum yields. Estimates of overall efficiency of applied fertilizer have been reported to be about or lower than 50% for N, less than 10% for P, and about 40% for K. Plants that are efficient in absorption and utilization of nutrients greatly enhance the efficiency of applied fertilizers, reducing cost of inputs, and preventing losses of nutrients to ecosystems. Inter- and intra-specific variation for plant growth and mineral nutrient use efficiency (NUE) are known to be under genetic and physiological control and are modified by plant interactions with environmental variables. There is need for breeding programs to focus on developing cultivars with high NUE. Identification of traits such as nutrient absorption, transport, utilization, and mobilization in plant cultivars should greatly enhance fertilizer use efficiency. The development of new cultivars with higher NUE, coupled with best management practice.

Increase nutrient use pattern by crop have five major sub-concepts, viz.

i. Plant nutrients stored in the soil for crop growth.

ii. Plant nutrients, those present in the crop residues, organic manure and domestic wastes.

iii. Plant nutrients purchased or obtained from outside the farm.

iv. Plant nutrient loosed e.g. those removed from the field in crop harvest and lost from the soil through volatilization (ammonia and nitrogen oxide gases and leaching (nitrate, sulphate etc.)

v. Plant nutrient outputs e.g. nutrient uptake by the crops at harvest time.

Goals of NUE

a. Optimum utilization of available nutrient for plant growth.

b. To maintain soil productivity through cycling of nutrient and balance nutrient budget of soil.

c. To ensure productive and sustainable agriculture.

d. To reduce expenditure on costs of purchased inputs by using farm manure and crop residue etc through better absorption capability.

e. To utilize the potential benefits of green manures, leguminous crops and biofertilizers.

f. To prevent degradation of the environment.

g. To meet the social and economic aspirations of the farmers without harming the natural resource base of the agricultural production.

Nutrient use efficiency (NUE) on the basis of agricultural farm is defined as:

NUE = tradable outputs / all inputs

= (exported crops + exported animals + exported milk, cheese & butter) / (imported animals + imported bedding + imported feed + mineral fertilisers + imported manure + deposition & biological fixation) (Fig. 1)

Zzz21q qaAzQSA wel` 1` NUE improvements should come from increased outputs and/or decreased inputs. This approach has many advantages, as well as some disadvantages, the most important of which are the following:

- NUE is an integrative indicator.

- NUE does not distinguish between the different losses and the different compartments the losses act on.
- NUE takes an integrated approach in which N losses to the atmosphere are as important as those to groundwater, etc.
- NUE can be a positive indicator for farm policy and farm management.
- Unlike the indicator of loss (input – output) and the rigid application standards, NUE can be used as a more positive indicator, as it focuses on increasing overall efficiency and makes use of 'end norms'.
- NUE takes a regional or even global approach. Overall, it focuses on increasing efficiency at any intensity level, optimising output for each input.

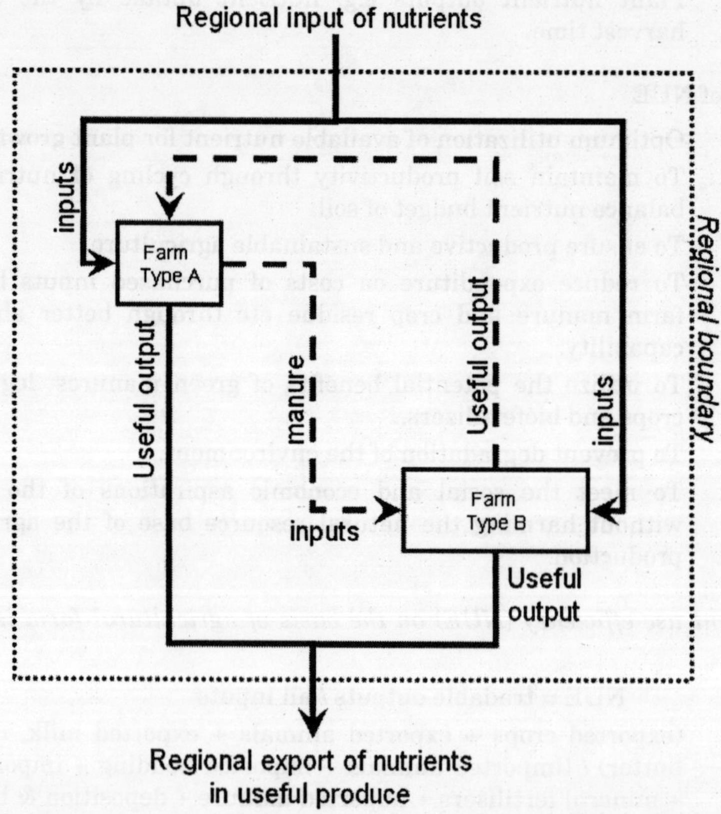

Fig. 1: Overview of the flows that are used to calculate NUE (Regional inputs of nutrients include imported animals, imported bedding, imported feed, mineral fertilizers, imported manure, deposition and biological fixation. Exports consist of exported crops and exported animal products. The dashed arrows indicate inter-farm nutrient recycling processes).

2. CHALLENGES TO AGRICULTURE

The major challenge facing plant biology is to improve crop production to feed an expanding world population. This is against a background of pressure on agricultural land use and climate change having negative impacts on growing conditions and limiting geographic regions for agriculture (Mukherjee, 2010). The adverse effects of agriculture, and specifically fertilizer use, include damage to the environment, a large carbon footprint for the manufacture and use of agrochemicals, and the utilization of non renewable resources. One solution is to increase the area of land for agriculture, as well as increasing production while maintaining the current rate of inputs; however, this is predicted to have substantial negative impacts on the environment, is unsustainable in terms of phosphate use, and would have a huge economic footprint in terms of energy demands for nitrogenous fertilizer production. The challenge is to increase yield, decrease inputs, and improve resistance to abiotic and biotic stresses. Improving crop nutrient use efficiency ideally requires an understanding of the whole system, from the macro - (agro - ecosystem) to the molecular level. While acknowledging the critical contribution of agronomy to improving efficient nutrient use, particularly in classically inefficient systems, there is a point at which crop genetic improvement becomes essential for further improvement. This may be achieved by conventional breeding, as indeed has been the case to the present day with ever increasing yields, albeit often in parallel with increased nutrient demands, with marker - assisted breeding utilizing genetic information derived from basic plant science, and by the utilization of this same information to produce genetically modified crops.

Nutrients, along with light, temperature, and water, are critical determinants of crop production, but fertilizers are costly inputs and inappropriate overuse can have many ecologically damaging effects, making efficient use of fertilizers a major issue for agriculture (Mukherjee and Sen, 2005). For example, excessive nitrogen use results in a major fraction of anthropogenic nitrous oxide and methane emissions, which contribute substantially to climate change and inefficient nutrient uptake may result in pollution of inland and coastal waters by leaching and runoff. Worldwide, it has been estimated that nitrogen use efficiency (NUE) for cereal production is only 33% (nitrogen removed in grain as a percentage of that applied) (Evans, 2009). Both agronomic practice and plant breeding have a responsibility to optimize efficient nutrient use, particularly nitrogen, in crop systems. Furthermore, crop improvements to anticipate changing patterns of rainfall and temperature must include an anticipation of nutritional demands influenced by changing cropping systems and crop ideotypes. Optimal plant growth demands a balanced nutrient supply.

2.1 Potential Environmental Concerns

Increase NUE is most important not only crop point of view but also to our biosphere. Excess use of nutrient deteriorates environment quality and sustainability too. Further, water quality problems can occur when nutrients are added to the soil at a time when they could be removed in surface runoff from rainfall or snow melt at rates exceeding that removed by the crop, or if applied at times that they cannot be utilized by the crop. When nitrogen or phosphorus are present in lakes or rivers at a high concentration, a condition called "eutrophication" or biological enrichment can occur. High nitrogen from the Mississippi River has been blamed for a hypoxic or "dead" zone in the Gulf of Mexico, where excess algae grow in response to the enriched nutrient solution and few fish can be found. When the algae die, their decomposition consumes enough dissolved oxygen to suffocate fish and other animal life. Sources of nitrogen contributing to the problem include agricultural runoff, sewage treatment plants, atmospheric nitrogen, and other sources. Excessive nitrate in ground water can present a direct health hazard to very young infants. Ingestion of nitrate (NO_3) can bind with hemoglobin in the infant's bloodstream and cause a condition called methemoglobinemia or "blue baby" syndrome. Nitrate does not bind to soil particles and is quite soluble, making it susceptible to leaching into groundwater if not used by the crop.

3. NUTRIENTS AND BALANCE

In crop production, plants synthesize nutrients in the soil such as nitrogen, phosphorus, and potassium (NPK) with air, sunlight, and water. Without proper management, continuous crop production can reduce nutrient reserves in the soil. On an average more than 30 net kilograms of NPK per hectare are re moved from soils each year (Sheldrick *et al.*, 2002). As reserves get depleted, crop growth and productivity can be compromised. Over time, cumulative depletion can decrease agricultural production, crop yields, and soil fertility, and lead to soil degradation. Techniques to conserve and add nutrients to the soil through the application of organic or in organic fertilizers can help to maintain and increase the nutrient reserves of the soil. But over supply of nutrients can also be a problem, causing economic in efficiency, damage to the environment, and, in certain situations, harm to the plants themselves, and to the animals and humans that consume them or products made from them. Whereas soil mining is primarily a problem in developing countries, over application of nutrients occurs chiefly in the developed world, where the relatively low cost of fertilizer leads some farmers to use it in amounts far in excess of plant needs and the capacity of soils to hold nutrients. Balance in the absolute and relative application of nutrients is a part of INM (Mukherjee and Chakraborty, 2010).

3.1 Achieving Balance: Nutrient Application

Achieving balance between the nutrient requirements of plants and the nutrient reserves in soils is essential for maintaining high yields and soil fertility, preventing environmental contamination and degradation, and sustaining agricultural production over the long term. In many cases, imbalances can be corrected through the application of appropriate inorganic and organic fertilizers. The application of nitrogenous fertilizer on nitrogen- poor soils increased maize yields from 43.8 to 5.9 metric tons per hectare, while application of a less appropriate fertilizer increased yields to only 3.8 tons per hectare (Mukherjee, 2010). Correcting nutrient imbalances not only leads to sustainable high crop yields, but it reduces the need to cultivate unsustainable marginal lands. Al though a relatively poor sub statute for commercial fertilizers, urban sludge improves soil structure, contains secondary and micronutrients as well as NPK, and has the potential to be a good source of nutrients for agricultural lands. From an economic perspective, the use of urban waste as fertilizer has an added benefit: it puts to good use material that would otherwise be costly to dispose off.

High NUE through some input modification of plant physiology system. Genetic engineering holds the promise of enabling plants themselves to generate some of the nutrients they require. By altering or creating microorganisms that can fix nitrogen in non-leguminous plants such as cereals, genetic engineering would reduce the amount of nitrogen farmers had to apply. While this strategy holds promise for the future, the process is quite complex and will be ex pensive to realize.

3.2 Conservation and Uptake of Nutrients

Boosting plant nutrient sources does not necessarily require the heavy application of in organic fertilizer or any kind of fertlizer. Planting practices such as terracing, alley cropping, and low- till farming prevent nutrient loss. Other practices, including use of cover crops and intercropping and application of organic manures, can improve nutrient reserves and soil characteristics and structure. The rotation of cereals and leguminous plants has been shown to reduce nutrient loss from soil system by up to 30 per cent, as cereals absorb the nitrates released from the decaying roots and nodules of leguminous plants. Through integrated approach of nutrient management also making nutrient uptake more efficient (Mukherjee and Roy, 2011). Most crops make in efficient use of nitrogen, often losing nitrogen into the atmosphere through volatilization. New techniques, such as deep placement of fertilizers and the use of inhibitors or urea coatings, have been developed to address this problem.

3.3 Nutrient Budget

The generally accepted approach to nutrient balance measures the difference between nutrient inputs and outputs in an agricultural system.

Nutrient or mineral balances establish a link between agricultural nutrient use, changes in environmental quality, and the sustainable use of soil nutrient resources. Depending on the data input, these budgets can be used at a variety of scales. Nutrient budgets are becoming increasingly common as a tool to describe nutrient flows within farming systems and to assist in the planning of the complex spatial and temporal management within rotational cropping and mixed farming systems. Budgets are the outcome of a nutrient accounting process, ranging from simple to complex, which details all the inputs and outputs to a given system over a fixed period of time. The underlying assumption of a nutrient budget is that of mass balance (i.e. nutrient inputs to the system minus any nutrient exports equal the change in storage within the system (Mukherjee, 2010). Many approaches have been used to estimate nutrient balances, depending on the intended use. For example, the technique for developing national, regional, or global estimates of efficiency may be much different from a field-scale or micro-plot approach. Additionally, a nutrient deficit or surplus over the short term does is not immediately indicative of undesirable consequences, but in fact may be beneficial and desirable for building overall soil fertility. Several basic techniques are used to measure nutrient balances- all of which have various limitations depending on the level of measurement and the availability of data. The usefulness and reliability of any type of budget depends on its completeness. The three main approaches are:

3.3.1 Soil surface balance

This approach measures the difference between the inputs (or the application) of nutrients and the output (or removal of nutrients) from the soil (Sheldrick *et al.*, 2002). While this budget provides the most detail for nutrient management planning, there is usually uncertainty associated with the data inputs and the partitioning of the components of the nutrient balance between air, soil, and water.

3.3.2 Farm gate balance

This type of balance simply measures the difference between the nutrient content of farm inputs and the nutrient content of farm outputs. This balance has been successfully used for P and K, but it ignores many of the complex on-farm transformations that N is subject to (e.g. NH_3 volatilization, denitrification, volatile losses during crop senescence, etc.). This method quantifies nutrients supplied to and removed from the farm, but does not quantify the nutrients circulating within the farm enterprise. This budget was based on crop production statistics, average nutrient content of harvest crops, recoverable animal manure, and fertilizer consumption. The contribution of legumes to the overall nutrient budget is an important N input in many area.

3.3.3 Soil system balance

This approach is commonly used where detailed information on inputs, outputs, and internal transformations is available for all the important components (Fig. 2). This type of balance requires much larger data inputs than the previous approaches, but the use of relevant computer models can help with parameter estimates when field observations are not available.

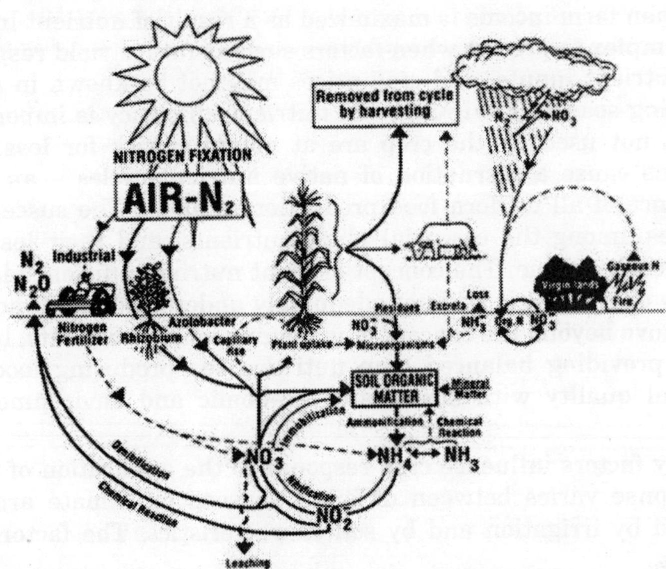

Fig. 2. Example of inputs required for a soil system balance based on mechanistic nutrient transformations (Brown and Johnson, 1996).

4. NUTRIENT USE EFFICIENCY

Efficient use of nutrients in agriculture may be defined differently when viewed from agronomic, economic, or environmental perspectives. Proper definition for the intended use is essential to understand published values and have meaningful discussion. For example, efficiency is frequently defined as the nutrient accumulated in the aboveground part of the plant. For N, this value frequently varies between 40 and 60%. Another common definition of efficiency is the nutrient recovered within the entire soil-crop-root system. For N, this value may be in the range of 65 to 85%, and even higher for P and K additions. It is a fallacy that the highest possible nutrient efficiencies should be the ultimate goal of fertilizer users. The highest “efficiency” occurs when small amounts of nutrients are applied on deficient soils (Mukherjee, 2012). While efficiency may be very high in this condition, crop growth in this region is generally stunted, profitability is low, water use efficiency is sub-optimal, and the potential for nitrate leaching is enhanced- compared with the situation where balanced and appropriate nutrition is provided.

Another example of inadequate understanding of "efficiency" is when an insufficient quantity of nutrients is regularly added to meet crop needs. In this condition, soil productivity will gradually decline as crop production continues to be increasingly reliant on nutrient stocks from soil reserves. Nominal nutrient efficiency may be very high under these circumstances, but it is clearly a non-sustainable scenario. Economic efficiency occurs when farm income is maximized as a result of nutrient inputs. This can be complex to predict when factors such as future yield responses, the cost of nutrient inputs, and crop prices may not be known in advance of the growing season. Environmental nutrient efficiency is important since nutrients not used by the crop are at potential risk for loss. All agro-ecosystems cause a disruption of native nutrient cycles – an inevitable consequence of all modern food production systems. The susceptibility of loss varies among the essential plant nutrients, and their loss mechanisms are each unique. The concept of plant nutrient efficiency is certainly not a new one- but it is still not adequately understood and practiced. It is time to move beyond the concept of managing single nutrients, but instead consider providing balanced crop nutrition for producing foods of high nutritional quality with sustainable economic and environmental yield levels.

Many factors influence crop response to the application of fertilizers. The response varies between different seasons or climate areas and is influenced by irrigation and by soil characteristics. The factors involved include:

- yield increase at different levels of the basic yield without fertilization;
- fertilizer-use management;
- changes in fertilizer response with time;
- the influence of price variations on the economics of fertilizer use;
- soil acidity;
- deficiency of other plant nutrients;
- cultivation techniques;
- sowing dates;
- different varieties;
- crop rotations.

4.1 Opportunities Abound for Improving Nutrient use Efficiency through Implementation of Fundamental best Management Practices are the following:

- Selection of genetics and management practices assuring maximum economic yields.

- Enhanced-efficiency fertilizer products using controlled release technologies and inhibitors of N transformation.
- Precision agriculture technologies to map fields and sense crop needs.
- Precision application equipment improving timing and placement of nutrient applications.
- Increased use of on-farm (field and sub-field specific) cropping system nutrient budgets.

A schematic metaphor is helpful in illustrating the potential role of enhanced efficiency fertilizers and at the same time places them in appropriate context (Fig. 3). The "nutrient management genome" pictured here started as a DNA double helix. One of the strands represents productivity while the other represents environmental impact. The base pairs are the fundamental components of nutrient best management practices, source, timing, placement, and rate. Two main points are illustrated: the strands (productivity and environmental impact) must be linked through the base pairs (a proper set of best management practice (BMP) components) before the genome (nutrient management system) can have the desired impact on the organism (agro-ecosystem); and the impact of any base pair (i.e., fertilizer source) is dependent on the other base pairs (i.e., timing, placement, rate) it occurs within the genome. The metaphor conveys the strong message that for research to be most meaningful it must measure impacts on both productivity and the environment and appropriately manage or measure the interacting influences of other management practices. More efficient fertilizers that better synchronize nutrient release and crop demand would be a great asset to crop producers as they strive to reduce nutrient losses while increasing productivity. However, they will only be effective if used in a system that recognizes the impact of enhanced efficiency fertilizers on rate, placement, and timing decisions.

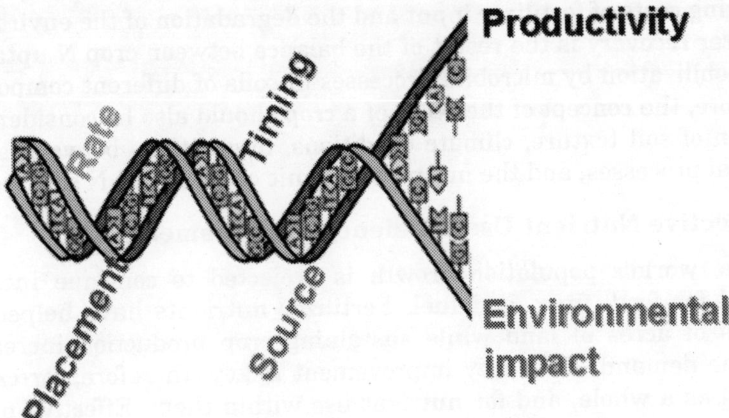

Fig. 3. The nutrient management genome. (Fixen, 2009)

4.2 Crop Nutrient Use Efficiency

Rice, wheat, maize, and, to a lesser extent, barley, coarse grains in legumes along with root crops are the most important crops cultivated in the world and account for the majority of end-products used or human diets (http://apps.fao.org/), and it is likely that they will still contribute to human nutrition in the next century. Moreover, the high yields of rice, wheat, and maize largely contributed to the total increase in the global supply of food production since 1967 (Cassman *et al.*, 2002). It is therefore of major importance to identify the critical steps controlling plant N use efficiency (NUE). This NUE can be divided into two processes: uptake efficiency (NupE; the ability of the plant to remove N from the soil as nitrate and ammonium ions) and the utilization efficiency (NutE; the ability to use N to produce grain yield). This challenge is particularly relevant to cereals for which large amounts of N fertilizers are required to attain maximum yield and for which NUE is estimated to be far less than 50%. In addition to the improvement of N fertilization, soil management, and irrigation practices here is still a significant margin to improve NUE in cereals by selecting new hybrids or cultivars from the available ancient and modern germplasm collections in both developed and developing countries. Consequently, the effective use of plant genetic resources will be required to meet the challenge posed by the world's expanding demand for food, the fight against hunger, and the protection of the environment. More recently, the production of oilseed rape (*Brassica napus* L.), an emerging oilseed crop, has been significantly increased, to become second only to soybean in the world supply. This increased interest in this crop is mostly due to the use of the oil in end-products, including biofuel. However, as for cereals, the ratio of plant N content to the N supplied does not exceed 50% whatever the level of N fertilization, which suggests that improvement of NUE in this species is also a possibility. Site specific (precision) technology in the future might help to develop sound management systems and lead to reduced fertilizer inputs, thereby improving costs of fertilizer input and the degradation of the environment. Fertilizer recovery is the result of the balance between crop N uptake and N immobilization by microbial processes in soils of different compositions. Therefore, the concept of the NUE of a crop should also be considered as a function of soil texture, climate conditions, interactions between soil and bacterial processes, and the nature of organic or inorganic N sources.

4.3 Effective Nutrient Use Efficiency Improvement

The world's population growth is projected to continue increasing demand for food, fiber, and fuel. Fertilizer nutrients have helped spare millions of acres of land while sustaining crop production increases to meet the demand. Efficiency improvement is key, therefore, for cropping systems as a whole, and for nutrient use within them. Effective nutrient use improves both, minimizing risks of losses that potentially harm air and water quality (Table 1).

Table 1. Plant and external factors that affect nutrient use efficiency in plants

Plant Factors	External Factors
Genetic control	Fertilizers Source
Species/cultivar/genotypes	• Ammonification, nitrification inhibitors • Time depth method of placement and application • Applying in combination • Reduce losses (NH3, NO3) • Use slow release form
Physiological • Roots: length, and density of main, lateral and root hair • Higher shoot yield, harvest index internal demand • Higher physiological efficiency • Higher nutrient uptake and utilization	Climatic • Adequate soil moisture • Extreme temperature
Biochemical enzymes: nitrate reductase (N), phosphatise (P), pyruvate kinase (K), arginine residue (N), phytic phosphate (P) rhodotorubic acid (Fe), Proline, aspharagine pinitol (salinity), Abscisic acid, proline (drought), Matallothionein (trace element), Root exudate (citric, malic, transaccionitic acid)	Elements toxicities: acidic soil (Al, Mn, pH), saline (Na, Mg, Cl, SO4) and alkaline (Na, Na2, CO3) soils Deficiencies (N, P, K micro) Others • Arbuscular mycorrhizae, beneficial soil microbes • Control of weeds, diseases and insects • Incorporate crop residue, cover crops, crop rotation

Source: Fageria (1992)

Efficiencies are calculated as ratios of inputs to outputs in a system. Different forms of nutrient use efficiency are related to crop production. Few of the important nutrient use efficiency patterns are as

a. ***Agronomic efficiency*** may be defined as the nutrients accumulated in the above-ground part of the plant or the nutrients recovered within the entire soil-crop-root system.

b. ***Economic efficiency*** occurs when farm income is maximized from proper use of nutrient inputs, but it is not easily predicted or always achieved because future yield increases, nutrient costs, and crop prices are not known in advance of the growing season.

c. ***Environmental efficiency*** is site-specific and can only be determined by studying local targets vulnerable to nutrient impact.

Nutrients not used by the crop are at risk of loss to the environment, but the susceptibility of loss varies with the nutrient, soil and climatic conditions, and landscape. In general, nutrient loss to the environment is only a concern when fertilizers or manures are applied at rates above agronomic need. Though perspectives vary, agronomic nutrient use efficiency is the basis for economic and environmental efficiency. As agronomic efficiency improves, economic and environmental efficiency will also benefit.

Agronomic indices commonly used to describe nutrient use efficiency: partial factor productivity (PFP, kg crop yield per kg nutrient applied); agronomic efficiency (AE, kg crop yield increase per kg nutrient applied); apparent recovery efficiency (RE, kg nutrient taken up per kg nutrient applied); and physiological efficiency (PE, kg yield increase per kg nutrient taken up). Crop removal efficiency (removal of nutrient in harvested crop as % of nutrient applied) is also commonly used to explain nutrient efficiency (Bruulsema, 2007).

Four of them are very commonly used, but are often misinterpreted:

(i) ***Partial factor productivity*** (crop yield per unit of nutrient applied) answers the question: "How productive is this cropping system in comparison to its nutrient input?"

(ii) ***Agronomic efficiency*** (yield increase per unit of nutrient applied) answers a more direct question: "How much productivity improvement was gained by the use of this nutrient?"

(iii) ***Partial nutrient budget*** (nutrient in harvested crop per unit of nutrient applied) answers the question: "How much nutrient is taken out of the system in relation to the amount put in?"

(iv) ***Recovery efficiency*** (increase in above-ground crop uptake per unit of nutrient applied) answers the question: "How much of the nutrient applied did the plant take up?"

In the short term, all four of these ratios increase as rates of fertilizer application are decreased, even to levels well below the economic optimum. This might cause one to falsely conclude that the lowest fertilizer rate results in the most efficient cropping system. This is untrue.

Cropping systems depend on multiple inputs, including land, labour, seed, plant protection, capital, and more. At the rate where the net return to the use of a plant nutrient peaks, it is making its best contribution to increasing the efficiency of all other inputs involved. This most economic rate is also often associated with minimal nutrient loss. Best management practices ensure effective use of fertilizers in improving the efficiency of all inputs used in cropping systems. The goal of their use is to apply the most appropriate sources at the right rate, time, and place. Opportunities abound for improving nutrient use efficiency effectively (Bruulsema, 2007):

- Genetics and management practices assuring maximum economic yields.
- Enhanced-efficiency fertilizer products using controlled-release technologies.
- Precision agriculture technologies to sense crop needs and improve application.
- Increased use of on-farm measures evaluating nutrient use efficiency.
- Decision support tools applying science at the farm level.

Nutrient use efficiency improvements must always be evaluated in terms of their effectiveness in supporting cropping systems with both increasing yields and diminishing nutrient losses.

Nutrient concentration and uptake by different plant genotypes are the most important criteria for identifying the existing genetic specificity of plant nutrition (Mukherjee and Sharma, 2012). The tolerance in a given plant species or genotype to nutrient stress is closely related to its nutrient use efficiency. Genotypic differences in nutrient use efficiency are linked with root nutrient acquisition capacity or with utilization by the plant, or both. Various plant species have developed morphological and/or physiological mechanisms to improve acquisition and use efficiency of mineral nutrients when grown on poor, infertile soils. By such mechanisms plants achieve nutrient deficiency stress tolerance. However, sometimes the ability to adapt to stress is limited, depending on the intensity and duration of stress. For example, some nutrient stress factors, when operating for a long period, can result in inhibited growth and root physiological processes responsible for nutrient uptake with the consequence of an impaired capability for adaptation and stress tolerance.

5. IMPROVING NUTRIENT USE EFFICIENCY

The evaluation of NUE is useful to differentiate plant species, genotypes and cultivars for their ability to absorb and utilize nutrients for maximum yields. The NUE is based on (a) uptake efficiency (acquire from soil, influx rate into roots, influx kinetics, radial transport in roots are based on root parameters per weight or length and uptake is also related to the amounts of the particular nutrient applied or present in soil), (b) incorporation efficiency (transports to shoot and leaves are based on shoot parameters) and (c) utilization efficiency (based on remobilization, whole plant i.e. root and shoot parameters).

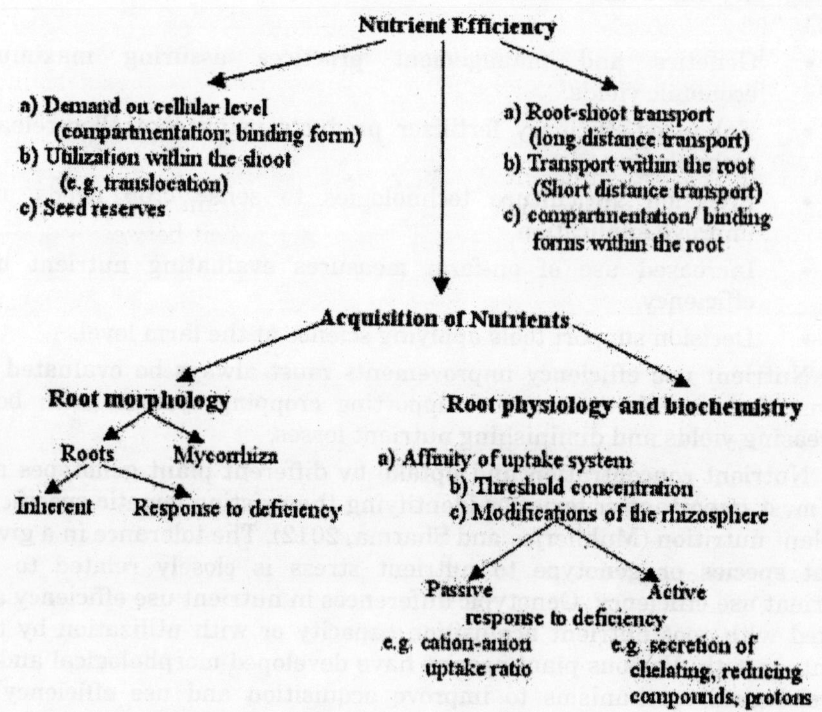

Fig. 4. Possible mechanisms in genotypic differences for nutrient efficiency

The fertilizer industry supports applying nutrients at the right rate, right time, and in the right place as a best management practice (BMP) for achieving optimum nutrient efficiency. However on the plant factor basis nutrient use efficiency increase by various ways. These are as follows:

5.1 Right Rate

Most crops are location and season specific — depending on cultivar, management practices, climate, etc., and so it is critical that realistic yield goals are established and that nutrients are applied to meet the target yield. Over- or under-application will result in reduced nutrient use efficiency or losses in yield and crop quality. Soil testing remains one of the most powerful tools available for determining the nutrient supplying capacity of the soil, but to be useful for making appropriate fertilizer recommendations good calibration data is also necessary. Unfortunately, soil testing is not available in all regions of the world because reliable laboratories using methodology appropriate to local soils and crops are inaccessible or calibration data relevant to current cropping systems and yields are lacking. Other techniques, such as omission plots, are proving useful in determining the amount of fertilizer required for attaining a yield target (Witt and Doberman, 2002). In this method, N, P, and K are applied at sufficiently high rates to ensure that yield is not limited by an insufficient supply of the added nutrients. Target yield can be determined from plots with unlimited NPK. One nutrient is omitted from the plots to determine a nutrient-limited yield. For example, an N omission plot receives no N, but sufficient P and K fertilizer to ensure that those nutrients are not limiting yield. The difference in grain yield between a fully fertilized plot and an N omission plot is the deficit between the crop demand for N and indigenous supply of N, which must be met by fertilizers. Nutrients removed in crops are also an important consideration. Unless nutrients removed in harvested grain and crop residues are replaced, soil fertility will be depleted.

5.2 Right Time

Greater synchrony between crop demands and nutrient supply is necessary to improve nutrient use efficiency, especially for N. Split applications of N during the growing season, rather than a single, large application prior to planting, are known to be effective in increasing N use efficiency (Cassman *et al.,* 2002). Tissue testing is a well known method used to assess N status of growing crops, but other diagnostic tools are also available. Chlorophyll meters have proven useful in fine-tuning in season N management (Francis and Piekielek, 1999) and leaf color charts have been highly successful in guiding split N applications in rice and now maize production in Asia (Witt *et al.*, 2005). Precision farming technologies have introduced, and now commercialized, on-the-go N sensors that can be coupled with variable rate fertilizer applicators to automatically correct crop N deficiencies on a site-specific basis. Another approach to synchronize release of N from fertilizers with crop need is the use of N stabilizers and controlled release fertilizers. Nitrogen stabilizers (e.g., nitrapyrin, DCD [dicyandiamide], NBPT [n-butylthiophosphoric triamide]) inhibit nitrification or urease activity, thereby slowing the conversion of the fertilizer to nitrate (Havlin *et al.,* 2005). When soil and environmental conditions are favorable for nitrate losses, treatment with a stabilizer will often increase fertilizer N efficiency. Controlled-release fertilizers can be grouped into compounds of low solubility and coated watersoluble fertilizers. Most slow-release fertilizers are more expensive than water-soluble N fertilizers and have traditionally been used for high-value horticulture crops and turf grass. However, technology improvements have reduced manufacturing costs where controlled-release fertilizers are available for use in corn, wheat, and other commodity grains (Blaylock *et al.*, 2005). The most promising for widespread agricultural use are polymer-coated products, which can be designed to release nutrients in a controlled manner. Nutrient release rates are controlled by manipulating the properties of the polymer coating and are generally predictable when average temperature and moisture conditions can be estimated.

5.3 Right Place

Application method has always been critical in ensuring fertilizer nutrients are used efficiently. Determining the right placement is as important as determining the right application rate. Numerous placements are available, but most generally involve surface or sub-surface applications before or after planting. Prior to planting, nutrients can be broadcast (i.e. applied uniformly on the soil surface and may or may not be incorporated), applied as a band on the surface, or applied as a subsurface band, usually 5 to 20 cm deep. Applied at planting, nutrients can be banded with the seed, below the seed, or below and to the side of the seed. After planting, application is usually restricted to N and placement can be as a top dress or a subsurface side dress. In general, nutrient recovery efficiency tends to be higher with banded applications because less contact with the soil lessens the opportunity for nutrient loss due to leaching or fixation reactions. Placement decisions depend on the crop and soil conditions, which interact to influence nutrient uptake and availability. Plant nutrients rarely work in isolation. Interactions among nutrients are important because a deficiency of one restricts the uptake and use of another. Numerous studies have demonstrated that interaction between N and other nutrients, primarily P and K, impact crop yields and N efficiency. For example, data from a large number of multi-location on-farm field experiments conducted in India show the importance of balanced fertilization in increasing crop yield and improving N efficiency. Adequate and balanced application of fertilizer nutrients is one of the most common practices for improving the efficiency of N fertilizer and is equally effective in both developing and developed countries.

For plant nutrition science to work well across disciplines, between public and private sectors, and across geographies, a common framework for viewing goals, practices, and performance is likely helpful. The seeds for such a framework were planted more than 20 years ago by Thorup and Stewart (1988) when they wrote: "This means using the right kind of fertilizer, in the right amount, in the right place, at the right time." Figure 5 is a schematic representation of the 4R nutrient stewardship framework based on the concepts described by Thorup and Stewart (Bruulsema *et al.*, 2008). At its core are the 4Rs – application of the right nutrient source at the right rate, right time, and right place. Best management practices are the in-field manifestation of these 4Rs. The 4Rs are shown within a cropping system circle because they integrate with agronomic BMPs selected to achieve crop management objectives. Those farm-level crop management objectives contribute toward the larger economic, social and environmental goals of sustainable development. Furthermore, the 4Rs cannot truly be realized if problems exist with other aspects of the cropping system.

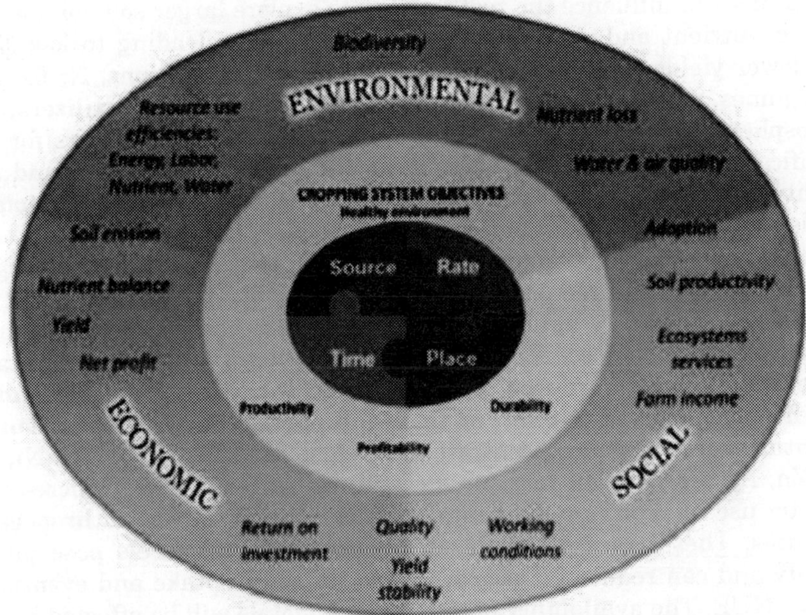

Fig. 5: The 4R nutrient stewardship framework (Bruulsema *et al.*, 2008)

5.4 Soil Factors

Production potential of many soils in the world are affected by the low supply of nutrients due to adverse soil physical and chemical constraints. In tropical regions the main soil problems in rain fed systems that affect crop production are low soil fertility, salinity, alkalinity, acidity, and Fe toxicity, and P and Zn deficiencies. Fertilizer efficiency in these soils is profoundly influenced by fertilizer placement and timing. Salinity, acidity, elemental deficiencies, and toxicities, and low organic matter content are some of the major chemical constraints. Physical constrains such as high bulk density layers or pans, poor structure and texture, surface sealing and crusting, high or low water holding capacity, water logging and extreme drying or poor aeration can also reduce NUE. Among other nutrient dynamics, these factors can affect mineralization and immobilization, fixation by adsorption and precipitation mechanisms, leaching, runoff, and gaseous losses via denitrification and ammonia volatilization.

Adverse soil physical properties affect the longitudinal and radial root growth, root distribution, morphology by stunting, thickening, reduction of second and third order lateral roots and root anatomical changes. High mechanical impedance leads to loss of root caps and reduction in radial thickening primarily due to shorter and wider cells with the same volume in the cortex and a thicker cortex. This may also cause changes in cell structure of the endodermis and pericycle. Such changes in the size and internal and external morphology of roots due to the adverse soil physical conditions will influence the root's ability to explore larger soil volume and reduce nutrient and water availability and uptake, leading to low NUE and lower yields. Leaching and crop removal of basic cations, N_2 fixation by legumes, use of heavy levels of organic and inorganic N fertilizers, and atmospheric deposition of N and sulfur oxides are major factors for soil acidification that leads to degradation and lower productivity and soil quality in temperate and tropical regions. Acidic soils have phytotoxic levels of Al, Mn, Fe and H and deficient levels of N, P, K, Ca, Mg, Mo and Zn to support good plant growth. Both of these factors are largely responsible for reduced growth and lower NUE (Mukherjee, 2012 a). Excess salt affects N uptake by plants and also contributes to reduced permeability of roots, consequently decreasing water and nutrient uptake. Increasing salinity reduced plant concentrations of Ca, Mg, and P significantly, but had no effect on the Na, K and Cl concentrations. During recent decades the soil concentrations of elements such as Cd, Cr, Ni, Pb, Cu, Zn, As, Co and Mn in some agricultural soils have been incre-asing due to use of soil amendments, pesticides and other anthropogenic activities. These trace elements, if present at excess levels pose phyto-toxicity and can reduce plant growth and nutrient uptake and eventually reduce NUE. The availability of these heavy metals will be affected by soil pH, temperature, redox potentials, anion ligand formation, and compo-sition and quantity of soil solution among other factors. Changes in the soil nutrient reserve and alteration in root systems under different tillage systems might have direct bearing on the nutrient availability and uptake by crops.

5.5 Plant Factor

Root morphology parameters such as length, thickness, surface areas, density, root hairs and root growth rate expressed as dry mass and/or root: shoot ratios are affected by deficiencies of essential minerals and/or excess of minerals. Clark (1970) reported that in solution culture studies with maize, reducing the supply of essential nutrients from full strength to none increased root: shoot ratio in P, Ca, S and Zn treatments; however, root: shoot ratios decreased in NO_3-N, Mg, Mn and Cu treatments. Plant genotypes respond differently in their uptake, translocation, accumu-lation, and use of mineral elements. Efficient plant cultivars could have better fertilizer use efficiency (Epstein and Bloom, 2005) and hence reduce input cost and conserve environment (Baligar *et al.*, 2001).

Selection of improved genotypes adaptable to a wide range of climatic changes has been a major contributor to the overall gain in crop productivity. Steady increase in the average yields of major crops during the second half of the 20th century has been achieved through genetic improvement coupled with improvement in best management practices. Modern genotypes of rice (*Oriza sativa* L), corn, wheat (*Triticum aestivum* L) and soybean (*Glycine max* L. Merrill) are more efficient in absorption and utilization of nutrients as compared to older cultivars (Witt *et al.*,

2005). Climatic stress such as water deficits and the low availability of seeds of improved cultivars are also affecting cropping systems and plant nutrient absorption capacity.

5.5.1 Soil and plant mechanisms and processes and other factors that influence genotypic differences in nutrient efficiency in plants grown under nutrient stress conditions:

A. Nutrient acquisition

i. Diffusion and mass flow (buffer capacity, ionic concentration, ionic properties, tortuosity, soil moisture, bulk density, temperature)
ii. Root morphological factors (number, length, root hair density, root extension, root density)
iii. Physiological [root:shoot, root microorganisms such as mycorrhizal fungi, nutrient status, water uptake, nutrient influx and efflux, rate of nutrient transport in roots and shoots, affinity to uptake (Km), threshold concentration Cmin]
iv. Biochemical (enzyme secretion as phosphatase, chelating compounds, phytosiderophore), proton exudate, organic acid production such as citric, transaconitic, malic acid exudates

B. Nutrient movement in root

i. Transfer across endodermis and transport within root
ii. Compartmentalization/binding within roots
iii. Rate of nutrient release to xylem

C. Nutrient accumulation and remobilization in shoot

i. Demand at cellular level and storage in vacuoles
ii. Retransport from older to younger leaves and from vegetative to reproductive parts
iii. Rate of chelates in xylem transport

D. Nutrient utilization and growth

i. Metabolism at reduced tissue concentration of nutrient
ii. Lower element concentration in supporting structure, particularly the stem
iii. Elemental substitution, e.g. Na for K function
iv. Biochemical (nitrate reductase for N-use efficiency, glutamate dehydrogenase for N metabolism, peroxidase for Fe efficiency, pyruvate kinase for K deficiency, metallothionein for metal toxicities)

E. Other factors

1. Soil factors

a. Soil solution (ionic equilibria, solubility precipitation, competing ions, organic ions, pH, phytotoxic ions)
b. Physico-chemical properties of soil (organic matter, pH, aeration, structure, texture, compaction, soil moisture)

2. Environmental effects

a. Intensity and quality of light (solar radiation)
b. Temperature
c. Moisture (rainfall, humidity, drought)
d. Plant diseases, insects, and allelopathy

6. ORGANIC MATTER AND OTHER SOIL AMENDMENT

The SOM helps to maintain good aggregation and increase water holding capacity and exchangeable K, Ca, and Mg. It also reduces P fixation, leaching of nutrients and decreases toxicities of Al and Mn. Best management practices such as addition of crop residues, green manure, compost, animal manure, use of cover crops, reduced tillage and avoiding burning of crop residues can significantly improve the level of SOM and contribute to the sustainability of the cropping systems and higher NUE. Liming is an effective way to correct soil chemical constraints. It improves the availability of Ca, Mg, Mo, P, soil structure, and CEC. The fixation of atmospheric dinitrogen (N_2) by free living and symbiosis organisms like *rhizobium* is increased. Lime has very low mobility in soil and when surface applied it does not reduce the acidity of sub-surface soil horizons. Contrary to lime, gypsum ($CaSO_4$) has a greater downward movement and when applied to the surface it can still impact and reduce the acidity of the subsoil. Downward movement of Ca in soil has resulted in increased rooting depth and in higher uptake rates of N, Ca, Cu, P and Mn by corn (*Zea mays* L) (Mukherjee, 2010). Reduction of subsoil acidity usually leads to deeper rooting and higher water and mineral uptake by plants. In such a case, micronutrients will have to also be monitored for these soils to ensure that higher yields and NUE are maintained (Yau and Ryan, 2008).

6.1 Fertilizer Factors

The fertilizer use efficiency is affected by several factors such as soil properties, efficiency of crops, climate, chemical species of the fertilizer used, mycorhiza, and others. The availability and recovery efficiencies of fertilizers, are greatly affected by amendments such as lime, organic materials and others, due to their effects in nutrient dynamics. Best management practices such as source, rate, method of application, and split application of nutrients should be optimized based on soil, plant, and climatic factors to reduce nutrient losses due to leaching, denitrification, ammonia volatilization, runoff, and fixation. Improvement and careful considerations to these several factors increases NUE of added fertilizers.

6.2 Agronomic Factor

Tillage practices such as conventional, conservation and no tillage are known to bring changes in SOM, nutrient concentrations, bulk density, water holding capacity and soil temperature among others. Higher contents of available P, Ca, K and organic C and N have been reported for no tillage than for conventional tillage. Minimum tillage increases root growth in the top 12 cm of soil for barley (*Hordeum vulgare* L) and oat (*Avena sativa* L) cropping systems (Evans, 2009). Minimum tillage has also been reported to increase root weight, length, and density, increasing the nutrient and water use efficiencies. Shoot dry matter yields and root length and densities of silage corn in no-till were significantly higher than in conventional tillage. Such improved root parameters contributed to higher yields and uptake efficiencies of N, P, Ca, S, Cu, Fe, and Zn (Mukheree and Roy, 2011). Improved tillage equipment and practices need to continue being developed to increase NUE across different agro-ecosystems. Slow and controlled release fertilizers have added advantages in increasing nutrient recovery by plants, lowering N_2O and NH_3 emissions and NO_3^- N leaching from cropping systems, while supplying a lasting nutrient source. Nitrification inhibitors that are widely used are N-serve [2-chloro-6 (trichloromethyl) pyridine], AM [2-amino-4chloro 6 methylpyrimidine], DCD (Dicyandiamide), and KN_3. Application of neem cake, PPD (phenly phosphorodiamidate), and NBPT [N-(nbutyl) thiophosphoric triamide] with urea has been suggested to reduce the rate of urea hydrolysis and improve its efficiency. Nitrificator inhibitors such as DCD have the potential to reduce N_2O emissions and increase NUE of irrigated systems such as barley.

Minimum tillage, no tillage, conservation tillage and traditional tillage can bring profound changes in soil quality, SOM and nutrients throughout different soil horizons (Mukherjee, 2012). Rooting pattern, water holding capacity, water penetration, aeration, soil compaction, and soil temperature are also influenced by type of tillage practices (Mukherjee, 2008). Crop rotation and use of cover crops and green manure crops are known to improve soil fertility and physical properties and to minimize pest and weed problems. Improved tillage practices and tillage equipment need to be developed to enhance NUE in crop plants.

6.2.1 Agronomic approaches for improving NUE

Crop production in most countries is highly dependent on the supply of exogenous fertilizers. Regardless of whether the source is chemical fertilizers, biofertilizers, animal dung, residue burning, or farm-waste manure, the adverse consequences of their accumulation in the environment are the same, barring the additional dependence of synthetic fertilizers on fossil fuels (Chien *et al.*, 2009). This necessitates consideration of the overall NUE in food grain production expressed as partial factor productivity of N (PFPN). PFPN is an aggregate efficiency index that includes contributions to crop yield derived from uptake of indigenous soil N, fertilizer N UpE, and the efficiency with which N acquired by the plant is converted to grain yield. Fertilizer NUE can be controlled by crop demand for N, supply of N from the soil, fertilizer and manure, and loss of N from the soil–plant system. NUE is interplay of several complex interdependent parameters that need close monitoring in order to arrive at a deterministic approach for its improvement. Crop responses also depend on weather and environmental conditions due to large and variable effects. Cropping practice-based or crop management-based methods to improve NUE, such as crop rotation with N-fixing crops and optimization of chemical, biofertilizers, manures, and other N sources, as well as selection of natural plant varieties with better NUE. Biological interventions are aimed at further improvement of the natural or genetically inherent NUE using plant breeding and molecular marker-assisted methods, and recently transgenic interventions for improving NUE.

6.2.2 Transgenic efforts to improve NUE

NUE is a multigenic trait spread across hundreds of genes that extends beyond primary nitrate assimilation and metabolism. Naturally, transgenic efforts have concentrated on diverse targets that include genes belonging to uptake, translocation, remobilization, and carbon metabolism. In addition, signaling targets and regulatory elements have recently emerged as prospective candidates for biotechnological interventions. The following transgenic attempts to improve NUE based on the processes targeted.

6.2.3. NUE – Transgenic approach

Biotechnological interventions to improve NUE have largely revolved around manipulation and over expression of many crucial candidate genes apart from using knockout mutations to assess its effects on biomass and plant N status and overall yield (Good and Muench, 2004). N assimilatory pathway is one of the most widely chosen targets for improvement of NUE. The primary nitrate assimilation pathway that involves nitrate transporters, nitrate reductase (NR), nitrite reductase (NiR), plastidic glutamine synthetase (GS2), and ferredoxin glutamate synthase (Fd-GOGAT) along with secondary N assimilation and remobilization that includes cytosolic glutamine synthetase (GS1), reduced nicotinamide adenine dinucleotide glutamate synthase (NADH-GOGAT), glutamate dehydrogenase (GDH), and various aminotransferases; offers myriad opportunities of intervention at the uptake, assimilation, translocation, and remobilization stages (Masclaux Daubresse *et al.*, 2010). Emerging evidences from these efforts reveal that attempts to improve NUE in plants need to target different organs for individual plants. For example, in cereals, the total grain biomass or grain N content would be most suitable indicators of NUE, rather than any other organ. Thus, areas where NUE can be targeted is to improve the distribution of N between canopy (leaves and stem) and roots, better photosynthetic rate/ unit leaf N, reduced leaf senescence, transgenic developing C4 options for rice and wheat, or otherwise increasing the efficiency of net photosynthesis in warmer environments by modifying Rubisco, Rubisco activase, and the enzymes that modulate photorespiration in C3 plants (Pathak *et al.*,2008).

6.2.4 Manipulating genes of N uptake

N uptake is the first step in nitrate assimilation, which can be manipulated to enhance NUE. Two types of nitrate uptake systems: low-affinity transport system encoded by NRT 1 gene family and high-affinity transport system encoded by NRT 2 gene family. In addition, a number of ammonium transporters and putative amino acid transporters have also been identified in Arabidopsis. In the last decade, several attempts of over expression of high-affinity nitrate/nitrite transporters in tobacco and Arabidopsis (Shrawat *et al.*, 2008) reported enhanced nitrate uptake, but concomitant increase in assimilation and NUE was not observed. Though the various nitrate and ammonium transporters in plants are very well characterized and their differential regulation mechanisms are well known, over expression studies involving their genes have not been conclusive.

7. BIOLOGICAL CONSIDERATION

Enhanced beneficial microbes such as rhizobia, diazotrophic bacteria, and mycorrhizae in the rhizosphere have improved root growth by fixing atmosphere N_2, suppressing pathogens, producing phytohormones, enhancing root surface area to facilitate uptake of less mobile nutrients such as P and micronutrients, and mobilization and solubilization of unavailable organic/inorganic nutrients. Fixation of N_2 by *Rhizobium* is very effective in humid and sub-humid regions and the reported N_2 fixed ranges from 24 to 267 kg/ha/year (Evans and Fischer, 1999). The amount of N_2 fixed varies, depending on crop species/cultivars, soil acidity, temperature, drainage, and the timing of harvest. Ladha *et al.* (1996) reported that free-living and/or associated phototrophs and heterotrophs in irrigated rice paddies can fix from 50 to 100 kg N /ha, contributing to the increased supply and efficiency of N. *Arbuscular mycorrhizal* (AM) fungi forms a beneficial symbiosis with roots, thereby increasing root surface area which assists roots in exploring larger soil volumes thereby bring more ions closer to roots and contributing to higher nutrient inflow (Zhang *et al.*, 2010). Primary benefits of AM are enhanced acquisition of mineral nutrients, plant tolerance to soil chemical constraints such as acidity, salinity, alkalinity, and increases the ability of the host plants to withstand or have reduced acquisition of elements toxic to plant growth. Other benefits for the host-plants are the improvement of water uptake and the withstanding of drought. The AM infection may act as general modifiers of NUE regardless of the extent to which the plant roots are infected and extent of infection appears to be under genetic control and shows considerable variability between crop species/cultivar and AM isolates. Such an interaction could be used to select crop cultivars and effective AM isolates to enhance nutrient uptake from nutrient deficit or low input systems. The P uptake in plant is most affected by AM interaction, but AM can also directly increase the uptake of Zn, Cu, N, and the cation/anion ratio. Weeds compete with crop plants for water, nutrients, and sunlight, thereby reducing crop yields and consequently NUE (Mukherjee, 2006). Allelopathic interactions of weed crop plants are quite common. Appropriate crop rotation is an effective way to alleviate weed and allelopathic problems. Infections of diseases and insects also reduce crop yields and consequently NUE. Soil borne pathogens such as actinomycetes, bacteria, fungi, nematodes, and viruses present in the soil around roots lead to pathogenic stress and bring profound changes in the morphology and physiology of roots and shoots that reduces plants ability to absorb and use nutrients effectively. Proper biological, chemical, physical, and cultural management practices can be used to alleviate the pathogenic stress. Diseases and insects that infect plant leaves, reduce photosynthesis activity resulting in lower utilization of absorbed nutrients. Plant diseases are greatly influenced by environmental factors, including deficiencies and /or toxicities of essential nutrients. Balanced nutrition has an important role in determining plant resistance or susceptibility to diseases. The severity of obligate and facultative parasites on plants is influenced by the level of nutrient available to plants.

8. ENHANCEMENT OF NITROGEN USE EFFICIENCY FOR CROP PRODUCTION

Plants need nitrogen to grow, develop and produce usable products. Since plants get their nitrogen from the soil, farmers must replenish nitrogen to ensure successful growth and to replace nitrogen removed in the harvested crop. Nitrogen is generally applied to the soil through various types of fertilizers. Due to the substantial yield increases resulting from fertilization, farmers have steadily increased the amount of fertilizer added per unit of land area. When plants are not provided with sufficient nitrogen, they become nitrogen deficient. Plants with nitrogen deficiency often have spindly stems and their growth is stunted. Nitrogen use efficiency (NUE) is the fraction of applied nitrogen that is absorbed and used by the plant. Improving a plant's ability to utilize nitrogen is a key component in enhancing environmental sustainability. Today, improved plant breeding through the use of genetic engineering has the greatest potential to produce plants that will utilize fertilizer more efficiently than conventional varieties.

8.1 Impact of Leaf N Distribution and Photosynthesis–N Relations on Crop Growth

The increase in leaf area of plants and canopies is brought about by a large effect of N supply on the expansion of individual leaves and on branching, or tillering in grasses. The rate of leaf appearance andthe duration of leaf expansion was little affected by N in many species. From the top to the bottom of a canopy, individual leaves require progressively less N to maximize their carbon assimilation, due to light attenuation within the canopy and due to the lower amount of N needed to maximize leaf photosynthesis under lower light intensities. Hence it has been suggested that leaf N gradients observed in canopies represent a way to maximize carbon assimilation by the canopy. In fact, it has been shown that the actual non-uniform N distribution between leaves in a canopy increases the daily canopy carbon assimilation by 20–40% (depending on species, daily radiation and model assumptions) compared with a theoretical uniform N distribution (Evans, 2009). The adaptation of leaf N content to light intensity during leaf growth, and subsequent N mobilization related to changes in light environment of the leaf during canopy development, may apparently interfere with other physiological aspects or constraints (leaf age; N mobilization capacity of mature or senescing leaves). Leaf N distribution within the canopy with its consequences for leaf and canopy photosynthesis, are major determinants of the radiation use efficiency (RUE) of a crop (Mosier *et al.*, 2004). Thus the relationships between N supply, leaf N distribution, and leaf and canopy photosynthesis, provide a fundamental basis to explain the almost systematic effect of crop N supply on RUE. Thus the effect of N on crop growth through its effect on leaf growth is generally as important as its effect on photosynthesis. However, the situation can arise where significant crop N shortage only occurs after canopy closure and thus after maximum light interception is achieved. In these situations, as illustrated by a study with maize where favourable soil N mineralization and rapid leaf area expansion occurred early in crop development despite a limited N fertilization rate, the effect of N on crop growth is predominantly through its effect on RUE.

Nitrogen (N) limitation reduces canopy carbon assimilation by directly reducing leaf photosynthesis, and by developmentally reducing the rate of new leaf area development and accelerating leaf senescence. Effective use of N for biomass production under N limitation may be considered to be a result of a trade-off between the use of N to maintain high levels of specific leaf nitrogen (SLN the amount of N per unit leaf area) for high photosynthetic rate versus using N to maintain leaf area development (leaf area index – LAI). Compare the effects of N supply on the dynamics of LAI and SLN for two crops, maize (*Zea mays* L.) and sunflower (*Helianthus annuus* L.) that contrast in the structure and development of their canopy. Three irrigated experiments imposed different levels of N and plant density. While LAI in both maize and sunflower was reduced under N limitation, leaf area development was more responsive to N supply in sunflower than maize. Observations near anthesis showed that sunflower tended to maintain SLN and adjust leaf area under reduced N supply, whereas maize tended to maintain leaf area and adjust SLN first, and, when this was not sufficient, SLN was also reduced. The two species responded differently to variation in N supply, and the implication of these different strategies for crop adaptation and management is discussed (Massignam *et al.*, 2011).

9. ENHANCEMENT OF PHOSPHORUS USE EFFICIENCY FOR CROP GROWTH

With increasing demand of agricultural production and as the peak in global production will occur in the next decades, phosphorus (P) is receiving more attention as a non renewable resource (Cordell *et al.*, 2009; Gilbert, 2009). One unique characteristic of P is its low availability due to slow diffusion and high fixation in soils. All of this means that P can be a major limiting factor for plant growth. Applications of chemical P fertilizers and animal manure to agricultural land have improved soil P fertility and crop production, but caused environmental damage in the past decades. Maintaining a proper P-supplying level at the root zone can maximize the efficiency of plant roots to mobilize and acquire P from the rhizosphere by an integration of root morphological and physiological adaptive strategies. Furthermore, P uptake and utilization by plants plays a vital role in the determination of final crop yield. A holistic understanding of P availability from soil to plant is necessary for optimizing P management and improving P-use efficiency. Further, maximizing exploitation of the biological potential of root/rhizosphere processes for efficient mobilization, and acquisition of soil P by plants as well as recycling P from manure and waste. Taken together, overall P dynamics in the soil plant system is a function of the integrative effects of P transformation, availability, and utilization caused by soil, rhizosphere, and plant processes (Tarkalson and Mikkelsen, 2004). The P nutrition of plants is predominantly controlled by P dynamics in the soil/rhizosphere-plant continuum. The distribution and dynamics of P in soil has a significant spatio-temporal variation. Root architecture that distributes more roots to the place where P resources are located plays an important role in efficiently exploiting these P resources. The integration of P dynamics from soil to plant via the rhizosphere provides a comprehensive picture of available P behaviour and efficient acquisition in association with plant adaptive strategies (Fig. 6).

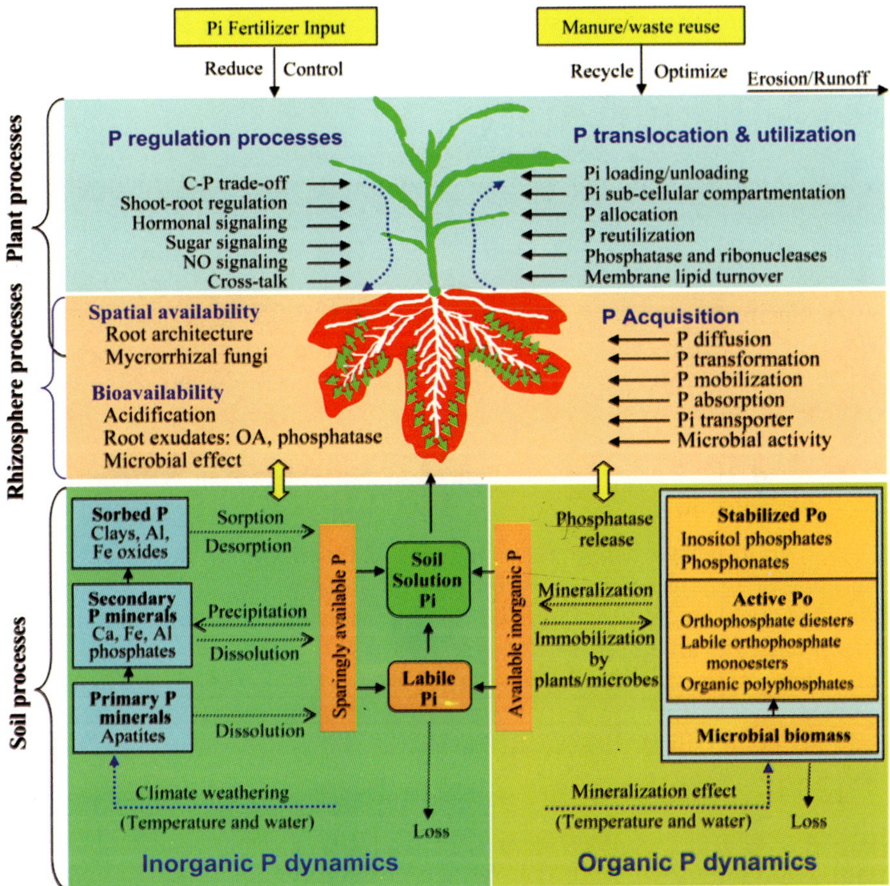

Fig. 6: P dynamics in the soil/rhizosphere- plant continuum. C-P, Carbon- P; NO, nitric oxide; OA, organic acids (Shen et al., 2011: Plant Physiol. Vol. 156, 2011)

9.1 Role of Increased Efficiency of Phosphorus and Potassium on Plant Growth

Phosphorus plays a critical role in photosynthesis, genetic coding, and energy transfer in all plants. Phosphorus, like potassium, is essential for nitrogen fixation in legumes. Top and root growth of virtually all crops is increased with balanced phosphorus and potassium nutrition. Corn research in Illinois and wheat research in Colorado, for example, have shown that response to nitrogen is reduced when phosphorus is deficient. Similar results have been reported for brome grass in Kansas research. With adequate phosphrous, response to nitrogen is improved through increased nitrogen and phosphorus uptake, hastened maturity, increased nitrogen-use efficiency, and increased yields. These effects work together for increased profit potential. Improved root growth with adequate phosphorus can also enhance the capture of soil and fertilizer nitrogen and can reduce the amount of residual nitrate-nitrogen, which may be prone to leaching or runoff losses (Synder, 2009). Adequate phosphorus increases crop yields and profit potential while minimizing the potential for environmental nitrogen loss.

Potassium plays a critical role in enzyme activation, water use, photosynthesis, transport of sugars, protein synthesis, and starch synthesis in plants. Potassium is especially important in its interaction with nitrogen throughout the growth cycle. It affects both nitrate absorption and reduction to amino acids and protein formation in plants. Adequate potassium in the field results in higher crop yields and higher nitrogen-use efficiency. Crops respond to higher potassium levels when nitrogen is sufficient, and greater yield response to nitrogen occurs when potassium is sufficient (Synder, 2009).

9.2 Role of Biofertilizers to Increase Nutrient Use Efficiency in Crop Production

Soil micro-organisms play a significant role in regulating the dynamics of organic matter decomposition and the availability of plant nutrients such as N, P and S. It is well-recognized that microbial inoculants constitute an important component of integrated nutrient management that leads to sustainable agriculture (Chen, 2006). In addition, microbial inoculants can be used as an economic input to increase crop productivity; fertilizer doses can be lowered and more nutrients can be harvested from the soil.

Biofertilizer is defined as a substance which contains living micro-organisms and is known to help with expansion of the root system and better seed germination. A healthy plant usually has a healthy rhizosphere which should be dominated by beneficial microbes. Conversely, in unhealthy soil, dominated by pathogenic microbes, optimum plant growth would not be possible. Biofertilizers differ from chemical and organic fertilizers in the sense that they do not directly supply any nutrients to crops and are cultures of special bacteria and fungi. The production technology for biofertilizers is relatively simple and installation cost is very low compared to chemical fertilizer plants.

9.2.1 The main micro-organisms used as biofertilizers and their functions to enhance NUE

i. Rhizobia

Rhizobia are symbiotic bacteria that fix atmospheric N_2 gas in plant root nodules and have a mutually helpful relationship with their host plants. The plant roots supply essential minerals and newly synthesized substances to the bacteria. Because of their N-fixing ability, legumes are less reliant on inorganic N fertilizer than many other non-legume crops such as cereals and pasture grasses. N fixation by legumes can also maintain soil fertility and can be of benefit to the following crop. *Rhizobium* inoculation is a well-known agronomic practice to ensure adequate N supply for legumes in place of N fertilizer. It is reported that *rhizobium* can fix 50-300 kg N/ha. Heavier application of inoculums mixed into peat granules trickled into soil as the seeds are planted is an alternative technique to encourage nodulation and enhance nutrient use efficiency of plant.

ii. Azotobacters and azospirillum

These are free-living bacteria that fix atmospheric nitrogen in cereal crops without any symbiosis and they do not need a specific host plant. Azotobacters are abundant in well-drained, neutral soil. They can fix 15 - 20 kg/ha N per year. *Azotobacter* sp. can also produce antifungal compounds to fight against many plant pathogens and help to enhance uptake of nutrient by plant. They also increase germination and vigour in young plants leading to improved crop stands.

iii. Phosphate-solubilizing bacteria (PSB)

Under acidic or calcareous soil conditions, large amounts of phosphorus are fixed in the soil but are unavailable to the plants. Phosphobacterins, mainly bacteria and fungi, can make insoluble phosphorus available to the plant. The solubilization effect of phosphobacterins is generally due to the production of organic acids that lower the soil pH and bring about the dissolution of bound forms of phosphate. It is reported that PSB culture increased yield up to 200-500 kg/ha and thus 30 to 50 kg of superphosphate can be saved.

iv. Vesicular arbuscular mycorrhiza (VAM)

Mycorrhizae are mutually beneficial (symbiotic) relationships between fungi and plant roots. VAM fungi infect and spread inside the root. They possess special structures known as vesicles and arbuscules. The plant roots transmit substances (some supplied by exudation) to the fungi, and the fungi aid in transmitting nutrients and water to the plant roots. The fungal hyphae may extend the root lengths 100-fold. The hyphae reach into additional and wetter soil areas and help plants absorb many nutrients, particularly the less available mineral nutrients such as phosphorus, zinc, molybdenum and copper. Some VAM fungi form a kind of sheath around the root, sometimes giving it a hairy, cottony appearance. Because they provide a protective cover, mycorrhizae increase seedling tolerance to drought, to high temperatures, to infection by disease fungi and even to extreme soil acidity. Application of VAM produces better root systems which combat root rotting and soil-borne pathogens. The greatest growth response to mycorrhizal fungi is probably in plants in highly weathered tropical acid soils that are low in basic cations and P, and may have toxic levels of aluminium. Plants that have coarse or limited root systems should benefit the most.

v. Plant growth promoting rhizobacteria (PGPR)

PGPR represent a wide variety of soil bacteria which, when grown in association with a host plant, result in stimulation of host growth. PGPR modes include fixing N2, increasing the availability of nutrients in the rhizosphere, positively influencing root growth and morphology and promoting other beneficial plant microbe symbioses (Adesemoye *et al.*, 2009). Some researche rs have indicated that PGPR will often have multiple modes of action. Ratti *et al.* (2001) found that a combination of the arbuscular mycorrhizal fungi *Glomus aggregatum*, the PGPR *Bacillus polymyxa* and *Azospirillum brasilense* maximized biomass and P content of the aromatic grass palmarosa (*Cymbopogon martinii*) when grown with an insoluble inorganic phosphate.

vi. Inoculation of biofertilizers

Biofertilizers are generally applied to soil, seeds or seedlings, with or without some carrier for the microorganisms, for example, peat, composts or stickers. Regardless of methods, the number of cells reaching the soil from commercial products is smaller than the existing numbers of soil or rhizosphere microorganisms; these added cells are unlikely to have a beneficial impact on the plant unless multiplication occurs. In addition, the population of introduced micro-organisms will decline and be eliminated in a very short time, often days or weeks. The formulation of inoculums, method of application and storage of the product are all critical to the success of a biological product. Short shelf life, lack of suitable carrier materials, susceptibility to high temperature, problems in transportation and storage are biofertilizer bottlenecks that still need to be solved in order to obtain effective inoculation.

vii. Seed inoculation

Seed inoculation uses a specific strain of microbe that can grow in association with plant roots; soil conditions have to be favourable for the inoculants to perform well. Selected strains of N-fixing Rhizobium bacteria have proven to be effective as seed inoculants for legumes. The seed treatment can be done with any of two or more bacteria without antagonistic effect. In the case of seed treatment with *Rhizobium*, *Azotobacter*, *Azospirillum* along with PSB, first the seeds must be coated with *Rhizobium* or *Azotobacter* or *Azospirillum*. When each seed has a layer of the aforesaid bacteria then the PSB inoculant has to be treated on the outer layer of the seeds. This method will provide maximum numbers of population of each bacterium to generate better results and ultimately help to more microbial activity, which enhance nutrient uptake pattern of plant.

viii. Soil inoculation

In soil inoculation, microbes are added directly to the soil where they have to compete with microbes already living in the soil that are already adapted to local conditions and greatly outnumber the inoculums. Inoculants of mixed cultures of beneficial micro-organisms have considerable potential for controlling the soil microbiological equilibrium and providing a more favourable environment for plant growth and protection. Therefore, adequate quality control and a high level of consistency in performance and benefits must be ensured. Although inoculations with PSBs have not been very effective, joint inoculation of PSBs with mycorrhizae and N_2-fixing bacteria have been successful.

9.2.2 Combined use of biofertilizers with chemical or organic fertilizers

The activity of soil organisms is very important for ensuring sufficient nutrient supply to the plant. If the micro-organisms find suitable conditions for their growth, they can be very efficient in dissolving nutrients and making them available to plants. Sundara *et al.* (2002) found that the application of PSB, *Bacillus megatherium* var. *phosphaticum*, increased the PSB population in the rhizosphere and P availability in the soil. It also enhanced sugarcane growth, yield and quality. When used in conjunction with P fertilizers, PSB reduced the required P dosage by 25%. In addition, 50% of costly superphosphate could be replaced by a cheap rock phosphate, when applied in combination with PSB. The effects of a combined treatment of multifunctional biofertilizer (mixture of Bacillus sp. *B. subtillis*, *B. erythropolis*, *B. pumilus* and *P. rubiacearum*) plus 50% chemical fertilizer (½ CF + biofertilizer) on a treatment of with chemical fertilizer (CF) and biofertilizer on the growth of lettuce were compared by Young *et al.* (2003). Results showed there was a 25% increase of lettuce yield for the treatment of ½ CF + biofertilizer compared to that of the CF treatment, indicating that at least 50% of chemical fertilizer can be saved as multifunctional biofertilizer was used along with chemical fertilizer. Young *et al.* (2004) evaluated the effects of multifunctional biofertilizer (mixture of *Bacillus* sp. *B. subtillis*, *B erythropolis*, *B. pumilus* and *P. rubiacearum*) on rhizosphere microbial activity and the growth of water celery in a field experiment. Results showed that the dry weight of water celery in the treatment with 50% organic compound fertilizer with multifunctional biofertilizer (MC + ½ OCF) was increased by 34% compared to the treatment with 100% organic compound fertilizer added (OCF). In addition, the beneficial bacterial counts including mineral PSB, cellulolytic bacteria and N_2-fixing bacteria in the rhizosphere of water celery in the MC + ½(OCF) treatment were increased 102~104 CFU/g.

REFERENCES

Adesemoye, A.O., Torbert, H.A. and Kloepper, J.W. 2009. Plant growth pomoting rhizobacteria: Allow reduced application rates of chemical fertilizers. *Microbial Ecology*, 58 : 921–929

Baligar, V.C. and Fageria. N.K. 1997. Nutrient use efficiency in acid soils: nutrient management and plant use efficiency pp. 75–93. In: A.C. Monitz, A.M.C. Furlani, N.K. Fageria., C.A. Rosolem, and H. Cantarells (eds.), Plant-Soil Interactions at Low pH: Sustainable Agriculture and Forestry Production. Brazilian Soil Science Society Companies, Brazil.

Baligar, V.C., Fageria, N.K and He, Z.L. 2001. Nutrient use efficiency in plants. Commonwealth Soil Science. *Plant Annals*, 32: 921-950.

Blaylock, A.D., J. Kaufmann and R.D. Dowbenko. 2005. Nitrogen fertilizer technologies. In: Proceedings of the Western Nutrient Management Conference, Salt Lake City, Utah, March 3-4, Vol. 6, pp. 8-13.

Brown, L.C. and J.W. Johnson. 1996. Nitrogen and the hydrologic cycle. Ohio State Univ. Ext. Fact Sheet AEX-463-96. Columbus, OH. Pp 66-78

Brown, L.C. and Johnson, J.W. 1996. Nitrogen and the hydrologic cycle. Ohio State Univ. Ext. Fact Sheet AEX-463-96. Columbus, OH.

Bruulsema, T. 2007. *Plant Nutrition TODAY* articles are available online at the IPNI website: www.ipni.net/pnt

Bruulsema, T.W., C. Witt, Fernando García, Shutian Li, T. Nagendra Rao, Fang Chen, and S. Ivanova. 2008. Better Crops 92(2):13-15.

Carangal, V.R and Dart, P.J. 1996. Legume productivity and soil nitrogen dynamics in lowland rice-based cropping systems. *Soil Science Society of America Journal*, 60: 183–192.

Cassman, K.G., A. Dobermann and D.T. Walters 2002. Agroecosystems, nitrogen use efficiency, and nitrogen management. *Ambions*, 31: 132-140

Chen, J.H. 2006. The combined use of chemical and organic fertilizer and or biofertilizer for crop growth and soil fertility. International Workshop on Sustained Management of the Soil-Rhizosphere System for Efficient Crop Production and Fertilizer Use 16 – 20 October 2006, Land Development Department, Bangkok.

Chien, S. H., Prochnow, L. I. and Cantarella, H. 2009: Recent developments of fertilizer production and use to improve nutrient efficiency and minimize environmental impacts. *Advances in Agronomy*, 102 : 267–306.

Clark, R. B. 1970. Mineral deficiencies of corn. *Agrichemical Age*. 13:4 –8.

Cordell, D., Drangert, J.O and White, S. 2009. The story of phosphorus: global food security and food for thought. *Global Environment Change* 19: 292–305

Duncan, R. R. 1994. Genetic manipulation. pp. 1–38. In:. R. E. Wilkinson. (ed.), Plant-Environment Interactions. Marcel Dekker Inc., New York, NY.

Epstein, E. and Bloom, A. 2005. Mineral Nutrition of Plants: Principles and Perspectives, 2, Sinauer Associates, Sunderland, MA.

Evans, L.T. and Fischer, R.A. 1999. Yield potential: its definition, measurement, and significance . *Crop Science* 39, 1544 – 1551 .

Evans, A. 2009. The Feeding of the Nine Billion: Global Food Security for the 21st Century (A Chatham House Report). Royal Institute of International Affairs, London.

Fageria, N.K. 1992. Maximizing Crop Yields. Marcel Dekker, New York, NY.

Fixen, P.E. 2009. Perspective on current and future agricultural and environmental need for enhanced efficiency fertilizers. Online. *Crop Management*, 12-16, doi:10.1094/CM-2009-0112-01-RV.

Francis, D.D. and W.P. Piekielek. 1999. Assessing Crop Nitrogen Needs with Chlorophyll Meters. Site-Specific Management Guidelines, Potash & Phosphate Institute. SSMG-12. Reference 99082/

Gerloff, G. C. 1987. Intact-plant screening for tolerance of nutrient deficiency stress. *Plant and Soil*. 99: 3-16.

Gilbert, N .2009. Environment: the disappearing nutrient. *Nature* 461: 716–718

Good A, Shrawat A, and Muench D. 2004. Can less yield more? Is reducing nutrient input into the environment compatible with maintaining crop production? *Trends in Plant Science*, 9: 597–605.

Havlin, J.L., J.D. Beaton, S.L. Tisdale and W.L. Nelson 2005. Soil Fertility and Fertilizers. An Introduction to Nutrient Management. Pearson Education, Inc., Upper Saddle River, New Jersey. Item#10-1012.

Jianbo, S., Lixing, Y., Junling, Z., Haigang, L, Zhaohai, B., Xinping, C., Weifeng, Z and Fusuo, Z.2011. Phosphorus Dynamics: From Soil to Plant. *Plant Physiology*, 156: 997–1005, www.plantphysiol.org _ 2011 American Society of Plant Biologists.

Khouri, N., Shideed, K and Kherallah, M. 2011. Food security: perspectives from the Arab World. Food Secur. 3 (Suppl 1), S1–S6.

Ladha, J. K., Kundu, D. K., Angelo- van Coppenolle, M. G., Peoples, M. B.,

Marschner, H. 1995. Mineral Nutrition of Higher Plants. Academic Press, San Diego, CA.

Marschner, H. 1997. Introduction, Definition, and Classification of Mineral Nutrients. In: *Mineral Nutrition of Higher Plants* 2nd Ed. Academic Press, London. pp. 3-5.

Masclaux-Daubresse, C, Daniel-Vedele, F, Dechorgnat, J. 2010. Nitrogen uptake, assimilation and remobilization in plants: Challenges for sustainable and productive agriculture. *Annals of Botany* 105: 1141–1157.

Massignam, A. M., Chapman, S. C., Hammer, G. L. and Fukai, S. 2011. Effects of nitrogen supply on canopy development of maize and sunflower. *Crop and Pasture Science*, 62(12) : 1045-1055 http://dx.doi.org/10.1071/CP11165

Meisinger, J.J., and G.W. Randall. 1991. Estimating nitrogen budgets for soil-crop systems. p. 85–124. *In* R.F. Follett et al. (ed.) Managing nitrogen for groundwater quality and farm profitability. ASA, CSSA, and SSSA, Madison, WI.

Mosier, A. R, Syers, J. K., and Freney, J. R. 2004. Nitrogen fertilizer: An essential component of increased food, feed, and fiber production. Pages 3-18 in: SCOPE 65: Agriculture and the Nitrogen Cycle: Assessing the Impacts of Fertilizer Use on Food Production and the Environment, Washington, DC.

Mukherje, D and Chakrabarty, S. 2010. Effect of hormonal treatment on seed germination of *Valeriana jatamansi*. *Journal of Medicinal and Aromatic Plant Sciences* 32 (2) : 145-147.

Mukherjee, D and Sen, A. 2005. Influence of rice husk and fertility levels on the nutrient content of rice (*Oryza sativa* L.). *Agricultural Science Digest* 25 (2) : 151-152

Mukherjee, D and Sharma, B.R. 2012. Studies on productivity, economics and disease incidence of Garden Pea (*Pisum sativum* L.) as influenced by sowing date and cultivars. *Journal of interacademicia* 16 (2) : 203-09.

Mukherjee, D. 2002. Food security coping with nutritional challenge ahead. *Agriculture Today* 8 : 30-32.

Mukherjee, D. 2006. Techniques of weed management in chickpea -A Review. *Agricultural Review* 28 (1) : 34-41.

Mukherjee, D. 2008. Effect of tillage practices and fertility levels on the performance of wheat (*Triticum aestivum*) under mid hill condition of West Bengal. *Indian Journal of Agricultural Sciences* 78 (12) : 1038-1041.

Mukherjee, D. 2010. Biodynamics of organic farming. (eds. Tiwari, V.N., Gupta, D.K., Maloo, S.R and L.L. Somani). Agrotech Publishing Academy, Udaipur, India. pp 31-47.

Mukherjee, D. 2010. Indigenous traditional knowledge in the context of conservation agriculture in Eastern Himalaya Range. *Asian Agri-History Foundation* (AAHF) 45 : 61-68.

Mukherjee, D. 2011. Organic farming for sustainable agricultural development. (eds. Trievedi, P.C). Aavishkar Publishers, Jaipur, India. pp 101-135.

Mukherjee, D. 2012a. Sustainable hill agriculture through Organic Farming. Himalayn Times Vol. 4 issue 5, May 2012 (1): 20-21.

Mukherjee, D. 2012. Resource conservation through indigenous farming system in hills of West Bengal. *Journal of Crop and Weed* 8 (1): 160-64.

Mukherjee, Dhiman and Roy, Amar 2011. "Organic practices in hill agriculture" (Book). pp 66-87

Mukherjee, Dhiman. 2010. Productivity, profitability and apparent nutrient balance under different crop sequence in mid hill condition. *Indian Journal of Agricultural Sciences* 80 (5) : 420-22.

Pathak RR, Ahmad A, Lochab S. and Raghuram N. 2008 Molecular physiology of plant N-use efficiency and biotechnological options for its enhancement. *Current Science,* 94(11): 1394–1403.

Ratti, N., Kumar, S., Verma, H.N. and Gautams, S.P. 2001. Improvement in bioavailability of tricalcium phosphate to *Cymbopogon martini* var. *motia* by rhizobacteria, AMF and azospirillum inoculation. *Microbiology Research,* 156: 145-149.

Richardson, A.E., Barea, J.M, McNeill, A.M. and Prigent-Combaret, C. 2009. Acquisition of phosphorus and nitrogen in the rhizosphere and plant growth promotion by microorganism. *Plant Soil,* 321: 305–339

Sheldrick, W.F., Syers, J.K. and Lingard, J. 2002. A conceptual model for conducting nutrient audits at national, regional, and global scales. *Nutrient Cycling Agroecosystems,* 62:61-72.

Shrawat, A.K., Carroll, R.T., DePauw, M. 2008. Genetic engineering of improved nitrogen use efficiency in rice by the tissue-specific expression of alanine aminotransferase. *Plant Biotechnology Journal* 6: 722–732.

Summer. 2007, No. 1 For more information, contact Dr., Northeast Director, IPNI, 18 Maplewood Drive, Guelph, Ontario N1G 1L8, Canada. Phone: (519) 821-5519. E-mail: Tom.Bruulsema@ipni.net.

Sundara, B., Natarajan, V. and Hari, K. 2002. Influence of phosphorus solubilizing bacteria on the change in soil available phosphorus and sugarcane and sugar yields. Field Crop Research, 77:43-49.

Synder, C.S. 2009. Phosphorus and potassium improve nitrogen use efficiency. Agronomic News Items, No. 6 (For more information, contact Dr. Cliff S. Snyder, Midsouth Director, PPI, P.O. Drawer 2440, Conway, AR 72033-2440. Phone: (501) 336-8110. E-mail: csnyder@ppi-far.org. Potash & Phosphate Institute).

Tarkalson, D.D., and Mikkelsen, R.L. 2004. Runoff phosphorus losses as related to phosphorus source, application method and application rate on a Piedmont soil. *Journal of Environment Quality*, 33:1424-1430.

Witt, C. and Dobermann, A. 2002. A site-specific nutrient management approach for irrigated, lowland rice in Asia. *Better Crops International*, 16: 20-24.

Yau, S.K. and Ryan, J. 2008: Boron toxicity tolerance in crops: a viable alternative to soil amelioration. *Crop Science*, 48 : 854–865.

Young, C.C., Lai, W.A., Shen, F.T., Huang, W.S. and Arun, A.B. 2004. Characterization of multifunctional biofertilizer from Taiwan and biosafety considerations. International Symposium on Future Development of Agricultural Biotechnology Park. The symposium series for celebrating the establishment of the Agricultural Biotechnology Park, Council of Agriculture, Executive Yuan, & the 80th Anniversary of National Pingtung University of Science and Technology. pp. 373-388.

Young, C.C., Lai, W.A., Shen, F.T., Hung, M.H., Hung, W.S. and Arun, A.B. 2003. Exploring the microbial potentially to augment soil fertility in Taiwan. In Proceedings of the 6th ESAFS International Conference: Soil Management Technology on Low Productivity and Degraded Soils, Taipei, Taiwan. pp. 25-27.

Zhang, F., Shen, J., Zhang, J., Zuo, Y., Li, L. and Chen, X. 2010. Rhizosphere processes and management for improving nutrient use efficiency and crop productivity: implications for China. *Advances in Agronomy* 107: 1–32.

6

IMPACT OF MINERAL TOXICITY ON CROP PRODUCTIVITY

A.K. Trivedi and Jyoti Sati

National Bureau of Plant Genetic Resources, Regional Station Bhowali – 263 132, Nainital, Uttarakhand (INDIA)

INTRODUCTION

Concentration of mineral elements in the soil and plant tissues significantly affects growth, development and productivity of crops. Plants require a minimum of 14 mineral elements, in addition to these carbon dioxide, oxygen and water are essential for proper growth and development of plants (Marschner, 1995; Mengel *et al.,* 2001). Mineral elements required in tissue concentrations of 1000 $\mu g.g^{-1}$ of dry matter or more are known as macronutrients and mineral elements required in tissue concentrations equal to or less than 100 $\mu g.g^{-1}$ of dry matter are known as micronutrients. Plants acquire these mineral elements from the soil solution. Although soil is reservoir of all these mineral elements but plants utilize the elements which are in available form. The concept of term 'available' refers to the existence of nutrients in the soil in a chemical form which may be absorbed by the plant roots or may be readily converted into such a condition.

Deficiency or toxicity of any one of these mineral elements reduces plant growth and crop yields. If phyto-availability of essential mineral elements is low these are externally supplied to the crops as fertilizers to achieve proper growth and development and more yield. Excess availability of mineral elements in the soil solution may reduce or inhibit plant growth as well as crop yields (Mengel *et al.,* 2001). High concentration of some mineral elements viz., aluminium (Al), boron (B), chlorine (Cl), iron (Fe), manganese (Mn) and sodium (Na) is a general problem of agricultural soils. Toxicities of these mineral elements depend on soil pH such as toxicity of Al and Mn occurs on acidic soils while toxicities of B and Na are a problem on sodic (Na-rich) soils. In saline soils toxicities of Na and Cl are global problem. In irrigated condition Na, B and Cl toxicities and imbalances of Ca, Mg and K also occur. In waterlogged or flooded soils, Mn and Fe toxicity is a problem for agricultural production. In serpentine soils nickel (Ni), cobalt (Co) and chromium (Cr) toxicities and in seleniferous soils selenium (Se) toxicity is common problem. Toxic concentrations of certain mineral elements viz., Cd, Cu, Hg, Pb and Zn due to anthropogenic activities are adversely affecting agricultural fields. Counter measures allowing crop production on such soils are usually expensive and success is either partial or temporary. Physiological mechanisms that allow crop plants to grow on soils containing toxic concentrations of mineral elements are based on their exclusion from the plant and/or tolerance through their sequestration as non-toxic compounds and/or in non-vital cellular compartments (Marschner, 1995). Plant breeders are therefore trying to develop crop genotypes tolerant to these soils.

Table 1. Available forms of mineral elements.

S. No.	Mineral element	Available form
1	Nitrogen	NO_3^-, NO_2^-, NH_4^+
2	Phosphorus	HPO_4^-, $H_2PO_4^{--}$
3	Potassium	K^+
4	Calcium	Ca^{++}
5	Magnesium	Mg^{++}
6	Sulphur	SO_3^{--}, SO_4^{--}
7	Iron	Fe^{++}, Fe^{+++}
8	Zinc	Zn^{++}
9	Manganese	Mn^{++}, Mn^{++++}
10	Copper	Cu^{++}
11	Boron	BO_3^{--}
12	Molybdenum	MoO_4^{--}
13	Chlorine	Cl^-

Due to vital role of mineral elements in plant processes their uptake by plant roots and subsequent distribution within the plant has been widely studied (Marschner, 1995; Mengel *et al.*, 2001; Karley and White, 2009; Miller *et al.*, 2009; Miwa *et al.*, 2009; White and Broadley, 2009). Mineral elements are acquired from the soil solution by plant roots. They traverse the root via apoplastic (extracellular) and/or symplastic (intracellular) pathways to the stele, where they are loaded into the xylem for transport to the shoot. Cytotoxic cations are transported either through the apoplast or through symplast in a chelated form. Some mineral elements such as Ca, Mo, Na, Cd and Al are retained in the roots of certain plant species (Marschner, 1995). The xylem delivers mineral elements to transpiring leaf tissues, where these are taken up from the apoplast by specific cell types. Recirculation of mineral elements within the plant, and their delivery to non-transpiring or xylem-deficient tissues, occurs via the phloem. Some mineral elements viz., K, Na, Mg, Cd, N, P, S, Se and Cl are readily mobile within the plant while Fe, Zn, Cu, Mo and iodine are less mobile, Mn and Ca are immobile in the phloem of most plant species (Marschner, 1995). The mobility of B in the phloem varies between species (Brown *et al.,* 2002). Mineral elements which have low phloem mobility accumulate in tissues with high transpiration rates, these are present at low concentrations in fruits, seeds and tubers (Karley and White, 2009; White and Broadley, 2009).

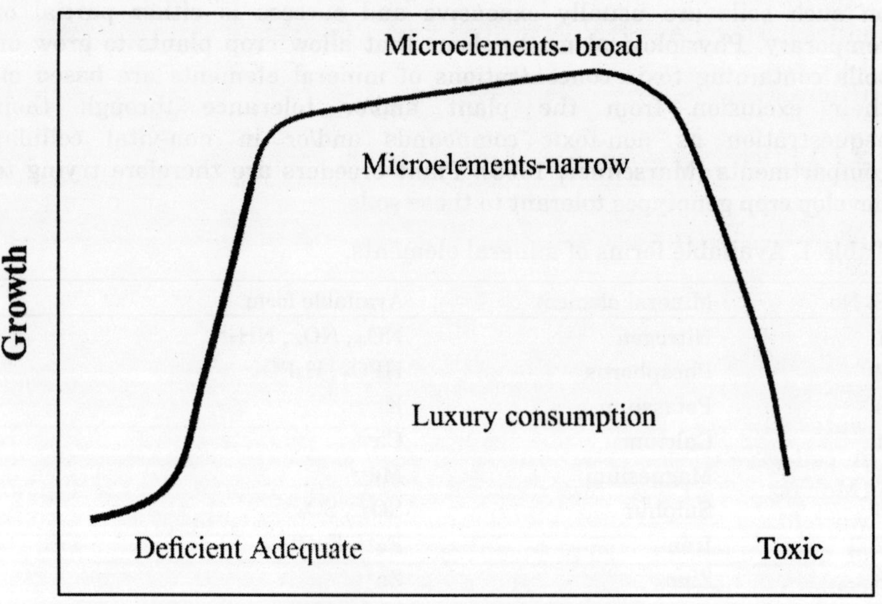

Fig. 1. Impact of mineral concentration on plants growth

2. ALUMINIUM TOXICITY

Although sometimes low concentrations of aluminium can increase plant growth or induce other desirable effects but it is not included in the list of essential nutrients (Foy and Fleming, 1982). In acidic soils below pH 5.0 aluminium toxicity is an important growth-limiting factor for plants. Al toxicity can be observed in soils having pH as high as 5.5 (Roy *et al.,* 1988). Aluminium interferes with cell division in root tips and lateral roots. By cross linking pectins Al increases cell wall rigidity. It reduces DNA replication by increasing the rigidity of the DNA double helix. It fixes phosphorous in less available forms in soils and on root surfaces, decreases root respiration, interferes with enzyme activity governing sugar phosphorylation and the deposition of cell wall polysaccharides, and the uptake, transport as well as use of several essential nutrients (Ca, Mg, K, P and Fe) (Foy, 1992). Al is present in all soils, but Al toxicity is visible only in acid conditions, in which the phytotoxic form Aluminium (Al^{3+}) predominates (Abdulla, *et, al.,*1985).

2.1 Effects of Aluminium (Al) on Root

Aluminium was found to help in new root development and seedling establishment (Nosko *et al.,* 1988). Inhibition of root growth is a characteristic symptom of Al toxicity in plants (Marschner, 1991 and Ryan 1993). The roots were found to be more affected at cellular level as compared to other parts of the plant (Wagatsuma *et al.,* 1987). Al toxicity affects root-tips and lateral roots; lateral roots become thickened and turn brown (Kinraide, 1988).

2.1.1 Effects on plant morphology and physiology

In general, young seedlings are more susceptible to Al than older plants (Thawornwong and van Diest,1974). Al has been found to affect, cell division in roots. It converts phosphorous in less available forms in the soil as well as in or on plant roots, decreases root respiration. It interferes with certain enzymes governing the deposition of polysaccharides in cell walls and increases cell wall rigidity (cross-linking pectins). Aluminium affects the uptake, transport of essential nutrients (Ca, Mg, K, P) and water supply to plants (Foy 1992., Nichol and Oliveira, 1995). Alteration in cell-wall Donnan free space (Jentschke *et al.*, 1991), plasma membrane (Meharg, 1993), membrane transport proteins (Suhayda and Haug 1986) were found to be obvious symptoms of Al toxicity. It regulates the activity of many enzymes (Wagatsuma *et al.,* 1987) and metabolic pathway for repair mechanism (Plucinska and Ziegler,1995).

2.2 Effect of Aluminium on Metabolism

Alteration of root membrane structure and function due to Al toxicity results in other toxic effects at biochemical level (Hechi-Buchholz and Foy 1981) including toxicity of Al on nitrogen metabolism. Aluminium was found to inhibit the influx of the cations of ammonium, calcium and potassium and enhances the influx of the anions of nitrate and phosphate. Aluminium interferes with the binding of the cations in the cell wall by the same order of magnitude as their respective influxes. Aluminium was found to be bound to the plasma membrane phospholipids, forming a positively charged layer that influences ion movement to the binding sites of the transport proteins. Al^{3+} induced inhibition of ion fluxes, particularly Ca^{2+} played an important role in mechanisms of Al^{3+} toxicity due to binding of cations or screening of the negative charges on the plasma membrane, thus reducing the activity of Al^{3+} close to the cell surface (Huang *et al.,* 1992).

2.3 Al Interaction with Other Mineral Elements

Al interferes with the uptake, transport and utilization efficiency of the other mineral elements. Calcium influx at the root apex was found to be strongly inhibited by Al^{3+} (Huang *et al.*, 1992). High concentration of Al in the nutrient solution was found to influence the uptake of minerals elements particularly divalent cations like Ca and Mg (Delhaize and Ryan, 1995). Aluminium affects P uptake which results in P deficiency in plants grown on acidic soils or in nutrient solutions. It has been noted that aluminium inhibits the activities of ATPase in plants (Woolhouse, 1969).

3. ARSENIC (As) TOXICITY

Aresenic is present in the environment in both inorganic and organic forms. Inorganic arsenic species are generally highly toxic to plants. In the soil solution arsenate [As (V)] and arsenite [As (III)] are the two phyto available forms of arsenic. Inorganic arsenic may be transformed into organic and vice versa by microbes which can methylate and demethylate arsenic species in soil (Turpeinen *et al.*, 1999). Arsenic in the environment is often associated with other elements such as gold (Au), silver (Ag), copper (Cu) and tin (Sn). World wide mining and processing of these ores cause arsenic pollution in the mining areas (Nriagu, 1994). The use of arsenic-based pesticides was found to cause contamination of agricultural as well as domestic land (Murphy and Aucott, 1998). Plants vary in their sensitivity or resistance to arsenic (Meharg, 1994). Plant species not resistant to arsenic suffer considerable stress upon exposure, with symptoms ranging from inhibition of root growth to death (Barrachina *et al.*, 1995). Arsenate acts as a phosphate analogue and is transported across the plasma membrane via phosphate cotransport systems. Inside the cytoplasm arsenate competes with phosphate such as replacing phosphate in ATP to form unstable ADP-As, and leads to the disruption of energy flows in cells (Meharg, 1994).

4. BORON (B) TOXICITY

Boron toxicity is a major problem in soils of arid and semi arid environments where it limits plant growth. High concentrations of B may occur naturally in the soil or in groundwater or be added to the soil from mining, fertilizers or irrigation water (Gupta *et al.* 1985; Leyshon and Jame, 1993). The typical visible symptoms of B toxicity on a wide variety of plant species is leaf burn – chlorotic and/or necrotic patches, often at the margins and tips of older leaves (Bennett, 1993; Bergmann, 1992). Lovatt and Bates (1984) found that in young squash plants (*Cucurbita pepo*) early effect of B toxicity development is decreased chlorophyll concentrations, followed by reduced growth, loss of leaf area and decreased CO_2 fixation. All of these effects occur well before the development of visible toxicity symptoms. Boron toxicity exerts different effects on diverse processes in vascular plants such as reduced root cell division, lower leaf chlorophyll contents, altered metabolism, decreased photosynthetic rate and less lignin and suberin content (Reid, 2007b). Reduced B accumulation follows a pattern from leaf base to tip in many plants. This leads to typical toxicity symptoms on older leaves which appear as marginal or tip chlorosis or both and necrosis (Roessner *et al.*, 2006). Although the physiological basis for B toxicity is not clear but taking into account B chemistry three main causes have been proposed, that is the ability of B to bind compounds with two hydroxyl groups in the cis-configuration: (1) alteration of cell wall structure, (2) metabolic disruption by binding to the ribose moieties of molecules such as ATP, NADH or NADPH and (3) disruption of cell division and development by binding to ribose either as the free sugar or within RNA (Reid *et al.*, 2004).

5. CADMIUM (Cd) TOXICITY

Industrial processes and farming practices are the main reason of Cd release in the agricultural fields. It has been ranked No. 7 among the top 20 toxins (Yang *et al.*, 2004). Cd is toxic for most of the plants even at low concentrations i.e., at concentrations greater than 5–10 μg $Cd.g^{-1}$ leaf dry weight (White and Brown, 2010). But Cd-hyperaccumulator plants can tolerate Cd concentrations of 100 μg Cd g^{-1} leaf dry weight (Verbruggen *et al.*, 2009). In spite of its high phytotoxicity, Cd is easily taken up by plant roots and transported to above-ground tissues (Lux *et al.*, 2010) and enters into the food chain where it may create serious threats to human health (Gill *et al.*, 2011). In 1993, The International Agency for Research on Cancer classified Cd as a human carcinogen (Gianazza *et al.*, 2007) and interestingly it has also been suggested that crops are the main source of Cd intake by humans (Satarug *et al.*, 2002). Being highly mobile in phloem, Cd can be accumulated in all plant parts which causes stunted growth, chlorosis and leaf epinasty. It alters chloroplast ultrastructure, inhibits photosynthesis, inactivates enzymes in CO_2 fixation, induces lipid peroxidation, inhibits pollen germination and pollen tube growth. It also disturbs the nitrogen (N) and sulfur (S) metabolism and antioxidant machinery (Khan *et al.*, 2007, Anjum *et al.*, 2008). Cd can also inhibit the activity of several groups of enzymes such as those of the Calvin cycle, carbohydrate metabolism and phosphorus metabolism (Shah and Dubey, 2006). The effect of Cd on nitrate and S assimilation has been studied in several plants showing an inhibition of the nitrate uptake rate and the activity of the enzymes involved in the N assimilation pathway (Hasan *et al.*, 2008). Alternatively, Cd-caused induction of enzymes of S assimilation pathway has been reported in many plants (Dominguez *et al.*, 2003). Therefore, the regulation of S assimilation may be necessary to ensure an adequate supply of S compounds required for heavy metal tolerance. However, Cd is a non-redox active metal, but it induces the generation of reactive oxygen species (ROS) including superoxide radical ($O_2^{\bullet-}$), hydrogen peroxide (H_2O_2) and hydroxyl radical ($OH^{\bullet}$) (Shah *et al.*, 2001), which has to be kept under tight control because the presence of Cd leads to excessive production of ROS causing cell death due to oxidative stress such as membrane lipid peroxidation, protein oxidation, enzyme inhibition and damage to nucleic acid (Mishra *et al.*, 2008).

6. CHROMIUM (Cr) TOXICITY

Chromium and its compounds are highly toxic to plants and are detrimental to their growth and development. Although some crops are not affected by low Cr concentration (3.8104 AM) (Huffman and Allaway, 1973a, b), but it is toxic to most of the higher plants (Davies *et al.*, 2002). Seed germination is the first physiological process affected by Cr, the ability of a seed to germinate in a medium containing Cr would be indicative of its level of tolerance to this metal (Peralta *et al.*, 2001). Seed germination of the weed *Echinochloa colona* was found to be reduced due to Cr exposure (Rout *et al.*, 2000). High levels (500 ppm) of hexavalent Cr in soil reduced germination of bush bean *Phaseolus vulgaris* up to 48% (Parr and Taylor, 1982). Zeid (2001) has suggested that the reduced germination of seeds under Cr stress could be a depressive effect of Cr on the activity of amylases and on the subsequent transport of sugars to the embryo axes. Protease activity, on the other hand, increases with the Cr treatment, which could also contribute to the reduction in germination of Cr-treated seeds (Zeid, 2001). General response of decreased root growth due to Cr toxicity could be due to inhibition of root cell division/ root elongation or to the extension of cell cycle in the roots. Under high concentrations of both the Cr species combination, the reduction in root growth could be due to the direct contact of seedlings roots with Cr in the medium causing a collapse and subsequent inability of the roots to absorb water from the medium (Barcelo *et al.*, 1986). Adverse effects of Cr on plant height and shoot growth have been reported (Rout *et al.*, 1997). Leaf growth, leaf area development and total leaf number decisively determines the yield of crops. Sharma and Sharma (1993) found 50% reduction in the leaf number per plant in wheat when 0.5 mM Cr was added in nutrient solution. Tripathi *et al.* (1999) found leaf area and biomass of *Albizia lebbek* seedlings was severely affected by a high concentration (200 ppm) of Cr (VI). It was found that dry matter production was severely affected by Cr(VI) concentrations above 2.5 $\mu g/ml^{-1}$ in nutrient medium (Vajpayee *et al.*, 2001).

6.1 Effects on Physiological Processes

Photosynthesis in terms of CO_2 fixation, electron transport, photophosphorylation and enzyme activities were found to be adversely affected by Cr stress (Clijsters and Van Assche, 1985). Due to structural similarity of chromium with some essential elements, it affects mineral nutrition of plants in a complex way. Interactions of Cr with uptake and accumulation of other inorganic nutrients have received considerable attention by researchers. Cr (III) and Cr (VI) are taken up by the plants by different mechanisms (Zaccheo *et al.*, 1985). It has been suggested that both the species can interfere with uptake of several other ionically similar elements like Fe and S (Skeffington *et al.*, 1976).

Nitrate reductase (NR) activity of leaves was significantly increased over control values and negatively correlated with root and shoot length, leaf area and biomass of the plants, indicating stress due to Cr (VI) in *Albizia lebbek* (Tripathi *et al.*, 1999). ATPase plays a significant role in the adaptation to heavy-metal conditions and it is regulated at the molecular and biochemical level (Dietz *et al.*, 2001). A toxic effect of Cr on the transport activities of plant cell plasma membrane was suggested by (Zaccheo *et al.*,1982). Induction and activation of superoxide dismutase (SOD) and of antioxidant catalase are some of the major metal detoxification mechanisms in plants (Shanker *et al.*, 2003a).

7. COPPER (Cu) TOXICITY

Copper is one of the essential micronutrients for plant growth. It is involved in numerous physiological functions as a component of several enzymes, mainly those which participate in electron flow, catalyze redox reactions in mitochondria and chloroplasts (Hansch and Mendel, 2009). Nonetheless, in excessive quantities copper becomes toxic as it interferes with photosynthetic and respiratory processes, protein synthesis and development of plant organelles (Upadhyay and Panda, 2009). Specifically excess copper can cause chlorosis, inhibition of root growth and damage to plasma membrane permeability, leading to ion leakage (Bouazizi *et al.*, 2010). Copper toxicity induces deficiency of other essential mineral elements which may be deleterious for growth and development of plant (Bouazizi *et al.*, 2010; Lequeux *et al.*, 2010).

Higher concentrations of copper show a decrease in the root and shoot length and leaf area (Xiong *et al.*, 2006). Moreover, in high concentrations of copper, the root and shoot elongation was found to be poor with a concomitant decrease in root and shoot length (Bouazizi *et al.*, 2008, 2010). This view was supported by previous finding, indicated that application of copper slightly increased the dry weight at lower concentration, while excess of copper reduced the biomass (Xiong *et al.*, 2006). At higher external Cu concentration growth was depressed, chlorophyll contents were found to be lower, photosynthetic capacity reduced, including an inhibitory effect on photosytem II (Baron *et al.*, 1995). It has been observed that the toxic action of Cu is substitution of Cu for Mg in chlorophyll molecules, thus reducing photosynthesis (Kupper *et al.*, 1996). A side effect of Cu inhibiting photosynthesis is an increase in the production of free radicals and therefore, an increase in the rate of leaf senescence due to oxidative damage (Luna *et al.*, 1994). However, in some species an increase in photosynthetic pigments occurs because of Cu induced reduction in CO_2 fixation and as such photosynthesis does not decreases, at least initially (Romeu-Moreno and Mas, 1999).

8. LEAD (Pb) TOXICITY

Lead (Pb) is one of the major heavy metals and has gained considerable importance as a potent environmental pollutant. Natural weathering processes, mining and smelting activities, Pb containing paints, gasoline and explosives as well as disposal of municipal sewage sludges enriched in Pb are the main source of Pb contamination of the environment (Chaney and Ryan, 1994). The visual non-specific symptoms of Pb toxicity are rapid inhibition of root growth, stunted growth of the plant and chlorosis (Burton *et al.*, 1984). Once entered inside the cells even in small amounts Pb produces a wide range of adverse effects on physiological and biochemical processes. Pb phytotoxicity leads to inhibition of enzyme activities, disruption of mineral nutrition, water imbalance, change in hormonal status and alteration in membrane permeability. Such phytotoxic effects upset normal plant processes at different levels. At the cellular level Pb inhibits the activities of enzymes containing sulphydryl (-SH) groups necessary for their activity (van Assche and Clijsters, 1990). Pb interacts with free –SH groups that are present in the active site of the enzyme and essential for enzyme activity as well as with –SH groups that are necessary for the stabilization of enzyme tertiary structure. Besides the reaction with –SH groups, blockage of –COOH groups with Pb ions also appears to play a key role in inhibition of enzyme activity under Pb treatment. Pb forms a mercaptide with the –SH group of cysteine and also forms complexes with phosphate groups. Inhibition of metalloenzymes under Pb treatment appears to be due to displacement of the essential metal by Pb. The key enzyme of chlorophyll biosynthesis, δ-amino laevulinate dehydrogenase was found to be strongly inhibited by Pb ions (Prassad and Prassad, 1987). Pb also inhibits the activities of enzymes of the reductive pentose phosphate pathway (Hampp *et al.*, 1973). Vallee and Ulmer (1972) has reported inhibition of ribulose-bis-phosphate carboxylase/ oxygenase activity in leaf homogenates of spinach even at a Pb nitrate concentration of 5 μM. A decrease in the activity of lactate dehydrogenase was also found but magnitude of decrease was less pronounced. Conversely, the activity of pyruvate kinase was enhanced by Pb nitrate (Vallee and Ulmer, 1972). Pb acetate has been shown to considerably decrease the activities of protease and amylase in rice endosperms after 4 days of germination, whereas RNase and DNase activities were found to be not much affected (Mukherji and Maitra, 1976). In the leaves of Pb-treated soybean plants a decrease in the activity of glutamate dehydrogenase and a low content of nitrate has been observed whereas no significant change was found in malate dehydrogenase activity (Lee *et al.*, 1976). Pb was found to be highly effective in inhibiting ATP synthetase/ ATPase (TuShu and Brouillette, 1987). At high concentrations Pb eventually may cause cell death (Seregin and Ivanov, 2001). Pb decreases germination percent, germination index, retards growth of seedlings, root/ shoot length, tolerance index and dry mass of roots and shoots (Mishra and Choudhari, 1998). High concentrations of Pb (1 mM) caused 14 to 30% decreased germination in rice seeds and reduced the growth of seedlings by 13 to 45% (Verma and Dubey, 2003).

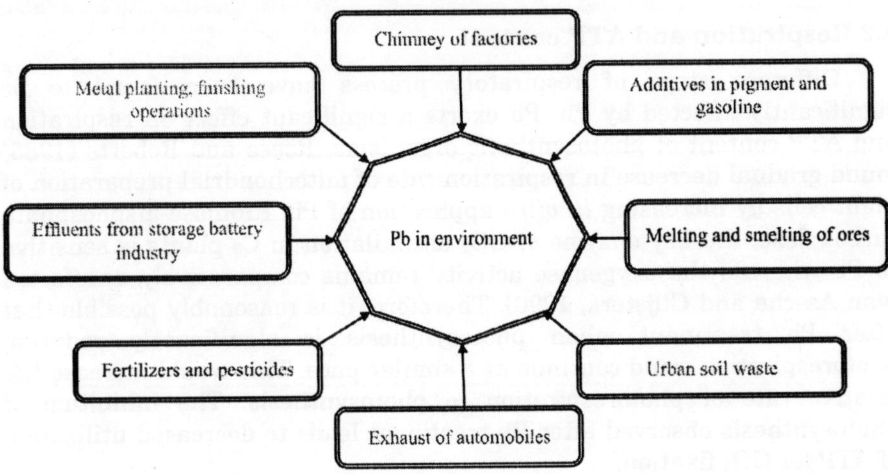

Fig. 2. Sources of lead

8.1 Photosynthesis

Different steps of photosynthetic process were found to be adversely affected by Pb toxicity which results decrease in yield. Plants exposed to Pb ions show a decline in photosynthetic rate which results from distorted chloroplast ultrastructure, restrained synthesis of chlorophyll, plastoquinone and carotenoids, obstructed electron transport, inhibited activities of Calvin cycle enzymes, as well as deficiency of CO_2 as a result of stomatal closure. Rebechini and Hanzely (1974) reported distinct changes in chloroplast fine structure of *Ceratophyllum demersum* plants when grown in aquatic medium containing $Pb(NO_3)_2$. In the leaf cells of such plants reduction in grana stacks together with a reduction in the amount of stroma in relation to the lamellar system as well as absence of starch grains was found. In plants exposed to Pb changes in the lipid composition of thylakoid membranes was also found (Stefanov *et al.*, 1995). Exposure of plants to Pb causes changes in various components of photosynthesis, irregularities in mitotic process, respiration, water regime and nutrient uptake. Pb inhibits chlorophyll synthesis by causing impaired uptake of essential elements such as Mg and Fe by plants (Burzynski, 1987). It damages the photosynthetic apparatus due to its affinity for protein N- and S- ligands (Ahmed and Tajmir-Riahi, 1993). An enhancement of chlorophyll degradation occurs in Pb-treated plants due to increased chlorophyllase activity (Drazkiewicz, 1994). Chlorophyll 'b' was found to be more affected than chlorophyll 'a' by Pb treatment (Vodnik *et al.*, 1999). An inhibition effect on electron transport has been reported (Rashid *et al.*, 1994). Pb effects have been reported for both donor and acceptor sites of PS II, the cytochrome b/ f complex and PS I. It has been found that PS I electron transport is less sensitive to inhibition by Pb than PS II (Sersen *et al.*, 1998).

8.2 Respiration and ATP content

Different steps of respiratory process have been found to be significantly affected by Pb. Pb exerts a significant effect on respiration and ATP content of photosynthetic organisms. Reese and Roberts (1985) found gradual decrease in respiration rate of mitochondrial preparation of plant cells by increasing *in vitro* application of Pb. Ribulose-bisphosphate carboxylase, the key enzyme of CO_2 assimilation in C_3 plants is sensitive to Pb whereas the oxygenase activity remains comparatively unaffected (van Assche and Clijsters, 1990). Therefore, it is reasonably possible that after Pb treatment when photosynthesis is significantly reduced, photorespiration could continue at a similar pace. This would increase the relative rate of photorespiration to photosynthesis. The inhibition of photosynthesis observed after Pb treatment leads to decreased utilization of ATP for CO_2 fixation.

8.3 Nutrient uptake

High concentration of Pb in the agricultural fields causes imbalance of mineral nutrients in the growing plants. Many of the observed effects of Pb appear to be mediated through imbalance of mineral elements within the tissues. Under Pb toxicity significant changes in nutrient contents as well as in internal ratios of nutrients occur in plants (Kabata-Pendias and Pendias, 1992). Pb blocks the entry of cations (K^+, Ca , Mg, Mn, Zn, Cu, Fe^{3+}) and anions (NO_3^-) in the root system and physically blocks the access of many ions from absorption sites of the roots (Godbold and Kettner, 1991). It induces production of ROS within plants, production of ROS depends on the intensity of the stress, repeated stress periods, species and plant age (Verma and Dubey, 2003). Lipid peroxidation, which is most easily ascribed symptom of oxidative damage, involves oxidative degradation of polyunsaturated fatty acyl residues of membranes (Girotti, 1990). Pb ions induce lipid peroxidation, decrease the level of saturated fatty acids and increase the content of unsaturated fatty acids of membrane in several plant species (Halliwell and Gutteridge, 1999). Although, ROS generating processes are slow under normal conditions, Pb exposure accelerates them (Verma and Dubey, 2003).

9. MANGANESE (Mn) TOXICITY

Mn toxicity causes necrotic brown spots on leaves, petioles and stems (Wu, 1994). This spotting starts on the lower leaves and gradually progresses with time towards the upper leaves (Elamin and Wilcox, 1986a; Elamin and Wilcox, 1986b; Horiguchi, 1988). Speckles increase with time both in number and size resulting in necrotic lesions, leaf browning and death (Elamin and Wilcox, 1986a; Elamin and Wilcox, 1986b). Crawford *et al.* (1989) has reported general leaf bronzing and shortening of internodes in *Cucumis sativus* (cucumber) (Crawford *et al.*, 1989). Wu (1994) reported another common symptom i.e., "crinkle-leaf" which occurs in the youngest leaf, stem and petiole tissue and is also associated with the chlorosis and browning of these tissues. Roots exhibiting Mn toxicity is usually brown in colour (Foy *et al.*, 1995) and sometimes cracks (Foy *et al.*, 1995). Macfie and Taylor (1992) and Macfie *et al.* (1994) have reported a drop in the photosynthetic rate of plants in the excess Mn condition. This lowering of photosynthetic rate occurred due to decrease in the chlorophyll and the photosynthesis per unit chlorophyll in a sensitive cultivar and only decrease in the chlorophyll content in a tolerant cultivar of *T. aestivum* (Macfie and Taylor, 1992). In *G. max* manganese toxicity has been found to be associated with swollen chloroplasts (Wu, 1994).

10. SELENIUM (Se) TOXICITY

In 1930s it was realized that selenium toxicity is responsible for certain livestock disorders i.e., alkali disease and blind staggers in animals grazing on soils with high selenium content such as seleniferous soils (Brown and Shrift, 1982). Plants exposed to high concentration of Se in their root medium exhibit symptoms of injury including stunting of growth, chlorosis, drying of leaves, decreased protein synthesis and premature death of the plant. In non accumulator plants the threshold toxicity concentration dependents on the form of Se accumulated. Selenium is taken up as selenates (SeO_4^{2-}) or selenites (SeO_3^{2-}) which are readily absorbed by the plants and assimilate to organic Se compounds. At the plasma membrane of root cells selenate and sulfate (SO_4^{2-}) compete for same binding sites. The uptake, transport and assimilation of selenate follow same pathway as sulfate. This leads to the formation of selenium analogues of cystein and methionine namely selenocysteine and selenomethionine. In accumulator plants, the seleno amino acids are transformed into nonprotein amino acids while in non-accumulator plants, seleno aminoacids are incorporated into proteins which are either nonfunctioning or at least much less capable of functioning as enzyme protein than the corresponding sulfur-containing proteins (Brown and Shrift, 1982). Some studies indicate that selenite is more toxic than selenate (Rosenfeld and Beath 1964). This may be due to the faster conversion of selenite to selenoaminoacids (Zayed, *et al.*, 1998), which may then be incorporated into plant proteins in replacement of S and cause toxicity to the plant. Wu *et al.,*(1988) observed that selenate caused a greater growth inhibition than selenite with respect to the growth of tall fescue.

11. ZINC (Zn) TOXICITY

Zinc toxicity in crops is far less widespread than Zn deficiency. However, Zn toxicity occurs in soils contaminated by mining and smelting activities, agricultural soils treated with sewage sludge and urban and peri-urban soil enriched by anthropogenic inputs of Zn especially in low-pH soils (Chaney, 1993). Zn toxicity symptoms include reduced yields and stunted growth, Fe-deficiency-induced chlorosis through reductions in chlorophyll synthesis and chloroplast degradation and interference with P (Mg and Mn) uptake (Foy *et al.*, 1978; Chaney, 1993). Initial symptom of Zn toxicity in most species is general chlorosis of the younger leaves (Harmens *et al.*, 1993; Ren *et al.*, 1993; Fontes and Cox, 1995). Depending on the degree of toxicity this chlorosis progresses to reddening due to anthocyanin production in younger leaves (Fontes and Cox, 1995; Lee *et al.*, 1996a). Leaves of plants exhibiting Zn toxicity were found to be smaller than control plants (Ren *et al.*, 1993). Necrotic lesions on leaves may be observed in severe cases and eventually entire leaf dies (Harmens *et al.*, 1993). In roots, Zn toxicity is apparent as a reduction in the growth of the main root, fewer and shorter lateral roots and yellowing of roots (Ren *et. al.*, 1993).

CONCLUSION

High concentration of some mineral elements viz., aluminium (Al), boron (B), chlorine (Cl), iron (Fe), manganese (Mn) and sodium (Na) is a general problem of agricultural soils. Toxicities of these mineral elements depend on soil pH such as toxicity of Al and Mn occurs on acidic soils. Toxic concentrations of certain mineral elements viz., Cd, Cu, Hg, Pb and Zn due to anthropogenic activities are adversely affecting agricultural fields. Mineral elements which have low phloem mobility accumulate in tissues with high transpiration rates, these are present at low concentrations in fruits, seeds and tubers Physiological mechanisms that allow crop plants to grow on soils containing toxic concentrations of mineral elements are based on their exclusion from the plant and/or tolerance through their sequestration as non-toxic compounds and/or in non-vital cellular compartments (Marschner, 1995).

The review on the meaningful problem prompts us to be aware with a number of alarming situation being arisen due to toxicities of over 10 elements including the heavy metals. In this regard aluminium affects the uptake, transport of essential nutrients (Ca, Mg, K, P) and water supply to plants. The use of arsenic-based pesticides was found to cause contamination of agricultural as well as domestic land. Inside the cytoplasm arsenate competes with phosphate such as replacing phosphate in ATP to form unstable ADP-As, and leads to the disruption of energy flows in cells. Boron toxicity exerts different effects on diverse processes in vascular plants such as reduced root cell division, lower leaf chlorophyll contents, altered metabolism, decreased photosynthetic rate and less lignin and suberin content, whereas Cd inhibits the activity of several groups of enzymes such as those of the Calvin cycle, carbohydrate metabolism and phosphorus metabolism. General response of decreased root growth due to Cr toxicity could be due to inhibition of root cell division/ root elongation or to the extension of cell cycle in the roots. Besides these, copper toxicity induces deficiency of other essential mineral elements which may be deleterious for growth and development of plant. Similarly, Pb phytotoxicity leads to inhibition of enzyme activities, disruption of mineral nutrition, water imbalance, change in hormonal status and alteration in membrane permeability. Mn toxicity is usually brown in colour and sometimes with cracks and also causing a drop in the photosynthetic rate of plants in the excess Mn condition. Interestingly, selenium toxicity is responsible for certain livestock disorders i.e., alkali disease and blind staggers in animals grazing on soils with high selenium content such as seleniferous soils. Last but not the least, Zn toxicity occurs in soils contaminated by mining and smelting activities, agricultural soils treated with sewage sludge and urban and peri-urban soil enriched by anthro-pogenic inputs of Zn especially in low-pH soils that creates necrotic lesions on leaves in severe cases and eventually entire leaf dies.

Plant breeders are suggested, therefore, for making ceaseless endeavour to develop crop genotypes tolerant to these soils in addition to a thorough survey of varying conditions supposed also to have been generating under the climate change.

REFERENCES

Abdulla, M., Nair, B.M. and Chandra, R.K. (1985). Health effects and interactions of essential and toxic elements. *Nutr. Res.* 1–751.

Ahmed, A. and Tajmir-Riahi, H.A. (1993). Interaction of toxic metal ions Cd^{2+}, Hg^{2+} and Pb with light-harvesting proteins of chloroplast thylakoid membranes. An FTIR spectroscopic study. *J. Inorg. Biochem.* **50**: 235 - 243.

Anjum, N.A., Umar, S., Ahmad, A., Iqbal, M. and Nafees, N.A. (2008). Sulphur protects mustard (*Brassica campestris* L.) from cadmium toxicity by improving leaf ascorbate and glutathione Sulphur protects mustard from cadmium toxicity. *Plant Growth Regul.* **54** : 271 – 279.

Barcelo, J., Poschenrieder, C. and Gunse, B. (1986). Water relations of chromium VI treated bush bean plants (*Phaseolus vulgaris* L cv Contender) under both normal and water stress conditions. *J Exp Bot.* **37**:178 – 87.

Baron, M., Arellano, B. and Lopez Gorge, J. (1995). Copper and photosystem II: A controversial relationship. *Physiol plant.* **94:** 174-180.

Barrachina, A.C., Carbonell, F.B. and Beneyto, J.M. (1995). Arsenic uptake, distribution, and accumulation in tomato plants – effect of arsenite on plant growth and yield. *Journal of Plant Nutrition* 18: 1237 – 1250.

Bennett, W. F. (1993). Nutrient Deficiencies and Toxicities in Crop Plants. APS Press, St Paul, MN, USA.

Bergmann,W. (1992). Colour Atlas : Nutritional Disorders of Plants. 204–239. Gustav Fischer, New York.

Bouazizi, H., Jouili, H., Geitmann, A. and El Ferjani, E. (2008). Effect of copper excess on H_2O_2 accumulation and peroxidase activities in bean roots*Acta. Biol. Hung.* **59(2)**: 233 – 45.

Bouazizi, H., Jouili, H., Geitmann, A. and Ferjani, E.E.I. (2010). Copper toxicity in expanding leaves of *Phaseolus vulgaris* L.: antioxidant enzyme response and nutrient element uptake. Ecotox. *Environ. Safe.* **73**: 1304 – 1308.

Brown, P.H., Bellaloui, N. and Wimmer, M.A. (2002). Boron in plant biology. *Plant Biol.* **4**: 205 - 223.

Brown, T.A. and Shrift, A. (1982). Selenium: Toxicity and tolerance in higher plants. *Biol. Rev. Cambridge Philos. Soc.* **57**: 59 – 84.

Burton, K.W., Morgan, E. and Roig, A. (1984). The influence of heavy metals on the growth of sitka-spruce in South Wales forests. II green house experiments. *Plant Soil* **78**: 271 - 282.

Burzynski, M. (1987).The influence of lead and cadmium on the absorption and distribution of potassium, calcium, magnesium and iron in cucumber seedlings. *Acta Physiol. Plant*. **9**: 229 - 238.

Chaney, R.L. and Ryan, J.A. (1994). Risk based standards for arsenic lead and cadmium in urban soils. Dechema, Frankfurt, Germany.

Chaney, R.L. (1993). Zinc phytotoxicity. In: Robson AD, ed. *Zinc in soil and plants.* Dordrecht, the Netherlands: Kluwer Academic Publishers, 135 – 150.

Clijsters, H. and Van Assche, F. (1985). Inhibition of photosynthesis by heavy metals. *Photosynth. Res.* **7**: 31 – 40

Crawford, T.W., Jr., Stroehlein, J.L. and Kuehl, R.O. (1989). Manganese and rates of growth and mineral accumulation in cucumber.*J. American Soc.Hort. Sci.* **114**: 300 - 306.

Davies, F, T., Puryear, J.D., Newton, R.J., Egilla, J.N. and Grossi, J.A.S. (2002). Mycorrhizal fungi increase chromium uptake by sunflower plants: influence on tissue mineral concentration, growth, and gas exchange. *J. Plant Nutr*. **25**: 2389 – 407.

Delhaize, E. and Ryan, P.R. (1995). Aluminium toxicity and tolerance in plants. *Plant Physiol.* **107**: 315 – 321.

Dietz, K.J., Tavakoli, N., Kluge, C., Mimura, T., Sharma, S.S. and Harris, G.C. (2001). Significance of the V-type ATPase for the adaptation to stressful growth conditions and its regulation on the molecular and biochemical level. *J Exp Bot*. **52**: 1969 – 80.

Dominguez, M.J., Gutierrez, F., Leon, R., Vilchez, C., Vega, J.M. and Vigara, J. (2003). Cadmium increases the activity levels of glutamate dehydrogenase and cysteine synthase in *Chlamydomonas reinhardtii*. *Plant Physiol. Biochem*. **41:** 828 – 832.

Drazkiewicz, M. (1994). Chlorophyll-occurrence, functions, mechanism of action, effects of internal and external factors. *Photosynthetica.* **30**: 321 - 331.

Elamin, O.M. and Wilcox, G.E. (1986a). Effect of magnesium and manganese nutrition on muskmelon growth and manganese toxicity. *J. American Soc. Hort. Sci.*. **111**: 582 - 587.

Elamin, O.M. and Wilcox, G.E. (1986b). Effect of magnesium and manganese nutrition on watermelon growth and manganese toxicity. *J. American Soc. Hort. Sci.* **111**: 588 - 593.

Fontes, R.L.F. and Cox, F.R. (1995). Effects of sulfur supply on soybean plants exposed to zinc toxicity. *J. Plant Nut.* **18**: 1893 - 1906.

Foy, C.D. (1992). Soil chemical factors limiting plant root growth, in: Hatfield J.L., Stewart B.A. (Eds.). *Advances in Soil Sciences.* Limitations to Plant Root Growth. Springer Verlag, New York. **19**: 97 – 149.

Foy, C.D. and Fleming, A.L. (1982). Aluminium tolerance of two wheat cultivars related to nitrate reductase activities. *J. Plant. Nut.* **5**: 1313–1333.

Foy, C.D., Chaney, R.L. and White, M.C. (1978). The physiology of metal toxicity in plants, *Annu. Rev. Plant Physiol.* **29**: 511 – 566.

Foy, C.D., Weil, R.R. and Coradetti, C.A. (1995). Differential manganese tolerances of cotton genotypes in nutrient solution. *J. Plant Nut.*. **18**: 685 - 706.

Gianazza, E., Wait, R., Sozzi, A., Regondi, S., Saco, D. and Labra, M. (2007). Growth and protein profile changes in *Lepidium sativum* L. plantlets exposed to cadmium. *Env. Exp Bot.* **59**: 179 – 187.

Gill, S.S., Khan, N.A., Anjum, N.A. and Tuteja, N. (2011). Amelioration of cadmium stress in crop plants by nutrients management : morphological, physiological and biochemical aspects. *Plant Stress.* **5** (Special Issue 1):1 – 23.

Girroti, A.W. (1990). Photodynamic lipid peroxidation in biological systems. *Photochem. Photobiol.* **51**: 497 - 509.

Godbold, D.L. and Kettner, C. (1991). Lead influences root growth and mineral nutrition of Piceaabies seedlings. *J. Plant Physiol.* **139**: 95 - 99.

Gupta, U.C., Jame, Y.W., Campbell, C.A., Leyshon, A.J. and Nicholaichuk, W. (1985). Boron toxicity and deficiency: a review. *Can. J. Soil Sci.* **65:** 381 – 409.

Halliwell, B. and Gutteridge, J.M.C. (1999). Free Radicals in Biology and Medicine. 3rd edn. Oxford University Press, New York.

Hampp, R., Ziegler, H. and Ziegler, I. (1973). Influence of lead ions on the activity of enzymes of reductive pentose phosphate pathway. *Biochem. Physiol. Pflanzen.* **164**: 588 - 595.

Hansch, R. and Mendel, R.R. (2009). Physiological functions of mineral micronutrients (Cu, Zn, Mn, Fe, Ni, Mo, B, Cl). *Curr. Opin. Plant. Biol.* **12**: 259 – 266.

Harmens, H., Den Hartog, P.R., Ten Bookum, W.M. and Verkleij, J.A.C. (1993). Increased zinc tolerance in *Silene vulgaris* (Moench) Garcke is not due to increased production of phytochelatins. *Plant Physiology* **103**: 1305–1309.

Hasan, S.A., Hayat, S., Ali, B. and Ahmad, A. (2008). 28-Homobrassinolide protects chickpea (*Cicer arietinum*) from cadmium toxicity by stimulating antioxidants. *Env. Pollut.* **151**: 60 – 66.

Hechi-Buchholz, C. and Foy, C.D. (1981). Effect of aluminium toxicity on root morphology of barley, in: Brouwer R. (Ed.), Structure and Function of Plant Roots, MartinusNijhoff / Dr. W. Junk Publishers, The Hague. pp. 343 – 345.

Horiguchi, T. (1988). Mechanism of manganese toxicity and tolerance of plants. IV. Effects of silicon on alleviation of manganese toxicity of rice plants. *Soil Sci. Plant Nut*. **34**: 65 - 73.

Huang, J.W., Grunes, D.L. and Kochian, L.V. (1992). Aluminium effects on the kinetics of calcium uptake into cells of the wheat root apex : Quantification of calcium fluxes using a calcium-selective vibrating microelectrode. *Planta.* **188**: 414 – 421.

Huffman, Jr. E.W.D. and Allaway, W.H. (1973b). Growth of plants in solution culture containing low levels of chromium. *Plant Physiol*. **52:**72– 75.

Huffman, Jr. E.W.D.andAllaway, H.W. (1973a). Chromium in plants: distribution in tissues, organelles, and extracts and availability of bean leaf Cr to animals. *J Agric Food Chem*. **21**: 982 – 6.

Jentschke, G., Schlegel, H. and Godbold, D.L. (1991). The effect of aluminium on uptake and distribution of magnesium and calcium in roots of mycorrhizal Norway spruce seedlings. *Physiol. Plant*. **82**: 266 – 270.

Kabata-Pendias, A. and Pendias, H. (1992). Trace elements in soils and plants. 2nd edn. CRC Press, Boca Raton, London.

Karley, A.J. and White, P.J. (2009). Moving cationic minerals to edible tissues: potassium, magnesium, calcium. *Current Opinion in Plant Biology*. **12:** 291 - 298.

Khan, N.A., Samiullah, Singh, S. and Nazar, R. (2007). Activities of antioxidative enzymes, sulphur assimilation, photosynthetic activity and growth of wheat (*Triticum aestivum*) cultivars differing in yield potential under cadmium stress. *J Agro Crop Sci*. **193**: 435 – 444.

Kinraide, T.B. (1988). Proton extrusion by wheat roots exhibiting severe aluminium toxicity symptoms. *Plant Physiol*. **88**: 418 – 423.

Kupper, H., Kupper F. and Spiller M. (1996). Environmental relevance of heavy metal substituted chlorophylls using the example of water plants. *J. Exp. Bot*. **47**: 259 – 266.

Lee, C.W., Jackson, M.B., Duysen, M.E., Freeman, T.P. and Self, J.R. (1996a). Induced micronutrient toxicity in 'Towndown' Kentucky bluegrass. *Crop Science*. **36**: 705 - 712.

Lee, K.C., Cunningham, B.A., Poulsen, G.M., Liang, J.M. and Moore, R.B. (1976). Effects of cadmium on respiration rate and activities of several enzymes in soybean seedlings. *Physiol. Plant*. **36**: 4 - 6.

Lequeux, H., Hermans, C., Lutts, S. and Verbruggen, N. (2010). Response to copper excess in Arabidopsis thaliana: Impact on the root system architecture, hormone distribution, lignin accumulation and mineral profile. *Plant Physiol. Biochem*. **48**: 673 – 682.

Leyshon, A. J. and Jame, Y. M. (1993). Boron toxicity and irrigation management. In Boron and It's Role in Crop Production. Ed. U C Gupta. 207–226. CRC Press, Boca Raton, FL, USA.

Lovatt, C. J. and Bates, L. M. (1984). Early effects of excess boron on photosynthesis and growth of *Cucurbita pepo*. *J. Exp. Bot.* **35**: 297 – 305.

Luna, C.M., Gonzalez, C.A. and Trippi, V.S. (1994). Oxidative damage caused by an excess of copper in oat leaves. *Plant Cell Physiol*. **35**:11–15

Lux, A., Martinka, M. and Vaculík, P.J. (2010). White Root responses to cadmium in the rhizosphere: A review. *J Exp Bot*. **62**: 21 – 37.

Macfie, S.M., Cossins, E.A. and Taylor, G.J. (1994). Effects of excess manganese on production of organic acids in Mn-tolerant and Mn-sensitive cultivars of *Triticum aestivum* L. (Wheat). *J. Plant Physiol.* **143**: 135-144.

Marschne, H. (1995). Mineral nutrition of higher plants. 2nd edn. London: Academic Press.

Marschner, H. (1991). Mechanisms of adaptation of plants to acid soils. *Plant and Soil*. **134**: 1–20.

Meharg, A.A. (1993). The role of the plasmalemma in metal tolerance in angiosperms. *Physiol. Plant*. **88**: 191–198.

Meharg, A.A. (1994). Integrated tolerance mechanisms – constitutive and adaptive plant- responses to elevated metal concentrations in the environment. *Plant, Cell & Environment* **17**: 989 – 993.

Mengel, K., Kirkby, EA., Kosegarten, H. and Appel, T. (2001). Principles of plant nutrition. Dordrecht: Kluwer Academic.

Miller, A.J., Shen, Q. and Xu, G. (2009). Freeways in the plant: transporters for N, P and S and their regulation. *Current Opinion in Plant Biology*. **12**: 284 - 290.

Mishra, A. and Choudhari, M.A. (1998). Amelioration of lead and mercury effects on germination and rice seedling growth by antioxidants. *Biol. Plant*. **41**: 469 - 473.

Mishra, S., Srivastava, S., Tripathi, R.D., Dwivedi, S. and Shukla, M.K. (2008). Responses of antioxidant enzymes in coontail (*Ceratophyllum demursum* L.) plants under cadmium stress. *Environ Toxicol*. **23**: 294 – 301.

Miwa, K., Kamiya, T. and Fujiwara, T. (2009). Homeostasis of the structurally important micronutrients. B and Si. *Current Opinion in Plant Biology*. **12**: 307 - 311.

Mukherji, S. and Maitra, P. (1976). Toxic effects of lead on growth and metabolism of germinating rice (*Oryza sativa* L.) root tip cells. *Ind. J. Exp. Biol*. **14**: 519 - 521.

Murphy, E.A. and Aucott, M. (1998). An assessment of the amounts of arsenical pesticides used historically in a geographical area. *Science of the Total Environment*. **218**: 89 – 101.

Nichol, B.E. and Oliveira, L.A. (1995). Effects of aluminium on the growth and distribution of calcium in roots of an aluminium-sensitive cultivar of barley (*Hordeum vulgare*). *Can. J. Bot*. **73**: 1849 – 1858.

Nosko, P., Brassard, P., Kramer, J.R. and Kershaw, K.A. (1988). The effect of aluminium on seed germination and early seedling establishment growth and respiration of white spruce (*Picea glauca*). *Can. J. Bot*. **66**: 2305 – 2310.

Nriagu, J.O. (1994). Arsenic in the environment: part 1 cycling and characterization. New York, USA: Wiley, 430.

Parr, P.D. and Taylor, Jr. FG. (1982). Germination and growth effects of hexavalent chromium in Orocol TL (a corrosion inhibitor) on *Phaseolus vulgaris*. *Environ Int* .**7**:197 – 202.

Peralta, J.R., GardeaTorresdey, J.L., Tiemann, K.J., Gomez, E., Arteaga, S. and Rascon, E. (2001). Uptake and effects of five heavy metals on seed germination and plant growth in alfalfa (*Medicago sativa*) L. *B Environ Contam Toxicol*. **66(6)**:727 – 34.

Plucinska, G.L. and Ziegler, H. (1995). The effect of aluminium on adenylate levels in Scots pine roots. *Acta Physiol. Plant*. **17**: 225–232.

Prassad, D.D.K. and Prassad, A.R.K. (1987). Altered d-aminolaevulinic acid metabolism by lead and mercury in germinating seedlings of Bajra (*Pennisetum typhoideum*). *J. Plant Physiol*. **127**: 241- 249.

Rashid, A., Camm, E.L., Ekramoddoullah, K.M. (1994). Molecular mechanism of action of Pb and Zn^{2+} on water oxidizing complex of photosystem II.*FEBS Lett*. **350**: 296 - 298.

Rebechini, H.M. and Hanzely, L. (1974). Lead-induced ultrastructural changes in chloroplasts of the hydrophyte *Ceratophyllum demersum*. *Z. Pflanzenphysiol*. **73**:377-386.

Reese, R.N. and Roberts, L.W. (1985). Effects of cadmium on whole cell and mitochondrial respiration in tobacco cell suspension cultures (*Nicotiana tobacum* L. var. *xanthi*). *J. Plant Physiol*. **120**: 123 - 130.

Reid, R. (2007b). Update on boron toxicity and tolerance in plants. In: Xu F, Goldbach HE, Brown PH, Bell RW, Fujiwara T, Hunt CD, Goldberg S, Shi L, eds. *Advances in Plant and Animal Boron Nutrition*. Springer, Dordrecht, The Netherlands, pp. 83 - 90.

Reid, R.J., Hayes, J.E., Post, A., Stangoulis, J.C.R. and Graham, R.D. (2004). A critical analysis of the causes of boron toxicity in plants. *Plant Cell Environ*. **25:** 1405 - 1414.

Ren, F., Liu, T., Liu, H. and Hu, B. (1993). Influence of zinc on the growth, distribution of elements, and metabolism of one-year old American ginseng plants. *J. Plant Nut.*. **16**: 393 - 405.

Roessner, U., Patterson, J.H., Forbes, M.G., Fincher, G.B., Langridge, P. and Bacic, A. (2006). An investigation of boron toxicity in barley using metabolomics. *Plant Physiol.* **142:** 1087 - 1101.

Romeu–Moreno, A. and Mas, A. (1999). Effect of copper exposure in tissue cultured *Vitis vinifera*. *J. Agric. Chem.* **47**: 2519–22

Rosenfeld, I. and Beath, O.A. (1964). Selenium, Geobotany, Biochemistry, Toxicity, and Nutrition. New York: Academic. 411 pp.

Rout, G.R., Samantaray, S. and Das, P. (1997). Differential chromium tolerance among eight mungbean cultivars grown in nutrient culture. *J Plant Nutr*. **20**: 473 – 83.

Rout, G.R., Sanghamitra, S. and Das, P. (2000). Effects of chromium and nickel on germination and growth in tolerant and non-tolerant populations of *Echinochloa colona* (L). Chemosphere. **40**: 855 – 9.

Roy, A.K., Sharma, A. and Talukder, G. (1988). Some aspects of aluminium toxicity in plants, *Bot. Rev*. **54**: 145 – 177.

Ryan, P.R., DiTomaso, J.M. and Kochian, L.V. (1993). Aluminium toxicity in roots. An investigation of spatial sensitivity and the role of root cap. *J. Exp. Bot.* **44**: 437 – 446.

Satarug, S., Baker, J.R., Reilly, P.E.B., Moore, M.R. and Williams, D.J. (2002). Cadmium levels in the lung, liver, kidney cortex and urine samples from Australians without occupational exposure to metals. *Arch Environ Health* **57:** 69 – 77.

Sersen, F., Kralova, K. and Bumbalova, A. (1998). Action of mercury on the photosynthetic apparatus of spinach chloroplasts. *Photosynthetica* **35**: 551 - 559.

Shah, K. and Dubey, R.S. (2006). A 18 kDa Cd inducible protein complex: its isolation and characterization from rice (*Oryza sativa* L.) seedlings. *J Plant Physiol.* **152**: 448 – 454.

Shah, K., Kumar, R.G., Verma, S. and Dubey, R.S. (2001). Effect of cadmium on lipid peroxidation, superoxide anion generation and activities of antioxidant enzymes in growing rice seedlings. *Plant Sci.* **161**:1135 – 1144.

Shanker, A.K., Sudhagar, R. and Pathmanabhan, G. (2003a). Growth, Phytochelatin SH and antioxidative response of Sunflower as affected by chromium speciation. 2nd International Congress of Plant Physiology on sustainable plant productivity under changing environment, New Delhi, India.

Sharma, D.C. and Sharma, C.P. (1993). Chromium uptake and its effects on growth and biological yield of wheat. *Cereal Res Commun.* **21**: 317 – 21.

Skeffington, R.A., Shewry, P.R. and Petersen, P.J. (1976). Chromium uptake and transport in barley seedlings Hordeumvulgare. *Planta* **132**: 09 – 14.

Stefanov, K., Popova, I., Kamburova, E., Pancheva, T., Kimenov, G., Kuleva, L. and Popov, S. (1993). Lipid and sterol changes in *Zea mays* caused by lead ions. *Phytochemistry.* **33**:47-51.

Suhayda, C.A. and Haug, A. (1986). Organic acids reduce aluminium toxicity in maize root membranes. *Physiol. Plant.* **68**: 189 – 195.

Thawornwong, N. and van Diest A. (1974). Influences of high acidity and aluminium on the growth of low land rice. *Plant and Soil.* **41**: 141–159.

Tripathi, A.K. and Tripathi, S. (1999). Changes in some physiological and biochemical characters in *Albizia lebbek* as bio-indicators of heavy metal toxicity. *J Environ Biol.* **20:** 93 – 8.

Turpeinen, R., Pantsar-Kallio, M., Haggblom, M. and Kairesalo, T. (1999). Influence of microbes on the mobilization, toxicity and biomethylation of arsenic in soil. *Science of the Total Environment* **236**: 173 – 180.

TuShu, I. and Brouillette, J.N. (1987). Metal ion inhibition of corn root plasma membrane ATPase. *Phytochem.* **26**: 65 - 69.

Upadhyay, R.K. and Panda, S.K. (2009). Copper-induced growth inhibition, oxidative stress and ultrastructural alterations in freshly grown water lettuce (*Pistia stratiotes* L.). *C. R. Biologies.* **332**: 623 – 632.

Vajpayee, P., Rai, U.N., Ali, M.B., Tripathi, R.D., Yadav, V. and Sinha, S. (2001). Chromium induced physiological changes in *Vallisneria spiralis* L and its role in phytoremediation of tannery effluent. *Bull. Environ. Contam. Toxicol.* **2:** 246 – 56.

Vallee, B.L. and Ulmer, D.D. (1972). Biochemical effects of mercury, cadmium and lead. *Annu. Rev. Biochem.* **41**: 91 - 128.

van Assche, F. and Clijsters, H. (1990). Effects of metal on enzyme activity in plants. *Plant Cell Environ.* **13:** 195 - 206.

Verbruggen, N., Hermans, C. and Schat, H. (2009). Mechanisms to cope with arsenic or cadmium excess in plants. *Curr.Opin. Plant Biol.* **12**: 364 – 372.

Verma, S. and Dubey, R.S. (2003). Lead toxicity induces lipid peroxidation and alters the activities of antioxidant enzymes in growing rice plants. *Plant Sci.* **164**: 645 - 655.

Vodnik, D., Jentschke, G., Fritz, E., Gogala, N. and Godbold, D.L. (1999). Root-applied cytokinin reduces lead uptake and affects its distribution in Norway spruce seedlings. *Physiol. Plant.* **106**: 75 - 81.

Wagatsuma, T., Kaneko, M. and Hayasaka, Y. (1987). Destruction process of plant root cells by aluminium. *Soil.Sci. Plant Nutr.* **33**: 161 – 175.

White, P.J. and Broadley, M.R. (2009). Biofortification of crops with seven mineral elements often lacking in human diets – iron, zinc, copper, calcium, magnesium, selenium and iodine. *New Phytol.* **182**: 49 - 84.

White, P.J. and Brown, P.H. (2010). Plant nutrition for sustainable evelopment and global health. *Ann Bot.* **105**:1073 – 1080.

Woolhouse, H.W. (1969). Differences in the properties of the acid phosphatases of plant roots and their significance in the evolution of edaphic ecotypes, in: Rorison I.H. (Ed.), Ecological aspects of the mineral nutrition of plants, Blackwell Publishers, Edinburg, Oxford. 357 – 380.

Wu, L., Huang, Z.Z. and Burau, R.G. (1988). Selenium accumulation and selenium-salt cotolerance

Wu, S. (1994). Effect of manganese excess on the soybean plant cultivated under various growth conditions. *J. Plant Nut.* **17**: 993 - 1003.

Xiong, Z., Liu, C. and Geng B. (2006). Phytotoxic effects of copper on nitrogen metabolism and plant growth in *Brassica pekinensis* Rupr. *Ecotox. Environ. Safe.* **64**: 273 – 280.

Yang, X.E., Long, X.X., Ye, H.B., He, Z.L., Calvert, D.V. and Stoffella, P.J. (2004). Cadmium tolerance and hyperaccumulation in a new Zn hyperaccumulating plant species (*Sedum alfredii* Hance). *Plant Soil.* **259**:181 – 189.

Zaccheo, P., Genevini, P.L. and Cocucci, S. (1982). Chromium ions toxicity on the membrane transport mechanism in segments of maize seedling roots. *J Plant Nutr.* **5**:1217– 1227.

Zayed, A., Lytle, C.M. and Terry, N. (1998). Accumulation and volatilization of different chemical species of selenium by plants. *Planta.* **206**: 284 – 292.

Zeid, I.M. (2001). Responses of *Phaseolus vulgaris* to chromium and cobalt treatments. *Biol Plant.* **44**: 111 – 115.

7

Significance of Nickel in Plant Metabolism

A. Hemantaranjan

Department of Plant Physiology, Institute of Agricultural Sciences, Banaras Hindu University, Varanasi, (U.P.), India
E-mail: hemantaranjan@gmail.com

INTRODUCTION

In 1987, Nickel (Ni) was first established as an essential nutrient for completion of plant life cycle. Ni deficiency decreased the capacity of barley to develop viable seeds because of hindered embryo growth. The embryonic root developed poorly or even stayed undeveloped; in addition, several anomalies were reported in endosperm development together with declined dehydrogenase activities (Brown *et al.,* 1987). In fact, plants require no less than 14 mineral elements for their nutrition. These include the macronutrients nitrogen (N), phosphorus (P), potassium (K), calcium (Ca), magnesium (Mg) and sulphur (S) and the micronutrients chlorine (Cl), boron (B), iron (Fe), manganese (Mn), copper (Cu), zinc (Zn), nickel (Ni) and molybdenum (Mo). These are generally obtained from the soil. Crop production is often limited by low phytoavailability of essential mineral elements and/or the presence of excessive concentrations of potentially toxic mineral elements, such as sodium (Na), Cl, B, Fe, Mn and aluminium (Al), in the soil solution. Nickel (Ni), comparatively the recently discovered essential element (Brown *et al.,* 1987), is unique among plant nutrients in that its metabolic function was determined well before it was confirmed that its deficiency could disrupt plant growth. Subsequent to the discovery of its essentiality in the laboratory, Ni deficiency has now been observed in field situations in several perennial species (Wood *et al.,* 2004).

Interestingly, nickel (Ni) is a ubiquitous trace metal and occurs in soil, water, air and the biosphere. Nickel ore deposits are accumulations of nickel sulfide minerals (mostly pentlandite) and laterites. Most nickel is used for production of stainless steel and other nickel alloys with high corrosion and temperature resistance. Nickel levels in terrestrial and aquatic organisms can vary over several orders of magnitude. Ni has gained considerable attention in recent years, because of its rapidly increasing concentrations in soil, air, and water in different parts of the world. Human use of Ni has been traced as far back as 3500 BC. However, it was first isolated and classified as a chemical element from *Kupfernickel* by the Swedish mineralogist and chemist, A. F. Cronstedt, in 1751. Nickel has five valences: 0, +1, +2, +3, and +4. Among these, only Ni^{2+} is considered an available form for plants. Nickel comprises approximately 3% of the earth's crust composition and is the twenty-fourth most abundant element. Total Ni concentration commonly ranges from 5 to 500 mg kg^{-1}, with an average of 50 mg kg^{-1} in soils. However, Ni concentrations in dried biosolids (also referred to as treated sewage sludge) or soil near metal refineries range between 24,000 and 53,000 mg kg^{-1}. Soils for crop production contain 3–1,000 mg kg^{-1}. Since Ni^{2+} is the available form of Ni for plants, total Ni concentration is not a useful measure for Ni bioavailability. Positive 2-valence Ni ion (Ni^{2+}) readily oxidizes and becomes unavailable.

Based on the criteria for essentiality of elements for plant growth and development established by Arnon and Scott (1939) and Meyer and Anderson (1939). Brown *et al.* (1987) determined in the late 1980s that Ni is an essential nutrient for both monocotyledons and dicotyledons. However, American Association of Plant Food Control officials did not recognize Ni as essential until 2004. Nickel is the seventeenth element recognized as essential for plant growth and development (Liu, 2001). Plants' Ni requirement is the lowest of all essential elements at < 0.5 mg per kg of dry weight, making it an essential plant micronutrient. Nickel is required for plants at such low concentrations that analytical technologies were not available until the mid-1970s. Relatively little is known about the role of nickel (Ni) in plant nutrition, physiology, and metabolism, especially in woody perennial species. The crop plants accumulated 2.5% urea dry weight in the leaves. Nickel represents the first of several likely new essential elements that will be shown to be critical for certain metabolic processes normally active in plants, but not necessarily essential for the completion of the species' life cycle under all conditions. Nickel is a functional constituent of seven enzymes. Among the seven, urease (EC 3.5.1.5) (i.e., urea amidohydrolase) is extremely important to N metabolism in plants.

Evidence for a key metabolic role was strengthened with observation of field-level growth responses to foliar Ni applications to crops as diverse as wheat (*Triticum aestivum*), potatoes (*Solanum tuberosum*), and broad beans (*Vicia faba*) (Roach and Barclay, 1946; Dobrolyubskii and Slavvo,

1957; Welch, 1981). Although Ni is a recognized essential mineral nutrient element for higher plants, its agricultural and biological significance is poorly understood. This is largely because of the low levels thought to be needed by plants (about 1–100 ng g^{-1} dry weight) in relation to the relative abundance of Ni in essentially all soils (>5 kg ha^{-1}). The existence of field-level Ni deficiency in crops was only recently discovered.

Therefore, based on existing substantial imperative reports on the availability and essentiality of nickel in plant growth and development, this makes vital to put emphasis on meaningful Ni researches with a view to extending judicious research with reference to metabolic studies in diverse plants and crops that might be of utmost significance in precise perceptions for nickel requirements at different growth stages in varied metabolic systems without leaving traces of toxicity as found with heavy metals. This brief review with the broad objective on significance of nickel in plant metabolism highlights importance of nickel, its uptake, roles in plant metabolism, disruptions of metabolic activities under nickel deficient situations along with certain examples on its direct and indirect influences on crop growth that have been subdivided into the following six major sections:

2. IMPORTANCE OF NICKEL

The putative mechanisms of nickel hyperaccumulation are considered in several representatives of angiosperm plant families. The existing evidence was used to outline the metabolic changes in plants affected by nickel. The comparison with other heavy metals is used to disclose the general mechanisms that disturb plant mineral nutrition, water regime, photosynthesis, and morphogenesis as well as the common cell responses aimed at detoxification of heavy metals. The numerous nonspecific effects of heavy metals depend on their direct and indirect action; in addition, some effects of nickel are specific. To illustrate, high Ni content in endoderm and pericycle cells blocks cell divisions in the pericycle and results in the inhibition of root branching. Nickel holds a special place among the heavy metals (Seregin and Kozhevnikova, 2006). Unlike Cd, Pb, Hg, Ag, and several other metals that are not the components of plant enzymes, Ni is a constituent of urease, and small quantities of Ni (0.01 to 5μg g^{-1} dry wt) are essential for some plant species. On the other hand, Ni is not as important for plant metabolism as Zn and Cu. However, same as with other heavy metals, high Ni concentrations may turn toxic to plants. The metal cations are classified into hard and soft Lewis acids (Brown *et al.*, 1987). According to this classification, the ions of hard acids, such as Li^+, Sr^{2+}, etc., would rather interact with hard bases, whereas the ions of soft acids, such as Pb^{2+}, Hg^{2+}, Ag^+, etc., with soft bases. Cd^{2+}, Cu^{2+}, Ni^{2+}, and several other ions come into an intermediate class, with Ni^{2+} shifting towards hard and Cd^{2+}, soft class. In the biological chemistry, there are four major classes of donor atoms in ligands. Among these, oxygen and aliphatic nitrogen are classified as hard, aromatic nitrogen as intermediate, and sulfur as soft. It follows that Ni would rather bind to aromatic nitrogen, and Cd^{2+}, Hg^{2+}, and Ag^{+}, to SH-groups. Ni^{2+} and Zn^{2+} practically do not differ in several physical and chemical properties, such as the size of the ion. However Zn^{2+} is a constituent of numerous enzymes, whereas Ni^{2+} is found in few plant enzymes, probably because in the former case the rate of ligand exchange is higher by three orders of magnitude. Thus, diverse effects of the metals of similar properties apparently depend on the different rates of ligand exchange, and their physiological properties are immediately derived from their physical and chemical parameters.

The interest of plant scientists in the role of nickel was initiated following the discovery in 1975 (Dixon *et al.,* 1975) that it was a critical constituent of the plant enzyme, urease. The ultimate determination that nickel was essential for plant growth (Brown *et al.,* 1987a) depended heavily on the development of new techniques to purify growth media and to measure extremely low concentrations of nickel in plants. The establishment of nickel as an essential element, however, highlights the limitations of the current definition of essentiality of nutrients as applied to plants (Arnon and Stout, 1939). It has been argued, for example, that even though nickel is clearly a normal and functional constituent of plants, it does not fulfill the definition of essentiality, since urease is not essential for plant growth and nickel deficiency apparently does not prevent the completion of the life cycle of all species, even though that criterion has not been explicitly satisfied for any element (Arnon and Stout, 1939). Several authors (Arnon and Stout, 1939; Epstein and Bloom, 2004) now suggest that the criteria for essentiality should be modified to include elements that are normal functional components of plants.

Nickel was established as an essential micronutrient for the growth of temperate cereal crops (Brown, 1987). Grain from barley (*Hordeum vulgare* L. cv `Onda'; containing 40 to 80 nanograms of Ni per gram dry weight) grown in solution culture with negligible Ni concentrations (< 30 nanograms of Ni per liter) exhibited greatly reduced germination rates (*i.e.* 50% less than grain from Ni-adequate plants) and seedling vigor of the viable grain was greatly depressed. Grain containing less than 30 nanograms per gram dry weight was inviable. Under Ni-deficient conditions, barley plants fail to produce viable grain because of a disruption of the maternal plant's normal grain-filling and maturation processes that occur following formation of the grain embryo. The observations that (a) barley plants fail to complete their life cycle in the absence of Ni and (b) addition of Ni to the growth medium completely alleviates deficiency symptoms in the maternal plants satisfies the essentiality criteria; thus, Ni should be considered a micronutrient for cereals. Because Ni is required by legumes, and is now established as essential for cereals, Ni was added to the list of micronutrients essential for all higher plant growth (Brown, 1987a).

Nickel, in low concentrations, fulfills a variety of essential roles in plants, bacteria, and fungi. Therefore, Ni deficiency produces an array of effects on growth and metabolism of plants, including reduced growth, and induction of senescence, leaf and meristem chlorosis, alterations in N metabolism, and reduced Fe uptake. In addition, Ni is a constituent of several metallo-enzymes such as urease, superoxide dismutase, NiFe hydrogenases, methyl coenzyme M reductase, carbon monoxide dehydrogenase, acetyl coenzyme-A synthase, hydrogenases, and RNase-A. Therefore, Ni deficiencies in plants reduce urease activity, disturb N assimilation, and reduce scavenging of superoxide free radical. In bacteria, Ni participates in several important metabolic reactions such as hydrogen metabolism, methane biogenesis, and acetogenesis. Although Ni is metabolically important in plants, it is toxic to most plant species when present at excessive amounts in soil and in nutrient solution.

3. NICKEL UPTAKE

Learning of the aspects related with uptake, transport and distributive localization of Ni is very important in various cellular metabolic processes particularly under increased nitrogen metabolism. Nickel even though recognized as a trace element, its metabolism is very decisive for certain enzyme activities, maintaining proper cellular redox state and various other biochemical, physiological and growth responses (Yusuf *et al.*, 2011). The pattern of Ni accumulation in cell walls and protoplasts of all root tissues presumes that Ni is translocated via both the apoplast and symplast. When reaching the xylem by radial transport, Ni will reach the shoots with the xylem sap. The amount and distribution of Ni-containing complexes would vary depending on plant age.

Plants have two transport systems (Brown, 2006): low affinity and high affinity. With the low-affinity transport system, plants can absorb Ni^{2+} ions at the low concentration of 4.4 ppb, which is approximately 0.6 ounces Ni per million gallons of water. With the high-affinity transport system, plants can take up 1.8 parts per million (ppm) of Ni^{2+} ions, which is 237.7 ounces per million gallons of water. Ni is transported from nonsenescent source leaves to sink tissues via the phloem of leguminous and nonleguminous plants. The main mechanisms by which Ni is taken up by plants are passive diffusion and active transport. Soluble Ni compounds are preferably absorbed by plants passively, through a cation transport system; chelated Ni compounds are taken up through secondary, active-transport-mediated means, using transport proteins such as permeases. Insoluble Ni compounds primarily enter plant root cells through endocytosis. Once absorbed by roots, Ni is easily transported to shoots via the xylem through the transpiration stream and can accumulate in neonatal parts such as buds, fruits, and seeds. The Ni transport and retranslocation processes are strongly regulated by metal-ligand complexes (such as nicotianamine, histidine, and organic acids) and by some proteins that specifically bind and transport Ni.

In bacterial systems, several families of nickel permeases and ATP-dependent nickel carriers have been characterized. No equivalent mechanism has yet been identified in animals or plants (Maroney, 1999). In plant systems, most studies have been conducted at unrealistically high soil-nickel concentrations and as such may be relevant for nickel toxicity, but are not relevant for nickel uptake under normal conditions. Cataldo *et al.* (1978) using 63Ni indicated that a high-affinity Ni^{2+} carrier functioned at 0.075 or 0.25μM Ni^{2+} with a K_m of 0.5 μM which approaches the nickel concentration in uncontaminated soils Brown *et al.* (1989). Either Cu^{2+} or Zn^{2+}competitively inhibits Ni^{2+} uptake suggesting that all the three elements share a common uptake system Kochian (1991). Uptake at higher nickel-supply levels (0.5 to 30 μM) was energy dependent and had a K_m of 12μM indicative of an active, low-affinity transport system. Nickel, unlike many other divalent cations, is readily re-translocated within the plant likely as a complex with organic acids and amino acids Cataldo *et al.* (1988). Nickel rapidly re-translocates from leaves to young tissues in the phloem, particularly during reproductive growth. Indeed, up to 70% of nickel in the shoots was transported to the seed of soybean (Tiffin, 1971). Nickel is associated primarily with organic acids and amino acids in the phloem. Above pH 6.5, histidine is the most significant chelator, whereas at pH <5, citrate is the most significant one Gerendas *et al.* (1999).

4. NICKEL IN PLANT METABOLISM

The field of nickel metallobiochemistry has seen tremendous growth over the preceding 10 years, and nickel is clearly a biologically important element in a diverse range of organisms. Indeed, it is highly likely that with the advent of molecular techniques to search for genetic and functional homology rapidly, the diversity of known functions of nickel in biology will increase substantially in the coming years. The discovery in 1975 that Ni is a component of the enzyme urease (Dixon *et al.,* 1975), which is present in a wide range of plant species (Welch, 1981), led to renewed scientific interest and research concerning the role of Ni in higher plants. Several researchers have since reported growth responses of plants to Ni fertilization under field conditions and in plants grown in nutrient solutions or in tissue culture media furnished with urea as the sole N source (Polacco, 1977).

As a Ni-metalloenzyme, urease assists in the hydrolysis of urea. Nickel works as a cofactor to enable urease to catalyze the conversion of urea into the ammonium ion, which plants can use as a source of N. Without the presence of Ni, urea conversion is impossible. Nickel is accumulated in plant organs or tissues, such as leaves.

So far, seven nickel-dependent enzymes have been identified. Two of these enzymes have nonredox function (urease and glyoxylase), and the remaining five involve oxidation–reduction reactions (Ni-superoxide dismutase, methyl coenzyme M reductase, carbon monoxide dehydrogenase, acetyl coenzyme A synthase, and hydrogenase). Of the established nickel enzymes and proteins, urease is the sole nickel-specific enzyme known to function in plants; however, nickel-dependent hydrogenase also indirectly influences plant productivity through its role in nitrogen-fixing symbionts (Cammack, 1995) and in leaf commensal bacteria (Holland and Pollacco, 1992) . Recent biochemical and molecular biological experiments have provided clear evidence for the existence of multiple auxiliary genes that facilitate nickel incorporation into urease and hydrogenase. Similarly, accessory factors are also likely to be required for the other two enzymes. One of the urease-related genes (ureE) encodes a cytoplasmic protein that has been purified and shown to bind nickel reversibly. The ureE protein serves as a nickel donor to urease apoprotein. A second urease-related auxiliary gene (ureG) possesses a sequence motif that is found in ATP- and GTP-binding proteins. Nickel incorporation into urease requires energy and speculate that the UreG protein may serve as an energy transducer, coupling the energy of NTP hydrolysis to metallocenter incorporation. The UreG protein is related in sequence to HypB, a protein that has been proposed to function in nickel processing in hydrogenases. Hence, the mechanisms for metallocenter biosynthesis in these two dissimilar enzymes may have evolved from a common nickel incorporation system.

Currently, none of the bacterial proteins involved in nickel uptake and assimilation (NikA, NixA, UreE, etc.) is known to be present in plants. Interestingly, the hydrogenase and urease activities of leaf-surface symbionts are clearly inhibited when they colonize urease-deficient soybean mutants (Holland and Pollacco, 1992). The mechanism by which this inhibition occurs is unknown but may suggest that the urease deficient mutants lack key nickel assimilatory proteins, thus preventing the transfer of nickel to the leaf-surface bacterial enzymes. This possibility would suggest that plants might contain nickel dependent assimilatory proteins.

Additional nickel-dependent enzymes and proteins are present in higher plants is supported by the observation that several of the known bacterial nickel-containing enzymes have analogs in plants and animals (including superoxide dismutase, glyoxylase, acetyl coenzyme A synthase, and hydrogenase). The similarity between the effects of nickel deficiency in animals and plants also provides evidence of a common biological role for nickel in all organisms.

4.1 Nickel-Dependent Oligomerization

After activation with $NiCl_2$, the recombinant α subunit of the Ni-containing $\alpha_2\beta_2$ acetyl-CoA synthase/carbon monoxide dehydrogenase (ACS/CODH) catalyzes the synthesis of acetyl-CoA from CO, CoA and a methyl group donated from the corrinoid-iron-sulfur protein (CoFeSP) (Tan *et al.,* 2007). The α subunit has two conformations (open and closed), and contains a novel [Fe_4S_4]-[Ni_p Ni_d] active site in which the proximal Ni_p ion is labile. Prior to Ni activation, recombinant apo-α contain only an Fe_4S_4 cluster. Ni-activated α subunits exhibit catalytic, spectroscopic and heterogeneity properties typical of α subunits contained in ACS/CODH. Evidence presented here indicates that apo-α is a monomer whereas Ni-treated α oligomerizes, forming dimers and higher molecular weight species including tetramers. No oligomerization occurred when apo-α was treated with Cu(II), Zn(II) or Co(II) ions, but oligomerization occurred when apo-α was treated with Pt(II) and Pd(II) ions. The dimer accepted only 0.5 methyl groups/α and exhibited, upon treatment with CO and under reducing conditions, the NiFeC EPR signal quantifying to 0.4 spins/α. Dimers appear to consist of two types of α subunits, including one responsible for catalytic activity and one that provides a structural scaffold. Higher molecular weight species may be similarly constituted. It is concluded that Ni binding to the A-cluster induces a conformational change in the α subunit, possibly to the open conformation, that promotes oligomerization. These interrelated events demonstrate earlier unrealized connections between: a) the conformation of the α subunit; b) the metal which occupies the proximal/distal sites of the A-cluster; and c) catalytic activity (Tan *et al.,* 2007).

4.2 Nickel-Binding Proteins

Nickel enzymes are a relatively new class of metalloenzymes. The seven known nickel enzymes are urease, hydrogenase, CO-dehydrogenase, methyl-coenzyme M reductase, Ni-superoxide dismutase, glyoxalase I and cis-trans isomerase. The requirement for nickel implies the presence of a nickel-processing system, since free transition metals are harmful to the cell. A nickel-processing system involves the recognition and transport of nickel into the cell and the handling of the nickel once it enters the cell until it is inserted into the nickel enzyme. Several mechanisms for nickel transport have been identified and will be reviewed here. Accessory proteins required for the biosynthesis of the nickel active site have been identified. Accessory proteins bind the nickel when it enters the cell and are proposed to assist with the insertion of nickel into the enzyme (Watt and Ludden, 1999).

Further, by inhibiting key enzyme activities of the Calvin cycle, such as rubisco, 3-phosphoglycerate kinase, fructose-1,6-bisphosphatase, aldolase, and NAD- and NADP-dependent phosphoglyceraldehide dehydrogenases, nickel can also hold back the dark reactions of photosynthesis. Such effects were demonstrated in *Cajanus cajan* leaves following several days of incubation on 1 mM $NiCl_2$ solution (Sheoran *et al.,* 1990).

5. NICKEL DEFICIENCY DISRUPTS METABOLISM

Nickel is a key component of selected enzymes involved in N metabolism and biological N fixation. Plants suffering from Ni deficiency show necrosis initiating from the tip of the leaf. This symptom can be reversed or corrected by applying a dilute Ni solution. In 1983, D. L. Eskew and his colleagues reported that legume plants grown in nutrient solutions containing ammonium nitrate as the N source developed necrotic lesions on the leaflet tips. The leaf tip necrosis symptom of Ni deficiency is even more recurrent and widespread when plants are highly dependent on biological N fixation for their N. This concentrated urea kills leaf cells, resulting in the development of necrotic lesions on the legume leaf tips. Ni deficiency also results in delayed nodulation and reduced efficiency of N fixation. Leguminous plants might have a unique requirement for Ni. Therefore, for leguminous crops such as green bean and cowpea, Ni fertilization might be needed, particularly for those soils with high Zn or copper (Cu) concentrations, or with pH > 6.7 (Brown, 2006). Plants grown in high pH soils are vulnerable to Ni deficiency. Additionally, excessive use of Zn and Cu may induce Ni deficiency in soil because these three elements share a common uptake system. Over-liming, which raises pH excessively, also causes soil to be deficient in plant-available Ni. Thus, in soils that have high pH, either naturally or artificially, Ni fertilization may be needed to ensure good crop quality and yield.

Many proteins contain Ni (Thomson, 1982), Ni nutrition of higher plants and its physiological significance, especially to woody perennials, have received little attention. There are several enzyme systems (NiFe-hydrogenase, carbon monoxide dehydrogenase, acetyl-CoA decarbonylase synthase, methyl-coenzyme M reductase, superoxide dismutase, Ni-dependent glyoxylase, aci-reductone dioxygenase, and methyleneurease) in bacteria and lower plants that are activated by Ni (Walsh and Orme-Johnson, 1987; Mulrooney and Hausinger, 2003); however, the activation of urease appears, to date, to be the only enzymatic function of Ni in higher plants (Gerendás *et al.*, 1999). Urease contains two Ni ions at the active site (Ciurli, 2001). Ni can also replace Zn or Fe, and other metal ions, in certain other metalloenzymes of lower plants (Mulrooney and Hausinger, 2003). Circumstantial evidence indicates that ureide-transporting species, such as pecan, possess a higher Ni requirement than amide-transporting species (Wood, 2006), thus raising the possibility that ureide transporters might possess enzymes, other than urease, that require Ni for activation or for enhanced activity. Likely candidates are one or more enzymes affecting ureide catabolism.

5.1 Ureides, Free Amino Acids and Organic Acids in Relation to Nickel

The existence of nickel (Ni) deficiency is becoming increasingly apparent in crops, especially for ureide-transporting woody perennials, but its physiological role is poorly understood. There are quantitative differences in ureides, free amino acids, organic acids, and probably other organic substances between Ni-S and Ni-deficient foliage. Presently three basic pathways of urea biosynthesis are known to function in higher plants (Welch, 1981; Walker *et al.,* 1985; Zonia *et al.,* 1995). First, it is the terminal step of the ornithine cycle catalyzed by the mitochondrial arginase; this step involves arginine hydrolysis to ornithine and urea. Second, it is purine catabolism leading to formation of ureides, such as allantoin and allantoinic acid; the latter is hydrolyzed to glyoxylate and urea in the reaction catalyzed by allantoicase. This pathway is characteristic of the Fabaceae plants, which accumulate ureides as the major reserve form for translocating nitrogen from roots into shoots. In addition, arginine and ureides are often the reserve forms of nitrogen in the seeds. Finally, an additional pathway of urea synthesis from a nonprotein amino acid canavanine was found in *Canavalia* and soybean.

5.1.1 Ureides

It is known that higher plants primarily transport nitrogen compounds as either amides or ureides (Schubert and Boland, 1990). Those primarily transporting ureides tend to be hydrophiles (e.g. pecan, river birch) or tropical legumes [e.g., soybean, cowpea, Phaseolus beans, pigeon pea (*Cajanus cajan*), and mungbean (*Vigna radiata*)]. Ureides appear to be primarily metabolized in foliage and fruit (Pate and Atkins, 1983). Ureide catabolism eventually results in cleavage into urea and glyoxylate (with CO_2 and NH_4 as by-products) as urease is ever-present in plant organs and tissues. Since Ni is a necessary cofactor, Ni deficiency would be expected to influence ureide catabolism and consequently influence the availability of nitrogen from nitrogen-storage compounds for growth and development processes. Ureide concentrations are relatively high in dormant roots of ureide-transporting perennials, where they are exported via uphill xylem sap to emerging foliage, catabolized, and assimilated into amino acids, peptides, polypeptides, proteins, purines, and nucleic acids (Schubert and Boland, 1990). The catabolic pathway from purine nucleotides and ureides to urea or glyoxylate and NH_4 and CO_2 via xanthine, uric acid, allantoin, allantoate, ureidoglycine, and ureidoglycolate is well known (Mathews *et al.,* 2000). Urea is further broken down by urease (which is Ni dependent) to form CO_2 and 2 NH_4 (*Gerendás et al.,* 1999). Ni-deficient foliage exhibited relatively high levels of three intermediates of the ureide catabolic chain. The accumulation of these intermediates (xanthine, allantoic acid, and ureidoglycolate, plus urea accumulation was greatly reduced) indicates reduced enzyme activity in at least three ureide catabolic steps. The enzymes catabolizing ureidoglycolate has a direct or indirect metabolic role for Ni. The accumulated ureidoglycolate is potentially degraded by two enzymes (Wells and Lees, 1991; Munoz *et al.,* 2001). One is via ureidoglycolate amidinohydrolase (UGL; EC 4.3.2.3) to produce urea and glyoxylate. The second is via ureidoglycolate amidohydrolase (UGAH; EC 3.5.3.19) to produce NH_4, CO_2, and glyoxylate (Wells and Lees, 1991; Munoz *et al.*, 2001), and requires manganese as a cofactor (Lukaszeski *et al.*, 1992; Vadez and Sinclair, 2000).

Ureide-transporting plants come into view to possess the potential to catabolize ureides by at least two pathways (i.e. UGL or UGAH), with one predominating over the other. For example, in soybean, ureidoglycolate catabolism was approximately 95% via UGAH and approximately 5% via UGL; thus, urease did not appear to be essential for ureide catabolism in this species (Stebbins and Polacco, 1995), whereas in chickpea (*Cicer arietinum*) UGL appears to be the predominant path (Munoz *et al.*, 2001). The accumulation of ureidoglycolate and quantitative reduction in urea from Ni-deficient foliage suggests that the UGL activity is being impaired. Ureidoglycolase is a metalloenzyme whose activity is inhibited by EDTA and enhanced by several different bivalent cations (Munoz *et al.*, 2001). It is possible that one of these metal components is Ni. It is also possible that UGAH is a metalloenzyme requiring Ni to stimulate activity because it is also inhibited by EDTA and activity is enhanced by Mn^{2+}(Wells and Lees, 1991), of which Ni^{2+} is likely present as a trace contaminant. An investigation of the relevance of Ni to both UGL and UGAH enzymes is therefore, defensible. The accumulation of urea and reduced urease activity in zucchini and cowpea is attributed to insufficient Ni for sufficient activation of urease.

The unfavorable pressure of Ni deficiency on ureide catabolism in pecan indicates that comparable effects can occur in other ureide-transporting species. Examples of ureide-transporting genera are Acer, Alnus, Annona, Betula, Carpinus, Carya, Cercis, Chamaecyparis, Cornus, Corylus, Diospyros, Juglans, Nothofagus, Ostrya, Platanus, Populus, Pterocarya, Salix, and Vitis (Schubert and Boland, 1990). In addition, tropical legumes (e.g. soybean, Phaseolus beans, mungbean, and cowpea) are also candidates for disrupted nitrogen metabolism by Ni deficiency. Because the equilibrium between production of ureides and amides in certain crops is influenced by factors such as shade and soil nitrogen sources (nitrate favors amides and ammonium favors ureides), the expression of Ni deficiency could also be partially dependent on orchard or nursery management strategies (Schubert and Boland, 1990).

5.1.2 Amino Acids

Amino acids serve a wide range of functions in plants and are also the structural units by which proteins are made; thus, any disruption in free amino acid pools potentially disrupts a multitude of growth and developmental processes, plus the production of secondary products that influence plant pest resistance, toxicity, appearance, and taste. Little is known about the influence of Ni deficiency on amino acid metabolism. Ni deficiency in barley increased the pool of total free amino acids and nonprotein nitrogen compounds in shoots and seeds by 20% to 40% (Brown *et al.,* 1990). For shoots, the increase in the various amino acid pools was attributed to Asp, Glu, Val, Lys, His, Arg, Tyr, and Phe, whereas in seeds Glu, Gly, Ala, Val, Leu, Lys, His, Tyr, and Phe increased when plants were supplied with mineral nitrogen (except that Ala was decreased). The same general net accumulation of free amino acids occurred in Ni-deficient pecan foliage, except the accumulation was much greater than that of Ni-deficient barley (20%–40% versus 210% in Ni-deficient pecan), with six amino acids (Gly, Val, Ile, Tyr, Trp, and Arg) increasing and two (His and Glu) decreasing. The Ala pool was not depressed in Ni- deficient pecan, as was previously observed in barley. On the contrary, urea-grown plants subjected to Ni deficiency accumulated urea and exhibited a generalized depletion of free amino acids (Gerendás and Sattelmacher, 1997; Gerendás *et al.*, 1999).

It becomes noticeable that for crops as diverse as barley and pecan, when the nitrogen source is either mineral or stored, most free amino acid pools increase under conditions of Ni deficiency. This disruption of the various amino acid pools and associated apparent disruption in usage is potentially serious in that these amino acids are the building blocks of proteins needed for canopy deployment and other key growth processes. While end-product inhibition usually appears to be the principal mechanism regulating amino acid pools, it appears that the normal metabolism of free amino acid pools of several such monomers is being blocked by Ni deficiency.

Amino acids glycine and serine are precursors for phospholipids and purine synthesis and are the main sources of one-carbon units in higher plants (Morot-Gaudry *et al.,* 2001). The free Gly pool is normally much lower than that of Ser in green plants, with the ratio being dependent on the degree of irradiance and photorespiration, although the ratio is typically in the 3- to 10-fold range. The relative abundance of these two interconvertible amino acids (via Ser hydroxymethyl transferase; EC 2.1.2.1) was reversed, with the Gly pool being about 10-fold greater in foliage than the Ser pool (Morot-Gaudry *et al.,* 2001). This drastic equilibrium shift suggests either a major influence of Ni on photorespiration processes that produce Gly or the conversion of Gly to Ser. The impressive accumulation of free Gly raises the possibility that Ni depredation is limiting the activity of either Gly decarboxylase or Ser hydroxymethyl transferase, which acts to convert Gly to Ser in the mitochondria. These two enzyme systems comprise about 40% to 50% of the soluble proteins in the mitochondria, with their buildup potentially being dependent on leaf development (Vauclare *et al.,* 1996). Alternatively, the incorporation of Gly into purines could also be blocked. The branched-chain amino acid family contains Ile, Val, and Leu. The Asp-derived Ile pool increased to about 3-fold in Ni- deficient foliage, whereas the pyruvate-derived Val and Leu pools increased to about 3- and 8-fold, respectively, in Ni- deficient foliage. These amino acids share several common enzymes in their synthesis and their pathways are reported to be localized in chloroplasts (Bryan, 1990). Furthermore, the interruption in the aromatic amino acid pool by Ni deficiency indicates a likely disruption of the shikimate pathway existing in chloroplasts and cytosol. Several secondary metabolites depend upon this pathway for precursors; also, other key metabolites, such as indole-3-acetic acid (IAA) and lignin, are similarly linked to the shikimate pathway (Heldt, 2005). This pathway produces aromatic amino acids via three different routes, with the synthesis of chorismate being the last common link for all three routes (Morot-Gaudry *et al.*, 2001). The increase of Trp and Tyr pools in Ni-deficient foliage indicates the possibility of a Ni-associated disruption of enzyme activity or substrate availability downstream of these two aromatic amino acids. Such a disruption likely explains certain prominent symptoms of Ni deficiency in which dwarfed foliage and shoots and loss of apical dominance and rosetting suggest an auxin deficiency and exceedingly brittle wood suggests poor lignification (Wood *et al.,* 2004a). In the case of IAA, conversion from Tyr to either indole-3-pyruvic acid (via Trp transaminase and indole pyruvate decarboxylase) or tryptamine (via Trp decarboxylase and amino oxidase) or indole-3-acetaldehyde (via indole aldehyde dehydrogenase) appears to be directly or indirectly disrupted by Ni deficiency, although it appears that plants have different pathways for IAA biosynthesis at different developmental stages, tissues, and environmental conditions (Bartel, 1997). While lignin biosynthesis is complex, it also appears likely that its synthesis or incorporation into cell walls is being influenced by Ni, perhaps in the shikimate pathway downstream of Phe ammonia lyase in the phenylpropanoid pathway because Phe did not accumulate in Ni- deficient foliage. This apparent influence of Ni deficiency on the shikimate pathway, from which many defense-related biochemicals are derived, indicates that disease resistance is likely to be influenced by Ni deficiency.

Although the Arg pool of Ni- deficient foliage was not big relative to that of many other amino acids, the percentage increase in Arg (21.4-fold), an increase of about 10-fold that of other amino acids, indicates that Ni deficiency substantially influences Arg catabolism and likely affects the Orn cycle because Arg is an intermediate. The citrulline pool, a nonprotein nitrogen compound, also increased, and is an upstream intermediate between Orn and Arg. A similar increase in Arg was reported by Shimada *et al.* (1980) for tomato (*Lycopersicon esculentum*) growing under Ni depredation, where it has been suggested that urea accumulation may have interfered with Orn cycle intermediates via feedback inhibition (Gerendás *et al.,* 1999).

The His pool was affected by Ni depredation more so than any other measured amino acid, with the pool being only about 4% of that in Ni-S foliage. Histidine synthesis is an extremely energy-consuming process and thus appears to take place in chloroplasts (Morot-Gaudry *et al.,* 2001). The notably low level of His raises the possibility that (1) chloroplasts were poorly developed at this stage in Ni-deficientfoliage; (2) Ni deficiency somehow reduces the availability of energy needed for synthesis (i.e. 41 ATPs/His molecule) because the necessary Gln and Glu substrates appeared to be unaffected by Ni deficiency; or (3) the activity of one or more enzymes of the His biosynthetic pathway was being reduced because of Ni deficiency.

The synthesis of amino acids is complex, with several amino acids being made by more than one synthetic pathway, as influenced by environment, age, tissue, etc. This study indicates that plant Ni nutritional physiology influences either directly or indirectly the pools of several essential amino acids as well as potentially several different amino acid pathways, but does not elucidate the nature of this Ni-amino acid interaction. However, results indicate a possible role of Ni in certain aspects of amino acid metabolism.

5.1.3 Organic Acids

There is a dearth of information on the influence of Ni deficiency on organic acids in plant tissues. In Ni-deprived barley, malate decreased in shoots but was unchanged in seeds (Brown *et al.,* 1990). In the case of Fe-deficient sugar beets (*Beta vulgaris*), certain TCA organic acids (i.e. citrate and malate) increased in moderately chlorotic leaves (López-Millán *et al.,* 2001). The increase in lactic acid is especially puzzling in that it is normally a product of glycolysis under anaerobic conditions and the Ni-deficient foliage of this study was not growing in water-saturated soil and did not appear to experience oxygen depredation. It is possible that the accumulation of these acids poisons cells and tissues of Ni-deficientfoliage rather than the toxic accumulation of urea (Walker *et al.,* 1985). The concentration drop in citric acid, a TCA cycle acid, is evidence that Ni deficiency may be influencing the availability of acetyl-CoA, which is key to the conversion of oxaloacetate to citric acid in the mitochondria. The accumulation of lactic acid, Val, Leu, Try, Trp, and Gly, all potentially derived from intermediates in glycolysis, is additional evidence of impaired respiration. Thus, Ni depredation can adversely influence carbohydrate oxidation via indirect effects on glycolysis and a direct effect on the TCA cycle. Further evidence for impairment of acetyl-CoA is the fact that the three nitrogen-containing components (adenine, panthothenic acid, and β-mercaptoethylamine) of CoA are likely to be influenced by low Ni-associated disruption of normal nitrogen metabolism as a consequence of ureide and amino acid metabolism. It is noteworthy that a slowdown of the TCA cycle results in considerable reduction in the ATP, NADH, and FADH2 available for the electron transfer/oxidative phosphorylation stage of respiration. Thus, the low vegetative growth vigor of Ni-deficient plants is likely due in part to a disruption of respiration by Ni deficiency. Because Ni deficiency symptoms in pecan are first apparent at bud break as poor growth vigor, when Suc and other carbohydrates are translocating from roots to the developing canopy (Wood, 1987), Ni-deprived cells might thus potentially experience a disruption in carbohydrate oxidation sufficient to alter growth. This raises the possibility that disrupted carbohydrate oxidation and metabolism, and not just nitrogen-associated metabolism, contribute to the morphological symptoms of Ni-deficient plants (Wood *et al.*, 2004a). The wide range of morphological and growth symptoms exhibited by Ni-deficient woody perennials suggests that Ni might affect plant metabolism by influencing a key metabolic intermediate. A viable candidate is acetyl-CoA, as it is probably the most central intermediate in cellular metabolism, provides a link between several pathways (e.g. glycolysis, fatty acid oxidation, amino acids, flavonoids, sterols, isoprenoid derivatives, TCA cycle), does not appear to cross membranes, and appears to be influenced by Ni. Influence of Ni-deficient on amino acid biosynthesis and organic acid metabolism in pecan leaves was studied by Buchanan *et al.* (2000).

Aerobic respiration is the most *common* form of energy release in organisms, with glycolysis of Glc resulting in pyruvate. The converse is anaerobic respiration, in which glycolysis yields lactic acid or ethanol instead of water. During anaerobic glycolysis, if glycolysis is proceeding at a high rate (or in anaerobic organisms), there is substantial oxidation of NADH. The large quantity of NADH produced is oxidized by reducing pyruvate to lactate, leading to a buildup of lactic acid (King, 2005). Thus, pyruvate undergoes oxidation with products entering the TCA cycle via acetyl-CoA, whereas lactic acid is a product of anaerobic glycolysis. If the energy level in these cells is sufficiently high, the carbons of pyruvate can be diverted back to Glc via the gluconeogenesis pathway (King, 2005). The high concentration of lactic acid is puzzling because it is typically associated with anaerobic respiration, yet the plants used in this study did not appear to be anaerobic, nor were soil conditions anaerobic. Although not measured in this study, it is possible that the activity of pyruvate dehydrogenase (EC 1.2.1.51) is being suppressed as a consequence of Ni deficiency. It is an iron-requiring enzyme (Inui *et al.,* 1987) and thus there is the possibility of it also being activated by Ni.

The occurrence of oxalic acid is typically greatest in leaves and lowest in roots. Oxalic acid content is related to photosynthesis and carbohydrate metabolism and varies according to tissue age, season of the year, climatic stresses, and soil type. Oxalic acid can be harmful to plants because it is produced by certain pathogenic fungi and often plays an essential role in pathogenicity (Caliskan, 2000). *Botrytis cinerea*, a phytopathogenic fungus, secretes oxalic acid as a pathogenicity factor suspected of facilitating degradation of cell walls and repressing plant defenses. Thus, degradation of oxalic acid during host-pathogen interaction potentially interferes with the infection process. *Sclerotinia minor* is the causal agent of Sclerotinia blight, a highly destructive disease of peanut (*Arachis hypogaea*). Evidence indicates that oxalic acid serves as a pathogenicity factor during Sclerotinia infection. Direct application of oxalic acid to stem or leaf tissue causes tissue injury and wilting, similar to plant responses to fungal infection by *Sclerotinia sclerotiorum* (Livingstone *et al.*, 2005). The double amount of oxalic acid present in Ni-deficient foliage compared to Ni-S foliage, in addition to lactic acid, is a potential marker for identifying Ni-deficientfoliage and is thus a potential indicator of cellular damage. It appears that the relatively high levels of lactic and oxalic acids in young foliage is likely a contributing factor to leaflet dwarfing (i.e. mouse-ear), a key identifying factor of Ni deficiency in woody perennials.

Metabolism of free amino acids and organic acids is linked to ureide metabolism. For example, ureidoglycolate can be converted to glyoxylate, which is then converted to Gly or urea and pyruvate. The pyruvate can then be converted to Val or lactic acid. In plants, two main reactions involving Gly are potential sources of C1 units (Mouillon *et al.*, 1999). The greater level of Val and lactic acid content in Ni-deficient foliage may possibly be linked to a shift in pyruvate metabolism.

6. NICKEL AND CROP GROWTH

The first evidence of a response of a field crop to application of a nickel fertilizer was demonstrated in 1945 for potato (*Solanum tuberosum* L.), wheat (*Triticum aestivum* L.), and bean (*Phaseolus vulgaris* L.) crops (Roach and Barclay, 1946). In these crops, the application of a dilute nickel spray resulted in a significant increase in yield. These experiments were conducted on the 'Romney Marshes' of England, a region that is well known for its trace mineral deficiencies, particularly of manganese and zinc. These experiments were conducted very carefully and excluded the possibility that the nickel applied was merely substituting for manganese, zinc, iron, copper, or boron, suggesting that the growth response was indeed due to the application of nickel. Interestingly, the soils of this region may be low in nickel since the conditions that limit manganese and zinc availability in these soils (acid sands of low mineral content) would also limit nickel availability to crops, and the concentrations of nickel provided were appropriate based on the current knowledge of nickel demand.

Many early reports of the role of nickel in agricultural productivity have been questioned since they did not adequately exclude the possibility that nickel was acting directly as a fungicidal element (Welch, 1981). The clearest agronomic responses to nickel have been observed when nitrogen is supplied as urea or by nitrogen fixation. The most illustrative example of the relationship between nickel and urea metabolism is provided from studies with foliar urea application and tissue-culture growth of plants. Plants without a supply of nickel have low urease activity in the leaves, and foliar application of urea leads to a large accumulation of urea and severe necrosis of the leaf tips (Eskew and Welch, 1982). Nicoulaud and Bloom (1998) observed that nickel, provided in the nutrient solution of tomato (*Lycopersicon esculentum* Mill.) seedlings growing with foliar urea as the only nitrogen source, significantly enhanced growth. The effect of nickel was more consistent with its role in urea translocation than that on urease activity directly (Nicoulaud and Bloom, 1998). Brown *et al.*(1987) suggested that nickel has a role in the transport of nitrogen to the seed thereby influencing plant senescence and seed viability. The first demonstration of an agricultural Ni deficiency did not occur until 2004 (Wood *et al.*, 2004), when it was observed in pecan (*Carya illinoinensis*). Nickel deficiency in pecan is associated with a physiological disorder 'mouse-ear' which occurs sporadically, but with increasing frequency, throughout the southeastern United States (portions of South Atlantic region) where it represents a substantial economic impact. In agreement with the results of Brown *et al.* (1987), Ni deficiency in pecan results in a disruption of nitrogen metabolism and altered amino acid profiles (Bai *et al.*, 2006).

The value of addition of nickel to Murashige and Skoog plant tissue-culture medium was shown by Witte *et al.* (2002). These authors suggested that the lack of nickel and urease activity may represent a stress factor in tissue culture and recommended that the addition of 100 nM Ni be adopted as a standard practice. The benefits of adding nickel to solution cultures was also demonstrated by Khan *et al.* (1997), who determined that a mixture of 0.05 mg Ni L^{-1} and 20% nitrogen as urea resulted in optimal growth of spinach (*Spinacia oleracea* L.) under hydroponic conditions. High Ni concentrations in growth media severely retards seed germinability of many crops. This effect of Ni is a direct one on the activities of amylases, proteases, and ribonucleases, thereby affecting the digestion and mobilization of food reserves in germinating seeds. At vegetative stages, high Ni concentrations retard shoot and root growth, affect branching development, deform various plant parts, produce abnormal flower shape, decrease biomass production, induce leaf spotting, disturb mitotic root tips, and produce Fe deficiency that leads to chlorosis and foliar necrosis. Additionally, excess Ni also affects nutrient absorption by roots, impairs plant metabolism, inhibits photosynthesis and transpiration, and causes ultrastructural modifications. Ultimately, all of these altered processes produce reduced yields of agricultural crops when such crops encounter excessive Ni exposures.

7. CONCLUSIONS

The establishment of nickel as an essential element highlights the limitations of the current definition of essentiality of nutrients as applied to plants (Arnon and Stout, 1939). Total Ni concentration commonly ranges from 5 to 500 mg kg^{-1}, with an average of 50 mg kg^{-1} in soils. Nickel levels in terrestrial and aquatic organisms can vary over several orders of magnitude. The amount and distribution of Ni-containing complexes would differ depending on plant age. Nickel (Ni), comparatively the recently discovered essential element, is unique among plant nutrients in that its metabolic function was determined well before it was confirmed that its deficiency could disrupt plant growth. Nickel is a functional constituent of seven enzymes. Among the seven, urease (EC 3.5.1.5) (i.e., urea amidohydrolase) is extremely important to N metabolism in plants.

Ni is a constituent of several metallo-enzymes such as urease, superoxide dismutase, NiFe hydrogenases, methyl coenzyme M reductase, carbon monoxide dehydrogenase, acetyl coenzyme-A synthase, hydrogenases, and RNase-A. Therefore, Ni deficiencies in plants reduce urease activity, disturb N assimilation, and reduce scavenging of superoxide free radical. Nickel clearly has a significant effect on the productivity of field-grown, nitrogen-fixing plants, those in which ureides are a significant form of nitrogen and those utilizing urea as a primary nitrogen source. The symptoms of nickel deficiency in barley, wheat, and oats observed by Brown *et al.* (1987) and Wood *et al.* (2004) are consistent with the observations made in nickel-deficient animals and are indicative of a role of nickel in nitrogen metabolism that cannot be easily explained through an exclusive role of nickel in urease. Under Ni-deficient conditions, barley plants fail to produce viable grain because of a disruption of the maternal plant's normal grain-filling and maturation processes that occur following formation of the grain embryo. Excess Ni, however, affects nutrient absorption by roots, impairs plant metabolism, inhibits photosynthesis and transpiration, and causes ultrastructural modifications.

The researchers still debate over the causes and mechanisms of hyperaccumulation, which may depend on the morphological and physiological characteristics of particular plant species, the capacity of Ni to overcome the physiological barriers, etc. Plants comprise several such barriers that curb the entry of heavy metals into the shoots, primarily the plasmalemma and endodermis at the cell and tissue levels, respectively.

In bacteria, Ni participates in several important metabolic reactions such as hydrogen metabolism, methane biogenesis, and acetogenesis. The diverse known functions of nickel in bacteria suggests that nickel may indeed play a role in many, yet undiscovered processes in plants. Although Ni is metabolically important in plants, it is toxic to most plant species when present at excessive amounts in soil and in nutrient solution.

The comparison with other heavy metals is used to disclose the general mechanisms that disturb plant mineral nutrition, water regime, photosynthesis, and morphogenesis as well as the common cell responses aimed at detoxification of heavy metals. Last but not the least, enough scope have been left for extending judicious research with nickel with reference to metabolic studies in diverse plant and crop species for precise understanding of nickel requirements as an essential element at different growth stages in varied metabolic systems without toxicity. Such studies have great relevance in crop improvement programme under changing environment too.

REFERENCES

Arnon, D.I., P.R. Stout. 1939. The essentiality of certain elements in minute quantity for plants with special reference to copper. *Plant Physiol*. 14:371–375.

Bai, C., C.C. Reilly, B.W. Wood. 2006. Nickel deficiency disrupts metabolism of ureides, amino acides, and organic acids of young pecan foliage. *Plant Physiol*. 140:433–443.

Bartel, B. 1997. Auxin biosynthesis. *Annu. Rev. Plant Physiol. Plant Mol. Biol*. 48: 51–66.

Brown, P.H., R.M. Welch, J.T. Madiso, 1990. Effect of nickel deficiency on soluble anion, amino acid and nitrogen levels in barley. *Plant Soil* 125: 19–27

Brown, P. H. 2006. "Nickel." In *Handbook of Plant Nutrition*, edited by A. V. Barker and D. J. Pilbeam, 395–410. Boca Raton, FL: CRC Press Taylor & Francis Group.

Brown, P.H., L. Dunemann, R. Schultz, H. Marschner. 1989. Influence of redox potential and plant species on the uptake of nickel and cadmium from soils. *Zh. Pflanzenernahr. Bodenkd*. 152:85–91.

Brown, P.H., R.M. Welch and E. E. Cary. 1987. Nickel: a micronutrient essential for higher plants. *Plant Physiol*. **85**, 801-803.

Brown, P.H., R.M. Welch, E.E. Cary, R.T. Checkai. 1987b. Beneficial effects of nickel on plant growth. *J. Plant Nutr*. 10:2125–2135.

Brown, P.H., R.M. Welch and E.E. Cary 1987a. Nickel: A micronutrient essential for higher plants. *Plant Physiol*. 85:801–803.

Bryan, J.K. 1990. Advances in the biochemistry of amino acid biosynthesis. *In*B.J. Miflin, P.J. Lea, eds., The Biochemistry of Plants, Vol 16. Academic Press, New York, pp 161–195.

Buchanan, B.B., W. Gruissem, R. Jones. 2000. Biochemistry and Molecular Biology of Plants. American Society of Plant Physiologists, Rockville, MD.

Caliskan, M. 2000. The metabolism of oxalic acid. *Turk. J. Zool*. 24: 103–106.

Cammack. R. 1995. Splitting molecular hydrogen. *Nature* 373:556–557.

Cataldo, D.A., K.M. McFadden, T.R. Garland, R.E. Wildung. 1988. Organic constituents and complexation of nickel(II), iron(III), cadmium(II) and plutonium(IV) in soybean xylem exudates. *Plant Physiol*. 86:734–739.

Cataldo, D.A., T.R. Garland, R.E. Wildung. 1978. Nickel in plants. I. Uptake kinetics using intact soybean seedlings. *Plant Physiol*. 62: 5636–5665.

Ciurli, S. 2001. Electronic structure of the nickel ions in the active site of urease. *Chemistry (Easton)* 2001: 99–100.

Dixon, N.E., C. Gazzola, R.L. Blakeley, B. Zerner. 1975. Jack bean urease (EC 3.5.1.5). A metalloenzyme. A simple biological role for nickel. *J. Amer. Chem. Soc*. 97:4131–4133.

Epstein, E., A.J. Bloom. 2004. Mineral Nutrition of Plants: Principles and Perspectives. 2nd edition. Sinauer Associates, Sunderland, MA, pp. 45.

Eskew, D.L., Welch, R.M., and Cary, E.E.1983. Nickel: An essential micronutrient for legumes and possibly all higher plants. *Science*, 222: 621–623.

Eskew, D.L., R.M. Welch. 1982. Nickel supplementation 1 $\mu g\ L^{-1}$ prevents leaflet tip necrosis in soybeans grown in nutrient solutions purified using 8 hydroxy quinoline-controlled pore glass chromatography. *Plant Physiol*. (Supp.) 69:134,.

Gerendas, J. and Sattelmacher, B. 1997. Significance of Ni supply for growth, urease activity and the concentrations of urea, amino acids and mineral nutrients of urea-grown plants. *Plant Soil*, 190: 153-162.

Gerendas, J., J.C. Polacco, S.K. Freyermuth, B. Sattelmacher. 1999. Significance of nickel for plant growth and metabolism. *J. Plant Nutr. Soil Sci.* 162:241–256.

Heldt, H.W. 2005. Plant Biochemistry, Ed 3. Elsevier Academic Press, Burlington, M.A.

Holland, M.A., J.C. Pollacco. 1992. Urease-null and hydrogenase-null phenotypes of a phylloplane bacterium reveal altered nickel metabolism in two soybean mutants. *Plant Physiol.* 98:942–948.

Inui, H., K. Ono, K. Miyatake, Y. Nakano, S. Kitaoka. 1987. Purification and characterization of pyruvate:NADP+ oxidoreductase in *Euglena gracilis*. *J. Biol. Chem.* 262: 9130–9135.

Khan, N.K., M. Wantanabe, Y. Wantanabe. 1997. Effect of different concentrations of urea with or without nickel on spinach (*Spinacia oleraceae* L.) under hydroponic culture. In: T. Ando, K. Fujita, T. Mae, S. Matsumoto, S. Mori, J. Sekiya, eds. Plant Nutrition for Sustainable Food Production and Environment. Dordrecht: Kluwer Academic Publishers, pp. 85–86.

King, M.W. 2005. Glycolysis. The Medical Biochemistry Page, Indiana University, Terre Haute, In, http://web.indstate.edu/thcme/mwking/glycolysis.html

Kochian, L.V. 1991. Mechanisms of micronutrient uptake and translocation in plants. In: J.J. Mortvedt, ed. Micronutrients in Agriculture, 2nd ed. SSSA book series #4. Madison, WI: Soil Science Society of America, pp. 229–296.

Livingstone, D.M., J.L. Hampton, P.M. Phipps, E.A. Grabau. 2005. Enhancing resistance to *Sclerotinia minor* in peanut by expressing a barley oxalate oxidase gene. *Plant Physiol*. 137:1354–1362.

López-Millán, A.F., F. Morales, A. Abadía, J. Abadía. 2001. Changes induced by Fe deficiency and Fe supply in the organic acid metabolism of sugar beet (*Beta vulgaris* L.) leaves. *Physiol. Plant.* 112: 31–38.

Lukaszeski, K.M., D.G. Blevins, D.D. Randall. 1992. Asparagine and boric acid cause allantoate accumulation in soybean leaves by inhibiting manganese-dependent allantoate amidohydrolase. *Plant Physiol.* 99: 1670–1690.

Maroney, M.J. 1999. Structure/function relationships in nickel metallobiochemistry. *Curr. Opin. Chem. Biol.* 3:188–199.

Mathews, C.K., K.E. van Holde, K.G. Ahern 2000. Biochemistry. Addison-Wesley, New York.

Morot-Gaudry J.F., D. Job, P.J. Lea 2001. Amino acid metabolism. *In* P.J. Lea, J.F. Morot-Gaudry, eds, Plant Nitrogen. Springer-Verlag, New York, pp 167–211.

Mouillon, J.M., S. Aubert, J. Bourguignon, E. Gout, R. Douce, F. Rébeillé 1999. Glycine and serine catabolism in non-photosynthetic higher plant cells: their role in C1 metabolism. *Plant J.* 20: 197–205.

Mulrooney, S.B. and R.P. Hausinger 2003. Nickel uptake and utilization by microorganisms. *FEMS Microbiol. Rev.* 27: 239–261.

Munoz, A., P. Piedras, M. Aguilar and M. Pineda 2001. Urea is a product of ureidoglycolate degradation in chickpea: purification and characterization of the ureidoglycolate urea-lyase. *Plant Physiol.* 125: 828–834.

Neumann, P.M., A. Chamel. 1986. Comparative phloem mobility of nickel in nonsenescent plants. *Plant Physiol.* 81, 689-691.

Nicoulaud, B.A.L., A.J. Bloom 1998. Nickel supplements improve growth when foliar urea is the sole nitrogen source for tomato. *J. Amer. Soc. Horti. Sci.* 123:556–559.

Pate, J.S. and C.A. Atkins. 1983. Nitrogen uptake, transport and utilization. *In:* W.J. Broughton, ed, Nitrogen Fixation. Oxford University Press, London, pp 245–298.

Roach, W.A. and C. Barclay 1946. Nickel and multiple trace deficiencies in agricultural crops. *Nature* 157: 696.

Schubert, K.R. and Boland, M.J. 1990. The ureides. *In:*B.J. Miflin, P.J. Lea, eds., The Biochemistry of Plants, Vol 16. Academic Press, San Diego, pp 197–283.

Seregin, I. V. and A. D. Kozhevnikova 2006. Physiological role of nickel and its toxic effects on higher plants. *Russian Journal of Plant Physiology,* 53: 257–277.

Sheoran, I.S., Singal, H.R., and Singh, R. 1990. Effect of cadmium and nickel on photosynthesis and the enzymes of the photosynthetic carbon reduction cycle in pigeonpea (*Cajanus cajan* L.). *Photosynth. Res.* 23: 345–351.

Shimada, N., T. Ando, M. Tomiyama, H. Kaku 1980. Role of nickel in plant nutrition. Effects of nickel on the growth of tomato and soybean. *Japanese J. Soil Sci. Plant Nutr.* 51: 493–496.

Stebbins, N., J.C. Polacco 1995. Urease is not essential for ureide degradation in soybean. *Plant Physiol.* 109: 169–175.

Tan, X, I. Kagiampakis, I. V. Surovtsev, B. Demeler and P. A. Lindahl 2007. Nickel-dependent oligomerization of the alpha subunit of acetyl-coenzyme a synthase/carbon monoxide dehydrogenase. *Biochemistry,* 46: 11606–11613.

Thomson, A.J. 1982. Proteins containing nickel. *Nature* 298: 602–603.

Tiffin, L.O. 1971. Translocation of nickel in xylem exudates of plants. *Plant Physiol.* 48:273–277.

Vadez, V. and T.R. Sinclair. 2000. Ureide degradation pathways in intact soybean leaves. *J. Exp. Bot.* 51: 1459–1465.

Vauclare, P., N. Diallo, J. Bourguignon, D. Macherel and R. Douce 1996. Regulation of the expression of the glycine decarboxylase complex during pea leaf development. *Plant Physiol.* 112: 1523–1530.

Walker, C.D., Graham, R.D., Madison, J.T., Cary, E.E. and Welch, R.M. 1985. Effects of Ni deficiency on some nitrogen metabolites in cowpeas (*Vigna unguiculata* L. Walp.), *Plant Physiol.* 79: 474–479.

Walsh, C.T. and W.H. Orme-Johnson. 1987. Nickel enzymes. Biochemistry 26: 4901–4906.

Watt R.K. and P.W. Ludden 1999. Nickel-binding proteins. *Cell Mol Life Sci.* 56: 604-625.

Welch, R.M. 1981. The biological significance of nickel, *J. Plant Nutr.*, 3, 345–356.

Wells, X.E. and E.M. Lees. 1991. Ureidoglycolate amidohydrolase from developing French bean fruits (*Phaseolus vulgaris* [L.]) *Arch. Biochem. Biophys.* 287: 151–159.

Witte, C.P., S.A. Tiller, M.A. Taylor and H.V. Davies. 2002. Addition of nickel to Murashige and Skoog medium in plant tissue culture activates urease and may reduce metabolic stress. *Plant Cell Tissue Organ Culture* 68:103–104.

Wood, B.W., C.C. Reilly and A.P. Nyczepir 2004a. Mouse-ear of pecan: I. Symptomology and occurrence. *Hort. Sci.* 38: 87–94.

Wood, B.W., C.C. Reilly and A.P. Nyczepir 2004. Nyczepir. Mouse-ear of Pecan: A nickel deficiency. *HortScience* 39 (6):1238–1242.

Wood, B.W. (2006) Field deficiency of nickel in trees: symptoms and causes. *Acta Hortic.*

Yusuf M, Fariduddin Q, Hayat S and Ahmad A. 2011. Nickel: an overview of uptake, essentiality and toxicity in plants. *Bull Environ Contam Toxicol.* 86(1):1-17.

Zonia, L.E., Stebbins, N.E. and Polacco, J.C. 1995. Essential role of urease in germination of nitrogen-limited *Arabidopsis thaliana* seeds, *Plant Physiol.* 107, 1097–1103.

Section II:

PLANT RESPONSES TO CHANGING ENVIRONMENT AND ENVIRONMENTAL STRESSES

8

CAPACITIES AND UNCERTAINTIES IN MONITORING THE CARBON DIOXIDE FLUXES AND CARBON POOLS AT DIFFERENT SCALES IN TERRESTRIAL ECOSYSTEMS

Rakesh Pandey[1]*, Vijay Paul[1], Vinay Sehgal[2], Madan Pal Singh[1] and Kalikinkar Bandyopadhyay[2]

[1] Division of Plant Physiology, [2] Division of Agricultural Physics, Indian Agricultural Research Institute (IARI), New Delhi-110 012, India
** Corresponding author :* r_pan_pdcsr@yahoo.co.in

INTRODUCTION

The IPCC reports have brought about a general and scientific awareness about the impending climate change and the problems associated with it. This awareness has lead to a global consensus that concerted efforts need to be made to reduce emission of the greenhouse gases in to the atmosphere so that mitigation of the effects of climate change can be achieved and food production is not threatened (IPCC report, 2007; Lal, 2007). Under Kyoto Protocol, nations therefore, agreed on three special mechanisms for emission reduction i.e. (i) Joint implementation (ii) Clean development mechanism and (iii) Emission trading. Also, under the United Nations Framework Convention on Climate Change (UNFCCC) and Kyoto protocol, the countries need to maintain and update the greenhouse gas emission inventories and category on land use change and forestry. The importance of accurate and sustained monitoring of carbon fluxes and pools over large areas in the present era of changing climate have a considerable importance and is not a simple and routine task. In fact, monitoring of CO_2 by renowned scientist Charles David Keeling (1928-2005) has been acknowledged as the single most important environmental data set taken in the 20th century.

The present review therefore hypothesizes that the procedures for monitoring and estimation of CO_2 and carbon pools in the vegetation and soil are adequate and their estimates have high degree of certainty. Moreover, this will lead to development of sound policy decisions on mitigation and adaptation strategies. The objectives of the present review is therefore to understand the monitoring and role of important CO_2 fluxes - photosynthesis and respiration in relation to the carbon pools present in the vegetation and soil and how they affect global carbon cycle. Also attempt has been made to review the monitoring methods at different scales because literature is seldom available that integrates the methods of monitoring the biotic and soil carbon, their partitioning and their interactions. Aspects related to the quantification of the aboveground and belowground carbon stocks and their functioning i.e. photosynthesis, respiration, growth of aboveground parts, net primary productivity, responses to elevated CO_2, root growth and soil organic matter carbon dynamics have been described highlighting the multidisciplinary approaches. Finally, the complexities and uncertainties in monitoring, understanding the ecosystem responses due to multiple processes to climate change and their implications in terms of implementation of policy issues have been analyzed.

2. SOURCES AND SINKS OF CARBON AND THEIR DYNAMICS

On of plant physiological basis, when photosynthesis is greater than respiration and fossil fuel combustion, the land biosphere serves as a sink for carbon. When carbon losses are greater than photosynthesis the land acts as carbon source. The annual uptake of CO_2 (photosynthesis) by the terrestrial biosphere approximates the loss due to respiration and therefore, it is in balance. However, this absorption is sensitive to climate as well as to atmospheric CO_2 concentration (Cox *et al.*, 2000).

The different ecosystem i.e. land, atmosphere, ocean human spheres can be integrated in a global carbon cycle. There are different forms of carbon in carbon cycle. The principal global carbon pools and fluxes between the terrestrial ecosystem and atmosphere are shown in (**Fig. 1**) linked with the exchange of CO_2 gas which also is a potent greenhouse gas. The current level of carbon dioxide in the atmosphere is about 387 μmol mol^{-1} and accounts for about 60% of the global warming potential. It will increase to nearly 450 μmol mol^{-1} by 2050 (Karl *et al.*, 2009). The components of an ecosystem absorb and release carbon from and to the atmosphere. If they show a net emission of carbon then they are called sources of carbon and if there is a net absorption then they are called as sinks of carbon. The sinks of carbon associated with carbon cycle consist of atmospheric absorption of 4.1 PgC yr^{-1}, land 1.5 PgC yr^{-1} and ocean 2.3

$PgC\ yr^{-1}$ and there is an unknown terrestrial sink of about 1.5 $PgC\ yr^{-1}$ (1 Pg = 10^{15} g = 1 Gt; where Pg refers to Petagram and Gt refers to Gigaton). The source of carbon into the carbon cycle consists of 7.9 $PgC\ yr^{-1}$ from fossil fuel combustion and cement manufacture and 1.0 $PgC\ yr^{-1}$ from deforestation, biomass burning and soil cultivation (based on decadal average 2000-2010 values from www.globalcarbonproject.org). Also, currently about half (45%) of the total emissions are being absorbed by atmosphere, 26% by ocean and about 29% by land ecosystems (Le Quere *et al.*, 2009). Presently, about 40% of the Earth's land surface is used as cropped land and pasture and is estimated to emit about 10-12% of global anthropogenic emissions (Foley *et al.*, 2005; Ortiz-Monasterio *et al.*, 2010) and combined with agricultural activities and land use changes contribute about one third of the total anthropogenic emissions (Lal, 2010). Globally, soil contain approximately 1500 Pg of organic carbon (Batjes, 1996), roughly three times the amount of carbon in vegetation and twice the amount in the atmosphere (IPCC, 2001). The CO_2 efflux from soil is an important component of global carbon cycle because the amount of carbon in soil organic matter is twice as large that in atmosphere. The soil CO_2 efflux represents the largest source of CO_2 from terrestrial ecosystems to the atmosphere. The total below-ground carbon flux i.e. the plant carbon allocation belowground is a critical determinant of soil carbon sequestration.

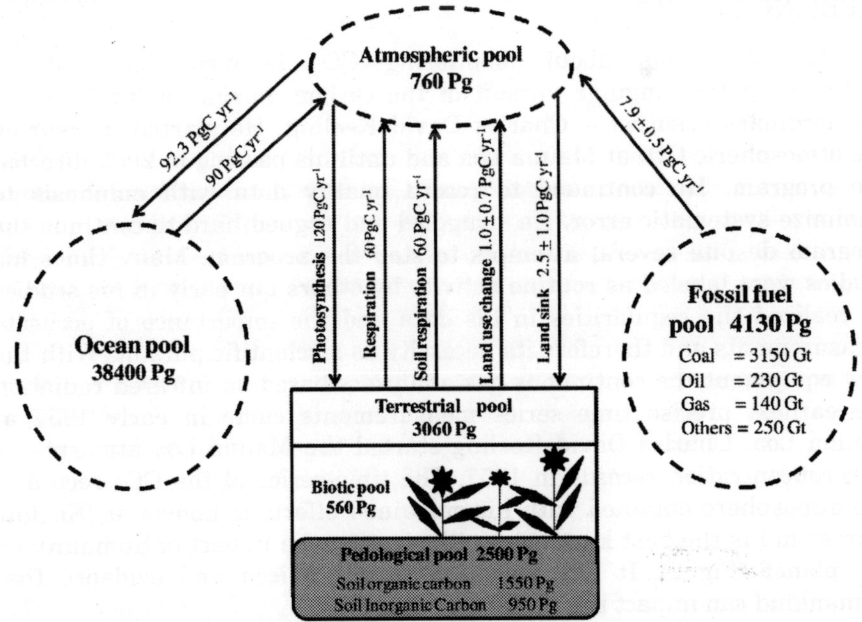

Fig. 1. The global carbon cycle showing different carbon pools and fluxes. The size of pools is represented from Lal (2008). The changes in net source or sink levels ($PgC\ yr^{-1}$) in different pools represent decadal average (2000-2010) taken from *www.globalcarbonproject.org.*

From the studies and extrapolation about earth climate in future, the terrestrial carbon sink will not be able to balance fossil fuel emissions. Therefore, it is important to understand the contemporary carbon cycle and the strength of carbon sources and sinks with their spatial and temporal variation (Canadell *et al.*, 2000). The source and sinks of carbon have changed over the part 200 years (Schimel, 2004) as shown in the Table 1. It indicates that 265 Gt + 30 Gt i.e., 295 Gt C has been released from fossil fuel burning and from land use change from terrestrial biosphere, respectively. This has gone to the atmosphere and oceans (180+115 = 295 Gt), respectively.

Table 1. The carbon cycle over the past two centuries

Reservoir	Source (GtC since 1800)	Sink (GtC since 1800)	Net change (GtC since 1800)
Fossil carbon	265	0	-265
Oceans	0	115	+115
Terrestrial biosphere	140	110	-30
Atmosphere	0	180	+180

(Source: Schimel, 2004)

3. THE LONGEST RECORD OF ATMOSPHERIC CO_2 AT MAUNA LOA: "KEELING CURVE"

The discussion about monitoring CO_2 is incomplete without highlighting the singular pursuit of the carbon dioxide molecule by an extraordinary scientist – Charles David Keeling. He started measuring the atmospheric CO_2 at Mauna Loa and until his passing in 2005 directed the program. He continued to record quality data with emphasis to minimize systematic error. He struggled and argued hard to continue the program despite several attempts to stop the program. Many times his studies were labeled as routine activity by others but early in his studies he realized the regularities in his data and the importance of accurate measurements and therefore its necessity as a scientific pursuit. With the new equipment the continuous gas analyzers based on infrared radiation the earliest precise time series measurements came in early 1957 at Mauna Loa. Charles David Keeling started the Mauna Loa atmospheric CO_2 concentration records in 1957. The time-series of the CO_2 record of the atmosphere obtained with his sustained efforts is known as 'Keeling Gurve' and is the best known icon illustrating the impact of humanity on the planet (Fig 2). It has become a standard icon and evidence that humankind can impact the global atmosphere.

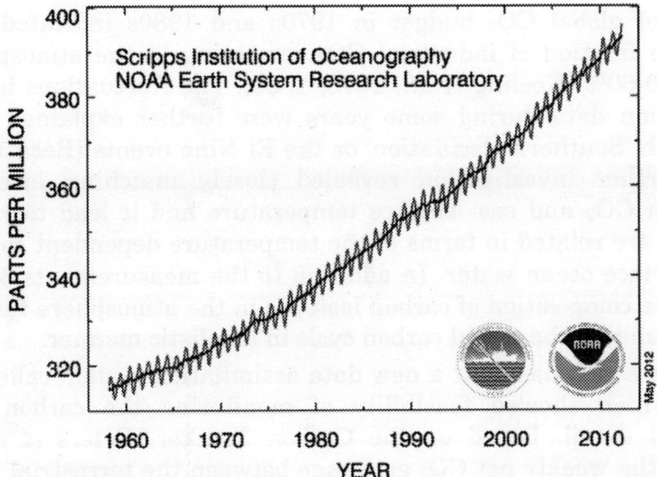

Fig. 2 Atmospheric CO_2 at Mauna Loa observatory - "The Keeling Curve" (Source: *http://www.esrl.noaa.gov/gmd/ccgg/trends/#mlo_full*).

Before his work, the published data pointed to high variability in CO_2 concentration in the atmosphere. Initially during his earlier studies (1953-56) he took air samples every few hours over a full day and night and found consistent diurnal pattern from several undisturbed locations. The air contained higher CO_2 concentration at night than during day. These measurements showed that air in the afternoon contained nearly same amount of CO_2 (about 310 ppm) whereas the concentration at night was higher than afternoon and also highly variable. Further sampling from several locations like mountains, barren land and Pacific Ocean the concentration of 310 ppm prevailed. On the basis of his findings, suggestions and enthusiasm for continuous monitoring CO_2 a program was started at island of Hawaii at Mauna Loa and other locations at Antarctica and South Pole. The Mauna Loa observatory is located 3,397 meters above sea level on the northern slopes of Earth's largest volcano. This place is located away from dense vegetation or human population that might skew the measurements. From these studies, within a year, the data showed a regular seasonal cycle in atmospheric CO_2. Maximum concentration at Mauna Loa occurred just before plants of these temperate climate started showing emergence of new leaves. He could witness for the first time absorption of CO_2 from air for plant growth during summer and release of CO_2 into air in the succeeding winter season. Another important finding came within few years in 1961-1963 (Bolin and Keeling, 1963) which showed a long term increase in the CO_2 concentration. The data also showed a distinct seasonal cycle in the northern hemisphere and a weaker seasonal cycle in the southern hemisphere. Further, attempts were made for a well explained model of global carbon cycle (Ekdahl and Keeling, 1973; Becastow and Keeling, 1973). More observations from modeling of global CO_2 budget in 1970s and 1980s indicated that the cumulative fraction of industrial CO_2 remaining in the atmosphere was between 50-60% (Keeling *et al.,* 1976; 1985). The fluctuations in the CO_2 concentration data during some years were further explained to match closely with 'Southern Oscillation' or the El Nino events (Becastow *et al.,* 1980). Further investigation revealed closely matching inter annual patterns in CO_2 and sea- surface temperature and it lead to realization that these are related in terms of the temperature dependent solubility of CO_2 in surface ocean water. In addition to the measurements of CO_2 the work on the composition of carbon isotopes in the atmosphere further lead to understanding the global carbon cycle in a holistic manner.

NOAA ESRL has built a new data assimilation system called Carbon Tracker which showed feasibility of monitoring the carbon cycle in substantial detail. Based on the Carbon Tracker, Peters *et al.* (2007) estimated the weekly net CO_2 exchange between the terrestrial biosphere and the atmosphere from 2000 to 2005. They concluded this initiative can be helpful in precisely finding the sink capacity and the sensitivity of the sink to climate extremes and to evaluate the effectiveness of efforts to reduce or store carbon emissions.

4. NET PRIMARY PRODUCTION

The estimation of net primary production (NPP) provides the inputs of carbon in a system. The NPP relates to the amount of photosynthetically fixed carbon and losses due to respiration. Ecosystem respiration consists of (i) leaf respiration (ii) respiration by woody stem or non photosynthetic parts of plants and (iii) soil respiration which further consists of root respiration, microbial respiration. Additionally, the CO_2 efflux from soil may also be from weathering of rocks. The relationship between gross primary production (GPP), photosynthesis (P), leaf maintenance respiration during day time (Rmd), net primary production (NPP), autotrophic respiration (Ra), net ecosystem exchange (NEE) or net ecosystem production (NEP) and heterotrophic respiration (Rh) is as follows-

$$GPP = P - Rmd \quad \text{(i)}$$

$$NPP = GPP - Ra \quad \text{(ii)}$$

$$NEE \text{ or } NEP = NPP - Rh \quad \text{(iii)}$$

The equation (iii) is the fundamental model of ecosystem carbon cycling (Woodwell and Whittaker, 1968) embedded in the models of global carbon cycle. Leaf maintenance respiration is a function of leaf nitrogen concentration (Ryan, 1991) and day length. The heterotrophic respiration depends on the decomposition activity as regulated by soil temperature, soil moisture as well as the amount of decomposition material. The decomposable material is governed by mortality rate of vegetation biomass. If Ra < NPP then ecosystem acts as carbon sink and if Rh > NPP then it acts as net C source (Pietsch and Hasenauer, 2009). Part of the carbon store of NEP is also lost and the final stored carbon is termed the net biome production (NBP). This is the difference between the amount of carbon fixed by photosynthetic assimilation and the amount lost by the oxidation or removal of plant material by all routes. The NPP and NEP in three major regions (Schimel *et al.*, 2001) are shown in **Table 2**.

Table 2 The NEP and NPP of major regions

Region	Vegetated area (10^{12} m^2)	NEP GtC yr^{-1}	NPP GtC yr^{-1}	NEP/NPP
North America	20	0.8	7	12%
Eurasia	36	1.7	11	15%
Tropics	47	0.4	36	1%

(Source: Schimel *et al.*, 2001)

Studies indicate that on an average about 80% of GPP is released to the atmosphere through respiration (Law *et al.*, 2000) and of the total ecosystem respiration the CO_2 released from the soil is a substantial part. Therefore respiration may be more important than photosynthesis in relation to the interannual variability in NEP (Valentini *et al.*, 2000). Soil respiration can be further divided into different components (**Fig. 3**) and these are root respiration, rhizomicrobial respiration, decomposition of plant residues, the priming effect induced by root exudation or by addition of plant residues and basal respiration by microbial decomposition of soil organic matter (SOM).

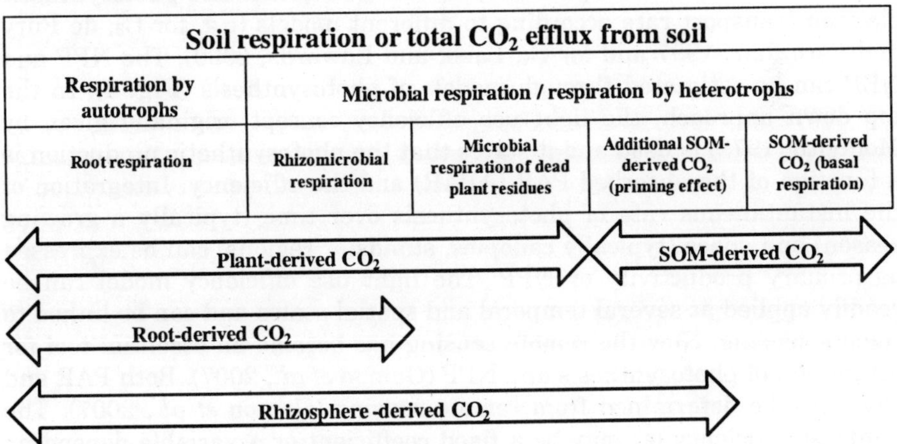

Fig. 3. Five main sources of biogenic CO_2 efflux from the soil (Adapted from Kuzyakov *et al.*, 2006).

Lieth and Whittaker (1975) estimated global NPP based on temperature data and actual evapotranspiration. The direct estimation of NPP is a time consuming and labour-intensive process. It also requires the information on accumulation and turnover of carbon pools in biomass, leaves, shoots, roots, exudates, volatile organic compounds over a period of time. In reality, estimation of all these is seldom done and therefore the accuracy of NPP assessments have been questioned (Geider *et al.*, 2001). With some data on cropping systems, the net primary productivity and annual carbon inputs to agro ecosystem can be estimated (Bolinder *et al.*, 2007). The approach uses a series of plant carbon allocation coefficients for each crop type, shoot to root ratios and yields. From these the below ground net primary production can be estimated. Further, estimation of source carbon from crop residues, root studies and rhizodeposits and crop yields can be done as suggested by Johnson *et al.* (2006). With the current interest to characterize, predict and mange the soil carbon dynamics, there is a requirement for more accurate and precise estimates of carbon inputs to the soil. In this context, Chapin III *et al.* (2009) have also indicated that recent advances in the soil carbon cycling have not been incorporated into current climate carbon cycle models. It seems rightly stated by Leonardo Da Vinci, "We know better the mechanics of celestial bodies than the functioning of the soil below our feet" (Dessaux *et al.*, 2009).

5. PHOTOSYNTHESIS IN RELATION TO NPP AND NEP

The rate of photosynthesis in C_3 plants depends upon the amount of the primary carbon fixing enzyme Ribulose 1, 5-bisphosphate carboxylase/oxygenase (Rubisco) and on Phosphenolpyruvate carboxylase (PEP carboxylase) in C_4 plants. The reaction rate is calculated using enzyme kinetics, leaf temperature, absorbed radiation and photosynthetic electron transport rate according to different models (e.g. for C_3; de Pury and Farquhar, 1997 and for C_4; Laisk and Edwards, 2000). The NPP and NEP can be estimated through models of photosynthesis (**Fig. 4**). In the top down approach, the light use efficiency concept originally given by Monteith (1972) is used and it states that the photosynthetic production is a function of the absorbed PAR (APAR) and the efficiency. Integration of the instantaneous rate of photosynthesis over time (typically a growing season) and space (typically canopies, stands or regions) can be expressed as primary productivity or NPP. The light use efficiency model can be readily applied at several temporal and spatial scales and can be linked to remote sensing. Now the remote sensing has become an essential tool for estimation of photosynthesis and NPP (Gamon *et al.*, 2007). Both PAR and F_{APAR} can be determined from remote sensing (Gamon *et al.*, 2007). The light use efficiency (ε) may be a fixed coefficient or a variable depending upon the environmental conditions particularly moisture and temperature. For example, in the widely used CASA carbon cycle model, LUE is estimated as a function of theoretical universal LUE, temperature and water regime (Friend *et al.*, 2009). The limited spatial coverage of ground based C flux measurements can be overcome through use of remote sensing. However, GPP cannot be directly estimated from the remote sensing measurements but estimated from LAI and fraction of absorbed PAR or through the greenness indices like NDVI. Some of the satellite based observations that can monitor the photosynthesis and primary productivity are listed in Table 3.

Table 3 Some important satellite based observations to monitor the photosynthesis and primary productivity

Satellite / Sensor	Application for vegetation and photosynthetic aspects
Advanced Very High Resolution Radiometer (AVHRR) with optical and near infrared (NIR) spectral channels	Normalized difference vegetation index or NDVI (Tucker, 1978) is expressed as ratio (NIR-R) / (NIR+R), gives a measure of the greenness of land cover. However, it is not a measure of photosynthesis (Grace *et al.*, 2007). Relationship of NDVI and atmospheric CO_2 variations has been observed (Tucker *et al.*, 1986). Also1-Km resolution global land-cover database from AVHRR was made (Belward, 1996)
Moderate Resolution Imaging Spectroradio-meter (MODIS) sensor	MODIS is radiometrically superior to previous AVHRR and it is also hyperspectral in nature (36 bands). It gives Landcover: 96 day, Vegetation indices: 16 day, LAI/FPAR: 8 day, GPP: 1 day NPP: 1 year and helps in estimation of growing season
Synthetic Aperture Radar (SAR)	To map above ground biomass based on microwaves (Kasischke *et al.*, 1997)
Light Detection and Ranging (Lidar)	Based on pulse of laser, Lidar sensors give measurements about the three-dimensional distribution of plant canopies, sub-canopy topography, high-resolution topographic maps, vegetation height and canopy structure (Lefsky *et al.*, 2002).
GOSAT satellite (Japan)	Chlorophyll fluorescence is a good signal about plant physiological state e.g. photosynthetic activity. Remote sensing of solar induced chlorophyll fluorescence (Fs) e.g. the raw fluorescence (from GOSAT) is linearly related to GPP at global scale (Frankenberg *et al.*, 2011).
Satellites with hyperspectral bands	Hyperspectral sensors on satellites can be used to estimate Photochemical Reflectance Index (PRI) which is expressed as ratio (R_{531} – R_{570}) / (R_{531} + R_{570}). It is strongly correlated with light use efficiency at leaf scale (Penuelas *et al.*, 1995), small canopy scale (Gamon *et al.*, 1992), at the ecosystem scale (Nichol *et al.*, 2000) and overall proportional relation between PRI and LUE is observed in different vegetation and forest types (Garbulsky *et al.*, 2011). It gives a measure of ecosystem CO_2 uptake and gross primary productivity (Penuelas *et al.*, 2011).

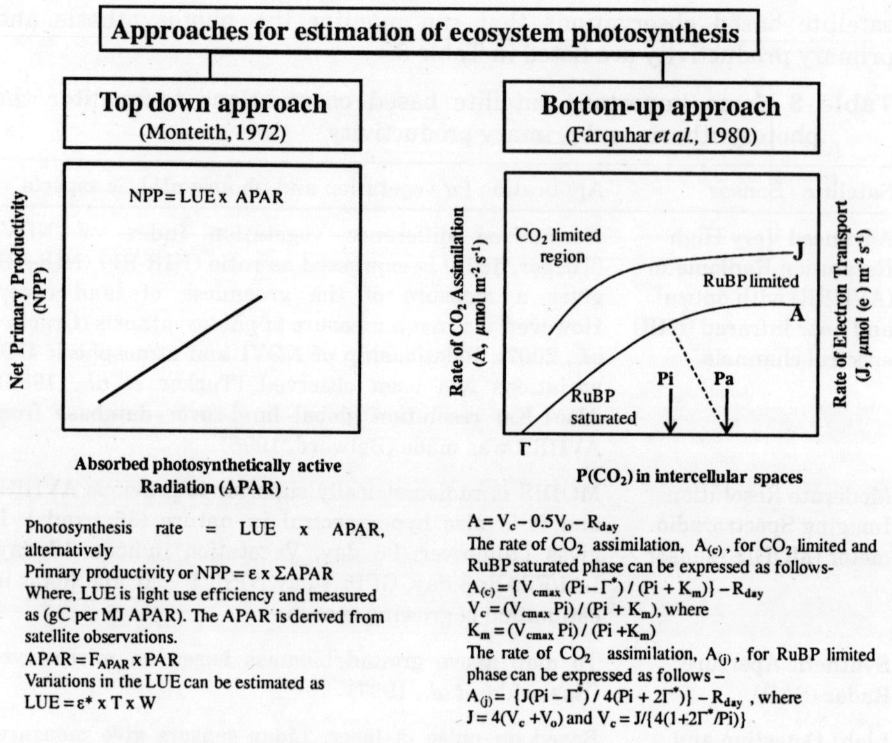

Fig. 4 Approaches to model photosynthesis and net carbon assimilation for ecosystem (The symbols are as follows: A = Net CO_2 assimilation, V_c = Rate of carboxylation, V_o = rate of oxygenation, R_{day} = day respiration, K_m = K_c (1 + O_i / K_o), K_c = Michaelis Menten constant for carboxylation reaction, K_o = Michaelis Menten constant for oxygenation reaction, O_i = Partial pressure for oxygen, V_{cmax} = rate of CO_2 assimilation at saturating intercellular partial pressure of CO_2, Pi = intercellular partial pressure of CO_2, Γ = CO_2 compensation point, Γ* = CO_2 compensation point in the absence of day respiration, J = Rate of electron transport, $A_{(c)}$ = Rate of CO_2 assimilation for CO_2 limited and RuBP saturated phase, $A_{(j)}$ = Rate of CO_2 assimilation for RuBP limited phase, MJ APAR = mega joule absorbed photosynthetically absorbed radiation).

6. MEASURING CO_2 FLUXES: LEAF TO ECOSYSTEMS

The gas exchange methods are of two types (i) Enclosure methods (Fig. 5), where portable gas analyzers or photosynthesis systems are used for measuring the photosynthesis of an enclosed leaf and (ii) non-enclosure methods or micrometeorological methods where the CO_2 exchange can be monitored over a crop or vegetation over a large area e.g. eddy covariance technique (Fig 6).

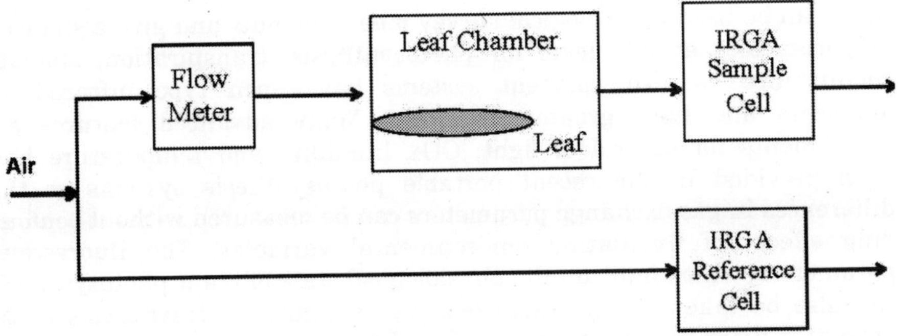

Fig. 5 Leaf enclosure method for measuring CO_2 and H_2O exchange in an open photosynthesis system.

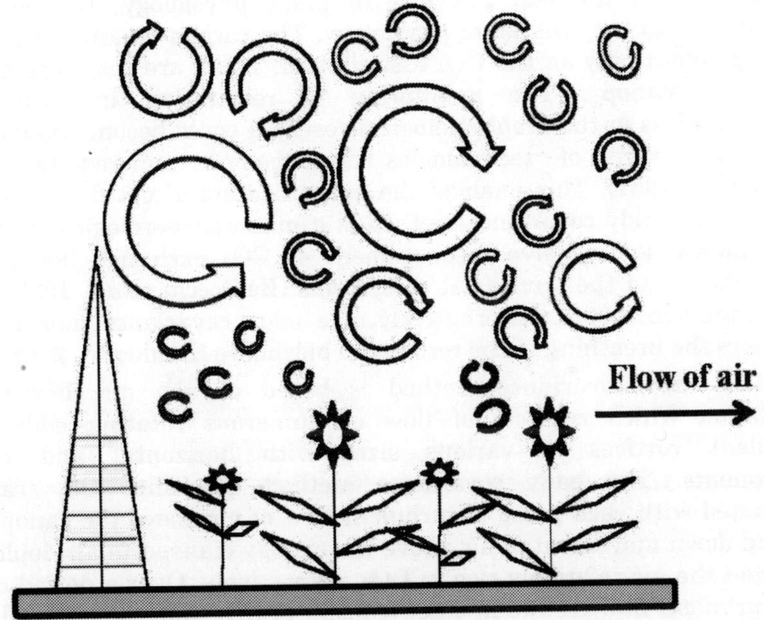

Fig. 6 Eddy covariance technique, a non-enclosure method for CO_2 exchange between the atmosphere and the vegetation measures the flow and concentration of CO_2 of numerous rotating eddies in the flowing air.

6.1 Leaf Enclosure Methods: Portable Photosynthesis Systems

The portable photosynthesis systems basically consist of the infrared gas analyzers (IRGA) for measuring CO_2/H_2O concentration, air flow pumps, flow control devices and leaf cuvette for monitoring the leaf gas exchange. The infrared gas analyzers are based on the principle that heteroatomic molecules (like CO_2 and H_2O) absorb infrared radiation and this absorption obeys Lambert-Beer's law. The major IR absorption band used for CO_2 and H_2O detection are 4.25 μm and 2.7 μm, respectively.

They can be used for ecological survey measurements and give a snapshot of parameters at leaf level for photosynthesis, transpiration, stomatal conductance etc. The present systems have open type infrared gas analyzers and have greater flexibility. Many advanced features and attachments for control of light, CO_2, humidity and temperature have been provided in the recent portable photosynthesis systems so that differences in gas exchange parameters can be measured without confounding effects of fluctuating environmental variables. The fluorescence parameters, which indicate the physiological state of plant photosynthesis, can also be taken on the same leaf while taking photosynthesis in the present instruments with more advanced features.

6.2 Micrometeorological Methods: Eddy Covariance Method

Similar to the leaf gas flux in plant physiology, the eddy flux corresponds to the ecosystem physiology. The carbon balance of plants is not dependent only on the CO_2 assimilation. There are respiratory losses from the canopy. The monitoring of respiration in addition to photosynthesis in the ecophysiological research could become possible due to breakthrough of technologies developed by micrometeorologists (Swinbank, 1951). This enabled the quantification of net flux of a whole canopy. The eddy covariance method is a micro-meteorological technique that allows non-invasive measurement of CO_2 exchange between the atmosphere and the terrestrial ecosystems (Baldocchi *et al.*, 1988; Burba and Anderson, 2005). Alternatively, the eddy covariance measurement monitors the breathing of the terrestrial biosphere (Baldocchi, 2008).

The eddy covariance method is based on the air flow from a vegetation which consists of flow of numerous rotating eddies, i.e., turbulent vortices of various sizes with horizontal and vertical components. The eddy covariance method quantifies the transport associated with each pulse of turbulent flow of air above the canopy. The up and down movement of air above the canopy transports air depleted of CO_2 and the air relatively rich in CO_2, respectively. Over a period of time this turbulent flow results in a net transfer of CO_2 from the atmosphere to the photosynthetic canopy and *vice-versa* during respiration. The net flux is equal to the mean covariance between fluctuations in vertical wind velocity and concentration of CO_2. This method is an effective way to monitor large areas where CO_2 may escape from the subsurface. In a typical setup, the sensors are mounted on a tower for measuring fluxes. The basic sensors include rapidly responding CO_2/H_2O open path analyzers and sonic anemometers for measuring the CO_2/H_2O concentrations associated with individual eddies of air and to measure the turbulent fluctuations in the vertical wind velocity. The eddy flux towers can measure stand fluxes of up to 1 Km^2.

Internationally, a network of eddy covariance sites has been named as FLUXNET which helps to understand the responses of NEE and its components to environment variables (Baldocchi *et al.*, 2001). FLUXNET started formally in 1997 and now a network of more than 300 sites are there worldwide in different regions and climatic zones which help in the measurement of the fluxes of different vegetation stand types. In contrast to the smaller size of eddy flux towers, very tall towers of height more than 400 m are being used for giving the carbon flux measurements at landscape level (Bakwin *et al.*, 1995).

The Net Ecosystem Exchange of CO_2 depends on the complex interaction of many factors including leaf area, CO_2 exchange capacity, photosynthetic and non photosynthetic tissue activity, canopy structure, stand biomass and meteorological conditions including light, air temperature, air humidity, wind speed and atmospheric CO_2 concentration. Therefore net canopy flux as measured by eddy covariance, however, is not equal to net flux of the ecosystem. This is because of lateral transports and losses of the dissolved organic carbon to the groundwater (Schulze, 2006). Therefore for the ecosystem studies and measurement of NEP the eddy covariance method can give more useful estimate of the net ecosystem CO_2 exchange (NEE) between the land surface and the atmosphere provided other carbon addition or losses are properly credited.

Continuous measurement of instantaneous net CO_2 exchange between the biosphere and the atmosphere and its integration over time can help in measurement of net ecosystem productivity (Baldocchi 2008). It can also help in understanding the effect of management practices on the soil carbon sequestration (Baker and Griffis, 2005; Verma *et al.*, 2005). Based on the global eddy covariance studies, Baldocchi (2008) have summarized the key findings as follows (1) greatest net carbon uptake in ecosystems is not necessarily related to the greatest canopy photosynthesis but to the longest growing season, (2) disturbance leads to loss of carbon from ecosystems (3) many old-growth forests act as carbon sinks and (4) Stresses decrease canopy photosynthesis and respiration in tandem and this leads to differences observed year to year in net annual exchange of CO_2.

7. THE CARBON COLUMN OBSERVATIONS

The CO_2 data from *in situ* measurements from the network of surface sites may be difficult to interpret due to atmospheric transport and mixing on diurnal and seasonal time scales and also proximity to local sources and sinks of CO_2. Therefore, column measurements which measure the column-averaged dry air mole fraction of CO_2 and other greenhouse gases are useful. The column vertically integrates the concentration of CO_2 above the surface. They are also insensitive to variations in surface pressure and atmospheric water vapour. The column measurements can be taken from reflected sun observations of the space-based sensors (e.g. OCO - Orbiting Carbon Observatory). But their accuracy is affected by aerosols, uncertainty in air mass variation and variations in land surface properties. The other methodology is ground based network of Fourier Transform Spectrometers (FTSs) which can retrieve precise column abundance of CO_2 and other greenhouse gases from near-infrared absorption spectra. This ground based network was established in 2004 and is called the Total Carbon Column Observing Network (TCCON). These can help in the primary validation of space based instruments (Wunch *et al.*, 2011). For proper validation, the column measurements need to be precise enough to 0.1 per cent to detect the inter-hemispheric flux difference of 1 GtC yr^{-1} (Olsen and Randerson, 2004).

8. PARTITIONING OF THE FLUXES: CARBON AND OXYGEN ISOTOPE DISCRIMINATION

The two naturally occurring stable isotopes of carbon in air are ^{12}C (98.9%) and ^{13}C (1.1%). The composition of these isotopes reveals information about the types of physical, chemical and metabolic process during different carbon transformations (Farquhar *et al.* 1989). Plant biomass shows lesser abundance of ^{13}C as compared to atmosphere due to carbon isotope discrimination during photosynthesis (Lloyd and Farquhar, 1994). Rubisco has higher ability to discriminate $^{13}CO_2$ (δ or carbon isotope discrimination values of about -27‰) as compared to C_4 (-5.7‰) plant where the first fixation of CO_2 occurs via PEP carboxylase enzyme. The study of ^{13}C has advantage that all the carbon pools of plant are labeled, it is non intrusive method and there is easy handling due to non radioactivity.

The initial studies on the plant carbon isotopic composition were done about fifty years back (Craig, 1953). Further, the extensive studies on the carbon isotope composition of atmosphere were pioneered by Charles David Keeling (Keeling, 1960; 1961). Keeling (1958) initiated the study of ecosystem carbon isotope discrimination and observed that during daytime, the concentration of $^{12}CO_2$ is reduced faster as compared to $^{13}CO_2$ due to isotopic discrimination by plants. This leads to enrichment of $^{13}CO_2$ in air. However, during night $^{12}CO_2$ concentration of air increased leading to reduced $^{13}CO_2$ concentration. He observed that that CO_2 concentration and isotope composition in forest air undergo a regular daily cycle that reflects the influence of photosynthesis, respiration and atmospheric mixing. This lead him to estimate the isotope composition of the respired air by using a classical graphical approach now called as 'Keeling Plot'. It has become a standard tool in ecosystem physiology. The carbon isotope discrimination studies have also provided evidence of the increased atmospheric CO_2 due to burning of fossil fuels. Study of Antarctic ice cores by Francey *et al.* (1999) showed a rapid decline of $\delta^{13}C$ in the modern era as compared to preindustrial area. The isotopic composition in air is being monitored in the NOAA/CMDL cooperative sampling network since 1991. There is global $\delta^{13}CO_2$ observation network e.g. NOAA/ESRL which helps in partitioning of the fluxes at finer spatial and temporal scales (Balantyne *et al.*, 2011).

The process of carbon isotope discrimination is affected by photosynthesis, stomatal conductance, respiration, air humidity, temperature, light etc. Thus, it can be stated that estimates of terrestrial carbon isotope discrimination is an important method to quantify the terrestrial carbon sink. It is an important and integrated signature of the plant physiology, metabolism and environment of the plant. Based on the application of $^{13}C/^{12}C$ ratio approach at global scale, Ciais *et al.* (1995) were able to locate a large northern hemisphere terrestrial CO_2 sink. Battle *et al.* (2000) also observed general agreement of the variability of global carbon sinks with atmospheric CO_2 and $\delta^{13}C$ values. Based on the studies it has also emerged that the global flux weighted estimate of discrimination by terrestrial biosphere is highly sensitive to total annual precipitation amount and to a lesser extent to mean annual temperature (Pataki *et al.*, 2003). In addition to ^{13}C, the oxygen isotopes are also important for ecosystem studies. The aboveground and belowground respiratory sources can lead to the changes in the oxygen isotopic ratio ($^{18}O/^{16}O$) of the atmospheric CO_2 (Farquhar *et al.*, 1993) and this knowledge has been used at ecosystem (Flanagen *et al.*, 1997) and global level (Cias *et al.*, 1997).

Carbon isotope discrimination studies have also been used to partition the net CO_2 uptake between land and oceans. This is based on the preferential uptake of ^{12}C by the land relative to oceans. The partitioning of sink activity on the earth surface into terrestrial and oceanic components is because plants discriminate against ^{13}C by about 18‰ whereas uptake by oceans discriminates by about only 2‰. However, earlier studies have also indicated that temporal variations in isotope discrimination may occur with a season and between years in a given biome and this dynamics may also need to be considered in the partitioning of atmospheric CO_2 between terrestrial biosphere and ocean (Ciais *et al.*, 1995b; Flangen *et al.*, 1996).

9. PLANT AND SOIL RESPONSES TO CLIMATE CHANGE AND EFFECT ON CARBON CYCLE

Early work on carbon budget showed that it was close to balance and carbon in land biosphere was stable or slightly increasing during the industrial era (Bacastow and Keeling, 1973). However, later it was observed that a large amount of CO_2 emissions due to land use change have not been accounted (Woodwell *et al.*, 1983) and was therefore, thought to be balanced by an unknown or 'missing sink'. It was therefore reasoned that the sink is due to CO_2 fertilization effect i.e. increased photosynthesis due to elevated CO_2 into the atmosphere. Hymus and Valentini (2007) have reviewed the potential of terrestrial vegetation as a carbon dioxide sink. A strong terrestrial carbon sink was found in the temperate latitudes of the Northern Hemisphere in 1992 and 1993, the magnitude of which is roughly half of the global fossil fuel emission of those years (Ciais *et al.* 1995). Sabine *et al.* (2004) have mentioned that the global excess of photosynthesis over respiration in the past few decades has resulted in an annual accumulation of 0.2-1.4 Pg carbon in the terrestrial biosphere which has slowed down the rate of rise in the atmospheric CO_2.

Free air CO_2 enrichment experiments have provided valuable insight about the cycling and storage of carbon in the terrestrial ecosystems (Ainsworth and Long, 2005; Norby and Zak, 2011). The study of response of elevated CO_2 as plants and ecosystems have been done with the help of open top chambers (OTC's) and FACE (Free Air CO_2 Enrichment). The OTC's have approx. diameter of 2-3 m whereas FACE experiments can have diameter as large as 30 m and can accommodate trees up to 20 m tall, much larger than OTC's. FACE has therefore, advantage of larger size, longer experimental duration and undesirable artifacts due to chamber walls (Hendry *et al.*, 1999).

The response of plants to elevated CO_2 has been studied based on several hypotheses related to-leaf area index, growth, root production, N-cycling, water use, stand composition, soil microbial responses and carbon cycling of the system (Norby and Zak, 2011). A slower turnover of root material grown in elevated CO_2 may result in a limited net increase in carbon storage (Groenigen *et al.,* 2005). Heath *et al.* (2005) found that CO_2 enrichment in European tree species lead to an increase in soil microbial respiration and a marked decline in sequestration of root derived carbon in the soil. They however, did not measure the respiration of roots. But recent literature indicates that there is marked increase in respiration also with elevated CO_2 (Leaky *et al.*, 2009b). Further, the indications of promotion of fine root growth with elevated CO_2 is also important aspect as its turnover represents a substantial proportion of annual net primary productivity in trees and forests (Norby *et al.,* 2004). The global biomass production has increased due to rising atmospheric CO_2 and it is expected that ecosystem will sequester CO_2. However, Sayer *et al.* (2011) observed that increased litter fall can stimulate release of soil organic carbon in tropical forest. They found that the soluble organic carbon leached from the litter enhances the soil organic matter turnover due priming effect. Creating a positive feedback that can lead to reduction in the carbon sequestration potential of tropical forest ecosystem. Not only the priming effect, but the major effect of elevated atmospheric CO_2 is to increase the plant biomass production and rhizodeposition even more than above ground production (DeGraff *et al.*, 2006). Therefore, it is necessary that both above and belowground studies need to be undertaken for a better understanding of the effects on the carbon cycle and the interactions (Kuzyakov, 2011).

10. ROOT GROWTH AND TURNOVER AS RELATED TO PHOTOSYNTHESIS, NPP AND SOIL RESPIRATION

Roots are a major sink for assimilates and their requirement to produce unit dry matter is about twice as compared to shoots (Passioura, 1983). Studies indicate that roughly 40% of the net fixed carbon is allocated belowground (Jones *et al.,* 2009). Of the belowground partitioned carbon, 50% resides in root biomass, 33% is used for respiration and 12% is recorded as soil residues. Illustrations of general scheme of assimilate portioning to the roots based on Lambers *et al.* (1998) is given in **Fig. 7.** Many studies indicate that recent photosynthates move through phloem to the roots and within 24 hours about half of them are respired by roots.

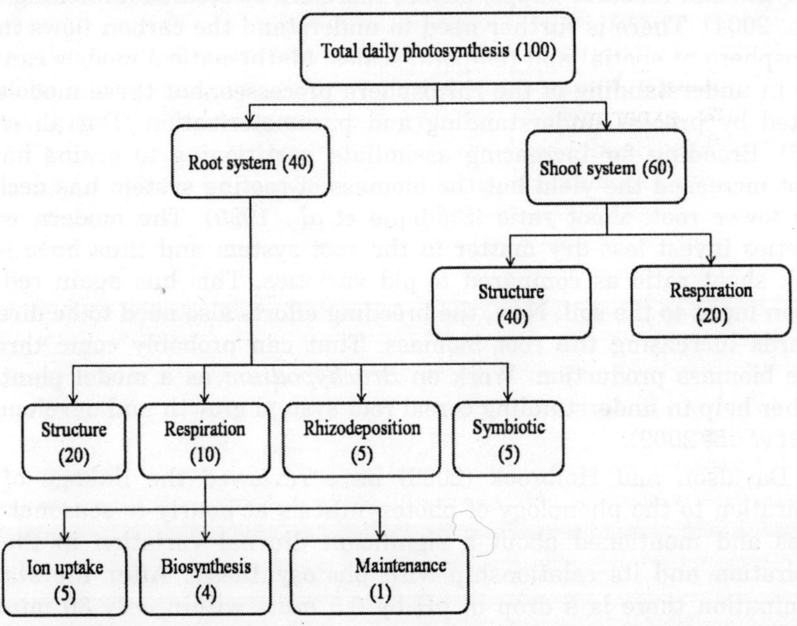

Fig 7. Approximate carbon costs i.e. percentage of daily carbon fixation (shown in parentheses) associated with shoot and root growth in fast growing herbs (Based on Lambers *et al.*, 1998).

The root turnover represents annual root production divided by maximum standing root biomass expressed on per year basis. Root turnover is important to the global carbon budget as well as to nutrient cycling in ecosystems and to the success of individual plants (Eissenstat *et al.*, 2000). In fact, the role of fine roots, which play a primary role in plant water and nutrient uptake, has been highlighted by Jackson *et al.* (1997). They estimated that global fine root carbon constitutes about 5% of all carbon in the atmosphere and represents about 33% of global annual NPP. A compilation of root turnover across different climates and vegetation types based on sequential root biomass estimates was overviewed by Gill and Jackson (2000). Some values of root turnover are as follows - whole tree root systems (0.10 per year), shrublands (0.34 per year), fine root system in grasslands, wet lands or forests (0.53-0.56 per year) and fine root in tropical vegetation (0.6-0.9 per year). The root turnover increased with mean annual temperatures and the Q_{10} values were 1.4 for fine roots (<5 mm dia.) in forests, 1.6 for fine roots in grasslands and 1.9 for shrub lands. The rapid root turnover seems to be beneficial. It is determined by the nutrient content and it in turn determines the mineral nutrient acquisition e.g. phosphorus. The physiological mechanism behind the higher root turnover is not clear (Noordwijk *et al.*, 2004).

Plant roots adapt to environment by placing roots in more suitable habitat patches. Complexity of root growth in terms of heterogeneities of soil types and resource supply should therefore be studied (Hutchings and John, 2004). There is further need to understand the carbon flows in the rhizosphere at spatial and temporal scales. Mathematical models can also help in understanding of the rhizosphere processes, but these models are limited by process understanding and parameterization (Darrah *et al.*, 2006). Breeding for increasing assimilate partitioning to grains has no doubt increased the yield but the biomass of rooting system has declined with lower root: shoot ratio (Siddique *et al.*, 1990). The modern wheat varieties invest less dry matter in the root system and thus have lower root : shoot ratio as compared to old varieties. This has again reduced carbon input to the soil. Now, the breeding efforts also need to be directed towards increasing the root biomass. That can probably come through more biomass production. Work on *Brachypodium* as a model plant can further help in understanding cereal root system growth and development (Watt *et al.*, 2009).

Davidson and Holbrook (2009) have reviewed the linkage of soil respiration to the phenology of photosynthesis at hourly to seasonal time scales and mentioned about a significant diurnal variation in the soil respiration and its relationship with photosynthesis. After the start of illumination there is a drop in pH by 0.5 units within only 30 minutes. Evidences indicate that canopy photosynthesis influence soil respiration (Kuzyakov and Gavrichkova, 2010) and its speed is based on phloem transport (Mencuccini and Holtta, 2010). The temporal scales of the influence of photosynthesis and soil respiration are from hours to weeks and at seasonal and annual intervals also (Vergas *et al.*, 2011). Soil respiration consists of root respiration and respiration by soil heterotrophs. It is very difficult to analyze root respiration from the soil. Hogberg *et al.* (2001) designed experiment in which the interruption of carbohydrate supply to roots was interrupted by stem-girdling. In their large scale stem-girdling of the boreal pine forests, Hogberg *et al.*, 2001 found that there was up to 37-50% reduction in the CO_2 fluxes from the soils. This established that a major portion of the soil respiration consists of soil respiration. Soil heterotrophs also have a major contribution in soil respiration. Chapin III *et al.* (2009) emphasized a close coupling between NPP and heterotrophic respiration (Rh) because NPP constrains the food available to heterotrophs. Study by Subke *et al.* (2011) using girdling experiments also showed that the total soil CO_2 efflux was stabilized within two weeks at 65% of control plot values indicating that heterotrophic respiration dominates soil respiration. They found significant contribution of mycorrhiza and other rhizospheric organisms to soil respiration and concluded a tight coupling between auxotroph activity and soil decomposition processes.

11. SOIL CARBON SEQUESTRATION: BENEFITS, CAPACITY AND COST

The future food demand of the growing population essentially requires a quantum jump in plant productivity and genetic manipulations in yield improvement will play an important role (Parry *et al* 2011). But at the same time it is important to realize that SOM also plays an equally or more important role in survival of humankind because without it, the genetic potential of a plant can't be realized. The soil organic matter functions as - source and sink of plant nutrients, ion exchange, water holding capacity, soil aggregation, high water infiltration, low surface runoff, sustenance for microorganisms, biodiversity, biological habitat and gene pool, formation of aggregates, high nutrient use efficiency, buffering capacity, moderation of soil temperature, reduction of sediment load, filtration, buffering emission of green house gases from soil to the atmosphere, storing, filtering and transformation of compounds (Lal 2004, 2008; Bouma, 2010). Lal (2006) emphasized the need for improving soil organic carbon as it is essential for increasing the agronomic yields. The increased in yields with improvement in SOM can help in meeting the food deficit of the world. Enhancing soil carbon storage in addition to climate mitigation potential has other benefits also like improvement in productivity, improving soil health and quality of environment and therefore it has been described as a 'win-win' strategy (Lal, 2001).

In future, the biotic and soil carbon storage will be important for reducing the emission. Soil carbon sequestration can play central solution of the measures to reduce atmospheric CO_2 concentration (Lal, 2004; Smith, 2004). Soil carbon sequestration is defined as the net removal of CO_2 from the atmosphere into long-lived pools of carbon in soil. The maximum yearly global soil carbon sequestration potential is 0.9±0.3 PgC yr^{-1} (Lal, 2004). However, the realistically achievable potential for carbon sequestration has been less than 20% of the biological/technical potential (Cannel 2003) because of policy and technological constraints (Smith *et al.* 2007). The soil carbon sequestration process comes to equilibrium in about 20-30 years after a change in management. Sequestration of carbon in long lived forms such as wood, substituting biomass for fossil energy and soil management is also important. However, the biotic sequestration occurs over a time period of 70-150 years. Degraded soils and ecosystems have higher sink capacity of soil organic matter. Abrol (1995) reviewed the gaseous emissions from agro-ecosystems in India and concluded that in Indian soils more than 50% of the carbon in surface and 10% of the carbon in the rest of the soil profile has been lost. The carbon pool present in Indian soils is 24.3 Pg whereas the estimated potential is 34.9 Pg. The difference i.e. 10.6 Pg represents the amount of additional soil carbon that can be sequestered. Lal (2004) has shown that ecosystem with high and attainable soil carbon sequestration potential are croplands, grazing lands, degraded lands and irrigated soils and estimated that world soils have potential of carbon sequestration of 0.4-1.2 GtC yr^{-1} **(Table 4).**

Table 4 Potential for C sequestration in major soils

Ecosystem	Area	Potential for C sequestration (GtC yr^{-1})
Cropland	1350 mha	0.4-0.8
Degraded and desertified soils	1.1 billion ha	0.2-0.4
Irrigated soils	275 mha	0.01-0.03
Range lands and grass lands	3.7 billon ha	0.01-0.3

(Source: Lal, 2004)

West *et al.* (2010) made a global comparative analysis of carbon stock versus crop yields on agricultural land and showed that for each unit of land cleared, the tropics lose nearly two times as much carbon (~120 t ha^{-1} yr^{-1} *vs.* ~63 t ha^{-1} yr^{-1}) and produce less than one-half the annual crop yield compared with temperate regions (1.71 t ha^{-1} yr^{-1} *vs.* 3.84 t ha^{-1} yr^{-1}). Besides the potential of the ecosystems, the management practices in agro ecosystems play an important role. Some good practices like green manuring and legume crop rotation need to be promoted for improving the soil quality (Drinkwater *et al.*, 1998). Green manures can increase the nitrogen available from the legume. There is increase in growth and yield of crops after green manuring. It helps in reducing the environmental impact of agriculture. Agroforestry ecosystem can have a high soil carbon stock. For example, the home gardens in Kerala state (India) have soil carbon of about 101-126 Mg ha^{-1} (Saha *et al.,* 2009). The effect of carbon sequestration can be visualized from long term soil experiments. Such a study by Pathak *et al.* (2011) indicated the potential and cost of carbon sequestration from long term experiments (average 16.9 years) in India. The cost of carbon sequestration is expressed in terms of monetary units per Mg of carbon sequestered. The average SOC concentration increased from the level of 0.54% in control to 0.82% in NPK + FYM treatments. Carbon sequestration level of 0.33 MgC ha^{-1} yr^{-1} was observed in NPK + FYM practice indicating a good potential compared to 0.16 MgC ha^{-1} yr^{-1} in NPK treatments. Therefore, NPK + FYM have good potential in carbon sequestration in Indian soils by mitigating GHG emission without any additional cost and in addition increasing the yields and net returns in majority of experiments.

12. METHODS FOR MONITORING ROOT GROWTH, SOIL ORGANIC CARBON AND SOIL CO_2 FLUXES

The vastness of the root system can be imagined with an example of a single rye plant consisting of more than ten thousand kilometers of root hairs (Dittmer, 1937). Despite the importance of roots the knowledge of the root system is limited due to the inadequacy of the techniques available and there is still lot to be learnt about the rhizhosphere processes. Further, the effects of human activities on the belowground zone of earth's "critical zone" have not been well quantified (Richter and Mobley, 2009). The contribution of plant roots to soil and atmospheric carbon is significant and difficult to survey accurately. Worldwide interest in reducing greenhouse gases has led to research activities for quantifying the root biomass and evaluating their critical role in space and time. There are several methods for root growth and biomass measurements. The traditional methods are destructive, labour intensive and limited by soil volume or surface area to be assessed (Bohm, 1979), however, several advances in the root studies have lead to a better understanding of root growth, architecture and behavior (Manusco, 2012). Some methods are described in **Table 5**.

Table 5 Some methods for measuring root biomass, length, architecture and functioning

Method	Description
Soil core sampling	Roots are taken from soil core samples and the root length and surface area parameters can be estimated
Trench profile wall method	Root distribution from the soil profile is traced
Minirhizotrons	Transparent tubes inserted into soil with digital camera allow frequent root observation and estimation of root production and root longevity
Multielectrode arrays	In plants, most cells are electrically excitable and active, releasing and propagating action potentials (Masi *et al.*, 2012)
High-throughput imaging systems	High-throughput digital imaging of *Arabidopsis thaliana* roots is followed by image analysis for studying the root architecture (French *et al.*, 2012)
Ground penetrating radar or microwaves	Microwaves or ground penetrating radar (GPR) can help in estimating the root length non-destructively (Hrusca *et al.*, 1999; Butnor *et al.*, 2001). It can help in detection of coarse woody roots, estimate biomass, root diameter, and spatial distribution of roots
Electrical impedence spectroscopy	This method is based on correlation between electrical capacitance and resistance of roots at a single low frequency with root biomass and morphology e.g., surface area (Repo *et al.*, 2012)

Planar optodes	The functioning of roots can be visualized non-invasively. Measurement of the rhizhosphere pH with optodes can help in quantitative visualization of pH dynamics of the rhizhosphere non-invasively (Blossfeld and Gansert., 2007)
Flat optical scanner	Flat optical scanners inserted into the ground can take wide images of root developing on a flat surface and can be automated for image acquisition at high frequency (Dannoura *et al.*, 2012)
Multi-electrode resistivity imaging	It is based on relationships between root biomass and electrical resistivity of soils. It helps in visualizing the distribution of soil volumes with different root densities rapidly and with minimum disturbance (Amato *et al.*, 2012)

Soils are the biggest terrestrial carbon store. Therefore, measurement of soil carbon is necessary to understand the long term changes. However, over longer periods, it is very difficult to generalize about linking the pool changes with environmental changes. Moreover, the measurements of soil carbon stock changes are difficult and require intensive sampling. Therefore, only a systematic periodic measurement of soil respiration can indicate the level of plant metabolism, plant activity and belowground allocation. This can help in relating to long term changes in soil carbon pools. Several methods have been provided for assessment of soil carbon and *in situ* soil carbon measurement (Lal *et al.*, 2001; Gehl and Rice, 2007). Some methods have potential to improve our understanding of soil C dynamics at a scale of space and time not feasible before. Some of these approaches are presented in Table 6.

Table 6 Some methods for measuring soil carbon

Method	Description
Wet chemical method	Chromic acid is used for measuring the oxidizable C (Walkley and Black, 1934)
Dry combustion method	Total combustion measures total carbon (Nelson and Sommer, 1982)
Laser induced breakdown spectroscopy	It is based on the principle that focusing laser pulse on a sample leads to formation of a microplasma that emits light characteristic of the elemental compositional of the sample (Ebinger *et al.*, 2003)
Soil reflectance spectra and remote sensing	Soil reflectance spectra and remote sensing has potential for predictions of soil organic carbon rapidly and at low cost (Chen *et al.*, 2000; Gomez *et al.*, 2008; Stenberg *et al.*, 2010; Rourke and Holden, 2011)
Inelastic neutron scattering	It is based on inelastic scattering of 14 MeV neutrons from C nuclei present in the soil and measurement of the resulting 4.44 MeV gamma ray emission (Wielopolski *et al.*, 2000; 2003)

The sources of CO_2 flux from soil and the partitioning methods have been reviewed by Kuzyakov (2006). Prediction of the soil respiration CO_2 flux remains limited because of multiple controlling mechanisms that interact over different temporal and spatial scales (Vergas *et al.*, 2011). Further, the fluxes do not reflect the turnover in the total carbon pools because most of the fluxes originate from small pools but having a very fast turnover (Kuzyakov, 2011). The total CO_2 efflux of soil do not provide sufficient information about the soil's capacity as a net source or net sink (Kuzyakov, 2006) because not all of the CO_2 released from soil is soil-derived. Therefore, experimentally separating soil respiration into autotrophic and heterotrophic sources is important challenge facing current soil respiration research (Kuzyakov and Gavrichkona, 2010) and in this regard the isotopic partitioning methods are advantageous over other non-isotopic techniques because they impart much less disturbance to plant-soil system (Table 7 and 8).

Table 7. Some non-isotopic methods for measuring soil respiration and its partitioning

Total or partitioned respiration	Method	Description
Total soil respiration	*In situ* automated chamber measurements (Open and closed type IRGA)	Automated chambers measure the CO_2 flux at the soil surface and therefore, integrate all biophysical processes that contribute to soil respiration
	In situ automated soil CO_2 concentration profiles (gas wells or solid state CO_2 sensors)	With soil CO_2 concentrations profiles, the location of production of CO_2 within the soil profile can be determined
Partitioning of soil respiration to root-derived CO_2 and SOM-derived CO_2	Root exclusion techniques (planted *vs.* unplanted soil and Trenching	Difference of CO_2 from planted and unplanted soil or without and with trenched roots gives only a crude estimate of root-derived CO_2 and SOM-derived CO_2
	Shading and clipping	Difference of CO_2 from shading and clipping occurs due to stopping the leaf photosynthesis and assimilate transport to the roots
	Regression	It is based on a linear relation between root mass and root-derived CO_2 at a given temperature
	Component integration	Manual separation of CO_2 sources and measuring CO_2 from each source
	Excised roots	Measuring of respiration of excised roots separated from soil by infrared gas analyzer (IRGA), oxygen electrode or capture of CO_2 in alkali

	In situ root respiration	Respiration of living roots without separation from plant and placed in a cuvette located in the soil
	Tree girdling	Interruption of C flows into roots by phloem girdling stops assimilates to roots
	Gap formation	Difference of CO_2 from soil with vegetation and from the soil under the gap
Partitioning of root-derived CO_2 into root respiration and rhizomicrobial respiration	Component integration	Manual separation of sources and measuring CO_2 from each source is done by incubation
	Substrate induced respiration	Differential response of microbial respiration and root respiration to glucose addition

(Source: Kuzyakov, 2006)

Isotopic or tracer approaches help in a direct and quantitative manner about the links of plant and microbial aspects of soil respiration (Table 8). These links include quantification of C transport within plants, the transfer of C from roots to microbes, SOM, and/or respiration, and the contributions to soil respiration from autotrophic and heterotrophic sources. Tracer studies include pulse and continuous labeling and also monitoring of changes occurring in the introduced samples (of CO_2, litter and carbohydrates) to the atmosphere or soil (Kuzyakov, 2006; Pendall *et al.*, 2004).

Table 8. Some isotopic methods for measuring soil respiration and its partitioning

Type of partitioned respiration	Method	Description
Partitioning of soil respiration to root-derived CO_2 and SOM-derived CO_2	Continuous labeling of shoots in $^{14}CO_2$ or $^{13}CO_2$ atmosphere	Growing plants continuously in $^{14}CO_2$ or $^{13}CO_2$ atmosphere and dilution of root-derived CO_2 by unlabeled CO_2 from SOM.
	Pulse labeling of shoots in $^{14}CO_2$ or $^{13}CO_2$ atmosphere	Short-time labeling of shoots in $^{14}CO_2$ or $^{13}CO_2$ atmosphere helps in estimation of belowground allocation, time lags between assimilation and soil respiration, the role of photosynthesis and substrate supply in modulating soil respiration and the transfer of C to microbial pathways.
	^{13}C natural abundance	Growing C_4 plants on a C_3 soil or vice versa. Rochette and Flanagen (1997) and Rochette *et al.* (1999) used the abundance of ^{13}C for estimating rhizo-

		sphere respiration (Rh) and this has advantage in handling as compared to method with ^{14}C.
	Free air carbon dioxide enrichment	Application of additional CO_2 with changed $\delta^{13}C$ signature compared to air, C tracing of $\delta^{13}C$ of CO_2 from the soil.
Partitioning of root-derived CO_2 into root respiration and rhizomicrobial respiration	Comparison between root-derived $^{14}CO_2$ and rhizomicrobial $^{14}CO_2$	Comparison of root-derived $^{14}CO_2$ with rhizomicrobial $^{14}CO_2$ during continuous labeling by decomposition of uniformly ^{14}C labeled samples of rhizodeposits of the same plants.
	$\delta^{13}C$ values of CO_2 and microbial biomass	Growing C_4 plant on a C_3 soil (or *vice versa* or growing plants under FACE) and calculation by $\delta^{13}C$ values of CO_2, microbial biomass, soil and roots.

(Source: Kuzyakov, 2006)

13. COMPLEXITIES AND UNCERTAINTIES ISSUES RELATED TO – MONITORING, PLANT RESPONSES, FEEDBACKS AND CARBON TRADING

The terrestrial CO_2 exchange processes depend upon multiple factors which can have direct effect or their interactive effects can lead to changes in the carbon balance (Fig. 8). This leads to uncertainties due to complexities in monitoring and disentangling their effects. The estimates of these uncertainties in the sources and sinks have been shown in Fig. 1 (represented by $\pm$ values). The highest degrees of uncertainty occur in the source and sink estimates of the land ecosystem as compared to the atmosphere, ocean and fossil fuel (Le Quere *et al.*, 2009). Houghton (2003) analyzed the differences in the estimates of terrestrial carbon balance. He observed that these differences occur due to uncertainties, errors and incomplete accounting in some methods. These can be resolved through detailed historical reconstructions where data are available and from systematic monitoring of land use changes with satellite data. A considerable amount of data has been acquired by measuring and modeling the carbon balance of the ecosystems. However, there is still much uncertainty about the strength of carbon sources and sinks due to discrepancies in the estimation methods and the complexity of the terrestrial ecosystems (Wang *et al.* 2009). Le quere *et al.* (2010) also mentioned role of sophisticated monitoring mechanisms and development of methods for full GHG accounting. The CO_2 concentration near the ground is affected by the asymmetric transport phenomenon in the atmosphere which is also called as atmospheric rectifier effect. This effect is diurnal as well as seasonal. Similarly, cyclones can also take up air modified by photosynthesis to other places changing the CO_2 concentration of that place. Therefore, coupled atmosphere-biosphere models are necessary for interpreting measurements of CO_2 concentration and estimation of whether a given area is a source or sink for CO_2 (Berry, 2012).

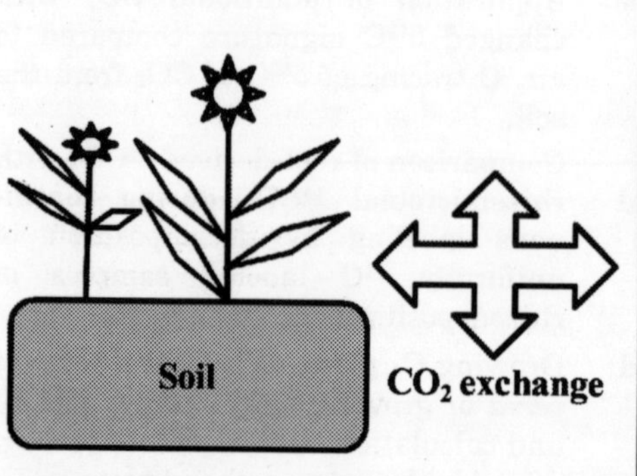

Factors affecting CO_2 exchange from vegetation and soil

- Light
- [CO_2]
- Water / precipitation
- Temperature
- Nutrients
- Age of vegetation
- Biodiversity
- Disturbances (insect pest invasions, floods etc.)
- Changes in species composition
- Land use change (deforestation, burning etc.)
- Soil physico-chemical properties

Fig. 8 Multifactorial effects on the ecosystem CO_2 exchange processes.

Bacastow and Keeling (1973) estimated the CO_2 fertilization effect of land as the natural logarithm of the ratio of current to pre industrial CO_2, suggesting accelerating plant growth with increasing atmospheric CO_2. It is however, not based on mechanistic approach. The mechanistic relation between photosynthesis and CO_2 concentration at single leaf level has been robustly shown by Farquhar *et al.* (1980). However, at higher levels of organization i.e. larger space and over longer time periods, mixed results of CO_2 fertilization emerged from the ecosystem experiments (FACE etc.) (Norby and Zak, 2011). Therefore, directly ascribing the missing sink to the CO_2 fertilization effect is not so simple. The magnitude of sink from CO_2 fertilization is still with large uncertainty. On one hand, Kohlmaier *et al.* (1989) estimated CO_2 fertilization level as equal to terrestrial sink whereas Thompson *et al.* (1996) concluded that CO_2 may not act as sole driver for the historical sink. Estimates indicated that CO_2 fertilization accounts for about one-half of the sink or less and it can modulate the rate of CO_2 increase in the atmosphere (Field, 2001).

In the changing climatic scenario there are uncertainties about how the biotic carbon (Friend 2010) and soil carbon (Heimann and Reichstein, 2008; Norby and Luo, 2004) will respond to the climate change. It is because of the fact that there will not only be increased incidences of changes in temperature, soil moisture, extreme events etc. but the plants also will acclimate to the elevated CO_2 i.e. their capacity for photosynthesis may be up-regulated or down-regulated (Vijai *et al.*, 2006). Canadell *et al.* (2007) have also pointed out that increase in the rate of airborne fractions of CO_2 emission in the recent years also implies a decline in the efficiency of CO_2 sinks (i.e., down regulation). They also observed that carbon cycle has changed faster than that simulated by models. There will be changes in the plant respiratory patterns (Leakey *et al.*, 2009b), nitrogen and water relations (Leakey *et al.*, 2009a), assimilate partitioning to roots (Pendall *et al.*, 2004). It can be hypothesized that the increased C : N ratio of biomass can lead to reduced decomposition. But the decomposition process did not get affected much at elevated CO_2 (Norby *et al.*, 2001). However, review by Torbert *et al.*, (2000) indicates that mineralization of N from decomposing plant material will be reduced. With global warming, the interactive effect of temperature, nutrient availability and soil moisture on the total belowground carbon flux needs to be further studied so that partitioning of global primary productivity to the total belowground carbon flux can be helpful in predicting carbon sequestration and these can then be incorporated into global scale carbon cycling models. The relationship of NPP and soil organic carbon and soil respiration is discussed further.

Plants are potential victims of climate change as well as major regulators of climate (Field 2001). Due to increasing atmospheric concentration and the fertilizing effect of CO_2 the ability of plants for CO_2 uptake (i.e. photosynthesis) has increased as compared to respiration and there is a sustained imbalance between photosynthesis and respiration (Field, 2001; Candell *et al.*, 2007 and Norby and Zak, 2011). This increased C storage in the terrestrial biosphere creates a negative feed-back on the rate of CO_2 accumulation in the atmosphere and can moderate the trajectory of climate change (Matthews, 2007). Therefore, the long term study of fertilization effects of CO_2 needs further investigations and also on the carbon cycling in that ecosystem. Another hypothesis is that increased CO_2 released from soil can accelerate global climate change, acting as positive feedback (IPCC, 2007; Wixon and Balser, 2009). It is however difficult to conclude on this hypothesis and it needs to be further investigated, particularly the complex relationship of organic matter decomposition and temperature (Agren and Wetterstelt, 2007). Not only the plants and soil have feedback to changing climate, but also there is a mutual feedback between plants and soil. A change in soil conditions causes change in the plant component which in turn causes further changes (positive or negative feedback) in the soil and *vice-versa*. Most of the studies show one-way effects of the plant and soil on each other. However, real feedback involves a reciprocating sequence of effects between plants and soil (Ehrenfeld *et al.*, 2005) which also makes the situation more complex.

In the Kyoto Protocol, the earlier discussions were related to cap-and-trade policy for fossil fuel carbon and the consideration of the carbon sinks came later. Therefore, the cap-and-trade architecture needed modification with inclusion of sinks. Among the above-mentioned mechanisms the carbon trading refers to a market based approach for reducing greenhouse gas emissions where a price is attached to the emission of carbon whereas incentive is given for cutting down the emission. The alternative land use practices which lead to loss of soil carbon (or sink reversal) results in loss of carbon credits. Lal (2004) also highlighted the commoditification of soil carbon and its importance for trading carbon credits. There are however a number of issues and complexities associated with trading in relation to sink in a climate mitigation policy. Some of these (based on Reilly and Asadoorian, 2007; Houghton, 2009; Neider and Benbi, 2008) are listed as follows - size of sinks, changes in sinks over time, types of sink-enhancement actions, the definitions of-'forest', 'management', 'difference between aforestation and reforestation' etc., identification of changes in carbon stocks due to human activities than from those due to natural causes, considerations of spatial and temporal leakage, measurement and monitoring, credits for sequestration versus avoided emissions, variability of ecosystem uptake, large releases due to fire, determining direct human responsibility, permanence and leakage, pre-existing distortions, enforcement, carbon stored in products, limits of allowable emissions, distribution of permits amongst participants, selection of projects, human rights, supportive issues, cost-effectiveness of approach, land prices, opportunity cost of land and cost of land use changes, variation due to regional/climatic factors etc.

The account of sink measurement and monitoring will help in reducing the uncertainties and improving the transparency and verifiability. Monitoring and scaling of carbon stocks and fluxes in agro-ecosystems is necessary for verification of carbon sequestration and trading schemes (Lal 2006). Durant *et al.* (2011) have also emphasized the need for enhancement of measurement infrastructure to reduce uncerta-inty in global carbon balance and estimated that even 5% of unreported emissions can cost US$ 20 billion each year by 2050. Therefore, it has significance in carbon-equivalent trading and reducing the volatility of carbon market. To implement this, a question arises that how much carbon is coming and how much is released from a system? Answer to such question can help us in better implementation of mitigation options. However, there emerge other important questions such as - what are the methods for estimation of the above ground and below ground carbon stocks and whether some methods are more rapid, applicable to larger scales, non invasive, cost effective and with a tolerable compromise on accuracy? The verifiability depends upon the stringency of the metho-dology. More stringency can cost prohibitively at national level and least stringency can lead to use of default values. Intermediate scheme can be of more practical use and require use of remote sensing, use of controlled experiments or benchmark sites and models (Smith, 2004). However, the estimation of components of the global carbon cycle are still not accurate as mentioned by Houghton (2005) and also accounting of sources and sinks of carbon becomes more complex with scaling. Presently there is emphasis for monitoring greenhouse gases which is still inadequate in terms of number of sites, large geographic gaps or spatial resolution (Manning, 2011; Levin, 2011) and therefore, high frequency atmospheric GHG measurements (Weiss and Prinn, 2010) are required.

14. CONCLUSIONS AND FUTURE SCOPE

The flux of carbon among various carbon pools in the carbon cycle occurs due to various CO_2 exchange processes such as photosynthesis, respiration, deposition of carbon into soil, anthropogenic emissions and other natural processes. Global efforts are being made to reduce the emissions to slow down the pace of climate change. Therefore, the measurements, monitoring and understanding of the controls of carbon inputs and outputs by the terrestrial ecosystems i.e. vegetation and soil have become increasingly important. The key measures of the input of carbon in a system are net primary production (NPP) and the net ecosystem productivity (NEP). Estimates of photosynthesis and NPP can be made through traditional approaches as well as other approaches e.g. bottom up approach (using mechanistic photosynthesis models) and the top down approach (using light use efficiency with satellite data). Advancements have taken place for monitoring the CO_2 exchange at different scales *viz.* leaf (portable photosynthesis systems), stand level (eddy covariance method), landscape level (tall towers) and vertical carbon column observations (Fourier transform spectrometers and satellites). These flux measurements along with the use of satellite data for land cover changes, growing season, phenology and their coupling with models at various scales can characterize the changes in the carbon cycle. There are also methods to partition the fluxes based on discrimination of isotopes of carbon and oxygen by terrestrial ecosystem processes. The CO_2 exchange dynamics in the vegetation and soil can also be investigated at smaller scale with Open Top Chambers (OTCs) and with larger ecosystem experimental set-ups e.g. Free Air CO_2 Enrichment (FACE). The soil is a vast reservoir of carbon – consisting of roots, organic matter, organic carbon and inorganic carbon and has a great potential for carbon sequestration. Monitoring of carbon and fluxes in soil is therefore an essential aspect in the era of changing climate. The root systems are monitored mostly in a destructive manner but many non-destructive methods have also been devised such as use of microwaves, electrical impedance spectroscopy, minirhizotrons etc. Similarly, soil carbon estimation with traditional chemical method can be replaced with soil reflectance spectroscopy for rapid and large area estimations. Measurement of soil respiration and its partitioning also helps in verifying the capacity of soil as a net source or net sink. The uncertainties in the estimates need understanding of the multi-factorial effects on carbon exchange processes, plant acclimation responses, feedback response from changing climate and carbon trading issues. Monitoring therefore, needs concerted efforts in terms of enhancement of measurement infrastructure at different scales, use of multi-techniques and multi-disciplinary approaches so that more constrained and better estimates can be made.

REFERENCES

Abrol, I.P. 1995. Gaseous emissions from agro-ecosystems in India. In: Lal, R., Kimble, J.M., Levine E. and Stewart B.A. (eds.). Soils and Global Change - Advances in Soil Science, CRC Press, Boca Raton, pp 143-151.

Agren, G.I. and Wetterstedt, J.A.M. 2007. What determines the temperature response of soil organic matter decomposition? *Soil Biol. Biochem.*, **39**:1794-1798

Ainsworth, E.A. and Long S.P. 2004. What have we learned from 15 years of free-air CO_2 enrichment (FACE)? A meta-analytic review of the responses of photosynthesis, canopy properties and plant production to rising CO_2. *New Phytol.*, **165**:351–372

Amato, M., Lapenna, V., Rossi, R. and Bitella, G. 2012. Multi-electrode resistivity imaging. In: Mancusa, S. (ed.). Measuring Roots an Updated Approach, Springer-Verlag, Berlin Heidelberg, pp 189-211

Bacastow, R. and Keeling, C.D. 1973. Atmospheric carbon dioxide and radiocarbon in the natural carbon cycle: II. Changes from AD 1700 to 2070 as deduced from a geochemical model. In: Woodwell G.M. and Pecan E.V. (eds.). Carbon and the Biosphere, Proceedings of the 24th Brookhaven Symposium in Biology, Upton, New York, pp 86–135

Bacastow, R.B., Adams, J.A., Keeling, C.D., Moss, D.J., Whorf, T.P. and Wong, C.S. 1980. Atmospheric carbon dioxide, the southern oscillation, and the weak 1975 El Nino. *Science,* **210**:66–68

Baker, J.M. and Griffis, T.J. 2005. Examining strategies to improve the carbon balance of corn/soybean Agriculture. *Agric. Forest. Meteorol.*, **128**:163-177

Bakwin, P.S., Zhao, C., Ussler III, W., Tans, P.P. and Quesnell, E. 1995. Measurements of carbon dioxide on a very tall tower. *Tellus,* **47B:** 535–549

Balantyne, A.P., Miller, J.B., Baker, I.T., Tans, P.P. and White, J.W.C. 2011. Novel applications of carbon isotopes in atmospheric CO_2: what can atmospheric measurements teach us about processes in the biosphere. *Biogeosciences Discuss.*, **8**:4603-4631

Baldocchi, D.D. 2008. Breathing' of the terrestrial biosphere: lessons learned from a global network of carbon dioxide flux measurement systems. *Austral. J. Bot.,* **56:** 1-26

Baldocchi, D.D. and Wihon, K.B. 2001. Modelling CO_2 and water vapour exchange over a temperature broad-leafed forest across hourly to decadal time scales. *Ecological Modelling,* **142**:155-184

Baldocchi, D.D., Falge, E., Gu, L.H., Olson, R., Hollinger, D., Running, S., Anthoni, P., Bernhofer, C., Davis, K., Evans, R., Fuentes, J., Goldstein, A., Katul, G., Law, B., Lee, X.H., Malhi, Y., Meyers, T., Munger, W., Oechel, W., Paw, U.K.T., Pilegaard, K., Schmid, H.P., Valentini, R., Verma, S., Vesala, T., Wilson, K. and Wofsy, S. 2001. FLUXNET: A new tool to study the temporal and spatial variability of ecosystem-scale carbon dioxide, water vapor, and energy flux densities. *Bull. Am. Meteorol. Soc.*, **82**:2415-2434

Baldocchi, D.D., Hicks, B.B. and Meyer, T.P. 1988. Measuring biosphere atmosphere exchange of biologically related gases with micrometerological methods. *Ecology,* **69**:1331-1340

Batjes, N.H. 1996. Total carbon and nitrogen in the soils of the world. *Eur. J. Soil Sci.* **47**:151-163

Battle, M., Bender, M.L., Tans, P.P., White, J.W.C., Ellis, J.T., Conway, T. and Francey, R.J. 2000. Global carbon sinks and their variability inferred from atmospheric O_2 and ^{13}C. *Science,* **287**:2467-2470

Belward, A. 1996. The IGBP-DIS global 1 km land cover data set (DISCover): Proposal and Implementation Plans. IGBP-DIS Working Paper 13, International Geosphere-Biosphere Programme Data and Information System Office. Toulouse, France

Berry, J.A. 2012. There ought to be an equation for that. *Annu. Rev. Plant Biol.*, **63**:1-17

Blossfeld, S. and Gansert, D, 2007. A novel non-invasive optical method for quantitative visualization of pH dynamics in the rhizhosphere of plants. *Plant Cell Environ.* **30**:176-186

Böhm, W. 1979. Methods of Studying Root Systems. Springer-Verlag, Berlin-Heidelberg-New York

Bolin, B. and Keeling, C.D. 1963. Large scale atmospheric mixing as deduced from the seasonal and meridional variations of carbon dioxide. *J. Geophys Res.*, **68:** 3899-3920

Bolinder, M.A., Janzen, H.H., Gregorich, E.G., Angers, D.A. and VandenByaart, A.J. 2007. An approach for estimating net primary productivity and annual carbon inputs to soil for common crops in Canada. *Agric. Ecosyst. Environ.*, **118**: 29-42

Bouma, J. 2010. Implications of the knowledge paradox for soil science. *Adv. Agron.*, **106**:143-171

Burba, G. and Anderson, D. 2005. A brief practical guide to eddy covariance flux measurements. LI-COR Inc., pp 213

Butnor, J.R., Doolittle, J.A., Kress, L., Cohen, S. and Johnsen, K.H. 2001. Use of ground-penetrating radar to study tree roots in the southeastern United States. *Tree Physiol.*, **21:**1269–1278

Canadell J.G., Mooney H.A., Baldocchi D.D., Berry J.A., Ehleringer J.R., Field C.B., Gower S.T., Hollinger D.Y., Hunt J.E., Jackson R.B., Running S.W., Shaver G.R., Steffen W., Trumbore S.E., Valentini R. and Bond, B.Y. 2000. Carbon metabolism of the terrestrial biosphere: a multitechnique approach for improved understanding. *Ecosystems,* **3:**115–130

Canadell, J.G., Le Quere, C., Raupach, M.R., Field, C.B., Buitenhuis, E.T., Ciais, P., Conway, T.J., Gillett, N.P., Houghton, R.A. and Marland G. 2007. Contributions to accelerating atmospheric CO_2 growth from economic activity, carbon intensity, and efficiency of natural sinks. *PNAS.*, **104:**18866–18870

Cannel, M.G.R. 2003. Carbon sequestration and biomass energy offset-theoretical, potential and achievable capacities globally, in Europe and the UK. *Biomass Bioenergy,* **24**:97-116

Chapin III, F.S., McFarland, J., McGuire, D.A., Euskirchen, E.S., Ruess, R.W. and Kielland, K. 2009. The changing global consequences. *J. Ecol.*, **97**:840-850

Chen, F., Kissel, D.E., West, L.T. and Adkins, W. 2000. Field-scale mapping of surface soil organic carbon using remotely sensed imagery. *Soil Sci. Soc. Am. J.*, **64:**746–753

Ciais, P., Tans, P.P., Trolier, M., White, J.W.C. and Francey, R.J. 1995. A large northern hemisphere terrestrial CO_2 sink indicated by the $^{13}C/^{12}C$ ratio of atmospheric CO_2. *Science,* **269**:1098-1102

Ciais, P., Tans, P.P., White, J.W.C., Trolier, M., Francey, R.J., Berry, J.A., Randall, D.R., Sellers, P.J., Collatz, J.G. and Schimel, D.S. 1995b. Partitioning of ocean and land uptake of CO_2 as inferred by $\delta^{13}C$ measurements from NOAA climate monitoring and diagnostics laboratory global air sampling network, *J. Geophys Res.*, **100**:5051-5070

Ciais. P, Denning, A.S., Tans, P.P., Berry, J.A., Randall, D.A., Collatz, G.J., Sellers, P.J., White, J.W.C., Trolier, M., Meijer, H.J., Francy, R.J., Monfray, P.M. and Heimann, M. 1997. A three dimensional synthesis study of ^{18}O in atmospheric CO_2. 1. Surface fluxes. *J. Geophys. Res.*, **102**:5857-5872

Cox, P.M., Betts, R.A., Jones, C.D., Spall, S.A. and Totterdell, I.J. 2000. Acceleration of global warming due to carbon cycle feedbacks in a coupled climate model. *Nature,* **408**:184-187

Craig, H. 1957. Isotopic standards for carbon and oxygen and correction factors for mass-spectrometric analysis of carbon dioxide. *Geochimica et Cosmochimica Acta,* **12**:133-149

Dannoura, M., Kominami, Y., Makita, N. and Oguma, H. 2012. Flat optical scanner method and root dynamics. In: Mancuso, S. (ed.). Measuring Roots an Updated Approach. Springer-Verlag, Berlin Heidelberg, pp 127-133

Darrah, P.R., Jones, D.L., Kirk, G.J.D. and Roose, T. 2006. Modelling the rhizosphere: a review of methods for 'upscaling' to the whole plant scale. *Eur. J. Soil Sci.*, **57**:13-25

Davidson, E.A. and Holbrook, N.M. 2009. Is temporal variation of soil respiration linked to phenology of photosynthesis? In: Noormets, A. (ed.). Phenology of Ecosystem Processes: Applications in Global Change Research. Springer, New York, pp 187-197

de Graaff, M.A., van Groeningen, J., Six, J., Hungate, B. and van Kessel, C. 2006. Interactions between plant growth and nutrient dynamics under elevated CO_2: a meta-analysis. *Global Change Biol.*, **12**:1-15

DePury, D.D.G. and Farquhar, G.D. 1997. Simple scaling of photosynthesis from leaves to canopies without the errors of big-leaf models. *Plant Cell Environ.*, **20**:537-557

Dessaux, Y., Hinsinger, P. and Lemanceau, P. 2009. Rhizosphere: so many achievements and even more challenges. *Plant Soil,* **321**:1-3

Dittmer, H.J. 1937. A quantitative study of the roots and root hairs of a winter rye plant (*Secale cereal*) *Am. J. Bot.*, **24**:417-420

Drinkwater, L.E., Wagoner, P. and Sarrantonio, M. 1998. Legume based cropping system have reduced carbon and nitrogen losses. *Nature,* **396**:262-265

Durant, A.J., Le Quere, C. Hope C. and Friend, A.D. 2011. Economic value of improved quantification in global sources and sinks of carbon dioxide. *Philos. Transact. A Math. Phys. Eng. Sci.*, **369**:1967-1979

Ebinger, M.H., Norfleet, M.L., Breshears, D.D., Cremers, D.A., Ferris, M.J., Unkefer, P.J., Lamb, M.S., Goddard, K.L. and Meyer, C.W. 2003. Extending the applicability of laser-induced breakdown spectroscopy for total soil carbon measurement. *Soil Sci. Soc. Am. J.*, **67**:1616–1619

Ehrenfeld, J.G., Ravit, B. and Elgersma, K. 2005. Feedback in the plant-soil system. *Annu. Rev. Environ. Resour.*, **30**:75-115

Eissenstat, D.M., Wells, C.E., Yanai, R.D. and Whitbeck, J.L. 2000. Building roots in a changing environment: implications for root longevity. *New Phytol.*, **147**: 33-42

Ekdahl, C.A. and Keeling, C.D. 1973. Atmospheric carbon dioxide and radiocarbon in the natural carbon cycle: I. Quantitative deductions from the records of Mauna Loa Observatory and at the South Pole. In: Woodwell, G.M. and Pecan, E.V. (eds.). Carbon and the Biosphere, Proceedings of the 24th Brookhaven Symposium in Biology. Upton, New York, pp 51-85

Farquhar, G.D., Ehleringer, J.R. and Hubick, K.T. 1989. Carbon isotope discrimination and photosynthesis. *Annu. Rev. Plant Physiol. Plant Mol. Biol.*, **40**: 503-537

Farquhar, G.D., Lloyd, J., Taylor, J.A., Flanagan, L.B., Syvertsen, J.P., Hubik, K.T., Wong, C.S. and Ehleringer J.R. 1993. Vegetation effects on the isotope composition of oxygen and atmospheric CO_2. *Nature,* **363**:439-443

Farquhar, G.D., von Caemmerer, S. and Berry, J.A. 1980. A biochemical model of photosynthetic CO_2 assimilation in leaves of C_3 species. *Planta,* **149**:78-90

Field, C.B. 2001. Plant physiology of the "missing" carbon sink. *Plant Physiol.*, **125**: 25–28

Flanagan, L.B., Brooks, D.R., Varney, G.T., Berry, S.C. and Ehleringer, J.R. 1996. Carbon isotope discrimination during photosynthesis and the isotc ratio of respired CO_2 in boreal forest ecosystems. *Global Biogeochem. Cycles,* **10**:629-640

Flanagen, L.B., Brooks, J.R., Varney, G.T. and Ehleringer, J.R. 1997. Discrimination against ($^{18}O/^{16}O$) during photosynthesis and oxygen isotope ratio of respired CO_2 in boreal ecosystems. *Global Biochem. Cycles* **18**:83-98

Foley, J.A., DeFries, R., Asner, G.P., Barford, C., Bonan, G., Carpenter, S.R., Chapin, F.S., Coe, M.T., Daily, G.C., Gibbs, H.K., Helkowski, J.H., Holloway, T., Howard, E.A., Kucharik, C.J., Monfreda, C., Patz, J.A., Prentice, I.C., Ramankutty, N. and Snyder, P.K. 2005. Global consequences of land use. Science, **309**: 570-574

Francey, R.J., Allison, C.E., Etheridge, D.M., Trudinger, C.M., Enting, I.G., Leuenberger, M., Langenfelds, R.L., Michel, E. and Steele, L.P. 1999. A 1000-year high precision record of $\delta^{13}C$ in atmospheric CO_2. *Tellus,* **51**:170–193

Frankenberg, C., Fisher, J.B., Worden, J., Badgley, G., Saatchi, S.S., Lee, J.E., Toon, G.C., Butz, A., Jung, M., Kuze, A. and Yokota T. 2011. New global observations of the terrestrial carbon cycle from GOSAT: Patterns of plant fluorescence with gross primary productivity. *Geophys. Res. Lett.*, **38**: L17706

French, A., Wells, D., Everitt, N. and Pridmore, T. 2012. High-throughput quantification of root growth. In: Stefano, M. (ed.). Measuring Roots an Updated Approach. Springer-Verlag, Berlin Heidelberg, pp 109-126

Friend, A.D. 2010. Terrestrial plant production and climate change. *J. Exp. Bot.*, **61**:1293-1309

Friend, A.D., Geider, R.J., Behrenfeld, M.J. and Still, C.J. 2009. Photosynthesis in global-scale models. In: Laisk, A., Nedbal, L. and Govindjee (eds.). Photosynthesis *In Silico*: Understanding Complexity from Molecules to Ecosystems. Springer Science, pp 465-497

Gamon, J.A., Penuelas, J. and Field, C.B. 1992. A narrow waveband spectral index that tracks diurnal changes in photosynthetic efficiency. *Remote Sens. Environ.*, **41**:35-44

Gamon, J.A., Qiu, H.L. and Sanchez-Azofeifa, A. 2007. Ecological applications of remote sensing at multiple scales. In: Pugnaine F.I. and Valladares F. (eds.). Functional Plant Ecology. CRC Press-Taylor and Francis, pp 655-683

Garbulsky, M.F., Penuelas, J., Gamon, J., Inoue, Y. and Filella, I. 2011. The photochemical reflectance index (PRI) and the remote sensing of leaf, canopy and ecosystem radiation use efficiencies: a review and meta analysis. *Remote Sens. Environ.*, **115**:281–297

Gaten, C.T. and Wullschleger, S.D. 1999. Soil carbon inventories under a bioenergy crop (Switch grass): measurement limitation. *J. Environ. Qual.*, **28**:1359-1365

Gehl, R.J. and Rice, C.W. 2007. Emerging technologies for *in situ* measurement of soil carbon. *Climatic Change,* **80**:43–54

Geider, R.J., Delucia, E.H., Falkowski, P.G., Finzi, A.C., Grime, J.P., Grace, J., Kama, T.M., La Roche, J., Long, S.P., Osborne, B.A., Platt, T., Prentice, I.C., Raven, J.A., Schlesinger, W.H., Smetacek, V., Stuart, V., Satyendranath, S., Thomas, R.B., Vogelmann, T.C., Williams, P. and Woodward, F.I. 2001. Primary productivity of planet earth biological determinations and physical constraints in terrestrial and aquatic habitats. *Global Change Biol.*, **7**:849-882

Gill, R. and Jackson, R.B. 2000. Global patterns of root turnover for terrestrial ecosystems. *New Phytol.*, **147**:13-31

Gomez, C., Viscarra Rossel, R.A. and McBratney, A.B. 2008. Soil organic carbon prediction by hyperspectral remote sensing and field vis-NIR spectroscopy: An Australian case study. *Geoderma* **146:** 403-411

Grace, J., Nichol, C., Disney, M., Lewis, P., Ouife, T. and Bowyer, P. 2007. Can we measure terrestrial photosynthesis from space directly, using spectral reflectance and fluorescence *Global Change Biol.*, **13**: 1484-1497

Groenigen, K.J., Gorissen, A., Six, J., Harris, D., Kuikman, P.J., Groenigen, J.W. and Kessel, C. 2005. Decomposition of ^{14}C labeled roots in a pasture soil exposed to 10 years of elevated CO_2. *Soil Biol. Biochem.*, **37**:497-506

Heath, J., Ayres, E., Possel, M., Bardgett, R.D., Black, H.I.J., Grant, H., Ineson, P. and Kersteins, G. 2005. Rising atmospheric CO_2 reduces sequestration of root derived soil carbon. *Science* **309**:1711-1713

Heimann, M. and Reichstein, M. 2008. Terrestrial ecosystem carbon dynamics and climate feedbacks. *Nature* **451**:289-292

Hendrey, G.R., Ellsworth, D.S., Lewin, K.F. and Nagy, J. 1999. A free-air enrichment system for exposing tall forest vegetation to elevated atmospheric CO_2. *Glob. Change Biol.*, **5**:293–309

Hogberg, P., Nordgren, A., Buchmann, N., Taylor, A.F.S., Ekblad, A., Hogberg, M.N., Nyberg, G., Ottossen-Lofvenium, M. and Read, D.J. 2001. Large scale forest girdling shows that current photosynthesis drives soil respiration. *Nature,* **411**:789-792

Houghton, J. 2009. Global Warming-The Complete Briefing (fourth edition). Cambridge University Press, Cambridge UK

Houghton, R.A. 2003. Why are estimates of the terrestrial carbon balance so different? *Global Change Biol.*, **9**:500-509

Houghton, R.A. 2005. Aboveground forest biomass and the global carbon balance. *Global Change Biol.*, **11**:945-958

Hruska, J., Cermak, J. and Sustek, S. 1999. Mapping tree root systems with ground-penetrating radar. *Tree Physiol.*, **19**:125–130

Hutchings M.J. and John, E.A. 2004. The effects of environmental heterogeneity on root growth and root/shoot partitioning. *Ann. Bot.*, **94**:1-8

Hymus, G. and Valentini, R. 2007. Terrestrial vegetation as a carbon dioxide sink. In: Reay, D.S. Hewitt, C.N., Smith, K.A. and Grace, J. (eds.). Greenhouse Gas Sinks. CABI, pp 11-30

IPCC (2001): Climate change 2001: The scientific basis. In: Houghton, J.T., Ding, Y., Griggs, D.J., Noguer, M., van der Linden, P.G., Dai, X., Maskell K. and Johnson C.A. (eds.). Contribution of Working Group I to the Third Assessment Report of the Intergovernmental Panel on Climate Change. Cambridge University Press, Cambridge UK

IPCC. 2007. Inter governmental Panel on Climate Change. Summary Report of the Working Group of IPCC, Paris, February, 2007

Jackson, R.B., Mooney, H.A. and Schulze, E.D. 1997. A global budget for fine root biomass, surface area, and nutrient contents. *PNAS* (USA) **94**:7362-7366

Johnson, J.M.F., Allmaras, R.R. and Reicosky, D.C. (2006). Estimating source carbon from crop residues, root and rhizodeposits using the national grain-yield database. *Agron. J.*, **98**: 622-636

Jones, D.L., Nguyen, C. and Finlay, R.D. 2009. Carbon flow in the rhizosphere: carbon trading at the soil-root interface. *Plant Soil,* **321**:5-3

Karl, T.R., J.M. Melillo and T.C. Peterson (eds.). 2009. Global Climate Change Impacts in the United States. Cambridge University Press, New York, NY

Kasischke, E.S., Melack, J.M. and Dobson, M.C. 1997. The use of imaging radars for ecological applications - A review. *Remote Sensing Environ.*, **59**:141-156

Keeling, C.D. 1958. The concentration and isotopic abundances of atmospheric carbon dioxide in rural areas. *Geochim Cosmochim. Acta,* **13**:322-334

Keeling, C.D. 1960. The concentration and isotopic abundances of carbon dioxide in the atmosphere. *Tellus,* **12**:200-203

Keeling, C.D. 1998. Rewards and penalties of monitoring the earth. *Annu. Rev. Energy Environ.*, **23**:25-82

Keeling, C.D., Bacastow, R.B., Bainbridge, A.E., Ekdahl Jr., C.A., Guenther, P.R., Waterman, L.S., Chin, J.F.S. 1976. Atmospheric carbon dioxide variations at Mauna Loa Observatory, Hawaii. *Tellus,* **8**:538–551

Keeling, C.D., Whorf, T.P., Wong, C.S. and Bellagay, R.D. 1985. The concentration of atmospheric carbon dioxide at ocean weather station P from 1969 to 1981. *J. Geophys. Res.*, **90:**10511–10528

Keeling, C.D. 1961. The concentration and isotopic abundances of carbon dioxide in rural and marine air. *Geochim. Cosmochim. Acta.*, **24:**277–298

Kohlmaier, G.H., Sire, E.O., Janecek, A., Keeling, C.D., Piper, S.C. and Revelle, R. 1989. Modelling the seasonal contribution of a CO_2-fertilization effect of the terrestrial vegetation to the amplitude increase in atmospheric CO_2 at Mauna Loa Observatory. *Tellus,* 41:487-510

Kravchenko, A.N. and Robertson, A.N. 2011. Whole-profile soil carbon stocks: the danger of assuming too much from analyses of too little. *Soil Sci. Soc. Am. J.*, **75:**235–240

Kuzyakov Y. 2006. Sources of CO_2 efflux from soil and review of partitioning methods. *Soil Biol. Biochem.* **38:**425–448

Kuzyakov, Y. 2011. How to link soil C pools with CO_2 fluxes? *Biogeosci.*, 8:1523–1537

Kuzyakov, Y. 2011. Prime time for microbes. *Nature Climate Change,* **1:**295-297

Kuzyakov, Y. and Gavrichkova, O. 2010. Time lag between photosynthesis and carbon dioxide efflux from soil: a review of mechanisms and controls. *Global Change Biol.*, **16:**3386–3406

Laisk, A. and Edwards, G.E. 2000. A mathematical model of C_4 photosynthesis. The mechanisms of concentrating CO_2 in NADP malic enzyme type species. *Photosynth. Res.*, **66**:199-224

Lal, R. 2001. World cropland soils as a source or sink for atmospheric carbon. *Adv. Agron.*, **71**:145-190

Lal, R. 2004a. Soil carbon sequestration impacts on global climate change and food security. *Science,* **304**:1623-1627

Lal, R. 2004b. Soil carbon sequestration to mitigate climate change. *Geoderma,* **123**:1-22

Lal, R. 2006. Enhancing crop yields in developing countries through restoration of soil organic carbon pool in agricultural lands. *Land Degrad. Dev.*, **17**:197-209

Lal, R. 2007. Anthropogenic influences on world soils and implications to global food security. *Adv. Agron.*, **93**: L 69-93

Lal, R. 2008. Carbon sequestration. *Philos. Trans. R. Soc. Lond. B. Biol. Sci.*, **363**: 815-830

Lal, R. 2010. Soil as source and sink of environmental carbon dioxide. In: Xu, J.M. and Huang, P.M. (eds.). Molecular Environmental Soil Science at the Interface in Earth's Critical Zone. Zhejiang University Press and Springer-Verlag, pp 11-12

Lal, R., Kimble, J.M., Follett, R.F. and Stewart, B.A. (Eds.) 2001. Assessment methods for soil carbon. CRC Press, Boca Raton

Lambers, H., Chapin III, F.S. and Pous, T.L. 1998. Plant Physiological Ecology. Springer, USA

Law, B.E., Waring, R.H., Anthoni, P.M. and Abers, J.D. 2000. Measurements of gross and net ecosystem productivity and water vapour exchange of a *Pinus ponderosa* ecosystem and an evaluation of two generalized models. *Global Change Biol.* **6:**155–168

Le Quéré, C., Canadell, J.G., Ciais, P., Dhakal S., Patwardhan A., Raupach, M.R. and Young O.R. 2010. An international carbon office to assist policy-based science. *Curr. Opinion Env. Sustainability,* **2:**1-4

Le Quéré, C., Raupach, M.R., Canadell, J.G., Marland, G., Bopp, L., Ciais, P., Conway, T.J., Doney, S.C., Feely, R.A., Foster, P., Friedlingstein, P., Gurney, K., Houghton, R.A., House, J.I., Huntingford, C., Levy, P.E., Lomas, M.R., Majkut, J., Metzl, N., Ometto, J.P., Peters, G.P., Prentice, I.C., Randerson, J.T., Running, S.W., Sarmiento, J.L., Schuster, U., Sitch, S., Takahashi, T., Viovy, N., van der Werf, G.R. and Woodward, F.I. 2009. Trends in the sources and sinks of carbon dioxide. *Nature geosci;* **2:**831–836

Leakey, A.D.B., Ainsworth, E.A., Bernacchi, C.J., Rogers, A., Long, S.P. and Ort, D.R. 2009a. Elevated CO_2 effects on plant carbon, nitrogen, and water relations: six important lessons from FACE. *J. Exp. Bot.*, **60**:2859–2876

Leakey, A.D.B., Xua, F., Gillespiea, K.M., McGratha, J.M., Ainsworth, E.A. and Ort, D.R. 2009b. Genomic basis for stimulated respiration by plants growing under elevated carbon dioxide. *PNAS.*, **106**:3597–3602

Lefsky, M.A., Cohen, W.B., Parker, G.G. and Harding, D.J. 2002. Lidar remote sensing for ecosystem studies. *BioSci.*, **52:**19-30

Leith, H. and Wittaker, R.H. 1975. Primary Productivity of the Biosphere. Springer-Verlag, New York

Levin, I., Hammer, S., Eichelmann, E. and Vogel, F.R. 2011. Verification of greenhouse gas emission reductions: the prospect of atmospheric monitoring in polluted areas. *Phil. Trans. R. Soc. Am.*, **369:**1906-1924

Lloyd, J. and Farquhar, G.D. 1994. ^{13}C discrimination during CO_2 assimilation by the terrestrial biosphere. *Oecologia.*, **99**:201-215

Mancuso, S. 2012. Measuring Roots an Updated Approach, Springer-Verlag, Berlin Heidelberg.

Manning, A.J. 2011. The challenge of estimating regional trace gas emissions from atmospheric observations. *Phil. Trans. R. Soc. Am.*, **369:** 1943-1954

Masi, E., Azzarello, E., Pandolfi, C., Pollastri, S., Mugnai, S. and Mancuso, S. 2012. Multi electrode arrays (MEAs) and the electrical network of the roots. In: Mancuso, S. (ed.). Measuring Roots an Updated Approach. Springer-Verlag, Berlin Heidelberg, pp 51-65

Matthews, H.D. 2007. Implications of CO_2 fertilization for future climate change in a coupled climate–carbon model. *Global Change Biol.*, 13:1068–1078

Mencuccini, M. and Hölttä, T. 2010. The significance of phloem transport for the speed with which canopy photosynthesis and belowground respiration are linked. *New Phytol.*, **185:**189-203

Monteith, J.L. 1972. Solar radiation and productivity in tropical ecosystems. *J. Appl. Ecol.*, **9:**747-766

Nelson, D.W., and. Sommer, L.E. 1982. Total carbon, organic carbon, and organic matter. In Page A.L. (ed.). Methods of Soil Analysis (second edition). ASA Monogr. 9(2). Amer. Soc. Agron. Madison, WI, pp 539-579

Nichol, C.J., Huemmrich, K.F., Black, T.A., Jarvis, P.G., Walthall, C.L., Grace, J., and Hall, F. G. 2000. Remote sensing of photosynthetic-light-use efficiency of boreal forest. *Agric. Forest Meteorol.*, **101**:131–142

Nieder, N. and Benbi, D.K. 2008. Carbon and Nitrogen in the Terrestrial Environment. Springer Verlag

Noormets, A. (ed.) 2009. Phenology of Ecosystem Processes: Application in Global Change Research. Springer Verlag

Norby, R.J. and Luo, Y. 2004. Evaluating ecosystem responses to rising atmospheric CO_2 and global warming in a multi-factor world. *New Phytol.*, **162:** 281–293

Norby, R.J. and Zak, D.R. 2011. Ecological lessons from free-air CO_2 enrichment (FACE) experiments. *Annu. Rev. Ecol. Evol. Syst.*, **42**:181-203

Norby, R.J., Cotrufo, M.F., Ineson, P., O'Neill, E.G. and Canadell, J.G. 2001. Elevated CO_2 litter chemistry and decomposition: a synthesis. *Oecologia.*, **127**: 153-165

Norby, R.J., Ledford, J., Rilley, C.D., Miller, E., O'Neill, E.G. 2004. Fine-root production dominates response of a deciduous forest to atmospheric CO_2 enrichment *PNAS (USA),* **101**: 9689-9693

Olsen, S.C. and Randerson, J.T. 2004. Differences between surface and column atmospheric CO_2 and implications for carbon cycle research. *J. Geophys. Res.*, **109**:D02301

Ortiz-Monasterio, I., Wassman, R., Govaertz, B., Hosen, Y., Katayanagi, N. and Verhulst, N. 2010. Greenhouse mitigation in main cereal systems: rice, wheat and maize. In: Reynolds, M.P. (ed.). Climate Change and Crop Production. CAB International, pp151-176.

Parry, M.A.J., Reynolds, M., Salucci, M.E., Raines, C., Andralojc, P.J., Zhu, X.G., Price, G.D., Condon, A.G. and Furbank, R.T. (2011). Raising yield potential of wheat II. Increasing photosynthetic capacity and efficiency. *J. Exp. Bot.*, **62**:453-467

Passioura, J.B. 1983. Roots and drought resistance. *Agric. Water Mgmt.*, **7**: 265-280

Pataki, D.E., Bowling, D.R. and Ehleringer, J.R. (2003). Seasonal cycle of carbon dioxide and its isotopic composition in an urban atmosphere: Anthropogenic and biogenic effects. *J. Geophys. Res.*, **108**: D23, 4735

Pathak, H., Byjesh, K., Chakrabarti, B. and Aggarwal, P.K. 2011. Potenital and cost of carbon sequestration in Indian agriculture: Estimates from long-term experiments. *Field Crops Res.*, **120**:102-111

Pendall, E., Bridgham, S., Hanson, P.J., Hungate, B., Kicklighter, D.W., Johnson, D.W., Law, B.E., Luo, Y., Megonigal, J.P., Olsrud, M., Ryan, M.G. and Wan, S. (2004). Below-ground process responses to elevated CO_2 and temperature: a discussion of observations, measurement methods and models *New Phytol.*, **162**:311–322

Penuelas, J., Filella, I. and Gamon, J.A. 1995. Assessment of photosynthetic radiation use efficiency with spectral reflectance. *New Phytol.*, **131**:291-296

Penuelas, J., Garbulsky, M.F. and Filessa, I. 2011. Photochemical reflectance index (PRI) and remote sensing of plant CO_2 uptake. *New Phytol.*, **191**:596-599

Peters, W., Jacobson, A.R., Sweeney, C., Andrews, A.E., Conway, T.J., Masarie, K., Miller, J.B., Bruhwiler, L.M.P., Petron, G., Hirsch, A.I., Worthy, D.E.J., van der Werf, G.R., Randerson, J.T., Wennberg, P.O., Krol, M.C. and Tans, P.P. 2007. An atmospheric perspective on North American carbon dioxide exchange: CarbonTracker. *PNAS (USA),* **104**:18925-18930

Pietsch, S.A. and Hasenauer, H. 2009. Photosynthesis with large scale ecosystem models. In: Laisk, A., Nedbal, L. and Govindjee (eds.). Photosynthesis *In Silico*: Understanding Complexity from Molecules to Ecosystems. Springer Science, pp 441-464

Pritchard, S.G. and Rogers, H.H. 2000. Spatial and temporal deployment of crop roots in CO_2 - enriched environments. *New Phytol.* **147**:55–71

Reilly, J.M. and Asadoorian, M.O. 2007. Mitigation of greenhouse gas emissions from land use: creating incentives within greenhouse gas emissions trading systems. *Climatic Change,* **80**:173–197

Repo T., Cao Y., Silvennoinen, R. and Ozier-Lafontaine, H. 2012. Electrical impedance spectroscopy and roots. In: Mancuso S. (ed.). Measuring Roots An Updated Approach. Springer-Verlag, Berlin Heidelberg, pp 25-49

Richter, D.D. and Mobley, M.L. 2009. Monitoring earth's critical zone. *Science,* **326**:1067-1068

Rochette, P. and Flanagen, L.B. 1997. Quantifying rhizosphere respiration in a corn crop under filed conditions. *Soil Sci. Soc. Am. J.*, **61**:466-474

Rochette, P., Flanagan, L.B. and Gregorich, E.G. 1999. Separating soil respiration into plant and soil components using analysis of the natural abundance of carbon-13. *Soil Sci. Soc. Am. J.,* **63**:1207-1213

Ryan, M.G. 1991. A simple method for estimating gross carbon budgets for vegetation in forest ecosystems. *Tree Physiol.*, **9**:255-266

Sabine, C.L., Heimann, M., Artaxo, P., Bakker, D.C.E., Chen, C.T.A., Field, C.B.N., Gniber, N., Le Quere, C., Prinn, R.G., Richey, J.E., Lankao, P.R., Sathaye, J.A. and Valentini, R. 2004. Current status and part trends of the global carbon cycle. In: Field C.B.and Raupach, M.R. (eds.). The Global Carbon Cycle. Islands Press, Washington, pp 17-44

Saha, S., Nair, P.K.R., Nair, V.D., Kumar, B.M. 2009. Soil carbon stock in relation to plant diversity of homegardens in Kerala, India. *Agroforestry Syst.*, **76**:53-65

Sayer, E.J., Heard, M.S., Grant, H.K., Marthews, T.R. and Tanner, E.V.J. 2011. Soil carbon release enhanced by increased tropical forest litter fall. *Nature Climate Change,* **1**:304-307

Schimel, D.S., House, J.I., Hibbard, K.A., Bousquet, P., Gais, P., Peylin, P., Braswell, B.H., Apps, M.J., Baker, D., Bondean, A., Canadell, J., Churkina, G., Cramer, W., Denning, A.S., Field, C.B., Friedlingstein, P., Goodale, C., Heimann, M., Houghton, R.A., Melillo, J.M., Moore, B., Murdiyarso, D., Noble, I., Pacala, S.W., Prentice, I.C., Raupach, M.R., Rayner, P.J., Scholes, R.J., Steffen, W.L. and Wirth, C. 2001. Recent patterns and mechanisms of carbon exchange by terrestrial ecosystems. *Nature,* **414**:169-172

Schmiel, D. 2004. Photosynthesis and the global carbon cycle. In: Archer, M.D. and Barber, J. (eds.). Molecular to Global Photosynthesis. World Scientific Publishing Company, pp 613-626

Schulze, E.D. 2006. Biological control of the terrestrial carbon sink. *Biogeosciences,* **3**:147-166

Siddique, K.H.M., Belford, R.K. and Tennant, D. 1990. Root : shoot ratio of old and modern tall and semi-dwarf wheats in Mediterranean environment. *Plant Soil,* **121**:89-98

Smith, P. 2004. Carbon sequestration in crop lands: the potential in Europe and the global context. *Eur. J. Agron.*, **20**:229-236

Smith, P., Martino, D., Cai, Z., Gwary, D., Janven, H., Kumar, P., McCarl, B., Ogle, S., Mara, F.O., Rice, C., Scholes, B., Sirotenko, O., Howden, M., McAllister, T., Pan, G., Romanenkov, V., Schneider, U. and Towprayoon, S. 2007. Policy and technological constraints to implementation of greenhouse gas mitigation options in agriculture. *Agric. Ecosyst. Env.* **118**:6-28

Stenberg, B., Rossel, R.A.V., Mouazen, A.M. and Wetterlind, J. 2010. Visible and near infrared spectroscopy in soil science. *Adv. Agron.* **107**:163-215

Subke, J.A., Voke, N.R., Leronni, V., Garnett, M.H. and Ineson, P. 2011. Dynamics and pathways of autotrophic and heterotrophic soil CO_2 efflux revealed by forest girdling. *J. Ecol.*, **99**:186–193.

Swinbank, W.C. 1951. The measurement of vertical transfer of heat and water vapour by eddies in the lower atmosphere. *J. Meteorol.*, **8**:135–145

Thompson, M.V., Randerson, J.T., Malmstrm, C.M. and Field, C.B. 1996. Change in net primary production and heterotrophic respiration: How much is necessary to sustain the terrestrial carbon sink? *Global Biogeochem. Cycles,* **10**:711-726

Torbert, H.A., Prior, S.A., Rogers, H.H. and Wood, C.W. 2000. Review of elevated atmospheric CO_2 effects on agro-ecosystems: residue decomposition processes and soil C storage. *Plant Soil,* **224**:59-73

Tucker, C.J. 1978. A comparison of satellite sensor bands for vegetation monitoring. *Photogrammetry Remote Sensing,* **44**:1369-1380

Valentini, R., Matteucci, G., Dolman, A.J., Schulze, E.-D., Rebmann, C., Moors, E.J., Granier, A., Gross, P., Jensen, N.O., Pilegaard, K., Lindroth, A., Grelle, A., Bernhofer, C., Grunwald, T., Aubinet, M., Ceulemans, R., Kowalski, A.S., Vesala, T., Rannik, U., Berbigier, P., Loustau, D., Guomundsson, J., Thorgeirsson, H., Ibrom, A., Morgenstern, K., Clement, R., Moncrieff, J., Montagnan, L., Minerbi, S. and Jarvis, P.G. 2000. Respiration as the main determinant of carbon balance in European forests. *Nature,* **404**:861–865

van Noordwijk, M., Rahayu, S., Williams, S.E., Hairiah, K., Khasanah, N. and Schroth, G. 2004. Crop and tree root system dynamics. In: van Noordwijk, M., Cadisch, G. and Ong C.K. (eds.). Below-ground Interaction in Tropical Agro-ecosystem: Concept and Models with Multiple Plant Components. CABI, pp 83-107

Vargas, R., Baldocchi, D.D., Bahn, M., Hanson, P.J., Hosman, K.P., Kulmala, L., Pumpanen, J. and Yang, B. 2011. On the multi-temporal correlation between photosynthesis and soil CO_2 efflux: reconciling lags and observations. *New Phytol.*, **191**:1006–1017

Verma, S.B., Dobermann, A., Carsman, K.G., Walters, D.T., Knops, J.M., Arkebauer, T.J., Suyker, A.E., Burba, G.G., Amos, B., Yang, H., Ginting, D.,

Hubbard, K.G., Gitelson, A.A. and Walter-Shea, E.A. 2005. Annual carbon dioxide exchange in irrigated and rainfed maize-based agroecosystem. *Agric. Forest Meteorol.*, **131**:77-96

Vijai, P., Sharma-Natu, Poonam., Sreekanth, B. and Ghildiyal, M.C. 2006. Photosynthetic response of wheat and sunflower varieties to long term exposure of carbon dioxide concentration. *Photosynthetica,* **44**:586-590

Walkley, A. and Black, I.A. 1934. An examination of Degtjareff method for determining soil organic matter and a proposed modification of the chromic acid titration method. *Soil Sci.,* **37**:29-37

Watt, M., Schneebeli, K., Dong, P. and Wilson, I.W. 2009. The shoot and root growth of Brachypodium nad its potential as a model for wheat and other cereal crops. *Funct. Plant Biol.*, **36**:960-969

Weiss, R.F. and Prinn, R.G. 2011. Quantifying greenhouse-gas emissions from atmospheric measurements: a critical reality check for climate legislation. *Phil. Trans. R. Soc. Am.* **369** 1925-1942

West, P.C., Gibbs, H.K., Monfreda, C., Waquer, J., Bradford, C.C., Carpenter, S.R. and Foley, J.A. (2010). Trading carbon for food: Global comparison of carbon stocks *vs.* crop yields on agricultural land. *PNAS,* **107**:19645-19648

West, T.O. and Six, J. 2007. Considering the influence of sequestration duration and carbon saturation on estimates of soil carbon capacity. *Climatic Change,* **80**:25-41

Wielopolski, L., Mitra, S., Hendrey, G., Rogers, H., Torbert, A. and Prior, S. 2003. Non-destructive *in situ* soil carbon analysis: Principles and results. Proceedings of 2nd National Conference on Carbon Sequestration: Developing and Validating the Technology base to reduce carbon intensity. 5–8 May, 2003, Alexandria, VA

Wielopolski, L., Orion, I., Hendrey, G. and Rogers, H. 2000. Soil carbon measurements using inelastic neutron scattering. *IEEE Trans. Nucl. Sci.*, **47**:914-917

Wixon, D.L. and Balser, T.C. 2009. Complexity, climate change and soil carbon: a systems approach to microbial temperature response. *Systems Res.*, **26**:601–620

Woodwell, G.M. and Whittaker, R.H. 1968. Primary production in terrestrial communities. *Am. Zool.*, **8**:19-30

Woodwell, G.M., Hobbie, J.E., Houghton, R.A., Melillo, J.M., Moore, B., Peterson, B.J. and Shaver, G.R. 1983. Global deforestation: contribution to atmospheric carbon dioxide. *Scienc,e* **222**:1081-1086

Wunch, D., Toon, G.C., Blavier, J.F.L., Washenfelder, R.A., Notholt, J., Connor, B.J., Griffith, D.W.T., Sherlock, V. and Wennberg, P.O. 2011. The total carbon column observing network. *Phil. Trans. R. Soc. Am.*, **369**:2087-2112

Zenone, T., Morelli, G., Teobaldelli, M., Fischanger, F., Matteucci, M., Sordini, M., Alessio, A., Chiara, F., Chiti, T. and Seufert, G. 2008. Preliminary use of tomography to study tree roots in pine forests and poplar plantation. *Funct. Plant Biol.*, **35**:1047–1058.

9

CROSS-TALK SIGNALING AND SURVIVAL STRATEGIES IN PLANTS UNDER CHANGING ENVIRONMENT

Pramod Kumar Singh[1], Sananda Mondal[2] and Bandana Bose[2]

[1]*Plant Physiology Laboratory, Department of Botany, Udai Pratap Autonomous College, University of M. G. Kashi Vidyapeeth, Varanasi-221002, UP, India (Corresponding author: drpksingh2001@yahoo.com)*

[2]*Department of Plant Physiology, Institute of Agricultural Sciences, Banaras Hindu University, Varanasi-221005, UP, India*

INTRODUCTION

Plants protect themselves against various biotic and abiotic stresses that damage their tissues in many ways. Hormone signaling is integrated at several levels during plant growth and development (Santner and Estelle, 2009; Orsini *et al.*, 2009). Crosstalk between phytohormone and defense signaling pathways should reveal new potential targets for the development of multiple stress tolerance mechanisms and higher plant productivity (Singh *et al.*, 2011, Grant and Jones, 2009). Phytohormonal interaction is specific and depends on type of attack such as the physical wound created by insects feeding may itself facilitate the entry of opportunistic pathogens. Conversely, some insects like aphids and thrips prefer the yellow color of diseased plants. Attack by one organism may be positively associated with attack by other organisms, plants should utilize strong defense systems that either effective against both type of attackers or develop a resistance against them (Pandey *et al.*, 2008; Thaller and Bostock, 2004). Similarly, crosstalk in cold, draught, and salt signaling components might be organized into a network action (Reyes *et al.*, 2009, Taylor and Mc Ainsh, 2004). This complexity of signaling makes the answer to the question how crosstalk signaling pathways are regulated in response to biotic and abiotic stresses. Ecological significance of signal interactions in plant defense system is depending on the specific metabolic defensive compounds. For example, Jasmonate controls defense gene expression, growth and fertility in *Arabidopsis thaliana* through conjugation with amino acid isoleucine to form the active hormone jasmonoyl-isoleucine (JA-Ile), a major biologically active jasmonate (Gfeller *et al.*, 2010) Whereas, salicylic acid is synthesized by plants in response to diverse range of biotic and abiotic factors to establish the resistance mechanisms (Loake and Grant, 2007; Bari and Jones, 2009). A major goal of phytohormonal ecology is to understand the role of plant hormones in determining plant responses to various environmental challenges. This includes the strategy that each hormone plays in response to individual stress as well as the coordination of multiple hormones in response to multiple stresses. However, complex defense signaling network in nature should join forces to place molecular mechanism of induced plant defences in an ecological perspective. Phytohormone interact together for defense signal networking to fine tune defense. SA, JA, ABA are recognized as key players in the regulation of the signaling pathway (Wang *et al.*, 2007). The primary immune response has evolved to recognise common features of organism that interact with the plant and to translate this recognition into a defense response that is especially directed against the attack encounter (Koornneef and Pieterse, 2008).

Plant growth is greatly affected by a combination of environmental stresses such as extreme temperatures, drought or high salinity. From agricultural perspective, such stresses are among the most significant factors responsible for sustainable agriculture and plant productivity. For example, salinity is a soil condition characterized by a high concentration of soluble salts and often impair the availability of water for its biological functions like its role as a solvent, transpirant, electron donor; transport medium etc. (Hasegawa *et al.*, 2000; Zhu, 2002; Sairam and Tyagi, 2004). To understand the physiological and molecular mechanisms for the survival strategies of plants, it is necessary to elucidate whether their growth is being limited by the osmotic effect of the salt in the soil, or the toxic effect of the salt within the plant (Munns and Tester, 2008). Plants have evolved several adaptive mechanisms to cope with the salinity in their environment, but the understanding of these mechanisms still remains incomplete. Thus development of methods for inducing stress tolerance in plants is vital and is a major focus of research over many decades.

Recently, some of the endogenous growth regulators like ABA, JA and SA have been implicated in imparting stress tolerance in plants (Overmyer *et al.*, 2003; Mittler *et al.*, 2004; Gunes *et al.*, 2007; Gautam and Singh, 2009; Hayat *et al.*, 2010). Investigations show the role of SA in modulating plant responses to a wide range of oxidative stresses (Dat *et al.*, 2007; Singh *et al.*, 2008,). The established effects of SA on stomata function, chlorophyll content, transpiration rate and respiratory pathways raise the assumption that SA might possess another physiological function, most probably involved in regulation of some photosynthetic reactions and other growth regulatory mechanisms under salinity stress in plants (Borsani *et al.*, 2001; Senaratna *et al.*, 2003; Zahra *et al.*, 2010). Thus studies suggest that exogenous application of SA in plants may bring some potential practical utilization, such as manipulating the tissue level of SA in plants may be a promising area for the importance of biotechnology to crop protection and high yield. The main objective of this article was to develop a new concept about the role of salicylic acid in growth regulation, gene induction and signal transduction pathways in plants grown various environmental stresses. In the present review, information is presented on SA induced protein kinases (SIPK), osmotolerance, antioxidative system, and ion homeostasis in plants exposed to biotic and abiotic stresses.

2. PATTERNS AND CORRELATION BETWEEN PLANT STRESSES

Plant hormonal response to one stress influences its response to additional stresses. This may be due to major role of water in the movement, germination, and leaf entry of many pathogens (Thaller and Bostock, 2004). Availability of water may also affect osmoticum balance and associated with the changes in the microenvironment (Temperature, Humidity). The logic of this argument is that stressed plants invest less in defense and may be of higher nutritional value. This hypothesis was tested in a meta-analysis of various experiments that manipulated abiotic stress and the results showed support for the plant stress hypothesis. For example, Loeb *et al.* (1997) demonstrated that increasing drought in tomato plants had nonlinear effects on plant chemistry and resistance to herbivores. There is also evidence for an ecological association between moisture and herbivory (Chase *et al.*, 2000). Plants response to natural damage and induction by elicitors in a dose dependent manner and develop biochemical link through signaling mechanism (Niki *et al.*, 1998). Signaling interaction between phytohormonal pathways was detected through either antagonistic or synergistic activity of phytohormones (Singh *et al.*, 2011). These ecological correlations between stresses can be used as a framework to examine phytohormonal interactions that regulate plant responses to the respective stresses Fig. 1.

3. MECHANISMS OF STRESS TOLERANCE

3.1 Stress Sensing and Homeostasis

Numerous signal or signal-like molecules have now been identified and presumed, through some evidence, to function in plants as mediators of osmotic adaptation. A signal or any environmental stimuli must be sensed by receptor/receptors present in the plasma membrane, cytosol or other cellular organelles. To protect actively growing and metabolizing cells, plant regulates ion movement into tissues. Study regarding salinity stress suggests that one mode by which plants control salt flux to the shoot is the entry of ions into the xylem stream, which may depend on symplastic ion transport through the epidermal and cortical cells, contributes to a reduction in Na^+ that is delivered to the xylem. However, at the endodermis, radial movement of solutes must be via a symplastic pathway, as the casparian strip constitutes a physical barrier to apoplastic transport (Flowers and Yeo, 1992).

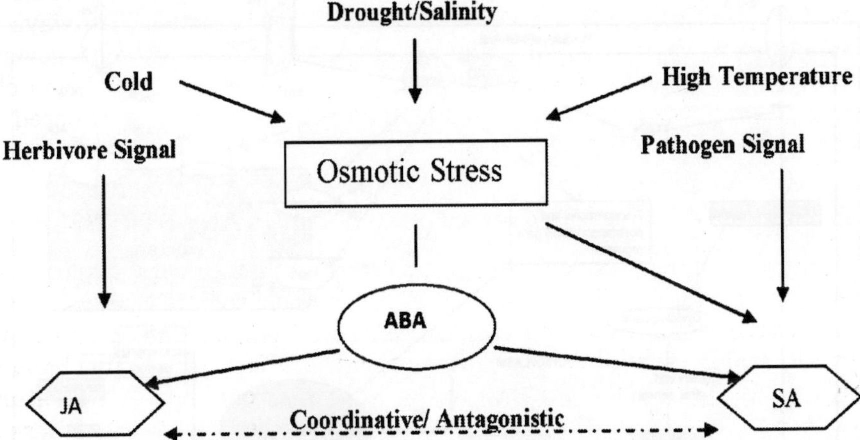

Fig. 1. Proposed model to show a relationship between ABA-JA-SA signaling pathways in plants in response to biotic and abiotic stresses.

Ion homeostasis in saline environments is dependent on trans-membrane transport proteins that mediate ion fluxes, including H^+ translocating ATPases and pyrophosphatases, Ca^{2+}-ATPases, secondary active transporters, and channels (Sze *et al.*, 1999). Actually, salt stress is sensed through SOS pathway, which comprises of three key components: SOS3, acting as Ca^{2+} sensor (Liu and Zhu, 1998); SOS2, a serine/threonine protein kinase (Liu *et al.*, 2000); and SOS1, a plasma membrane Na^+/H^+ antiporter (Shi *et al.*, 2000). Compartmentalization of Na^+ into the vacuoles or its efflux across the plasma membrane are mediated by the expression and activity of Na^+/H^+ antiporters as well as of V-type H^+-ATPase and H^+-PPase (Chinnusamy *et al.*, 2005).

The SOS1 protein has transmembrane domains to sense Na^+ and is supposed to exist in the cytoplasm (Shi *et al.*, 2000; Zhu 2003; Zhang *et al.*, 2004; Shabala *et al.*, 2005). Thus, Na^+/H^+ antiport activity of SOS1 results in efflux of excess Na^+ ions and thus contributes to Na^+ ion homeostasis (Liu and Zhu, 1998; Halfter *et al.*, 2000; Ishitani *et al.*, 2000; Liu *et al.*, 2000; Quintero *et al.*, 2002). SOS1 has also been reported to protect plasma membrane K^+ transport during salinity stress (Shabala *et al.*, 2005). SOS3/SOS2 protein kinase complex appears to down-regulate the activity of AtHKT1 (low-affinity Na^+ transporter) under salt stress (Mahajan and Tuteja, 2005), which mediates Na^+ entry into the root cells of *Arabidopsis* (Uozumi *et al.*, 2000; Zhu, 2002). SOS2 is shown to interact with vacuolar Na^+/H^+ antiporter (NHX) influencing its Na^+/H^+ exchange activity, resulting in sequestration of excess Na^+ ions into vacuolar compartment and thus further contributing ion homeostasis (Qiu *et al.*, 2002; Munns and Tester, 2008) (Fig. 2).

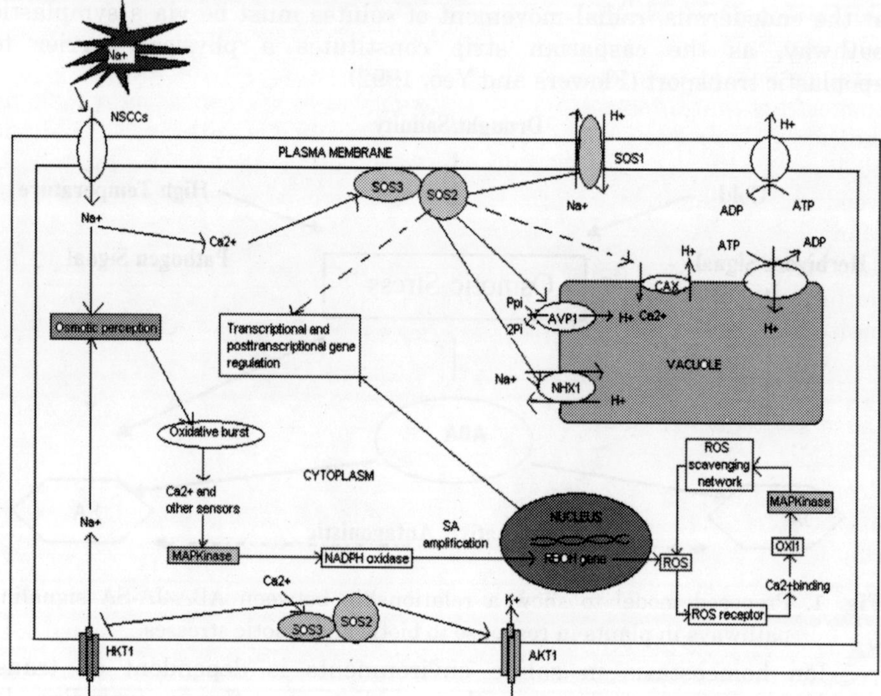

Fig. 2. Proposed model showing the possible survival strategies of plants under adverse environmental conditions. Ion homeostasis and probable signal transduction pathways in plants in response to phytohormones: Na^+ is passively transported into the cytosol by non-selective cation channels (NSCCs). For cellular homeostasis, two mechanisms operate; one is exclusion of Na^+ across the membrane (SOS1, a Na^+-H^+ exchanger located at plasma membrane). Second is the compartmentalization of Na^+ into vacuoles (NHX1), where Na^+ is less toxic. Salinity induced osmotic stress is perceived through both ABA-dependent and ABA independent signaling. Na^+ stressed gene expression of ABA biosynthesis is stimulated through SA amplification of AAO protein (Szepesi et al. 2009). ABA induced gene expression involve Ca^{2+} coupled signal transduction cascade(s), may regulate genes for required osmolytes and HSPs biosynthesis (Leung and Giraudat 1998).

Kaydan *et al.* (2007) observed that pre-sowing soaking treatment of seeds with SA positively affected the osmotic potential, shoot and root dry mass, K^+/Na^+ ratio and contents of photosynthetic pigments (chlorophyll a, b and carotenoids) in wheat seedlings, under both saline and non-saline conditions.Salicylic acid is used for regulation of oxidative stress in plants subjected to unfavorable environmental conditions. SA treatments enhanced H_2O_2 production, lipid peroxidation, and oxidative damage to proteins, and resulted in the formation of chlorophyll and carotene isomers. SA-enhanced H_2O_2 levels were related to increased activities of Cu, Zn-superoxide dismutase and were independent of changes in catalase and ascorbate peroxidase activities. Prolonging SA treatments inactivated catalase and ascorbate peroxidase and resulted in phytotoxic symptoms, suggesting that inactivation of H_2O_2-degrading enzymes serves as an indicator of hypersensitive cell death as reported in Arabidopsis thaliana by Rao *et al.* (1997). Again application of herbicides for the purpose of weed killing can affect not only the weeds but also the main crop as well. Many herbicides have the ability to cause oxidative stress and further degradation of cell components. SA has the ability to regulate clethodim herbicide detoxification through regulation of the antioxidant status of maize leaf cells by Radwan (2012). Habibi (2012) reported that the effects of 500 µM salicylic acid (SA) application on drought stress acclimation of barley (*Hordeum vulgare* L. *cv* Nosrat) plants grown in soil culture. In these experiments the following treatments were used: CK (control), DR (drought), SA (500 µM) and DSA (SA+drought). The results showed that drought stress decreased the dry mass and net CO_2 assimilation rate (*A*) of plants, which were all increased by the addition of SA. Under drought conditions, the improvement of photosynthesis of barley plants treated with SA was associated with an increase in g_s (stomatal conductance), whereas the maximal quantum yield of PSII (Fv/Fm) did not change with SA treatment. Malondialdehyde (MDA) content remained unchanged in DSA plants because of an efficient scavenging of reactive oxygen species (ROS) following a significant enhancement of some antioxidative enzyme activities. The improvement of SA on drought tolerance of barley plants was associated with the increase of antioxidant defense abilities and maintenance of photosynthesis under drought, which may elucidate the physiological mechanism of SA in improvement of drought tolerance of barley plants.

3.2 Phytohormonal Interactions in Response to Stress Conditions

Biochemical signaling interactions between phytohormones are known to have strong influences on the metabolic profile of plants in response to various environmental stresses (Achard *et al.*, 2009). Current biochemical and molecular signaling approved the involvement of phytohormones and their predominant role for plant resistance to biotic and abiotic stresses (SA, Deloney *et al.*, 1994; Alonso-Ramı´rez *et al.*, 2009, JA, Creelman, 1998; ABA, Lin *et al.* 2009). These lines of action have also been supported by microarray to analyze expression of many ABA-JA and SA regulated genes emphasizes the coordination responses mediated by these hormones. As Wei *et al.* (2000) stated that ABA involved in the changes of a specific portion of metabolic activity in response to salt stress, while JA has been shown to be involved in an antagonistically with the ABA regulated response to salt stress. Exogenous SA interfered with various physiological processes (Vicente and Plasencia, 2011). Thus complex interactions occur between ABA, JA, and SA phytohormones that are important regulators of biotic and abiotic stress responses in plants.

The ability of environmental stresses (biotic and abiotic) to switch on signaling cross-talk in plants may be presented through models suggested that ABA had a synergistic role with jasmonate – mediated responses to herbivory (Westerneck and Parthier, 1997). Up-regulation of several environmental stimuli induced osmotic stress but not all (Fig. 1), wound inducible genes are also responsible for interaction of phytohormones to generate a machinery of defense molecule (Chao *et al*, 1999).

Three signaling molecules, SA, JA and ET (ethylene), are known to play key roles in various aspects of plant defense. These include defense against abiotic stresses, such as wounding and exposure to ozone, as well as defense against insect and microbial attack (Creelman and Mullet, 1997; Ecker, 1995). The JA, SA and ET defense signaling pathways do not function independently. They are involved in a complex signaling network in which the different pathways influence each other through positive and negative regulatory interactions. Both JA and ET signaling are required for the expression of the defense-related gene *PDF1.2* in response to infection by *A. brassicicola* (Penninckx *et al.*, 1996), and for the expression of *PDF1.2, HEL*, and *CHIB* in response to treatment with *E. carotovora* culture filtrates (Norman-Setterblad *et al.*, 2000). In tomato, the development of disease symptoms following infection by *X. campestris* pv. *vesicatoria* requires both SA and ET, and the accumulation of SA in infected plants is dependent on ET synthesis (O'Donnell *et al.*, 2001). Constitutive SA signaling in the *mpk4* and *ssi2* mutants is likely due to loss of an antagonistic effect of JA signaling on the SA pathway. *MPK4* is predicted to encode a mitogenactivated (MAP) kinase that is required for JA-dependent gene expression. The *SSI2* gene encodes a steroyl-ACP fatty-acid desaturase, which is hypothesized to catalyze the synthesis of a fatty-acid-derived signal that is involved in mediating both JA signaling and negative cross talk between the JA and SA pathways (Kachroo *et al.*, 2001).There are several ways in which the jasmonate and salicylate pathways are known to interact and regulate growth and development of plants (Singh *et al.*, 2008; Mori *et al.*, 2004). SA itself interferes with JA signaling in different sites, first conversion of 13 S-hydroperoxidate dihydrase and, the second is conversion of 12-oxy-phytodienoic acid to JA. JA may interfere with salicylate biosynthesis in tobacco plants via a G-protein switch that is required for appropriate response to external stimuli (Seo *et al.*, 1995). Interaction between phytohormonal mediated signaling pathways depends on contraction and timing of elicitation or nature of stresses (Taylor and Mc Ainsh, 2004). Both jasmonate and salicylate pathways are two of the key biochemical response mechanism that can be triggered by various biotic and abiotic stresses and function as necessary signaling molecules responsible for defensive responses in plants. Recent studies suggested that SA-dependent inhibition of JA action could be manifested via activation of NPRI (non-repressor of pathogenesis-related genes) (Kawano *et al.*, 2004, Shah, 2003). The SA dependent increase in SA production could be explained by crosstalk between SA and JA. Cooperative and antagonistic interaction of SA and JA is key events in the signal transduction mechanism of defense (Koorneef and Pieterse, 2008). Phytohormones interact together for defense signal networking to fine-tune defense. SA, JA and ABA are recognized as key players in the regulation of the signaling pathway (Wang *et al.*, 2007).

3.3 Regulation of ROS Signaling

Genetics involved in managing the level of ROS is highly dynamic and redundant network encodes both ROS-producing and ROS-scavenging proteins (Mittler *et al.*, 2004). ROS produced by NADPH oxidase regulate plant cell growth and development under adverse environmental conditions. However, being toxic levels are capable of injuring cells (Foreman *et al.*, 2004). Plasma membrane NADPH-dependent oxidase and mammalian calcium regulated NADPH oxidase play an important role in biosignaling of ROS (Apel and Hirt, 2004). ROS signals interact directly or indirectly with other signaling pathways such as nitric oxide and the stress hormones especially SA, JA and ABA. Crosstalk between these pathways somehow, determines the survival or death of the cells (Singh *et al.*, 2011). Overmyer *et al.* (2003) have also proposed that SA might be involved in this positive amplification loop and amplifies signals leading to defense responses through induction of rboh genes. Recently, Lin *et al.* (2009) isolated and characterized maze NADPH oxidase (rboh) genes regulated by MAPK cascade in ABA signaling. He reported that ABA treatment induced a biphasic response in the expression of zmrbohA-D and the activity of NADPH oxidase in maize plants. The MAPK cascade is one of the major pathways by which extracellular stimuli are transduced in to intracellular responses in plants through phytohormonal crosstalk and ROS signaling (Takeda *et al.,* 2008, Singh *et al.,* 2011). Recently identified serine/threonine protein kinase (OX1) has been shown to play a central role in ROS sensing and the activation of mitogen activated protein kinases (Rentel, 2004). The accumulation of ROS in cells might activate the ROS scavenging pathways and result in the suppression of ROS in specific cellular locations or the entire cells.

3.3.1 ROS scavenging enzymes

Plants have evolved efficient systems for ROS removal, which include specific ROS-scavenging enzymes like superoxide dismutase (SOD; EC

1.15.1.1), ascorbate peroxidase (APX; EC 1.11.1.11), catalase (CAT; E.C 1.11.1.6), glutathione reductase (GR; EC 1.6.4.2), glutathione peroxidase (GPX; EC 1.11.1.9) and some small non-enzymatic molecules such as ascorbate, glutathione, tocopherol and some phenylpropanoids (Turkan & Demiral, 2009). Therefore, it is seems that steady- state level of ROS in cells needs to be tightly regulated and number of genes are involved in to the encoding of ROS scavenging and ROS-producing proteins. Recently, Sakhabutdinova *et al.* (2004) observed that exogenously applied salicylic acid might be involved in the down regulation of active oxygen species in the roots of young wheat seedlings. These findings indicate that the activities of these antioxidant enzymes are directly or indirectly regulated by salicylic acid, thereby providing protection against salinity stress. The possible explanation of involvement of SA in ROS regulation under saline condition could be an enhancement of amplification loops involving NADPH oxidases in ROS signaling in several studies (Dat *et al.*, 2003, Rizhsky *et al.*, 2004). These loops might be activated by low levels of ROS and result in enhanced production and amplification of the ROS signals in specific cellular locations for the homeostasis of cellular redox (Mittler *et al.*, 2004).

The genes upregulated by SA under stress conditions include glutathione S-transferase (*GST*), glycosyl transferase (*GT*), peroxiredoxin (*PRX*), peroxidase (*PX*), thioredoxin (*TRX*), glutaredoxin (*GRX*) and pathogen resistance (*PR*) genes (Vanderauwera *et al.* 2005; Wang *et al.*, 2005). It is also documented that SA activates defense gene expression by triggering redox changes in components of the signal transduction pathway (Durrant and Dong, 2004; Fobert and Despres, 2005). There are at least two components of SA signal transduction pathway involved in the activation of *PR* genes, NPR1 and the transcription factors, TGAs. Also, one of the promoter elements responsive to SA in defense genes (*as-1*-like element) is responsive to oxidative signals (Garreton *et al.*, 2002). One piece of concrete evidence in this regard is the use of SA-deficient rice that has higher ROS levels and reduced antioxidant capacity. They also exhibit spontaneous lesion formation. However, the symptoms developed were suppressed by exogenous application of SA-analog, benzothiadiazole (Yang *et al.*, 2004), which clearly indicates that SA scavenges hydroxyl radicals and protects plant catalase inactivation by H_2O_2 (Durner and Klessig, 1996). Hence, SA can play a critical role in modulating the cell redox balance, thereby protecting the plants against the oxidative damage (Yang *et al.*, 2004). According to Yusuf *et al.* (2008) the activity of POX in the leaves of the plant treated with SA was higher than those of plants grown under varied salinity levels also added that SOD activity in the leaves of stressed plants was higher than that of the control (Singh and Usha, 2003).

3.4 Phytohormone-induced Protein Kinase pathways

Protein kinases are central signal transduction links and integration point in plants and mammals (Jonak *et al.*, 2002). It has been shown that

MAPKs are involved in plant signal transduction in response to pathogens, drought, salinity, cold, wounding, heavy metals, O_3, ROS, UV-B and hormone stimuli (Lin *et al.*, 2009). Several plant protein kinases have now been activated by osmotic stress due to its involvement in osmosensing pathways (Zhu, 2002). Recent studies using in-gel kinase assays indentified some specific protein kinase that are activated by osmotic stress in plants, but whether any of the pathways is directly activated by osmotic stress or by a derived stress needs to be addressed. SA induced protein kinases (SAIPK) may be involved in induction of gene expression modulating multiple stress responses (Miko-Iajczyk *et al.*, 2000). They demonstrated that immunological experiment showed induction of 48 KD and 42 KD protein kinases (SAIPK) under osmotic stress imposed by exogenous NaCl in tobacco cells. Osmotic stress based rapid activation of these protein kinases by protein phosphorylation suggests their involvement in abiotic stress signaling pathways in plants. Mahajan *et al.* (2006) have reported that the CBL and CIPK genes from pea were induced in response to SA and wounding. These SA-responsive genes are known to be regulated by basic/leucine zipper-type transcription factors of the TGA family. NPR1 (non-expressor of PR genes 1) is one of the important regulators of SA signaling, which interacts with TGA transcription factors that are involved in the activation of SA-responsive PR genes (Dong, 2004). These findings suggest that SA-induced NPR1 activates GRX480, which forms a complex with TGA factors and suppresses the expression of JA-responsive genes in defense system and cross talk between SA and JA signaling (Miao and Zentgraf, 2007). Similarly, Mitogen activated protein kinases (MPKs) have been identified as another key component involved in stress tolerance signaling in *Arabidopsis* (Brodersen and Petersen, 2006). WRKY53 has been shown to interact with the JA-inducible protein epithiospecifying senescence regulator (ESR). More importantly, the expression of these genes is antagonistically regulated in response to JA and SA suggesting that WRKY53 and ESR mediate negative cross-talk between pathogen resistance and senescence in Arabidopsis (Miao and Zentgraf, 2007). During the last few years the information on the nature of chemicals and the signaling pathways that they are involved has been generated. Earlier, it was thought that calcium, phosphoinositides and ABA were the key molecules that participated in abitoic stress signaling. Recently, calcium signaling is also reported to be involved in the regulation of plant cell cycle progression in response to abiotic stress. However, as it has been described previously chemicals like SA, JA, NO, which were thought to respond to biotic stresses, have now been implicated in abiotic stresses also.

4. CONCLUSION AND FUTURE PROSPECTS

Phytohormones are critical for plant growth and development and play an important role in integrating various stress signals controlling downstream stress responses. It is now well established that the stress-induced gene products are also involved in the generation of regulatory molecules like ABA, JA, and SA. However, in recent years regulation and involvement of these hormones in acclimation/adaptation mechanisms to adverse environmental conditions is very vital. Phytohormones involved in ion exclusion and/or compartmentalization, osmotic adjustments, reduction in lipid peroxidation, stress induced synthesis of protein kinases (SIPK), regulation of oxidative system and also enhancement of antioxidant activity to combat the lethal oxidative events. Discovery of phytohormonal interaction for the regulation of upstream/downstream targets under adverse environmental conditions will be a future challenge to disentangle the sophisticated network of plant metabolism and signaling pathways. The molecular analysis of stress responses has arrived at a stage where research can facilitate a large collection of characterized genes. Identification and characterization of genetic loci for abiotic stress resistance and its phytohormonal control could be an effective analytical tool. The use of novel approaches combining genetic, biochemical and molecular techniques should provide results in near future for food security.

REFERENCE

Achard P, Cheng H, De Grauwe L, Decat J, Schoutteten H,Moritz T, Van Der Straeten D, Peng J, Harberd NP (2006) Integration of plant responses to environmentally activated phytohormonal signals. Science 311: 91–93

Alonso-Ramırez A, Rodrı´guez D, Reyes D, Jime´ nez JA,Nicola´s G, Lo´ pez-Climent M, Go´ mez-Cadenas A, Nicola´s C (2009) Evidence for a role of gibberellins in salicylic acid-modulated early plant responses to abiotic stress in Arabidopsis seeds. Plant Physiol 150: 1335–1344

Bari R, Jones JDG, Innes J (2009) Role of plant hormones in plant defence responses. Plant Mol Biol 69(4): 473-488

Bohnert HJ, Nelson DE, Jensen RG (1995) Adaptations to environmental stresses. Plant Cell 7: 1099-1100

Borsani O, Valpuesta V, Botella MA (2001) Evidence for a role of salicylic acid in the oxidative damage generated by NaCl and osmotic stress in *Arabidopsis* Seedlings. Plant Physiol 126: 1024-1030

Badawi GH, Kawano N, Yamauchi Y, Shimada E, Sasaki R, Kubo A, Tanaka K (2004) Over-expression of ascorbate peroxidase in tobacco chloroplasts enhances the tolerance to salt stress and water deficit. Physiol Plantarum 121: 231–238

Bartels D, Engelhardt K, Roncarati R, Schneider K, Rotter M, Salamini F (1991) An ABA and GA modulated gene expressed in the barley encodes an aldose reductase related protein. EMBO Journal 10: 1037-1043

Binzel ML, Hess FD, Bressan RA, Hasegawa PM (1988) Intracellular compartmentation of ions in salt adapted tobacco cells. Plant Physiol 86: 607-614

Bowler C, Van Montagu M, Inze D (1992) Superoxide dismutase and stress tolerance. Annu Rev Plant Physiol Plant Mol Biol 43: 83-116

Bowler C, VanCamp W, Van Montagu M, Inze D (1994) Superoxide dismutases in plants. Crit Res Plant Sci 13: 199-218

Breusegem FV, Slooten L, Stassart JM, Moens T, Botterman J, Van Montagu M, Inze D (1999) Over production of *Arabidopsis thaliana* FeSOD confers oxidative stress tolerance to transgenic maize. Plant Cell Physiol 40: 515–523

Chen C, Dickman MB (2005) Proline suppresses apoptosis in the fungal pathogen *Colletotrichum trifolii*. Proc Natl Acad Sci USA 102: 3459–3464

Creelman RA, Mullet JE (1997) Oligosaccharins, brassinolides, and jasmonates: nontraditional regulators of plant growth, development, and gene expression. Plant Cell 9:1211-1223

Der-Ent SV, Wees SCMV, Pieterse CMJ (2009) Jasmonate signaling in plant interactions with resistance-inducing beneficial microbes *Phytochemistry* 70: 1581–1588

Dat JF, Pellinen R, Cotte BVD, Langerbartels C, Kangasjarvi J, Inze D, Van Breusegem F (2003) Changes in hydrogen peroxide homeostasis trigger an active cell death process in tobacco. Plant Journal 33: 621–632

Deef HE (2007) Influence of Salicylic Acid on Stress Tolerance During Seed Germination of *Triticum aestivum* and *Hordeum vulgare*. Advances in Biological Research 1: 40-48

Demiral T, Türkan I (2005) Comparative lipid peroxidation, antioxidant defense systems and proline content in roots of two rice cultivars differing in salt tolerance. Env Exp Bot 53: 247–257

Ecker JR (1995) The ethylene signal transduction pathway in plants. Science 268:667-675

El Tayeb MA (2005) Response of barley grains to the interactive effect of salinity and salicylic acid. Plant Growth Regul 45: 215–224

Flowers TJ, Colmer TD (2008) Salinity tolerance of halophytes. New Phytol 179: 945–963

Foyer CH, Souriau N, Perret S, Lelandais M, Kunert KJ, Pruvost C, Jouanin L (1995) Overexpression of glutathione reductase but not glutathione synthetase leads to increases in antioxidant capacity and resistance to photoinhibition in poplar trees. Plant Physiol 109: 1047–1057

Gautam S, Singh PK (2009) Salicylic acid-induced salinity tolerance in corn grown under NaCl stress. Acta Physiol Plantarum 31: 1185-1190

Gibon I, Bessieres MA, Larher F (1997) Is glycine betaine a non-compatible solute in higher plants that do not accumulate it? Plant Cell Environ 20: 329-340

Grant MR, Jones JDG (2009) Hormone Disharmony Moulds Plant Health and Disease. *Science* 324(5928): 750-752

Gfeller A, Liechti R, Farmer EE (2010) Arabidopsis jasmonate signaling pathway *Sci Signal* 3(109), cm4.

Gunes A, Inal A, Alpaslan M, Eraslan F, Bagci EG, Cicek N (2007) Salicylic acid induced changes on some physiological parameters symptomatic for oxidative stress and mineral nutrition in maize *(Zea mays* L.) grown under salinity. J Plant Physiol 164: 728-736

Habibi G (2012) Exogenous salicylic acid alleviates oxidative damage of barley plants under drought stress. Acta Biologica Szegediensis 56(1): 57-63

Halfter U, Ishitani M, Zhu JK (2000) The Arabidopsis SOS2 protein kinase physically interacts with and is activated by the calcium-binding protein SOS3. Proc Natl Acad Sci U.S.A. 97: 3735–3740

Hamada AM, Al-Hakimi AMA (2001) Salicylic acid versus salinity-drought induced stress on wheat seedlings. Rostl Vyr 47: 444–450

Hare PD, Cress WA (1997) Metabolic implications of stress-induced proline accumulation in plants. Plant Growth Regul 21: 79–102

Hasegawa PM, Bressan RA, Jhu JK, Bohnert HJ (2000) Plant cellular and molecular responses to high salinity. Annu Rev Plant Physiol Plant Mol Biol 51: 463-499

Hernández JA, Jimenez A, Mullineaux P, Sevilla F (2000) Tolerance of pea (*Pisum sativum* L.) to long-term salt stress is associated with induction of antioxidant defenses. Plant Cell Env 23: 853–862

Hayat Q, Hayat S, Irfan M, Aqil (2010) Effect of exogenous salicylic acid under changing environment: A review Environmental and Experimental Botany 68, 14–25

Ishitani M, Liu J, Halfter U, Kim CS, Shi W, Zhu JK (2000) SOS3 function in plant salt tolerance requires N-myristoylation and calcium-binding. Plant Cell 12: 1667–1677

Jagendorf AT, Takabe T (2001) Inducers of glycinebetaine synthesis in barley. Plant Physiol 127: 1827–1835

Kachroo P, Shanklin J, Shah J, Whittle EJ, Klessig DF (2001) A fatty acid desaturase modulates the activation of defense signaling pathways in plants. Proc Natl Acad Sci USA 98:9448-9453

Kavi Kishor PB, Sangam S, Amrutha RN, SriLaxmi P, Naidu KR, Rao KRSS, Rao S, Reddy KJ, Theriappan P, Sreenivasulu N (2005) Regulation of proline biosynthesis, degradation, uptake and transport in higher plants : Its implications in plant growth and abiotic stress tolerance. Curr Sci 88: 424-438

Kaydan D, Yagmur M, Okut N (2007) Effects of salicylic acid on the growth and some physiological characters in salt stressed wheat (*Triticum aestivum* L.). Tarim Bilimleri Dergisi 13: 114–119

Klessig DF, Malamy J (1994) The salicylic acid signal in plants. Plant Mol Biol 26: 1439–1458

Koca H, Ozdemir F, Turkan I (2006) Effect of salt stress on lipid peroxidation and superoxide dismutase and peroxidase activities of *Lycopersicon esculentum* and *L. pennellii*. Biol Plant 50: 745–748

Koca H, Bor M, Ozdemir F, Türkan I (2007) The effect of salt stress on lipid peroxidation, antioxidative enzymes and proline content of sesame cultivars. Env Exp Bot 60: 344–351

Koornneef A, Pieterse C M J (2008) Cross talk in defense signaling. Plant Physiol 146: 839–814

Leung J, Giraudat J (1998) Abscisic acid signal transduction. Annu Rev Plant Physiol Plant Mol Biol 49: 199-222

Loake G, Grant M (2007) Salicylic acid in plant defence—the players and protagonists. Curr Opin Plant Biol 10(5): 466-472

Lindquist Y, Branden CL, Mathews FS, Lederer F (1991) Spinach glycolate oxidase and yeast flavocytochrome b2 are structurally homologous and evolutionarily related enzymes with distinctly different function and flavin mononucleotide binding. J Biol Chem 266: 3198–3207

Liu J, Ishitani M, Halfter U, Kim CS, Zhu JK (2000) The *Arabidopsis thaliana* SOS2 gene encodes a protein kinase that is required for salt tolerance. Proc Natl Acad Sci USA 97: 3730–3734

Liu J, Zhu JK (1998) A calcium sensor homolog required for plant salt tolerance. Sci 280:1943–1945

Ma QQ, Wang W, Li YH, Li DQ, Zou Q (2006) Alleviation of photoinhibition in drought-stressed wheat (*Triticum aestivum*) by foliar-applied glycinebetaine. J Plant Physiol 163: 165–175

Madhusudhan R, Ishikawa T, Sawa Y, Shigeoka S, Shibata H (2003) Characterization of an ascorbate peroxidase in plastids of tobacco BY-2 cells. Physiol Plant 117: 550-557

Mahajan S, Tuteja N (2005) Cold, salinity and drought stresses: an overview. Arch Biochem Biophys 444: 139–158

Mikolajczyk M, Awotunde OS, Muszynska G, Klessig DF, Dobrowolska G (2000) Osmotic stress induces rapid activation of a salicylic acid-induced protein kinase and a homolog of protein kinase ASK1in tobacco cells. The Plant Cell 12: 165-178

Mittler R, Vanderauwera S, Gollery M, Breusegem FV (2004) Reactive oxygen gene network of plants. Trends Plant Sci 9: 490-498

Mittova V, Tal M, Volokita M, Guy M (2003) Up-regulation of the leaf mitochondrial and peroxisomal antioxidative systems in responses to salt-induced oxidative stress in the wild salt-tolerant tomato species *Lycopersicon pennellii*. Plant Cell Env 26: 845–856

Mohamed EA, Iwaki T, Munir I, Tamoi M, Shigeoka S, Wadano A (2003) Overexpression of bacterial catalase in tomato leaf chloroplasts enhances photooxidative stress tolerance. Plant Cell Env 26: 2037–2046

More SD, Hangarge DS, Raghavaiah CV, Joshi BM (2004) Performance of different safflower, *Carthamus tinctorious* L. genotypes with varied soil salinity levels. J Oilseeds Res 21: 196-197

Munns R, Tester M (2008) Mechanisms of salinity tolerance. Annu Rev Plant Biol 59: 651-681

Norman-Setterblad C, Vidal S, Palva ET (2000) Interacting signal pathways control defense gene expression in *Arabidopsis* in response to cell wall-degrading enzymes from *Erwinia carotovora*. *Mol* Plant Microbe Interact 13:430-438

O'Donnell PJ, Jones JB, Antoine FR, Ciardi J, Klee HJ (2001) Ethylene dependent salicylic acid regulates an expanded cell death response to a plant pathogen. Plant J 25:315-323

Overmyer K, Brosche M, Kangasjarvi J (2003) Reactive oxygen species and hormonal control of cell death. Trends in Plant Sci 8: 335-342

Orsini F , Santacroce M , Arosio P , Castagna M , Lenardi C , Poletti G , Sacchi FV (2009) Intermittent contact mode AFM investigation of native plasma membrane of Xenopus laevis. Eur Biophys J 38 (7): 903-10.

Parida A, Das AB, Das P (2002) NaCl stress causes changes in photosynthetic pigments, protein and other metabolic components in the leaves of a true mangrove, *Bruguieva parviflora* in hydroponic cultures. J Plant Biol 45: 28-36

Pandey SP, Shahi P, GaseK, Baldwin IT (2008) Herbivory-induced changes in the small-RNA transcriptome and phytohormone signaling in Nicotiana attenuata. *Proc. Natl. Acad. Sci.USA* 105(12): 4559-4564.

Parida AK, Das AB (2005) Salt tolerance and salinity effects on plants: a review. Ecotoxic Env Safety 60: 324-349

Park EJ, Jeknic Z, Chen THH (2006) Exogenous application of glycinebetaine increases chilling tolerance in tomato plants. Plant Cell Physiol 47: 706–714

Penninckx IAMA, Eggermont K, Terras FRG, Thomma GPHJ, De Samblanx GW, Buchala A, Metraux J, Manners JM, Broekaert WF (1996) Pathogen-induced systemic activation of a plant defensin gene in *Arabidopsis* follows a salicylic acid-independent pathway. Plant Cell 8:2309-2323

Qiu QS, Guo Y, Dietrich MA, Schumaker KS, Zhu JK (2002) Regulation of SOS1, a plasmamembrane Na^+/H^+ exchanger in *Arabidopsis thaliana*, by SOS2 and SOS3. Proc Natl Acad Sci USA 99: 8436–8441

Quintero FJ, Ohta M, Shi H, Zhu JK, Pardo JM (2002) Reconstitution in yeast of the *Arabidopsis* SOS signaling pathway for Na^+ homeostasis. Proc Natl Acad Sci USA 99: 9061–9066

Radwan DEM (2012) Salicylic acid induced alleviation of oxidative stress caused by clethodim in maize (*Zea mays* L.) leaves. Pesticide Biochemistry and Physiology 102: 182-188

Rai VK (2002) Role of amino acids in plant responses to stress. Biol Plant 45: 481–487

Ramani B, Reeck T, Debez A, Stelzer R, Huchzermeyer B, Schmidt A (2006) *Aster tripolium* L. and *Sesuvium portulacastrum* L.: two halophytes, two strategies to survive in saline habitats. Plant Physiol Biochem 44: 395-408

Rao MV et al. (1997) Influence of salicylic acid on H_2O_2 production, oxidative stress, and H_2O_2-metabolizing enzymes. Plant Physiol 115: 137-149

Raza SH, Athar HR, Ashraf M, Hameed A (2007) Glycinebetaine-induced modulation of antioxidant enzymes activities and ion accumulation in two wheat cultivars differing in salt tolerance. Env Exp Bot 60: 368–376

Rich PR, Bonner WDJr (1978) The sites of superoxide anion generation in higher plant mitochondria. Arch Biochem Biophys 188: 206–213

Rizhsky L, Davletova S, Hongjian L, Mittler R (2004) The zinc finger protein Zat12 is required for cytosolic ascorbate peroxidase 1 expression during oxidative stress in *Arabidopsis*. J Biol Chem 279: 11736–11743

Rontein D, Basset G, Hanson AD (2002) Metabolic engineering of osmoprotectant accumulation in plants. Metabol Eng 4: 49–56

Roxas VP, Smith Jr RK, Allen ER, Allen RD (1997) Overexpression of glutathiones-transferase/glutathione peroxidase enhances the growth of transgenic tobacco seedlings during stress. Nature Biotech 15: 988–991

Reyes AL *et al*. (2009) Ethylene modulates the role of npr1 in cross talk between salicylate and jasmonate signaling. Plant Physiol 149: 1797–1809

Roxas VP, Sundus AL, Garrett DK, Mahan JR, Allen RD (2000) Stress tolerance in transgenic tobacco seedlings that over express glutathione-S-transferase/glutathione peroxidase. Plant Cell Physiol 41: 1229-1234

Russell BL, Rathinasabapathi B, Hanson AD (1998) Osmotic stress induces expression of choline monooxygenase in sugar beet and amaranth. Plant Physiol 116: 859–865

Sairam RK, Tyagi A (2004) Physiology and molecular biology of salinity stress tolerance in plants. Curr Sci 86: 407-421

Sakhabutdinova AR, Fatkhutdinova DR, Shakirova FM (2004) Effect of salicylic acid on the activity of antioxidant enzymes in wheat under conditions of salination. Appl Biochem Microbiol 40: 501–505

Santner A, Estelle, M (2009) Recent advances and emerging trends in plant hormone signalling Nature 459: 1071-1078

Sekmen AH, Turkan I, Takio S (2007) Differential responses of antioxidative enzymes and lipid peroxidation to salt stress in salt-tolerant *Plantago maritime* and salt-sensitive *Plantago media*. Physiol Plant 131: 399–411

Senaratna T, Merritt D, Dixon K, Bunn E, Touchell D, Sivasithamparam K (2003) Benzoic acid may act as the functional group in salicylic acid and derivatives in the induction of multiple stress tolerance in plants. Plant Growth Regul 39: 77-81

Shabala L, Cuin TA, Newman IA, Shabala S (2005) Salinity-induced ion flux patterns from the excised roots of *Arabidopsis* SOS mutants. Planta 222: 1041–1050

Shakirova FM, Sakhabutdinova AR, Bezrukova MV, Fatkhutdinova RA, Fatkhutdinova DR (2003) Changes in the hormonal status of wheat seedlings induced by salicylic acid and salinity. Plant Sci 164: 317–322

Shi H, Ishitani M, Kim C, Zhu JK (2000) The *Arabidopsis thaliana* salt tolerance gene SOS1 encodes a putative Na^+/H^+ antiporter. Proc Natl Acad Sci USA 97: 6896–6901

Sreenivasula N, Grimn B, Wobus U, Weschke W (2000) Differential response of antioxidant compounds to salinity stress in salt tolerant and salt sensitive seedlings of foxtail millet (*Setaria italica*). Physiol Plant 109: 435–442

Singh NK, La Rosa PC, Handa AK, Hasegawa PM, Bressan RA (1987b). Hormonal regulation of protein synthesis associated with salt tolerance in plant cells. Proc Natl Acad Sci USA 84: 739-743

Singh PK, Bose B, Sharma MK, Singh A (2008) Physiological and Molecular actions of salicylate in plants. In:Development in Physiology, Biochemistry and Molecular Biology of Plants Vol 2 (Eds. Bose B and Hemantaranjan A) pp 135-155

Singh PK, Singh HB, Chaturvedi VK (2011) Cross talk signaling: an emerging defense strategy in plants. Current Science 100: 1-2

Singla SL, Pareek A, Grover A (1997) In Plant Ecophysiology Ed. Prasad MNV, John Wiley and Sons, pp 101-127

Storey R, Ahmad N, Wyn Jones RG (1977) Taxonomic and ecological aspects of distribution of glycine-betaine and related compounds in plants. Oecologia (Berl.) 27: 319-332

Sze H, Li X, Palmgren MG (1999) Enegrization of plant cell membranes by H^+-pumping ATPases: regulation and biosynthesis. Plant Cell 11: 677-689

Szepesi A, Csiszar J, Bajkan Sz, Gemes K, Horvath F, Erdei L, Deer A, Simon LM, Tari I (2005) Role of salicylic aicd pre-treatment on the acclimation of tomato plants to salt- and osmotic stress. Acta Biol Szegediensis 49: 123–125

Szepesi A, Csiszar J, Gemes K, Horvath E, Horvath F, Simon ML, Tari I (2009) Salicylic acid improves acclimation to salt stress by stimulating abscisic aldehyde oxidase activity and abscisic acid accumulation, and increases Na^+ content in leaves without toxicity symptoms in *Solanum lycopersicum* L. J Plant Physiol 166: 914-925

Turkan I, Demiral T (2009) Recent developments in understanding salinity tolerance. Environmental Experimental Botany 67: 2-9

Thaller JS, Bostock RM (2004) The role of the jasmonate response in plant susceptibility to diverse pathogens with a range of lifestyles. Ecology 85(1): 48–58

Tari I, Csiszar J, Szalai G, Horvath F, Pecsvaradi A, Kiss G, Szepesi A, Szabo M, Erdei L (2002) Acclimation of tomato plants to salinity stress after a salicylic acid pre-treatment. Acta Biol Szegediensis 46: 55–56

Tari I, Simon LM, Deer KA, Csiszar J, Bajkan Sz, Kis Gy, Szepesi A (2004) Influence of salicylic acid on salt stress acclimation of tomato plants: oxidative stress responses and osmotic adaptation. Acta Physiol Plantarum 26S, 237.

Timasheff S (1992) Stabilization of protein structure by solvent additives. In: Ahern, T.J., Manning, M.C. (Eds.), Stability of protein pharmaceuticals part b: *in vivo* pathways of degradation and strategies for protein stabilization. Plenum, New York, pp. 265–285

Türkan I, Bor M, Ozdemir F, Koca H (2005) Differential responses of lipid peroxidation and antioxidants in the leaves of drought-tolerant *P. acutifolius* Gray and drought-sensitive *P. vulgaris* L. subjected to polyethylene glycol mediated water stress. Plant Sci 168: 223–231

Taylor JE, McAinsh MR (2004) Signalling crosstalk in plants: emerging issues. J Exp Bot 55(395): 147–149

Uozumi N, Kim EJ, Rubio F, Yamaguchi T, Muto S, Tsuboi A, Bakker EP, Nakamura T, Schroeder JI (2000) The *Arabidopsis* HKT1 gene homolog mediates inward Na^+ currents in *Xenopus laevis* oocytes and Na^+ uptake in *Saccharomyces cerevisiae*. Plant Physiol 122: 1249–1260

Verma VS, Raghavaiah CV, Kumar V, Prasad R (2005) Identification of linseed, *Linum usilatissimum* L. genotypes for high salinity tolerance. J Oilseeds Res 22: 75-78

Vicente MRS, Plasencia J (2011) Salicylic acid beyond defense: its role in plant growth and development. J of Exp Botany 1-8 (dio:10 1093/jxb/err 031)

Vierling E (1991) The roles of heat shock proteins in plants. Annu Rev Plant Physiol Plant Mol Biol 42: 579-620

Wang J, Zhang H, Allen RD (1999) Overexpression of an *Arabidopsis* peroxisomal ascorbate peroxidase gene in tobacco increases protection against oxidative stress. Plant Cell Physiol 40: 725–732

Wyn Jones RG, Storey R, Leigh RA, Ahmad N, Pollard A (1977) A hypothesis on cytoplasmic osmoregulation. In: Marre, E., Cifferi, O. (Eds.), Regulation of Cell Membrane Activities in Plants. Elsevier, Amsterdam, pp. 121–136

Wang D, Pajerowska-Mukhtar K, Culler AH, Dong X, (2007) Systemic Signalling in Plant Defence. Curr Biol 17: 1784–1790

Xiong L, Schumaker KS, Zhu JK (2002) Cell Signaling during cold, drought, and salt stress. The Plant Cell 14: S165-S183

Yang B, Verkman AS (1997) Water and glycerol permeabilities of aquaporin 1-5 and MIP determined quantitatively by expression of epitope-tagged constructs in *Xenopus* oocytes. J Biol Chem 272: 16140-16146

Yazici I, Türkan I, Sekmen AH, Demiral T (2007) Salinity tolerance of purslane (*Portulaca oleracea* L.) is achieved by enhanced antioxidative system, lower level of lipid peroxidation and proline accumulation. Env Exp Bot 61: 49–57

Yeo AR (1998) Molecular biology of salt tolerance in the context of whole-plant physiology. J Exp Bot 49: 915-929

Yoshimura K, Yabuta Y, Ishikawa T, Shigeoka S (2000) Expression of spinach ascorbate peroxidase isozymes in response to oxidative stress. Plant Physiol 123: 223-234

Yusuf M, Hasan SA, Ali B, Hayat S, Fariduddin Q, Ahmad A (2008) Effect of salicylic acid on salinity induced changes in *Brassica juncea*. J Int Plant Biol 50: 1–4

Zhang J, Kirkham MB (1994) Drought stress induced changes in activities of superoxide dismutase, catalase and peroxidase in wheat species. Plant Cell Physiol 35: 785-791

Zhang JZ, Creelman RA, Zhu JK (2004) From laboratory to field. Using information from *Arabidopsis* to engineer salt, cold, and drought tolerance in crops. Plant Physiol 135: 615-621

Zhu JK (2003) Regulation of ion homeostasis under salt stress. Curr Opin Plant Biol 6: 441-445

Zhu JK (2002) Salt and drought stress signal transduction in plants. Annu Rev Plant Biol 53: 247–273.

10

PLANTS: INDICATORS OF AIR POLLUTION

A. K. Trivedi

National Bureau of Plant Genetic Resources,
Regional Station Bhowali – 263132 Nainital, Uttarakhand (INDIA)

INTRODUCTION

Plant which indicate some specific conditions of environment are called indicator plants. Plant species or plant community acts as a measure of environmental conditions, it is referred to as biological indicators/bioindicators/phytoindicators. For example evergreen forests indicate high rainfall in winter as well as summer; Sclerophyllous vegetation indicate heavy rainfall in winter and low during summer; Grassland indicate heavy rains during summer and low during winter; xerophytic vegetation indicate a very low or no rainfall in the year. Plants like *Utricularia, Chara, Wolffia* to grow in polluted waters. Bacteria like *E. coli* also indicate water pollution. Presence of diatoms in water indicates pollution by sewage. Many plants indicate the presence of characteristics minerals in the soils, theses plant are called metallocoles/ metallophytes e.g., Gold- *Equisetum arvense*, Diamond- *Vallozia candida,* Silver - *Eriogonium ovalifolium*. Crop plants like *Hordeum vulgare, Phaseolus vulgaris, Raphanus sativus, Pisum sativum, Medicago sativa* are used as indicators of air pollution.

Crops require certain critical levels of various environmental factors for optimum physiological functioning. Even a slight deviation in critical levels results in abnormal crop growth and performance of plant. Usually three components of environment (viz., lithosphere, hydrosphere and atmosphere) affect plant growth and development. The atmosphere is an insulating blanket like layer around the earth; it is source of essential gasses, it maintains a narrow difference of day and night temperatures. It also acts as shield around the earth against fatal UV radiation. Ambient air is composed of mixture of different gases and pollutants. Concentration of individual pollutant varies with time and location. Based on their physical and chemical properties, air pollutants exhibit varying levels of toxicity to different environmental receptors. Depending on the nature of the exposure, a particular air pollutant can be dominant in exerting adverse effects on a receptor sensitive to that particular pollutant.

Major contributors of air pollution are rapid urbanization, industrialization, expansion of cities, increases in traffic load, rapid economic development and higher level of energy consumption. In urban areas the main sources of air pollution are motor vehicles, industries, power plants and domestic sources. In rural areas use of fossil fuel, kerosene oil, cow dung, and wood for cooking and heating during winter are the main source of in door air pollution. Use of motor vehicles is growing fast particularly in the developing countries. The share of motor vehicles in the pollution load is thus set to increase in the developing countries. In the absence of the introduction of stringent control measures for traffic-related pollutants particularly in developing countries, air quality in urban areas is deteriorating.

The atmosphere of the earth can remain within the limits of its homeostasis or equilibrium if there is no significant alteration due to anthropogenic activities (Lovelock, 1987). However, depending on their physical and chemical properties, addition of air pollutants due to human interventions can cause adverse effect on appropriate receptors such as crops (Krupa *et al.*, 1994). Furthermore, the original air pollutants emitted can be transformed in the atmosphere to form secondary pollutants and can also be involved in regional, continental or global-scale phenomena such as 'acidic precipitation' and 'climate change'.

2. RESPONSES OF PLANTS TO AIR POLLUTION

Responses of plants to air pollution exposures are of two types - acute and chronic. Acute responses involve rapid changes in the physiological and biochemical processes of the plant (Schulte-Hostede *et al.,* 1988). These changes cause foliar injury (Krupa and Manning, 1988). Acute responses are usually caused by relatively high hourly pollutant concentrations from a few to several hours on a particular day. Manifestation of injury symptoms appear within a few to several days after such exposures. Evidence suggest that acute and chronic responses to ozone appear to be due to moderate or intermediate but not the highest hourly pollutant concentrations (Grünhage *et al.*, 1993; Krupa *et al.*, 1993, 1995). The reason may be due to lower atmospheric gaseous pollutant (O_3) flux or stomatal closure by the plant during periods of high pollutant (O_3) concentration (US EPA, 1986) as a protective mechanism to stress condition, since the stomata are predominant route for entry of gases into the leaves.

In contrast to the acute response, chronic effects are due to the exposures to gaseous pollutants for long period such as weeks, months, whole growth seasons and the entire life cycles. Acute responses may or may not lead reduction in crop yield but chronic responses can manifest in the form of foliar chlorosis, retarded growth, retarded flowering, foliar/ flower abscission, reduced biomass and economic yield and changes in nutritional quality of the plant product (Kickert and Krupa, 1990). In agriculture, chronic effects are of much concern, since growth and biomass effects can occur without appearance of visible foliar symptoms.

3. CHARACTERISTICS OF AIR POLLUTION EXPOSURE

Ambient air is composed of different gases and mixture of pollutants having different physical and chemical properties (Finlayson-Pitts and Pitts, 1986). Normally single air pollutants have been viewed as causal agents of adverse effects on a particular biological receptor (Mudd and Kozlowski, 1975). But joint effects of these pollutants in the mixture may differ from individual effects, it can be additive, antagonistic or synergistic (Torn *et al.,* 1987). Investigations show that, profiles exhibit latitudinal, altitudinal, seasonal, intra seasonal as well as daily variation (Lefohn *et al.*, 1990; Legge and Krupa, 1990; Lovett and Kinsman, 1990). In most of the biological systems rapid response to acute pollutant exposures occur. Receptors in biological system may respond after one or a series of such exposures. Therefore, in biological system the study of effect of air pollutants on frequency of pollutant episodes is a major consideration. Some prediction models using weather data are available which can predict with accuracy the potential or possible frequency of acute exposures of primary air pollutants at specific geographical areas in the vicinity of point sources (Lyons and Scott, 1990). Some models can predict pollutant inputs at large scale like regional or continental scales (Venkatram, 1986; Dennis *et al.,* 1990). The uncertainties attached to many such models increase with geographical scaling and with the increasing complexity of stochastic atmospheric processes included in the model. Any newly developed model should be properly validated prior to application for the first time. Subsequent modifications of the original model for new application should also be validated.

Coarse particulate air pollutants (>2.0 μm) are rapidly removed from the atmosphere by sedimentation (Whitbby, 1978). But fine particles (<2.0 µm) produced by atmospheric processes, remain suspended in the atmosphere for long time due to their low deposition velocity and tend to accumulate (Lee *et al.*, 1986). Accumulation of fine particle acid aerosols in the atmosphere can increase the atmospheric residence times of O_3 and the potential for adverse environmental effects (Chevone *et al.,* 1986). Photo chemically generated fine particles accumulate in the atmosphere at night-time (Stevens *et al.*, 1978). Effects of fine particle aerosols on vegetation are less studied as compared to other air pollutants (Gmur *et al.*, 1983). These affect plant processes at physiological and molecular level.

4. ROLE OF DEVELOPMENTAL STAGES IN AIR POLLUTION TOLERANCE

Maximum visible injury in a leaf due to O_3 exposure occurs when the leaf is just beyond half way to its maximum size (Health, 1994)). The variation in response to pattern of plant at different developmental stages is related to general metabolic demand. Higher level of soluble sugar and amino acids were found to be responsible for affording protection against O_3 exposure (Dugger and Ting 1970). Reich and Lassoie (1985) have shown that in young leaves of poplar, pigment breakdown and decline in net photosynthesis are delayed up to some extent due to repair mechanism such as an increase in dark respiration, which also compensate for the shortened life span of the leaves by increasing leaf production. In the pollutant- tolerant plants leaf turnover was found to be more rapid (Pandey and Agrawal, 1994)

It was found that young leaves take up more SO_2 than the mature leaves (Bressan *et al.*, 1978a) but in young leaves injury was found to be far less than mature leaves on the same plant (Bressan *et al.*, 1978). Thus the young leaves have to have a biochemical based resistive mechanism which function after SO_2 metabolism in resistant versus sensitive plants which are otherwise closely similar.

On the other hand, there was a vast difference between the emission of volatile sulphur by young, resistant leaves and mature, sensitive leaves on the same plant. The young leaves emitted a lot of H_2S, while the mature leaves emitted very little (Filner *et al.*, 1984). Thus H_2S emitted by young leaves was equal to more than 10% of the SO_2 absorbed. The positive correction of H_2S emission with SO_2 resistance is a meaningful clue to the developmentally regulated, biochemically based resistance mechanism (Filner *et al.*, 1984). The level of SO_2 is higher inside young leaves than inside mature leaves throughout an acute exposure to SO_2 but the mature leaves are injured far more severely (Wilson *et al.*, 1979; Filner *et al.*, 1984). This indicates that SO_2 itself is not toxic but it might be transformed to a toxic compound able to cause injury. Conversely, it could be mean that young leaves do not have sites prone to injury accessible to the absorbed SO_2.

5. BIOCHEMISTRY OF AIR POLLUTION TOLERANCE

Gaseous pollutants after entering through the stomatal cavity dissolve in extra cellular fluid and react with biological compounds to produce ionic species or free radicals (Mehlhorn *et al.*, 1990; Heath, 1994).

As a result of free radicals interaction with proteins and lipids in the cell wall and membranes, oxidation, lipid peroxidation and initiation of chain reaction takes place giving rise to more free radicals followed by increases in cell permeability (Mehlhorn *et at.*, 1986; Pell and Dann, 1991). This leads to severe metabolic disorder and damage to genetic material (Brawn and Fridovich, 1981). Damage to genetic material may cause mutations and change in characters of species or development of a new species.

The gaseous air pollutants viz., O_3, SO_2, NO_x and HF destroy the ultrastructural organization of leaf cells, thylakoids become swell gradually curl and finally disintegrate, cytoplasmic and plastid matrix becomes granular, mitochondria injured and number of ribosome decreases (Mlodzianowski and Biolobok, 1977). Fluoride inhibits enzyme action by binding with the calcium and/or magnesium essential for enzyme activity and by the possible formation of fluoride ion of toxic metabolic analogues.

As a result of disintegration organelles release the enzymes contained in them or on their surfaces. This may account for an increase in the activity of some enzymes and also for changes in isozyme patterns sometimes observed in extracts obtained from leaves of trees growing in regions polluted by industrial emissions (Mejnartowicz, 1984). In some species, the sub cellular distribution of some enzymes differs in trees of varying tolerance (Mlodzianowski and Mlodzianowska, 1980). Horsman and Wellburn (1975) have used isoenzymes of peroxidase enzyme as useful markers of pollution sensitivity. In crops such as soybean, wheat and rice crops exposed to elevated O_3 in the field decrease in net assimilation is a common feature which takes place due to both decreased leaf level photosynthesis rates and significantly decreased leaf area (Feng *et al.*, 2011).

Moreover, detoxification of toxic air pollutants is another strategy in plants to overcome deleterious effects. Plants have an array of mechanisms to detoxify the toxic molecules (Larson, 1988; Foyer *et al.*, 1994). These include enzyme systems (catalase, peroxidase, superoxide dismutase, etc.), ascorbic acid, vitamin E (α-tocopherol), peptides (glutathione), carotenoids (β-carotene), polyamines as well as organic buffering systems. An understanding of the biochemical markers providing protection against oxidative damage is essential for the development of plants more resistant to the growing number of environmental stresses (Monk *et al.*, 1989). There is also limited evidence to suggest that O_3 can affect CH_4 emissions from peat lands, possibly through O_3 causing plants to alter substrate availability to soil microbes or causing changes in transport of CH_4 through vascular plants with aerenchymatous tissue (Toet *et al.*, 2011).

6. AIR POLLUTION TOLERANCE MECHANISMS IN PLANTS

Plants adopt different strategies to cope up the problem of air pollution. Resistance or susceptibility of plants to air pollution depends on functioning of these strategies. Strategies adopted by plants for building resistance against air pollutants include stomatal movements, enzyme action, detoxifying processes as well as genetic and developmental factors (Malhotra and Khan, 1984; Wolfenden and Mansfield, 1991).

6.1 Stomatal Movement

The extent of damage that air pollutant(s) can cause in plants depends upon the pollutant influx into the leaf and the reactivity of the pollutant and/or reactivity of its reaction products with cellular constituents. Primary route of the pollutant uptake in a leaf is controlled by the stomatal aperture and conductance to gas diffusion (Health, 1994; Kangasjarvi *et al.,* 1994). Movement of CO_2 inwards and O_2 outwards in leaves is controlled by diurnal variation in stomatal aperture. These apertures influence the pollutant entry also. The opening and closing of stomata depend on turgidity/flaccidity of the guard cells and their neighboring cells due to massive intercellular fluxes of potassium (Mac Robbie, 1988).

Stomatal movement is controlled by hormonal level. Two hormones Abscisic acid (ABA) and indol-3-acetic acid (IAA) play crucial role in this regard (Snaith and Mansfield, 1982). Stomatal responses vary with the variation in water status of the plant (Raschke, 1975). An ideal mechanism for avoiding stress by air pollutants involves a balancing of the leaf's priorities between import of CO_2 for photosynthesis and the prevention of excessive water loss by transpiration (Mansfield and Freer-Smith, 1984; Darrall, 1989).

There are two types of action of air pollutants on cell of the epidermis (Mansfield and Freer-Smith, 1984). In the first type molecules diffuse in through the stomata to reach the sub stomatal cavity and then enter other cell of the epidermis. Inner surfaces of guard cells are more tolerant to the strain imposed by the pollutant than the subsidiary cells due to their anatomical feature. Hence the latter become injured and lose turgor while the former retain turgidity. Stomatal opening enlarges purely as a result of the release of surrounding turgor pressure. In the second type molecules enter the epidermis as in the first type and cause a loss of turgor of the guard cells, this brings about stomatal closure.

Air pollutants affect both stomata and other parts of the epidermal layer (Mansfield and Freer-Smith, 1984). Cells surrounding the sub stomatal cavity are most exposed to air pollutants. Effect of air pollutant on plant leaves depends on the number and size of stomata and the degree of their opening (Yunus *et al.*, 1979). Stomata have central role in determining resistance or susceptibility of plants to air pollution because these regulate pollutant entry to plant cells. There are two types of air pollution stress avoidance mechanisms in plants at the stomatal level (i) active avoidance reducing gas exchange during exposure period and (ii) passive avoidance through maintenance of low gas exchange capacity under favorable and unfavorable condition. Active avoidance mechanism is a metabolically active process or manifestation of injury to guard cells, leading to stomatal closure. Species with higher stomatal conductance are more susceptible to air pollutant (Reich and Amundson, 1985).

The foliar cuticle restricts water loss from leaf surface and protects the plant from the effects of air pollutants. Amongst the most important properties of the cuticular membranes are the thickness and tortuosity of the pathway, the wettability, the chemical composition and the arrangement of the membrane constituents (Merida *et al.*, 1981; Price, 1982). Molecular radius of the solute, its affinity for the cuticular membrane (the partition coefficient) and the concentrations inside and outside the tissues are important factors in determining effect of air pollutants (Schonherr and Schmidt, 1979; Price, 1982). In soybean, the coupling of lower stomatal conductance and reduced leaf area index at elevated O_3 was found to result in a 10% decrease in canopy evapotranspiration, which has implications for the terrestrial hydrological cycle (Bernacchi *et al.*, 2011). In a recent modeling analysis it was concluded that despite substantially increased future O_3 in central and southern Europe, the flux-based risk of O_3 damage to vegetation was found to be unchanged or decreased at sites across Europe, mainly as a result of projected reductions in stomatal conductance under rising CO_2 (Klingberg *et al.*, 2011).

6.2 Genetically Controlled Tolerance

The variation in responses of various species and cultivars to air pollutants suggest phenotypic expressions of genotypes evolved naturally or introduced by breeding (Pitelka, 1988). The genetic resistance enables plants to grow, survive and reproduce in polluted environments either through avoidance or tolerance mechanisms (Pitelka, 1988).

In the family cucurbitaceae genetic differences exist between genera in susceptibility to acute injury (Bressan *et al.*, 1981). For example, cultivars of *Cucumis sativus* (cucumber) were found to be more sensitive than cultivars of *Cucurbita pepo* (squash and pumpkin). Differences in susceptibility occur among cultivars of the same species also. Differences in resistance due to differences in uptake of a gas are presumed to reflect differences in stomatal characters viz., number, size and conductance. Levit (1972) suggested that there is a common denominator of stress tolerance – probably some property of a membrane. The significance of the plasma membrane is decisive in this regard, if the membrane fails, nothing inside can save the cell from damage (Filner *et al.*, 1984).

6.3 Antioxidants and Growth Regulators in Air Pollution Tolerance

Plant varieties naturally sensitive to oxidant injury can be transformed into tolerant ones by some chemical treatments that alter plant metabolic processes (Rao and Dubey, 1990; Lee *et al.*, 1992; Miller *et al.*, 1994). EDU is one of such chemical which substantially enhance plant tolerance to oxidants. The enhancement in the foliar tolerance to O_3 was found to be up to 30-fold in plant treated with EDU systemically (Brennan *et al.*, 1987). No statistical difference was found between EDU treated and untreated plant in stomatal resistance on the abaxial side of the leaves (Lee and Bennett, 1979). Findings suggest that EDU-induced tolerance to O_3 exposure is probably a biochemical process at leaf mesophyll sites rather than physical exclusion by stomatal closure and restricted O_3 diffusion (Lee *et al.*, 1992). It was found that neither membrane phospholipids nor polyamines within the leaf tissue are responsible for the protective nature of EDU in O_3 treated plants. EDU does not change the level of chlorophyll and cartotenoid pigments in leaf, but reduces the loss of pigments due to O_3 treatment (Whitaker *et al.*, 1990). Furthermore, EDU plays an important role in maintaining integrity of the foliar membranes.

Plant tolerance to environmental stresses may be increased by means of plant growth regulators and antioxidant treatments (Hausladen and Kunert, 1990). Tolerance to air pollutant exposure induced by gibberellic acid (GA) inhibitor is biochemical at leaf mesophyll sites as stomatal closure is not involved in GA inhibitor-induced resistance to SO_2 (Lee *et al.*, 1992). GA_3 inhibitors modify the biosynthesis and action of GA by affecting RNA and protein synthesis, which in turn may regulate the growth and SO_2 sensitivity.

Ascorbic acid, an antioxidant found in plants, play an important role in pollution tolerance (Wolfenden *et al.*, 1992; Pandey and Agrawal, 1994). A direct relationship between endogenous levels of ascorbic acid and plant tolerance to pollutants has been established (Polle *et al.*, 1991). Varshney and Varshney (1984) reported that SO_2-resistant species have higher ascorbic acid concentration in leaves compared to sensitive ones and a slower decline of ascorbic acid in the former. Ascorbic acid, being a powerful reductant, maintains the stability of cell membranes during pollution stress (Dhindsa *et al.*, 1982a) and scavenges cytotoxic free radicals (Haliwell and Gulteridge, 1989). In plants exposed to relatively low concentrations of air pollutants for long duration, biochemical changes can lead to air pollution tolerance possibly through two ways i.e., (i) plant species with high antioxidant activities are comparatively more tolerant even if the activity increase a little in polluted areas and (ii) some plant species develop tolerance by raising the antioxidant activities substantially in polluted areas (Rao and Dubey, 1990).

Sugars and alcohols are among cellular products in large quantities that readily scavenge the initiating OH radicals. Carbohydrate in plant tissues including free sugars, starch and other cellular components may be important stabilizers of cell and organelles exposed to OH radicals. Polyamines have also been shown to be antioxidants and protect membranes from peroxidation (Bores *et al.*, 1989). Exogenous application of polyamines was found to provide protection against O_3 pollution (Bores *et al.,* 1989). Ameliorative effect of polyamines on O_3 injury has been linked to their ability to stabilize cell membranes, scavenge free radicals (Drolet *et al.,* 1986) and prevent lipid peroxidation (Tadolini, 1988). Polyamines in plants increases in response to O_3 which is an adoptive response to stress (Rowland - Bamford *et al.,* 1989). A possible mechanism for preventing O_3 injury could be the reduction of ethylene formation by polyamines (Langebartels *et al.,* 1991). High levels of polyamines are known to reduce 1-aminocyclopropane-1-carboxylic acid (=ACC) synthase and ACC oxidase activity (Apelbaum, 1990), which would in turn reduce the ethylene production.

Recognition of air pollution damage or injury is subject to a differential diagnosis, which includes physiological and biochemical parameters. If possible it is more appropriate to avoid pollution damage than to repair it. Physiological and biochemical parameters have been used for preventive measures such as evaluation of acceptable pollution levels (Bucher and Keller, 1978) and selection of resistant cultivars (Bressan *et al.,* 1978). In this regard, metabolic products in plants could receive more attention than substances emitted by plants.

6.4 Role of Enzymes

*6.4.1. Superoxide dismutase (*EC *1.15.1.1)*

Superoxide dismutase (SOD) metalloproteins are classified on the basis of specific metal co-factor (iron, manganese or copper) essential for their activity. This family of oxyradical-scacenging enzymes is of universal occurrence in aerobic and aero tolerant cells. Zinc in the Cu-Zn-containing SOD can be replaced by a variety of other metals without loss of enzymatic activity (Fridovich, 1981). The widespread occurrence of superoxide dismutases in plants as well as other aerobic organisms has brought out SODs as important cellular protective enzymes. Since the enzyme substrate, O_2^-, is less toxic than the other potential secondary oxyradicals that may be formed in the cells, the protective action is believed to be largely indirect, i.e., by removing O_2^- radicals which may cause generation of more toxic chemical species. The low reactivity of the superoxide anion radical makes it capable of transmitting radical properties over wide distances in the cell. Superoxide dismutases along with catalase and the peroxidases that act on the end product (H_2O_2) of SOD activity can interact to regulate injurious oxyradical and peroxyl concentrations in cells and organelles as well as determine equilibrium rates (Fridovich, 1986).

Superoxide dismutase is found to be the most efficient known scavenger for superoxide radicals. The different isozymes of SOD occur in different subcellular compartments. Rennenberg and Polle (1994) have estimated that chloroplasts contain SOD, ascorbate and glutahione at concentrations of 10µM, 10-50 mM and 3 mM, respectively and contribute in scavenging of $O_2.^-$ in the order of 10:1-5:1.

6.4.2. Peroxidase (EC 1.11.1.7)

Under various stresses plant cells show increased peroxidase activity, such as mechanical injury, attack by pathogen or influence of environmental pollution (Nandi *et al.*, 1990). Under stress condition an increase in the activity of enzymes is associated with destruction of cellular membranes and the simultaneous release of previously immobilized proteins. In plant cells exposed to SO_2, peroxidase activity increases when sulphite is oxidized to sulphate in the presence of hydrogen peroxide and peroxidase enzyme. This increase in peroxidase activity varies with the species of the plant, the season and concentration of the gas (Nikolayevskii, 1986). The changes in peroxidase activity are accompanied changes in catalase activity in reverse direction (Nandi *et al.*, 1984).

The increase in peroxidase activity was found to be accompanied by an increase in the accumulation of sulphur in leaves, decline in CO_2 assimilation and reduced ascorbic acid content (Khan and Malhotra, 1982). Preceding the appearance of symptoms of leaf injury, increase in peroxidase activity may occur in the leaves exposed to toxic gases such as SO_2 (Pierre and Queiroz, 1981) and O_3 (Endress *et al.*, 1980).

6.4.3. Acid phosphatase (EC 3.1.3.2)

Acid Phosphatase (APH) enzyme is produced in cell walls and occurs in the cytosol of higher plants also. In the leaves, it occurs in the spongy parenchyma and in minor veins, but it is much scarce in the palisade parenchyma (Besford and Syred, 1979). APH is a glycoprotein, it has been separated into two components: an "acid" one and an "alkalin" one (Jonsson, 1979). APH metabolically decomposes phosphate esters (Banasik *et al.*, 1980) and releases inorganic phosphate in the cells (Heredia *et al.*, 1963). If the level of phosphorus declines below 0.25% in tissue APH activity rises substantially (Besford, 1979). Some ions, including those found in the ambient air polluted with industrial fumes, such as Al^{3+}, Fe^{3+}, arsenate, fluoride, molybdate and phosphate have an inhibitory effect on APH activity.

At the cellular level, specific metabolic changes particularly in enzyme concentration and activity resulting from rise in the level of air pollution or elevated O_3 need to be elucidated. For example negative correlation between O_3 and transcript abundance of cytosolic ATP-citrate lyase and mitochondrial alternative oxidase 2b (AOX_2b) was observed in soybean (Gillespie *et al.*, 2012).

CONCLUSIONS

The epidermal cells are most vulnerable and susceptible to gaseous air pollutants as they are the first to interact with the pollutants and these cells lack chloroplasts. The cells without chloroplasts cannot remove excess sulphur through production and release of H_2S for want of energy that comes from photosynthetic electron transport. Plants adapt to avoid or tolerate the harmful effect of air pollution include changes at biochemical, physiological and morphological levels. Susceptibility to acute injury by gaseous air pollutants may differ among different genera/species or among cultivars of the same species. This could be due to differences in pollutant absorption from the air which in turn may depend on stomatal number per unit area, stomatal pore size and/or permeability of plasma membrane. The pathways for infusion of different gaseous air pollutants to the leaf tissues are not similar. Low SO_2 concentrations normally widen the stomatal opening without affecting the diurnal functioning of the stomata. The higher concentrations of SO_2 induce stomatal closure in two ways: (i) directly by causing injury to the guard cells and (ii) indirectly by modulating the internal CO_2 concentration in the leaf through its inhibitory effect on photosynthetic CO_2 fixation. In a high vapour pressure deficit condition, the stomata may, however, tend to have smaller apertures which work as an avoidance mechanism. Young leaves are generally less injured than mature leaves on the same plant. The gaseous pollutants destroy the ultrastructural organization of leaf cells and produce oxygen-free radicals that lead to the oxidation of biomolecules such as proteins, lipids and DNA. Although, plants posses defense mechanisms consisting of oxyradical scavenging enzyme (e.g., SOD, catalase and peroxidase) and antioxidants. SOD acts upon O_2 and converts it into H_2O_2, which is then converted into H_2O and O_2 by the enzymes catalase and peroxidase. Fluoride inhibits enzyme action by binding with the calcium and/or magnesium needed for enzyme activity and by the possible formation with the fluoride ion of poisonous metabolic analogues. Distinctly it reduces the content of free phosphates, and causes precipitation of calcium. Peroxidase activity increases invariably against stresses but it remains insensitive to gaseous pollutants with heavy metals. The enzyme activity is usually higher in mesophyll tissue, while in resistant plants the activity is higher in the epidermis, hypodermis, endodermis and in the resin canals. Therefore, these tissues must be involved in the process of detoxification. Some common metabolic reductants, such as ascorbic acid, glutathione, α-tocopherol, starch and sugars etc., protect plants from the pollution injury. These antioxidants serve as a free oxyradical scavenging system in plants exposed to gaseous air pollutants and protect membranes from peroxidation. Plants emit volatile compounds such as ethylene, ethane, monoterpense, acetaldehyde and ethanol in response to cellular disturbance and injury due to air pollutants. These are not only the consequence of a pollution stress, but may be pollutant themselves. Stress ethylene may be a good indicator of pollution stress under standardized conditions with high pollutant concentrations but this is not so in the field condition because ethylene emission fluctuate strongly under persistent stress. Ethylene is produced by living cells, both normal and stress ethylene are produced in the plants through the same biochemical pathway. Ethane production increases with increasing membrane damage and is mostly associated with the dying cells, a visible injury. Sensitive plants emit more ethane than the resistant ones. Ethane/ethylene ratio is one of the most reliable assays for assessing the injury level of a tissue exposed to air pollutants. Terpenes are released into the atmosphere through the stomatal pathway but cuticle acts as a barrier. Occurrence of necrosis indicates the breakopen of this barrier due to high emissions of air pollution induced monoterpenes. Monoterpenes also act as a sink for ozone and, thus, provide a preventive effect.

REFERENCES

Apelbaum, A. (1990). Interrelationship between polyamines and ethylene and its implication for plant growth and fruit ripening. In Flores, H.E., Arteca, R.N, and Shannon, J.C. (eds) *Polyamines and ethylene: Biochemistry, Physiology and interactions*: 278-294. American Society of Plant Physiologists, Rockville, MD.

Banasik, L., Rudnicki, R.M. and Saniewski, M. (1980). Studies on the physiology of hyacinth bulb (*Hyacinthus orientalis* L.). XIII. The distribution of amylase and acid phosphatase activities and starch grain in hyacinth bulbs. *Acta Physiol. Plant.* **2:** 145-156.

Bernacchi C.J., Leakey A.D.B., Kimball B. A. and Ort D.R. (2011). Growth of soybean at future tropospheric ozoneconcentrations decreases canopy evapotranspiration and soil water depletion. *Environ. Pollut.* **159**:1464–72.

Besford, R.T. (1979). Quantitative aspects of leaf acid phosphatase activity and phosphorus status of tomato plants. *Ann. Bot.* **44:** 153-161.

Besford, R.T. and Syred, A.D. (1979). Effect of phosphorus nutrition on the cellular distribution of acid phosphatase in the leaves of *Lyvopersicon* L. *Ann. Bot.* **43:** 431-453.

Bors, W., Langebartels, C., Michel, C. and Sandermann, H. Jr. (1989). Polyamines as radical scavengers and protectants against ozone damage. *Phytochem.* **28**: 1589-1595.

Brawn, K. and Fridovich, I. (1981). DNA strand scission by enzymatically generated oxygen radicals. *Arch. Biochem. Biophys.* **206**: 414-419.

Brennan, E., Leone, I. and Clarke, B. (1987). EDU a chemical for evaluating ozone foliar injury and yield reduction in field-grown crops. *Int. J. Trop. Dis.* **5**: 35-42.

Bressan, R. A. Wilson, L. G. and Filner, P. (1978). Mechanisms of resistance to sulphur dioxide in Cucurbitaceae. *Plant Physiol.* **61**: 761 – 767.

Bressan, R.A., Lecureux, L., Wilson, L.G., Filner, P. and Baker, L.R. (1981). Inheritance of resistance to sulphur dioxide in cucumber. *Hort. Sci.* **16**: 332-333.

Bucher, J.B. and Keller, Th. (1978). Einwirkungen niedriger SO_2-Konzentrationen in mehrwochigen Begassungsversuch auf Waldbaume. *VDI-Bericht* **341**: 237-242.

Chevone, B.I., Herzfeld, D.E., Krupa, S.V. and Chappelka, A.H. (1986). Direct effect of atmospheric sulphate deposition on vegetation. *JAPCA* **36**: 813-816.

Darrall, N.M. (1989). The effect of air pollutants on physiological processes in plants. Plant, *Cell Environ.* **12**: 1-30.

Dennis, R., Clark, T., Barchet, R., Roth, P., Reynolds, S., Seilkop, S., Laulainen, N., Clark, J., Chang, J. and Samson, P. (1990). SOS/T 5: *Evaluation of Regional Acidic Deposition Model.* US NAPAP, Washington, DC.

Dhindsa, R.S., Plumb-Dhindsa, P.L. and Reid, D.M. (1982a). Leaf senescence and lipid peroxidation: Effects of some Phytohormones and free radical scavengers. *Plant Physiol.* 69 (Suppl.) **10**: Abstract No.48.

Drolet, G., Dumbroff, E.B., Legge, R.L. and Thompson. J.E. (1986). Radical scavenging properties of polyamines. *Phytochem.* **25**: 367-371.

Dugger, W.M. and Ting, I.V. (1970). Air pollution oxidants - Their effect on metabolic processes in plants. *Annu. Rev. Plant. Physiol.* **21:** 215-235.

Endress, A. G., Saurez, S. J. and Taylor, O. C. (1980). Peroxidase activity in plant leaves exposed to gaseous HCl or ozone. *Environ. Pollut*. (Ser. A) **22:** 47 – 58.

Feng Z. Z., Pang J., Kobayashi K., Zhu J. G. and Ort D. R. (2011). Differential responses in two varieties of winterwheat to elevated ozone concentration under fully open-air field conditions.Glob. *Change Biol.* **17**:580–91.

Filner, P., Rennenberg H., Sekiya, J., Bressan, R.A., Wilson, L.G., Le Cureux, L. and Shimei, T. (1984). Biosynthesis and emission of hydrogen sulfide by higher plants. In Koziol, M.J. and Whatley, F.R. (eds) Gaseous Air Pollutants and Pant Metabolism : 291-312 Butterworths, London.

Finlayson- Pitts, B.J. and Pitts, J.N. (1986). *Atmospheric Chemistry: Fundamentals and Experimental Techniques.* Wiley, New York, 1098 pp.

Foyer, C.H., Descourvieres, P. and Kunert, K. (1994). Protection against oxygen radicals: an important defense mechanism studied in transgenic plants. *Plant, Cell Environ*. **17**: 507-523.

Fridovich, I. (1981). Role and toxicity of superoxide in cellular systems. In Rodgers M.A.J. and Powers E.L. (eds) Oxygen and Oxy-Radicals in Chemistry and Biology : 197-239. Academic Press, New York.

Fridovich, I. (1986). Superoxide dismutases. *Adv. Enzymol.* **58:** 62-97.

Gillespie K.M., Xu F., Richter K.T., Mc Grath J.M. and Markelz R.J.C. (2012). Greater antioxidant andrespiratory metabolism in field-grown soybean exposed to elevated O_3 and two CO_2 concentrations.*Plant Cell Environ.* **35**:169–84.

Gmur, N. F., Evans, L.S. and Cunningham, E.A. (1983). Effect of ammonium sulfate aerosols on vegetation. II. Mode of entry and responses of vegetation. *Atmos. Environ.* **17**: 715- 721.

Grunhage, I., Dammgen, U., Haenel, H.D., Jarger, H.J., Holl, A., Schmitt, J. and Hanewald, K. (1993). A new potential air a quality criterion derived from vertical flux densities of ozone and from plant response. *Angew, Bot.* **67**: 9-13.

Halliwell, B. and Gutteridge, J.M.C. (1989). *Free Radicals in Medicine and Biology*, 2nd edition. Clarendon Press, Oxford.

Hausladen, A. and Kunert, K.J. (1990). Effects of artificially enhanced levels of ascorbate and glutathione on the enzymes monohydroascorbate reductase and glutathione reductase in spinach. *Physiol. Plant.* **79**: 384-388.

Heath, R.L. (1994). Alterations of plant metabolism by ozone exposure. In Alscher, R.G. and Wellburn, A.R. (eds) *Plant Response to the Gaeous Environment*: 121-145. Chapman & Hall Londen.

Heredia, C.F., Yen, F. & Sols, A. (1963). Role and formation of the acid Phosphatase in yeast. *Biochem. Biophys. Res. Commun.* **10:** 14-18.

Horseman, D.C. and Wellburn, A.R. (1975). Synergistic effects of SO_2 and NO_2 polluted air upon enzyme activity in pea seedlings. *Environ. Pollut.* **8:** 123-133

Jonsson, J. (1979). Isolation and partial characterization isoenzymes of acid Phosphatase from needles of Pinus sylvestris L. In Rudin, D. (ed.) *Proceedings of the Conference on Biochemical Genetics of Forest Trees*: 5-14. Report 1. Swedish University of Agricultural Sciences, Umea, Sweden.

Kangasjarvi, J., Talvinen, J., Utriainen, M. and Karjalainen, R. (1994). Plant defense systems induced by ozone. *Plant, Cell Environ.* **17**: 783-794.

Khan, A.A. and Malthotra, S.S. (1982). Peroxidase activity as an indicator of SO2 injury in jack pine and white birch. *Biochem. Physiol. Pflanzen.* **177:** 643-650.

Kickert, R.N. and Krupa, S.V. (1990). Forest responses to tropospheric ozone and global climate change: An analysis. *Environ. Pollut.* **70**: 271-383.

Klingberg, J., Engardt, M., Uddling, J., Karlsson, P. E. And Pleijel, H. (2011). Ozone risk for vegetation in the future climate of Europe based on stomatal ozone uptake calculations. *Tellus* **63A**: 174 – 184.

Krupa, S. V., Nosal, M. And Legge, A. H. (1994). Ambient ozone and crop loss: Establishing a cause-effect relationship. *Environ. Pollut.* **83**: 269 – 276.

Krupa, S.V. and Manning, W.J. (1988). Atmospheric ozone: Formation and effects on vegetation. *Environ. Pollut.* **50**: 101-137.

Krupa, S.V., Grunhage, L., Jager, H.J., Nosal, M., Manning, W. J., Legge, A.H. and Hanewald, K. (1995). Ambient ozone and adverse crop response: A unified view of cause and effect. *Environ Pollut.* **87**: 119-126.

Krupa, S.V., Manning, W.J. and Nosal, M. (1993). Use of tobacco cultivars as biological indicator of ambient ozone pollution: an analysis of exposure-response relationships. *Environ. Pollut.* **81**: 137-146.

Langebartels, C., Kerner, K., Loenardi, S., Schraudner, M., Trost, M., Heller, W. and Sandermann, H.J. (1991). Biochemical plant responses to ozone I. Differential induction of polyamine and ethylene biosynthesis in tobacco. *Plant Physiol.* **95**: 882-889.

Larson, R. A. (1988). The antioxidants of higher plants. *Phytochem.* **27**: 969 – 978.

Lee, E.H. and Bennett, J.H. (1979). Comparative studies of Foliar protection from ozone induced by EDU in greenhouse and growth chamber preconditioned plants. In Abdel-Rahman, M. (ed.) *Proceedings of the Sixth Annual Meeting of the Plant Growth Regulator Working Group*: 218. Agway, Inc., Syracuse, NY.

Lee, E.H., Kramer, G.F., Rowland, R.A. and Agrawal, M. (1992). Antioxidants and growth regulators counter the effects of O_3 and SO_2 in crop plants. *Agric. Ecosyst. Environ.* **30**: 99-106.

Lee, S. D., Schneider, T., Grant, L. D. and Verkerk, P. J. (1986). Aerosols. Lewis Publishers, Inc. Chelsea, MI, p. 1221.

Lefohn, A.S., Krupa, S.V. and Winstanley, D. (1990). Surface ozone exposure measured at clean location around the world. *Environ. Pollut*. **63**: 189-224.

Legge, A.H. and Krupa, S.V. (1990). *Acidic Diposition*. Sulpher and Nitrogen Oxides. Lewis Publication, Inc., Chelsea, MI, 659 pp.

Levitt, J. (1972). *Response of Plants to Environmental Stresses.* Academic Press, New York. Liebermann, M. 1979. Biosynthesis and action of ethylene. *Annu. Rev. Plant Physiol.* **30**: 533-591.

Lovelock, J.E. (1987). *Gaia: A New Look at Life on Earth*. Oxford University Press, Oxford, 157pp.

Lovett, G.M. and Kinsman, J.D. (1990). Atmospheric pollutant deposition to high elevation ecosystems. *Atmos. Environ.* **24A**: 2767-2786.

Lyons, T.J. and Scott, W.D. (1990). *Principles of Air Pollution Meteorology*. CRC Press, Boca Raton, FL, 224 pp.

Mac Robbie, E.A.C. (1988). Control of ion fluxes in stomatal guard cells. *Bot. Acta.* **101**: 140-148.

Malhotra, S.S. and Khan, A.A. (1984). Biochemical and Physiological impact of major pollutants. In Treshow, M. (ed.) *Air Pollution and Plant Life*: 113-157. Wiley, Chichester.

Mansfield, T.A. and Freer-Smith, R.H. (1984). The role of stomata in resistance mechanism. In Koziol, M.J. and Whatley, F.R. (eds) *Gaseous Air Pollutants and Plant Metabolism*: 131-146. Butterworths, London.

Mehlhorn, H., Gunther, S., Schmidt, A. and Kunert, K.J. (1986). Effect of SO_2 and O_3 on production of antioxidants in conifers. *Plant Physiol*. **82**: 336-338.

Mehlhorn, H., Tabner, B. and Wellburn, A.R. (1990). Electron spin resonance evidence for the formation of free radicals in plants exposed to ozone. *Physiol. Plant*. **79**: 377-383.

Mejnartowicz, L.E. (1984). Enzymatic investigations on tolerance in forest trees. In Koziol, M.J. and Whatley, F.R. (eds) *Gaseous Air Pollutants and Plant Metabolism*: 381-398 Butterwoeths, London.

Merida, T., Schonherr, J. and Schmidt, H.W. (1981). Fine structure of plant cuticles in relation to water permeability: The fine structure of the cuticle of Clivia miniata Reg. leaves. *Planta.* **152**: 259-267.

Miller, J.E., Pursley, W.A. and Heagle, A.S. (1994). Effects of ethylenediurea on snap bean at a range of ozone concentrations. J. *Environ. Qual.* **23**: 1082-1089.

Mlodzianowski, F. and Bialobok. S. (1977). The effect of sulphur dioxide on ultrastructural organization of larch needles. *Acta Soc. Bot. Pol.* **46**: 629-634.

Mlodzianowski, F. and Mlodzianowska, L. (1980). Cytochemical localization of enzyme activity in Scot pine needles treated with SO_2. In Bialobok, S. (ed.) Studies on the Effect of Sulphur Dioxide and Ozone on the Respiration of Trees and Shrubs in Order to Select Individuals Resistant to the Action of these Gases: 79-81. Polish Academy of Sciences, Institute of Dendrology, Kornik, Poland.

Monk, L.S., Fagerstedt, K.V. and Crawford, R.M.M. (1989). Oxygen toxicity and superoxide dismutase as an antioxidant in physiological stress. *Physiol. Plan.* **76**: 456-459.

Mudd, J.B. and Kozlowski, T.T. (1975). *Responses of Plants to Air Pollution.* Academic Press, New York, 383 pp.

Nandi, P.K. Agrawal, M., Agrawal, S.B. and Rao, D.N. (1990). Physiological reponses of *Vicia faba* plants to sulphur dioxide. *Ecotoxic. Environ. Safety.* **19:** 64-71.

Nandi, P.K., Agrawal, M. and Rao, D.N. (1984). SO_2-induced enzymatic changes and ascorbic acid oxidation in *Oryaza sativa. Water Air Soil Pollut.* **21:** 25-32.

Nikolayevskii, V. S. (1968). Aktivnost nekotorikh fyermyentovi gazoustoichivosti rastyenii. [Plant resistance to gaseous emission and activity of some enzymes.] *Trudy Inst. Ekologii Rastyenii I Zhyvotnykh. Uralskii Filial AN SSR* **52:** 208-211.

Pandey, J. and Agrawal, M. (1994). Evaluation of air pollution phytotoxicity in a seasonally dry tropical urban environment using three woody-perennials. *New Phyto.***126:** 53-61.

Pell, E.J. and Dann, M.S. (1991). Multiple stress induced foliar senescence and implications for whole plant longevity. In Monney, H.A., Winner, W.E. & Pell, E.J. (eds*) Response of Plants to Multiple stresses:* 189-204. Academic Press, San Diego.

Pierre, M. and Qyeiroz, O. (1981). Enzymic and metabolic changes in bean leaves during continuous pollution by subnecrotic levels of SO_2. *Environ. Pollut. (Ser. A)* **25:** 41-53.

Pitelka, P.S. (1988). Evolutionary responses of plants to anthropogenic pollutants. *Trends Ecol.* **3**: 233-236.

Polle, A., Chakrabarti, K., Schurmann, W. and Rennenberg, H. (1990). Composition and properties of hydrogen decomposing systems in extracellular and total extracts from needles of Norway spruce *(Picea abies* L.) *Plant Physiol* **94**: 312-319.

Price, C.E. (1982). A review of the factors influencing the penetration of pesticides through plant leaves. In Cutler, D.F., Alvin, K.L. and Price, C.E. (eds). *The Plant Cuticle*: 237-252. Academic Press London.

Rao, M.V. and Dubey, P. S. (1990). Biochemical aspects (antioxidants) for development of tolerance in plants growing at different low levels of ambient air pollutants. *Environ. Pollut.* **64**: 55 – 66.

Raschke, K. (1975). Simultaneous requirement for carbon dioxide and abscisic acid for stomatal closing in *Xanthium strumarium* L. *Planta.* **125**: 243-259.

Reich, P. B. and Lassoie, J. P. (1985). Influence of low concentrations of ozone on growth, biomass partitioning and leaf senescence in young hybrid poplar plants. *Environ. Pollut.* **39**: 39 – 51.

Reich, P.B. and Amundson, R.G. (1985). Ambient levels of ozone reduce net photosynthesis in tree and crop species. *Science* **230**: 566-570.

Rennenberg, H. & Polle, A. 1994. Metabolic consequences of atmospheric sulphur influx into plants. In Alscher, R.G. & Wellbern, A.R. (eds) *Plant Response to the Gaseous environment*: 165-180. Chapman & Hall, London.

Rowland- Bamford, A.J., Borland, A.M., Lea, P.J. and Mansfield, T.A. (1989). The role of arginine decarboxylase in modulating the sensitivity of barley to ozone. *Environ. Pollut*. **61**: 95-106.

Schonherr, J. and Schmidt, H.W. (1979). Water permeability of plant cuticle. *Planta*. **144**: 391-400.

Schulte-Hostede, S., Darrall, N.M., Blank, L.W. and Wellburn, A.R. (eds.) (1988). Air Pollution and Plant Metabolism. Elsevier Applied Science, London, 379 pp.

Snaith, P. J. and Mansfield, T. A. (1982). Control of the CO2 responses of stomata by indol-3ylacetic acid and abscisic acid. *J. Exp. Bot*. **33**: 360-365.

Stevens, R.K., Dzubay, T.G., Russwurm, G. and Rickel, D. (1978). Sampling and analysis of atmospheric sulfates and related species. Atmos. Environ. **12**: 55-68.

Tadolini, B. (1988). Polymine inhibition of lipoperoxidation. *Biochem. J*. **249**: 33-36.

Toet S., Ineson P., Peacock S., and Ashmore M. (2011). Elevated ozone reduces methane emissions from peatlandmesocosms. *Glob. Change Biol*. **17**: 288-96.

Torn, M.S., Degrange, J.E. and Shinn, J. H. (1987). The Effects of Acidic Deposition on Alberta Agriculture. Alberta Acid Deposition Research Programme, ADRP-B-08/87, The University of Calgary, Alberta, 160 pp.

US EPA, (1986). Effects of ozone and other photochemical oxidants on vegetation. In Air Quality Criteria for Ozone and other Photochemical Oxidants, **Vol.3**, pp 6-1-6-298; EPA-600/8-84-020cF, Environmental Criteria and Assessment Office, Research Triangle Park, NC.

Varshney, S.R.K. and Varshney, C.K. (1984). Effect of SO_2 on ascorbic acid in crop plants. *Environ. Pollut, (Ser. A)* **35**: 285-290.

Venkatram, A. (1986). Statistical long – range transport models. *Atmos. Environ*. **20:** 1317 – 1324.

Whitaker, B.D., Lee, E.H. and Rowland, R.A. (1990). EDU and ozone protection: Foliar glycerolipids and steryl lipids in snap bean exposed to O_3. *Physiol. Plant*. **80**: 286-293.

Whitby, K.T. (1978). The physical characteristics of sulfur aerosols. Atmos. Environ. **12**: 135-159.

Wilson, L.G., Bressan, R., Lecureux, L., Ream, J.E. & Filner, P. (1979). Destruction of bisulfite by young and mature leaves. *Plant Physiol*. **63:** Abstr. 59.

Wolfenden, J., Wookey, P.A., Lucas, P.W. and Mansfield, T.A. (1992). Action of pollutants individually and in combination. In Barker, J.R. and Tingey, D.T. (eds) *Air Pollution Effects o Biodiversity*. 72-92. Van Nostrand Reinhold, New York.

Wolfenden, L. and Mansfield, T. A. (1991). Physiological disturbances in plants caused by air pollutants. *Proc. Royal Soc. Edinb*. **97B**: 117 – 138.

Yunus, M., Ahmad, K.J. and Gale, R. (1979). Air pollutants and epidermal traits in *Ricinus communis* L. *Environ. Pollut*. **20**: 189-198.

Advances in Plant Physiology, Vol. 14
Ed. A. Hemantaranjan
Scientific Publishers (India), Jodhpur, 2013 pp. 331-347.
E-mail: **info@scientificpub.com** www.scientificpub.com

11

ROS MODULATED RESPONSES DURING PATHOGEN INFESTATION IN PLANTS

[1]Ashutosh Sharma, [2]P.K. Nagar and [1]Pratap Kumar Pati*

[1]Department of Biotechnology, Guru Nanak Dev University, Amritsar, India
[2]Former Head, Biotechnology Division,
Institute of Himalayan Bioresource Technology, Palampur, (H.P.), India
**Corresponding author e-mail: pkpati@yahoo.com*

INTRODUCTION

Reactive Oxygen Species (ROS) are a variety of molecules and free radicals derived from the molecular oxygen (Turrens, 2003). They are partially reduced or activated derivatives of oxygen like superoxide anion, H_2O_2 and hydroxyl radical (Mittler *et al.*, 2004) and are very reactive. In normal conditions, the Reactive Oxygen Species (ROS) are generated as the by-products of main aerobic metabolism, like electron transport systems in chloroplasts and mitochondria and photores-piration in peroxisomes (Dat *et al.*, 2000; Asada and Takahashi, 1987; Mittler, 2002) and are metabolized rapidly into other relatively non toxic metabolites by antioxidants (Noctor and Foyer, 1998) and antioxidant enzymes (Mittler, 2004).

Plants in their natural environment are subjected to various abiotic and biotic stresses. Both these abiotic and biotic stresses may disrupt this delicate balance between ROS production and scavenging, leading to rapid accumulation of ROS (Figure 1). During biotic stress, when the plant is challenged with the pathogen, a defense response is evinced with the rapid increased production of ROS (Wojtaszek, 1997). Such a response of enhanced production of ROS (oxidative burst) was initially noticed in potato tuber protoplasts after *Phytophthora infestans* elicitor treatment (Doke, 1983). Besides causing direct toxicity by oxidative damage in the plant cells (Dat *et al.*, 2000), ROS is also found to be associated with various signaling networks during plant immune responses (Alvarez *et al.* 1998). When the avirulent pathogens are successfully recognized by the plants, a biphasic ROS accumulation is observed. In the first phase a weak, transient, low-magnitude of,non-specific ROS accumulation and in the second phase a sustained and higher magnitude of ROS production has been observed. The biphasic ROS kinetics has been implicated as typical incompatible plant-pathogen interaction and corresponds to disease resistance (Lamb and Dixon, 1997; Gara *et al.*, 2003). Whereas a virulent pathogen induces only transient, low-amplitude first phase of oxidative burst (Torres *et al.*, 2006) (Figure 2).

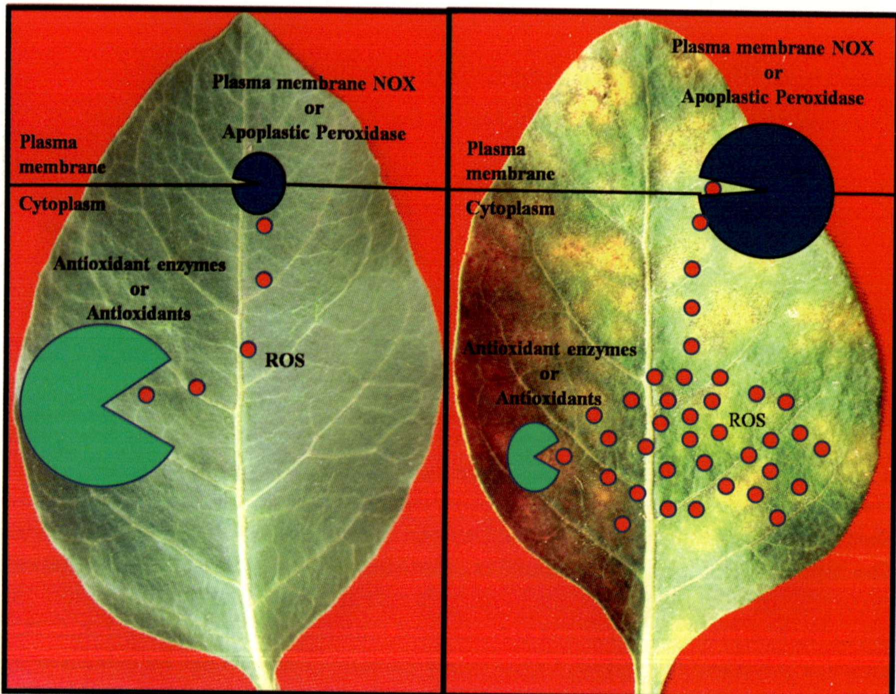

Fig. 1. Balance in ROS generating systems and ROS scavenging systems precisely regulates ROS accumulation and redox status of the cell and thereby controls HR in response to pathogen attack. Two different redox statuses of plant cells are described, In the first case enhanced activity of ROS generator (during pathogen infestation) and oxidative burst are described, which may lead to HR. In the later case healthy ROS scavenging mechanism and healthy steady state level of ROS of the plant cell is described.

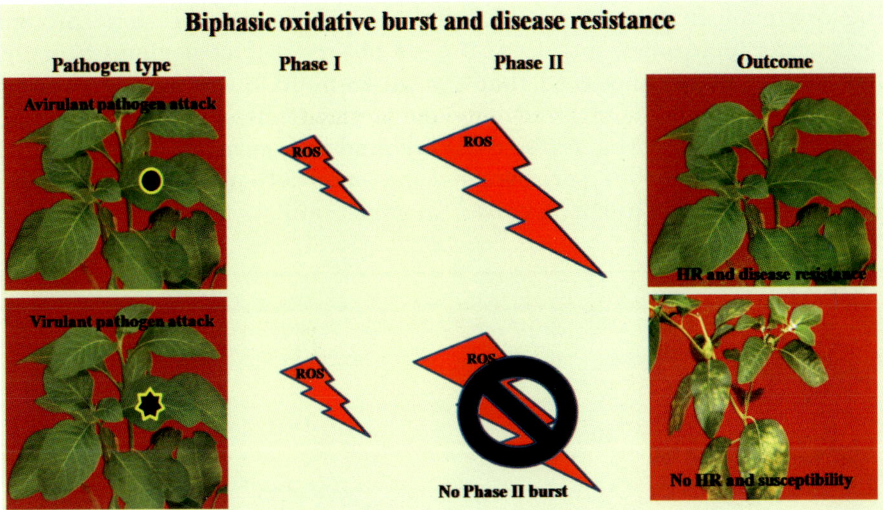

Fig. 2. Role of biphasic oxidative burst in disease resistance. Avirulent pathogen durin incompatible host-pathogen interaction causes a biphasic oxidative burst that leads to HR mediated defense, whereas virulent pathogen during a compatible reaction causes only first non-specific phase and second more intense specific phase of oxidative burst do not occur leading to susceptibility.

This clearly points towards a critical role of massive and prolonged production of ROS in HR response and plant defenses. A low-frequency, systemic mico-HR was observed in discrete cells of *Arabidopsis* away from site of inoculation of *Pseudomonas syringae* (Alvarez *et al.*, 1998), which suggests that ROS not only protect the plant at the penetration site, but it may also generate a systemic signal for resistance response. Further, the role of ROS during different plant-pathogen interactions also varies with the pathogen lifestyle. Pathogen induced cell death mediated by ROS, provides resistance to the biotrophic pathogen, whereas it increases susceptibility towards the necrotrophic pathogen (Barna *et al.*, 2012) (Figure 3). Hence elevation of antioxidant capacity of plants could enhance their tolerance to development of necrosis caused by nacrotropic pathogens.

The steady state levels of cellular ROS are tightly regulated by ROS generating and scavenging systems. The generation of various ROS as the by-products of aerobic metabolic metabolism in various organelles has been discussed earlier as water-water cycle, Ascorbate-glutathione cycle, glutathione peroxidase cycle, Haber-weiss reaction and Fenton reaction *etc.* (Arora *et al.*, 2004; Mittler, 2002; Wojtaszek, 1997). The production of ROS is site specific. Although, the main sites of ROS production has been reported in chloroplast (Asada, 2006) and Mitochondria (Rhoads *et al.*, 2006), yet the apoplastic generation of ROS in response to pathogens is also documented (Bindschedler *et al.*, 2006). Peroxisomes also generate ROS by glycolate oxidase and fatty acid β-oxidation (Corpas *et al.*, 2001).

The origin of ROS in a specific site as well as its diffusion through aquaporins (peroxoporins) has also been observed (Henzier and Steudle, 2000). It is further observed that the different ROS-scavanging systems which regulate the ROS levels are compartmentalized differently in the plant cells (Mittler *et al*, 2004). SOD is found in almost all compartments of the cell, catalase in peroxisomes and ascorbate-glutathione cycle in chloroplast, mitochon-dria, cytosol, apoplast and peroxisomes (Mittler, 2002).

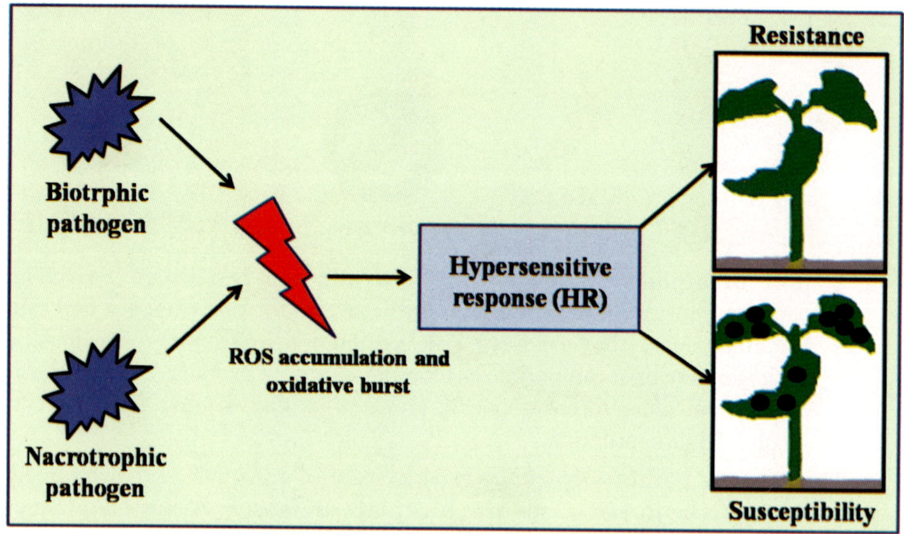

Fig. 3. Differential role of ROS in biotrophic vs nactrotrophic pathogens. Oxidative burst induced HR leads to cell death which aids the growth of necrotrophic pathogen

The higher levels of ROS generation in response to pathogens (Oxidative burst) has three major functions in the plant tissue (Figure 4). Firstly it may directly kill the pathogen (being toxic) (Chen and Schopfer, 1999), secondly it is involved in the establishment of physical barriers by oxidative cross linking of cell wall (Bradley *et al.*, 1992; Huckelhoven, 2007) and finally ROS alone or in combination with Nitrous oxide (NO) may act as signaling molecules regulating the defense related genes (Mou *et al.*, 2003; Kovtun *et al.*, 2000; Delledonne *et al.*, 1998; Delledonne *et al.*, 2001). ROS also lead to the accumulation of phytoprostanes (cyclic oxylipins) which may lead to defense gene activation and subsequent enhancement of phytoalexins (Thoma *et al.*, 2003). It has been estimated that a network of at least 152 genes is involved in the management of ROS in *Arabidopsis* (Mittler *et al.,* 2004). Transgenic approaches indicate the involvement of several genes are associated with ROS generating and scavenging systems, during plant-pathogen interactions. In addition to this, the role of ROS has also been implicated in the enhancement of the expression of some key transcription factors of the family WRKY, Myb, Zat, GRAS and RAV (Mittler at al., 2004). For the assessment of oxidative stress and to calculate the changes in concentrations and fluxes of various oxidants and antioxidants, computer aided simulations have also been employed (Polle, 2001). In addition to above, the present review also highlights the generation and scavenging of ROS during pathogen infestations and ROS modulated responses during plants defence.

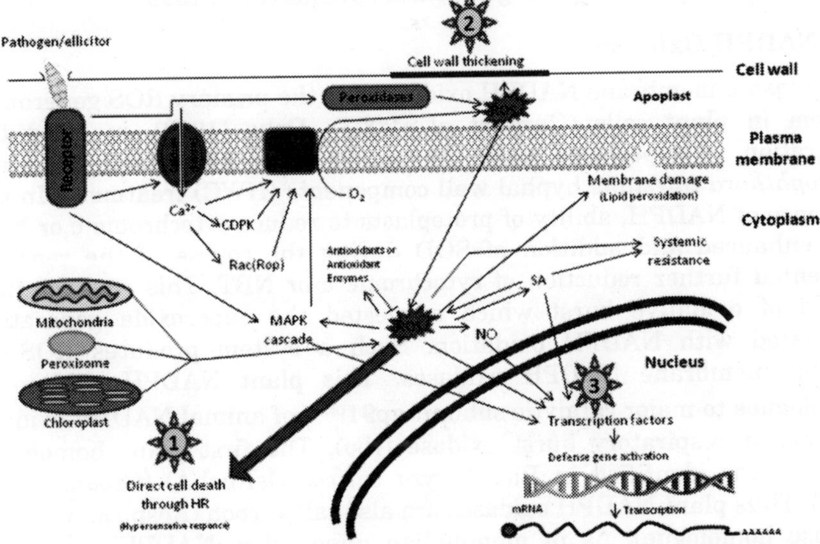

Fig. 4: ROS production, their scavenging and function in response to pathogen attack. Pathogen invasion/elicitation and MAMP recognition by the receptors triggers ROS accumulation through apoplastic Peroxidases or NADPH oxidases or both. The pathogen attack leads to Ca^{2+} influx that may activate plasma membrane NADPH oxidase directly by interacting with its EF hands or through CDPKs, or through Rac. The apoplastic ROS may come in cytoplasm through aquaporins and cause oxidative burst. The high ROS levels may alter antioxidants/antioxidant enzyme metabolism, which checks them. Besides this, Chloroplasts, Mitochondria and peroxisomes also generate ROS as their metabolic by-products, which also contributes to cellular steady state ROS pool. ROS may then trigger the three major responses (Marked by numbers), including HR, Cell wall thickening and defense gene activation. However ROS may also damage the cell membranes by lipid peroxidation. ROS itself or upon its interaction of SA may activate systemic resistance response (Refer text for detailed description of each step).

2. ROS GENERATION, SCAVANGING AND SENSING IN RESPONSE TO PATHOGEN

To keep the steady state levels of ROS under sub-lethal doses, excess of ROS is detoxified by antioxidants/antioxidant enzymes. An attempt of the infection or the elicitors of microbial origin causes the sudden increase in ROS production to facilitate HR (Hypersensetive Response) like responses. To understand the mechanism of HR and related response, the understanding of ROS generators and ROS scavengers is essential. During oxidative burst in response to pathogens, the major source of ROS are apoplastic peroxidases or NADPH oxidases, or both (Daudi *et al.*, 2012) and may differ in different plant-pathogen combinations (Bolwel *et al.*, 1998). However, oxalate oxidase has also been observed as a ROS producer in response to pathogen attack (Dumas *et al.*, 1995).

2.1. NADPH Oxidase

Plasma membrane NADPH oxidases are the primary ROS generating system in plant cells (Gara *et al.*, 2003). Doke (1983), reported the generation of superoxide anion by potato tuber protoplasts following *Phytophthora infestans* hyphal wall components (HWC) treatment. In the presence of NADPH, ability of protoplasts to reduce cytochrome c or NBT was enhanced and addition of SOD during the course of the reaction prevented further reduction of cytochrome c or NBT. This was the first record of oxidative burst which suggested the superoxide generation associated with NADPH oxidation. Such a system produces ROS by plasma membrane NADPH oxidases. This plant NADPH oxidase is homologues to major catalytic subunit gp91^{phox} of animal NADPH oxidase also called respiratory burst oxidase (rbo), The first plant homologue gp91^{phox} was identified in Rice (*Oryza sativa*, OsRbohA) (Groom *et al.*, 1996). Thus plant NADPH oxidases are also called rboh (Respiratory burst oxidase homologue). As in mammalian phagocytes, NADPH oxidase is involved in microbial killing (Lambeth, 2004), a similar function of this ROS generator was also anticipated in plants. Rbohs are membrane proteins with six trans-membrane domains and their N-terminal hydrophilic of rbohs contains EF-hands, calcium binding motifs suggesting calcium as a major regulatory component. This N-terminal extension is absent in phagocyte oxidase gp91^{phox}, but exists in other animal NADPH oxidases (NOX5 and Duox), and contains some phosphorylation sites (Suzuki *et al.*, 2011) and EF hand motifs (Wong *et al.*, 2007), responsible for its regulation. The major regulatory components of Rboh are Ca^{2+} (Ogasawara *et al.*, 2008), calcium-dependent protein kinases (CDPKs) (Kobayashi *et al.*, 2007) and Rop (Rac) (Wong *et al.*, 2007). Rac proteins may either up-regulate or down-regulate the ROS production. Ntrac5 is a down-regulator of NtrbohD generated ROS in elicited tobacco (Morel *et al.*, 2004), whereas Osrac1 is a positive regulator (Ono *et al.*, 2001). Plant Rac protein are however also shown to induce superoxide production in heterologous mammalian system (Hassanain *et al.*, 2000).

A. thaliana contains 10 homologues of this NADPH oxidase gene (AtrbohA-J) (Torres and Dangl, 2005). *A. thaliana* mutants lacking functional AtRbohD and AtRbohF, Infected with *Pseudomons syringae* and *Perconospora parasitica*, demonstrated that AtRbohD was mainly responsible for ROS production in response to avirulent pathogens and AtRbohF have a role in the regulation of HR (Torres *et al.*, 2002). Whereas the AtRbohD overexpression line, following infection by *Botrytis cinerea*, the negative regulation of cell death was observed by triggering salicylic acid (SA) accumulation (Torres *et al.*, 2005). The endogenous elicitors of plant defence response, Oligogalacturonides (OGs) elicit defense response against *B. cinerea* through AtRbohD generated oxidative burst, which is independent of SA, ethylene and jasmonic acid (Galletti *et al.*, 2009). AtRbohD is known to plays a dual role in *Arabidopsis-Alternaria* phytopathosystem. While it triggers cell-death in infected cells, simultaneously it inhibits cell-death in neighbouring healthy cells through the suppression of free salicylic acid and ethylene levels (Pogány *et al.*, 2009).

Besides HR, some homologues of NADPH oxidase are also involved in other physiological responses in response to pathogens. Involvement of NbRbohA and NbRbohB has been reported in elicitor-induced stomatal closure but not in elicitor-induced HR using virus-induced gene silencing (Zhang *et al.*, 2009). Some NADPH oxidase homologue genes show temporal differences in their activation upon pathogen recognition and show differential activation of other defense related genes. Studies in RNAi knockdown lines of OsRbohA and OsRbohE, showed that that OsRbohA is involved in rapid H_2O_2 generation at early stage and OsRbohE is involved in H_2O_2 production at later stages of plant immune response against an incompatible strain of *Acidovorax avenae*. OsRbohE is involved in the regulation of defense related genes El2 and Lox, whereas Cht-1 gene is regulated by both OsRbohA and OsRbohE during this plant-pathogen interaction (Yoshie *et al.*, 2005). NADPH oxidase mediated ROS generation may affect the penetration efficiency of fungal pathogens. (Trujillo *et al.*, 2006). NADPH oxidase is also involved in ROS production in elicited tobacco cells. The antisense NtRbohD plants are not capable of ROS production in response to cryptogein (a fungal elicitor) (Simon-Plas *et al.*, 2002). Induction of AtrbohD mediated ROS production was observed during free sphingobases increase in *Arabidopsis* in response to its pathogen *P. syringae*. The specific sphingobases have been suggested to play role in Rboh mediated cell death (Peer *et al.*, 2011).

2.2. Peroxidases

Besides NADPH oxidase (rboh), peroxidases are also known to produce ROS during plant-pathogen interaction. $CaPO_2$ (a *Capsicum annuum* peroxidase) silenced plants of pepper were highly susceptible to *Xanthomonas campestris* infection, in which H_2O_2 accumulation was compromised and hypersensitive cell death was noticed during avirulent pathogen infection. Whereas in $CaPO_2$ over-expressed lines in *A. thaliana*, showed enhanced disease resistance accompanied by cell death, PR gene induction and H_2O_2 accumulation. Moreover, this H_2O_2 generation was insensitive to DPI but sensitive to potassium cyanide (a peroxidase inhibitor) suggesting peroxidase dependent H_2O_2 generation (Choi *et al.*, 2007).

Oxidative burst in *A. thaliana* in response to *Fusarium oxysporum* cell wall preparations was found to be azide sensitive. The hetrologous anti-sense expression of *Phaseolus vulgaris* (French been) peroxidase (type

III peroxidase) cDNA in *A. thaliana* impaired the oxidative burst and reduced level of two class III peroxidase (PCa/PRX33 and PCb/PRX34) of *A. thaliana*. Blocking of oxidative burst by reducing peroxidase levels led to the enhanced susceptibility of these plants to both bacterial and fungal pathogens (Bindschedler *et al.*, 2006). Among 73 peroxidase genes in *Arabidopsis* (Bakalovic *et al.*, 2006), levels of only two peroxidases (PRX33 and PRX34) were reduced due to their high degree of homology with French been peroxidase. The anti-sense lines show reduced ROS production and callose deposition in response to Flg22 and Elf36 (MAMPs; Microbe associated molecular patterns). Down-regulation of MPMP-activated genes was observed in unchallenged leaves of these anti-sense lines suggesting the possible requirement of apoplastic ROS to preprime the basal resistance (Daudi *et al.*, 2012).

2.3. Germin-Like Oxalate Oxidase System

Erysiphe graminis f. sp. *hordei* induces a germin-like Oxalate oxidase in barley during infection and has been proposed to have a role in plant defense through H_2O_2 production (Dumas *et al.*, 1995).

Besides these three major ROS generators, alternative enzymes like Polyamine oxidase have also been recorded as the contributors of oxidative burst. Upon the incompatible interaction among Barley and powdery mildew, the polyamine levels are increased, which is also accompanied with the increased in the activity of their catabolic enzymes di-amine oxidase and polyamine oxidase (Cowley and Walters 2002). Virus-induced gene silencing of polyamine oxidase led to the decrease in H_2O_2 production and the disappearance of visible HR related cell death with an increase in bacterial growth during the interaction between Tobacco and its non-host pathogen (Yoda *et al*, 2009).

2.4. ROS Scavangers

Steady state levels of ROS are maintained in the plants by ROS scavenging system, which is mainly consists of antioxidants (Noctor and Foyer, 1998) and antioxidant enzymes (Mittler, 2004). These antioxidant enzymes work in co-ordination (Arora *et al.*, 2004) to keep cellular ROS levels under check. Major ROS-scavanging system in plants include Ascobate peroxidase (APX), Catalase (CAT) and Superoxide dismutase (SOD) (Mittler, 2002). Since SOD converts superoxide ions to H_2O_2, and APX and CAT scavenge H_2O_2, the balance between SOD and CAT/APX is critical for maintaining the steady state levels of these two types of ROS.

2.4.1. Enzymatic ROS scavangers

The post-transcriptional suppression of Ascobate peroxidase (APX) was observed in TMV induced HR in Tabacco (Mittler *et al.*, 1998). Statistically significant increase in APX activity was observed at the advance stages of leaf spot disease progression in *Withania somnifera* (Sharma *et al.*, 2011), that corresponds to increase in the ion leakage (a measure of cell death). Transgenic anti-sense Tobacco plants for APX and CAT (Catalase) were hyperresponsive against bacterial pathogen *Pseudomonas syringae* pv. *Phaseolicola*. PCD (Programmed cell death) was activated in these anti-sense lines in response to relatively low amount of pathogens, which was unable to trigger PCD in control plants (Mittler *et al.*, 1999). However the double antisense plants of tobacco lacking APX and CAT both are relatively less sensitive to oxidative stress than the single antisense plants lacking APX or CAT (Rizhsky *et al.*, 2002). Zat12, a zinc finger transcription factor is required for Apx1 expression and protects *Arabidopsis* against oxidative stress (Rizhsky *et al.*, 2004). Overexpression of CAT (*Cat2* gene from Maize) in Tobacco plants were more sensitive to the bacterial pathogen *P. syringae* (Polidoros *et al.*, 2001). The three independent CAT antisense lines of Tobacco, that showed the most severe reduction in CAT activity and developed chlorosis/necrosis symptoms on some of the lower leaves, which accumulated very high levels of PR-1 proteins. Further, their resistance towards Tobacco Mosaic Virus (TMV) was also enhanced. Necrosis and elevated salicylic acid (SA), appeared as a result from severe reduction of catalase levels (Takahashi *et al.*, 1997). In a model proposed by Klessig group, intercellular H_2O_2 act as a signal to induce PR responce (Chen *et al.*, 1993). Further they showed that SA produced upon pathogen infection binds to CAT and inactivates it (Chamnongpol *et al.*, 1996). *Botrytis cinerea* triggered changes in peroxisomal antioxidant system in tomato, that led to collapse of protective mechanisms at advanced stage of its infection, which appeared partly as the effect of leaf senescence promoted by the pathogen (Kuźniak and Skłodowska, 2005).

2.4.2. Non-enzymatic ROS scavangers

The antioxidants like Ascorbate (ASC) and a tri-peptide, Glutathione (GSH) protect the plant cells against ROS (Noctor and Foyer, 1998). Increase in GSH content by avirulrnt biotrophic pathogens and decrease in GSH content by virulent necrotrophic pathogens has been highlighted by Gara *et al.* (2003), aiding spread of nactotic area (HR) to promote the spread of necrotrophic pathogen.

2.5. ROS Sensing

Besides the ROS generators and scavangers, there are three main ROS sensing mechanisms in plants *viz.* unidentified receptor proteins, Redox sensitive transcription factors (e.g. NPR1) or heat shock proteins and direct inhibition of phosphatases (Barna *et al.*, 2012; Mittler *et al.*, 2004) in plants.

3. ROS AND PLANT DEFENCE

Generally the plant defense is elevated by ROS accumulation on incompatible plant pathogen interaction, which is coupled with HR and restricts the pathogen. However recent increase in our understandings towards the necrotrophic pathogen highlighted the opposite role of ROS in plant defense. Further, SA signaling is generally employed in resistance to biotrophs, whereas JA signaling against nacroteophs (Barna *et al*, 2012). StCDPK5 (a potato calcium dependent protein kinase), that phosphorylates the N terminus of RBOH, the driven by a pathogen inducible promoter of potato that activates NADPH oxidase showed a high degree of resistance to late blight pathogen *P. infestans* whereas extreme susceptibility to early blight pathogen *Alternaria solani* (Yoshioka *et al*, 2009). A constitutively active form of StCDPK5, called StCDPK5VK, was targeted to plasma membrane and its conditional expression activated StRBOH A-D. Transgenic potato line expressing StCDPK5VK under pathogen inducible promorter, showed resistance against near-obligate hemibiotrophic pathogen, *P. infestance*, whereas showed succesptibility to necrotrophic pathogen *A. solani* (Kobayashi *et al.*, 2012). Cryptogein (*Phytophthora* elicitor) is also known to stimulate clathrin-mediated endocytosis in Tobacco Bright Yellow-2 (BY-2) cells, that requires the ROS production by NtrbohD (Leborgne-Castel *et al*. 2008).

Besides the various role of ROS in response to pathogen attack, ROS may interact with other grown regulators (SA and Ethylene) and signaling molecules (NO and Ca^{2+}). SA levels may shoot up in cells near infection cites during HR in response to TMV (Enyedi *et al.*, 1992). The Avirulent pathogen *P. syringae* Induces of H_2O_2 accumulation and hypersensitive cell death, which is blocked by the phenylpropanoid synthesis inhibitor α-aminooxy-β-phenylpropionic acid in *Glycine max*. Since these responses were rescued by exogenous SA, and existence of antagonist-dependent gain control at physiological levels of SA, a rapid fine control signal amplification making an important contribution to SA function in the induction of disease resistance mechanisms was proposed (Shirasu *et al.*, 1997). ROS may act along with SA in a synergestic amplification loop to initiate HR and establish systemic defence (Draper, 1997). Accumulation of SA may lead to down-regulation of ROS scavenging system, and contribute to oxidative burst upon pathogen recognition (Klessig *et al.*, 2000).

Wi *et al*. (2012) noticed a biphasic rise in ethylene and ROS in tobacco plants in response to *P. parasitica*. This biphasic production of ROS was inhibited in antisence ACC synthase (an ethylene metabolism gene) and also in transgenic plant impaired in ROS generation. A correlation in ROS and ethylene was hence observed with the pathogen proliferation in susceptible plants. Ethylene can induce programmed cell death (PCD) in tomato suspension cell cultures through ROS (de Jong *et al.*, 2002). Function of ethylene receptor, ETR1 as a ROS sensor, mediating stromatal closure in response to H_2O_2 has already known (Desikan *et al.*, 2005).

HR related cell death is activated upon interaction of NO with H_2O_2 (Delledonne *et al.*, 2001). Increase in cytosolic Ca^{2+} concentration is the most rapid response upon pathogen infection (Blume *et al.*, 2000), which was placed upstream to NADPH oxidase mediated oxidative bust by Grant *et al.*, (2000) and Rboh activation by Ca^{2+} was reported (Sagi and Fluhr, 2001). The calcium binding EF-hands motifs are present on hydrophilic N-terminal extension of Rbohs, which may act as a bridge between Rboh and Rop (Wong *et al.*, 2007). A serine/thrionine protein kinase OXI1 is an essential component of oxidative burst mediated signaling and activation of MAP kinases by Ca^{2+} (Rentel *et al.*, 2004). The MEKK1-MKK1/MKK2-MPK4 cascade negatively regulates SUMM2 (a nucleotide-binding leucine-rich repeat, NB-LRR protein) -mediated immunity (Zhang *et al.*, 2012).

However, there is a different perspective of fungal pathogen generated ROS during plant pathogen interaction. Studies on Nox1 and Nox2 mutants in *Magnaporthe grisea*, the rice blast fungi, revealed that the generation of ROS by NADPH oxidase of fungi is required for appressoria development and in the initiation of rice blast disease (Egan *et al.*, 2007). Whereas, the activation of ROS scavenging mechanisms by *Ustilago maydis* Yap1 gene, are essential in overcoming the plant defense mechanisms and to allow the infection (Molina and Kahmann, 2007).

4. FUTURE PERSPECTIVE

Hypersensitive reaction (HR) mediated cell death in plants in response to plant pathogens is a well known defense mechanism in plants, however the signaling events leading to HR have not been fully understood. Further, there is functional overlap between various ROS producers (e.g. various Atrbohs). The precise spatial and temporal regulation of such genes in response to pathogen attack is to be determined. However, the studies using pharmacological inhibitors (DPI and Sodium azided etc.) have suggested the role of NADPH oxidase and peroxidases in pathogen induced ROS generation, there precise role in ROS generation in various phyto-pathosystems and their relative contributions in oxidative burst are to be understood. Further, the differential kind of ROS defense mechanisms in biotrophic and necrotrophic pathogens, which are to be worked out.

Since ROS metabolism regulates various defense responses in plants, the generation of transgenic lines with altered ROS metabolism, and the study of other interacting partners of ROS metabolism is required to design the viable strategies for better disease resistance. Translating this knowledge in crop improving programs will help us to engineer the future crops with better disease resistance.

ACKNOWLEDGEMENT

Financial assistance for this work from Council of Scientific and Industrial Research, Government of India is duly acknowledged.

REFERENCES

Alvarez ME, Pennell RI, Meijer PJ, Ishikawa A, Dixon RA, Lamb C (1998) Reactive oxygen intermediates mediate a systemic signal network in the establishment of plant immunity. *Cell* **92**: 773–784

Arora A., Sairam R.K., Srivastava G.C. (2004) Oxidative stress and antioxidative system in plants. *Current Science* **82**: 1227-1238

Asada K (2006). Production and scavenging of reactive oxygen species in chloroplasts and their functions. *Plant Physiol.* **141**: 391-396

Asada, K. and Takahashi, M. (1987) Production and scavenging of active oxygen in photosynthesis. In: *Photoinhibition (Topics in Photosynthesis)* (**Vol. 9**) (eds. Kyle, D.J. *et al.*,), pp. 227–287, Elsevier

Bakalovic N, Passardi F, Ioannidis V, Cosio C, Penel C, Falquet L, Dunand C (2006) PeroxiBase: a class III plant peroxidase database. *Phytochemistry*. **67**: 534-539

Barna B., Fodor J., Harrach B.D., Pogány M. and Király Z (2012). The Janus face of reactive oxygen species in resistance and susceptibility of plants to necrotrophic and biotrophic pathogens. *Plant Physiology and Biochemistry*. doi:10.1016/j.plaphy.2012.01.014

Bindschedler LV, Dewdney J, Blee KA, Stone JM, Asai T, Plotnikov J, Denoux C, Hayes T, Gerrish C, Davies DR, Ausubel FM, Bolwell GP (2006) Peroxidase-dependent apoplastic oxidative burst in Arabidopsis required for pathogen resistance. *Plant J.* **47**: 851–863

Blume B, Nürnberger T, Nass N, Scheel D (2000) Receptor-mediated increase in cytoplasmic free calcium required for activation of pathogen defense in parsley. *Plant Cell.* **12**: 1425–1440

Bolwell GP, Davies DR, Gerrish C, Auh C-K, Murphy TM (1998) Comparative biochemistry of the oxidative burst produced by rose and French bean cells reveals two distinct mechanisms *Plant Physiol.* **116**: 1379–1385

Bradley D, Kjellbom P, Lamb C (1992) Elicitor- and wound-induced oxidative cross-linking of a proline-rich plant cell wall protein: a novel, rapid defense response. *Cell.* **70**: 21–30

Chamnongpol, S., Willekens, H., Langebartels, C., Van Montagu, M., Inzé, D., and Van Camp, W. (1996). Transgenic tobacco with a reduced catalase activity develops necrotic lesions and induces pathogenesis related expression under high light. *Plant J.* **10**: 491– 503.

Chen S-X, Schopfer P (1999) Hydroxyl-radical production in physiological reactions. *Eur J Biochem.* **260**: 726–735

Chen, Z., Silva, H., and Klessig, D.F. (1993). Active oxygen species in the induction of plant systemic acquired resistance by salicylic acid. *Science.* **262**: 1883–1886

Choi HW, Kim YJ, Lee SC, Hong JK, Hwang BK (2007) Hydrogen peroxide generation by the pepper extracellular peroxidase $CaPO_2$ activates local and systemic cell death and defense response to bacterial pathogens. *Plant Physiol.* **145:** 890–904

Corpas, F.J. Barroso JB, del Río LA (2001) Peroxisomes as a source of reactive oxygen species and nitric oxide signal molecules in plant cells. *Trends Plant Sci.* **6**: 145–150

Cowley T, Walters DR (2002) Polyamine metabolism in barley reacting hypersensitively to the powdery mildew fungus Blumeria graminis f. sp. hordei. *Plant Cell Environ.* **25**: 461–468.

Dat, J. Vandenbeele S, Vranova E, Van Montagu M, Inzé D, Van Breusegem F (2000) Dual action of the active oxygen species during plant stress responses. *Cell Mol Life Sci.* **57**: 779–795

Daudi A, Cheng Z, O'Brien JA, Mammarella N, Khan S, Ausubel FM, Bolwell GP (2012). The apoplastic oxidative burst peroxidase in Arabidopsis is a major component of pattern-triggered immunity. *Plant Cell.* **24**: 275-87

de Jong AJ, Yakimova ET, Kapchina VM, Woltering EJ (2002) A critical role for ethylene in hydrogen peroxide release during programmed cell death in tomato suspension cells. *Planta.* **214**: 537–545

Delledonne M., Xia Y., Dixon R.A., Lamb C. (1998). Nitric oxide functions as a signal in plant disease resistance. *Nature*. **394**: 585-588.

Delledonne M., Zeier J., Marocco A., Lamb C. (2001). Signal interactions between nitric oxide and reactive oxygen intermediates in the plant hypersensitive disease resistance response. *Proc Natl Acad Sci USA*. **98**: 13454-13459.

Desikan R, Hancock JT, Bright J, Harrison J, Weir I, Hooley R, Neill SJ (2005) A role for ETR1 in hydrogen peroxide signaling in stomatal guard cells. *Plant Physiol*. **137**: 831–834

Doke N (1983) Involvement of superoxide anion generation in hypersensitive response of potato tuber tissues to infection with an incompatible race of *Phytophthora infestans* and to the hyphal cell wall components. *Physiol Plant Pathol.* **23**: 345–357

Draper J (1997) Salicylate, superoxide synthesis and cell suicide in plant defense. *Trends Plant Sci*. **2**: 162–165

Dumas B, Freyssinet G, Pallett KE (1995). Tissue-Specific Expression of Germin-Like Oxalate Oxidase during Development and Fungal Infection of Barley Seedlings. *Plant Physiol.* 107:1091-1096

Egan MJ, Wang ZY, Jones MA, Smirnoff N, Talbot NJ (2007). Generation of reactive oxygen species by fungal NADPH oxidases is required for Rice blast disease. *Proc Natl Acad Sci USA*. **104**: 11772-11777.

Enyedi AJ, Yalpani N, Silverman P, Raskin I (1992) Localization, conjugation and function of salicylic acid in tobacco during the hypersensitive reaction to tobacco mosaic virus. *Proc Natl Acad Sci USA.* **89**: 2480–2484

Galletti R, De Lorenzo G, Ferrari S (2009). Host-derived signals activate plant innate immunity. *Plant Signal Behav.* **4**: 33-34.

Gara L D, Pinto M C d, Tommasi F (2003) The antioxidant systems vis-à-vis reactive oxygen species during plant–pathogen interaction. *Plant Physiology and Biochemistry* **41**: 863-870

Grant M, Brown I,Adams S, KnightM, AinslieA, Mansfield J (2000) The RPM1 plant disease resistance gene facilitates a rapid and sustained increase in cytosolic calcium that is necessary for the oxidative burst and hypersensitive cell death. *Plant J.* **23**: 441–450

Groom QJ, Torres MA, Fordham-Skelton AP, Hammond-Kosack KE, Robinson NJ, Jones JD. (1996) RbohA, a Rice homologue of the mammalian gp91phox respiratory burst oxidase gene. *Plant J.* **10**: 515-522.

Hassanain HH, Sharma YK, Moldovan L, Khramtsov V, Berliner LJ, Duvick JP, Goldschmidt-Clermont PJ (2000). Plant rac proteins induce superoxide production in mammalian cells. *Biochem Biophys Res Comm.* **272**: 783-788

Henzler T, Steudle E (2000). Transport and metabolic degradation of hydrogen peroxide in *Chara corallina*: model calculations and measurements with the pressure probe suggest transport of H(2)O(2) across water channels. *J Exp Bot.* **51**: 2053-2066

Huckelhoven R (2007) Cell wall-associated mechanisms of disease resistance and susceptibility. *Annu Rev Phytopathol.* **45**: 101–127

Klessig DF, Durner J, Noad R, Navarre DA, Wendehenne D, Kumar D, Zhou JM, Shah J, Zhang S, Kachroo P, Trifa Y, Pontier D, Lam E, Silva H (2000) Nitric oxide and salicylic acid signaling in plant defense. *Proc Natl Acad Sci USA* **97**: 8849–8855

Kobayashi M, Ohura I, Kawakita K, Yokota N, Fujiwara M, Shimamoto K, Doke N, Yoshioka H. (2007) Calcium-dependent protein kinases regulate the production of reactive oxygen species by potato NADPH oxidase. *Plant Cell.* **19**: 1065-1080.

Kobayashi M, Yoshioka M, Asai S, Nomura H, Kuchimura K, Mori H, Doke N, Yoshioka H. (2012). StCDPK5 confers resistance to late blight pathogen but increases susceptibility to early blight pathogen in potato via reactive oxygen species burst. *New Phytol.* doi: 10.1111/j.1469-8137.2012.04226.x.

Kovtun Y, Chiu W-L, Tena G, Sheen J (2000) Functional analysis of oxidative stress-activated mitogen-activated protein kinase cascade in plants. *Proc Natl Acad Sci USA.* **97**: 2940–2945

Kuźniak E and Skłodowska M (2005) Fungal pathogen-induced changes in the antioxidant systems of leaf peroxisomes from infected tomato plants. *Planta* **222**: 192–200

Lamb C, Dixon RA (1997) The oxidative burst in plant disease resistance. *Annu Rev Plant Physiol Plant Mol Biol.* **48**: 251–275

Lambeth JD (2004) NOX enzymes and the biology of reactive oxygen. *Nature Rev Immunol.* **4**: 181–189

Leborgne-Castel N, Lherminier J, Der C, Fromentin J, Houot V, Simon-Plas F (2008) The plant defense elicitor cryptogein stimulates clathrin-mediated endocytosis correlated with reactive oxygen species Production in Bright Yellow-2 tobacco cells. *Plant Physiol.* **146**: 1255–1266

Mittler R, Herr EH, Orvar BL, van Camp W, Wilikens H, Inzé D, Ellis BE (1999) Transgenic tobacco plants with reduced capability to detoxify reactive oxygen intermediates are hyperresponsive to pathogen infection. *Proc Natl Acad Sci USA* **96**: 14165–14170

Mittler R, Vanderauwera S, Gollery M, Van Breusegem F (2004) Reactive oxygen gene network of plants. *Trends Plant Sci.* **9**: 490–498

Mittler, R. (2002) Oxidative stress, antioxidants and stress tolerance. *Trends Plant Sci.* **7**: 405–410

Mittler, R., Feng, X. and Cohen, M. (1998) Post-transcriptional suppression of cytosolic ascorbate eroxidase expression during pathogen-induced programmed cell death in tobacco. *Plant Cell* **10**: 461-474

Molina L, Kahmann R (2007) An *Ustilago maydis* gene involved in H2O2 detoxification is required for virulence. *Plant Cell* **19**: 2293-2309

Morel J, Fromentin J, Blein JP, Simon-Plas F, Elmayan T. (2004) Rac regulation of NtRbohD, the oxidase responsible for the oxidative burst in elicited tobacco cell. *Plant J.* **37**: 282-293

Mou Z, Fan W, Dong X (2003) Inducers of plant systemic acquired resistance regulate NPR1 function through redox changes. *Cell* **113**: 935–944

Noctor, G. and Foyer, C. (1998) Ascorbate and glutathione: keeping active oxygen under control. *Annu. Rev. Plant Physiol. Plant Mol. Biol.* **49**: 249–279

Ogasawara Y, Kaya H, Hiraoka G, Yumoto F, Kimura S, Kadota Y, Hishinuma H, Senzaki E, Yamagoe S, Nagata K, Nara M, Suzuki K, Tanokura M, Kuchitsu K (2008) Synergistic activation of the Arabidopsis NADPH oxidase AtRbohD by Ca2+ and phosphorylation. *J Biol Chem.* **283**: 8885-8892.

Ono E, Wong HL, Kawasaki T, Hasegawa M, Kodama O, Shimamoto K (2001) Essential role of the small GTPase Rac in disease resistance of rice. *Proc Natl Acad Sci USA* **98**: 759–764

Peer M, Bach M, Mueller MJ, Waller F (2011). Free sphingobases induce RBOHD-dependent reactive oxygen species production in Arabidopsis leaves. *FEBS Lett.* **585**: 3006-10

Pogány M, von Rad U, Grün S, Dongó A, Pintye A, Simoneau P, Bahnweg G, Kiss L, Barna B, Durner J (2009). Dual roles of reactive oxygen species and NADPH oxidase RBOHD in an *Arabidopsis-Alternaria* pathosystem. *Plant Physiol.* **151**: 1459-1475.

Polle A. (2001) Dissecting the superoxide dismutase–ascorbate peroxidase–glutathione pathway in chloroplasts by metabolic modeling. Computer simulations as a step towards flux analysis *Plant Physiol.* **126**: 445–462

Rentel MC, Lecourieux D, Ouaked F, Usher SL, Petersen L, Okamoto H, Knight H, Peck SC, Grierson CS, Hirt H, Knight MR (2004). OXI1 kinase is necessary for oxidative burst-mediated signalling in Arabidopsis. *Nature.* **427**: 858-861

Rhoads DM, Umbach AL, Subbaiah CC, Siedow JN (2006). Mitochondrial reactive oxygen species. Contribution to oxidative stress and interorganellar signaling. *Plant Physiol.* **141**: 357-366.

Rizhsky L., Davletova S., Liang H. and Mittler R. (2004) The zinc-finger protein Zat12 is required for cytosoli ascorbate peroxidase 1 expression during oxidative stress in Arabidopsis. *J. Biol. Chem.* **279**: 11736-11743

Rizhsky, L. Hallak-Herr, E., Van Breusegem, F., Rachmilevitch, S., Rodermel, S., Inzé, D., Mittler, R. (2002) Double antisense plants with suppressed expression of ascorbate peroxidase and catalase are less sensitive to oxidative stress than single antisense plants with suppressed expression of ascorbate peroxidase or catalase. *Plant J.* **32**: 329-342

Sagi M, Fluhr R (2001) Superoxide production by plant homologues of the gp91(phox) NADPH oxidase. Modulation of activity by calcium and by tobacco mosaic virus infection. *Plant Physiol.* **126**: 1281–1290

Sharma A, Sharma I, Pati PK (2011) Post-infectional changes associated with the progression of leaf spot disease in *Withania somnifera* (L.) Dunal. *J Plant Pathol.* **93**: 397-405

Shirasu K, Nakajima H, Rajashekar K, Dixon RA, Lamb C (1997) Salicylic acid potentiates an agonist-dependent gain control that amplifies pathogen signal in the activation of defense mechanisms. *Plant Cell* **9**: 261–270

Simon-Plas F, Elmayan T, Blein JP (2002). The plasma membrane oxidase NtRbohD is responsible for AOS production in elicited tobacco cells. *Plant J.* **31**: 137-147.

Suzuki N, Miller G, Morales J, Shulaev V, Torres MA, Mittler R (2011). Respiratory burst oxidases: the engines of ROS signaling. *Curr Opin Plant Biol.* **14**: 691-699

Takahashi H, Chen Z, Du H, Liu Y, Klessig DF (1997). Development of necrosis and activation of disease resistance in transgenic tobacco plants with severely reduced catalase levels. *Plant J.* **11**: 993-1005

Thoma I, Loeffler C, Sinha AK, Gupta M, Krischke M, Steffan B, Roitsch T, Mueller MJ (2003) Cyclopentenone isoprostanes induced by reactive oxygen species trigger defense gene activation and phytoalexin accumulation in plants. *Plant J* **34**: 363–375

Torres MA, Jones JDG., Dangl JL (2006). Reactive Oxygen Species Signaling in Response to Pathogens. *Plant Physiology* **141**: 373–378

Torres MA, Dangl JL (2005) Functions of the respiratory burst oxidase in biotic interactions, abiotic stress and development. *Curr Opin Plant Biol* **8**: 397-403.

Torres MA, Dangl JL, Jones JD (2002). Arabidopsis gp91phox homologues AtRbohD and AtRbohF are required for accumulation of reactive oxygen intermediates in the plant defence response. *Proc Natl Acad Sci USA* **99**: 517-522.

Torres MA, Jones JD, Dangl JL (2005). Pathogen-induced, NADPH oxidase-derived reactive oxygen intermediates suppress spread of cell death in *Arabidopsis thaliana*. *Nat Genet.* **37**:1130-1134

Trujillo M, Altschmied L, Schweizer P, Kogel KH, Hückelhoven R (2006). Respiratory burst oxidase homologue A of barley contributes to penetration by the powdery mildew fungus *Blumeria graminis* f. sp. hordei. *J Exp Bot* **57**: 3781-3791.

Turrens JF (2003). Mitochondrial formation of reactive oxygen species. *J Physiol.* **552**: 335-44

Wi SJ, Ji NR, Park KY (2012). Synergistic biosynthesis of biphasic ethylene and reactive oxygen species in response to hemibiotrophic *Phytophthora parasitica* in tobacco plants. *Plant Physiol.* **159**: 251-65

Wojtaszek P (1997) Oxidative burst: an early plant response to pathogen. *Biochem J* **322**: 681–692

Wong HL, Pinontoan R, Hayashi K, Tabata R, Yaeno T, Hasegawa K, Kojima C, Yoshioka H, Iba K, Kawasaki T, Shimamoto K (2007). Regulation of rice NADPH oxidase by binding of Rac GTPase to its N-terminal extension. *Plant Cell*. **19**: 4022-4034

Yoda H, Fujimura K, Takahashi H, Munemura I, Uchimiya H, Sano H (2009) Polyamines as a common source of hydrogen peroxide in host- and nonhost hypersensitive response during pathogen infection. *Plant Mol Biol* **70**: 103–112

Yoshie Y, Goto K, Takai R, Iwano M, Takayama S, Isogai S, Che FS (2005). Function of the Rice gp91phox homologs OsrbohA and OsrbohE genes in ROS-dependent plant immune responses. *Plant Biotechnol* **22**: 127-135.

Yoshioka H, Asai S, Yoshioka M, Kobayashi M (2009). Molecular mechanisms of generation for nitric oxide and reactive oxygen species, and role of the radical burst in plant immunity. *Mol Cells*. **28**: 321-329

Zhang H, Fang Q, Zhang Z, Wang Y, Zheng X (2009). The role of respiratory burst oxidase homologues in elicitor-induced stomatal closure and hypersensitive response in *Nicotiana benthamiana*. *J Exp Bot* **60**: 3109-3122.

Zhang Z, Wu Y, Gao M, Zhang J, Kong Q, Liu Y, Ba H, Zhou J, Zhang Y (2012). Disruption of PAMP-induced MAP kinase cascade by a Pseudomonas syringae effector activates plant immunity mediated by the NB-LRR protein SUMM2. *Cell Host Microbe*. **11**: 253-63.

12

CALCIUM SIGNALING DURING ABIOTIC STRESS IN PLANTS: ROLES OF CALCIUM DEPENDENT PROTEIN KINASES

Ajay Kumar Dixit and Chelliah Jayabaskaran*

Department of Biochemistry, Indian Institute of Science, Bangalore, Karnataka PIN-560012 India

**Corresponding author : Department of Biochemistry, Indian Institute of Science, Bangalore, Karnataka -560012 India*

E-mail: cjb@biochem.iisc.ernet.in

INTRODUCTION

Plants being immobile are rapidly exposed to various biotic and abiotic stress conditions like drought, salinity, cold, anoxia (caused by water logging), UV radiations, pathogen attack and mechanical injury by strong wind and gazing. Most of the losses in agriculture productivity are caused due to these drastic environmental conditions throughout the world. Thus understanding plant signaling and response to abiotic and biotic stresses would have far reaching implications in agriculture, biotechnology and forestry. How plants perceive, transduce and respond to different stress conditions have fascinated many plant biologists and remain an area of intense investigation in recent years.

In order to survive under these adverse conditions, sessile plants have developed sophisticated signaling network, however not much is known about plant signaling network and most of our understanding about plant signaling is derived from animal signaling systems. Most of the living system responds to different kind of stresses by altering the physiology/ gene expression, through systematic actions of myriad signaling components like receptors, secondary messengers and various transcription factors etc. In plants, perception of most of the biotic and abiotic signals via receptors result in production of non –protein molecules called secondary messengers. Most common secondary messengers are cyclic nucleotides, lipid molecules, reactive oxygen species, H_2O_2, Nitric oxide and Ca^{2+} (Demidchik *et al.*, 2002b; Guo *et al.*, 2001; Levine *et al.*, 1994; Sanders *et al.*, 2002; Scrase-Field and Knight, 2003; Talke *et al.*, 2003; Trewavas and Malho, 1997). These secondary messengers generated during particular stress which in turn regulate various cellular processes by regulating various downstream targets.

In this current chapter we will focus on calcium and how it regulates abiotic signaling pathways through Calcium dependent protein kinases.

1.2 Calcium as a Secondary Messenger

Calcium ions play major roles as a second messenger in coupling physiological responses to external and developmental signals (Reddy and Reddy, 2004). Changes in cytosolic free calcium ion concentration are well documented during transduction of most of abiotic stimuli including light, low and high temperature, touch, hyper osmotic and oxidative stresses as well as in the case of biotic stimuli like fungal elicitors and nodulation (Nod) factors (Rudd and Franklin-Tong, 2001; Sanders *et al.*, 2002).

1.3 Why Calcium?

Out of many inorganic ions why nature has chosen only calcium as a secondary messenger? Though there are no definite answers to this question, but we will discuss about the possibility about what makes calcium better suited for this purpose.

In the cell, Intracellular calcium levels are primarily maintained by various channels, pumps and transporters. In plants, calcium ions are basically stored in endoplasmic reticulum, vacuoles, plastids, mitochondria. The sequestering of calcium in these organelles could have originated for the purpose of defense against "calcium holocaust" (Campbell, 1983).

Calcium was much abundant on earth's crust during the origin of life and high concentrations of calcium was lethal as it could causes cell death by damaging organelles, making proteins and nucleic acids to aggregate and leads to precipitation of the phosphates. In order to circumvent this problem, cells have to develop various pumps and transporters to pump out the Ca^{2+} or store the calcium in specific organelles.

Though "Calcium holocaust" hypothesis argued the possible mechanism of evolution of calcium signaling , but it failed to explain why other ions like Mg^{2+}, Zn^{2+}, Na^{+}, and K^{+}, though present in abundance are not preferred.

The answer to the above question lies in the chemical properties of these ions. In order to act as signaling molecule the ions/ molecule must able to interact with various biomolecules like protein, lipids and carbohydrates. Calcium has high affinity for carboxylate oxygen which is present in acidic amino acids like aspartic acid, and glutamic acid and where as Zn^{2+} and Mg^{2+} has high affinity for nitrogen ligands which is mainly present in amino acids like asparagine, histidine and glutamine. Aspartic acid and glutamic acid are more abundant amino acid over asparagine, histidine and glutamine.

Other parameter which is requisite for a good secondary messenger is its binding kinetics. To measure these rates of water exchange of metal aquo complexes can be utilized. Such a rate (in units of s^{-1} at 25°C) has been found to be 10^8 for Ca^{2+}, 2×10^7 for Zn^{2+} and Mn^{2+}, 7×10^5 for Mg^{2+} (Ochiai, 1991). As Ca^{2+} binds to and dissociates much faster than any other metal ions, it can regulates biomolecules much more efficiently than any other metal ions. Apart from this, an irregular geometry and ability to display high coordination numbers (6-9) also puts calcium at an advantage over other ions like Mg^{2+}, Mn^{2+}, and Zn^{2+}.

Taking all the above parameters, it could be said that Ca^{2+} is better suited as a secondary messenger than any other metal ions.

1.4 Calcium Signaling during Abiotic Stresses

Change in cytosolic calcium level have been well documented in abiotic stresses like salinity, drought, cold, light , etc. (Table 1) indicates direct involvement of calcium during these stresses.

Table 1. Abiotic stress induced Ca^{2+}- mediated cellular processes in plants

Abiotic stress	Response	Reference
NaCl	Gene expression, osmolyte synthesis, ion transport, K^+uptake, stomatal regulation, salt tolerance	Knight *et al.* (1997), Liu and Zhu (1998), Lynch *et al.* (1989)
Cold	COR gene expression, proline synthesis, changes in membrane lipid profile and cold acclimation	Campbell *et al.* (1996), Knight *et al.* (1999), Knight *et al.* (1991), Thomashow (1998)
Anoxia	Anaerobic adaptation	Subbaiah *et al.* (1994, 1998)
Drought and hypoosmotic	Gene expression, synthesis of osmoprotectants, stomatal regulation and osmotolerance	Knight *et al.* (1997), Takahashi *et al.* (1997)
Oxidative stress	Production of ROSs, HR	Cessana & Low (2001), Levine *et al.* (1994),
Heat shock	Thermotolerance	Gong *et al.* (1998), Larkindale & Knight (2002)

The information encrypted in the stimulus induces change in cytosolic calcium levels. The stimulus-specific change in free cytosolic Ca^{2+} is called calcium signature. Any given Ca^{2+}-mediated cellular process begins with the generation of such signal-specific calcium signature in the cytoplasm (or nucleus) by a synchronized activity of channels, pumps and transporters. Change in cytosolic free calcium ion levels are sensed by a specific set of proteins named calcium sensors. There are two groups of calcium sensors. The first group of sensors called sensor relay- are without any responder domains. They bind to Ca^{2+} and undergo conformational changes and in turn regulate the activities/functions of a variety of target proteins (e.g., calmodulin and calcineurin B-like proteins). Sensors of the second group are called "responders" as they have kinase domains (eg. CDPKs and CaMKs) through which they relay the message to their downstream targets.

Most of the calcium binding proteins have highly conserved, 29 amino acids helix-loop-helix structure which resembles that of a hand having the index finger and thumb (the two helices) extended at roughly right angles from the rest of the fingers, which are curled towards the palm (the loop). The loop in each EF-hand motif consists of 12 residues with a pattern of conserved residues that coordinate the calcium ion (Strynadka and James, 1989; Goodman *et al.,* 1990). The EF-hand modules bind Ca^{2+} ions in the loops and undergo conformational changes that result in the exposure of hydrophobic pockets, which in turn facilitates interactions of the protein with a variety of protein partners. The presence of pairs of EF-hand motifs in proteins increases their stability and affinity for calcium (Reddy and Reddy, 2004).

2. PLANT KINASES AS CALCIUM DECODING MACHINERY

Plant's CDPK/SnRK protein kinases are a large and differentiated group of calcium sensors. After analyzing *Arabidopsis* genome sequence, a CDPKs/SnRK superfamily of kinases was formed which includes seven different types of serine/threonine kinases. All members of the CDPK/SnRK family have a similar length kinase domain and sequence and also have similar structural organization with kinase domain followed by a junction domain and C- terminal regulatory domain. The distinguishing feature in all the nine subfamilies is the differences in the primary structure of the terminal domains and the mode of regulation of the kinase domain (Figure 1) (Hrabak *et al.*, 2003).

CDPK/SnRK family of protein kinases contain nine subfamilies of protein kinases among these only five subfamilies i.e., Calcium-dependant protein kinase, (CDPK), Calcium/Calmodulin dependant protein kinase (CCaMK), Calmodulin dependant protein kinase (CaMK), CDPK- related kinase (CRK), and SnF1 related kinase 3 (SnRK3) are regulated by calcium directly or indirectly. Two subfamilies CDPK and CCaMK are regulated by direct binding of calcium to their EF hands in their C terminal domain. Three of them viz. CCaMK, CaMK and CRF are regulated by Calmodulin while SnRK3 is activated by binding another EF hand protein SOS3/calcinuerin B- like proteins (CBLs) .The remaining 4 subfamilies; PPCK, PEPRK, SnRK1 and SnRK2 are either known or predicted not to be activated by calcium.

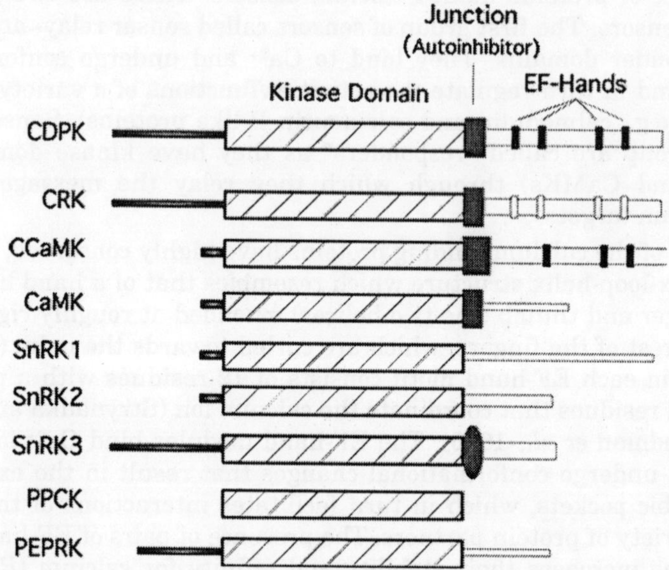

Fig. 1. Diagrammatic representation of domain organization in CDPK/SnRK superfamily (adopted from Hrabak *et al.,* 2003).

CRKs: CDPK related kinase

The CRKs have a degenerate Calmodulin like domain and hence cannot bind calcium. Out of the 8 CRKs identified in model plant *Arabidopsis* only 3 are predicted to have EF hands but are non- functional. But recent studies on two isoforms from tobacco plant suggested that some CRKs may be regulated by calcium through interaction with calmodulin (Harmon *et al* 2003).

CCaMKs: Ca^{2+}/ Calmodulin dependant protein kinase

The CCaMKs have a calcium binding domain which resembles visinin rather than Calmodulin. They are absent in *Arabidopsis)* but are reported in lily and tobacco plants (Harmon, 2001). The C-terminal side of the auto inhibitory domain contains the CaM binding site. Binding of CaM to this region regulates activities of these protein kinases (Ramachandiran *et al.*, 1997).

CaMKs : Calmodulin dependent protein kianse

The CaMKs subfamily of kinases is found mainly in animals while the exception in plants is a putative CaMK from apple (Watillion *et al.*, 1993). These kinases lack EF hands and are ~100 amino acids shorter than CCaMKs (Harmon, *et al.*, 2003). Recombinant CaMK was shown to bind Calmodulin *in vitro*, but the effect of calcium/CaM on its activity was not well studied (Watillion *et al.*, 1993).

SnRK3: SnF1 related Kinase

SnRK3 (SnF1 related protein kinase, type 3) subfamily of kinases are also called CIPKs (Calcinerin B-like interacting protein kinase) or PKS (Protein Kinase related to SOS2). They have typical catalytic domain, auto inhibitory domain and C terminal domains. *Arabidopsis* have 25 CIPKs and may have shown to interact with atleast one of the 10 CBLs. The C-terminal domain is different from most of the other CDPK/SnRK family of proteins (Hrabak *et al* 2003)

Apparently, in plant cells the above mentioned four types of calcium binding protein kinases (CCaMKs, CaMKs, CRKs and SnRK3) identified in plants may not efficiently perform the multifunctional signal integration. It also seems unlikely that the myriads of Ca^{2+}-mediated signal response-coupling in plant occur via the few CaM-regulated proteins thus far identified in plants. It can therefore be presumed that the majority of calcium signal propagation may occur through a large family of calcium sensor protein called calcium-dependent, calmodulin-independent protein kinases (CDPKs). CDPKs being ubiquitous have been found to be involved in plethora of physiological responses to abiotic and biotic stresses like drought, salt, pathogen attacks phytohormones signaling, etc. Besides plants, CDPKs are reported from few protozoans *like Plasmodium falciparam*, the malarial causative agent and some algae. The following section will discuss about CDPKs in much detail.

2.2 Calcium Dependent Protein Kinases

CDPKs are encoded by multigene families; there are 34 isoforms of CDPKs in *Arabidopsis* and 29 isforms in rice. The significance of this multiplicity can be explained by the differences in Ca^{2+} activation thresholds, substrate recognition, expression patterns and sub-cellular localization. This suggests that they are involved in multiple signaling pathways and they regulate myriads of proteins. There are only few *invivo* substrates reported but different potential sites have been identified using several synthetic peptides.

2.2.1. Structure

CDPKs are monomeric serine/threonine protein kinases ranging from 40 -90 KDa, consisting of four domains; an amino-terminal variable domain, a kinase domain, an auto-inhibitory domain and a regulatory calmodulin-like domain (CaM-LD). They represent a unique class of Ca^{2+} sensors, having protein kinase as well as calmodulin-like domain in a single polypeptide chain, enabling them to couple the calcium sensor directly to its responder (kinase).

The kinase domain and the auto inhibitory domain are well conserved among CDPKs from different plant source however the length of N and C terminal varies among CDPKs. N -terminal variable domain varies in length from 40-180 amino acids and highly variable. Many CDPKs have N - terminal myristoylation and palmitoylation motif which tether them to the membrane. In the absence of calcium signature CDPKs activity is inhibited by auto-regulatory domain which acts as a pseudo substrate and blocks the active site of the enzyme. In the presence of calcium, the CaM like domain binds to junction domain and displaces auto inhibition from the active site (Roberts and Harmon, 1992).

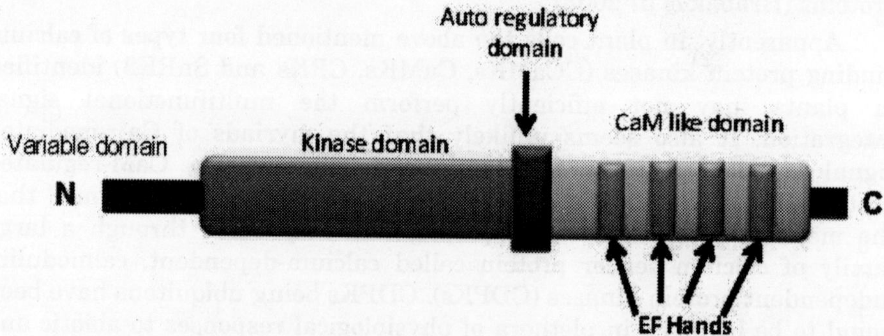

Fig. 2. Graphical representation of CDPK. CDPKs consist of four domains namely a variable domain, a kinase domain, an auto-regulatory domain and a C-terminal CaM-like domain which contain four EF hand motifs.

2.2.2 Autophosphorylation of CDPKs

Apart from calcium CDPKs are also regulated by autophosphory-lation. Occurring in many proteins, phosphorylation is used as an elegant regulatory phenomenon in many signaling pathways. Phosphorylation of CDPK proteins is known to have multiple effects on the kinase activity *viz* stimulatory (Chaudhuri *et al* 1999), phosphory-lation of *Nt*CDPK2 resulted in 10-200 fold increase in maximum Ca^{2+} stimulated activity. Saha and Singh 1995 showed that auotphosphorylation has inhibitory effect on Wingbean CDPK and according to Anil *et al* 2001 autopho-sphorylation did not show any effect on *Sw*CDPK. Auto-phosphorylation was found to be a prerequisite for activation of peanut CDPK (Dasgupta *et al.* 1999). In contrast to above report, inhibition was observed in Wingbean CDPK (Saha and Singh, 1995). Autophosphorylation in almost all CDPKs is Ca^{2+}-dependent with an exception of one CDPK isoform from Wingbean, wherein the second isoform of CDPK does not even exhibit autophos-phorylation activity (Ganguly and Singh 1998).

Autophosphorylation causes the change in sub-cellular localization for example *Mt*CDPK, upon autophosphorylation it re-localizes itself from plasma membrane to cytosol. Physiological or mechanistic role of autophosphorylation *in vivo* still remains obscure (Hegeman *et al.* 2006).

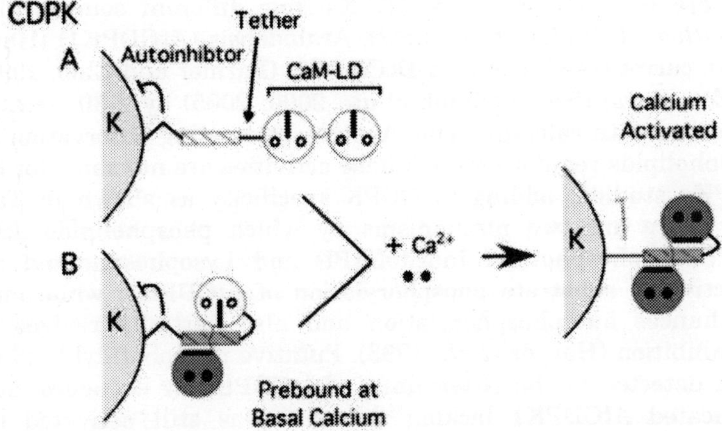

Fig. 3. Calcium induced conformational change in CDPKs. The junction domain contain auto inhibitory domain which act as pseudo-substrate and blocks the active site of enzyme. Calcium binds to the CaM-LD domain and induces the conformational change which removes the block from active site of CDPK. (Adopted from Harmon *et al.*, 1992)

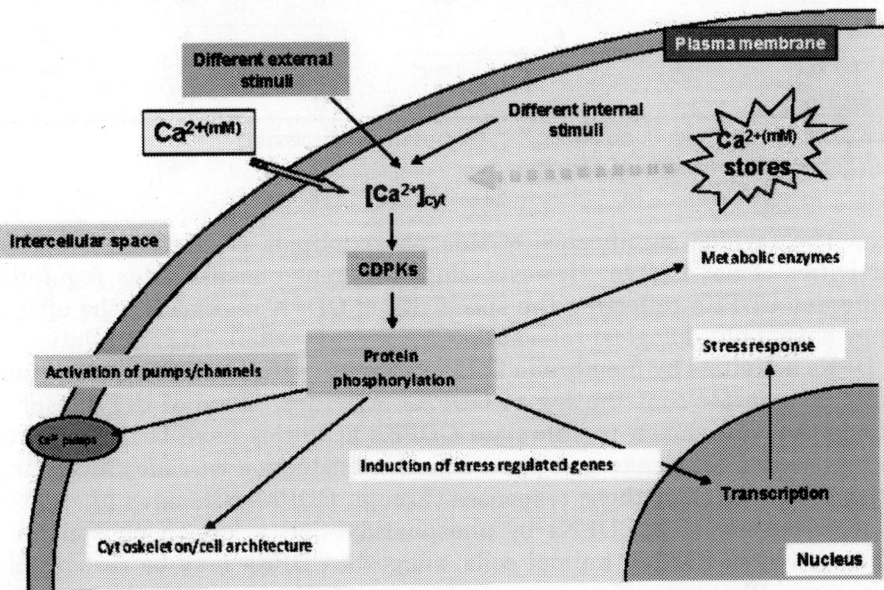

Fig. 4. Physiological functions of CDPKs. CDPK regulate a wide variety of cellular processes like transcription, transport, metabolism and cellular architecture in response to various abiotic stress. (Discussed in text in detail)

2.2.3 Regulation of CDPKs by phospholipids

Various biochemical analyses have revealed that in presence of Ca^{2+}, specific phospholipid enhances *in vitro* substrate phosphorylation. This findings are reported for many CDPKs from different sources like Oat (*Avena sativa*) (Schaller *et al.*, 1992), Arabidopsis (AtCDPK1) (Harper *et al.*, 1993), carrot (*Daucus carota* DcCDPK1) (Farmer and Choi, 1999) and maize (*Zea mays*) (Szczegielniak *et al.*, 2000; 2005) by 2-30 times above that observed with calcium alone. Another interesting observation is that the phospholipids regulating the kinase activities are not same for each of the CDPKs studied, adding to CDPK specificity as shown in Table 2. Perhaps there are two mechanisms by which phospholipids activates CDPKs Both phosphotidyl Inositol (PI) and Lysophosphotidyl choline (LPC) activates substrate phosphorylation of *At*CDPK1, while only the prior enhances autophosphorylation and also partially revives poly-L Lysine inhibition (Harper *et al.*, 1993). Putative phosphotidyl binding site has been detected in the N-terminus of *At*CDPK1 by sequence analysis, but truncated *At*CDPK1 lacking this site was still activated by the phospholipid indicating other PI-binding sites in the protein (Harper *et al.*, 1993).

Table 2.

CDPK	PA	PS	PI	PC	LysoPC	PE
AtCPK1	0	0	++	0	++	0
Oat CDPK	n.d.	+	+	+	+	n.d.
DcCPK1	++	+	+	n.d.	0	++
ZmCPK11	++	++	+	0	0	n.d.

n.d., not determined; 0, no effect; "+", stimulation of activity;
"++" strong stimulation of activity.

The *in vivo* significance of this phospholipids regulation on CDPK activities is not known. However since different phospholipids regulates different CDPKs reflecting the specificity of CDPK regulation, the effects may have a physiological relevance (Cheng *et al.*, 2002). The stimulation of CDPKs activities by membrane lipids suggests that membrane association may be a factor contributing to CDPKs activities. Some of the phospholipids that are shown to stimulate CDPKs activities have been shown to act as second messengers in plant signal transduction cascades indicating that they may elicit those responses through CDPKs (Cheng *et al.*, 2002). The activation of DcCDPK1 by phosphatidyl-Ser, a known phospholipid that activates PKC in animal cells, suggests CDPKs may be responsible for some PKC-like responses in plant cells (Farmer and Choi, 1999). Whether there is a plant PKC homolog is still not known, but this observation may also help explain reports of kinases that seems to have properties of both CDPKs and PKC (Abo-el-Saad and Wu, 1995; Lino *et al.*, 1998). Contrastingly, recombinant *Dc*CDPK1 is stimulated by phosphatidic acid (Farmer and Choi, 1999), a cleavage product of phospholipids by phospholipase D, but is not stimulated by diacylglycerol, a Protein kinase C regulator in animal system. The question of regulation by lipids is important because this could represent a possible point of crosstalk between signaling pathways, or a means of targeting activity to specific locations by reversible membrane association. It remains to be determined if CDPKs are subject to cross-regulation by components of other pathways such as Jasmonic acid or brassinosteroids.

Apart from DcCDPK1, *Zm*CDPK11 activity is shown to be regulated by phosphatidic acid (Klimecka *et al* 2011). Maize leaves treated with n-butanol, a potent inhibitor of phosphatidic acid (PA) synthesis catalyzed by phospholipase D, a significant decrease of *Zm*CDPK11 activity was observed, indicating contribution of PA in the kinase activation and involvement of *Zm*CDPK11 in Phosphatidic acid signaling.

2.3.5 CDPKs and 14-3-3 protein

*At*CDPK1 have shown to be regulated by three different isoformes of 14-3-3 *in vitro* in calcium dependent manner (Camoni *et al.*, 1998). Therefore, 14-3-3 does not merely mimic Ca^{2+} but rather provides an additional stimulation once the enzyme is activated by calcium. It is speculated that 14-3-3 proteins might recognize the phosphorylated residues of CDPKs. N-terminal domain of AtCDPK1 contains one site similar to the most common 14-3-3 consensus binding site, R-S/T-X-S-X-P, where the underlined Ser is phosphorylated. Such putative 14-3-3 binding sites are also there in two other *Arabidopsis* CDPKs, AtCDPK24 and AtCDPK28 (Cheng *et al.*, 2002). Apart from binding to CDPKs, 14-3-3 reported to interact and regulate other proteins like nitrate reductase, (which have been phosphorylated by CDPK) (Chung *et al.*, 1999), sucrose phosphate synthase, trehalouse-6-phosphate synthase and glutamine synthetases in a phosphorylation-dependant manner (Moorhead *et al.*, 1999). It has been anticipated that 14-3-3 proteins may dock other proteins that are involved in two consecutive metabolic steps. How does 14-3-3 proteins regulate CDPK remains unclear and remains an interesting area of further investigation?

2.2.4 Localization of CDPKs

In plants, CDPKs are present in both soluble and membrane associate forms. CDPKs have been reported to be localized to the cytosol (Putnam-Evans *et al.*, 1990; Klimczak and Hind, 1990; DasGupta, 1994; MacIntosh *et al.*, 1996; Frylinck and Dubey, 1998), nucleus (Dammann *et al.*, 2003), cytoskeleton (Putnam-Evans *et al.*, 1989; Dammann *et al.*, 2003), oil bodies of endosperm cells (Anil *et al.*, 2003; Kumar and Jayabaskaran, 2004), endoplasmic reticulum (Lu and Hrabak, 2002; Dammann *et al.*, 2003), plasma membrane (Schaller *et al.*, 1992; Yoon *et al.*, 1999; Dammann *et al.*, 2003), and peroxisomes (Dammann *et al.*, 2003).

In addition to their diverse sub-cellular localization, some of them reported to change their localization during stress conditions. Patharkar and Cushman, 2000 showed that GFP tagged *Mc*CDPK1 relocate itself from plasma membrane to nucleus under dehydration and salt stress. The groundnut*Ah*CDPK2 is translocated to the nucleus under hyper-osmotic conditions through an interaction with importins (Raichaudhuri *et al.* 2006). Thus understanding of the cellular and subcellular localization in conjunction with identification of potential substrates and regulatory proteins would provide insight into the physiological roles of individual CDPKs during abiotic stress conditions. It will also provide an understanding of the molecular mechanisms by which multiple cellular processes are regulated by calcium-dependent phosphorylation in plants.

3. FUNCTIONAL AND EVOLUTIONARY DIVERSITY OF CDPKS

As discussed in earlier sections CDPKs exhibit remarkable diversity in terms of their function, localization as well as in calcium sensing. For example three CDPK isoforms α, β and γ from soybean are responsive to calcium concentrations over different ranges (Lee *et al.*, 1998). The calcium concentration required for half maximal activity ($K_{0.5}$) for CDPKα, β and γ varied over two orders of magnitude in presence of syntide-2, as substrate. *Pf*CDPK1 from *P. falciparum* required higher concentrations of calcium when compared to the soybean CDPKs (Zhao *et al.*, 1994b). Calcium-binding properties have been experimentally determined for only a few CDPKs, and it is difficult to predict the calcium-binding properties of each isoform the sequence information available. Some CDPKs appear to have defects in one or more of their EF hands that would affect their calcium-binding properties (Hrabak *et al.*, 1996). Zhao *et al.*, 1994 has shown that *Pf*CDPK1, in which each EF hand was disabled by mutating critical glutamate residue, showed that only the first EF-hand is critical for the enzyme to be stimulated by calcium. The above mentioned finding therefore suggests that CDPKs with defects in their EF hands can still be regulated by calcium. CDPKα binds to Ca^{2+} with a K_d of 50 μM in absence of substrate. However, it has been observed that calcium sensitivity can increase tenfold or more in the presence of a substrate (Zhao *et al.*, 1994b; Lee *et al.*, 1999). Thus for different CDPKs, Ca^{2+} required for half maximal stimulation ($K_{0.5}$ or apparent binding constant) of substrate phosphorylation activity falls in the sub-micromolar or micromolar level. However, considerable differences exist in the Ca^{2+}-binding properties among the CDPKs for which binding studies have been carried out (Zhao *et al.*, 1994b; Lee *et al.*, 1998). These differences in sensitivity towards calcium might denote that each isoform of CDPK responds to a specific calcium signature, which differ in frequency of oscillation, magnitude and duration depending on the stimulus (McAnish and Hetherington, 1998; Sanders *et al.*, 1999).

CDPKs display a wide range of variations in their properties such as Ca^{2+}-binding, substrate specificity, enzyme kinetics, and other biochemical properties indicating that the members of this family of kinases are evolved to respond to a multitude of environmental and hormonal stimuli.

This belief is corroborated by various studies that show regulation of a wide range of developmental and cellular functions by different CDPKs.

CDPKs show evidence of a wide range of diversity. So far CDPK isolated from in a soluble enzyme as well as in with membrane associated forms have been characterized in a number of plant species (Bassarab and Werner, 1987; Harmon *et al.*, 1992; Polya *et al.*, 1987; Klucis and Polya, 1988; Olah *et al.*, 1989; Klimczak and Hind, 1990; Weaver *et al.*, 1991; Johnson and Chrispeels., 1992; Schaller *et al.*, 1992; Yuasa and Muto, 1992; Pical *et al.*, 1993; Verhey *et al.*, 1993; DasGupta, 1994; Zhao *et al.*, 1994b; Abo-el-Saad and Wu, 1995; Ohto and Nakamura, 1995; Saha and Singh, 1995; Barker *et al.*, 1998; Ganguly and Singh, 1998; Iwata *et al.*, 1998; Satterlee and Sussman, 1998; Ellard-Ivey *et al.*, 1999; Yoon *et al.*, 1999, Hrabak *et al.*, 2003; Chehab *et al.*, 2004).

The hypothesis that CDPKs have evolved by the fusion of a CaM-like domain with the catalytic domain of divergent kinases (Harper *et al.*, 2003), explains to some extent, the similarity and diversity observed among members of this family. The diversity among CDPKs also reflects the specialization of different isoforms to regulate diverse physiological functions. CDPKs exist as multiple isoforms in single species (Hrabak *et al.*, 1996; Harmon *et al.*, 2000; Asano *et al.*, 2005). One reason for the multiplicity of CDPKs in a given plant species might be related to the specialization of different isoforms with respect to calcium binding and activation. Thus, it is conceivable that different isoforms of CDPKs perform different functions in various signal transduction pathways.

4. ROLES OF CALCIUM DEPENDENT PROTEIN KINASES DURING ABIOTIC STRESSES.

4.1 Dehydration and Salinity

Changes in cytosolic calcium level are seen in response to various abiotic stresses like high salinity, drought, cold and mechanical stress. These calcium signatures are decoded by CDPKs.

High transcript levels of CDPK as well activities are reported in response to various abiotic stresses.

Many studies have been performed with many different plant systems like rice, *Arabidopsis*, chickpea, tobacco etc. In recent past indicate the direct involvement of CDPKs during dehydration and salinity. Cold or salt treatment causes the increase transcript level of *Os*CDPK7 in roots and shoots of 10 day old rice seedlings. Moreover over-expression of *Os*CDPK7 made transgenic rice showed enhanced tolerance to cold, drought and salt stress (Saijo *et al.*, 2000).

In *Arabidopsis* northern blot analysis of two genes *At*CDPK1 and AtCDPK2 revealed that high –salt and drought causes induction of either AtCDPK1 and AtCDPK2 genes but low temperature or heat stress fails to do so as shown by northern blot analysis (Urao *et al.*, 1994).

In tobacco leaves, the levels of *Nt*CDPK1 mRNA increases rapidly after 1-2 hrs of treatment with chitosan, or NaCl or fungal elicitors. But the duration of the elevation of the NtCDPK1 mRNA levels differ depending on the treatment (Yoon *et al.*, 1999). Another isoform *Nt*CDPK4 was induced by both phytohormone (Gibberellic acid) as well as NaCl treatment. The maximal transcript level was observed 30 min after NaCl treatment and returned to basal level after 2 h (Zhang *et al.*, 2005).

In Chickpea, *Ca*CDPK1 expression both mRNA and protein kinase activity are rapidly induced by salt (NaCl) stress or by fungal spore treatment but not by dehydration whereas *Ca*CDPK2 by dehydration and not by salt or by fungal spore. These findings suggest the possible involvement of *Ca*CDPK1 in the salt- and fungal spore induced signal transduction pathways and *Ca*CDPK2 in the dehydration-induced signal transduction pathway.

Dehydration-inducible gene AtDi19 encodes a nuclear zinc finger transcription factor specifically interact with CDPK 4,11 and 12 from *Arabidopsis*, which further support the involvements of these isoforms during dehydration (Rodriguez Milla *et al.* 2006).

CDPK may reduce water loss during dehydration conditions by regulating diverse channel activities, for example in spinach CDPK regulates the activity of aquaporin PM28A (Johansson *et al.* 1996). In faba bean guard cells, a CDPK phosphorylates and inhibits the K^+ inward channel KAT1 in vitro (Li *et al.* 1998), which results in stomatal closure (Berkowitz *et al.* 2000). In contrast, *At*CDPK1 activates a vacuolar Cl^- channel, resulting in Cl^- uptake into the vacuole and stomatal opening (Pei *et al.* 1996).

Drought responses and stomatal movements are regulated by ABA. ABA-induced stomata closure is reduced in *Arabidopsis CDPK3*, *CDPK6* double mutant associated with an impaired ABA activation of slow-type anion channels and calcium permeable channels (Mori *et al.* 2006). Therefore, *At*CDPK3 and *At*CDPK6 are both positive regulators of stomata ABA signaling system. However, their functions may not be redundant since they belong to distinct CDPK subfamilies (Cheng *et al.* 2002). ABA considerably stimulates the activity of the grape berry CDPK1 (Yu *et al.* 2006), which positively regulates ABA-induced stomatal closure and the expression of stress-responsive genes (Yu *et al.* 2007).

In *Arabidopsis*, *At*CDPK32 phosphorylates and activates the ABA-responsive transcription factor ABF4, causing enhanced expression of ABF4 (Choi *et al.* 2005a). It is likely that multiple CDPKs act through ABF4 and related transcription factors to activate ABA and stress signaling (Sheen 1996; Choi *et al.* 2005a).

Over-expression of OsCDPK13/OsCDPK7 in rice plants showed enhanced resistance to drought and salinity (Saijo *et al.* 2000). A recent study has shown that *At*CDPK23 appears to have a negative function in drought and salt responses (Ma and Wu 2007). In addition to this, CDPK also regulates Ca^{2+} signatures by inhibiting the Ca^{2+} pump ACA2 located in the ER (Hwang *et al.* 2000).

4.2 Wounding and cold

Wounding causes a constant threat to plant survival since it physically destroys plant tissues, which could provide a way for various pathogens.

Wounding commonly occurs as a result wind, rain, and hail or is caused by phytopathogens and herbivores. NtCDPK1 from tobacco is a CDPK which is induced transiently during wounding (Ludwig *et al.*, 2004). In tobacco plants, the *NtCDPK1* transcript appeared within 2 h after wounding and went down at 11 h (Yoon *et al.*, 1999). An increase in the transcript level of *Vr*CDPK1was seen after mechanical strain treatment for mins in 10-day-old mung bean plants, the expression of VrCDPK1 reaches a maximum after 60 min and then remains unchanged up to 3 h (Botella *et al.*, 1996). In tomato plants mechanical wounding, causes a transient increase in the *LeCDPK1* mRNA levels. The increase in mRNA levels were seen both at the site of the injury as well as distant non-wounded leaves. *LeCDPK1* mRNA in whole plants subjected to wounding showed that the *LeCDPK1* expression in wounded leaves reaches a maximum at 4 h whereas in neighboring leaflets or distal leaves its mRNA appeared later (after 8–12 h or 18 h, respectively). Similar responses were also observed in maize, where wounding causes accumulation of the *ZmCDPK11* transcript in wounded and as well as in neighboring leaves (after 3 h and 6 h, respectively). These results indicate that monocots, like dicot plants are able to respond to wounding in a systemic way (Szczegielniak *et al.*, 2005).

In rice plant three CDPKs are reported to be involved in cold, salt and drought stress. The activity of a membrane-bound 56- kDa CDPK from rice plant was increased by cold treatment (Martin & Busconi, 2000). An accumulation of OsCDPK13 transcript was observed in 2-week-old rice plant upon cold treatment (Abbasi *et al.*, 2004). An increase in the activity of a 51 kDa cytoplasmic CDPK was observed in rice plants treated with cold, salt or drought stresses (Li & Komatsu, 2000). Apart from cold, some CDPKs transcript levels are reported to increase by heat shock. For example an induction of expression of MsCDPK3 from alfalfa was observed after thirty minutes of heat shock (37°C). This is a unique example of such induction, not observed in other plant species studied (Davletova *et al.*, 2001).

4.3 CDPKs Hormone Responses

CDPKs have been shown to modulate Ca^{2+} dependant plant responses to phyto-hormones mainly gibberlins, cytokinins and auxins. Rice seeds when treated with gibberlic acid (GA) showed 10 fold increased in activity of partially purified CDPK (Abo-El-Saad and Wu, 1995). Additionally, another CDPK from rice seedling, *Os*CDPK13 transcript level was increased by GA treatment (Yang *et al.*, 2003). In tobacco, GA treatment caused accumulation of *Nt*CDPK4 reaching to a maximum after 1 hr and returning to basal levels in two hours (Zhang *et al.*, 2005). An influence of GA was seen in aleurone cells of barley (*Hordeum vulgare*) expressing an inactivated (D140N) *Hv*CDPK1. Secretion, cell vacuolation and vacuolar acidification were inhibited while GA-induced gene expression was not affected by the inactivated *Hv*CDPK1 suggesting the role of *Hv*CDPK1 in regulation of post-transcriptional events in protein secretion (McCubbin *et al.*, 2004). A different regulation by phyto-hormones was observed in rice roots, a 45kDa CDPK's activity previously known to be decreased by cold stress was recovered by exogenous abscisic acid (ABA). This indicates that the 45kDa kinase activity modulated by ABA may play a role in cold stress response in rice (Komatsu *et al.*, 2001). Studies of CDPK1 from grape barrier (*Vitis vinifera x V. labrusca*) mesocarp revealed that ABA stimulates activity of kinase as well as causes increase in transcript level of ACDPK1, but both not at the same time. The rapid stimulation of the ACDPK1 activity 15 minutes after ABA treatment but without the increase of ACDPK1 protein level indicates the involvement of enzymes mediating some rapid ABA-responsive events (ABA may act on post translational modification of ACDPK1). A slower de novo ABA-induced synthesis of ACDPK1 may signify its role in regulating development related ABA-signaling pathway (Yu. 2006). Besides, in tobacco upon GA and ABA treatment *Nt*CDPK1 mRNA accumulated after 12 h of treatment of phytohormones (Yoon *et al.*, 1999)

Phytohormone auxin also has been shown to influence the expression and activities of CDPKs. *Nt*CDPK1 transcript was slightly induced upon activation of tobacco leaves with indole-3-acetic acid (IAA) (Yoon *et al.*, 1999) while in mung bean (*Vigna radiata*) the transcript level increased significantly reaching a maximum after 6 hrs of treatment (Botella *et al.*, 1996). Another example of transcript accumulation after auxin (2,4D) treatment was in the microcallus suspension of alfalfa (*Medicago sativa)* during somatic embryo induction. The transcript level of *Ms*CDPK3 peaked at about 2 hours after the treatment (Davletova *et al.*, 2001). Study in cucumber cotyledons have shown that a 50 kDa CDPK activity was stimulated IAA after first day of treatment and extended up to 3 days of continuous treatment. The same CDPK has been induced by NO, suggesting the role of IAA together with NO in Ca^{2+}-mediated signaling pathway regulating the CDPK activity leading to adventitious root formation (Lanteri *et al.*, 2006).

Cytokinins have been shown to affect the transcript level of *Cs*CDPK3 in cucumber, but in a organ specific manner. Treatment with benzyl adenine (BA) (synthetic cytokinin) increased *Cs*CDPK3 transcript level in cotyledons but contrastingly down-regulated in roots and had no significant effect in hypocotyls (Kumar *et al.* 2004). Studies with chickpea isoforms *Ca*CDPK1 and *Ca*CDPK2 reveal that the activity of *Ca*CDPK1 increased 3-fold in response to treatment with BA; however, IAA and GA_3

did not show any significant increase or decrease in activity. Furthermore, the activity of *Ca*CDPK2 was increased 3-fold in response to treatments with both GA_3 and BA, but did not change by IAA treatment (Prakash and Jayabaskaran, 2006b). It is also known that CDPKs phosphorylation regulate gaseous hormone ethylene mediated plant responses. Ethylene has been reported to play significant roles in physiological process like fruit ripening, senescence and leaf abscission and also responses to stress like wounding and pathogen attack (Kwak and Li., 1997; Ludwig *et al.*, 2005).

5. FUTURE DIRECTIONS

One of the major challenges in future will be to elucidate the functions of CDPKs in particular signaling pathways and also in the cross-talk in the complex signaling network. It will be interesting to study the regulation CDPK by specific phospholipid during abiotic stress signaling. Indentifying isoform specific *in vivo* substrate will give a better insight in abiotic signaling transduction pathways regulated by different CDPKs.

Acknowledgement

We would like to thank Prof. S.K. Podder his suggestions.

References

Abbasi, F; Onodera, H; Toki, S; Tanaka, H; and Komatsu, S (2004). OsCDPK13, a calcium-dependent protein kinase gene from rice, is induced by cold and gibberellin in rice leaf sheath. *Plant Mol. Biol* 55: 541-52.

Abo-El-Saad, M; Wu, R (1995). A rice membrane calcium-dependent protein kinase is induced by gibberellin. *Plant Physiol* 108: 787–793.

Anil, VS; Harmon, AC; and Rao, SK (2003).Temporal association of Calcium-dependent protein kinase with oil bodies during seed development in *Santalum album* L.: Its biochemical characterization and significance. *Plant and Cell Physiology* 44:367-376.

Anil, VS and Rao, SK (2001). Purification and characterization of a Ca^{2+}-dependent protein kinase from sandalwood (*Santalum album* L.): evidence for Ca^{2+}-induced conformational changes. *Phytochemistry* 58: 203-12.

Asano, T; Tanaka, N; Yang, G; Hayashi, N and Komatsu, S (2005). A rice calcium-dependent protein kinase OsCPK12 oppositely modulates salt-stress tolerance and blast disease resistance. *Plant Cell Physiol* 46: 356-66.

Barker, L; Templeton, M; and Ferguson, I; (1998) A 67-kDa plasma-membrane-bound Ca2+-stimulated protein kinase active in sink tissue of higher plants. *Planta* 205: 197-204.

Bassarab, S and Werner, D (1987). Ca^{2+}-dependent protein kinase activity in the peribacteroid membrane from soybean root nodules. *J. Plant Physiol* 130: 233-241

Berkowitz, G; Zhang, X; Mercie, R; Leng, Q and Lawton, M (2000) Co-expression of calcium-dependent protein kinase with the inward rectified guard cell K+ channel KAT1 alters current parameters in Xenopus laevis oocytes. *Plant Cell Physiol* 41:785–790

Botella, J R; Arteca, JM; Somodevilla, M; and Arteca, RN (1996). Calcium-dependent protein kinase gene expression in response to physical and chemical stimuli in mungbean (*Vigna radiata*) *Plant Mol. Biol* 30: 1129-1137.

Camoni, L; Harper, JF; and Palmgren, MG(1998) .14 3 3 proteins activate a plant calcium-dependent protein kinase (CDPK). *FEBS Lett* 430: 381 384.

Campbell, A K (1983). Intracellular calcium its universal role as regulator (Chicheester: John Wiley).

Chaudhuri, S; Seal, A and Dasgupta, M (1999) Autophosphorylation- dependent activation of calcium-dependent protein kinase from groundnut. *Plant Physiol* 120: 859 866.

Chehab, EW; Patharkar, OR; Hegeman, AD; Taybi, T and Cushman, JC (2004). Autophosphorylation and subcellular localization dynamics of a salt- and water deficit-induced calcium-dependent protein kinase from ice plant. *Plant Physiol* 135: 1430-46.

Cheng, S-H; Willmann, MR; Chen, H-Ch; Sheen, J (2002). Calcium signaling through protein kinases. The *Arabidopsis* calcium-dependent kinase gene family. *Plant Physiol* 129: 469–485.

Choi,HI; Park HJ; Park, JH; Kim, S; Im, MY; Seo, HH; Kim, YW, Hwang, I; Kim, SY (2005a). Arabidopsis calciumdependent protein kinase AtCPK32 interacts with ABF4, a transcriptional regulator of abscisic acid-responsive gene expression, and modulates its activity. *Plant Physiol* 139:1750–1761.

Chung, HJ; Sehnke, PC and Ferl, RJ (1999). The 14-3-3 proteins: cellular regulators of plant metabolism. *Trends Plant Sci* 4:367–371.

Dammann, Ch; Ichida, A; Hong, B; Romanowsky, SM; Hrabak, EM; Harmon, AC; Pickard, BG and Harper, JF (2003). Subcellular targeting of nine calcium-dependent protein kinase isoforms from *Arabidopsis*. *Plant Physiol* 132: 1840–1848.

DasGupta, M. (1994). Characterization of a Calcium-Dependent Protein Kinase from *Arachis hypogea* (Groundnut) Seeds. *Plant Physiol* 104: 961-969.

Davletova, S; Mészáros, T; Miskolczi, P; Oberschall, A; Török, K; Maryar, Z; Dudits, D and Deák, M (2001). Auxin and heat shock activation of a novel member of the calmodulin like domain protein kinase gene family in cultured alfalfa cells. *J Exp Bot* 52: 215–221.

Demidchik, V; Bowen, HC; Maathuis, FJ; Shabala, SN; Tester, MA; White, PJ; Davies, JM (2002a). Arabidopsis thaliana root non-selective cation channels mediate calcium uptake and are involved in growth. *Plant J* 32: 799–808.

Ellard-Ivey, M; Hopkins, RB; White, TJ and Lomax, TL (1999). Cloning, expression and N-terminal myristoylation of CpCPK1, a calcium-dependent protein kinase from zucchini (*Cucurbita pepo* L.). *Plant Mol Biol* 39: 199-208.

Farmer, PK and Choi, JM (1999) Calcium and phospholipid activation of recombinant calcium-dependent protein kinase (DcCPK1) from carrot (*Daucus carota* L.) *Biochim Biophys Acta* 1434: 6–17.

Frylinck, L and Dubery, IA (1998). Protein kinase activities in ripening mango (*Mangifera indica*) fruit tissue. II. Purification and characterization of a casein kinase I. *Biochem. Biophys. Acta* 1387: 342-354.

Ganguly, S and Singh, M (1998). Characterization of a second calcium-dependent protein kinase from winged bean, *Phytochemistry* 48: 61-70.

Guo, H; Mockler, T; Duong, H; Lin, C (2001) SUB1, an Arabidopsis Ca^{2+}-binding protein involved in cryptochrome and phytochrome coaction. *Science* 291:487–490.

Harmon, AC; Gribskov, M and Harper, JF (2000). CDPKs— a kinase for every Ca^{2+} signals? *Trends Plant Sci* 5: 154-159.

Harmon, AC (2003). Calcium-regulated protein kinases of plants. *Gravit Space Biol Bull* **16**: 83–90.

Harper, JF; Binder, BM; and Sussman, MR (1993) Calcium and lipid regulation of an Arabidopsis protein kinase expressed in Escherichia coli. *Biochemistry* 32: 3282-3290.

Hegeman, AD; Rodriguez, M; Han, BW; Uno, Y; Phillips, GN; Jr., Hrabak, EM; Cushman, JC; Harper, JF; Harmon, AC; Sussman, MR (2006). A phyloproteomic characterization of *in vitro* autophosphorylation in calcium-dependent protein kinases, *Proteomics* **6**: 3649-3664.

Hwang, I; Sze, H and Harper, JF (2000). A calcium-dependent protein kinase can inhibit a calmodulin-stimulated Ca2+ pump (ACA2) located in the endoplasmic reticulum of Arabidopsis. *Proc Natl Acad Sci* U.S.A. 97: 6224-6229.

Iwata, Y; Kuriyama, M; Nakamura, M; Kojima, H; Ohto, M and Nakamura, K (1998). Characterization of a calcium-dependent protein kinase of tobacco [*Nicotiana tabacum*] leaves that is associated with the plasma membrane and its inducible by sucrose. *Plant Cell Physiol* 39: 1176-1183.

Johansson, I; Larsson, C; Ek, B. and Kjellbom, P (1996) The major integral proteins of spinach leaf plasma membranes are putative aquaporins and are phosphorylated in response to Ca^{2+} and apoplastic water potential. *Plant Cell* 8:1181–1191.

Johnson, KD and Chrispeels, MJ (1992). Tonoplast-Bound Protein Kinase Phosphorylates Tonoplast Intrinsic Protein. *Plant Physiol* 100: 1787-1795.

Johnson, LN; Lowe, ED; Noble, MEM and Owen, DJ (1998). The structural basis for substrate recognition and control by protein kinases *FEBS Lett* 430: 1-11.

Klimczak, LJ and Hind, G (1990). Biochemical similarities between soluble and membrane-bound protein kinases of barley. *Plant Physiol* 92: 919-923.

Klimecka, M; Szczegielniak, J; Godecka, L; Lewandowska-Gnatowska, E; Dobrowolska, G and Muszyńska, G (2011), Regulation of wound-responsive Calcium-Dependent Protein Kinase from maize (ZmCPK11) by phosphatidic acid.

Klucis, E and Polya, GM (1988). Localization, Solubilization and Characterization of Plant Membrane-Associated Calcium-Dependent Protein Kinases. *Plant Physiol* 88: 164-171.

Komatsu, S; Li, W; Konishi ,H; Yoshikawa, M; Konishi, T and Yang, G(2001). Protein kinase from rice root: differential response to cold and regulation by abscisic acid. *Biol Pharm Bull* 24: 1316–1319.

Kumar, KGS; Ullanat, R and Jayabaskaran, C (2004). Molecular cloning, characterization, tissue-specific and phytohormone-induced expression of calcium-dependent protein kinase gene in cucumber (*Cucumis sativus* L.). *J Plant Physiol* 161:1061-71.

Kwak ,SH; Lee, SH (1997) The requirements for Ca2+, protein phosphorylation, and dephosphorylation for ethylene signal transduction in *Pisum sativum* L. *Plant Cell Physiol* 38: 1142–1149.

Lanteri, ML; Pagnussat, GC; Lamattina, L (2006). Calcium and calcium-dependent protein kinases are involved in nitric oxide- and auxin-induced adventitious root formation in cucumber. *J Exp. Bot* 57: 1341–1351.

Lee, J-Y; Yoo, BC and Harmon, AC (1998). Kinetic and calciumbinding properties of three calcium-dependent protein kinase isoenzymes from soybean. *Biochemistry* **37**: 6801–6809.

Levine, A; Tenhaken, R; Dixon, R and Lamb, C (1994). H_2O_2 from the oxidative burst orchestrates the plant hypersensitive disease resistance response. *Cell* 79, 583–593.

Li, J; Lee YR and Assmann SM (1998). Guard cells possess a calcium-dependent protein kinase that phosphorylates the KAT1 potassium channel. *Plant Physiol* 116:785–795.

Li, WG and Komatsu, S (2000). Cold stress-induced calcium-dependent protein kinase (s) in rice (*Oryza sativa* L.) seedling stems tissues. *Theor Appl Genet* 101: 355-363

Lino, B; Baizabal-Aguirre, VM and Gonzalez de la Vara, LE (1998). Purification and characterization of a calcium-dependent protein kinase from beetroot plasma membranes .*Planta* 204: 352-359.

Lu,SX; Hrabak, EM (2002). An *Arabidopsis* calcium-dependent protein kinase is associated with the endoplasmic reticulum. *Plant Physiol* 128: 1008–1021.

Ludwig, AA; Romeis, T and Jones JDG (2004). CDPK-mediated signalling pathways: specificity and cross-talk. *J Exp Bot* **55**: 181–188.

Ma, SY and Wu, WH (2007) AtCPK23 functions in Arabidopsis responses to drought and salt stresses. *Plant Mol Biol* 65: 511–518.

MacIntosh, GC; Ulloa, RM; Raices, M and Tellez-Inon, MT (1996). Changes in Calcium-Dependent Protein Kinase Activity during in Vitro Tuberization in Potato. *Plant Physiol* 112: 1541-1550.

Martin, ML and Busconi, L (2000). Membrane localization of a rice calcium dependent protein kinase (CDPK) is mediated by myristoylation and palmitoylation. *Plant J* 24: 429–435.

McAinsh, M R and Hetherington, AM (1998). Calcium ions as second messengers in guard cell signal transduction. *Trends Plant Sci* 3: 32-36.

McCubbin, AG; Ritchie, SM; Swanson, SJ and Gilroy, S (2004) the calcium-dependent protein kinase HvCDPK1 mediates the gibberellic acid response of the barley aleurone through regulation of vacuolar functions. *Plant J* 39(2):206-18.

Moncrief, ND; Kretsinger, RH and Goodman, M (1990). Evolution of EF hand Calcium-modulated proteins. Relationships based on amino acid sequences, *J Mol Evol* **30**:522-562.

Moorhead, G; Douglas, P; Cotelle, V; Harthill, J; Morrice, N; Meek, S; Deiting, U; Stitt, M; Scarabel, M; Aitken, A; MacKintosh C (1999). Phosphorylation-dependent interactions between enzymes of plant metabolism and 14-3-3 proteins. *Plant J* 18: 1-12

Mori, IC; Murata, Y; Yang, Y; Munemasa, S; Wang, YF; Andreoli, S; Tiriac, H.; Alonso, JM; Harper, JF; Ecker, JR; Kwak, JM and Schroeder, JI (2006). CDPKs CPK6 and CPK3 function in ABA regulation of guard cell S-type anion- and Ca^{2+}- permeable channels and stomatal closure. *PLoS Biol* 4: 1749 – 1762.

Ochiai El-Ichiro (1991) Why Calcium? *J Chem Edu* 68 : 10–12.

Ohto, MA and Nakamura, K (1995). Sugar-lnduced lncrease of Calcium-Dependent Protein Kinases Associated with the Plasma Membrane in Leaf Tissues of Tobacco. *Plant Physiol* 109: 973-981.

Oláh, Z; Bogre, L; Lehel, C; Farago, A; Seprodi, J and Dudits, D (1989). The phosphorylation site of Ca^{2+}-dependent protein kinase from alfalfa. *Plant Mol Biol* 12: 453-461.

Patharkar, OR and Cushman, JC (2000). A stress-induced calcium- dependent protein kinase from *Mesembryanthemum crystallinum* phosphorylates a two-component pseudoresponse regulator. *Plant J* 24: 679–691.

Paul, Cj; Van der Hoeven; Marco Siderius; Henrie, A.A; Korthout, j; Artem, V; Drabkin, and De Boer, AH (1996). A Calcium and Free Fatty Acid-Modulated Protein Kinase as Putative Effector of the Fusicoccin 14-3-3 Receptor. *Plant Physiol* 111: 857-865.

Pei, ZM; Ward, JM; Harper, JF and Schroeder, JI (1996) A novel chloride channel in Vicia faba guard cell vacuoles activated by the serine/threonine kinase, CDPK. *EMBO J* 15:6564–6574.

Pical, C; Fredlund, KM and Petit, PX (1993). The outer membrane of plant mitochondria contains a calcium-dependent protein kinase and multiple phosphoproteins. *FEBS Lett* 336, 347-351.

Polya, GM; Klucis, E and Haritou, M (1987). Resolution and characterization of two soluble calcium-dependent protein kinases from silver beet leaves. *Biochim Biophys Acta* 931: 68-77.

Prakash, SRS and Jayabaskaran, C (2006b) Expression and localization of calcium -dependent protein kinase isoforms in chickpea. *J Plant Physiol* 163:1135-49.

Putnam-Evans CL; Harmon AC and Cormier MJ (1990), Purification and characterization of a novel calcium-dependent protein kinase from soybean, *Biochemistry* 29 : 2488–2495.

Raichaudhuri, A; Bhattacharyya, R; Chaudhuri, S; Chakrabarti, P and DasGupta M (2006). Domain Analysis of a Groundnut Calcium-dependent Protein Kinase: nuclear localization sequence in the junction domain is coupled with nonconsensus calcium binding domains. *JBC* 281: 10399-10409.

Ramachandiran, S; Takezawa, D; Wang, W and Poovaiah BW (1997). Functional domains of plant chimeric calcium/calmodulin-dependent protein kinase: regulation by autoinhibitory and visinin-like domains. *Biochem J* 121: 984–990.

Reddy, VS and Reddy, ASN (2004). Proteomics of calcium-signaling components in plants *Phytochemistry* 65: 1745–1776.

Roberts, DM and Harmon, AC (1992) Calcium-modulated proteins- targets of intracellular calcium signals in higher plants. *Annu Rev Plant Physiol Mol Biol* 43: 375–414.

Roberts, DM and Harmon, AC (1992) Calcium-Modulated Proteins: Targets of Intracellular Calcium Signals in Higher Plants. *Annual Review of Plant Physiology and Plant Molecular Biology* 43: 375-414.

Rodriguez Milla ,MA; Uno ,Y; Chang, IF; Townsend, J; Maher, EA; Quilici D and Cushman JC (2006) A novel yeast two hybrid approach to identify CDPK substrates: characterization of the interaction between AtCPK11 and AtDi19, a nuclear zinc finger protein. *FEBS Lett* 580:904–911

Romeis, T; Ludwig, AA; Martin, R and Jones, JDG (2001). Calcium-dependent protein kinases play an essential role in a plant defence response. *EMBO J* 20: 5556–5567.

Rudd, JJ and Franklin-Tong,VE (2001). Unravelling response-specificity in Ca2þsignalling pathways in plant cells. *New Phytol* 151: 7–33.

Saha, P and Singh, M (1995). Characterization of a winged bean (Psophocarpus tetragonolobus) protein kinase with calmodulin-like domain: regulation by autophosphorylation. *Biochem J* 305: 205-210.

Saijo, Y; Hata, S.; Kyozuka, J; Shimamoto, K and Izui, K (2000). Over-expression of a single Ca^{2+}-dependent protein kinase confers both cold and salt/drought tolerance on rice plants. *Plant J* 23: 319–327.

Sanders, D; Brownlee, C and Harper, JF (1999). Communicating with Calcium. *Plant Cell* 11: 691-706.

Sanders, D; Pelloux, J; Brownlee, C and Harper, JF (2002). Calcium at the crossroads of signaling. *Plant Cell* 14: S401–S417.

Satterlee, JS and Sussman, MR (1998). Unusual protein kinases in the plasma membrane of plants. *J Membr Biol* 164: 205-213.

Schaller, GE; Harmon, AC and Sussman MR (1992). Characterization of a calcium- and lipid-dependent protein kinase associated with the plasma membrane of oat. *Biochemistry* 31: 1721–1727.

Scrase-Field, SA and Knight, MR (2003). Calcium: just a chemical switch? *Curr. Opin. Plant Biol* 6: 500–506.

Sheen, J (1996). Ca^{2+}-Dependent Protein Kinases and Stress Signal Transduction in Plants, *Science* 274: 1900-1902.

Strynadka NC and James, MNG(1989). Crystal structure of the helix-loop-helix calcium binding proteins, *Annu Rev Biochem* 58: 951-998.

Szczegielniak, J; Liwosz, A; Jurkowski, I; Loog, M; Dobrowolska, G; Ek, P; Harmon AC and Muszyńska G (2000) Calcium-dependent protein kinase from maize seedlings activated by phospholipids. *Eur J Biochem* 267: 3818–3827.

Szczegielniak, J; Klimecka, M; Liwosz, A; Ciesielski, A; Kaczanowski, S; Dobrowolska, G; Hormon, AC and Muszyńska, G (2005). A Wound-Responsive and Phospholipid-Regulated Maize Calcium-Dependent Protein Kinase, *Plant Physiol* 139: 1970–1983.

Talke, IN; Blaudez, D; Maathuis, FJ and Sanders, D (2003). CNGCs: prime targets of plant cyclic nucleotide signalling? *Trends Plant Sci* 8: 286–293.

Trewavas, AJ and Knight, MR (1996). Calcium imaging shows differential sensitivity to cooling and communication in luminous transgenic plants. *Cell Calcium* 19:211–218

Trewavas, AJ and Malho, R (1997). Signal perception and transduction: The origin of the phenotype. *Plant Cell* 9: 1181– 1195.

Urao, T; Katagiri, T; Mizoguchi, T; Yamaguchi-Shinozaki, K; Hayashida, N and Shinozaki, K (1994). An Arabidopsis thaliana cDNA Encoding Ca^{2+}-Dependent Protein Kinase. *Plant Physiol* 105: 1461-1462.

Verhey, SD; Gaiser, JC and Lomax, TL (1993). Protein Kinases in Zucchini (Characterization of Calcium-Requiring Plasma Membrane Kinases. *Plant Physiol* 103: 413-419.

Warren, DW and Knight, MR (2003). The sfr6 mutant of Arabidopsis is defective in transcriptional activation via CBF/DREB1 and DREB2 and shows sensitivity to osmotic stress. *Plant J* 34: 395–406.

Watillon, B; Kettman, R; Boxus, P and Burny A (1993). A calcium/ calmodulin-binding serine/threonine protein kinase homologous to the mammalian type II calcium/calmodulin protein kinase is expressed in plant cells. *Plant Physiol* 101: 1381 1384.

Weaver, CD; Crombie, B; Stacey, G and Roberts, DM (1991). Calcium-Dependent Phosphorylation of Symbiosome Membrane Proteins from Nitrogen-Fixing Soybean Nodules. *Plant Physiol* 95: 222-227.

Xiang-Chun Yu; Mei-Jun Li; Gui-Feng Gao; Hai-Zhong Feng; Xue-Qing Geng; Chang-Cao Peng; Sai-Yong Zhu; Xiao-Jing Wang; Yuan-Yue Shen and Da-Peng Zhang(2006).Abscisic acid stimulates a calcium-dependent protein kinase in grape berry. *Plant Physiol* 140: 558–579.

Yang, G; Shen, S; Yang S and Komatsu S (2003) *OsCDPK13*, a calcium-dependent protein kinase gene from rice, is induced in response to cold and gibberellin. *Plant Physiol* Biochem 41: 369–374.

Yoon, GM; Cho, HS; Ha, HJ; Liu, JR and Lee, HP (1999) Characterization of NtCDPK1, a calcium-dependent protein kinase gene in *Nicotiana tabacum*, and the activity of its encoded protein. *Plant Mol Biol* 39: 991–1001.

Yu, XC; Li, MJ; Gao, GF; Feng, HZ; Geng, XQ; Peng, CC; Zhu, SY; Wang, XJ; Shen, YY and Zhang, DP (2006) Abscisic acid stimulates a calcium-dependent protein kinase in grape berry. *Plant Physiol* 140:558–579.

Yu ,XC; Zhu, SY; Gao GF, Wang, XJ; Zhao, R; Zou, KQ, Wang, XF; Zhang ,XY; Wu, FQ; Peng, CC and Zhang DP (2007). Expression of a grape calcium-dependent protein kinase ACPK1 in Arabidopsis thaliana promotes plant growth and confers abscisic acid-hypersensitivity in germination, post-germination growth, and stomatal movement. *Plant Mol Biol* 64:531–538.

Zhang, M; Liang, S and Lu, YT (2005). Cloning and functional characterization of NtCPK4, a new tobacco calcium-dependent protein kinase. *Biochim Biophys Acta* 1729: 174-185.

Zhao, Y; Franklin, RM and Kappes, B (1994a). Plasmodium falciparum calcium-dependent protein kinase phosphorylates proteins of the host erythrocytic membrane. *Mol. Biochem. Parasitol* 66: 329-343.

Zhao, Y; Pokutta, S; Maurer, P; Lindt, M; Franklin, RM and Kappes, B (1994b). Calcium-Binding Properties of a Calcium-Dependent Protein Kinase from Plasmodium falciparum and the Significance of Individual Calcium-Binding Sites for Kinase Activation. *Biochemistry* 33: 3714-3721.

13

WATER DEFICIT STRESS AND ITS MANAGEMENT IN GROUNDNUT

A.L. Singh, R.N. Nakar, Nisha Goswami, K.A. Kalariya, K. Chakraborty and Man Singh*

Directorate of Groundnut Research, P.B. 5, Junagadh 362 001, Gujarat, India
Email: alsingh@nrcg.res.in and alsingh16@gmail.com
**WTC, Indian Agricultural Research Institute, New Delhi 110012, India*

INTRODUCTION

The groundnut (*Arachis hypogaea* L), is an important food legume of tropical and subtropical world presently grown on about 24 million hectare (ha) of land in about 120 countries under different agro-climatic zones between latitudes 40ºS and 40ºN. It is native of South America and was disseminated through colonial sea board by Spanish and Portuguese to other countries and presently cultivated mainly in Asian (11.5 m ha), African (11.5 m ha) and American (1.1 m ha) countries mainly in semi-arid regions and India, China, Nigeria, USA, Myanmar, Senegal, Sudan, Indonesia, Argentina and Vietnam are the major groundnut producing countries. Groundnut is an energy rich crop, but mostly grown under energy starved conditions across wide range of environments where frequent drought is one of the limiting factors adversely affecting its productivity in rainfed area. As a result the groundnut productivity is less than 1000 kg ha^{-1} in more than 50% of the groundnut growing countries in the world, between 1000-2000 kg ha^{-1} in 35-40% of the countries and only 10-15% of the countries had the productivity above 2000 kg ha^{-1} (FAO, 2012). However, the world average yield is around 1600 kg ha^{-1} and about 70% of the world groundnut production occurs in the semi-arid to arid tropics where the average yield is still around 1000 kg ha^{-1}.

Groundnut production fluctuates considerably as a result of rainfall variability. There is wide range of groundnut productivity varying from about 500 kg ha^{-1} (poor) in Angola and Mozambique (extremely low about 300 kg ha^{-1}), Madagasker, Namibia, Niger, Uruguay and Zimbabwe, about 3000 kg ha^{-1} (high) in China, Egypt, Syrian Arab Republic, about 4000 in USA, Malaysia, Saudi Arabia, Palestine and Nicaragua, as much as 6400 kg ha^{-1} (very high) in Israel and extremely high (more than 12000 kg ha^{-1}) in Cyprus (FAO, 2012).

Abiotic stresses are the major challenge in sustainable food production, with a potential yield reduction of 70% in crop plants. Of all the abiotic stresses, drought is regarded as the most damaging. In India the groundnut is grown on an area of about 6 million hectare, producing about 8 million tonne (mt) of pod and is the most important oilseed crop of the country. Presently, India has the largest groundnut area (24% of the world) accounting for only 20% of the world groundnut production, but China with only 18% area contributes 39% of the world production due to high productivity and better management practices. Though the average groundnut yield in India is around 1400 kg ha^{-1}, combination of improved varieties and better agronomic practices recorded more than 6000 kg ha^{-1} pod yield frequently and occasionally 8000 kg ha^{-1} (Singh 2004a, 2011). This clearly indicates that the potential of groundnut has not been exploited even by one-third and there is tremendous scope to increase the yield through understanding of its physiology and water relation.

Characterization of agricultural drought is essential before undertaking a yield improvement programme in semi-arid zones. A simplified model combining evapotranspiration (ET) and water balance concepts with basic data on plant responses to drought for groundnut, the applicability of the same for diagnosing drought types. The physiological studies and drought tolerance of this crop started in 80s and by now ample of studies have been conducted. Due to underground fruiting, indeterminate growth habit and different botanical types still certain aspects of physiology of this crop are not very clear. To develop a water stress response function in groundnut, research works have been done to improve the performance under varying degrees of stress at various physiological stages of crop growth. The water stress affects the vegetative, root and reproductive growth and a proper scheduling of irrigation is required.

The complex nature of drought tolerance limits its management through conventional breeding methods. Innovative biotechnological approaches have enhanced our understanding of the processes underlying plant responses to drought at the molecular and whole plant levels. Hundreds of drought stress-induced genes have been identified and some of these have been cloned. Plant genetic engineering and molecular marker approaches allow the development of drought-tolerant germplasm.

In groundnut drought-stress effects depend primarily on the stress pattern because genotypic variation is usually of secondary significance. The different responses of groundnut cv. to drought when assessed relative to the mean response of all genotypes to drought as two major aspects of drought (duration, intensity, and timing relative to crop phenophases) may vary independently. The timing of drought has a large impact on the variation about the mean response. The sensitivity of a genotype to drought increases with yield potential, increasing the closer the drought ends to final harvest. Genotypic variation in response to drought exists in the water-use ratio of genotypes, with some being able to accumulate up to 30% more shoot DM than others with the same total transpiration.

The knowledge of crop physiology and management under water-deficit stress is important for achieving optimal crop yield under limited water availability as introduction of an improved genotypes into new region is largely determined by temperature and phenology, which are essential component of whole crop simulation model and can be used to specify the most appropriate rate and time of specific developmental process to maximize yield. In this chapter an attempt was made to combine all the knowledge of water deficit stress and its impact on groundnut agronomy and physiology and management practices to grow high yielding groundnut varieties with targeted yield and recommend the same to groundnut workers to increase the productivity. Effects of water stress on these components are discussed separately; however all of them are interrelated.

2. EXTENT OF PROBLEM AREAS AND SOILS

The cultivation of groundnut, due to its wide adaptability has been spread on almost all soils in all the tropical and subtropical countries throughout the world. However, on large scale it is mainly grown in India, China, Nigeria, USA, Myanmar, Indonesia, Sudan, Senegal, Argentina and Vietnam producing more than 0.5 million tonne (mt) and in Ghana, Chad, Congo Republic, Mali, Guinea, Niger, Argentina, Brazil, Tanzania, Burkino Faso, and Malawi producing in between 0.25-0.5 mt. Though it is a energy rich crop, more than 85% of the world groundnut production come from low income food deficit countries with an average productivity of about 1500 kg ha^{-1} and having more than 90% of the groundnut growing areas of the world.

Though native of south America, the groundnut cultivation is settled in southern, eastern and south-eastern part of Asia, western Africa and northern and south America due to favourable soils and climates. The commercial groundnut cultivation is mainly in Asian (48% of the world groundnut area contributing 64% of the total world production), African (47% area, 27% production) and American (4.2% area and 8% production) countries due to suitable environment and matching growing season. Though grown in limited area, the groundnut productivity of Cyprus (<100 ha area) is highest (>12000 kg ha^{-1}) followed by Israel (2600 ha and 6440 kg ha^{-1}) in the word mainly due to favourable season and high cultivation practices. On the other hand the productivity of many African countries, with significant areas, are still around 400 kg ha^{-1} because of poor resources and scanty rainfall. However, on large scale cultivation the productivity of USA, China, Egypt, Turkey, Argentina and Nicaragua are very high (>3000 kg ha^{-1}).

In India, the groundnut is grown in about 260 districts mostly as rainfed dry lands, crop on well drained sandy soils in low (<750 mm) and medium (750-1000 mm) annual rainfall areas, often subject to the vagaries of the weather and only 20% of groundnut area is under irrigation. Between the decades of 60s and 70s, there is practically little difference in productivity (700-800 kg ha^{-1}) and the increase in production was largely due to the expansion in areas. But during 1988-89 due to favorable season and transfer of available technologies first time the productivity crossed one tonne (1132 kg ha^{-1}) and it was 1357 kg ha^{-1} during 2005-06 and 1459 kg ha^{-1} during 2007-08. In India generally the groundnut is grown as rainfed crop during rainy season (Kharif) with one or two protective irrigation and also during Rabi, summer and spring season as a irrigated crop with higher yield potential than in kharif. Presently, in India the average productivity of rabi-summer groundnut is about 1850 kg ha^{-1}, much higher than kharif season (1410 kg ha^{-1}) indicating more production potential during this season.

Presently, Gujarat (30% total area and 36-40% of production), Andhra Pradesh (28% area and 20-28% of production), Tamil Nadu (7% area and 11% of production), Karnataka (14.5% area and 10% of production), Rajasthan (6% area and 8.2% of production) and Maharashtra (6.1% area and 5.5% of production) are the main groundnut growing states. The Madhya Pradesh, Orissa, Uttar Pradesh and West Bengal are the other groundnut growing states with 7% of area contributing 7% of the total production of the country. If we critically analyse the situation of groundnut production in India, though the productivity (average of both the season) of groundnut during the year 2001 to 2010 ranged from 700-1460 kg ha^{-1}, the productivity in three major groundnut growing states, accounting for about 75% of the total productivity of the country, was in between 1473-2390 kg ha^{-1} in Gujarat, 1400-2130 kg ha^{-1} in AP and 2100-3730 kg ha^{-1} in Tamil Nadu during rabi-summer season, but fluctuated in between 510-2270, 300-1360, 1150-1880 kg ha^{-1}, respectively in these states during kharif season.

Drought is an insidious hazard of nature and is considered to be the most complex but least understood of all natural hazards. Large historical datasets are required to study drought and these involve complex interrelationships between climatological and meteorological data. Rainfall is an important meteorological parameter, the amount and distribution influence the type of vegetation in a region. Drought being very complex phenomenon, there is no universally accepted definition. It is a meteorological term and is commonly defined as a period without significant rainfall denoting scarcity of water in a region. Prolonged deficiencies of soil moisture adversely affect crop growth indicating incidence of agricultural drought. It is the result of imbalance between soil moisture and evapo-transpiration needs of an area over a fairly long period so as to cause damage to standing crops and to reduce the yields. The irrigation commission of India defines drought as a situation occurring in any area where the annual rainfall is less than 75% of normal rainfall. The book "Drought: Assessment, Monitoring, Management and Resources Conservation" by Nagarajan, listed chronically affected districts by drought conditions, of these groundnut growing one in various states are mentioned in Table 1.

Table 1. Groundnut growing districts chronically affected by drought conditions

Andhra Pradesh	Anantpur, Chittoor, Cuddapah, Karnool, Prakasam, Nalgonda, Mehboobnagar, Hyderabad
Gujarat	Ahmedabad, Amrely, Banaskantha, Bhavnagar, Jamnagar, Kheda, Kutch, Mehsana, Panchmahal, Rajkot, Surendranagar
Haryana	Bhiwani, Mahendranagar, Rohtak
Karnataka	Bangalore, Belgaum, Bellary, Bijapur, Chitradurga, Chickmagalur, Dharwad, Gulbarga, Hassan, Kolar, Mandya, Mysore, Raichur, Tumkur
Madhya Pradesh	Betul, Jhabhua, Khandak, Shahdol, Shahjapur, Sidhi, Ujjain
Maharashtra	Ahmednagar, Aurangabad, Beed, Nanded, Nashik, Osmanabad, Pune, Parbhani, Sangli, Satara, Solapur
Orissa	Phulbani, Kalahandi, Bolangir, Kendrapada
Rajasthan	Ajmer, Banswara, Barmer, Churu, Dungarpur, Jaisalmer, Jalore, Jhunjhunu, Jodhpur, Nagaur, Pali, Udaipur
Tamil Nadu	Coimbatore, Dharmapuri, Madurai, Ramanathapuram, Salem, Tiruchirapali, Tirunelveli, Kanyakumari
Uttar Pradesh	Allahabad, Banda, Hamirpur, Jalan, Mirzapur, Varanasi, Mainpuri
West Bengal	Bankura, Midnapore, Purulia
Jharkhand	Palamau
Chhattisgarh	Khargaon

Important causes for agricultural drought are inadequate precipitation, erratic distribution, long dry spells in the monsoon, late onset of monsoon, early withdrawal of monsoon along with lack of proper soil and crop management. Drought stress tolerance is seen in almost all plants but its extent varies from species to species and even within species.

Water deficit and salt stresses are global issues to ensure survival of agricultural crops and sustainable food production. Stress resistance may involve following mechanisms

- Avoidance mechanisms prevents exposure to stress
- Tolerance mechanisms permits the plant to withstand stress through osmotic adjustment

- Acclimation alter their physiology in response to stress

At whole plant level the effect of stress is usually perceived as a decrease in photosynthesis and growth, and is associated with alteration in carbon and nitrogen metabolism (Mwanamwenge *et al.*, 1999). Drought stress affects the growth, dry matter and harvestable yield, but the tolerance of genotypes to this menace varies remarkably.

Drought is the major abiotic constraint affecting groundnut productivity and quality worldwide. Groundnut plants are drought tolerant because of deep rooting and a water supply-related flexibility in time of flowering and fruiting. Various conservative water management treatments over a 3-yr period when used to determine water use and yield response of groundnuts growing on deep well-drained sandy soils yields were not reduced by droughts of short duration unless the seasonal water use was below about 50 cm. The pod yields were 2.26, 3.0 and 3.82 t/ha with approximately 33, 40 and 46 cm water, respectively (Hammond *et al.*, 1978).

Groundnut drought adaptation mechanisms when surveyed (with a view to developing selection criteria for breeding) under the headings (1) drought evasion (ability to complete the development cycle before water deficits occur), (2) drought avoidance (mechanisms such as modified root and leaf morphology which allow the plant to keep its tissues at a high water potential during drought) and (3) tolerance to drought (maintenance of potential turgidity by osmotic adjustments, and tolerance of desiccation due to properties of the cell membrane) (Annerose, 1988).

In a study the joint regression approach of stability analysis is briefly discussed and the economic concepts of risk and utility maximization and alternative approaches are considered. For groundnut improvement in India at 3 regions (Hyderabad, Anantapur and Gujarat) the response relationships between groundnut yields and relative available water are estimated and empirical yield distributions were simulated and the results of alternative risk analysis approaches when compared with the traditional stability analysis of the experimental data and recommendations for future plant breeding methodology and risk analysis are presented (Bailey and Boisvert, 1989).

To analyse the changes in vegetation cover due to variation in rainfall and identify the land-use areas facing drought risk, rainfall data from 1981 to 2003 were categorized into excess, normal, deficit and drought years. The Advanced Very High Resolution Radiometer (AVHRR) sensor's composite dataset was used for analysing the temporal and interannual behaviour of surface vegetation. The various land-use classes - crop land (annual, perennial crops), scrub land, barren land, forest land, degraded pasture and grassland were identified using satellite data for excess, normal, deficit and drought years. Normalized Difference Vegetation Indices (NDVIs) were derived from satellite data for each land-use class and the highest NDVI mean values were 0.515, 0.436 and 0.385 for the tapioca crop in excess, normal and deficit years, respectively, whereas in the drought year, the groundnut crop (0.267) showed the maximum. Grassland recorded the lowest value of NDVI in all years except for the excess year. Annual crops, such as groundnut (0.398), pulses (0.313), sorghum (0.120), tapioca (0.436) and horse gram (0.259), registered comparatively higher NDVI values than the perennial crops for the normal year. The Vegetation Condition Index (VCI) was used to estimate vegetation health and monitor drought. Among land-use classes, the maximum groundnut witnessed the maximum values of 78.2, 64.5 and 55.2% for normal, deficit and drought years, respectively. Based on the VCI classification, all land-use classes fall into the optimal or normal vegetation category in excess and normal years, whereas in drought years most of the land-use classes fall into the drought category except for sorghum, groundnut, pulses and grasses. These crops (sorghum 39.7%, groundnut 55.2%, pulses 38.5% and grassland 38.6%) registered maximum VCI values, with sustained under drought conditions. It is suggested that the existing crop pattern be modified in drought periods by selecting the suitable crops of sorghum, groundnut and pulses and avoiding the cultivation of onion, rice and tapioca (Muthumanickam *et al.*, 2011).

Most breeding programmes in groundnut follow an empirical approach to drought resistance breeding, largely based on kernel yield and traits of local adaptation, resulting in slow progress. Recent advances in the use of easily measurable surrogates for complex physiological traits associated with drought tolerance encouraged breeders to integrate these in their selection schemes. However, there has been no direct comparison of the relative efficiency of a physiological trait-based selection approach (Tr) vis-a-vis an empirical approach (E) to ascertain the benefits of the former.

In field studies, for 5 years at ICRISAT Patancheru, groundnuts advanced breeding lines produced greater pod yields on Vertisols (2.02-3.81 t/ha) than on Alfisols (0.61-1.56 t) and there was a strong soil type x genotype interaction. In another study, 4 cultivars irrigated or water stressed during flowering, pod-set or pod-filling showed that while CGR were greater on Alfisols, they were linearly related to those measured on Vertisols. However, pod growth rates and partitioning of DM to pods showed a strong soil type x genotype interaction and the genotypes developed on the Alfisol maintain relative ranking for total DM on Vertisol, but not necessarily for pod yields (Rao *et al.,* 1992).

A study was carried out to quantify the impact of drought on production of five major kharif crops (rice, groundnut, cotton, bajra, soyabeans) and a rabi crop (wheat) using the standardized precipitation index (SPI) that captures cumulative rainfall deviations at various time scales, computed for 36 meteorological sub-divisions at monthly (SPI1), bimonthly (SPI2) and tri-monthly (SPI3) scales using monthly rainfall data for the period of 1971-2002. July was identified as the most drought affected, followed by September, while June and August were near normal. In July, out of 36 meteorological sub-divisions, 26 showed rainfall below normal. Amongst these 26 drought affected sub-divisions, six sub-divisions were very severely affected. Correlation coefficients were computed between production of major kharif crops (1980-2001) and SPI values and September was the crucial month for defining the crop yield for most of the kharif crops throughout the country. Production forecast using SPI3 (July to September) showed good agreement with statistics provided by state department of agriculture for kharif crops (Chaudhari and Dadhwal, 2004).

The potential (no-water stress) and the lowest (no irrigation) yields for maize, soyabean and groundnut were calculated using three crop growth and water use models - CERES-Maize, SOYGRO, and PNUTGRO where rainfall, temperature and solar radiation records were used with these models to identify the 15 most severe drought years in the 53 year record in a 36-county region of Georgia, USA, that contains 75% of Georgia's irrigated land. In the 15 driest years, simulated yield losses averaged 75% for maize, 73% for soyabean, and 64% for groundnut. Irrigation amount and timing needed to provide 90% of the no-stress yields when calculated the irrigation needs of maize in these drought years occurred before that of groundnut or soyabean. For the reported irrigated crop acreage of the study area, simulated water withdrawals exceeded 3 million m^3/day, on an average, for most of the 130 days between late May and late September (Hook, 1994).

3. QUANTIFICATION OF DROUGHT AND YIELD LOSSES

Groundnut is an important crop of the semi-arid tropics where potential yields are frequently reduced by heat and water stress. Studies on occurrence and intensity of the drought during crop growing season revealed the effect of moisture stress on the groundnut yields in the dry lands. The groundnut is relatively drought resistant and important crop of semi-arid regions where evaporation exceeds precipitation for 5-10 months of the year. The plant water-status is the result of a balance between water uptake and loss which has been less understood in groundnut. Groundnut plant contains about 80% of water on fresh weight basis and reduction of the plant water status much below this level causes wilting and affects the rate of several plant functions. Though different stages have different sensitivity to water deficit, none of these can proceed normally below some minimum water. Increasing moisture stress from 0 to 2, 4 and 6 atm decrease the leaf RWC, increased water saturation deficit and relative saturation deficit, decreased DM accumulation in the shoot and increased it in the root, decreased RGR, increased specific leaf weight, decreased non-reducing sugar and increased reducing sugar in groundnut seedlings (Sharma *et al.*, 1985).

A study was conducted to characterize the plant extractable water pattern at four locations in India (Tirupati, ICRISAT, Jalgaon and Junagadh) and one location in Queensland, Australia (Kingaroy) and explore the possibility of clustering the multi-location trial environments based on similar water stress patterns. The APSIM groundnut model was used to compute daily changes in plant extractable soil water (P_{esw}) at each site, by using climate parameters (ambient temperature, radiation, rainfall or irrigation amounts), soil hydraulic parameters and crop parameters (planting and harvest dates). Results from the P_{esw} characterization of experimental sites clearly demonstrated that the crops grown at the multi-location have experienced a wide variation in timing, intensity and duration of crop water deficits during the growing season and that quantification of the P_{esw} during the growing season and clustering of environments based on P_{esw} patterns can assist in understanding the basis of G x E interactions for yield between clusters, and to examine the effect of breeding methods on yield variation within each of the clusters (Rachaputi, 2003).

Most part of India, shortage of water is caused by uneven distribution of rains, gaps between rain events and field water losses rather than from low seasonal or annual rainfall totals. The groundnut grown in the micro catchment during the rainy season, utilized 364-733 mm water in evapotranspiration (ET) and deep percolation (P) (Rathore *et al.*, 1996). In a water balance studies in M.P. in a 1.05 ha field, with a 0.09 ha farm pond (which stored excess water from the wet season) 28-37% of seasonal rainfall was available as surface runoff from a microcatchment (0.66 ha growing groundnut) for collection in the pond and is sufficient to prevent drought stress (Rathore *et al.*, 1996). Analysis of 20 years of rainfall data of Tirupati for drought classification using aridity index on annual and monthly basis the correlated groundnut yields were low due to uneven distribution of rainfall during crop growing season and moisture stress during July and September coincided with moisture critical periods (Sumathi and Subramanyam, 2007).

The crop coefficient curve facilitates the prediction of groundnut ET in preparation of planting at a new site from estimates of reference crop ET. On a sandy loam soil of Hyderabad, A.P. the crop coefficient (Kc) values of groundnut at different crop-growth subperiods were influenced by evapotranspiration deficits and leaf area development of the crop (Devi and Rao, 2003). The fully irrigated control (W-W-W) showed higher Kc values at all the crop-growth subperiods than other treatments. The crop coefficient curve for W-W-W showed that the Kc value was low (0.564) during the establishment of plant (0-10 DAS), increased linearly through vegetative period and remained constant at 1.024 from flowering to start of the pod filling period (35-80 DAS), then decreased through pod filling period and reached a lowest value of 0.547 during the final 10 days of the crop period. The crop coefficient curve facilitates the prediction of groundnut ET and for field application the net (39.9 cm) and gross irrigation requirements, both at field inlet (5563 m^3/ha) and headwork (9850 m^3/ha) were determined (Devi and Rao, 2003). The reported yield losses in groundnut due to soil water deficit stress under different conditions are summarised in Table 2.

Table 2. Yield losses due to water stresses in groundnut at various locations

SN	Groundnut varieties	Soil type and places	% Yield losses	Conditions	References
1	ICGV 86031, TMV 2 NLM	Andhra Pradesh	17-18% losses in total dry matter production	Under various soil water stress	Reddy, 1999
2	-	36-county region of Georgia, USA	64% pod yield	15 most severe drought years in 53 year simulated yield losses, PNUTGRO model	Hook, 1994
3	-	-	22, 18, 47 and 47%, Yield reductions with drought at 10-30, 30-50, 50-80 and 80-120 DAS, respectively.	Imposing drought at various stages	Billaz and Ochos, 1961
4	Six cultivars	Akola, Maharashtra	47% yield reduction in six cultivars under stressed conditions	Kharif on shallow soil (<20 cm) with 30% water holding capacity	Dhopte *et al.*, 1992
5	JL-24		32% in JL-24 a water stress tolerant	Kharif	Dhopte *et al.*, 1992
6	TAG-24		67% in TAG-24 the most susceptible cultivar.	Kharif	Dhopte *et al.*, 1992
7	GAUG-10 and J-11	Junagadh, Gujarat	Pod yield by 27, 45, 56 and 6%, respectively in J-11, and 13, 15, 38 and 6% , in GAUG-10 with water stress at FL, Pg I, pod development and pod maturation stages.	Summer, water stress at FL (28-48 DAE), Pg I (40-60 DAE), PD (55-75 DAE) and maturation (75-95 DAE)	Golakiya, 1993
8	SB-XI	-	24% yield reduction at IW:CPE ratio of 0.5 than with 0.75 IW:CPE throughout.	Irrigated at (IW): (CPE) ratios of 0.75 or 0.5 (40 mm/ irrigation)	Patil and Gangavane, 1990
9	Robut 33-1	-	28-96% reduction in seed yield at stress during the start of seed growth to maturity.	water stress during post-rainy	Rao *et al.*, 1985
10	GG 6	Clay soil at Junagadh	18% and 31% reduction due to water stress at FL (25-47 DAS) and P D (50-72 DAS) stages, respectively.	Soil moisture stress at various growth stages, summer season	Vaghasia *et al.*, 2010a

11	-	-	5% decrease in pod yields at soil moisture stress 30-45 DAS	Soil moisture stress at 30-45 DAS, caused drying of first flush of flowers up to 45 DAS	Gowda and Hegde, 1986
12	Robut-33 and McCubbin	Redland Bay, Queensland	Reduced pod yield by **30%** in the Virginia Bunch Robut-33 and by **45%** in a Spanish McCubbin	water was withheld from 84 d after sowing to maturity	Chapman *et al.*, 1993
13	-	-	20 and 59%, reduction in seed yields with drought for 3 weeks starting week 5 and 6 respectively.	Drought for 3 weeks at 7 growth stages starting 4-10 weeks after sowing in field	Zaharah, 1986
14	-	-	83 - 97% , yield reduc-tions due to water stress	Rainfed conditions	Huang and Ketring,1985
15	-	18 different locations in India	49% reduction in yield, water limiting yield 2750 kg ha^{-1} as against simulated water non-limiting yield of 5440 kg ha^{-1}	CROPGRO-Peanut model	Bhatia *et al.*, 2009
16	TG-17	-	Water stress during early and late VG, FL, Pg I and PD stages reduced pod yields by 20, 26, 33, 45 and 56%.	Water stress of -14 bar soil water potential during the various growth stages	Parmar *et al.*, 1989
17	GG-2	Junagadh, medium clay soil	36, 48, 52 and 56% pod yields losses due to water stress at FL, PgI, pod formation and pod development , respectively	Water stress at various growth stages	Sakarvadia and Yadav, 1994

Where, FL= Flowering, FR= Fruiting, PI= Pod initiation, PD= Pod development, VG= Vegetative, RP= Reproductive, M= maturity, MI= Maturity initiation stages, FI= Flower initiation, Pg I=Peg initiation

In central India Azam Ali, (1984) studied the interaction between population and water stress in 4 populations of groundnut by estimating, transpiration, stomatal resistance (rs), boundary layer resistance (ra), vapour concentration difference between leaf and air (deltachi) and LAI and the frequency distributions of rs, ra, deltachi and seasonal changes in LAI when plotted to analyse the dependence of transpiration rate on each variable both per unit area of leaf surface (E1) and per unit land surface (Ee), for estimates of E1, both rs and deltachi were of similar importance, exerting a far greater influence than changes in ra. However, in terms of Ee, changes in LAI were far more important than in any other variable, particularly late in the season when water was scarce. The ability of this technique to describe temporal and spatial variations as well as the dominant environmental and physiological influences on transpiration may outweigh any small loss in accuracy of estimates.

In Eastern India at Bhubaneswar, the crop coefficient of groundnut cv. AK12-24, for use in estimation of crop Et during the dry season (November-March) using different irrigation cycles, the daily moisture use rate increased gradually and reached the peak value (4.10-4.94 mm) during 55 to 60 days. The crop coefficient values followed the same trend as that of crop Et, which were lower at the initial stage of growth (0.61-0.80), increased gradually and attained the maximum value of 0.94-1.33 towards the peak period of crop growth and declined thereafter. The crop coefficient value approached unity or slightly exceeded it during the maximum growth stage of the crop. Plants stressed at the early vegetative stage showed lowest crop coefficient value (0.61) at initial stage and the highest value (1.33) at peak crop growth stage, and witholding irrigation at an early stage (14 DAS) resulted in lesser evaporation of water from the soil surface (Kar *et al.*, 2001).

Estimation of surface sensible and latent heat flux is the most important to appraise energy and mass exchanges among atmosphere, hydrosphere and biosphere and the surface energy fluxes were measured over irrigated groundnut during winter (dry) season using Bowen ratio (beta) micrometeorological method in a representative groundnut growing areas of eastern India, at Dhenkanal, Orissa by growing the crop with four irrigations based on phenological stages viz., (i) branching, (ii) pegging, (iii) pod development and (iv) seed filling and assessed what the crop stress was at those times to see if irrigation scheduling could be optimized further. The net radiation R_n varied from 393-437 to 555-612 W m^{-2} during two crop seasons. The soil heat flux (G) was higher (37-68 W m^{-2}) during initial and senescence growth stages as compared to peak crop growth stages (1.3-17.9 W m^{-2}). The latent heat flux (LE) showed apparent correspondence with the growth which varied between 250 and 434 W m^{-2} in different growth stages. The diurnal variation of Bowen ratio (beta) revealed that there was a peak in the morning (9.00-10.00 a.m.) followed by a sharp fall with the mean values varied between 0.24 and 0.28. The intercepted photosynthetic photon flux density or photosynthetically active radiation (IPAR) by the crop was measured and relationship between IPAR and leaf area index (LAI) was established with DAS, which will be useful in developing algorithm of crop simulation model for predicting LAI or IPAR. The stressed and non-stressed base lines were also developed by establishing relationship between canopy temperature and vapour pressure deficit (VPD). With the help of base line equation, [$(T_c-T_a) = -1.32VPD+2.513$], crop water stress index (CWSI) was derived on canopy-air temperature data collected frequently throughout the growing season. The soil moisture depletion during the crop period when plotted with CWSI at different stages the values of CWSI varied between 0.45 and 0.64 just before the irrigations were applied and at two stages (branching and pegging), CWSI were much lower (0.46-0.49) than that of recommended CWSI (0.60) for irrigation scheduling. Therefore, more research is required to optimize the phenology based irrigation scheduling further in the region (Kar and Kumar, 2007).

Field data on pod yield and seasonal ET as influenced by irrigation schedules during the summer predicted that the pod initiation and development stage (70 d to harvest) was the most sensitive stage for moisture stress with a yield response factor of 2.10. Water stress during 10 to 40 d was beneficial in enhancing pod yield with a yield response factor of 2.10 (Ramachandrappa and Nanjappa, 1994).During winter at Rajendranagar, Hyderabad, groundnut cv. ICGS 44 irrigated to give moderate to severe ET deficits at various stages showed that reproductive stage (35-115 DAS) was the most sensitive to a reduction in water supply, whereas water stress in the vegetative stage (10-35 days) had the least effect (Reddy *et al.*, 1996). At Coimbatore, TN, during summer 1994, water stress in groundnuts cv. Co 2 and VRI 2 at flowering, pegging, pod development or pod maturation when compared water stress at pod development was most detrimental on yield (Velu, 1998).

In Western India, a lysimeter experiment on black calcareous vertic Inceptisol at Junagadh, two groundnut cultivars subjected to water stress from the seedling to flowering (24-48 DAE), flowering to pegging (40-60 DAE), pegging to pod development (55-75 DAE) or pod development to maturation (75-95 DAE) decreased pod yields compared with plants given normal irrigations however yield reductions were greatest with stress imposed during the period between pegging and pod development and lowest with stress imposed from pod development to maturation (Patel and Golakiya, 1988). Further lysimeter trials on Spanish bunch groundnuts cv. J11 and GG 2 revealed that water stress from pegging to pod development gave the lowest pod yields with increased leaf temperature (35°C) markedly lowering photosynthesis. In all stress treatments, GG 2 out yielded J 11 mainly due to lower fluctuations in leaf temperature, stomatal resistance and lower vegetative growth (Patel and Golakiya, 1993). Also in another lysimeter studies, groundnut subjected to soil moisture tensions (SMT) of 330, 530 or 730 mbar, maximum daily water consumption occurred at 50-80 and 50-65 days in groundnuts grown at the 2 lower and the highest SMT, respectively and increase in SMT decreased total DM yield, but increased unshelled nut yields (Vivekanandan and Gunasena, 1976).

The total dry matter at harvest had positive correlation with TE, leaflet size was negatively correlated with TE under drought stress, the N content in leaves at 80 DAP and the chlorophyll content in leaves during moisture stress (28 days after imposing stress) showed positive relationship with TE. The leaf temperature 28 days after imposition of moisture stress had significant negative relationship with TE under adequately irrigated and simulated drought treatments. The mineral ash content of leaves 80 DAS in Spanish cultivars (ICG 476, ICG 221, ICG 1697, ICGV 86031 and TAG 24) had significant positive correlation with

TE in simulated drought treatment (Babitha and Reddy, 2001). Sharma *et al.* (1987) studied the performance of two groundnut cv. under soil moisture stress during rainy season where number of gynophores and pods/plant, 100-seed wt, pod yield and shelling percentage were highest with two irrigations at 50 and 80 DAS and were lowest under rainfed conditions. Irrigation at 80 DAS was more effective than irrigation at 50 DAS. The moisture stress suppressed pod setting more in cv. M13 than in cv. M37. Oil content in seeds was not affected by moisture stress in both cultivars.

Studies on radiation and energy budgets over a cropped surface in the Sabarmati river basin, Gujarat, India by recording continuous data on temperature, humidity, wind speed and direction at 1 and 4 m on a 9-m tower, soil heat flux sensible and latent heat fluxes from March to August1997, a polynomial relationship between residual flux and biomass under different phenological phases of the crop was observed, a linear relationship was found between residual flux and plant height under different phenophases and the biomass of crops increased exponentially with increasing AE: PE ratio and a polynomial trend was observed in water deficit, biomass and height (Padmanabhamurty *et al.*, 2001). A linear relationship for each month was observed in the AE: PE ratio parameterized for three months.

The water requirement of groundnut varies with the stages and is lowest from germination to flower formation and reaches maximum during pod formation. However, the utilization of available moisture is greatest during flowering and pod formation and the crop receiving adequate water during these stages only can give equal yield to the well watered crop. During these stages if stress is given and later on water supply is resumed only the vegetative growth is benefited not the reproductive growth of crop. Thus the period of maximum sensitivity to drought occurs between 50-80 DAS, the period of maximum flowering and vegetative growth. The groundnut production was directly proportional to light interception and to the ratio between water lost and the vapour pressure deficit from leaf to air. Root growth and development was favoured under limited water supply and high water demand. Leaf conductances to gas exchange were similar at different combinations of soil water content and atmospheric saturation deficit (Goncalves de Abreu, 1988).The groundnut cv. SB 11 grown at 3 levels of water stress applied at 4 growth stages, water stress of 0.8 (ratio of IW: CPE) applied at any growth stage reduced pod yield, Maximum pod yield obtainable (3.06 t/ha) was predicted to be obtained with 1131 mm irrigation water (Shinde and Pawar, 1982).

The water–yield relationship in groundnut when studied in Bhavanisagar under deficit irrigation, the yield response factor (k_y) ranges from 0.45 and 0.42 (normal irrigation) to 1.72 and 1.70 (full deficit irrigation) for summer and Rabi seasons, respectively and the pod formation and flowering stages were more sensitive to moisture stress and irrigation during these stages is more important to overcome the yield reduction in groundnut (Thiyagarajan *et al.*, 2010). Response of groundnut cv. Robut 33-1 to drought stress imposed at (a) emergence to maturity, (b) emergence to peg initiation, (c) from the start of flowering to the start of seed growth, and (d) from the start of seed growth to maturity when studied during the post-rainy seasons, the amount of water applied during these phases varied and the greatest reduction in seed yield (28-96%) occurred when stress was imposed during (d)(Rao *et al.*, 1985). Decreased irrigation during (b) increased pod yield relative to the fully irrigated control treatment by 13-19%. The evapotranspiration-yield relationships showed a strong interaction with timing of drought (Rao *et al.*, 1985).

4. FACTORS INFLUENCING WATER DEFICIT STRESS

4.1. Heat Stress

Groundnut is an important crop of the semi-arid tropics where potential yields are frequently reduced by heat and water stress. Craufurd (2000) studied 8 groundnut genotypes varying in heat tolerance in controlled environments at high (40/28°C) and near-optimum (30/24° C) temperatures from 32 DAS to maturity where significant variation among genotypes in main stem leaf number and total flower number at 30/24° and 40/28° and rates of appearance were faster at 40/28° than at 30/24°. Days from sowing to first flowering varied among genotypes from 28 to 41 days and therefore the time of plants were exposed to high temperature relative to first flowering ranged from -4 to 9 days. Fruit number at 40/28° was linearly and negatively related to the time of first flowering relative to the onset of high temperature (r^2=0.93; n=7; $P < 0.001$), indicating that 'escape' was an important component of heat tolerance. Further the fruit number in all genotypes at 40/28°C was closely associated with the cumulative number of flowers that had opened between first flowering and 3 days after the onset of the high temperature regime (r^2=0.95; n=8; $P < 0.001$). Variation in fruit number was due both to the timing of flowering and the initial rate of flower production. The most sensitive stage of development to high temperature occurred around 3 days before flowers opened and hence, it was the timing of flowering, rather than heat tolerance or susceptibility that was the dominant attribute determining fruit number (Craufurd, 2000). The DM partitioning to stems, leaves and pods in groundnut cv. Robut 33-1 investigated at mean air temperature ranging 19-31°C and water stress growing plants at variable levels of saturation vapour pressure deficit, pod: shoot wt. ratio (PWR) was max. at 22°C and decreased from 0.28 to 0.04 as temperature increased to 31°C, mild water stress promoted peg and pod production and increased PWR (Ong, 1984).

Groundnut cv. Florunner subjected to drought and high temperature stress (28-30°C) for periods of 20, 30, 40 or 50 d, and the seeds from these plants when separated into Jumbo, Medium and No. 1 market size categories, the soluble and total carbohydrate content in seed of Jumbo and Medium categories increased due to drought and temperature stress. The alpha -amino N content of the Jumbo category decreased while that of the No. 1 category increased following a 30-d stress exposure. Protein and oil content of all the categories were not affected after exposure to drought and temp stress. However, the protein and polypeptide profiles showed that a polypeptide with a MW of 70 000 and a pI between 6.2 and 7.0 increased with increasing periods of drought and temperature stress. Thus, Drought and temperature stresses increase accumulation and/or synthesis of carbohydrates and certain polypeptides may enhance *Aspergillus* invasion and aflatoxin production (Musingo *et al.*, 1989).

Groundnut germplasm evaluated for heat- and drought-tolerance traits, genotypic differences in tolerance to temperature above 35°C and heat tolerance, indicated by membrane thermostability (in vitro leaf-disc method with leaf tissue) were observed (Ketring, 1986). The means to improvement of hydration maintenance of this crop under soil-moisture deficits have been sought through genotypic diversity in rooting traits and water-potential components. Genotypes differed in rooting habit and ability to maintain plant-water and water-potential, differences in rate of decrease in water-potential components, osmotic adjustment, and apoplastic water fraction indicating potential for improved heat and drought tolerance (Ketring, 1986). The heat tolerance of ten groundnut genotypes studied under irrigated and rain-fed conditions in the field evaluated on four dates between 23 June and 14 August using an electrolyte leakage technique which measures the thermostability of cellular membranes observed genotype x date interactions, and in general Pearl Early Runner, X 537B, Florunner and X 487A were the most heat tolerant and water stress resulted in increased heat tolerance among four of the five genotypes (Bennett and Hammond, 1982).

Differential canopy/ambient temperature was used by Schubert and Sanders, (1985) to calculate a stress degree day index (the sum of the numbers of °C by which canopy temperature exceeded ambient temperature) for scheduling irrigation in groundnut cv. Florunner at Yoakum, Texas and observed that the plots irrigated after 5-25 stress degree days (SDD) showed declined yield and linearly as SDD level increased. The most irrigated plot, 5 SDD, yielded 2.98 t/ha while 10, 15, 20, 25 SDD and the unirrigated control yielded 2.67, 2.57, 2.45, 1.70 and 1.51 t/ha, respectively. Increased water stress increased the proportion of small seeds (Schubert and Sanders, 1985). Babu *et al.* (1983) studied the mechanism of alleviating water stress by leaflet angle variation in groundnut cv. Tindivanam 2 and a simple method of measuring leaflet angle was devised. Radiation avoidance by leaf closure was exhibited to different degrees by water stressed and non-stressed plants and leaflets under continuous water stress had reduced areas and showed greater leaflet movement (leaf closure) compared to the amount of movement of leaflets on plants without water stress (Babu *et al.*, 1983).

At Tirupati, India, Babitha *et al.* , (2006) screened several genotypes and found that Spanish genotype TIR 21 showed less reduction in Fv/ Fm ratio when exposed to 45°C (1.13% reduction) and 55°C (22% reduction), while the Virginia genotype TIR 34 maintained high Fv/ Fm ratio at temperature >50°C. Genotypes TIR 20 and JAL 31 showed a higher reduction of 84 and 82% respectively in Fv/ Fm ratio when exposed to 55°C. However, these genotypes showed more reduction (62%) in Fv/ Fm ratio at 55°C. Virginia groundnut CSMG 84-1 showed low membrane injury (34%). The genotypes TIR 21, TIR 34, JAL 07 and CSMG 84-1 are better under higher temperatures and hence can be recommended for the specified situation in order to increase the yield potential under high temperature conditions (Babitha, 2006).

4.2. Root Growth and Water Extraction

The groundnut is deep-rooted plant and its root can penetrate a depth ranging from 1.5-2.0 m, but rarely goes beyond 1 m. The groundnut roots extract most of the moisture from upper layer (36% in 0-30 cm depth and only 7% in the region of 120-150 cm depth). However, under moisture stress the crop extract water from greater depth. The root system is normally concentrated at a depth of 5 to 35 cm, and spread is confined to a radius of 12-14 cm. The spreading types are generally more vigorous than bunch types. Most of the roots are in shallower region having root densities of 1.5 cm cm^{-3} in 0-30 cm zone and only 0.1-0.4 cm cm^{-3} at higher depth. The epidermis sloughs off as the root extends destroying the basis for root-hair production resulting in no root hairs. However, Meisner and Karnok (1991) observed the existence of both lateral (0.3 mm) and rosette type (4 mm) root hairs on groundnut root under varying soil and soil-water conditions. In some wild species the hypocotyl and root may be modified to form tubers. Tap roots may vary from a few millimeters in diameter in annual species to 10 cm in perennial species.

Meisner (1991) studied groundnut root growth under 30 d water stress period beginning from 20, 50, 80 and 110 DAS using two non-destructive methods, a rhizotron and minirhizotron where root growth was reduced significantly by stress during 20-50 DAS in the rhizotron but was not affected by stress in the minirhizotron. Groundnut roots grew rapidly, representing a considerable portion of the C partitioned early in the growing season and by 80 DAS, >80% of the total root system was established, while flowering, peg and foliage production had peaked. Yield was significantly reduced by stress treatments 50-80 and 80-110 DAS in both studies. Most of the C partitioned after 80 DAS went into pod formation and pod filling, thus explaining yield reductions during the 50-80 and 80-110 DAS, stress treatments. Two types of root hairs were observed and quantified: rosette-type hairs surrounding lateral initiates emerging from primary roots and shorter, more profuse hairs on root lengths. While root hairs were not significantly affected by water stress, soil type and rooting depth significantly affected hair occurrance. Drought tolerance contributing factors in groundnuts was: an extensive root system established before maximum leaf area and consequent peak transpirational demand was reached; flowering, delayed when under water stress, recurred once stress was relieved; water storage cells in the abaxial side of the leaves provided a source of water when transpiration was greater than the roots to extract soil moisture; leaf folding during stress reduced solar incidence; and transpiration was regulated by high stomatal resistance during stress (Meisner, 1991).

The maintenance of higher leaf-water status, by few genotypes of groundnut during soil water deficits is due to greater density of roots in the lower depth of the soil profile, however, this does not account for the major variation in the HI associated with drought (Matthews *et al.*, 1988a; De Vries *et al.*, 1989). Meisner and Karnok (1992) in a Rhizotron study observed reduction in root growth in upper 40 cm depth during moisture stress from 20 to 50 DAS compared to well watered control, but the root growth was not affected in the lower depth due to adequate moisture. In groundnut more than 60% of the root growth is established by 50 DAS and 80% by 80 DAS, water stress imposed after 50 DAP reduces only root growth in upper depths where root density is highest and soil moisture readily extracted. NCAc 17090 is an efficient in extracting water from the top 40 cm of soil and also got greater WUE (ICRISAT, 1986).

A rainfed and two irrigated treatments of groundnut in Malaysia when compared in terms of yield and water use by taking rainfed treatment (A) as the control, (B) irrigation at 7-d intervals, (C) irrigations made whenever readings of tensiometers at the 20-cm soil depth were equal to or less than -30 kPa, where there was no significant difference in yield between the two irrigation treatments. The average groundnut yields obtained from treatments A, B and C were 1.9, 3.1 and 3.2 t ha^{-1}, respectively. The crop in the rainfed plot was exposed to water stress during its flowering stage owing to limited rainfall. The total water use of groundnut for a 30-d period beginning 30 d after planting (DAP) was 64.5, 124.5 and 152 mm for rainfed, irrigated B and irrigated C, respectively, and the low water use in the rainfed plot resulted in a low yield. As indicated by a continuously decreasing value of the soil water hydraulic head (< -70 kPa) in the rainfed plot from 35-54 DAP, the soil at 20 cm depth was continuously dry. With the yield response factor (ky) during flowering stage being 0.74, therefore the decrease in groundnut yield due to water deficit was relatively large (Ahmad *et al.*, 1999).

The effect of water deficit on phenology (vegetative and reproductive growth) of two groundnut cultivars (Tatu and PI 165317) reveals that, water stress reduced both root and shoot dry matter production, and root/shoot ratio and flower production, but number of secondary and tertiary ramifications was not affected, the peg and fruits were reduced by water stress affecting both yield and HI (Oliveira *et al.,* 2004). The soil moisture extraction patterns and root growth parameters when examined in five groundnut cultivars under various levels of soil water stress variation in soil moisture extraction, root length density and total dry matter production were observed among the cultivars and the differences in water extraction in deeper soil profiles were related to the variation in root length density and the cv. ICGV 86031 and TMV 2 NLM were efficient in soil water extraction and suffered least losses (17-18%) in total dry matter production under moisture stress (Reddy *et al.*, 1999).

Among many factors that are associated with drought tolerance in legume crops, root traits have been considered to be the most important attributes enabling the plant to mine water efficiently from deeper soil layer under dry environments. Most of the methods used to evaluate roots are time consuming that provide valuable information about the root morphology but they do not reflect the dynamic characteristics of roots and root systems. Also the morphological variation in roots which has specific significance of adaptation, their functional aspects involving direct water uptake and their related kinetics are equally important. Vadez *et al.* (2007) reviewed root structure, root hydraulics, and modes of water and nutrient absorption, mainly focusing on how inter- and intra-specific variations in these aspects can modify the way roots respond to a range of abiotic stresses and summarized the contribution of roots to stress tolerance including research on the role of roots in near isogenic lines containing terminal drought tolerance quantitative trait loci, and on the role of DREB1A gene in root growth in transgenic groundnut under drought conditions. Chemical and hydraulic signalling between roots and shoots, and its role in drought and salt tolerance also mentioned (Vadez *et al.*, 2007). Rooting depths were of the order of 200 cm with a density of 1.5 cm/cm^3 in the 0-30 cm zone and 0.1 to 0.40 cm/cm^3 at greater depths. Tensiometers and neutron meters showed that water extraction continued during prolonged drought at depths below the shallow irrigated surface soil layer (Hammond *et al.*, 1978).Considerable amount of genetic variability with respect to root traits involving length, dry weight and root length density (RLD) were observed and a large lysimetric system has been developed at ICRISAT to make progress in this direction. In groundnut, DREB1A triggers native genes that might be involved in root development. In this review, the progress made so far on roots in legume crops has been elucidated which might explore possibilities of breeding genotypes to inherit efficient root system in legumes (Vadez *et al*, 2008).

In a root block study in groundnut at DGR (DGR, 2010), the primary root length (cm) was directly associated with secondary root length (r=0.85**), root weight density between 61-75 cm (r= 0.41**), total biomass (r=0.21*) and inversely with degree of leaf folding (r =-0.19), and primary root length seems to be a desirable trait to increase biomass productivity under rain-dependent condition. The secondary root length was directly associated with secondary dense root length (r= 0.51**), root volume (r=0.35**), root weight density (r= 0.53**) in deeper soil layers, total biomass (r = 0.37**) and inversely with degree of leaf folding (r =-0.24*) and higher secondary root length seems to be a desirable trait in groundnut. Secondary dense root length (cm) was positively associated with root volume (r=0.52**), root weight density (r= 0.51**), total biomass (r=0.44**) and inversely with degree of leaf folding. Root volume was directly associated with number of branches (r=0.28*), root weight density in upper soil layers (r= 0.83**), shoot weight (r= 0.46**), total biomass (r=0.52**) and root shoot ratio (r=0.41**) but non-significant with degree of leaf folding. These desirable root traits associated with drought tolerance need to be utilised for developing water use efficient groundnut.

Soil compaction makes the soil denser, decreases permeability of gas and water exchange as well as alterations in thermal relations, and increases mechanical strength of the soil. Compacted soil can restrict normal root development. Simulations of the root restricting layers in a greenhouse are necessary to develop a mechanism to alleviate soil compaction problems in these soils. Duruoha *et al.* (2008) in an experiment, in Alabama, USA, assessed groundnut root volume and root dry matter in three distinct bulk densities (1.2, 1.4, and 1.6 g cm^{-3}) and two levels of soil water content (70 and 90% of FC) in a sandy loam soil (Plinthic Kandiudults) where groundnut yield responded favourably to subsurface compaction in the presence of high mechanical impedance clearly indicating the ability of its root to penetrate the hardpan with less stress. The root volume was not affected by the increase in soil bulk density and this mechanical impedance increased root volume when roots penetrated the barrier with less energy. Root growth below the compacted layer (hardpan), was impaired by the imposed barrier. This stress made it impossible for roots to grow well even in the presence of optimum soil water content. Generally, soil water content of 70% FC ($P<0.0001$) enhanced greater root proliferation. Nonetheless, soil water content of 90% FC in some occasions proved better for root growth. The mechanical impedance is not a good indicator for measuring root growth restriction in greenhouse and future research required using more levels of water to determine the lowest soil water level, which can inhibit plant growth (Duruoha *et al.*, 2008).

4.3. Water Use Efficiency

The water use efficiency (WUE) is g of total dry matter produced kg^{-1} water used. The transpiration efficiency of the leaf is the ratio of CO_2 assimilation rate to transpiration rate. The quantity of water transpired is proportional to the percentage of soil covered by the crop (KCOV), evaporative demand (EVPAN) and potential transpiration (CROPET) is estimated as CROPET = KCOV*KM*KCROP*EVPAN, where KCROP is equivalent of a crop coefficient. For well irrigated groundnut crop it is estimated as: KCROP=Crop's water requirement/ standardized class A EVPAN (Dan cette, 1981). The actual water used (E_t) from emergence onwards increased to potential use evapotranspiration (E_0) until both become equal about midway when canopy has closed cover and remains equal until harvest. The ICRISAT developed a technique for measuring WUE, partitioning of dry matter to pod and efficient root systems which was successfully used in groundnut by Nageswara Rao *et al.* (1985) who also computed the seasonal evapotranspiration using the water balance equation:

$$ET = (M_i - M_f) + (I + P) - (R + D)$$

where ET is evapotranspiration, M_i is initial moisture in 0 to 127 cm profile, M_f is final moisture in 0 to 127 cm profile, I is irrigation, P is precipitation, R is run-off, D is deep drainage (deeper than 127 cm) considered negligible.

The WUE was 19-68 kg pods/cm of water depending on the stage at which the crop was exposed to moisture stress (Raju *et al.*, 1981). In Junagadh, highest water use (84 cm) and benefit: cost ratio (2.42) were obtained under no moisture stress, but maximum water-use efficiency (WUE) was achieved under water stress imposed at flowering stage. Among the genotypes GG 6 recorded higher water-use efficiency (WUE) and benefit: cost ratio (Vaghasia *et al.*, 2010a and 2010b). In a multilocation trial, the WUE was high 3.71 g kg^{-1} in Virginia type and low 2.46 g kg^{-1} in Spanish type groundnut (Wright *et al.*, 1988). In Maharashtra the WUE was highest in (5.23 kg ha-1 mm) groundnuts with irrigation at 0.5 IW: CPE 0-40 d after sowing and 0.75 IW: CPE thereafter (Patil and Gangavane, 1990). Irrigation based on stress day index at Bhubaneswar, had the highest water consumptive use (WCU) and water requirement. WCU and water requirement were lowest when moisture stress was imposed at pod development stage. The WUE was highest in plants stressed at pod development stage and lowest in plants stressed at peg penetration stage. Irrespective of treatment, the extraction of soil moisture decreased with the increase in soil profile depth (Kar *et al.*, 2002).

A field method for evaluating the sensitivity of groundnut genotypes under various patterns of drought using line source sprinkler technique was developed by ICRISAT (Singh *et al.*, 1991) where water deficit (W_D) is estimated using the amount of water applied during the period of drought and the cumulative class `A' pan evaporation for the same period as: W_D=100 x (E -I/E), Where W_D is water deficient %, E is CPE for the period of drought and I is cumulative irrigation applied for the period of drought.

Collino (2000) compared the water extraction capability and WUE of two Argentinian groundnut varieties Florman INTA (a drought-sensitive) and Manfredi 393 INTA (a drought-tolerant) at two different water supply regimes (irrigated between 47-113 DAS, and no water) by measured soil water contents, canopy temperature, dry matter and root length density (RLD) periodically during the drought period, and calculating the water use (WU), stress degree days (SDD), extraction front velocity (EFV), uptake coefficient, shoot and pod WUE corrected (WUEc) by vapour pressure deficit, and the soil resistance, where Manfredi 393 INTA had a higher WU than Florman INTA in both irrigated (IRR) and water-stressed (WS) environment mainly due to higher transpiration rate as demonstrated by SDD time course values and higher uptake coefficient values. However, the EFV from lineal and logistic fitted models showed a similar pattern for both varieties. Uptake coefficient differences between varieties were not associated with RLD indicating that it was not useful as an indicator of the genotypic ability to extract soil water. Florman INTA possessed higher WUEc than Manfredi 393 INTA in both IRR and WS treatments. Both varieties showed similar values under the irrigated regime because the higher WU on Manfredi 393 INTA was compensated for a higher pod production due to an enhanced partitioning of assimilates to pods, but under WS regime, pod WUEc was significantly reduced in both varieties. The mechanical impedance in the soil upper layer contributed to this reduction, and although critical soil resistance was similar for the two varieties, its effect was remarkable on Florman INTA due to a delay in the onset of the beginning pod stage and non-synchronous reproductive development.

4.4. Carbon Isotope Discrimination and its Relationship with WUE and SLA

The groundnut plant, during carbon accumulation, discriminates against ^{13}C which changes the composition of isotopes of CO_2 the $^{13}C/^{12}C$ ratio in dry matter (Hubick *et al.*, 1986). Carbon isotope discrimination (CID) of plant dry matter is linearly related in a negative manner to leaf transpiration efficiency via Pi/Pa, the ratio of intercellular CO_2 pressure Pi, to ambient CO_2 pressure Pa (Hubick *et al.*, 1988). There is a strong negative correlation between transpiration efficiency (the ratio of dry matter produced to water used, W) and carbon isotope discrimination, Δ (Hubick, 1990) and also between Δ and total dry matter (Wright *et al.*, 1988). The discrimination against ^{13}C (Δ) in leaf dry matter is calculated as:

$$\Delta = (\delta a - \delta p) / (1 + \delta p)$$

Where, δa and δp being the isotope composition of the air and plant materials, respectively relative to PDB (Pee Dee Bele minite).

A high heritability of Δ and its strong relationship with W indicate that breeding programme which includes selection for W based on differences in Δ could lead to increase dry matter production and yield of groundnut in water limiting environments. Thus measurement of Δ may prove a useful trait for selecting cultivars with improved W and total dry matter yield. In a mini-lysimeters study the WUE, ranged 1.81 in Chico to 3.15 g kg^{-1} in Tifton -8, was negatively correlated with Δ (19.1 to 21.8%) and thus Δ is a useful trait for selecting groundnut genotypes with improved WUE under drought conditions in the field (Wright *et al*, 1994). A strong negative relationship also existed between WUE and SLA (cm^3

g^{-1}) and between Δ and SLA, indicating that genotypes with thicker leaves had greater WUE. Significant correlations amongst WUE, CID in leaf and SLA suggested that CID and SLA could be used to identify genotypes with high WUE (Rao *et al*, 1994). SLA could therefore be used as a rapid and inexpensive selection index for high W where mass spectrometry facilities are not available.

Nageswar Rao *et al.* (1993) reported that in groundnut the WUE was 1.38-2.5 g/kg and were inversely related to discrimination against $^{13}CO_2$ fixed in leaves (DELTA) in 8 of the 10 genotypes, but WUE and transpiration were not significantly correlated. The CGR were negatively related to DELTA under irrigated conditions, but not under drought with CGR value 12-17 g/m^2 per d with irrigation and 2-8 g/m^2 in stressed crops. Partitioning of DM to pods in drought conditions ranged from 0.56 in cv. ICGV 86707 to >0.95 in 3 early maturing genotypes.

Jayalakshmi, *et al.* (2002) studied CID in 21 F_1 hybrids of groundnut in Andhra Pradesh during the post rainy season observed lower values of CID in TMV2-NLM, ICG 2716, Tirupati 1 and ICGV 86031, highest significant heterosis in ICG 2716 x TAG 24 and positive heterosis over better parent in TAG 24 x TMV2-NLM. The ribulose-1, 5-bisphosphate carboxylase-oxygenase (Rubisco) content increase under water deficit and top leaves had a higher Rubisco content and lower DELTA, than bottom leaves (Rao *et al*, 1995). Cultivar x leaf position interaction observed for DELTA and Rubisco, indicate the importance of leaf position in selecting for WUE, using leaf traits in groundnut. Rubisco content and DELTA were negatively related ($r^2 = 0.65$, $P<0.01$). There is a positive correlation between Rubisco content and leaf weight per unit leaf area (rhoL) in the upper leaves ($r^2 = 0.60$, $P<0.01$). And the basis of genotypic variation in DELTA was mostly (>60%) attributable to Rubisco content. In view of the leaf positional effects on DELTA and Rubisco, the upper leaves in the canopy should be used for selecting genotypes for W based on leaf traits like rhoL or DELTA.

Under adequately irrigated and simulated drought treatments at Tirupati, Andhra Pradesh with 20 groundnut genotypes differing in their transpiration efficiency and repeated with 7 genotypes in the second season wher SLA and CID exhibited significant positive relationships indicating that SLA can be utilised as a surrogate to CID (Asalatha, *et al.*, 1999). Further SLA was negatively related to transpiration efficiency, while it was positively related to partitioning, suggesting that selection for low SLA might result in production of more dry matter with minimal influence on pod weight (Asalatha *et al* 1999). The mineral ash and total chlorophyll contents of leaves were strongly correlated with SLA and due to the simplicity in measurement these have merit considering screening tools in selection and breeding programmes for higher WUE under limited water environments Reddy, *et al.* (2000).

Krishnamurthy *et al.* (2007) evaluated the variation for Transpiration efficiency (TE) in a set of 318 recombinant inbred lines (RILs) of groundnut at F_8 generation, derived from a cross between a high TE (ICGV 86031) and a low TE (TAG 24) parent, and the value of SLA, SPAD chlorophyll meter readings (SCMR) and carbon isotope discrimination (Delta ^{13}C) as surrogates of TE were measured (on the dried tissue after harvest) and the study reveals that overall distribution of TE among the RILs indicated that TE, governed by dominant and additive genes, was negatively associated with SLA after the completion of stress treatment (r^2=0.15) and Delta ^{13}C in leaves (r^2=0.13) positively associated with SCMR during stress (r^2=0.17). Although the heritability of SCMR was relatively higher than that of TE, the stress-dependence of the relationship with TE, and the poor regression coefficients (r^2) with that RIL population, do not confer that these surrogates are adequately robust enough in that population (Krishnamurthy, *et al.*, 2007). Chuni Lal *et al.* (2009) evaluated 9 groundnut genotypes to investigate the influence of water stress on some phenological, morpho-physiological, and yield traits and found that water saturation deficit (WSD) and epicuticular wax load (EWL) increased in response to water stress and age of the crop, while SLA decreased with water stress and age of the crop. Though, the correlations of WSD, EWL, and SLA with yield traits were fairly weak, WSD in the early stage was positively associated with pod yield and EWL in the early stage was negatively associated with HI under stress. Genotypes that accumulated flowers sooner after initiation showed less yield reduction and the negative association between HI under stress and its reduction deems HI under moisture stress an important criterion of selection for drought tolerance in groundnut (Chuni Lal *et al.*, 2009).

The desirable traits such as WUE, partitioning of dry matter to pods and efficient root systems vary with genotypes and are heritable. A time-integrated approach based on stable isotope ratios of carbon and oxygen (Delta ^{13}C/Delta ^{18}O) were described by Bindumadhava *et al.* (2005) using groundnut (NCAC 17090, VRI 4, ICGS 11 and Sen Nghean) genotypes to identify crop genotypes with high mesophyll capacity for carbon assimilation as it has specific advantage in crop improvement, since such genotypes besides sustaining productivity under water-limited conditions can also save substantial amounts of irrigation water, this approach would provide a strong impetus to plant breeding efforts with assured success to improve productivity. Experimental evidence is presented to show that the ^{18}O enrichment in the leaf biomass and the mean (time-averaged) transpiration rate are positively correlated in groundnut genotypes (Sheshshayee, 2005). The relationship between oxygen isotope enrichment and stomatal conductance (g_s) was determined by altering g_s through ABA (abscisic acid), and subsequently using contrasting genotypes of groundnut. The Peclet model for the ^{18}O enrichment of leaf water relative to the source water is able to predict the mean observed values well, while it cannot reproduce the full range of measured isotopic values. As all the genotypes of both species experienced similar environmental conditions, the differences in transpiration rate could mostly be dependent on intrinsic g_s and hence, Delta ^{18}O of leaf biomass can be used as an effective surrogate for mean transpiration rate, further, at a given vapour pressure difference, Delta ^{18}O can serve as a measure of stomatal conductance as well (Sheshshayee, 2005).

4.5. Gene Action, Genotype and Environment Interactions

In an investigation to determine the gene action controlling the inheritance of SCMR and SLA in two crosses, ICG 7243 x ICG 9418 and ICG 6766 x Chico, and their reciprocals, 6 generations of each cross (P1, P2, F1, F2, BC1P1, and BC1P2) were evaluated for SCMR and SLA at two stages of the crop growth viz., 60 and 80 days after sowing (DAS) by Upadhyaya, (2011) where for SCMR at 80 DAS, additive effects were important in both the crosses whereas predominance of dominance effects with duplicate epistasis was observed for SCMR at 60 DAS and SLA at both stages in both the crosses. Predominance of additive effect for SCMR at 80 DAS suggested effective selection could be practiced even in early generations whereas for SCMR at 60 DAS and SLA at both stages in both crosses, it would be better to defer selection to later generations. Further, recording of SCMR and SLA should be done between 60 and 80 DAS for screening the germplasm lines for drought tolerance (Upadhyaya, 2011).

The genotype x environment (G x E) interaction for the relationship between SLA and DELTA were examined by Rao and Wright, (1994) in four groundnut genotypes (Chico, McCubbin, Shulamit and Tifton 8) with contrasting carbon isotope discriminating characteristics where the values of DELTA and SLA were significantly influenced by the location, genotype and irrigation treatments, but genotype x location interaction effects on the relationship between DELTA and SLA were not observed, however positive relationship between SLA and DELTA was maintained when data were combined over sites and treatments ($r^2 = 0.87$, $P < 0.01$). Further the SLA was negatively correlated with nitrogen content per unit leaf area (SLN) which in turn was negatively correlated with DELTA. The genotypic and environmental variation in transpiration efficiency (W) and its correlation with CID (DELTA) further investigated by Wright *et al.* (1993) in 7 groundnut cultivars and *Arachis villosa* and *A. glabrata* where the W was highly correlated with DELTA in leaves, SLA and leaf thickness and G x E interaction for W, DELTA and SLA was very low, while heritability of DELTA was high, indicating that these traits could be used for selecting high W in groundnut breeding programmes.

In Botswana, the breeding lines selected for adaptation to drought stress under rainfed tested in multilocation trials to compare their performance with that of locally grown cultivars showed largest variation across locations and years due to the environment, with minimal variation due to genotype and G x E interaction. All genotypes responded to changes in environmental conditions, with an indication that seasonal rainfall patterns were important in determining genotypic performance. Selection for drought adaptation under rainfed conditions, though commonly practiced, could be misleading, since it may not reflect the ability of the genotype if the stress occurs during the critical stages of plant development. More efficient selection would require simulated drought conditions, and the use of other indirect selection methods that give a good indication of drought adaptation (Maphanyane *et al.*, 1994).

Molecular markers and genetic linkage maps are pre-requisites for molecular breeding in any crop species. In groundnut, an amphidiploid (4X) species, not a single genetic map is available based on a mapping population derived from cultivated genotypes. In order to develop a genetic linkage map for tetraploid cultivated groundnut, a total of 1,145 microsatellite or simple sequence repeat (SSR) markers available in public domain as well as unpublished markers from several sources were screened by Varshney *et al,* (2009) on two genotypes, TAG 24 and ICGV 86031 that are parents of a recombinant inbred line mapping population and reported the construction of the first genetic map for cultivated groundnut and demonstrated its utility for molecular mapping of QTLs controlling drought tolerance related traits as well as establishing relationships with diploid AA genome of groundnut and model legume genome species. As a result, 144 (12.6%) polymorphic markers were identified and these amplified a total of 150 loci. A total of 135 SSR loci could be mapped into 22 linkage groups (LGs). While six LGs had only two SSR loci, the other LGs contained 3 (LG_AhXV) to 15 (LG_AhVIII) loci. As the mapping population used for developing the genetic map segregates for drought tolerance traits, phenotyping data obtained for transpiration, TE, SLA and SCMR for 2 years when analyzed together with genotyping data, the phenotypic variation explained by these QTLs was in the range of 3.5-14.1% and 2-5 QTLs for each trait were identified (Varshney *et al*, 2009). In addition, alignment of two linkage groups (LG_AhIII and LG_AhVI) of the developed genetic map was shown with available genetic maps of AA diploid genome of groundnut and Lotus and Medicago (Varshney *et al*, 2009). A biochemical test measuring pH of 0.1M EDTA leaf extract was developed by Dwivedi *et al.* (1986c) taking 22 cultivars and 3 wild species of *Arachis* grown under moisture stress and cultivars showing less than 2.77 p. 100 decline in the pH of leaves extract in EDTA, at -10 bar water stress were considered as drought resistant. A significant correlation co-efficient of r = 0.706 between percent decline in the pH of leaves extract and pod yield and r = 0.893 between percent decline in the pH of leaves extract and pH water deficit index (PWDI) were noted. The correlations of this test with drought resistance capacity of cultivars were found to be superior to that of proline accumulation, stomatal resistance, transpiration rate and saturated water deficit.

4.6 Minerals and Salinity

In field trials during summer season two groundnut cultivars subjected to a water stress of -14 bar soil water potential during the early vegetative, late vegetative, flowering, pegging or pod development stages gave average pod yields of 2.0, 1.8, 1.6, 1.3 and 1.0 t/ha, respectively, compared with 2.39 t in crops without water stress. Water stress increased P, K, Ca and Mg contents in both haulms and pods, the N contents increased in pods but decreased in haulms with water stress. Cv. TG 17 gave higher yields than cv. GAUG 1 both with and without water stress (Parmar *et al.*, 1989).

The influence of several pod characteristics on Ca accumulation and Ca concentration in groundnut fruit was studied in 8 genotypes with diverse fruit characteristics for 2 seasons under 5 water stress treatments (drought at 20-50, 50-80, 80-110, 110-140 DAS and an irrigated control) by Kvien *et al.* (1988) where 80-110 DAS drought period had the greatest negative impact on seed Ca concentration. Total Ca accumulation in the pod (hull + seed) was positively correlated (0.97) with pod surface area and five characteristics days required to mature a pod, specific hull wt, pod surface area, hull thickness and pod volume significantly influenced seed and hull Ca concentration. These characteristics were under genetic control, but their absolute value was modified by water stress. Thin light hulls and long pod maturity periods promoted high Ca concentration in the seed and thick dense hulls, short maturity periods and small pod vol. promoted high Ca concentration in the hull (Kvien *et al.*, 1988).

Potassium regulates water in plants, and is important under rainfed or moisture stress conditions have much greater importance than irrigated agriculture. Its application reduced the incidence of tikka disease of groundnut and improves resistance to environmental stresses by osmoregulation, proline accumulation, induction of phenolics, betaines, and phytoalaxins and by morpho-phenological changes in plants. Potassium besides increasing crop yields, seems to have the greatest effect on oil biosynthesis which often results in higher oil yields, its effect is more pronounced at higher rate of application. The current interest in the role of polyunsaturated fatty acids in the heart disease and cancer justifies studies of the effect of fertilizers on the fatty acid composition of vegetable oils. The relative concentrations of the various fatty acids are also important from an economic standpoint because they affect the susceptibility of the oil to flavour reversion and to oxidative rancidity. It is possible to increase production if we go by the attainable yields demonstrated by adoption of the full package of improved practices and if these best management practices can be adopted by the farmers on millions of hectares. Vertical growth in productivity demand higher inputs. Besides increasing fertilizer use, particularly potash, it would also be necessary to improve input use efficiency through proper harvesting of the interaction effects of plant nutrients by balanced fertilizer application (Tiwari, 2005).

Patel and Padalia (1980) studied the soil-water potential on nutrient uptake by groundnut with progressive increase in sequences of soil moisture stress from 1/3 to 5, 10 and 14 bar the root dry wt., haulm and pod yields and N, P, K, Ca, Mg, Fe, Mn and Zn uptake by different parts of plants decreased. Nutrient use efficiency was maximum with 1/3 bar and decreased with increase in soil moisture stress. A single stress of about 10 bar during flowering, pegging and pod development stages significantly decreased the nutrient uptake and shoot and pod yields compared to 1/3 bar. When continuous water was supplied by upward flow from the bottom of the pot keeping the surface soil dry (D), there was excessive vegetative growth yielding less pods; however, nutrient uptake was comparable with 1/3 bar. Pots irrigated according to plant requirement (3 bar) gave intermediate yields of pods and shoots and nutrient uptake ranged between those of 1/3 bar and D condition. Nutrient uptake with saturation (0 bar) treatment was significantly lower than that at 1/3 bar (Patel and Padalia, 1980).

In Oxisol during rabi (winter) at Wakawali, Maharashtra, groundnuts irrigated to 40 or 60 mm depth or at IW:CPE ratios of 0.50, 0.75 or 1.00 and given no fertilizers, 25 kg N + 50 kg P_2O_5 or 50 kg N + 50 kg P_2O_5/ha, where, N uptake in seeds, shell and haulm decreased by 31, 28 and 29% with irrigation at IW:CPE ratio of 0.5, while P uptake was decreased by 35, 45 and 36%, respectively and N and P uptakes were highest with 50 kg N + 50 kg P_2O_5 compared with irrigating at IW:CPE ratio 1.00 (Patil and Kadam , 1993).

In a study, AtNHX1, a vacuolar type Na^+/H^+ antiporter gene driven by 35S promoter was introduced into groundnut using Agrobacterium tumefaciens transformation system. The stable integration of the AtNHX1 gene was confirmed by polymerase chain reaction (PCR) and southern blot analysis. It was found that transgenic plants having AtNHX1 gene are more resistant to high concentration of salt and water deprivation than the wild type plants. Salt and proline level in the leaves of the transgenic plants were also much higher than that of wild type plants. The results showed that overexpression of AtNHX1 gene not only improved salt tolerance but also drought tolerance in transgenic groundnut. Our results suggest that these plants could be cultivated in salt and drought-affected soils (Asif *et al.*, 2011).

5. MANAGEMENT OPTIONS AND STRATEGIES

In most groundnut growing areas of the world, shortage of water is caused by uneven distribution of rains, gaps between rain events and field water losses rather than from low seasonal or annual rainfall totals. The groundnut requires 400-450 mm of water, however in sandy soil the water requirement goes to 600-700 mm, but the distinction between *Kharif* and Rabi are not very clear. The water requirement of groundnut varies with the stages and is lowest from germination to flower formation and reaches maximum during pod formation. However, the utilization of available moisture is greatest during flowering and pod formation and the crop receiving adequate water during these stages only can give equal yield to the well watered crop. During these stages if stress is given and later on water supply is resumed only the vegetative growth is benefited not the reproductive growth of crop. However, the period of maximum sensitivity to drought occurs between 50-80 DAS, the period of maximum flowering and vegetative growth. Under water scarcity condition, plant extractable soil water depletion of more than 45% of available soil water (ASW) must be avoided even during non-critical growth stages to obtain high WUE and net return (Panda and Behera, 2005).

The levels of irrigation have a major effect on the amount of water consumed, and it is 300-350 mm in non irrigated field and 500-600 mm in field irrigated at 40-60% moisture availability at 30 cm depth. The irrigation given to a depth of 100 mm showed highest yield. More than 70% of the kharif groundnut in India is rainfed where one or two life saving irrigations are required. The rabi groundnut is grown either in residual moisture or with 3-9 irrigations depending upon the soil and climate. But the summer groundnut is mainly irrigated and 6-15 irrigations are required depending upon the soil types, locations and varieties. The use of drip saves 30-50% water and produces up to 20-30% more yield than flood irrigation (Singh *et al*, 2000; 2001). Most of the scheduling of irrigation is based on the depletion of available soil moisture, IW/CPE ratio and maximum yield is obtained at IW/CPE ratio 0.6-0.9, but in practice mostly the frequency of interval is opted.

In last four decades enough progress has been made on drought research in groundnut and application of knowledge into practice in a systematic manner can lead to significant gains in yield and yield stability of the world's groundnut production, with transferable technology to help farmers of arid and semi-arid regions. Increasing soil moisture storage by soil profile management and nutrient management for quick recovery from drought are some of the areas that need to be explored further (Reddy *et al.*, 2003). The various water management practices and their principles are discussed here separately.

5.1. Evapotranspiration and Irrigation Scheduling

Water use by groundnut in different cropping seasons in different parts of the world varies between 250 mm under rain-fed conditions to 831 mm under irrigated conditions. The total water use of a groundnut crop may be affected by scheduling irrigations based on requirements at the various growth stages. The evapotranspiration (ET) in groundnut varies with crop duration and is nearly 400 mm for 100-110 days crop, 500-600 mm in 120-140 days crop and about 700 mm in 150 days crops. However, the evaporation from the bare soil is 350 mm for the same period. Further, the yield varied with soil types, it is maximum with 75% available soil moisture (ASM) in red loamy soils and only at 50% ASM in black loams, however, in general, the maximum yields are obtained under 50-60% FC and -0.3 to -0.4 bars of water tension (Reddy, 1988). The field experiment during rabi season has shown that crop coefficient (Kc) value was low initially, increased linearly with advancement of crop growth, attained peak value at reproductive growth period from flowering through pod initiation, and decreased towards maturity which is useful in developing a methodology for the determination of periodic and peak irrigation requirements (Hemalatha and Rao, 2006). The crop coefficient curve facilitates prediction of groundnut ET, the seasonal net, gross and peak irrigation requirement of groundnut were determined.

The scheduling of irrigation is based on the depletion of available soil moisture IW/CPE ratio. Using non-weighing lysimeters, a value of 1 times pan evaporation for the area covered by the canopy was determined to be an appropriate rate for the groundnut production of unstressed plants. The irrigation schedule recommended at various places is given in Table 3.

Table 3. Irrigation scheduling in groundnut at various locations

SN	Groundnut varieties	Soil type and Places	Critical stages of water stress and irrigations schedule	Conditions	References
1	TMV 2	Sandy loam soils AP, India	Supply adequate water during moisture-sensitive FL & PD by scheduling irrigation at 4 growth stages	In summer, moderate stress at VG and maturity stages, produced optimum yield (2.82 t/ha) and WUE (7.73 kg ha^{-1} mm^{-1})	Reddy and Reddy, 1993
2	GAUG-1	Medium black soil at Junagadh	Irrigations at 7-11 intervals	40 kg K2O/ha	Shahid Umar *et al.*, 1997
3	GG 2	Medium black soil at Junagadh	13 irrigations	Transient moisture-deficit stress during vegetative phase for 25 days, followed by weekly irrigations	Nautiyal *et al.*, 1999
4	GG-2 and Jl-24	—	8-days intervals	RWC, leaf water potential (psi), leaf diffusive resistance and transpiration	YC Joshi *et al.*, 1988
5	-	Parbhani	988 and 930 mm water consumption at IW:CPE 0.8 and 0.6, irrigating at IW:CPE ratio 1.0 throughout growth.	Irrigated at IW:CPE ratios of 0.4, 0.6 or 0.8 at different growth stages	Shinde and Pawar, 1984
6	SB XI	Akola, Maharashtra	Irrigation at IW:CPE 0.50 up to flowering, 0.65 FL to PD and 0.80 during PD to maturity	Pod yields increased by applying antitranspirants	Patil and Morey, 1993
7	-	—	Water stress for 20-25 days during the VG, relief with 2 irriga-	In summer crop increased the pod yields by 20% compared	Ravindra *et al.*, 1989

			tions at 5-d interval then normal irrigation at 10-d intervals	with yields of unstressed crops, besides savings of 2 irrigations.	
8	SB 11	—	Max. pod yield (3.06 t/ha) predicted with 1131 mm water	Using IW:CPE, 3 levels of water stress applied at 4 growth stages	Shinde and Pawar, 1982
9	Konkan Gaurav	—	11 irrigations at 10-day intervals.	The highest yield and net returns	Chavan *et al.*, 1999

A comprehensive field investigation was done on groundnut crop over a period of three years with five different irrigation treatments, to determine an efficient strategy for management of irrigation water under water stressed conditions in a sub-tropical region. Layer-wise soil moisture status was continuously monitored to determine the crop water extraction pattern and thereby the irrigation management depth. Five irrigation treatments maintained based on predefined levels of maximum allowable depletion (MAD) of ASW were 10% (T1), 30% (T2), 45% (T3), 60% (T4) and 75% (T5) maximum allowable depletion of ASW (measured periodically in 15-30, 30-45, 45-60, 60-90 and 90-120 cm soil profiles using a neutron probe). CROPGRO-Groundnut growth simulation model was calibrated and validated for further use. Thus, the plants extracted most of the soil moisture from 0-30 cm soil layer. The study shows that only 0-30 cm of soil profile be considered for scheduling of irrigation in case of groundnut grown in sandy loam soil in the sub-tropical regions. Measured and simulated results revealed that under water scarcity condition, plant extractable soil water depletion of more than 45% of ASW must be avoided even during non-critical growth stages to obtain high WUE and net return. The calibrated CROPGRO-Groundnut model was found to be quite efficient in simulation of yield parameters and layer-wise soil moisture extraction pattern (Panda and Behera, 2005).

In red sandy loam soils in Andhra Pradesh, a good relation was established by Reddy (1984) with class 'A' pan using a simple can evaporimeter and that irrigation of groundnut when cumulative can evaporation reached 2 cm with a depth of water equal to that lost in evaporation from the evaporimeter gave the highest yield of pods. In sandy loam soil of Tirupati, India, groundnut irrigated at IW/CPE ratio of 0.5 at 7 cm resulted in early flowering; however irrigation at IW/CPE ratio of 1.0 at 5 cm resulted in more filled pods, a greater volume weight of pods, a higher shelling percentage, and significantly higher pod and kernel weights, highest pod yield and net returns (Sree and Rao, 1998). The summer groundnut cv. TMV 2 grown by Reddy and Reddy (1993) with adequate soil moisture, 60% (moderate water stress) or 80% depletion (severe water stress) throughout the growing period, or 1 of 6 combinations of these irrigation schedules at 4 growth stages (vegetative, flowering, yield formation and maturity) found that scheduling irrigation to supply adequate water during the moisture-sensitive flowering and yield formation stages, yet allowing moderate stress in the vegetative and maturity stages, produced the optimum yield (2.82 t/ha) with maximum water-use efficiency (7.73 kg ha^{-1} mm^{-1}) and water economy. At times of water deficit, irrigations can be scheduled at 60% depletion of ASM throughout the growing period without reducing the cropping area for want of irrigation water. Yield response factors (k_y) which relate relative yield decrease to relative ET deficit were calculated. Lower k_y values due to no soil moisture stress at flowering and yield formation stages stress the need for an adequate water supply at these stages. For a relative ET of 0.71-0.74, the relative yield levels varied between 0.5 and 0.6, while for relative seasonal ET of 0.76-0.84, the groundnut yields varied from 0.74 to 0.86 (Reddy and Reddy, 1993). During summer, in Tamil Nadu, irrigation schedule at 0.75 IW/CPE led to highest yield (Lourduraj, 2000).

In Maharashtra, application of 80 mm depth of irrigation water at 0.40 IW/CPE ratio for first 40 days, at 0.90 ratio from 40-70 DAS and at 0.60 ratio from 70 DAS and onwards produced maximum pod yield in SB XI groundnut where the average consumptive use of 387 mm of water with consumptive use efficiency of 5.0 kg ha^{-1} mm^{-1} and ET/PE ratio of 0.71 (Dhonde *et al.*, 1985). However, Patil *et al.* (1984) reported 5.48 kg ha^{-1} mm^{-1} as the comsumptive use efficiency of JL 24 (Phule Pragati) at 3.5 mm day ET, 0.85 ET/PE ratio and 416 mm PE in lysimeter experiment. Irrigation at 50% ASM depletion or at 1-week intervals, during summer season, gave pod yields of 4.98 and 4.83 t ha^{-1}, respectively, compared with 4.65, 4.13 and 3.47 t ha^{-1} with irrigation at 50, 75 and 100 mm CPE, respectively (Babalad and Kulkarni, 1988).

At Parbani, Shinde and Pawar (1984) observed that irrigation at 0.8 IW/CPE ratio at all the growth stages with 930-998 mm of water in 14 irrigations increased pod yield, but was detrimental at the ratio of 0.4. The crop sown in Feb the maximum pod yield was obtained by irrigation at IW: CPE ratio of 1.0 throughout growth. Irrigation at ratio 0.8 and 0.6 was most economic, with 3.07 and 3.16 kg dry pods/ha per mm water applied (Shinde and Pawar, 1984). Response of groundnut to different levels of IW: CPE ratio (treated as a stress day index) was observed during the summer seasons. The relationship between stress day index (x) and pod yield (y) was negative and quadratic in nature and was expressed by y = 31.1692 - 0.13133x + 0.0000721x2. The maximum yield obtainable was predicted to be 3.12 t/ha with an opt. irrigation (zero stress day index) (Shinde and Pawar, 1983). At Palghar, groundnut cv. Konkan Gaurav, grown with 7, 9 or 11 irrigations gave higher pod yield than 5 irrigations and the highest yield and net returns were with 11 irrigations at 10-day intervals (Chavan *et al.*, 1999).

In medium black calcareous soil of Saurashtra, the FC and permanent wilting point were 24.8 and 9.9% soil moisture level in top 0-15 cm layer and irrigation at 1.0 IW:CPE ratio was suitable (Kachot *et al.*, 1984). However, Gajera and Patel (1984) reported that irrigation at IW/CPE ratio of 0.8 yielded at par with than that of ratio 1.0 with total water applied 550 mm, and to maintain this ratio a total of 11 irrigations were provided to summer groundnut variety GAUG 1. Thorat *et al.* (1984) obtained highest yield when crop was irrigated at 50 mm at, IW: CPE ratio1.0 and the total number of irrigations were 14 (700 mm-ha water) during rabi season. The irrigation at 15 days intervals during germination to pegging and 10 days during pegging to pod formation and maturity gave highest yield in calcareous soils of Saurashtra, Gujarat during summer season (Dwivedi, 1986b).

In India, Khan and Datta (1990) reported that the I/E (irrigation/ CPE) ratio of 0.75 was optimum irrigation index for potential yield and WUE in SB XI cv. of groundnut, and to ensure adequate moisture in the effective root zone, 6 cm depth of irrigation is considered to be the best irrigation depth. In Eastern India, during summer, pod yield and DM increased with rate of K application (up to 30 kg K_2O ha^{-1}) and frequency of irrigation at 0.3, 0.55 and 0.8 atm. soil water tension increased pod yield by 112, 150 and 159%, respectively, compared with the rainfed treatment (Ghatak *et al.*, 1997). At Kharagpur, Jain *et al.* (1997) developed a water stress response function by imposing various degrees of stress at various physiological stages of crop growth in groundnut crop using 22 treatments of different combinations of IW: CPE ratio (depth of water applied per irrigation to the CPE during kharif which can be used for optimum allocation of water resources for maximum benefit of the groundnut crop. The study concluded that in case of limited water supply, water saving should be made during periods other than the flowering and pod formation stages, however WUE increased at a decreasing rate when deficiency in IW: CPE ratio was made at one stage only (Jain *et al.,* 1997).

Under semi-arid tropics often with low and erratic rainfall the groundnut yield and yield attributes, growth and development are affected by soil moisture deficit or water stress and maintenance of optimum soil moisture at critical growth stages is the key factor for achieving higher yields. The knowledge on physiological characters like photosynthesis, stomatal conductance, leaf water potential and water use efficiencies would certainly be helpful in making irrigation scheduling of groundnut and decision support systems that aim at higher productivity (Thiyagarajan *et al.*, 2009). Combinations of N, P and K fertilizers on growth and production of groundnut growing in the absence of water stress and at a modest rate of deficit irrigation (0.75 times pan evaporation for the canopy area) when studied for the high irrigation rate, crop water stress index (CWSI) values did not exceed 0.2. For the low rate water stress values frequently ranged from 0.5 to 0.7 (Worthington *et al.*, 1995).

Nautiyal *et al.* (2001) synthesize the response of groundnut to various aspects of deficit irrigation practices during vegetative phase. The transient soil-moisture-deficit stress for 25 days, at the vegetative phase (20-45DAS) followed by two relief irrigations at an interval of 5 days, resulted in synchronized flowering, greater conversion of flowers to pods and higher pod yield and total biomass accumulation, but water deficit stress was highly detrimental when imposed at flowering (40-65 DAS) and pod development at (60-85 DAS) (Nautiyal *et al.*, 1999a)

Pressurized irrigation system found to be quite effective under less water availability not only in achieving higher productivity but also economizing other inputs such as fertilizers, pesticides, labour etc. compared to traditional irrigation methods (Keller and Bliesner, 1990). Drip irrigation system is a convenient and effective means of supplying water directly to soil and nearer to the plant without much loss of water resulting in higher water productivity (Parihar *et al.*, 2008; Singh *et al.*, 200, 2001; Man Singh *et al.*, 2012). Planting of groundnut on raised bed enhances pod yield and helps in achieving higher irrigation efficiency through higher pod yield compared to normal flat bed planting (Parihar *et al.*, 2008; Man Singh *et al.*, 2012).

The estimation of the water requirements (WR) of crop is one of the basic needs for crop planning on the farm and design of any irrigation system. The WR can be defined as the quantity of water, regardless of its source, required by crop or diversified pattern of crops in a given period of time for its normal growth at a given field conditions. The WR also includes the evapotranspiration (ET) losses or consumptive use (C_u) of the crop, unavoidable losses during irrigation, and the quantity of water required for special operations such as land preparation, transplanting, leaching, etc. Thus WR may be formulated as:

$$WR = ET / Cu + Unavoidable..losses + special.needs. \quad (1)$$

Water requirement is, therefore, a 'demand' and the 'supply' would consist of contribution from any of the source of water, the major source being the irrigation water (IR), effective rainfall (ER) and soil profile contributions (S) including that from shallow water tables. Therefore, water requirement is:

$$WR = IR + ER + S \quad (2)$$

The field irrigation requirement of a crop refers to the water requirement of crops, exclusively of effective rainfall and contribution from soil profile, and it may be given as:

$$IR = WR - (ER + S) \quad (3)$$

The plant canopies of young and wide-spaced crops shade only a portion of the soil surface area and intercept only a portion of incoming radiation. Conventional estimates of water requirement of young crops assume part of the applied water will be lost to non beneficial consumptive use. This loss is through evaporation from the wetted soil surface and through transpiration from undesirable vegetation. The plant canopies of young and medium spaced crops like groundnut shade only a portion of the soil surface area and gradually the canopy increases to full surface area as the crops attain its full vegetation stage. The daily crop water use rate under trickle irrigation is a function of the conventionally computed consumptive use rate and the extent of the plant canopy.

Trickle (drip) irrigation reduces evaporation losses to a minimum, so the transpiration by crop account for practically all the water consumed. Thus, estimates of consumptive use that assume the wetting of entire field surface should be modified for trickle irrigation. The transpiration rate under trickle irrigation is a function of conventionally computed consumptive rate and the extent of the plant canopy. A simple equation for estimating the average peak daily transpiration rate is:

$$T_d = U_d\left(0.1(P_d)^{0.5}\right) \quad (4)$$

Where, T_d = Average daily transpiration rate during the peak – use month for a crop under trickle irrigation mm/day, U_d = Conventionally estimated average daily consumptive use rate during the peak–use month for the mature crop with a full canopy, mm/day, P_d = Percentage of soil surface area shaded by crop canopies at midday.

Equation is based on the observation that even when the plant canopy is very small and P_d is 1% or greater, the minimum $T_d > (0.1\ U_d)$. This is because there is an oasis effect and some additional vegetation usually grows in the area wetted by the emitters. Furthermore, as the canopies of the plants increase toward full converge, T_d approaches U_d, and at full coverage when P_d= 100%, $T_d = U_d$

While calculating depth of irrigation water required, the percentage wetted area was 100 based on the plant spacing, number of emitters per plant, and optimal emitters spacing. The following formula can be used for calculating the percentage wetted area:

$$P_w = \frac{N_p S_e W}{S_p S_r}$$

Where, P_w = percentage of soil area wetted along a horizontal plane (%), N_p = Number of emitters per plant, S_e = Optimum emitter spacing (m), W = Wetted width (m), S_p = Plant spacing (m), S_r = Row spacing (m)

Man Singh *et al.* (2012), using same amount of irrigation water for entire season, but with different amount and duration at each irrigation, reported an innovative method of estimation of water requirement of irrigated groundnut crop for two distinct seasons *kharif* (July-November) and summer (March-July) in the semi-arid region under the trickle system of irrigation using 5 irrigation schedules (irrigation depth: as per the estimated crop evapotranspiration; irrigation interval (s): 1, 2, 3, 4 and 5 days) where the estimated seasonal water requirements were 304-310 mm during *kharif* and 380-450 and mm during summer seasons at Delhi, India. During summer the irrigation were applied through drip at one day interval for 15 minutes duration (T_1), two days interval with 30 minutes duration (T_2), three days interval with 45 minutes duration (T_3), four days interval with 60 minutes duration (T_4), five days interval with 75 minutes duration (T_5) where the pod yields were significantly higher in case of daily irrigation, but it was reduced if irrigation were scheduled beyond 3 days interval. Daily irrigation based on crop water use rate maintained moist soil surface throughout the irrigation season and thus promoted penetration of more number of pegs into the soil easily resulting in more pods per plant.

The reference evapotranspiration is determined using the Penman-Montith equation for the crop growth period. These ETo determined for 1-25 days, 25-75 days and 75-105 days and estimated daily water requirements of per plant for the entire crop growth season (*Kharif*) for the semi arid region is presented in the Table 4.

Table 4. Daily water requirement of groundnut plant under drip irrigation (Man Singh *et al.*, 2012)

Crop growth stage	Duration of crop (days)	Percentage shaded area	Water requirement (L $day^{-1}plant^{-1}$)
Initial	0-25	1-33	0.10-0.25
Vegetative and flowering	25-75	33- 67	0.25-0.45
Pegging and pod formation	75-105	67- 100	0.45-0.60

Using the daily value of U_d and P_d in equation (1) the T_d was estimated for trickle/drip irrigation. The values of U_d are the reference evapotranspiration estimated from the local weather data.

The water requirements of the crop under trickle irrigation for the two consecutive seasons varied with prevailing temperatures and rainfall events which led to the variation in the crop ET from 450 mm in the year 2006 to 380 mm in 2007. Also a comparison of water requirements estimate of ground nut crop between conventional and new method is given in Table 5.

Table 5. Water requirements estimate of groundnut crop, a *comparison* (Man Singh *et al.*, 2012)

Season	Conventional Method, mm	New Method, mm
Monsoon *(Kharif)*	400 - 450	304 -310
Summer	650 -750	380 - 450

Leaf growth is a sensitive indicator of water deficit stress and LAI recorded at three different phases of reproductive development; early pod formation (76 DAS), pod filling (104 DAS) and at physiological maturity (120 DAS) was highest in T1 treatment and T2 to T5 experienced water deficit. Canopy Temperature Depression (CTD) of the crop canopy recorded at 60, 76 and 101 DAS showed maximum CTD as 7 at 60 DAS and mininmum of 1.5 at 101 DAS. As soil water becomes limiting, transpiration is reduced and leaf temperature increase, effect of irrigation scheduling on CTD was clearly observed among the treatments with stress condition. The pod and haulm yields were significantly higher in T1 treatment where irrigation for 15 minute was provided daily, but decreases with increasing interval and was lowest with T5 treatment (interval 5 days) due to reduced transpiration and low CTD, and LAI (Man

Singh *et al* 2012). Thus water requirement of a particular crop is not constant or fixed regardless to any irrigation method, it depends on the weather/climatic fluctuations during the crop growth season.

5.2 Early Season Stress

Early-season moisture deficit greatly reduced leaf area expansion, although the rate recovered following irrigation. The early moisture stress (EMS), imposed from emergence to initiation of pegs, reduced ET with no apparent reduction in LAI, pod and seed growth during the rainy as well as the post-rainy seasons on a medium deep Alfisol at Patancheru, India and the WUE was substantially higher for the EMS treatment compared with the control and despite the contrasting climatic conditions during the rainy and post-rainy seasons, groundnut response to EMS was fairly similar (Sarma and Sivakumar, 1990). Stirling *et al.* (1990) in controlled-environment, withholding water in groundnut between sowing to pod initiation showed the pod yield 5 times higher than withholding water between pod initiation and harvest concluded that field management should aim to optimize water availability at pegging. Osmotic adjustment occurred in expanding but not in mature leaves; the latter often lost turgor around midday and pegs retained turgor during water stress in both treatments.

The groundnut cv. Yueyou 5, grown in a growth chamber with irrigation suspended after 11 d to create water stress, when sprayed with 3000 ppm triadimefon, SEM examination revealed that triadimefon treatment induced stomatal closure, reduced transpiration and increased drought resistance of the seedlings. From the 4th day after the suspension of irrigation, proline contents decreased and water contents increased in treated seedlings compared with the untreated controls. Treated seedlings withered on the 5th day and the control on the 4th day. Triadimefon also reduced plant height and increased photosynthetic rate and improved photosynthate distribution in favour of the sink, indicating that triadimefon, usually a systemic fungicide, may be a potential antitrans-pirant for groundnuts (Guo and Pan, 1989).

In greenhouse, groundnuts irrigated from sowing to pod intiation or from pod intiation to final harvest, when compared, the shoot DM yields were largely unaffected, but pod yields were 4-fold lower in early- than in late-irrigated crops. The insensitivity of pod yield to early moisture deficits reflected the extreme plasticity of growth and development in groundnut (Stirling, 1989). The effect of water stress on Tainan 9 groundnut cultivar studied on Warin soil series of the Agricultural Development Research Center (ADRC), Khon Kaen Province in Thailand during dry seasons by providing various treatments indicated that water stress plays an important role on the yield of groundnut and yield decreases due to water stress at different growth stages follows the following sequence: water stress at seed development > at early pod filling

> at early growth > at early pegging (Uthai-Arromratana *et al.*, 1993).The groundnut is the most important oil and cash crop in the sub-Sahelian tropics and traits expressed at the early stages of the cycle that could reveal cv differences in drought adaptation in the field are useful as plant adaptation to drought, i.e. cultivars (cvs) that can maintain yield when water is limited, is a complex phenomenon and not yet fully understood.

The yield advantages due to moderate water deficit during the pre-flowering phase are associated with greater pod synchrony after the release of water stress, resulting in production of more mature pods (Nageswar Rao *et al.*, 1988). When stress is released, the plant try to set more fruiting sites with the existing assimilates as the vegetative site demanding assimilate supply are reduced. To improve the conventional irrigation management practices to enhance yield and water use efficiency in groundnut during summer seasons a field experiment was conducted by Nautiyal *et al.* (2002) where dry matter partitioning among various plant parts, and leaf area index (LAI) varied significantly under water deficit and more dry matter accumulated in petiole and stem under stress. Transient water-deficit in vegetative phase resulted in higher dry matter accumulation in reproductive parts (peg+pod). Water use efficiency (Ef) was, however, higher under prolonged stress during vegetative phase; though cultivars response varied. Per cent reduction in total biomass under stress during flowering (F), and pod-development (P) ranged between 6 and 25%, this reduction in total biomass was mainly due to the reduction in the pod mass rather than in the vegetative mass. Water-deficit occurring during the vegetative stage, seedling stage until flowering, was most beneficial for the crop, and need to be utilized in irrigation scheduling (Nautiyal *et al.*, 2002).

5.3 Mid Season Stress

The mid-season drought (MSD) of common occurrence in groundnut in central India and AP in particular causing severe yield losses and reduction in yield attributes and quality, as against crop receiving full irrigation during the whole crop duration, revealed genotype differences, and genotype environment interaction for pod yield, shelling percentage and HI. There was maximum reduction in the number of mature pods (47%) under MSD, followed by pod yield (29.7%) while under ESD for pod yield (41%), followed by the number of mature pods (33%), indicating that number of mature pods was most sensitive to MSD and pod yield to ESD (Suvarna *et al.*, 2002).

Different drought management practices for groundnut were compared on shallow (15 cm depth) sandy loam soil with a loamy subsoil where pod yield was 1.36 t/ha in control plots, and with drought management practices it was 1.53 t with a 5% kaolin spray during mid-season stress and 1.68 t when mulched with 5 t groundnut shells/ha 10 d after sowing (Reddy, 1994). In a field at ICRISAT, Patancheru, during the rabi/summer season (Dec to April) screening of 20 groundnut genotypes under three drought conditions normal (with regular irrigation); mid-season drought (irrigation withheld 50-100 DAS); and end-season drought (irrigation withheld between 100 DAS to harvest) reveals that the genotypes, ICGV 86031 performed best under both normal and stress conditions, genotypes ICGV 93261, 93269, 93277 and KRG 1 had high yields and less percent yield reductions under the stress conditions (Suvarna *et al.*, 2006). By adopting moisture conservative methods during July and September months enhanced the groundnut yields in the drylands of Tirupati (Sumathi and Subramanyam, 2007).

At Tirupati, with groundnut cv. JL 24 moisture stress induced for 45 d in 1986 and 35 d in 1987, after irrigation at 25 d after sowing in summer by Selvam *et al.* (1989) reported that stress did not affect plant population, but reduced number of filled pods/plant. The pod yields in treatment (a) control, (b) water spray twice during stress, (c) irrigation twice to prevent moisture stress, (d) 2% urea spray once after relieving stress, (e) 2% urea spray twice during stress, and (f) 2% urea + trace element compound Tracel-1 (13.2% S, 6.6% Zn, 5.3% Fe and 3.7% Mn) sprayed twice during stress, followed (f) > (e) > (d) = (c) > (b) > (a) and ranged from 0.80 t/ha in (a) to 1.82 t ha^{-1} in (f). Thus, foliar application of 2% urea with trace elements is a promising technique to overcome the adverse effects of mid-season moisture stress (Selvam *et al.*, 1989).

Variations exist in the proportion of DM used for pod growth and large variations in the response of genotypes to midseason are due to recovery differences after the drought is relieved and a 3-factor interaction of genotype, gypsum, and drought exists because the gypsum may increase early pod development, thus providing escape effects (Williams *et al.*, 1985). In a field trial in Rahuri, Maharashtra on mulching with sugarcane trash or white polyethylene in groundnut water stress at different growth stages (flowering, pegging or pod development stage) during summer, indicated that mulching with white polyethylene along with water stress at the pod development stage resulted in the greatest plant height (38 cm), plant spread (25 cm), number of functional leaves per plant (73), leaf area per plant (21 dm^{-2}), dry matter per plant (18 g), dry pod yield (2896 kg ha^{-1}), haulm yield (4741 kg ha^{-1}), dry kernel yield (2083 kg ha^{-1}), oil yield (989 kg ha^{-1}), net monetary returns (34 097 rupees/ha) and benefit: cost ratio (2.44) (Bodare and Dhonde, 2011).The oil content with polyethylene and sugarcane trash mulch in water stress at the pod development stage was 47.7% and 47.3%, respectively (Bodare and Dhonde, 2011).

5.4 Late Season Stress

A large spectrum of genotype duration is now available, from long to short and extra-short duration and matching genotype duration with likely period of soil water availability is the first strategy used against terminal-drought stress. Reddy *et al.* (2003) in a review summarized recent information on the drought resistance characteristics of groundnut with a view to developing appropriate genetic enhancement strategies for water-limited environments and suggested that a considerable gain has to be made in increasing and stabilizing yield in environments characterized by terminal drought stress and further exploiting drought escape strategy by shortening crop duration. Many traits conferring dehydration avoidance and dehydration tolerance are available, but integrated traits, expressed at a high level of organization, are likely to be more useful in crop improvement programmes. Possible genetic improvement strategies are outlined, right from empirical selection for yield in drought environments to a physiological-genetic approach. Increasing soil moisture storage by soil profile management and nutrient management for quick recovery from drought are some of the areas which need to be explored (Reddy *et al.*, 2003).

Groundnut in the Sahel is often exposed to end-of-season drought and studies taking five new selected Spanish cultivars along with the control cultivar, 55-437 to identify traits associated with yield variation, indicated that earliness and general adaptation of the cultivars did not impair the expression of significant genetic variation for some traits relative to flowering, productivity and physiology (Clavel *et al.,* 2004). The partitioning coefficient and yield of the 5 cultivars under water stress were higher than those of cultivar 55-437, water deficit affected leaf area index, RWC and transpiration at approximately 2 weeks after the occurrence of water deficit at the soil level. Since genotypic differences seemed to be greatest at this time, measuring physiological traits during this period may be useful for breeding early groundnut for end-of-season water deficit conditions (Clavel *et al.,* 2004). In field during summer, 0.5% KCl spray reduced the effects of water stress imposed during seed growth stage (60-90 d after emergence) in groundnut cv. CO_2 suggesting that this may be useful in reducing water demand at critical growth stages (Mohandass *et al.*, 1989).

Four groundnut lines (H 2030, H 2060, H 2063 and H 2095) showing resistance to aflatoxin contamination (caused by *Aspergillus flavus* and *A. parasiticus*) when evaluated for pre-harvest contamination by aflatoxins under end-of-season drought stress in field alongwith susceptible line 88-1202 by controling irrigation at 80 DAS when plants showed slight wilting symptoms due to water deficit. The plants were harvested at 120 DAS, and the seeds were analysed for aflatoxin contamination within 30 days after harvesting. In 88-1202, drought stress significantly increased the aflatoxin content of seeds, which was higher than that of the 4 genotypes. Drought treatment did not significantly enhance the aflatoxin content in the resistant lines. H 2030 had the lowest aflatoxin content ($0.073\mu g\ g^{-1}$) among the lines. These results confirmed that aflatoxin resistance in vitro could reduce pre-harvest contamination of seeds by aflatoxins (Yu *et al.*, 2004).

Spanish type groundnut, due to relatively short duration, is preferred under lesser water availability environment. Field trials at DGR for three contrasting seasons with 30 Spanish groundnut cultivars showed wide genetic variability in physiological and yield components. The associations between pod yield and leaf area index during early crop growth stages, P_N and number of reproductive sink, and P_N and difference between T_{leaf} and T_{air} during pegging stage indicated that source is not a limiting factor. The lower SLA cultivars could be utilized under less water availability environment, the higher SLA types could be utilized in increasing early leaf area index and biomass, especially under irrigated condition. The P_N was higher during full pod (R4) (18 μmol $m^{-2}s^{-1}$) followed by beginning seed (R7) (17 μmol $m^{-2}s^{-1}$) and lower during full seed (R6) (8 μmol $m^{-2}s^{-1}$) and harvest maturity (R8) (8 μmol $m^{-2}s^{-1}$) showing enhanced P_N during beginning pod, to meet the additional requirement of photosynthates to developing reproductive sink (Nautiyal *et al.*, 2012).

5.5 Drought Tolerant Cultivars, QTLs and Breeding

Cultivated groundnut (*Arachis hypogaea* L.), an allotetraploid (2n=4x=40), is a self pollinated and widely grown crop in the semi-arid regions of the world. Improvement of drought tolerance is an important area of research for groundnut breeding programmes. The DGR, AICRP on groundnut and ICRISAT's research achievements for the past four decades in the domain of drought tolerance and present future perspectives in the genetic enhancement of crop water use and drought adaptation in the semiarid and arid tropics reveals that exploration of crop genetic variability and genotype-environment interactions have contributed significantly for developing suitable screening methods for specific drought-tolerant traits. Genetic sources of drought tolerance were also identified for groundnut, and some of the associated traits have been well characterized. A number of genotype with varied duration is now available, and matching genotype duration with likely period of soil water availability is the first strategy used against drought stress. Identification and genetic mapping of quantitative trait loci for specific drought-tolerant traits using molecular markers are currently receiving greater research focus. This approach provides a powerful tool for dissecting the genetic basis of drought tolerance. If validated with accurate phenotyping and properly integrated in marker-assisted breeding programmes, this approach will accelerate the development of drought-tolerant genotypes. Overall, the progress made at ICRISAT during the last three decades proves that it is realistic to develop varieties that have increased yield under drought-prone conditions, however further multidisciplinary research integrating plant breeding, simulation modelling, physiology and molecular genetics will realize the potential of these approaches and increase the efficiency of crop improvement in drought-prone environments (Serraj *et al.*, 2003).

Breeding for drought resistance in groundnut has been done using empirical selection for yield under drought stress conditions. Selection for physiological traits contributing to superior performance of the crop under drought stress conditions, SLA and HI are physiological traits to be used for this purpose. The identified groundnut cultivars showing drought tolerance in various climatic conditions and soil types are summarized in Table 6 and the sensitive one in Table 7. The sensitivity of a genotype to drought increases with yield potential, increasing the closer the drought ends to final harvest. Genotypic variation in response to drought exists in the water-use ratio of genotypes, with some being able to accumulate up to 30% more shoot DM than others with the same total transpiration (Williams *et al.*, 1985). Reddy *et al.* (2003) outlined possible genetic improvement strategies ranging from empirical selection for yield in drought environments to a physiological-genetic approach. For the choice of an efficient breeding procedure, a good knowledge on the types of gene action controlling the expression of these traits is needed. Field and laboratory studies conducted in Khon Kaen, Thailand, during the rainy season, to examine the various gene effects for SLA and HI in the groundnut crosses ICGV 86388 x IC 10, ICGV 86388 x KK 60-1 and IC10 x KK 60-1 showed that additive gene effects were predominant in determining the expression of SLA and HI in all 3 crosses, accounting for 80-95% of the total genetic variation for SLA and 63-73% for HI. The dominant gene effect for SLA was significant in one cross but its contribution was very small. Significant additive x dominant epistatic effects were also observed for SLA in all crosses, but additive x additive and dominant x dominant gene effects were significant in one cross each. Significant epistatic gene effects for HI were also detected in two crosses but were largely additive x additive, which is fixable. The predominance of additive gene effects for SLA and HI suggested that selection for the two traits in these crosses would be effective even in early segregating generations (Suriharn *et al.*, 2005).

The choice of a genotype for production under irrigated or moisture stress condition should be based on the yield potential and moisture stress tolerance of the genotype. Izge and Olorunju (2000) evaluated 16 groundnut cultivars in greenhouse and found most of them sensitive to moisture stress, as evident in most of the parameters (plants at emergence, days to first flowering, maturity, pods per plant, pod and haulm yields, shelling percentage, 100-seed weight, protein and oil contents). Pod yield and other attributes, except protein and oil contents, were significantly affected by moisture stress. SAMNUT 18 (RRB) had the highest pod yield under both normal moisture and moisture stress conditions. Two Argentinian groundnut varieties Florman INTA (a drought-sensitive) and Manfredi 393 INTA (a drought-tolerant) when compared at two different water supply regimes the Manfredi 393 INTA demonstrated an adaptative advantage to drought, expressed in its higher pod production; due to either an earlier onset of beginning peg stage that affected differentially peg penetration into the dry soil upper layer, and an enhanced partitioning of assimilates to pods (Collino *et al.*, 2000).

Table 6. Drought tolerant groundnut varieties/cultivars reported in various soils and climate

SN	Tolerant cultivars	Soil type, Place	Crop Stage	Criteria used	Details of Drought tolerance	References
1.	GG-2	Black calcareous, Junagadh	VG, PD	Transpi-ration rates and low leaf psi at rela-tively higher RWC.	High transpiration in GG-2 at 1800 h under water stress a mechanism ada-ptive to compensate for reduced CO_2 fixation during day. GG-2 with low leaf psi at relatively higher RWC than JL-24.	Joshi *et al.*, 1988
2.	J11, GG 2	Black calca-reous, Junagadh	VG, FL or PD stages	Yield	Water stress during vegetative stage had less effect on pod yield, nodules, plant DM and N uptake than at later stages.	Kulkarni *et al.*, 1988
3.	GG-2, G-13 and G-20	-	-	Germination in field and PEGstress	Putrescine prevented the fall in tissue moisture in water-deficit seedlings.	Vakharia *et al.*, 2003
4.	Ex-Dakar, RRB12, RMP12, 91	-	Early growth stage	Germination in field	The germination in PEG, glucose or NaCl solutions at 1.8 Mpa a quick proc-edure for drought screening	Mensah and Okpere, 2000
5.	ICGV8603 1, TMV-2, TCGS-41	-	-	Germination in labora-tory	Germination and seedling growth in water stress situation (-1.0 MPa)	Prathap *et al.*, 2006
6.	RS 218	-	Seedling stage	better perfor-mance of seedlings	0.25% $CaCl_2$ for 8 h was ascribed to higher accu-mulation of proline and K in seedlings soil moisture stress conditions	Sashidhar *et al.*, 1981
7.	Kadiri	-	Sowing to PI and PI to harvest	Growth stages	The insensitivity of pod yield to early moisture deficits reflected the extreme plasticity of growth and development.	Stirling,198 9
8.	TMV 2, Acc 847, 55-437, GNP 1157	-	FI to PI (14 to 80 DAS)	Yield in pot study	Interactions between cul-tivar and water stress on stomatal resistance, leaf water potential & root DW.	DelRosario and Fajardo, 1988
9.	GAUG-1	Medium black soil Junagadh	-	growth and yield para-meters	K application decreased these negative effects of water stress by increasing yield parameters, oil and protein content.	Shahid-Umar *et al.*, 1997

10.	Dh-3-30 TMV-2	-	FL peg and pod formation	Leaf diffusive resistance, proline transpiration	Leaf diffusive resistance and proline accumulation highest and transpiration rate lowest with water stress at 70 DAS.	Koti *et al.*, 1994
11.	ICGV 86031 TMV2NLM	Andhra Pradesh	-	Soil moisture extraction, root length density	Efficient in soil water extraction and least losses (17-18%) in total DM production under moisture stress	Reddy, 1999
12.	CO 2, VRI2	Coimbatore, TN	FL, PG, PD, PM stages	yield components and crop yield	Water stress at pod development had most effect on yield	Velu, 1998
13.	JL 24	-	5th sub-culture stage	Stress tolerance capacity of selected tissues	Tissues of JL 24 performed better than TMV 2 for growth, solute accumulation and water relation under PEG stress	Purushotham *et al.*, 1998
14.	AK-12-24	-	All growth stages	Yield and water use efficiency	Highest WUE in plants stressed at pod development stage & lowest when stressed at peg penetration.	Kar *et al.*, 2002
15.	SAMNUT-18 (RRB)	Samaru, Zaria, Nigeria	EMI, FLI, 50% FL	Pod yield and attributes	Pod yield and other attributes were affected by moisture stress.	Izge and Olorunju, 2000
16.	Tatu	-	VG and RG stages like FL, PI	In Green house, flowering, pegging	Lesser Flower production by water stress.	Oliveira-Junior *et al.*, 2004
17.	JL 24	Kalyani, IND	FL, PI	In field, pod, haulm yield	Water stress at the pod initiation and development stages reduced yield by 13.4 & 44.2%, respectively.	Patra *et al.*, 1999
18.	ICGV 86707	Medium deep Alfisol at ICRISAT	PI or PF (83-113 DAS	Partitioning to DM to pods in drought	Partitioning to pods in drought ranged 0.56 in cv. ICGV 86707 to >0.95 in early maturing genotypes.	Rao *et al.*, 1993
19.	TMV 2	Sandy loam, Andhra Pradesh	VG, FL, MS throughout growing	Yield	Produced optimum yield (2.82 t/ha) with maximum WUE (7.73 kg ha^{-1} mm^{-1}) and water economy	Reddy and Reddy, 1993
20.	JL-24	Akola, deep soil	VG and RP stages	Root respiration and yield	Cv. JL-24 was the most tolerant of water stress with 32.1% yield reduction compared with 47-67% yield reduction in TAG-24 the most susceptible cultivar.	Dhopte *et al.*, 1992

21.	GG-2	Junagadh, Gujarat	FL, PI, PD, PM stages	Pod yield, seed formation efficiency	Pod yield was reduced by reductions in fertility index, seed formation coefficient and seed formation efficiency and GG-2 was more drought tolerant than J-11.	Golakiya and Patel, 1992
22.	Q18801	Redland Bay, Queensland, Australia	Early RP stage	Harvest index	Q18801, a Virginia cultivar with high HI yielded higher than Virginia Bunch and McCubbin (a Spanish).	Chapman *et al.*, 1993
23.	SB XI	-	0-40, 40-80, 80-120 d after sowing	IW: CPE ratios	WUE highest (5.23 kg ha-1 mm) with irrigation at IW:CPE ratio 0.5 0-40 DAS, 0.75 during 50-80 DAS and adequate moisture 80-110 DAS for obtaining maximum yields.	Patil and Gangavane, 1990
24.	GG2	DGR, Junagadh	All pheno-phases	Leaf transpiration rate, RWC, pod yield,	GG 2 showed developmental plasticity and gave the best pod yields even under stress at any growth phase,	Ravindra *et al.*, 1990
25.	TMV-2		VG, FL, PI, PD stages	Pod yield	High pod yield of 2.62 t/ha was unde moisture stress during the pod development and maturation stages.	Raju *et al.*, 1981
26.	Baisha 1016	Shandong, China	FL, FR and ripening stages	Protein content of seeds,	Drought at the flowering reduced protein content in seeds, while drought at seed development and ripening stages it gave seeds with higher protein than control.	Yao *et al.*, 1982
27.	Robut 33-1-		EM to MI, Em to FL, start of SG to MI	Seed yield , evapotranspiration	Evapotranspiration- yield relationships showed a strong interaction with timing of drought and greatest reduction in seed yield (28-96%) when stress imposed during seed growth	Rao *et al.*, 1985
28.	Robut 33-1-		-	Pod:shoot wt. ratio (PWR), pod yield	Mild water stress promoted peg and pod production and increased PWR.	Ong, 1984

29.	GG-2	Medium black calcareous soil	FL, PI, PD, PF stages	Pod yield and nutrients (N, P, K) accumu-lation	Pod yield most affected due to water stress during pod development, nutrient accumulation increased with crop growth except K which showed maximum accumulation at 78 DAS	Sakarvadia *et al.*, 2010
30.	GG 6	Clay soil Junagadh	FL, PI, PD, PF stages	benefit: cost ratio, WUE	GG 6 recorded higher WUE and benefit: cost ratio.	Vaghasia *et al.*, 2010
31.	ICGS-11, ICGS-44 and Girnar-1	-	Fl stage	Stomatal conductance, proline content	The better performance under water stress was related to good stomatal conductance and increased proline level for osmotic adjustment.	Patil and Patil, 1993
32.	Florunner	Tifton, Georgia	Pre-flowe-ring, pod formation matu-ration	Fatty acid composition, O:L ratio, iodine value	Under water stress, long chain saturated fatty acids, eicosenoic acid [gadoleic acid], O:L ratio and alpha - tocopherol decreased in groundnuts and stress at maturity decreased O:L ratio, but increased iodine value.	Hashim *et al.*, 1993
33.	Giza 5	Egypt	Pod initiation and pod developm ent	Yield and yield attributes oil and protein content and yield	Water stress (skipping one irrigation) and potassium fertilizer application	Ali, 2001
34.	TAG 24	Jhargram, West Bengal, India, acid lateritic soils	VG and FL stages	WUE, Benefit: cost ratio	Maximum WUE under moisture stress during vegetative stage. Benefit: cost ratio was higher with moisture stress at vegetative stage.	Dutta and Mondal, 2006
35.	JL 24	Madurai, Tamil Nadu	At early PI, late PD stages and VG stges	Pod yield, seed quality, oil and protein content, nutrient uptake	Kaolinite (3%) spray reduced yield loss due to water stress.	Naveen *et al.*, 1992
36.	PI 165-3176	Ceara, Brazil, sandy soil	FL	Pod yield and transpiration	Pod yield was highest for PI 165-3176 under all water regimes.	Ferreira *et al.*, 1992

37.	Robut-33	Redland Bay, Queens land	-	Pod yield, harvest index (HI)	Robut-33 had the highest CGR and early and rapid pod growth and high HI more important in yield under water deficit.	Chapman *et al.,* 1993
38.	TMV2	-	-	Flowering , pod yield or oil content	Moisture stress reduced the number of gynophores formed from the 1st flush of flowers	Gowda, 1977
39.	Kadiri-3	Green-houses	At 28 °C (±5°) under CO_2 levels of 350 and 700 ppmv)	leaf water potential	Elevated CO2 affected growth and yield mediated by an increase in the conversion coefficient for intercepted radiation and maintenance of higher leaf water potentials during drought stress.	Clifford *et al.,* 1993
40.	J11, GG 2 and Robert, GAUG 1	-	Early growth period	Yield	Withdrawal of 2 irrigations in early crop growth followed by spraying of 50 ppm IBA at 40 and 60 DAS was economical.	Patel *et al.,* 1988
41.	JL 24	Vertisol and silty loam	-	Total DM, Pod yield	In water-stressed conditions, the cv. JL-24 partitioned more total DM to pods than susceptible cultivars	Dhopte and Ramteke,19 94
42.	JB 223 and 224	Junagadh, Gujarat	-	Yield	Recorded consistently superior and stable yield for the three years at all the locations.	Mandavia *et al.,* 2007
43.	ICGV8601 5	Sri Lanka	VG stage	Seed yield	highest seed yield under water-stressed condition	Costa *et al.,* 2001
44.	DVR50	-	PI stage	Drought index, Pod yields	highest yield under both natural and moisture-stress conditions.	Chavan *et al.,* 1992
45.	Robut-33	University of Queens land, Australia	Early RP stage	Harvest index	Robut-33 showed highly synchronous development and high HI and an ability to tolerate drought.	Chapman *et al.,* 1993
46.	GG 2	Junagadh, Gujarat	FL, PI, PD, PM stages	Pod yield	GG 2 was more tolerant to drought than J 11	Patel and Golakiya, 1991
47.	ICGV 87123	Niamey, Niger	-	pod yield	ICGV 87123 and 55-437 gave the highest yields over all irrigation treatments.	Greenberg and Ndunguru, 1989

48.	TG 17	-	FL, PI, PD stages	leaf area, pod yield, RWC	TG 17 tolerated water stress better than cv. GAUG 1.	Patel *et al.*, 1983
49.	Kadiri 3	-	PI stage	Filled - Pod wt/plant	Kadiri 3 intercropped with sorghum hybrid CSH-8, increased pod yield, in droughted stands.	Harris and Natarajan,1 987
50.	TG26	Southern Telangana, Alfisols, AP	PI to PD stage	pod yields	In rainy season, pod yields ranged from 0.58 (TCGS88) to 2.41 t/ha.	Thatikunta Ramesh *et al.*, 1996
51.	ICGV 86031	ICRISAT, Patan-cheru	VG and RP stages	yield	Performed best under both normal and water stress conditions .	Suvarna *et al.*, 2006
52.	Yueyou 5	-	11 d old seedling stage	Stomatal closure, tran-spiration, seedling growth	Triadimefon induced stomatal closure, reducing transpiration and increasing drought resistance of the seedlings.	Guo and Pan, 1989
53.	M 37	-	50 and 80 DAS	pod setting, oil content in seeds	The moisture stress suppressed pod setting, but not the oil content	Sharma *et al.*, 1987
54.	ICG-4504 and GPHY-35	-	-	Yield	The percentage of reduction in their yield under increasing moisture stress was less.	Shashikuma r *et al.*, 1988
55.	Robut 33-1	ICRISAT Patan-cheru, India	Emer-gence to maturity	Emergence, seedling vigour, pod and seed yields	Seeds from groundnut grown under stress from emergence to initiation of pegs gave higher field emergence, better seedling vigour pod yields.	Sarma and Sivakumar, 1987
56.	GG-2	Calcareous - vertic Inceptisol		Vegetative growth	Lower fluctuations in leaf temperature, stomatal resistance and vegetative growth.	Patel and Golakiya, 1993
57.	PBS 11049	Junagadh	95 DAS after sowing	Epicuticular wax load	Wax load ranged from 0.91 mg dm^{-2} in Chico to 1.74 mg dm-2 in PBS 11049, with a mean of 1.27 mg dm-2.	Samdur *et al.*, 2003
58.	RD 14 , RD 22 and RD 25	-	19, 28 DAS etc.	Symptoms	Transgenic lines RD 14, RD 22 and RD 25 showed lesser symptoms than JL 24. RD 14 reached the end point in 29 days, and RD 4 in 52 days.	Mathur *et al.*, 2004

59.	Manfredi 393 INTA	-	47 and 113 days after sowing	Critical soil resistance and Peg and pod development	Manfredi 393 INTA had an adaptative advantage to drought, due to earlier onset of peg and enhanced partitioning of assimilates to pods.	Collino *et al.*, 2000
60.	Q18801	Redland Bay, Queensland		LAI, transpiration efficiency	Q18801 was able to maintain a higher LAI and a greater crop transpiration efficiency than McCubbin.	Chapman *et al.*, 1993
61.	ICGVs 88369, 88371, 88381, -82 , 88403	ICRISAT Patancheru	40 and 80 DAS, mid-season drought	Oil content,	In ICGVs 88369, 88371, 88381, 88382 and 88403, total oil content remained unaffected while oleic fatty acid content increased under end-of-season drought	Dwivedi *et al.*, 1996
62.	K-134	-	-	The protease activity	Protease activity of K-134 makes it a more drought-tolerant cultivar	Madhusudhan *et al.*, 2002
63.	GG-2	-	Growth stages	Carbohydrate concentration	After 30 days without irrigation, increase in total carbohydrate.	Kandoliya *et al.*, 2000
64.	GAUG- 1	Gujarat	Different growth stages	Palmitic acid	Increase in palmitic acid due to stress during pod development phase was observed only in GAUG 1.	Misra and Nautiyal, 2005
65.	K 3	-		Yield components	Cv. K 3, performed better than JL 24 and Gangapuri if sowing was delayed until July.	Padma *et al.*, 1991
66.	Florunner	-	-	Soluble and total carbohydrate, total protein and oil contents	Soluble and total carbohydrate increased in Jumbo showing the highest increase under drought and temperature stress.	Musingo *et al.*, 1989
67.	TPT-4	-	-	CaM, proline and activity proline oxidase	The CaM in cotyledons of water-stressed seedlings decreased, but $CaCl_2$ treated maintained higher CaM and proline oxidase activity.	Sulochana and Savithramma, 2001
68.	TPT-4	-	-	Acid phosphatase activity	External Ca^{2+} maintained higher levels of ACPH activity in the seedlings of TPT-4 than TPT-1.	Sulochana and Savithramma, 2003

69.	Dragon - type cultivars	-	-	SOD activity, protein content, water potential	Under drought, water potential was higher in the dragon-type cultivars than in the other types cultivar.	Jiang and XiaoPing, 2004
70.	Baipi No.1	-	-	RPMP, oxygen radical, MDA content, and SOD, POD, CAT activity and AsA.	Resistant to drought as evident from smaller increase in RPMP, oxygen radical generation and MDA content, and in slower decrease in SOD, POD and CAT activity and AsA content.	Chen-YouQiang *et al.*, 2000
71.	Jun-40 and GG-2	-	-	SOD, SOR, APX, lipid peroxidation	Jun-40 accumulated SOR, higher SOD activity and GG-2 showed less lipid peroxidation in water stress.	Mittal *et al.*, 2006
72.	M-13	-	10-day-old seed lings	Relative turgidity, protein and NR activity	M-13, during stress, was able to preserve its protein conc. and nitrate reductase activity	Saini and Srivastava, 1981
73.	TAG 24 and Somnath	Mumbai	-	Leaflet area, stomatal frequency on adaxial	Reduced leaf area, increase in stomatal frequency on adaxial surface are related to WUE in TAG 24 and Somnath	Badigannavar *et al.*, 1999
74.	ICGV 99029 and ICR 48	-	-	SCMR, SLA	ICGV 99029 and ICR 48, with higher SCMR and lower SLA values indicated higher WUE.	Nigam and Aruna, 2008
75.	Vemana and K 1375	-	-	Water stress-regulated proteins,	Vemana, K 1375 are able to maintain expression of certain proteins (MW 14 kDa and 70 kDa)	Ramesh-Katam *et al.*, 2007
76.	Kadiri-3	-	-	CMS, K value, membrane stabilty	Kadiri-3, maintained higher CMS and K values greater cell membrane stability than JL-24, the drought susceptible one.	Venkateswarlu and Ramesh, 1993
77.	Manfredi 420	-	-	Stomatal opening, membrane stability	Manfedi 420, could keep their stomata most open during drought stress, without major alterations in membrane stability.	Collino *et al.*, 1994
78.	TMV2 , Kadiri 3	Sandy loam soil (Alfisol)	-	assimilation and water loss	At peak flowering these had lower canopy temperature.	Subramanian *et al.*, 1993

79.	GG-2	-	-	RWC, proline, ascorbic acid and reducing sugars	GG-2 is characterized by high RWC & greater accumulation of proline, ascorbic acid and reducing sugars.	Dhruve and Vakharia, 2007
80.	TMV-2	-	-	Pod yield	The lowest water stress was experienced by TMV2 and highest by TAG-24.	Mukherjee *et al.*, 2010
81.	TMV2 and J11	-	-	Gypsum requirement	Drought susceptible varieties has more requirement of gypsum.	Rajendrudu & Williams, 1987
82.	ICG221	Vriddhachalam, TN	-	-	ICG221 performed best in the simulated stress environment (rain out shelter).	Arjunan *et al.*, 1997
83.	Nigeria 55437	-	-	Leaf water potential, transpiration and DR	Nigeria 55437 showed high diffusion resistance and proline and low leaf water potential and transpiration under stress.	Nogueira-RJMC *et al.*, 1998
84.	Manfredi 393 INTA	-	-	RUE	RUE values higher in Manfredi 393 INTA than in Florman INTA,	Collino *et al.*, 2001
85.	JUG-37, JUG-43 and TIR-46	Tirupati, AP, India	-	Pod yield	Parents, JUG-37, JUG-43 and TIR-46 were desirable general combiners for pod yield, drought tolerance.	Venkateswarlu *et al*, 2007
86.	Spancross and NC 3033	-	-	Pod yield	The increase in pod yield was 21 and 23% for the drought resistant cv. Spancross and NC 3033	Womble and Garren, 1978
87.	TMV2 (Tindivanam 2)	-	-	leaflet area, leaf closure	Leaflets under water stress reduced areas and showed greater movement (leaf closure) than control.	Babu *et al.*, 1983
88.	TIR-21,34, JAL-07 & CSMG 84-1	-	-	Yield	Genotypes TIR-21, TIR-34, JAL-07 and CSMG 84-1 are better under higher temperatures.	Babitha *et al*, 2006
89.	Yuhua 13 and FDR-S10 dragon-type				Under drought stress, water potential was higher in the dragon-type cultivars than other.	Jiang and XiaoPing, 2004
90.	X537B, Florunner & X487A			Thermo stability of cellular membranes	Genotype x date interactions observed, but Pearl Early Runner, X537B, Florunner and X487A were the most heat tolerant	Bennett and Hammond, 1982

Where, FL= Flowering, FR= Fruiting, PI= Pod initiation, PD= Pod development, VG= Vegetatitive, RP= Reproductive, M= maturity, MI= Maturity initiation stages, FI= Flower initiation, Pg I=Peg initiation

Under limited rainfall conditions, the crosses of GG 2 x NCAC 17135, GG 2 x PI 259747, J 11 x PI 259747 and S 206 x FESR 8, kisan x FESR-S-PI-B1-B and the genotypes JB 223 and 224 recorded consistently superior and stable yield for the three years and could be grown under regions of limited rainfall and may be used as parents in breeding programmes for developing drought tolerant groundnut cultivars (Mandavia *et al.,* 2007). Six groundnut cultivars grown on a deep Vertisol or a shallow silty loam in the kharif season showed highest pod and total DM yield/plant in cv. TAG 24 in deep soil, and in cv. JL 24 on shallow soil. In water-stressed conditions, the tolerant cv. JL 24 partitioned more of the total DM to pods than did drought susceptible cultivars (Dhopte and Ramteke, 1994).

The transpiration response of 14 transgenic groundnut genotypes to water deficit was studied under greenhouse along with JL 24 as control by exposing the plants to drought stress (absence of irrigation) at 19 DAS where JL 24 started showing wilting symptoms (loss of turgor) after 21 days of stress, with severe symptoms later on. JL 24 reached the stage III (normalized transpiration rate or NTR <0.1) after 27 days. Wilting symptoms were not observed in the transgenic lines even after 21 DAS but later on these lines exhibited different levels of wilting symptoms, with a few transgenic lines showing no symptoms, and lines RD 14, RD 22 and RD 25 showing reduced levels of symptoms compared to JL 24. The transgenic lines varied largely in the number of days to reach the end point. RD 14 reached the end point in 29 days, whereas RD 4 reached the end point in 52 days. Data on NTR, fraction of transpirable soil water, and number of days to end point were subjected to average linkage cluster analysis for the development of a dendrogram indicating that the lines could be classified into 4 groups, as the transgenic lines varied in their stomatal response to water deficit (Mathur *et al.*, 2004).

The mechanism of water stress by leaflet angle variation when studied for cv. Tindivanam 2 (TMV 2) groundnut plants, radiation avoidance by leaf closure was exhibited to different degrees by water stressed plants and leaflets in plants under continuous water stress reduced areas and showed greater leaflet movement (leaf closure) compared to the amount of movement of leaflets on plants without water stress (Ramesh Babu *et al.*, 1983). At Vriddhachalam, evaluation of 68 groundnut genotypes for agronomic characteristics during rainy season the ICGS 76 and TAG 24 performed better under irrigated conditions, ICG 221 performed best in the simulated stress environment (rain out shelter), and ICGV 86635, DH 43 and ICG 2716 performed best under rainfed conditions (Arjunan *et al.*, 1997).

Relative drought tolerance of 17 groundnut genotypes when evaluated under six gradients of moisture stress using line-source irrigation, the JL 24 and ICG 5266 had high yield potential at all the moisture regimes but yield reduction with increasing moisture stress was higher than in other genotypes. On the other hand the ICG 4504 and GPHY 35 showed low yield potential under control as well as lesser yield reduction under increasing moisture stress (Shashikumar *et al.*, 1988). In groundnut cv. JL 24 grown in pot, water stress reduced number of nodules, leaf area, total DW and total chlorophyll and leaf protein contents and Ca application increased growth parameters and chlorophyll and protein contents (Mohan and Rao, 1989).

During February-May, the variation of micrometeorological parameters when assessed within different groundnut cultivars TG 51, ICGS 44, TAG 24, TMV 2 and AK 12-24 and their yield potentiality the ICGS 44 recorded the highest net radiant energy (490-540 w/m^2) irrespective of the dates of observation and TG 51 showed the lowest value. Based on the average canopy temperature value the cultivars are arranged in the following order TAG 24> ICGS 44> AK 12-24> TG 51> TMV 2. The lowest water stress was experienced by TMV 2 and highest by TAG 24. The highest and lowest pod yield was produced by TAG 24 and AK 12-24 cultivar, respectively (Mukherjee *et al.*, 2010).

Table 7. Drought sensitive groundnut cultivars/varieties reported in various soils and climate

SN	Sensitive varieties/ cultivars	Soil type, Place	Phenophase	Criteria used	Details of Drought tolerance	References
1	JL 24	Medium black, Junagadh	VG, PD	Transpiration and leaf psi and RWC.	Drought sensitive variety	Joshi *et al.*, 1988
2	Florman INTA	-	-	yield and oleic/linoleic ratio,	Nine variables along with yield and its compone-nts, and oleic/ lino-leic ratio, showed drought sensitive	Collino, 2000
3	M-13	-	FL stage (30 DAS)	RuBP carboxylase, PEP carboxylase	PEP carboxylase activities increased with stress RuBP carboxylase & NADP-glyceral-dehyde-3-phospha-te dehydrogenase decreased gradually.	Sharma *et al.*, 1993
4	Cultivar 55-437	Sahel	flowering	Partitioning coefficient, yield, LAI RWC and T	The partitioning coefficient & yield under water stress reduced, affected LAI, RWC & tran-spiration at 2 wee-ks of water deficit at the soil level	Clavel *et al.*, 2004
5	SB XI and UF 70103	-	12 d after sowing	Yield	UF 70103 and SB XI gave av. yields of 3.28 and 2.43 t/ha, resp.	Jadhao, *et al.*, 1989

6	AK-12-24	Bhubaneswar, Orissa, India	15 DAS of sowing	Crop coefficient value	Plants stressed at vegetative stage exhibited the lowest crop coefficient value (0.61) at initial stage and the highest value (1.33) at peak growth stage and lesser water evaporation from the soil surface.	Kar *et al.*, 2001
7	Chitra	-	30 DAS, 60 DAS	Proline, peroxidase, seed yield	Proline content and peroxidase activity increased with the simulated drought	Neelam-Yadav *et al.*, 2007
8	Ak 12-24, J-11, GAUG-1 and GG-3		pod initiation/ development stage	Viability, vigour, seed membrane integrity, dehydrogenase	Water stress at the PI, PD stages reduced germinability, vigour, seed membrane integrity, embryo RNA content, and dehydrogenase activity in cotyledons during germination	Nautiyal *et al.*, 1991
9	Florunner	Yoakum, Texas	-	Yield	Water stress increased proportion of small seeds.	Schubert and Sanders, 1985
10	Florigiant and Florunner	-	Germination	Seed weight of sound mature seed	Drought reduced the av. seed wt. of sound mature seed in Florigiant and Florunner but not in Tifspan	Pallas JE Jr *et al.*, 1977
11	JL-24	-	-	Proline , K content	Proline accumulation and tissue K content.	Venkateswarlu *et al.*, 1993
12	JL-24	-	-	Cell membrane stability	Kadiri-3, the drought tolerant cultivar maintained higher CMS than JL-24, the drought susceptible one.	Venkateswarlu and Ramesh, 1993

13 ICGS 44	-	Reproductive stage		The reproductive stage (35-115 days after sowing) was the most sensitive.	Reddy *et al.*, 1996
14 GG 11	Gujarat	-	Yield	Water stress during the late crop growth stages resulted in low yields	Sahu *et al.*, 2004
15 Konkan Gaurav	Palghar	-	Pod yield	Applying 7-11 irrigations gave higher pod yield than 5 irrigations.	Chavan *et al.*, 1999

Where, FL= Flowering, FR= Fruiting, PI= Pod initiation, PD= Pod development, VG= Vegetatitive, RP= Reproductive, M= maturity, MI= Maturity initiation stages, FI= Flower initiation, Pg I=Peg initiation

A combining ability analysis for drought tolerance and yield traits was conducted during rabi in Tirupati, AP, involving 8 parents (TIR 46, JUG 37, ICR 45, TIR 10, K 134, JAL 6, JUG 43 and JL 24) of groundnut in all possible combinations excluding reciprocals, where mean squares for both general combining ability (GCA) and specific combining ability (SCA) were significant for all the characters indicating the involvement of both additive and non-additive gene action in the inheritance of characters (Venkateswarlu *et al*, 2007). In general, high magnitude of GCA variance depicted the greater importance of additive gene action in the inheritance of traits and the parents, JUG 37, JUG 43 and TIR -46 emerged as desirable general combiners and the crosses K 134 x JUG 43, K 134 x JL 24 and TIR 46 x JAL 6 were the most desirable specific combinations for pod yield, drought tolerance and yield attributes (Venkateswarlu *et al.*, 2007).

For the identification of candidate QTLs for drought tolerance, a comprehensive and refined genetic map containing 191 SSR loci based on a single mapping population (TAG 24 x ICGV 86031), segregating for drought and surrogate traits was developed by Ravi *et al*, (2011) and genotyping and phenotyping data for more than ten drought related traits in 2-3 seasons were analyzed in detail for identification of main effect QTLs (M-QTLs) and epistatic QTLs (E-QTLs) using QTL Cartographer, QTL Network and Genotype Matrix Mapping (GMM) programmes. A total of 105 M-QTLs with 3.48-33.36% phenotypic variation explained (PVE) were identified using QTL Cartographer, while only 65 M-QTLs with 1.3-15.01% PVE were identified using QTL Network. A total of 53 M-QTLs were such which were identified using both programmes. On the other hand, GMM identified 186 (8.54-44.72% PVE) and 63 (7.11-21.13% PVE), three and two loci interactions, whereas only 8 E-QTL interactions with 1.7-8.34% PVE were identified through QTL Network. Interestingly a number of co-localized QTLs controlling 2-9 traits were also identified.

The identification of few major, many minor M-QTLs and QTL x QTL interactions in the study confirmed the complex and quantitative nature of drought tolerance in groundnut and study suggests deployment of modern approaches like marker-assisted recurrent selection or genomic selection instead of marker-assisted backcrossing approach for breeding for drought tolerance in groundnut (Ravi *et al*, 2011).

Five years of study on drought at DGR during 2005-2010 developed a concept on ideotype for water use efficiency and drought tolerance in groundnut and efficient root systems for maximum productivity under both irrigated and rain-dependent systems (DGR, 2010, 2009; Nautiyal *et al.*, 2012). In groundnut, compact canopy showed lesser leaf and canopy temperatures. The SLA, distance between two leaflets and petiole length were inversely associated with P_N and WUE. Number of branches and LAI during vegetative and early reproductive stages were directly associated with higher productivity (r = 0.62**), and JAL 42 an advance breeding line was ideal plant type under irrigated condition of summer season. The groundnut is characterized with degree of leaf folding to avoid the incidental radiations, especially under water deficit stress to avoid the water loss through leaf surfaces, i.e., higher the degree of leaf folding more drought tolerance (DGR, 2010). The large water storage cells, high stomatal density with small stomatal cell on both upper and lower surfaces of the leaf, higher palisade and spongy parenchyma cell thickness, higher number of xylem rows and number of cells are associated with drought tolerance, in groundnut. The epicuticular wax load accumulated under water deficit conditions cause drought avoidance mechanism. Plant height under water stress was associated positively with thickness of epidermis and inversely with thickness of palisade cell and number of xylem rows (DGR, 2010).

Uniform field emergence, early ground cover and synchronized flowering are desirable. The crop maturity period is positively associated with total biomass production, but inversely with harvest index (HI). The increase in pod yield, has been achieved is mainly due to increased HI. The groundnut cultivar with lower SLA and higher leaf area index and higher rate of biomass production during vegetative and early reproductive stage is a desirable trait (DGR, 2010). Water deficit stress is also associated with high temperature and in groundnut, a tolerant type is efficient in maintaining cell membrane integrity under high temperature stress, and accumulates less reserve food in stem but higher proline contents under water-deficit stress.

During rainy season under moisture-deficit, the low SLA groundnut cultivars were able to maintain higher RWC, P_N and stomatal conductance (g_s) with significant relationship between RWC and P_N (r= 0.91, P<0.01), and RWC and g_s (r=0.65, P<0.01) and a significant inverse association between SLA and RWC (DGR, 2010). The low SLA types (water use efficient) were drought tolerant in terms of total dry mass production and maintenance of higher RWC under water-deficit. Drought tolerant cultivars with higher biomass and pod yield could be made by combining high HI and WUE in terms of lower SLA. A superior rate of biomass accumulation and actual yield accumulation in order to acquire a higher HI is needed (DGR, 2010). The cultivar TAG 24 is most promising under summer season as it has higher P_N, g_s, HI and lower ΔT and higher stability in pod yield, stand close to the ideotype suitable for cultivation during summer season (DGR, 2010, 2009; Nautiyal *et al.*, 2012).

In a recent field study with six Spanish groundnut cultivars ('SG 99', 'ICGS 44', 'ICGV 86031', 'AK 159' and 'DRG 1') during summer season, Kalariya *et al.* (2012a) reported that water deficit stress at 30-60 DAS (WS1) and 60-85 DAS (WS2) reduced leaf RWC, Membrane Stability Index (MSI), but increased chlorophyll content, as compared to control plot irrigated at weekly interval. Under water deficit stress the leaf RWC reduction in 'ICGV 86031' (6.4% under WS1) and 'DRG 1' (10% in WS2) and reduction in MSI was highest in 'SG 99' and 'DRG 1' (22%). Under water deficit condition at Junagadh, the minimum fluorescence (F_0) and NPQ increased but the maximum quantum yield of PS II (Fv/Fm) and photosynthesis decreased (Kalariya *et al.*, 2012b). The rate of photosynthesis which was 29 and 36 μ mol m^{-2} s^{-1} in control plots decreased to 26 and 28 μ mol m^{-2} s^{-1} by imposing water deficit condition between 31-61 DAS and 61-87 DAS, respectively. The variety TAG 24 showed better stress recovery capacity with high photosynthesis under well watered as well as under water deficit condition whereas, data on chlorophyll fluorescence showed that variety ICGS 44 was least affected to damage via photoinhibitory action. Stress susceptibility index (STI) revealed that 'ICGS 44', 'TAG 24' and 'SG 99' were superior varieties compared to others.

In groundnut, WUE is correlated with SPAD chlorophyll meter reading (SCMR) and specific leaf area (SLA). These two traits, SCMR and SLA, can be used as surrogate traits for selecting for WUE. In order to improve SCMR and SLA, and in turn WUE in groundnut, a good knowledge of the genetic system controlling the expressions of these traits is essential for the selection of the most appropriate and efficient breeding procedure.

The Photosynthesis, Chlorophyll fluorescence and SPAD Chlorophyll meter reading (SCMR) were studied in 30 groundnut minicore germplasms during 70-90 Days after sowing during *Rabi-summer* 2012 where photosynthesis rate ranged from minimum of 18.3 μmol m^{-2} s^{-1} in NRCG 14327 to maximum of 34.2 μmol m^{-2} s^{-1} in NRCG 14338, while SCMR readings ranged from lowest value of 31 in NRCG 14332 to highest value of 41 in NRCG 14331, and the chlorophyll fluorescence highest Fv/Fm among the genotypes was 0.85 for NRCG 14347 (Nakar, *et al.*, 2012). Averages for Fv/Fm, SCMR and Photosynthetic rate in these groundnut genotypes were, 0.9, 37.3 and 28.1 respectively. On the basis of three physiological parameters studied, the physiologically efficient groundnut genotypes indentified were NRCG 14329, 14328, 14324, 14338, 14331,

14337, 14354 and 14348 accessions while the genotypes with NRCG accessions, 14351, 14347, 14332, 14343, 14344, 14335, 14327, 14351 and 14350 were physiologically less efficient. A positive correlations with r value of 0.28 between Fv/Fm and Photosynthetic rate, 0.67 between SCMR and Fv/Fm and 0.36 between Photosynthetic rate and temperature was observed indicating that these parameters account for physiological efficiency in groundnut.

5.6. Land configuration, Spacing and Intercropping

The land configuration, plant spacing and crop intercropping play an important role in increasing water use efficiency and finally pod yield as decreasing irrigation rates increases stomatal resistance, soil moisture tension and PAR penetration, and decreased transpiration rate, LAI, Pod yield and total DM (Dwivedi *et al.,* 1986d). Four type of canopy orientations S-N, E-W, radial and cross, at full, two-thirds or half the conventional irrigation rate (2.5 cm water applied at intervals of 10, 15 and 20 days, resp, giving totals of 30, 20 and 15 cm) were studied taking a Spanish bunch groundnuts cv. J 11 in Gujarat, The cross orientation favoured soil moisture storage compared with others (Dwivedi *et al.,* 1986d). Energy harvesting efficiency decreased with increasing water stress but the adverse effect was lower in cross than in the other treatments. The highest harvesting efficiency occurred in radial at the high irrigation rate. concluded that cris-cross was the most favourable orientation in terms of pod yields and energy harvesting by groundnuts (Dwivedi *et al.,* 1986d).

In an alfisol at Hyderabad the groundnut cv. TMV 2 grown in rows 35, 70 or 120 cm apart, the transpiration of ground cover (which varied 3-fold between spacings) and root: shoot ratio, was substantially greater in the wider row spacings; when the soil was wet, both the transpiration rate (T) and the canopy conductance (gc) were approximately proportional to the fraction (f) of incident radiation intercepted by foliage, but when the soil water content decreased below a threshold value, T/f and gc/f decreased because of an increase in stomatal resistance (Simmonds and Azam-Ali, 1989). Stomatal closure in response to soil water stress occurred sooner in the denser stands, because of more rapid depletion of soil water, also the sparser stands (which had a relatively large root: shoot ratio) had a greater capacity to keep stomata open as the soil water deficit increased (Simmonds and Azam-Ali, 1989).

In Japan, Runkulatile *et al.* (1998) characterized intercropping advantages in groundnut-finger millet (*Eleusine coracana*) intercrops in relation to crop combination ratios, soil moisture and nitrogen (N) availability taking three intercrops in 1:2, 1:1 and 2:1 alternating rows of groundnut and finger millet, their growth and yield in comparison with single crop and also the effect of adequately watered (W) and water stressed (D) conditions on the intercropping advantage for 1:1 intercrops revealed that the total aboveground biomass (DM) and its land equivalent ratio (LER) were highest in the 1:1 combination ratio. Intercropped groundnut exhibited significantly higher DM production after finger millet harvesting. The LERs were consistently higher under D than W conditions. Water stress severely reduced the leaf area index (LAI) of finger millet at the lower N rate, especially in the later stages, whereas higher N alleviated the water stress effect. A close linear relationship was observed between LAI and leaf area (LA) per unit leaf N both for groundnut and finger millet, with intercrops producing larger LA per unit leaf N than single crops. Intercropping maintained leaf net photosynthesis and transpiration of groundnut up to the later stages, and reduced water evaporation from soil surface compared with single cropping of finger millet (Runkulatile *et al.*, 1998).

An intercrop consisting of 1 row of sorghum between 2 rows of groundnut grown under two irrigation treatments a 'wet' (water stress was kept to a minimum by frequent irrigation) and a 'dry' treatment which received less water reveals that total crop performance ratio in wet intercrop was only 3% more total DM than the 2 crops separately, whereas in the dry treatment the advantage was 21%, the reproductive yield advantages were 14% and 88% in the wet and dry treatments, respectively, showing larger harvest indices in the intercrops (Harris *et al.*, 1987). Intercropped sorghum produced more TDM, but intercropped groundnut produced less, while leaf area indices were lesser than expected in all intercrop components. Sorghum intercepted more radiation in the intercrop than in the sole crop, but used it to produce DM less efficiently when water was plentiful. Groundnut intercepted less radiation than expected, but used it with greater efficiency in both wet and dry treatments. As well as intercepting more radiation, intercropped sorghum also used it more efficiently when water was limited, suggesting that sorghum was able to compete more successfully for soil water with groundnut in the intercrop than with itself in the sole crop (Harris *et al.*, 1987).

Further studies in a replacement series, intercrop of two rows of groundnut cv. Kadiri 3 alternating with one row of sorghum hybrid CSH-8, increased grain and filled-pod wt/plant due to intercropping were large, especially in drought stands sorghum, grain yields were 38 and 93% higher per unit row in the irrigated and drought treatments, respectively, while groundnut produced 81% more filled-pod wt per unit row than did sole stands during drought (Harris and Natarajan, 1987). Harvest index was larger for both species in the intercrops, by 8% and 33% in sorghum, and by 12% and 68% in groundnut in irrigated and drought treatments, respectively. In groundnut, HI was increased in the irrigated intercrop because individual pods were heavier, whereas the intercrop subjected to drought produced twice as many pods per plant in comparison with the sole crop. Large differences in plant temperature and water status between irrigated and drought stands throughout the post-rainy season, but mean differences between sole crops and intercrops within each water regime were small. Shading of groundnut by sorghum in the intercrop ameliorated to some extent the effects of high temperature and water stress, in the droughted stands. This was particularly important during peg production. It is suggested that less damage to flowers in the drought intercrop resulted in more pegs forming pods than in the sole crop, leading to the observed advantage in HI in groundnuts (Harris and Natarajan, 1987).

Using a line-source irrigation Natarajan and Willey, (1986) tested the range of moisture regimes (S1 to S5 in order of increasing moisture stress) on sole crops of sorghum cv. CSH 8R, *Pennisetum americanum* cv. BK 560 and groundnut cv. Robut 33-1, and intercrops of sorghum:groundnut at 1:2-3 rows, and *P. americanum*:groundnuts with 1:1-3 rows showed DM yield advantages of intercropping over sole cropping 8 to 30% for the *P. americanum*: groundnut systems and 0-19% for the sorghum/groundnut systems; moisture stress had no consistent effect on these DM advantages. For reproductive yields, all the intercropping systems showed some increase in relative advantages with increase in stress because of higher harvest indices in intercropping than in sole cropping. Largest advantages were 93% for sorghum + 2 rows of groundnut at S5 moisture regime and 78% for *P. americanum* + 2 rows groundnuts at S4 moisture regime, both of these being greater than at S1. The level of stress giving peak advantages depended on crop combinations and crop proportions (Natarajan and Willey, 1986).

Herbaceous legumes are becoming increasingly important for the crop-livestock farming systems in the moist and semi-arid regions of West Africa as these crops cover the ground quickly, check erosion, contribute to soil fertility and provide nutritious food and fodder to human beings and livestock. However, one of the major constraints in this region is the long dry season, which limits the productivity and duration of crop growth. A concerted effort was made to identify most suitable species and varieties with desirable agronomic traits including drought tolerance and high yield potential taking 72 accessions/varieties of relevant herbaceous legumes along with 3 cereals-millet, sorghum and maize for their relative drought tolerance in the wooden box method (130 cm long, 65 cm wide and 15 cm deep filled with soil of loamy composition). The number of days taken to first, 50% and 100% plant deaths as a measure of drought tolerance for different accessions/varieties showed that soybean variety TGX 1445-1D was the most susceptible as all plants were dead in 13 days while the lablab variety TLN 13 was the most drought tolerant which survived up to 46 days after stopping water. Based on the number of days taken to attain 100% plant death, the most drought tolerant group comprised of lablab, horse gram, centrosema and cowpea followed by chamaecrista and pearl millet as the second group; velvet bean, joint vetch, crotolaria, stylosanthes, sorghum and groundnut formed the third group and blue pea and soybean as the most drought susceptible group (Ewansiha and Singh, 2006).

On a well-drained Millhopper fine sand, maize, sorghum, groundnuts and sorghum-groundnuts intercropping subjected to four water treatments (optimum irrigation, irrigation allowing 2 d of wilting on sorghum or on groundnuts, and rainfed and visible crop water stress symptoms and daily soil water budgets using soil water depletion method) reveals that yields of maize, sorghum and groundnut increased linearly with seasonal irrigation and ET, with the respective slopes of the irrigation production functions 511, 204, and 160 kg ha-1 cm for DM and 341, 177, and 67 kg ha-1 cm for grain yields respectively and slopes of ET functions were 627, 486, and 383 for DM and 417, 397, and 198 for grain yields respectively (Omoko, 1990). The Irrigation-use efficiencies (IUE) were 82, 45, and 34% for maize, sorghum and groundnut, respectively. Yield levels in the irrigated treatments reached their near optimum potentials for all 3 crops and using LER concept, sorghum-groundnut intercrop yielded 15-36% more than pure stands, and the yield advantages increased with increasing irrigation, however, these advantages partially or completely disappeared when the analysis was done on the land water-use equivalency ratio (LWUER) concept. Similarly, intercropping did not provide any WUE superiority when its conjugate water production functions were compared with those of its equivalent sole crop and also intercropping did not improve yield stability (Omoko, 1990).

The WUE of groundnut sole (GG), sorghum sole (SS) and sorghum-groundnut intercrop (SG) were compared by Omoko and Hammond (2010) for two consecutive years with sorghum (*Sorghum bicolor* L. Moench) and groundnut on a loamy, Grossarenic Paleudult, using four water managements (T_1 Optimum irrigation, T_2 deficit irrigation allowing stress on sorghum, or T_3 on groundnut, T_4 rainfed, and crops seeded in rows at a density of (256000 (SS), 160000 (GG), 256000 + 160000 (SG, year 1), 157000+102000 (SG, year 2) plants/ha), sorghum grain yield (GY) range 3.55 (T_4) to 8.03 (T_1) Mg/ha in sole crop, and 2.71 to 6.27 Mg/ha in intercrop whereas the groundnut grain yield was 3.76 to 6.54 Mg/ha sole crop, but decreased in intercrop (0.13-3.26 Mg/ha). The mean Total Land Equivalent Ratio (TLER) was 1.14 for DM and 1.11 for grain yield (GY), showing a 14 and 11% advantages over sole cropping. But these advantages disappeared when the amount of water used was taken into account in the Total Land Water Use Equivalency Ratio (TLWUER). The overall mean TLWUER were 1.01 (irrigation) and 0.99 (seasonal ET) for DM, 0.98 and 0.96 for GY, indicating no advantage of intercropping over sole cropping. Nevertheless, based on water use ratios, intercropping was more water use efficient than sole crops. The contrasting results between the TLER and TLWUER may imply that the yield advantage of intercropping was not attributable to its overall improved water use ratio but rather to its higher seasonal water use (Omoko and Hammond, 2010).

5.7. Crop Modeling and its Simulation

A highly flexible and user-friendly method, with provisions to modify each input variable based on actual site and management requirements, was developed by Parekh and Shete (2008) using data on agro meteorological parameters (max and min temperature, dry and wet bulb temperature, sunshine h, pan evaporation, wind velocity, RH and rainfall) for 1991-2004 to estimate the daily water balance for summer groundnut and determine the optimum irrigation scheduling based on the daily crop water requirement (Reference crop ET calculated using meteorological data and daily crop water requirement calculated using the dual crop coefficient approach), where the crop coefficient varied with the crop growth stage and climatic condition in the area, and accordingly the irrigation scheduling can be planned using a computer programme based on MS Excel to avoid soil water stress.

The phenotype models utilized by plant breeders are partition traits reproductive yield (Y), and components of genetic (G), environmental (E), and GE interaction as yield commonly has large GE interaction. Breeders often have little information concerning the physiological basis of this GE interaction, without a clear idea of how to exploit the material further. Better knowledge of the physiological basis for the differential responses of genotypes to specific environments should improve the efficiency with which the breeder can characterize material for its G, and GE interaction, and hence increase the speed at which superior genotypes can be identified. Wright *et al.* (1996) described a simple physiological model to improve the understanding of the basis of GE interactions in groundnut under drought conditions. The physiological model, proposed by Passioura (1977) was used to define the yield (Y) as the product T x TE x HI, where T = amount of water transpired, TE = transpiration efficiency and HI = harvest index. Past and current studies have attempted to quantify these components in easily measurable ways. TE in groundnut was measured via carbon isotope discrimination and specific leaf area. T was estimated by substituting estimates of Y, TE, and HI in the identity above. The components analysed from an experiment consisting of 50 genotypes grown across multiple environments (seven locations) indicated selection of parents/genotypes with specific adaptive traits, highlight negative associations between yield determining traits. The sp. assumptions were further verified in an international collaborative project involving Indian and Australian scientists (Wright *et al.*, 1996).

Increased atmospheric CO_2 concentration will benefit the yield of most crops as two free air CO_2 enrichment (FACE) meta-analyses have shown increases in yield between 0 and 73% for C_3 crops. A novel perturbed-parameter method of crop model simulation based on peanut version of the general large-area model for annual crops (GLAM) was proposed by Challinor and Wheeler, (2008) where increases in yield simulated by GLAM for doubled CO_2 were between 16 and 62%. The difference in percentage increase between well-watered and water-stressed simulations was 6.8. These results were compared using CROPGRO and the groundnut model of Hammer *et al.*, (1995). The relationship between CO_2 and water stress in the models when examined, study shows that from a physiological perspective water-stressed crops are expected to show greater CO_2 stimulation than well-watered crops. However, this result is not seen consistently in either the FACE studies or in the crop models. In contrast, leaf-level models of assimilation do consistently show this result. The evidence from these models and from the data suggests that scale (canopy versus leaf), model calibration and model complexity are factors in determining the sign and magnitude of the interaction between CO_2 and water stress and the statement that 'water-stressed crops show greater CO_2 stimulation than well-watered crops' cannot be held to be universally true. Further the relationship between water stress and assimilation varies with scale (Challinor and Wheeler, 2008).

Deficit irrigation provides a means of reducing water consumption while minimizing adverse effects on yield. Models can play a major role in developing practical recommendations for optimizing crop production under conditions of scarce water supply. The applicability of FAO CROPWAT model on deficit irrigation scheduling for groundnut, was assessed at Bhavanisagar, Tamil Nadu during summer and Rabi where moisture stress imposed during flowering and pod formation stages were more sensitive than other stages and CROPWAT model can effectively simulate yield reduction as a result of moisture stress imposed by deficit irrigation at various growth stages (Thiyagarajan, and Ranghaswami, 2010).

In crop modelling, one of the most important problems is the model calibration for different cultivars (Gauch, 1988). In order to make a predictive model reliable it is necessary to find ways to calibrate it efficiently for different cultivars. Significant variability of photosynthesis rates between groundnut cultivars suggests that gas exchange parameters should be analyzed. Farquhar's model is often used to simulate photosynthesis on both levels of organization, single leaf and canopy. Ferreyra *et al.* (2000) showed that the simulated differences in both transpiration and photosynthesis qualitatively agree with observations of biomass vs. LAI and water consumption vs. LAI of experimental plots. With the inclusion of additional relevant phenomena such as heliotropism and canopy light distribution, the 2-dimensional model of leaf gas exchange 2D LEAF may be used to explain differences in water stress tolerance between different cultivars, and serve as a fine-tuning tool for other applications such as crop-scale peanut simulation models. A Theoretical model of water stress tolerance in groundnut proposed by Kameswara Rao *et al.* (2009) is given in Figure 1.

Representative model depicting the pathways implicated in water-deficit tolerance in groundnut leaves. Proteins identified in this study are displayed on the corresponding metabolic pathways. Solid arrows indicate induction of a particular protein, and dashed lines represent reduction. Dark, thick arrows show the signal molecules generated downstream of a pathway. ABA, NDK1 (nucleoside diphosphate kinase 1), TPR (tetratricopeptide repeat), DREB (water-deficit stress response element-binding protein).

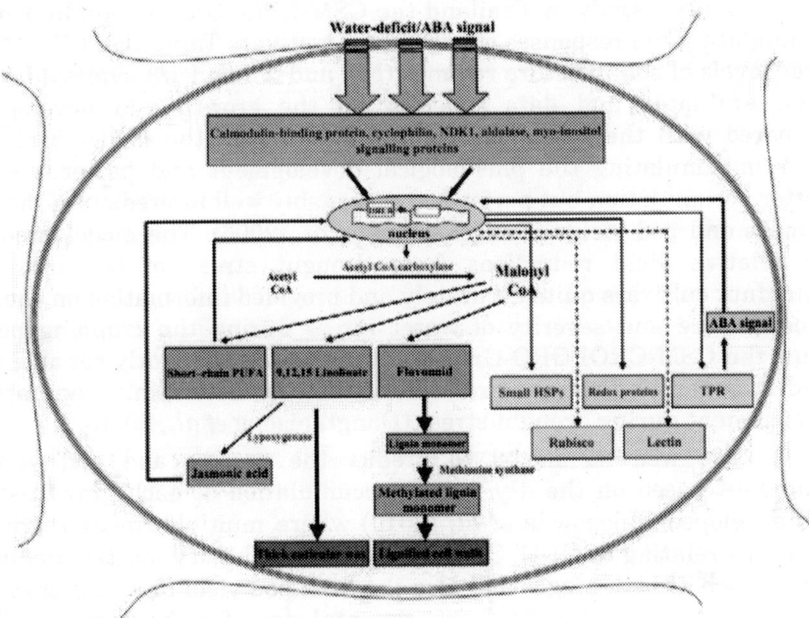

Fig. 1. Theoretical model of water stress tolerance in groundnut (Kameswara Rao *et al.*, 2009).

The PNUTGRO model calibrated and validated for a subhumid climate and sandy loam soil of Kharagpur, West Bengal by performing multiple year simulation to recommend a set of alternate decisions with different levels of assurance for effective management of water in groundnut crop, using actual weather data for 10 years and the study reveals that under limited water availability, a soil water depletion of more than 60% of plant extractable soil water should be allowed during the non-critical stages, but, 30% of ASW depletion can be allowed during critical stages of crop growth, at a moderate risk level of 50%, without affecting the crop yield where full irrigation is recommended for risk levels up to 30% (Panda *et al.*, 2003).

In NE Thailand the MACROS crop model evaluated for its utility to generate information on land suitability for dry season groundnut cropping based on water availability at the regional scale indicated that the model was specific for the condition where crop growth is limited by water stress, and evaluated using both calibration and validation phases in sequence (Katawatin *et al.*, 1996). The dynamics of observed and corresponding simulated values of shoot dry weight agreed in every condition involved in this validation study also the simulated pod yields agreed with the field data. For the validation B, where model was further validated using data from 35 farm trials conducted at 5 different test sites, a high positive correlation (r=0.91) existed between observed and simulated pod yields (Katawatin *et al.*, 1996).

In another study in Thailand the CSM-CROPGRO-Groundnut model in simulating the responses of groundnut cultivars Tainan and KK 60-3 to three levels of soil moisture regimes (FC, and 2/3 and 1/3 available water) when evaluated and data collected on the growth and development compared with the corresponding simulated data, the model performed fairly in simulating the phenological development and patterns of dry matter accumulation but performed reasonably well in predicting the final biomass and pod yields (Dangthaisong *et al.*, 2006). The model predicted the relative yield reductions from drought stress of the individual groundnut cultivars quite accurately and provided information on the time of occurrence and severity of water stress during the cropping period. Thus, the CSM-CROPGRO-Groundnut model is sufficiently capable to be used in generating the required information for determining appropriate management during drought stress (Dangthaisong *et al.*, 2006).

A crop weather model to predict the growth and pod yield of groundnut based on the dry matter accumulation at each growth stages was developed Rajegowda *et al.* (2010) where multiple linear regression equations relating to GDD, SSH and AET with the dry matter production during each growth stages and also the final pod yield of kharif crop were generated by using the field experimental data for the period of 2000-2008. The coefficient of determinants indicate that the climatic parameters and the initial TDM used to estimate the final TDM in each stage could predict an extent of 77 to 98 per cent (coefficients of determinants) in different growth stages. Comparison of the observed and the predicted yields indicates the close agreement between them in all the stages. Considering the observed TDM up to the first four stages and predicted the Total Dry Matter at the harvesting stage when model was validated for the year 2009, and there was a good agreement between the observed and the predicted crop yield. The favourable influence of AET at the beginning of peg initiation and peg formation stage, and higher GDD during pod formation and harvest stages were noticed. The increase in AET during pod filling stage did not favour to the pod yield (Rajegowda *et al*, 2010).

Improved management (high yielding cultivars, balance crop nutrition and control of pest and diseases) in high rainfall regimes and rainfall conservation and supplemental irrigations in low rainfall regimes are essential components of the improved technologies aimed at bridging the yield gaps of groundnut (Bhatia *et al.*, 2009). To assess the scope for enhancing productivity of groundnut in India, well-calibrated and validated CROPGRO-Peanut model used to assess potential yields (water non-limiting and water limiting) and yield gaps of groundnut for 18 locations representing major groundnut growing regions of India where the average simulated water non-limiting pod yield of groundnut was 5440 kg ha^{-1}, and the water limiting yield was 2750 kg ha^{-1} indicating a 49% reduction in yield because of deficit soil moisture conditions (Bhatia *et al.*, 2009). As against this, the actual pod yields of the locations averaged 1020 kg ha^{-1}, which was 4420 and 1730 kg ha^{-1} less than the simulated water non-limiting and water limiting yields, respectively. Across locations, the simulated water non-limiting yields were less variable than water limited and actual yields, and strongly correlated with solar radiation during the crop season (R^2=0.62, P<=0.01). Simulated water limiting yield showed a positive, but curvilinear relationship (R^2=0.73, P<=0.01) with mean crop season rainfall across locations. The relationship between actual yield and the mean crop season rainfall across locations was not significant, whereas across seasons for some of the locations, the association was found to be significant. Total yield gap (water non-limiting minus actual yields) ranged 3100-5570 kg ha^{-1}, and remained more or less unaffected by the quantity of rainfall received across locations. The gap between simulated water non-limiting and water limiting yields, which ranged from 710-5430 kg ha^{-1}, was large at locations with low crop season rainfall, and narrowed down at locations with increasing quantum of crop season rainfall. On the other hand, the gap between simulated water limiting yield and actual farmers yield ranged from 0 to 3150 kg ha^{-1}. It was narrow at locations with low crop season rainfall and increased considerably at locations with increasing amounts of rainfall indicating that type of interventions to abridge the yield gap will vary with the rainfall regimes (Bhatia *et al.*, 2009).

5.8. Pre and Post Season Sowing

The effects of sowing time and rainfall distribution on the yield of groundnut cv. GG 11 were studied in Junagadh, Gujarat, India, during 1975-2003 by sowing the crop after sufficient amount of rainfall received during or after the 25th standard meteorological week (SMW) by Sahu *et al.* (2004) dividing the sowing time into two periods, i.e., timely sowing (25th and 26th SMW) and late sowing (after 27th SMW) and classifying each year as low rainfall year (<590 mm), moderate rainfall year (590-740 mm) or high rainfall year (>940 mm) where the yield varied from 100 kg ha^{-1} during dry year of 1987 to 1784 kg ha^{-1} during moderate rain fall of 1975. The correlation between rainfall during vegetative period and pod formation period was negative, indicating that moderate rainfall had synergistic effects on the yield under timely sowing. Under late sowing, rainfall during the vegetative and flowering stages had positive effects on the yield. It was concluded that the timely onset of monsoon and commencement of sowing in the 25th SMW resulted in moderate and high rainfall and good and moderate yields, whereas the late commencement of sowing resulted in low rainfall and low yields and water stress during the late crop growth stages (Sahu *et al.*, 2004).

The effects of moisture stress on yield components of 4 groundnut cultivars when studied at 2 irrigation levels (rainfed and irrigated) and 3 sowing dates the cv. Kadiri 3 followed by M 13, performed better under rainfed conditions than JL 24 and Gangapuri if sowing was delayed until July. If delayed further, to August, JL 24 and K 3 gave the best performances. Irrigation improved yields in both the seasons, the values for K 3, M 13, JL 24 and Gangapuri being 4.74, 3.53, 3.01 and 2.85 t ha^{-1}, respectively, during kharif and 5.17, 7.62, 3.85 and 5.8 t ha^{-1}, respectively, during rabi (Padma *et al.*, 1991). In summer withholding irrigation for up to 10, 15, 20, 25 or 30 DAE gave av. yields of 2.58, 3.02, 3.10, 3.15 and 2.41 t ha^{-1}, respectively in two groundnut cultivars. Yields increased linearly with a delay in applying irrigation up to 25 and 20 DAE, resp in SB XI and UF 70103 with av. yields of 2.43 and 3.28 t ha^{-1}, with the highest yields of 2.68 and 3.65 t, respectively (Jadhao *et al.*,1989).

In peninsular India at Hyderabad, four groundnut genotypes studied across a drought-stress gradient to determine the influence of insect distribution in the post-rainy season reveals that gelechiid leaf-miner *Aproaerema modicella* was most abundant on the most stressed plants, *Cicadellid empoasca* kerri had the reverse distribution, the thrips *Frankliniella schultzei* and *Scirtothrips dorsalis* were at first densest where drought stress was least and their distribution subsequently became reversed and, as the condition of their hosts worsened, they again became most abundant at the wetter end of the gradient (Wheatley *et al.*, 1989). Bud necrosis disease caused most mortality where drought stress was highest (Wheatley *et al.*, 1989).

5.9. Fertilizer, Organic Manures, Mulching and Others

The effects of moisture stress and gypsum on pod development and yield of groundnut when examined, the gypsum increased the pod and kernel yields of Samaru 38 variety as moisture stress at 9-13 weeks after sowing coincide with the period of peg and pod development, lowered the uptake of nitrogen and increased the proportion of unfilled pods, drastically reducing yield (Balasubramanian and Yayock, 1981). As gypsum increases early pod development, it provides an escape mechanism from drought (Singh and Chaudhari, 1995; Williams *et al.*, 1986). Gypsum applied at flowering increased yield of genotype subjected to drought but there was no response if there was no drought since soil contained adequate amount of available calcium of about 600 ppm (Rajendrudu and Williams, 1987).

The cyclic dry spells in calcareous Vertic Inceptisol, caused up to 75% reduction in pod yield and potassium at 60 kg K_2O ha^{-1} enhanced the level of production and could also restore the loss in pod yield to a noticeable extent. A marked increase in the diffusive resistance of leaves with K fertilization supports the contention that potassium plays an important physiological role in counteracting adverse conditions caused by drought (Golakiya and Patel, 1988). In a field experiment in Gujarat, groundnut cv. GAUG 1, the solar energy harvesting efficiency was highest with 20 kg K and decreased with increasing water stress (Umar and Umar, 1997). Excised leaves of groundnut cv. J 11 exposed to normal and moisture stress conditions showed proline accumulation and KCl enhanced proline accumulation. The KCl pretreatment enhanced conversion of arginine to proline under non-stress conditions. In groundnut proline accumulation was high in the leaves and levels of all the precursors were also high (Rao, 1979).

In acid lateritic soils at Jhargram, India, the response of groundnut cv. TAG 24 to moisture stress and application of organic manure (FYM) and fertilizer with and without gypsum studied during summer reveals that moisture stress at vegetative stage (10-30 DAS) gave 34% higher pod yield than that at flowering stage (30-50 DAS), but this moisture stress at vegetative stage was on a par with no moisture stress for pod yield, yield attributes, oil content and nutrient uptake (Dutta and Mondal, 2006). The highest use of water was recorded with no moisture stress, as a result maximum WUE was obtained under moisture stress during vegetative stage. Farmyard manure at 7.5 tonnes ha^{-1} resulted in better yield attributes, yield, oil content, nutrient uptake and WUE than the control. Recommended dose of fertilizer (RDF), i.e., 30 kg N ha^{-1}, 60 kg P_2O_5 and 40 kg K_2O when applied with gypsum 500 kg ha^{-1} increased the pod yield by 17 and 11.5% over 100 and 125% RDF alone respectively. However nutrient uptake and oil content were also influenced with increase in fertilizer level in combination with gypsum. Benefit:cost ratio was higher with moisture stress at vegetative stage and application of 7.5 tonnes FYM/ha or 100% RDF+500 kg gypsum/ha respectively (Dutta and Mondal, 2006).

On the sandy soils at Ismailia Research Station, two long-term field trials were conducted to investigate the effect of sulfate of potash (SOP) and muriate of potash (MOP) at 70 and 140 kg K_2O ha^{-1}. Sprinkler irrigation was used in one trial with rotations of wheat-groundnut, berseem (*Trifolium alexandrinum*)-sesame and berseem-groundnut; and drip irrigation in the other with sesame, faba bean-sesame, onion-groundnut, and faba bean-fodder maize rotations. In sandy soils, the amounts of total soluble salts and chloride in the surface layer of the soil slightly increased with sprinkler irrigation irrespective of whether K was added as SOP or MOP. However, with drip irrigation addition of MOP at 140 kg K_2O ha^{-1} increased the total soluble salts under the dripper area by 18 times and chloride by 35 times compared to the concentration at the beginning of the experiment. In sandy soils using sprinkler irrigation, increased yields 140 kg K_2O ha^{-1} by SOP, were approximately 14% for wheat, 40-43% for groundnut, 17-18% for berseem, and 23% for sesame (Hadi *et al.*, 2003).

In Rahuri, Maharashtra, during the summer the effects of mulching and water stress at different growth stages was studied on the performance of groundnut by mulching with sugarcane trash or white polyethylene along with water stress at the flowering, pegging or pod development stage. Mulching with white polyethylene along with water stress at the pod development stage resulted in the greatest plant height (37.77 cm), plant spread (25.23 cm), number of functional leaves per plant (72.59), leaf area per plant (21.17 dm^{-2}), dry matter per plant (17.77 g), dry pod yield (2896 kg ha^{-1}), haulm yield (4741 kg ha^{-1}), dry kernel yield (2083 kg ha^{-1}), oil yield (989 kg ha^{-1}), net monetary returns (34 097 rupees ha^{-1}) and benefit:cost ratio (2.44). The highest oil contents were obtained with polyethylene mulch along with water stress at the pod development stage (47.79%) and mulching with sugarcane trash along with water stress at the pod development stage (47.31%) (Bodare and Dhonde, 2011).

5.10. Hormones and Growth Regulators

Plant hormones are involved in multiple processes. Phytohormones are essential for the ability of plants to adapt to abiotic stresses by mediating a wide range of adaptive responses as it play central roles in the ability of plants to adapt to changing environments, by mediating growth, development, nutrient allocation, and source/sink transitions (Santner and Estelle, 2009). Plant growth substances have key role in different physiological processes related to growth and development of crops. The changes in the level of endogenous hormones due to biotic and abiotic stress alter the crop growth and any sort of manipulation including exogenous application of growth substances would help for yield improvement or at least sustenance of the crop. Hormones usually move within plant from a site of production to site of action. Phytohormones are physiological intercellular messengers that are needed to control the complete plant lifecycle, including germination, rooting, growth, flowering, fruit ripening, foliage and death. In addition, plant hormones are secreted in response to environmental factors such as abundance of nutrients, drought conditions, light, temperature, chemical or physical stress. Hence, levels of hormones will change over the lifespan of a plant and are dependent upon season and environment. Cross-talk between the different plant hormones results in synergetic or antagonic interactions that play crucial roles in response of plants to abiotic stress (Peleng and Blumwald, 2011). Although ABA is the most studied stress-responsive hormone, the role of cytokinins, brassinosteroids, and auxins during environmental stress is now being emphasised.

The key role of ABA, jasmonic acid (JA) and salicylic acid (SA) as primary signals in the regulation of plant defense has been well established (Bari and Jones 2009). These hormones generate a signal transduction network that leads to a cascade of events responsible for the physiological adaptation of the plant to stress. The degree of drought tolerance varies with developmental stages in most plant species (Reddy *et al.*, 2004; Rassaa *et al.*, 2008). ABA is well known hormone for its regulatory role in integrating environmental adversity with the developmental programs of plants. Thus, it affects a wide range of processes at different developmental stages such as embryo and seed development, acquisition of desiccation tolerance and dormancy, flowering and organogenesis (De Smet *et al.*, 2006; Liang *et al.*, 2007). ABA also promotes plant growth under non stressful condition and has shown to be essential for vegetative growth in several organs (Sharp *et al.*, 2000; Chen *et al.*, 2002).

Similarly, SA is an endogenous regulator of growth involved in a broad range of physiologic and metabolic responses in plants (Hayat *et al.*, 2010). During the last few years, SA has been intensively studied as a signal molecule mediating local and systemic defense responses against pathogens. Currently, it has been reported that this compound plays also a role in plants responses to abiotic stresses, such as drought, low and high temperatures, heavy metals, and osmotic stress (Nemeth *et al.*, 2002; Munne-Bosch and Peñuelas, 2003; Shi and Zhu, 2008; Rivas-San Vicente and Plasencia, 2011). SA was also shown to influence a number of physiological processes, including seed germination, seedling growth, fruit ripening, flowering, ion uptake and transport, photosynthesis rate, stomata conductance, biogenesis of chloroplast (Fariduddin *et al.*, 2003; Khodary, 2004; Hayat *et al.*, 2005; Shakirova, 2007).

JA, and its cyclic precursors and derivatives constitute a family of bioactive oxylipins that regulate plant development and responses to environmental cues (Turner *et al.*, 2002; Devoto and Turner, 2003). This family of compounds is formed by 12-oxophytodienoic acid (OPDA), methyl jasmonate (Me-JA) are collectively receive the name of jasmonates (JAs). These molecules are involved in a variety of processes related to plant development and survival, including direct and indirect defense responses (e.g., defense against insects and necrotrophic pathogens, abiotic stresses viz. drought, salinity etc.), secondary metabolism, reproductive processes (e.g., pollen maturation and anther dehiscence, ovule development), and fruit development, among others (Seo *et al.*, 2001; Wasternack and Hause, 2002; Liechti and Farmer, 2006; Wasternack, 2007). The participation of JA in response to abiotic stress, such as drought and salinity, has been reported in several species. For instance, the treatment of barley leaves with sorbitol or mannitol (compatibles solutes to simulate water stress) increased JAs endogenous contents, followed by synthesis of jasmonate-induced proteins (Lehmann *et al.*, 1995).

The yield and quality of GAUG 1 groundnut under water stress during summer by withdrawing two irrigations in the first 30 DAE and various growth regulator treatments (50 ppm GA, 40 ppm NAA, 50 ppm IBA, 250 ppm daminozide, 500 ppm chlormequat chloride) applied alone or in combination and the effects of 12 irrigations + 50 ppm IBA on pod yields of cv. J11, GG 2 and Robot study reveals maximum pod yield (2.17 t/ha) by 250 ppm daminozide + 50 ppm GA treatment followed by 12 irrigations + 50 ppm IBA (2.15 t) and water stress + 50 ppm IBA (2.06 t), spraying with 50 ppm IBA at 40 and 60 DAS increased pod yield by 11-29% and gave the highest net return, but least cost:benefit ratio. Withdrawal of two irrigations in early growth followed by spraying of 50 ppm IBA at 40 and 60 DAS was economical (Patel *et al*., 1988).

Groundnut seeds soaked in 100 ppm GA_3 for 1 h and PEG for 24 h and subjected to drying for 0, 3, 5, 7 or 9 d when sown in soil containing 70, 40, 30, 21 and 14% moisture (w/w) to give 0, -0.1, -0.3, -0.6 and -15.0 bar, respectively, drying seeds for 9 d and growing in soil at 0 bar soil moisture decreased percentage seed germination to 80% compared with 91% with no drying (Golakiya, 1992). At -0.3 bar soil moisture, seed germination was 86% with no drying and 47% with 9 d drying. Pelleting seeds with polyvenyl alcohol, polyacryliamide and polyacrelic acid improved seed germination at low soil water potential the most effective germination was 78% with 7 d drying and -6.0 bar soil moisture when pelleted with polyacreliamide (Golakiya, 1992). Four groundnut varieties during summer and ten varieties during the rainy season grown, using seed pretreated with calcium chloride (1%) or ascorbic acid (50 ppm) the pod yield increased by calcium chloride in four varieties, more particularly in RS 218 which showed high proline accumulation under stress and then effect of seed hardening results into the ability of plants from treated seeds to produce proline when under stress (Sashidhar *et al*., 1977).

In Vertisol at Dharwad, groundnut cv. Dh 3-30 sprayed with different antitranspirants at 45DAS and withholding irrigation for 17 d during summer, among the antitranspirants, proline content was highest with 1500 ppm B-9 (daminozide), 20 ppm PMA (phenyl mercury acetate), 6% silica powder and 5% China clay, the reducing sugar was lowest in irrigated plants and highest with 0.2% Sunguard and 20 ppm alachlor, whereas non-reducing sugars were highest in irrigated crops, the leaf K content was lowest with irrigation and highest with Sunguard, alachlor and 100 ppm Cycocel (chlormequat), and among the antitranspirant, yield was highest with alachlor, Sunguard and Rallidhan (long chain fatty alcohol derivative), while spraying with PMA decreased yield (Amaregouda *et al*., 1994a). The pod yield increased by all antitranspirants except PMA and B-9, with Sunguard and alachlor giving the highest yields (1.71 and 1.70 t/ha, respectively) (Amaregouda *et al*., 1994). Also the antitranspirants and plant water relations when studied, stomatal resistance was lowest and greatest in plants treated with B-9 (1500 ppm) and PMA (20 ppm), respectively and Alachlor (20 ppm), Sunguard (0.2%), China clay (6% w/v) and silica powder (6% w/v) maintained moderate stomatal resistance compared with the control (Amaregouda *et al*., 1994b). In another field study on a Vertisol at Akola, groundnut cv. SB XI irrigated at IW: CPE ratios of 0.50, 0.65 or 0.80 up to flowering, flowering to pod development or pod development to maturity stages, stress up to flowering and pod formation stages decreased pod yield which were increased by foliar application of 400 ppm aspirin and/or 8% kaolin at 62 DAS as antitranspirants (Patil and Morey, 1993).

In summer two groundnut cultivars, irrigated at 12 or 8 days intervals (water stressed and non-stressed conditions, respectively), no mulch and mulched with 5 t wheat straw/ha, with and without 5 sprays of kaolin treatment reveal that, pod yields under non-stressed and stressed conditions were 3.23 and 2.76 t ha^{-1}, respectively. Yields without mulch, with mulch and mulch + kaolin sprays were 2.66, 2.99 and 3.34 t ha^{-1}, respectively. The WUE was highest with mulch + kaolin spray (Joshi *et al*, 1987). At Bambey the water balance in fallow soils and soils cultivated with groundnuts when examined groundnut gave pod yields at 2.26 and 3.43 t ha^{-1} in first year and 1.22 and 3.70 t ha^{-1} in second year, without and with irrigation, respectively (Dancette *et al.*, 1979). In plot groundnut cv. Nonghua No. 5 protected from rain during the early flower bud stage decreased root and shoot dry matter, main stem length, total root length and leaf area (Yao *et al.*, 1999).

Water logging is one of the most serious ecological restricting factors for groundnut in southern China. By exploring the correlation between main characteristics and pod yield in groundnut under natural water-logging stress, this study aimed to provide a theoretical basis for higher tolerance breeding in waterlogging conditions. Twenty characteristics including yield of 21 Spanish type germplasm lines were recorded. The bivariate correlation showed that the high-yielding groundnut lines had lower plant height, more sub-branches, less physiological defoliation, higher root biomass and lower shoot biomass (higher root/shoot ratio, i.e. R/S ratio), more fully developed pods $plant^{-1}$ (FDPP) and larger pods with bigger kernel, but lower fully developed pod ratio (FDPR) and fully developed kernel ratio, higher harvesting indexes (HI), and the correlation coefficients were in the rank of total pod number > HI > less developed pod plant-1 (LDPP) ~ FDPP > 100-seed mass (Ms) > 100-pod mass > R/S ratio. With multiple-factor stepwise regression analysis, however, only 5 characteristics were significantly related to pod yield, and they were in the rank of FDPP > Ms > LDPP >> FDPR > R/S ratio in standardized regression coefficients; meanwhile the partial correlation coefficients were in the rank of Ms > FDPP > LDPP > FDPR > R/S ratio. Path analysis indicated that FDPP and Ms directly affected pod yield, R/S ratio affected less on yield but had considerable indirect effect on yield, and FDPR had less direct effect on yield but strong indirect negative effect on yield. Thus, Ms, FDPP and LDPP can be used as criteria for screening high-yielding waterlogging-tolerant groundnut; apart from these parameters, R/S ratio may also be used as a reference (Li *et al.*, 2008).

During rabi seasons, presoaking of TAG 24 groundnut seeds in solutions of CFL and CCC (both 10^{-6} M) for 6 h resulted in higher yield under drought conditions and drought indices like RWC, proline accumulation and transpiration supported the antitranspirant action of these chemicals (Mathew and Pandey, 2006). During kharif seasons, at Madurai, India, water stress in groundnut cv. JL 24, imposed during flowering and pegging stages produced the greatest reduction in pod yield followed by water stresses at early pod stage, late pod stage and vegetative stages where Kaolinite (3%) spray reduced yield loss due to water stress. Seed quality, oil and protein contents and N, P and K uptake were correlated with pod yields (Naveen *et al*., 1992).

6. CONCLUSIONS AND FUTURE RESEARCH STRATEGIES

Drought is the major abiotic constraint affecting productivity and quality of groundnut worldwide. There are three major aspects of drought, duration, intensity and timing relative to crop phenophases which vary independently. Water deficit stress delay pod initiation, and the major cause of variability in pod yield and HI is the delay between peg initiation and onset of rapid pod growth. The period of reproductive growth stages occurs over a period of nearly two months and moisture stress has a depressing effect on flowering, stem growth and nodulation. No flowering occurs during the stress, but once the stress is relieved, there is a flush of flowering depending on the growth stages and sometimes it results in more flowers. The Virginia type groundnuts, due to their longer duration, are more tolerant to drought than Spanish and Valencia, however, the Spanish and Valencia due to short duration escape the late season drought. The flush of late flowers, following mid season drought, delay maturity and hence late harvesting, where late season rain helps. The fruiting occurs once the gynophores enter into the soil and soil physical condition is important and must be wet during the gynophore entering the soil as the gynophore can exert a force equivalent to 3-4 g only.

The water flow in intact plant under high soil moisture condition is for growth and transpiration and two concepts are expressed about the driving force for transpiration water flow, the water potential differences between the root and leaf as the primary force and hydrostatic and osmotic pressure differences, as the factors determining water flow. The management practices should aim to optimize the availability of resources at the time of pegging to ensure timely pod initiation. In order to sustain plant growth and hydration, water must be continuously supplied to the leaves as it is lost by transpiration which becomes difficult under low soil moisture condition.

The groundnut is relatively drought tolerant and an important crop of the semi-arid regions, however the plant water-status the balance between water uptake and loss has been less understood in groundnut. Though different growth stages have different sensitivity to water deficit, none of these can proceed normally below some minimum water. The water requirement of groundnut is lowest from germination to flower formation and reaches maximum during pod formation. However, the utilization of available moisture is greatest during flowering and pod formation and the crop receiving adequate water during these stages only can give equal yield to the well watered crop. During these stages if stress is given and later on water supply is resumed only the vegetative growth is benefited not the reproductive growth of crop. The period of maximum sensitivity to drought occurs baetween 50-80 days after sowing, the period of maximum flowering and vegetative growth.

The timing of drought has a large impact on the variation. The sensitivity of a genotype to drought increases with yield potential, increasing the closer the drought ends to final harvest. Genotypic variation to drought exists in the water-use ratio with some, being able to accumulate up to 30% more shoot DM with the same total transpiration and HI, and large variations in genotypes to midseason are due to recovery differences after the drought is relieved.

The yield is a function of many plant and environmental factors and moisture stress play an important role particularly the stage at which stress occurs. The water stress affects the vegetative, root and reproductive growth and a proper scheduling of irrigation is required. Moisture stress at flowering reduced phytobiomass and pod yield by limiting the number of mature pods per unit area as compared to stress at pegging and pod formation stages. The variation in HI account for the large proportion of variation in yield, and hence recommended to make selection for high HI. As reproductive development is sensitive to drought resulting to poor yield, the strategies to combat drought in groundnut genotypes are (i) early production of flowers pegs, and pods, with subsequent filling of the pods at a moderate, but essentially at constant rate despite the drought, (ii) faster development of later developed pegs into pods once water become available after drought late in the season.

Drought stress effects on groundnut depends primarily on the stress pattern because genotypic variation is usually of secondary significance. In a 110-120 days crop water deficit stress at 45-70 DAS (flowering) and pod development (60-90 DAS) phases was highly detrimental to leaf area development, dry matter production, pod formation causing 40-60 and 50-70% yield reductions, respectively. However, in a 140-150 days crop maximum reduction in kernel yield was when stress was imposed during seed filling phase, i.e., 93 DAS onwards. The early and continuous availability of water until the start of pod filling result in large canopy and which increases transpirational demand. The transient soil-moisture-deficit stress for 20-25 days as pre-flowering drought during vegetative phase (20-45 DAS) results in synchronized flowering, increases 10-20% pod yield and save 10-15% water mainly due to promotion of root growth during water stress and inhibition of number of vegetative sites (leaves and branches).

Pod yield is a function of transpired water (T), transpiration efficiency (TE) and harvest index (H) and the TE derived from measurements of carbon isotope discrimination in leaves indicated only small variation. The yield losses (%) due to drought are estimated as: Yield loss (%) = 100 (1 - D_y/W_y), where, W_y is the pod yield under adequate irrigation and D_y is the pod yield under drought. The reported yield reductions ranged 10-15, 15-30, 40-50 and 50-70%, when drought was imposed from 10-30, 30-50, 50-

80 and 80-120 DAS, respectively. The greatest yield reduction corresponds to peak flowering to early pod filling stage and adequate moisture during this period is critical for obtaining maximum yield. Under water stress there is poor pod filling that reduced kernel size, shelling, SMK% and lipid content of kernel.

Plant population, planting pattern, land configurations and minerals influence both the temporal and spatial patterns of water use, with high density crops extracting water from lower depths sooner than low density crop. High water use prior to early pod filling in high density crop was associated with more rapid leaf area development. The more rapid water extraction in a high, compared with a low, population density groundnut crop is associated with greater root production at depth. Minerals Ca and K play important role in the moisture stress tolerance. Gypsum increase early pod development and provides an escape mechanism from drought and hence it must be applied at flowering to increase yield of groundnut subjected to drought. Potassium also improve input use efficiency through interaction and quality of produce.

The groundnut crop of about 120 days duration, requires 400-450 mm of water in normal soil and 600-700 mm in sandy soil. The levels of irrigation have a major effect on the amount of water consumed, and it is 300-350 mm in non irrigated field and 500-600 mm in field irrigated at 40-60% moisture availability at 30 cm depth. The irrigation at a depth of 100 mm showed highest yield. The yield decreases due to water stress at different growth stages are in order of water stress at seed development > at early pod filling > at early growth > at early pegging. The pod yield and quality of groundnut are reduced when less than 30 cm water was received by the crop. Water deficit during seed production affected C_2H_2 and CO_2 production during subsequent germination. Water stress at pod initiation and development phase reduced germinability, vigour, seed membrane integrity and affects subsequent growth of seedlings and could pose a problem in establishment for the succeeding crop.

Seasonal water requirement for groundnut crop were 300-350 mm during *kharif* and 380-450 and mm during summer seasons for semi arid region. The best irrigation scheduling criteria would be to irrigate as per the daily use rate of the crop. A minimum of 350 and 500 mm of water in 120 and 140 days duration crops, respectively is necessary to produce seeds with high potential for germination and high proportion of vigorous seedlings. Many a time the superiority of genotype with high yield, in water stress is not reflected in their ancillary characters. As the technology is likely to increase pod yield more than two folds in addition to water saving farmers may adopt drip irrigation method for groundnut crop at large field scale.

Genetic improvement of crop resistance to drought stress is one component and will provide a good perspective on the efficacy of control strategy through genetic improvement. Selection for drought adaptation under rainfed conditions, though commonly practiced, could be misleading, since it may not reflect the ability of the genotype if the stress occurs during the critical stages of plant development. More efficient selection would require simulated drought conditions, and the use of other indirect selection methods that give a good indication of drought adaptation. When water deficit during seed filling phase, genotypic yield potential accounted for approximately 90% of the variation in pod yield sensitivity to water deficit, and it is unlikely that breeders will be able to combine high yield potential with low sensitivity to drought spanning the seed filling phase, therefore other important strategies are necessary. The pod yield potential accounted for less of the variation in drought sensitivity (15-64%) in the early and mid-season droughts. For these circumstances it may be possible to identify genotypes with both high yield potential and relatively low drought sensitivity.

Agrometeorological studies must include an awareness of the relationship between environment, crop phenology, maturity, and postharvest quality. The seed composition changes dramatically as the crop matures and also has relation with environment, postharvest quality of groundnut is the resultant of the particular set of environmental and cultural practices during pod growth and maturation. A biochemical basis exists for inferior quality in immature groundnut. Drought stress and soil temperature influence maturation rate and thus had an indirect effect on postharvest quality. *Aspergillus flavus* invasion and aflatoxin contamination in groundnuts are related to drought stress, soil temperature and maturity and small, immature seeds are more likely to be contaminated with *A. flavus* than larger, mature seeds.

Most breeding programmes in groundnut follow an empirical approach to drought resistance breeding, largely based on kernel yield and traits of local adaptation, resulting in slow progress. Recent advances in the use of easily measurable surrogates for complex physiological traits associated with drought tolerance encouraged breeders to integrate these in their selection schemes. However, there is no direct comparison of the relative efficiency of a physiological trait-based selection approach *vis-a-vis* an empirical approach to ascertain the benefits of the former. The drought tolerance contributing factors in groundnuts are, an extensive root system established before maximum leaf area to meet the transpirational demand, recurred and synchronized flowering once stress was relieved, water storage cells in the abaxial side of the leaves to provide water when transpiration was greater than the roots extraction of soil moisture; leaf folding during stress to reduce solar incidence; and transpiration regulated by high stomatal resistance during stress.

REFERENCES

Ahmad, A., Kirda, C., Moutonnet, P., Hera, C. and Nielsen, D.R. (1999). Yield response of groundnut grown under rainfed and irrigated conditions. *In:* Crop Yield Response to Deficit Irrigation, 128-135.

Ali, E.A. (2001). Effect of water stress and potassium fertilization on yield and yield components of peanut (*Arachis hypogaea* L.). *Annals of Agricultural Science*. 39: 1425-1434.

Amaregouda A., Chetti, M.B., Salimath, P.M. and Gowda, M.V.C. (1994a). Influence of antitranspirants on osmoregulants in summer groundnut (*Arachis hypogaea* L.). *Annals of Plant Physiology*. 8: 114-120.

Amaregouda, A., Chetti, M.B., Salimath, P.M. and Kulkarni, S.S. (1994b). Effect of antitranspirants on stomatal resistance and frequency, relative water content and pod yield in summer groundnut (*Arachis hypogaea* L.). *Annals of Plant Physiology*. 8: 18-23.

Annerose, D.J. (1988). Physiological criteria for improving groundnut adaptation to drought. *Oleagineux Paris*. 43: 217-222.

Arjunan, A., Senthil, N. and Dharmalingam, V. (1997). Selection of groundnut genotypes for irrigated and rainfed environments. *Indian Journal of Plant Physiology*. 2: 118-122.

Asalatha, M., Reddy, P.V. and Ramana Rao, D.V. (1999). Relationship of specific leaf area and carbon isotope discrimination with dry matter production and pod yield in groundnut genotype under moisture stress. *Indian Journal of Plant Physiology*. 4: 161-166.

Asif, M. A., Yusuf Zafar, Javaid Iqbal, Iqbal, M. M., Umer Rashid, Ali, G. M., Arif A., Nazir, F. (2011). Enhanced expression of AtNHX1, in transgenic groundnut (*Arachis hypogaea* L.) improves salt and drought tolerence. *Molecular Biotechnology*. 49(3): 250-256.

Babalad H.B. and Kulkarni G.N. (1988). Effects of irrigation frequencies on growth and yield of summer groundnut. *Journal of Maharashtra Agricultural Universities*. 13: 189-191.

Babitha, M. and Reddy, P.V. (2001). Physiological traits related to water use efficiency in groundnut under stress and non-stress conditions. Journal of Research ANGRAU. 29:9-15.

Babitha, M., Sudhakar, P., Latha, P., Reddy, P.V. and Vasanthi, R.P. (2006). Screening of groundnut genotypes for high water use efficiency and temperature tolerance. *Indian Journal of Plant Physiology*. 11: 63-74.

Babu, V.R., Murty, P.S.S., Reddi, G.H.S. and Reddy, T.Y. (1983). Stress day index and pod yield relationship in groundnut. *Environmental and Experimental Botany*. 23(2): 183-188.

Badigannavar, M., Kale, D.M., Bhagwat, S.G. and Murty, G.S.S. (1999). Genotypic and seasonal variation in stomatal characters in Trombay groundnut varieties. *Tropical Agricultural Research and Extension*., 2: 10-12.

Bailey, E and Boisvert, R N (1989) A risk evaluation of groundnut genotypes in drought prone areas of India. A. E. Research, Department of Agricultural Economics, Cornell University. 89 (11,iv): 95.

Balasubramanian, V. and Yayock, J.Y. (1981). Effect of gypsum and moisture stress on growth and pod-fill of groundnut (*Arachis hypogaea* L.). *Plant and Soil*. 62: 209-219.

Bari R. and Jones J. D. G. (2009). Role of hormones in plant defense responses. *Plant Molecular Biology*. 69: 473-488.

Bennett, J.M. and Hammond, L.C. (1982) Soil moisture and genotypic effects on the heat tolerance of peanut. *Proceedings: Soil and Crop Science Society of Florida*. 41: 97-100.

Bhatia, V.S., Singh, P., Rao, A.V.R.K., Srinivas, K. and Wani, S.P. (2009). Analysis of water non-limiting and water limiting yields and yield gaps of groundnut in India using CROPGRO-Peanut model. *Journal of Agronomy and Crop Science*. 195: 455-463

Billaz, R. and Ochos, R. (1961). The stage of susceptibility to drought in groundnut. *Oleagineux* 16: 605-609.

Bindumadhava, H., Sheshshayee, M.S., Shashidhar, G., Prasad, T.G. and Udayakumar, M. (2005). Ratio of stable carbon and oxygen isotope discrimination (Delta 13C/ Delta 18O) reflects variability in leaf intrinsic carboxylation efficiency in plants. *Curr. Sci*. 89: 1256-1258.

Bodare, C.A. and Dhonde, M.B. (2011). Effect of water stress on growth, yield and economics of summer groundnut. *Journal of Maharashtra Agricultural Universities*. 36: 152-155.

Challinor, A.J. and Wheeler, T.R. (2008). Use of a crop model ensemble to quantify CO_2 stimulation of water-stressed and well-watered crops. *Agricultural and Forest Meteorology*; 148: 1062-1077.

Chapman, S.C., Ludlow, M.M. and Blamey, F.P.C. (1993a). Effect of drought during early reproductive development on the dynamics of yield development of cultivars of groundnut (*Arachis hypogaea* L.). *Field Crops Research*. 32: 227-242.

Chaudhari, K.N. and Dadhwal, V.K. (2004). Assessment of impact of drought - 2002 on the production of major kharif and rabi crops using standardized precipitation index. *Journal of Agrometeorology*.; 6 (Special Issue): 10-15.

Chavan, A.A., Dhoble, M.V. and Khating, E.A. (1992). Effect of artificial water stress on different genotypes of groundnut (*Arachis hypogaea*) in dryland. *Indian Journal of Agricultural Sciences*. 62: 376-381.

Chavan, L.S., Patil, K.D. and Patil, B.P. (1999). Effect of water stress and nitrogen levels on yield of groundnut. *Journal of Maharashtra Agricultural Universities publ*. 24:309-310.

Chen S., Li J., Wang T., Polle A. and Hüttermann A. (2002). Osmotic stress and ion-specific effects on xylem abscisic acid and the relevance to salinity tolerance in poplar. *J. Plant Growth Reg*. 21: 224-233.

Chen YouQiang, Ye BingYing, Zhu-JinMao, Zhuang WeiJian, Chen, Y.Q., Ye, B.Y., Zhu, J.M. and Zhuang, W.J. (2000). Effects of osmotic stress on active oxygen damage and membrane peroxidation in young leaves of groundnut (*Arachis hypogaea*). *Chinese Journal of Oil Crop Sciences*. 22: 53-56.

Chung, SiYin., Vercellotti, J.R., Sanders, T.H. and Chung, S.Y. (1997). Increase of glycolytic enzymes in peanuts during peanut maturation and curing: evidence of anaerobic metabolism. *Journal of Agricultural and Food Chemistry*. 1997, 45: 4516-4521

Chuni Lal, Hariprasanna, K., Rathnakumar, A.L., Misra, J.B., Samdur, M.Y., Gor, H.K., Chikani, B.M. and Jain, V.K. (2009). Response of peanut genotypes to mid-season moisture stress: phenological, morpho-physiological, and yield traits. *Crop and Pasture Science*. 60: 339-347.

Clavel, D., Sarr, B., Marone, E., and Ortiz, R. (2004). Potential agronomic and physiological traits of Spanish groundnut varieties (*Arachis hypogaea* L.) as selection criteria under end-of-cycle drought conditions. *Agronomie*. 24: 101-111.

Clifford, S.C., Stronach, I.M., Mohamed, A.D., Azam Ali, S.N. and Crout, N.M.J. (1993). The effects of elevated atmospheric carbon dioxide and water stress on light interception, dry matter production and yield in stands of groundnut (*Arachis hypogaea* L.). *Journal of Experimental Botany*. 44: 1763-1770.

Collino, D., Biderbost, E. and Racca, R. (1994). Identification and characterization of physiological variables for the selection of drought-resistant material in groundnut (*Arachis hypogaea* L.). *Avances en Investigacion INTA Estacion Experimental Agropecuaria Manfredi*.1: 25-31.

Collino, D.J., Dardanelli, J.L., Sereno, R. and Racca, R.W. (2001). Physiological responses of argentine peanut varieties to water stress. Light interception, radiation use efficiency and partitioning of assimilates. *Field Crops Research*., 70: 177-184.

Collino, D.J., Dardanelli, J.L., Sereno, R. and Racca, R.W. (2000). Physiological responses of argentine peanut varieties to water stress. Water uptake and water use efficiency. *Field Crops Research*. 68:133-142.

Costa, W.A.J.M.de. and Nayakarathne, N.M.R.S. (2001). Effect of two different water regimes on growth and yield of different groundnut (*Arachis hypogaea* L.) genotypes in Sri Lanka. *Tropical Agricultural Research and Extension*. 4: 29-35

Craufurd, P.Q., Wheeler, T. R., Ellis, R.H., Summerfield, R.J. and Prasad, P. V. V. (2000). Escape and tolerance to high temperature at flowering in groundnut (*Arachis hypogaea* L.). *Journal of Agricultural Science*. 135: 371-378.

Dancette, C., Hamon, G. and Vachaud, G. (1979). Comparative study of water dynamics in fallow and cultivated sandy soils: modes of extraction of water by millet and groundnut crops in conditions of insufficient rainfall, in Senegal. Isotopes and radiation in research on soil plant relationships International Atomic Energy Agency, Vienna. 213-234.

Dangthaisong,, P., Banterng, P., Jogloy, S., Vorasoot, N., Patanothai, A. and Hoogenboom, G. (2006). Evaluation of the CSM-CROPGRO-Peanut model in simulating responses of two peanut cultivars to different moisture regimes. *Asian Journal of Plant Sciences*. 5: 913-922.

De Smet I., Zhang H., Inze D. and Beeckman T. (2006). A novel role for abscisic acid emerges from underground. *Trends Plant Sci*. 11: 434–439.

De Vries, J. D., Bennett, J. M., Albrecht, S. L. and Boote, K. J. (1989). Water relations, nitrogenase activity and root development of three grain legumes in response to soil water deficits. *Field Crop Res*. 21: 215-226.

Devi, K.B.S. and Rao, V.P. (2003). Crop coefficients for prediction of irrigation requirements of groundnut. *Journal of Research ANGRAU*. 31:1-8.

Devoto A. and Turner J.G. (2003). Regulation of Jasmonate mediated plant responses in Arabidopsis. Ann. Bot. 92: 329-337.

DGR, 2010. Annual report 2010-11. Directorate of groundnut Research, ICAR, Junagadh.

DGR, 2011. Annual report 2011-12. Directorate of groundnut Research, ICAR, Junagadh.

Dhonde, P.W., Shinde, S.H., Patil, B.B. and Pawar, A.D. (1985). Irrigation scheduling for kharif groundnut. *J. Maharashtra Agric. Univ*. 10: 28-30.

Dhopte, A.M. and Ramteke, S.D. (1994). Relative variation in dry matter partitioning of peanut genotypes under moisture stress conditions. *Annals of Plant Physiology*. 8: 174-178.

Dhopte, A.M., Ramteke, S.D. and Thote, S.G. (1992). Effect of soil moisture deficit on root growth and respiration, nodulation and yield stability of field grown peanut genotypes. *Annals of Plant Physiology*. 6:188-197.

Dhruve, J.J. and Vakharia, D.N. (2007). Amelioration of polyethylene glycol simulated water deficit stress by benzyladenine in groundnut. *Indian Journal of Agricultural Biochemistry*. 20: 53-58.

Duruoha, C., Piffer, C.R. and Silva, P.R.A. (2008). Root volume and dry matter of peanut plants as a function of soil bulk density and soil water stress. *IRRIGA*.13: 170-181.

Dutta, D. and Mondal, S.S. (2006).Response of summer groundnut (*Arachis hypogaea*) to moisture stress, organic manure and fertilizer with and without gypsum under lateritic soil of West Bengal. *Indian Journal of Agronomy*. 51: 145-148.

Dwivedi, R. S., Joshi, Y.C., Nautiyal, P. C., Singh, A. L., Ravindra, V., Thakkar, A. N., Koradia, V.G., and Dhapwal, G.S. (1986a). Developing parameters to predict drought resistance in groundnut genotypes. Transactions of Indian Society of Desert Technology and University Centre of Desert Studies. 11: 59-65.

Dwivedi, R.S. (1986b). Drought resistant plant attributes and breeding for drought resistance in groundnut. Ibid. pp. 66-67.

Dwivedi, R.S., Joshi, Y.C., P.C. Nautiyal, A.L. Singh, V. Ravindra, A.N. Thakkar, V.G. Koradia and G.S. Dhapwal, 1986c. Detection of drought resistance in peanut (*Arachis hypogaea* L.). Oleagineux, 41(6): 281-285.

Dwivedi, R.S., Joshi, Y.C., Saha, S.N., Thakkar, A.N., and Koradia, V.G.(1986d). Canopy orientations and energy harvests in groundnut (*Arachis hypogaea* L.) under water stress. *Oleagineux*, 41: 451-455

El Hadi, A.H.A., Khadr, M.S. and Johnston, A.E. (2003). Efficiency of potassium fertilization under saline and drought conditions in Egyptian soils. Potassium and water management in West Asia and North Africa. Proceedings of the Regional Workshop of the International Potash Institute, Amman, Jordan, 5-6 November 2001. 85-96.

Ewansiha, S.U. and Singh, B.B. (2006). Relative drought tolerance of important herbaceous legumes and cereals in the moist and semi-arid regions of West Africa. *Journal of Food Agriculture and Environment*. 4: 188-190.

FAO, 2012. FAOSTAT, FAO Statistics Division 2012, 26 September 2012.

Fariduddin Q., Hayat S. and Ahmad A. (2003). Salicylic acid influences net photosynthetic rate, carboxylation efficiency, nitrate reductase activity and seed yield in Brassica juncea. *Photosynthetica*. 41: 281-284.

Ferreira, L.G.R., Santos, I.F. dos, Tavora, F.J.F., Silva, J.V.da., Dos Santos, I.F., Da Silva, J.V. (1992). Effect of water deficit on groundnut (*Arachis hypogaea* L.) cultivars. Physiological reactions and yields. *Oleagineux Paris*. 47: 523-530.

Ferreyra, R.A., Ludmila B., Pachepsky, b., Collino, D., Acock B., (2000). Modeling peanut leaf gas exchange for the calibration of crop models for different cultivars. *Ecological Modelling.* 131: 285–298.

Gajera, M.S., Patel, J.C. (1984). Response of summer groundnut to irrigation scheduling based on cumulative irrigation under varying fertility levels. *Ind. J. Agron*. 29: 394-395.

Gauch, H.G. Jr, (1988). Model selection and validation for yield trials with interaction. *Biometrics.* 44: 705–715.

Ghatak, S., Sounda G,. Maitra S., Saren B. K., Roy D. K., Panda P. K. (1997). Effect of irrigation and potassium on growth and productivity of summer groundnut. *Environ and Ecology*. 15(2): 469 473.

Golakiya, B.A. (1992). Drought response of groundnut: VI. Hormonal treatment & seed pelleting to combat the drought. *Advances in Plant Sciences*. 5: 511-514.

Golakiya, B.A. (1993).Drought response of groundnut: VII. Identification of the critical growth stages most susceptible to water stress. *Advances in Plant Sciences*. 6: 20-27.

Golakiya, B.A. and Patel, M.S. (1988). Role of potassium in counteracting the effect of cyclic drought on groundnut. *Journal of Potassium Research*. 4: 163-167.

Golakiya, B.A. and Patel, M.S. (1992). Growth dynamics and reproductive efficiency of groundnut under water stress at different phenophases. Indian Journal of Agricultural Research. 26:179-186.

Gowda, A. (1977). Moisture stress and hormonal influence on the flowering and yield in groundnut (*Arachis hypogaea* Linn.).

Gowda, A. and Hegde, B.R. (1986). Moisture stress and hormonal influence on the flowering behaviour and yield of groundnut (*Arachis hypogaea* L.). *Madras Agricultural Journal*. 73: 82-86.

Greenberg, D.C. and Ndunguru, B.J. (1989). Groundnut drought-simulation studies at ICRISAT Sahelian Centre. *International Arachis Newsletter*. 6: 10-12.

Guo, Z.F. and Pan, R.Z. (1989). Effects of triadimefon on increasing drought resistance in groundnut seedlings. *Oil Crops of China.* 2: 14-18.

Hammer *et al.*, 1995. A peanut simulation model. I. Model development and testing. Agronomy Journal. 87, 1085-1093.

Hammond, L.C., Boote, K.J., Varnell, R.J. and Robertson, W.K. (1978). Water use and yield of peanuts on a well-drained sandy soil. *Proceedings of American Peanut Research and Education Association, Inc*., 10: 73.

Harris, D. and Natarajan, M. (1987). Physiological basis for yield advantage in a sorghum/groundnut intercrop exposed to drought. *Field Crops Research*. 17: 273-288.

Harris, D. and Natarajan, M. and Willey, R.W. (1987). Physiological basis for yield advantage in a sorghum/groundnut intercrop exposed to drought. I. Dry-matter production, yield, and light interception. *Field Crops Research*. 17: 259-272.

Hashim, I.B., Koehler, P.E., Eitenmiller, R.R. and Kvien, C.K. (1993). Fatty acid composition and tocopherol content of drought stressed Florunner peanuts. *Peanut Sci.* **20**: 21-24.

Hayat S., Fariduddin Q., Ali B. and Ahmad A. (2005) Effect of salicylic acid on growth and enzyme activities of wheat seedlings. *Acta Agron Hŭng.* 53:433–437.

Hook, J.E. (1994). Using crop models to plan water withdrawals for irrigation in drought years. *Agricultural Systems*. 45: 271-289.

Huang, M.T. and Ketring, D.L. (1985). Studies of water relations of peanut under rainfed and irrigated conditions. Proceedings, American Peanut Research and Education Society, Inc., 17: 65.

Hubick, K.T. (1990). Effect of nitrogen sources and water limitation on growth, transpiration efficiency and carbon isotope discrimination in peanut cultivars. *Aust. J. Plant. Physiol.* 17: 413-430.

Hubick, K.T., Farquhar, G.D. and Shorter, R. (1986). Correlation between water use efficiency and carbon isotope discrimination in diverse peanut (*Arachis*) germplasm. *Aust. J. Plant Physiol.* 13: 803-816.

Hubick, K.T., Shorter, R. and Farquhar, G.D. (1988). Heretibility and genotype environment interaction of carbon isotope discriminate and transpiration efficiency in peanut (*Arachis hypogaea* L.). *Aust. J. Plant Physiol.* 15: 799-813.

ICRISAT. (1986). Annual Report, 1985. ICRISAT, Patancheru, A.P. 502 324, India.

Izge, A.U. and Olorunju, P.E. (2000). Response of groundnut genotypes to moisture stress in the greenhouse. *Journal of Sustainable Agriculture and the Environment.* 2: 182-187.

Jadhao, P.N., Bhalerao, P.D., Fulzele, G.R., and Thorve, P.V. (1989). Response of peanut to water stress during seedling stage under summer season. *Annals of Plant Physiology*. 3: 146-151.

Jain, L.L., Panda, R.K., and Sharma, C.P. (1997). Water stress response function for groundnut (*Arachis hypogaea* L.). *Agricultural Water Management*. 32: 197-209

Jiang HuiFang and Ren-XiaoPing (2004). SOD activity and protein content in groundnut leaves as affected by drought stress. *Acta Agronomica Sinica*; 30: 169-174

Joshi, A.C., Patil, J.R. and Umrani, N.K. (1987). Use of mulch and anti-transpirant on groundnut under water stressed and non-stressed conditions. J. of Maharashtra Agricultural Universities. 12: 247-249.

Joshi, Y.C., Nautiyal, P.C., Ravindra, V., Dwivedi, R.S. (1988). Water relations in two cultivars of groundnut (*Arachis hypogaea* L.) under soil water deficit. *Tropical Agriculture*. 65: 182-184

Kachot, M. A., J. C. Patel, and Malviya, D.D. (1984). Response of Summer groundnut to irrigation scheduling based on IW/CPE ratio under varying levels of nitrogen and phosphorus. *Indian J. Agronomy*. 29:411-412.

Kalariya, K.A. Singh A.L. and K. Chakraborty, 2012a. Influence of water deficit stress on chlorophyll fluorescence in groundnut. In: National Seminar on New Frontiers of Plant Science for Sustainable development, 25-26 Feb 2012, PN College Khurda, Orissa p 20.

Kalariya, K.A., Singh, A.L, Chakraborty, K. Zala, P.V. and Patel, C.B. 2012b. Physiology of Groundnut (*Arachis hypogaea* L.) under water stress: I Chlorophyll Fluorescence and Photosynthesis, *Indian Journal of Plant Physiology (in press)*.

Kameswara Rao Kottapalli, Rakwal, R., Shibato, J., Gloria Burow David Tissue, John Burke, Puppala, N., Burow, M. And Payton, P. (2009). Physiology and proteomics of the water-deficit stress response in three contrasting peanut genotypes. *Plant, Cell and Environment*. 32: 380–407.

Kandoliya, U.K., Parameswaran, M., and Vakharia, D.N. (2000). Effect of drought on kernel carbohydrates and lipids in groundnut (*Arachis hypogaea* L.). *Advances in Plant Sciences*, 13: 573-579.

Kar, D.N., Ray, M. and Garnayak, L.M. (2002). Effect of moisture stress on yield & water use by rabi groundnut. *Annals of Agricultural Research*, 23:159-161.

Kar, D.N., Ray, M. and Garnayak, L.M. (2001). Determination of crop coefficient of groundnut for modified Penman method. *Environment and Ecology*. 19: 474-476.

Kar, G. and Kumar, A. (2007). Surface energy fluxes and crop water stress index in groundnut under irrigated ecosystem. *Agricultural and Forest Meteorology*. 146: 94-106.

Katawatin, R., Crown, P.H. and Grant, R.F. (1996). Simulation modelling of land suitability evaluation for dry season peanut cropping based on water availability in Northeast Thailand: evaluation of the MACROS crop model. *Soil Use and Management*., 12: 25-32.

Keller, J. and Bliesner, R.D. 1990. *Sprinkle and Trickle Irrigation*, Van Nostrand Reinhold, New York.

Ketring, D.L. (1986). Physiological response of groundnut to temperature and water deficits - breeding implications. Agrometeorology of groundnut. Proceedings of an international symposium, ICRISAT Sahelian Center, 21-26 August 1985: 135-143.

Khan, A.R. and Datta, B. (1990). Response of peanut to moisture stress and irrigation. Agrochimica 24: 15-23.

Khodary S.F.A. (2004). Effect of salicylic acid on the growth, photosynthesis and carbohydrate metabolism in salt stressed maize plants. *Int. J. Agric. Biol.* 6: 5-8.

Koti, R.V., Chetti, M.B., Manjunath, T.V.and Amaregowda, A. (1994). Effect of water stress at different growth stages on biophysical characters and yield in groundnut (*Arachis hypogaea* L.) genotypes. *Karnataka Journal of Agricultural Sciences*. 7: 158-162.

Krishnamurthy, L., Vadez, V., Devi, M.J., Serraj, R., Nigam, S.N., Sheshshayee, M.S., Chandra, S., Aruna, R. (2007). Variation in transpiration efficiency and its related traits in a groundnut (*Arachis hypogaea* L.) mapping population. *Field Crops Res*. 103: 189-197.

Kulkarni, J.H., Joshi, Y.C., and Bhatt, D.M. (1988) Effect of water stress on nodulation and nitrogen content of groundnut (*Arachis hypogaea* L.). *Zentralblatt fur Mikrobiologie*. 143: 4, 299-302.

Kulkarni, J.H., Ravindra, V., Sojitra, V.K. and Bhatt, D.M. (1988). Growth, nodulation and N-uptake of groundnut (*Arachis hypogaea* L.) as influenced by water deficit stress at different phenophases. *Oleagineux Paris*. 43: 415-419

Kvien, C.S., Branch, W.D., Sumner, M.E. and Csinos, A.S. (1988). Pod characteristics influencing calcium concentrations in the seed and hull of peanut. *Crop Science*. 28: 666-671.

Lehmann J., Atzorn R., Brückner C., Reinbothe S., Leopold J., Wasternack C. and Parthier B. (1995). Accumulation of Jasmonate, abscisic acid, specific transcripts and proteins in osmotically stressed barley leaf segments. *Planta.* 197: 156-162.

Li Lin, Liu DengWang, Zou Dong Sheng and Liu Fei (2008). The correlation between main characteristics and pod yield in peanut genotypes under natural waterlogging stress. *Chinese Journal of Oil Crop Sciences*. 30: 62-70.

Liang Y., Mitchell D.M. and Harris J.M. (2007). Abscisic acid rescues the root meristem defects of the Medicago truncatula latd mutant. *Dev Biol.* 304: 297–307.

Liechti R. and Farmer E.E. (2006). Jasmonate biochemical pathway. *Sci. STKE* 322: 1-3.

Lourduraj C. A. (2000). Effect of irrigation and manure application on the growth and yield of groundnut. *Acta Agronomica Hungarica*. 48(1): 83-88.

Madhusudhan, K.V., Giridarakumar, S., Ranganayakulu, G.S., Reddy, P.C., Sudhakar, C. (2002). Effect of water stress on some physiological responses in two groundnut (*Arachis hypogaea* L.) cultivars with contrasting drought tolerance. *Journal of Plant Biology*, 29: 199-202.

Mandavia, C., Dhruj, I.U., Chattrabhuji, B.J., Rajani, J.C. and Bharodia, P.S. (2007). Identification of drought tolerant groundnut (*Arachis hypogaea* L.) genotypes. *Indian Journal of Agricultural Research*. 41: 17-22.

Man Singh, S.S. Parihar Manoj Khanna, A.K. Singh and A.L. Singh 2012. Seasonal Water Requirements and Irrigation Scheduling of Groundnut (*Arachis hypogaea* L.) Crop under trickle irrigation *Journal of water management* (inpress).

Mante, E.F.G. (1976). Suitable soils for grain legume production. Proceedings of the joint University of Ghana Council for Scientific and Industrial Research symposium on grain legumes in Ghana [Doku, E.V. (Editor)]. 7-23.

Maphanyane, G.S., Ndunguru B.J .(ed.), Hildebrand, G.L. (ed.); Subrahmanyam, P. (1994). Selecting groundnuts for adaptation to drought under rainfed conditions in Botswana. Sustainable groundnut production in southern and eastern Africa: proceedings of a Workshop, 5-7 July 1994, Mbabane, Swaziland: 32-36.

Mathew, T. and Pandey, D.P. (2006). Protection of groundnut plants from water stress by chlorflurenol and cycocel. *Indian Journal of Plant Physiology*. 11: 209-212.

Mathur, P.B., Devi, M.J., Serraj, R., Yamaguchi Shinozaki, K., Vadez, V. and Sharma, K. K. (2004). Evaluation of transgenic groundnut lines under water limited conditions. *International Arachis Newsletter*. 24: 33-35.

Matthews, R. B., Harris, D., Nageswara Rao, R. C., Williams, J. H. and Wadia, K. D.R. (1988a). The physiological basis for yield differences between four genotypes of groundnut (*Arachis hypogaea*) in response to drought I. Dry matter production and water use. *Exp. Agric.* 24: 191-202.

Matthews, R. B., Williams, J. H., Nageswara Rao, R. C. (1988b). The physiological basis for yield differences between four genotypes of groundnut (*Arachis hypogaea*) in response to drought II. Solar radiation interception and leaf movement. *Expl. Agric*. 24: 203-213.

Meisner, C.A. (1991). Peanut roots, shoot, and yield under water stress. Dissertation-Abstracts-International-B,-*Sciences and Engineering*. 52: 3B-4B.

Meisner, C.A. and Karnok, K.J. (1992). Peanut root response to drought stress. *Agronomy Journal*. 84: 159-165.

Mensah, J.K. and Okpere, V.E. (2000). Screening of four groundnut cultivars from Nigeria for drought resistance. *Legume Research*. 23: 37-41.

Misra, J.B. and Nautiyal, P.C. (2005). Influence of imposition of soil moisture-deficit stress on some quality components of groundnut, *Arachis hypogaea* L. kernel. *J. of Oilseeds Res.* 22: 119-124

Mittal, G.K., Arunabh Joshi, Rajamani, G., Mathur, P.N. and Sharma (2006). A Water-deficit induced generation of reactive oxygen species and antioxidants in two Spanish groundnut cultivars. *National Journal of Plant Improvement*. 8: 7-10.

Mohan, C.V.K. and Rao, G.R. (1989). Influence of foliar application of calcium on growth performance, chlorophyll and protein contents in groundnut under water stress. *Narendra Deva Journal of Agricultural Research*. 4: 7-11.

Mohandass, S., Radhakrishnan, R., Panchanathan, R.M., Kandaswamy, P. (1989). Diurnal pattern of plant water relations of groundnut under water stress and potassium nutrition. *Journal of Agronomy and Crop Science*. 163: 18-20

Mukherjee, A., Moitra, R. and Chakravarti, A.K. (2010). Variation of micrometeo-rological environment within different groundnut cultivars. *Environment and Ecology*. 28(3A): 1842-1844.

Munne-Bosch S. and. Peñuelas J. (2003). Photo- and antioxidative protection, and a role for salicylic acid during drought and recovery in field-grown Phillyrea angustifolia plants. *Planta*. 217: 758–766.

Musingo, M.N., Basha, S.M., Sanders, T.H., Cole, R.J. and Blankenship, P.D. (1989). Effect of drought and temperaturestress on peanut (*Arachis hypogaea* L.) seed composition. *J. Plant Physiol.* **134:** 710-715.

Muthumanickam, D., Kannan, P., Kumaraperumal, R., Natarajan, S., Sivasamy and R., Poongodi, C. (2011) Drought assessment and monitoring through remote sensing and GIS in western tracts of Tamil Nadu, India. *International Journal of Remote Sensing*. 32(17/18): 5157-5176.

Nageshwar Rao, R.C. and Wright, G.C. (1994). Stability of the relationship between specific leaf area and carbon isotope discrimination across environment in peanut. *Crop Sci.* 34:98-103.

Nageshwar Rao, R.C. (2003). Environmental characterisation of experimental sites in India and Australia. Breeding of drought resistant peanuts Proceedings of a Collaborative Review Meeting Hyderabad, Andhra Pradesh, India, 25-27-February, 2002.; 61-66.

Nageshwar Rao, R.C., Reddy, L.J., Nigam, S.N. and Subrahmanyam, P. (1992). Influence of soil type on the adaptation of groundnut genotypes. Proceedings of the fifth regional groundnut workshop for Southern Africa, 9-12 March, 1992, Lilongwe, Malawi , 15-17.

Nageshwar Rao, R.C., Sardar Singh, Sivakumar, M.V.K., Srivastava, K.L. and Williams, J.H. (1985). Effect of water deficit at different growth phases of peanut. I. Yield responses. *Agronomy Journal*. 77: 782-786.

Nageshwar Rao, R.C., Williams, J.H., Wadia, K.D.R., Hubick, K.T. and Farquhar, G.D. (1993). Crop growth, water-use efficiency and carbon isotope discrimination in groundnut (*Arachis hypogaea* L.) genotypes under end-of-season drought conditions. *Annals of Applied Biology*. 122: 357-367.

Nakar R.N., A.L. Singh, Nisha Goswami, K.A. Kalariya, K. Chakraborty 2012. Physiological efficiencies in minicore groundnut germplasm accessions. In: Proceeding of the National Conference on "Research in Plant Sciences: Changing pathways, changing lives" Oct 29-30, 2012 at Talod, Sabarkantha, Gujarat P

Natarajan, M. and Willey, R.W. (1986). The effects of water stress on yield advantages of intercropping systems. *Field Crops Research*. 13:117-131.

Nautiyal, P.C., Ravindra, S. and Joshi, Y.C. (1991). Moisture stress and subsequent seed viability. Physiological and biochemical basis for viability differences in Spanish groundnut in response to soil moisture stress. *Oleagineux*. 46: 153-158.

Nautiyal, P.C., Ravindra, V., Zala, P.V. and Joshi, Y.C. (1999). Enhancement of yield in groundnut following the imposition of transient soil-moisture-deficit stress during the vegetative phase. *Exp Agric*. 35:371-385.

Nautiyal, P.C. Rajgopal, K. P.V. Zala, Dipti S. Pujari, Manojit Basu, Bhuwaneshwariba Dhadhal and Bharat M. Nandre, 2008. Evaluation of Wild *Arachis* species for abiotic stress tolerance: I Thermal stress and leaf water relations. *Euphytica*, 159:43-57.

Nautiyal, P.C., P.V. Zala, Rukam Singh, Paresh Sodavadiya, and Bhavesh Tavethia, 2011. A Evaluation of water use efficient new developed varieties of groundnut in on-farm trials in two different rain-fall areas in Gujarat, *Journal of SAT Agricultural Research, Dec, 9: 1-6.*

Nautiyal, P.C., Ravindra V, Rathnakumar, A.L., Ajay, B.C. and Zala, P.V. 2012. Genetic variation in photosynthetic rate, pod yield and yield components in Spanish groundnut cultivars during three cropping seasons. *Field Crop Research, 125:83-89.*

Naveen, P., Daniel, K.V., Subramanian, P. and Kumar, P.S. (1992). Response of irrigated groundnut (*Arachis hypogaea*) to moisture stress and its management. *Indian Journal of Agronomy*. 37: 82-85.

Neelam Yadav, Yadav, V.K. and Yadava, C.P.S. (2007). Influence of simulated whitegrub damage of roots on water relation parameters in groundnut. *Indian Journal of Plant Physiology*.12: 173-177

Nemeth M., Janda T., Horvath E. , Paldi E. and Szalai G. (2002). Exogenous salicylic acid increases polyamine content but may decrease drought tolerance in maize. *Plant Sci*. 162: 569-574.

Nigam, S.N. and Rupakula Aruna (2008). Stability of soil plant analytical development (SPAD) chlorophyll meter reading (SCMR) and specific leaf area

(SLA) and their association across varying soil moisture stress conditions in groundnut (*Arachis hypogaea* L.). *Euphytica*. 160: 111-117.

Nogueira, R.J.M.C., Santos, R.C.dos, Bezerra Neto, E., Santos, V.F. dos., dos Santos R.C. dos and Santos, V.F. (1998). Physiological response of two groundnut cultivars subjected to different water regimes. *Pesquisa Agropecuaria Brasileira.*, 33: 1963-1969.

Oliveira Junior, J. O. L. de., Tavora, F.J.A.F. and Barbosa Filho,-M (2004). Phenology of peanuts cultivars submitted to water stress. *Cientifica Jaboticabal*. 32: 25-29.

Omoko, M. (1990). Influence of crop-water managements on yield-water use relationships. Dissertation Abstracts International B, *Sciences and Engineering*. 51: 1050B

Omoko, M. and Hammond, L.C. (2010). Biological and water-use efficiencies of sorghum-groundnut intercrop. *Cameroon Journal of Experimental Biology*. 6: 1-10.

Ong, C.K. (1984). The influence of temperature and water deficits on the partitioning of dry matter in groundnut (*Arachis hypogaea* L.). *Journal of Experimental Botany*. 35: 746-755.

Padma, V., Rao, D.V.M. and Rao, I.V.S. (1991). Response of groundnut cultivars to time of sowing in different seasons. *Journal of Oilseeds Research*. 8: 275-279.

Padmanabhamurty, B., Saxena, N,. Sastry, P.S.N. (ed.), Rao, P.S. (ed.), Shekh, A.M. (ed.) and Pandey, V. (2001). Variation of residual energy fluxes in relation to plant height and biomass in groundnut. National workshop on preliminary results of Land Surface Processes Experiment (LASPEX-97) over Sabarmati River Basin, Anand, India, 18-20 November 1999. *J. Agrometeorology*. 3:119-126.

Pallas, JE Jr., Stansell, JR and Bruce, RR (1977). Peanut seed germination as related to soil water regime during pod development. *Agronomy Journal*. 69: 381-383

Panda, P.K. and Behera, S.K. (2005). Irrigation water management strategy for peanut under deficit conditions. *Zeitschrift-fur-Bewasserungswirtschaft*. 40: 91-114.

Panda, R.K., Kumar, P. and Kashyap, P.S. (2003). Effective management of irrigation water in a sub-humid region using PNUTGRO model. *Zeitschrift-fur-Bewasserungswirtschaft*. 38: 41-55.

Parekh, F.P. and Shete, D.T. (2008). Irrigation scheduling for summer groundnut considering agrometeorological parameters and dual crop coefficient approach. *Journal of Agrometeorology*. 10(Special Issue 2): 309-313.

Parmar, J.V., Patel, C.L. and Polara, K.B. (1989). Influence of soil-moisture stress at different stages of growth on yield response and nutrients in groundnut. *Annals of Arid Zone*. 28: 267-270.

Parihar, S.S., Man Singh, Manoj Khanna and A.K. Singh. 2008. Response of Summer Groundnut (*Arachis hypogaea* L.) to Method of Planting and Irrigation Regime under Drip Irrigation in Semi Arid Region. *Journal of water management.* 16(2): 101-113.

Passioura, J.B. (1977) Grain Yield, harvest index and water use of wheat. *J. Aust. Inst. Agric. Sci.* 43:117-121.

Patel, C.L. and Padalia, M.R. (1980). Nutrient accumulation and nutrient-use-efficiency in groundnut under different sequences of soil-water potential. *Legume Research*. 3: 45-50.

Patel, C.L., Padalia, M.R. and Babaria, N.B. (1983). Growth and plant-water relation in groundnut grown under different soil-moisture stresses. *Indian Journal of Agricultural Sciences*. 53: 340-345.

Patel, M.S and Golakiya, B.A. (1988). Effect of water stress on yield attributes and yield of groundnut (*Arachis hypogaea*). *Indian Journal of Agricultural Sciences*. 58: 701-703

Patel, M.S. and Golakiya, B.A. (1991). Effect of water stress on yield and yield attributes of groundnut (*Arachis hypogaea* L.). *Madras Agricultural Journal*. 78: 178-181.

Patel, M.S. and Golakiya, B.A. (1993). Moisture stress effects at different phenophases on biophysical parameters of groundnut cultivars. *Indian Journal of Agricultural Research*. 27: 126-130.

Patel, M.S., Gundalia, J.D., Davaria, R.L. and Vadher, P.G. (1988). Yield and quality of summer groundnut (*Arachis hypogaea*) as influenced by growth retardants and promoters. *Indian Journal of Agricultural Sciences*. 58: 773-775.

Patil, B.P. and Gangavane, S.B. (1990). Effects of water stress imposed at various growth stages on yield of groundnut and sunflower. *Journal of Maharashtra Agricultural Universities*. 15: 322-324.

Patil, N.A. and Patil, T.M. (1993). Physiological responses of groundnut genotypes to water stress. *Legume Research*.16: 23-30.

Patil, R.G., Radder, G.D., Patil V.C. and Ramachandrappa B.K. (1984). Dry matter production and distribution in bunch groundnut varieties as influenced by the time of sowing and foliar application of nitrogen and iron. *Madras Agricultural Journal*. 71: 177-181.

Patil, S.M. and Morey, D.K. (1993). Growth and yield of summer groundnut as influenced by antitranspirants and water management practices. *Journal of Soils and Crops*. 3: 18-22.

Patra, A.K., Tripathy, S.K. and Samui, R.C. (1999). Effect of sowing date, irrigation and spacing on nodulation, dry matter and yield of summer groundnut (*Arachis hypogaea*). *Indian Journal of Agronomy*., 44: 800-804

Peleg Z. and Bluwald E. (2011) Hormone balance and abiotic stress tolerance in crop plants *Current Opinion in Plant Biology*. 14:290–295.

Prathap, S., Reddy, S.S. and Swamy, G.N. (2006). Effect of water stress on germination and seedling growth of groundnut genotypes. *Crop Research Hisar*. 31: 58-60.

Purushotham, M.G., Patil, V., Raddey, P.C., Prasad, T.G., Vajranabhaiah, S.N. and Patil, V. (1998). Development of *in vitro* PEG stress tolerant cell lines in two groundnut (*Arachis hypogaea* L.) genotypes. *Indian Journal of Plant Physiology*. 3:49-51.

Rajegowda, M.B., Janardhangowda, N. A., Ravindrababu, B. T., Girish, J. (2010) Crop weather model for sustainable groundnut production under dry land condition. *J. of Agrometeorology*. 12(2): 221-224.

Rajendrudu, G and Williams, J.H. (1987). Effect of gypsum and drought on pod initiation and crop yield in early maturing groundnut (*Arachis hypogaea*) genotypes. *Experimental Agriculture*. 23: 259-271.

Raju, A.P., Reddi, G.H.S. and Reddy, T.B. (1981). Studies on the productivity of rainfed groundnut under moisture stress using neutron moisture probe. *J of Nuclear Agriculture and Biology*. 10: 131-133.

Ramachandrappa, B.K. and Nanjappa, H.V. (1994). Prediction of pod yield and quantification of stress effect to delineate sensitive stages in groundnut - A mathematical model. *Crop Research Hisar*. 8: 437-441.

Ramesh Babu, V., Murty, P.S.S., Sankara Reddi, G.H. and Reddy, T.Y. (1983) . Leaflet angle and radiation avoidance by water stressed groundnut (*Arachis hypogaea* L.) plants. *Environmental and Experimental Botany*. 23:183-188.

Ramesh Katam, Basha, S.M., Vasanthaiah, H.K.N. and Naik, K.S.S. (2007). Identification of drought tolerant groundnut genotypes employing proteomics approach. *Journal-of-SAT-Agricultural-Research*. 5: 1-2.

Ramesh Thatikunta, Durgaprasad, M.M.K. and Thatikunta, R. (1996). Identifying groundnut genotypes for the Southern Telangana Zone in India. International Arachis Newsletter., No. 16: 19-20.

Rassaa N., Ben Haj Salah H. and Latiri, K. (2008). Thermal responses of Durum wheat *Triticum durum* to early water stress. Consequence on leaf and flower development. *C. R. Biologies* 331: 363–371.

Rathore, A.L., Pal, A.R., Sahu, R.K. and Chaudhary, J.L. (1996). On-farm rainwater and crop management for improving productivity of rainfed areas. *Agricultural Water Management*. 31: 253-267.

Ravi, K., Vadez, V., Isobe, S., Mir, R.R., Guo, Y., Nigam, S.N., Gowda,M.V.C., Radhakrishnan, T., Bertioli, D.J., Knapp, S.J., Varshney, R.K. (2011) Identification of several small main-effect QTLs and a large number of epistatic QTLs for drought tolerance related traits in groundnut (*Arachis hypogaea* L.). *Theoretical and Applied Genetics*. 122(6): 1119-1132.

Ravindra, V., Nautiyal, P.C. and Joshi, Y.C. (1990). Physiological analysis of drought resistance and yield in groundnut (*Arachis hypogaea* L.). *Tropical Agriculture*. 67: 290-296.

Ravindra, V., Nautiyal, P.C., Koradia, V.G. and Joshi, Y.C. (1989). Manipulation of water stress to increase productivity in groundnut. *Groundnut News*. 2: 5.

Reddy A.R., Chaitanya K.V. and Vivekanandan M. (2004). Drought-induced responses of photosynthesis and antioxidant metabolism in higher plants. *J. Plant Physiol.* 161: 1189–1202.

Reddy, C.R. and Reddy, S.R. (1993). Scheduling irrigation for peanuts with variable amounts of available water. *Agricultural Water Management*. 23: 1-9.

Reddy, P.R.R., Reddy, B.B., Rao, V.P. and Sarma, P.S. (1996). Effect of water deficits at different crop growth periods on growth and yield of groundnut. *Journal of Research APAU*., 24: 147-149

Reddy, P.V., Asalatha, M., Babhita, M. (1999). Soil water extraction and root growth in five groundnut genotypes under different soil moisture regimes. *Legume Res*. 22: 94-98.

Reddy, P.V., Asalatha, M. and Babitha, M. (2000). Relationship of mineral ash and chlorophyll content with transpiration efficiency in groundnut under different moisture regimes. *Indian Journal of Plant Physiology*. 5: 59-63.

Reddy, S.R. (1984). Scheduling irrigation with can evaporation for groundnut. *Ind. J. Agron*. 29: 51-54.

Reddy, T.Y. (1994). Management practices for amelioration of moisture stress in groundnut (*Arachis hypogaea*) in arid tropics. *Indian Journal of Agronomy*. 39: 489-490.

Reddy, T.Y., Reddy, V.R. and Anbumozhi, V. (2003). Physiological responses of groundnut (*Arachis hypogaea* L.) to drought stress and its amelioration: a review. *Acta Agronomica Hungarica*., 51: 205-227

Reddy, T.Y., Reddy, V.R. and Anbumozhi, V. (2003). Physiological responses of groundnut (*Arachis hypogea* L.) to drought stress and its amelioration: a critical review. *Plant Growth Regulation*. 41: 75-88

Rivas-San Vicente M. and Plasencia J. (2011). Salicylic acid beyond defence: its role in plant growth and development. *J. Exp Bot*. doi: 10.1093/jxb/err 031.

Rosario, D.A.del. and Fajardo, F.F.(1988). Morphophysiological responses of ten peanut (*Arachis hypogaea* L.) varieties to drought stress. *Philippine Agriculturist*. 71: 447-459

Runkulatile, H., Homma, K., Horie, T., Kurusu, T. and Inamura, T. (1998). Land equivalent ratio of groundnut-finger millet intercrops as affected by plant combination ratio, and nitrogen and water availability. *Plant Production Science*., 1: 39-46

Sahu, D.D., Patra,B. K. and Patoliya, B.M. (2004). Effect of sowing time and rainfall distribution on groundnut yield. *International Arachis.Newsletter*.24: 39-42.

Saini, H.S. and Srivastava, A.K. (1981). Osmotic stress and the nitrogen metabolism of two groundnut (*Arachis hypogaea* L.) cultivars. *Irrigation Science*. 2: 185-192.

Sakarvadia, H.L. and Yadav, B.S. (1994). Effect of water stress at different growth phases of groundnut on yield and nutrient absorption. *Journal of the Indian Society of Soil Science*. 42: 147-149.

Sakarvadia, H.L., Parmar, K.B., Jetpara, P.I. and Kalola, A.D. (2010). Effect of cyclic water stress and potassium on dry matter and nutrients accumulation in summer groundnut.*Advances in Plant Sciences*. 23: 193-196.

Samdur, M.Y., Manivel, P., Jain, V.K., Chikani, B.M., Gor, H.K., Desai, S. and Misra, J.B. (2003). Genotypic differences and water-deficit induced enhancement in epicuticular wax load in peanut. *Crop Science*, 43:1294-1299.

Santner A, and Estelle M (2009). Recent advances and emerging trends in plant hormone signalling. *Nature*. 459:1071-1078.

Sarma, P. S. and Sivakumar, M.V.K. (1990). Evaluation of groundnut response to early moisture stress during the rainy and the post-rainy seasons. *Agricultural and Forest Meteorology*. 49: 123-133

Sarma, P.S. and Sivakumar, M.V.K. (1987). Productivity of groundnut as influenced by use of seed from a crop with moisture stress history. *Field Crops Research*. 15: 207-213.

Sashidhar, V.R., Mekhri, A.A., and Krishna Sastry, K.S. (1977). Proline accumulation in relation to seed hardening in groundnut genotypes. *Indian Journal of Agricultural Sciences*. 47: 595-598

Sashidhar, V.R., Mekhri, A.A. and Sastry, K.S.K. (1981). Potassium content and proline accumulation following seed treatment with calcium chloride in groundnut varieties. *Indian Journal of Plant Physiology*. 24: 89-92.

Schubert, A.M. and Sanders, T.H. (1985). Irrigation scheduling using a canopy temperature stress degree day index to induce variable water stress in field-grown Florunner peanut. *Proceedings, American Peanut Research and Education Society*. 17: 65.

Selvam, V.S., Rao, G.N.S. and Rajan, M.S.S. (1989). Mid-season correction for groundnut (*Arachis hypogaea*) during moisture-stress period. *Indian Journal of Agricultural Sciences*. 59: 795-796.

Seo H.S., Song J.T., Cheong J.J., Lee Y.-H., Lee Y.-W., Hwang I., Lee J.S. and Choi Y.D. (2001) Jasmonic acid carboxyl methyltransferase: a key enzyme for jasmonate-regulated plant responses. *Proc. Natl. Acad. Sci.* USA. 98:4788-93.

Serraj, R., Bidinger, F.R., Chauhan, Y.S., Seetharama, N., Nigam, S.N., Saxena, N.P., Kijne, J.W. (ed.), Barker-R (ed.), and Molden, D. (2003). Management of drought in ICRISAT cereal and legume mandate crops. Water productivity in agriculture: limits and opportunities for improvement. 127-144

Shahid Umar and Umar, S (1997). Influence of potassium on solar energy harvesting efficiency of groundnut under different soil moisture stress. *Indian Journal of Plant Physiology*. 2: 315-317.

Shahid Umar, Bansal, S.K. and Umar, S. (1997). Effect of potassium application on yield and quality of water stressed groundnut. *Fertiliser News*. 42: 27-29.

Shakirova F.M. (2007). Role of hormonal system in the manifestation of growth promoting and anti-stress action of salicylic acid. In: Salicylic Acid. A Plant Hormone (Eds.) Hayat S., Ahman A. Springer. ISBN 978-1-4020-5183-8. Dordrecht, Netherlands.

Sharma, K., Gurbaksh Singh and Sharma, H.L. (1987).Effect of simulated water stress by polyethylene glycol (PEG) on germination and some enzymes of groundnut (*Arachis hypogaea* L.). *Annals of Biology*. 3: 77-82

Sharma, K., Gurbaksh Singh and Singh, G. (1985). Seedling growth and sugar metabolism of groundnut under polyethylene glycol induced water stress. *Indian Journal of Ecology*. 12:252-256.

Sharma, K., Gurbaksh Singh and Singh, G. (1987). Effect of water stress on yield and quality of groundnut. *Environment and Ecology*. 5: 647-650

Sharma, P., Nisha., Malik, C.P. (1993). Photosynthetic responses of groundnut to moisture stress. *Photosynthetica*., 29:157-160.

Sharp R. E., LeNoble M. E., Else M. A., Thorne E. T. and Gherardi F. (2000). Endogenous ABA maintains shoot growth in tomato independently of effects on plant water balance evidence for an interaction with ethylene. *J. Exp. Bot.* 51: 1575-1584.

Shashikumar, M.R., Gowda, T.C.C., Prasad, T.G., and Kumar, M.U. (1988). Assessment of relative drought tolerance in groundnut genotypes using line-source technique. *Narendra Deva Journal of Agricultural Research*. 3: 48-54.

Sheshshayee, M.S., Bindumadhava, H., Ramesh, R., Prasad, T.G., Lakshminarayana, M.R. and Udayakumar, M. (2005). Oxygen isotope enrichment (Delta^{18}O) as a measure of time-averaged transpiration rate. *J. Exp. Bot.* 56: 3033-3039.

Shi Q. and Zhu Z. (2008). Effects of exogenous salicylic acid on manganese toxicity, element contents and antioxidative system in cucumber. *Environ. Exper. Bot.* 63: 317-326.

Shinde, G.G. and Pawar, K.R. (1984). Effects of water stress at critical growth stages on growth and yield of groundnut in summer season. *Journal of Maharashtra Agricultural Universities*. 9: 26-28.

Shinde, G.G. and Pawar, K.R. (1982). Effect of irrigation on the pod yield of groundnut. *Indian Journal of Agricultural Sciences*. 52: 576-578.

Shinde, G.G. and Pawar, K.R. (1983). Growth and plant-water relation in groundnut grown under different soil-moisture stresses. *Journal of Maharashtra Agricultural Universities*. 8: 32-34.

Shinde, G.G. and Pawar, K.R. (1984). Effect of water stress at critical growth stages on growth and yield of groundnut in summer season. *J. Maharashtra. Agric. Univ*. 9: 26-28.

Simmonds L.P. and Azam Ali S.N. (1989). Population, growth and water use of groundnut maintained on stored water. IV. The influence of population on water supply and demand. *Experimental Agriculture*. 25: 87-98.

Singh, A.L. 2004a. Growth and physiology of groundnut. pp 178-212. In: Groundnut Research in India (Eds. M.S. Basu and N.B. Singh) National Research center for groundnut (ICAR), Junagadh, India.

Singh, A.L. (2004b). Mineral nutrient requirement, their disorders and remedies in groundnut. pp 137-159. In: Groundnut Research in India (Eds. M.S. Basu and N.B. Singh) National Research center for groundnut (ICAR), Junagadh, India.

Singh, A.L. (2011). Physiological basis for realizing yield potentials in Groundnut. *In*: Advances in Plant Physiology (Ed. A. Hemantaranjan) Vol. 12 pp. 131-242. Scientific Publishers (India), Jodhpur, India.

Singh, A.L. and Vidya Chaudhari, 1995. Source and mode of sulphur application on groundnut productivity. *Journal of Plant Nutrition*, 18: 2739-2759

Singh, A.L., Ajai and Vidya Chaudhari. (2001). Drip irrigation- a potential system for micronutrient application in groundnut in semi-arid region. pp 501-507. *In*: Micro-irrigation (Eds. H.P. Singh, S.P. Kaushish, Ashwani Kumar, T.S. Murthy and J.C. Samuel), proceeding International Conference on Micro and Sprinkler Irrigation Systems, 8-10 February 2000, Jain irrigation Hills Jalgaon, India. Published by Central Board of Irrigation and Power, New Delhi, India.

Singh, A.L., Ajay, Vidya Chaudhari and J.B. Misra, 2000. Drip irrigation-an efficient system for micronutrient application in groundnut in calcareous soils of semi-arid region. In: Proc. GAU-PRII-IPI National Symposium on Balanced Nutrition of Groundnut and other Field Crops in Calcareous Soils of India (B.A. Golakiya, J.D. Gundalia, S.K. Bansal. and Patricia Imas eds) pp 194-198.

Sree, P.S.S. and Rao P.G. (1998). Response of groundnut to irrigation schedules based on IW/CPE ratios. *Journal of Oilseeds Research*. 15(2): 315 319.

Stirling, C.M. (1989). Environmental effects on partitioning and development in groundnut. Dissertation Abstracts International B, *Sciences and Engineering*. 50: 1726B

Stirling, C.M., Black, C.R. and Ong, C.K. (1990). Response of groundnut to timing of irrigation. *Monograph British Society for Plant Growth Regulation*. 21: 374-380.

Subramanian, V.B., Reddy, G.J. and Maheswari, M. (1993). Photosynthesis and plant water status of irrigated and dryland cultivars of groundnut. *Indian Journal of Plant Physiology*. 36: 236-238.

Sulochana, C. and Savithramma, N. (2003). Effect of calcium on acid phosphatase activity in groundnut (Arachis hypogaea L.) seedlings during water stress. *Indian Journal of Plant Physiology*. 8:186-188

Sulochana, C. and Savithramma, N. (2001). Effect of calcium in amelioration of PEG (6000) induced water stress in groundnut (*Arachis hypogaea* L.) cultivars during seedling growth. *Journal of Plant Biology*. 28: 257-263.

Sulochana, C.H., Rao, T.J.V.S. and Savithramma, N. (2002). Effect of calcium on water stress amelioration through calmodulin and scavenging enzymes in groundnut. *Indian Journal of Plant Physiology* (India). (Apl-Jun. v. 7(2) p. 152-158.

Sumathi, V. and Subramanyam, D. (2007). Analytical studies on drought categorization and its effects on rainfed groundnut yields. *Crop Research Hisar* .34: 88-91.

Suriharn, B., Patanothai, A. and Jogloy, S. (2005). Gene effects for specific leaf area and harvest index in peanut (*Arachis hypogaea* L.). *Asian Journal of Plant Sciences*. 4: 667-672.

Suvarna, Kenchanagoudar, P.V., Nigam, S.N. and Gowda, M.V. C (2006). Genotypic evaluation for yield and its attributes under imposed moisture stress conditions in groundnut (*Arachis hypogaea* L) *Research on Crops*. 7: 221-228.

Suvarna, Kenchanagoudar, P.V., Nigam, S.N. and Chennabyregowda, M.V. (2002). Effect of drought on yield and yield attributes of groundnut. *Karnataka Journal of Agricultural Sciences*. 15: 364-366.

Thiyagarajan, G. and Ranghaswami, M. V. (2010) Modeling yield response of groundnut to deficit irrigation at different growth stages by FAO CROPWAT. *Madras Agricultural J*. 97(4/6): 145-148.

Thiyagarajan, G., Rajakumar, D., Kumaraperumal, R., Manikandan, M. (2009) Physiological responses of groundnut (*Arachis hypogaea* L.) to moisture stress: a review. *Agricultural Reviews*. 30(3): 192-198.

Thiyagarajan, G., Ranghaswami, M.V., Rajakumar, D., Kumaraperumal, R. (2010). Deficit irrigation effects on groundnut (*Arachis hypogaea* L.) with micro sprinklers. *Madras Agricultural J*. 97: 40-42.

Thorat, S.T., Kalra, G.S., Pawar, A.B. (1984). Responses of groundnut varieties to different irrigation schedules. *Indian Journal of Agronomy*. 29: 43-45.

Tiwari, K.N. (2005). Meeting the challenges of oilseeds production through balanced use of potash. *Fertiliser Marketing News*. 36: 5-14, 19, 21-23.

Turner J.G., Ellis Ch. and Devoto A. (2002). The jasmonate signal pathway. *Plant Cell*. 14: 153-164.

Umar, S., Afridi, M.M.R.K. and Dwivedi, R.S. (1991). Influence of added potassium on the drought resistance of groundnut. *Journal of Potassium Research*. 7: 53-61.

Upadhyaya, H. D., Sharma S., Singh S. and Singh M. (2011) Inheritance of drought resistance related traits in two crosses of groundnut (*Arachis hypogaea* L.). *Euphytica*. 177(1): 55-66.

Uthai Arromratana, Tawatchai Na Nakara, Seri Sukakit, Boonlert Boonyoung (1993). Effect of water stress at different growth stages on yield of groundnut. Department of Agricultural Extension, Bangkok (Thailand). *Proceedings of the eleventh Thailand national groundnut meeting. Raingan kan sammana thua-lisong haeng chat khrang thi 11 na rongraem Chansomthara, Ranong 17-21 Phrusaphakhom 2536. Bangkok (Thailand)*. 568: 273-278.

Vadez, V., Krishnamurthy, L., Kashiwagi, J., Kholova, J., Devi, J.M., Sharma, K.K., Bhatnagar Mathur, P., Hoisington, D.A., Hash, C.T., Bidinger, F.R., Keatinge, J.D.H. (2007). Exploiting the functionality of root systems for dry, saline, and nutrient deficient environments in a changing climate. *Journal of SAT Agricultural Research*. 4: 1-61.

Vaghasia, P.M., Jadav, K.V. and Nadiyadhara, M.V. (2010a). Effect of soil moisture stress at various growth stages on growth and productivity of summer groundnut (*Arachis hypogaea* L.) genotypes. *International Journal of Agricultural Sciences*. 6: 141-143.

Vaghasia, P.M., Jadav, K.V., Jivani, L.L. and Kachhadiya, V.H. (2010b). Impact of water stress at different growth phases of summer groundnut (*Arachis hypogaea* L.) on growth and yield. *Research on Crops*. 11: 693-696.

Vakharia, D.N., Kukadia, A.D. and Parameswaran, M. (2003). Polyamines in response to artificial water stress in groundnut seedlings. *Indian Journal of Plant Physiology*. 8: 383-387

Varshney, R.K., Bertioli, D.J., Moretzsohn, M.C., Vadez, V., Krishnamurthy, L., Aruna, R., Nigam, S.N., Moss, B.J., Seetha, K., Ravi, K., He, G., Knapp, S.J., Hoisington, D.A. (2009) The first SSR-based genetic linkage map for cultivated groundnut (*Arachis hypogaea* L.). *Theoretical Applied Genetics*. **118**: 729-739.

Velu, G. (1998). Effect of water stress on yield potential of groundnut (*Arachis hypogaea*). *Indian Journal of Agricultural Sciences*. 68: 284-286.

Venkateswarlu, B. and Ramesh, K. (1993). Cell membrane stability and biochemical response of cultured cells of groundnut under polyethylene glycol-induced water stress *Plant Science Limerick*. 90: 179-185.

Venkateswarlu, B., Mukhopadhyay, K. and Ramesh, K. (1993). Effect of PEG induced water stress on cell cultures of groundnut in relation to the response of whole plants. *Oleagineux*, 48: 463-468.

Venkateswarlu, O., Reddy, K.R., Reddy, P.V., Vasanthi, R. P., Reddy, K.H.P and Reddy, N.P.E. (2007). Identification of superior donor parents for drought tolerance and yield through combining ability analysis in groundnut. *Legume Research*. 30: 128-132.

Vivekanandan, A. S. and Gunasena, H. P. M. (1976) Lysimetric studies on the effect of soil moisture tension on the growth and yield of maize (*Zea mays* L.) and groundnut (*Arachis hypogaea* L.). *Beitrage zur Tropischen Landwirtschaft und Veterinarmedizin*. 14:269-378

Wang Sheng Yu, Liao BoShou, Lei Yong; Li Dong and Xiao DaRen (2004). Role of in vitro resistance to aflatoxin contamination in reducing pre-harvest aflatoxin contamination under drought stress in groundnut. *International Arachis Newsletter*. 24: 38-39.

Wasternack C. and Hause B. (2002). Jasmonates and octadecanoids: signals in plant stress responses and development. *Prog. Nucleic Acid Research Molecular Biology*. 72: 165-221.

Wasternack C. (2007). Jasmonates: an update on biosynthesis, signal transduction and action in plant stress response, growth and development. *Annals of Botany*. 100: 681-697.

Wheatley, A. R. D., Wightman, J. A., Williams, J. H., and Wheatley, S. J. (1989). The influence of drought stress on the distribution of insects on four groundnut genotypes grown near Hyderabad, India. *Bulletin of Entomological Research*. 79: 567-577.

Williams, J.H., Rao, R.C.N., Matthews, R and Harris, D. (1986). Responses of groundnut genotypes to drought. Agrometeorology of groundnut. Proceedings of an international symposium, ICRISAT Sahelian Center, 21-26 August 1985. 99-106.

Womble, B.A. and Garren, K.H. (1978). Sodium azide offers some promise for control of Cylindrocladium black rot (CBR) of peanuts. *Proceedings of American Peanut Research and Education Association, Inc.*, 10:52.

Worthington, J. W., Schmidt, J. R., and Lamm, F. R. (1995). A microirrigation system for peanut research plots. Microirrigation for a changing world: conserving resources-preserving the environment. Proceedings of Fifth International Microirrigation Congress, Orlando, Florida, USA, 2-6 April 1995. 664-668.

Wright G. C., Rao, R. C. N., and Farquhar, G. D. (1994). Water use efficiency and carbon isotope discrimination in peanut under water deficit conditions. *Crop Science*. 34: 92- 97.

Wright, G.C., Hubick, K.T. and Farquhar, G.D. (1988). Discrimination in carbon isotopes of leaves correlates with water use efficiency of field grown peanut cultivars. *Aust. J. Plant Physiol*. 15: 815-825.

Wright, G.C., Hubick, K.T., Farquhar, G.D. and Nageswara, R.C. (1993). Genetic and environmental variation in transpiration efficiency and its correlation with carbon isotope discrimination and specific leaf area in peanut. *Stable Isotopes And Plant Carbon Water Relations*. 247-267.

Wright, G.C., Rahmianna, A., and Hatfield, P.M. (1988). A comparison of thermocouple psychrometer and pressure chamber measurements of leaf water potential in peanuts. *Exerimental Agriculture*. 24: 355-359.

Wright, G.C., Rao, R.C.N., Basu, M.S., Passioura, J.B., Cooper, M (ed.) and Hammer, G.L. (1996). A physiological approach to the understanding of genotype by environment interactions - a case study on improvement of drought adaptation in groundnut. *Plant Adaptation and Crop Improvement*: 365-381.

Yao ChunMei, Yang PeiLing, Hao ZhongYong, Wang QuanHai, Yao, C.M., Yang, P.L., Hao, Z.Y. and Wang, Q.H. (1999). Dynamic analysis of root/shoot development in groundnut under different water regimes. *Journal of China Agricultural University*. 4: 40-44.

Yao, J.P., Luo, Y.N. and Yang, X.D. (1982). Preliminary report on the effects of drought on seed development and quality of early groundnut. *Chinese Oil Crops Zhongguo Youliao*., 3:50-52.

Zaharah, H. (1986). The effect of water deficit at different growth phases on the performance of groundnut. Miscellaneous Crops Res. Div., MARDI, Serdang, Selangor, Malaysia. *MARDI Research Bulletin*. 14: 126-131.

14

MECHANISM OF SALINITY STRESS TOLERANCE IN CROP PLANTS AND RECENT DEVELOPMENTS

K. Chakraborty[1], A.L. Singh[1], D. Bhaduri[1] and R.K. Sairam[2]

[1]Directorate of Groundnut Research, P.B. No. 5, Junagadh – 362 001, Gujarat, India
[2]Indian Agricultural Research Institute, New Delhi – 110 012, India

INTRODUCTION

The last three decades has witnessed substantial increase in productivity of cereals, pulses, oilseeds and cash crops through adoption of high yielding varieties and intensive agriculture. There has been a shift from cereals to cash crops with development of high yielding varieties and increase in irrigated acreage.

Salts are component of soil, and originate from mineral weathering, inorganic fertilizers, soil amendments (e.g., gypsum, composts and manures), and irrigation waters. But when salts are present in relatively high amounts, plant growth is adversely affected. Soil salinity is a measure of the total amount of soluble salt in soil. As salinity levels increase, plants extract water less easily from soil, aggravating water stress conditions. High soil salinity afflicts about 95 million hectares of land worldwide (Szabolcs, 1994) can also cause nutrient imbalances, result in the accumulation of elements toxic to plants, and reduce water infiltration. In India, soil salinity, spread in almost 8.5 m ha area is the factor limiting plant growth and productivity.

There are two types of salinity, dry land and irrigation salinity. The dry land salinity is classified as either primary or secondary. Primary salting occurs naturally while secondary salting is induced by human activities such as agriculture. Secondary salinity is an insidious problem that may be undetected for years until saline discharge is discovered at the soil surface. Salinity through irrigation resembles dry land salinity, except that groundwater accession is induced through irrigation water rather than rainfall alone. Irrigation salinity refers to an accumulation of salt in the plant root zone or on the soil surface, commonly as a result of saline groundwater rising within two meters of the ground surface. Classification of salt affected soil as per ICAR (ICAR 2011) are given in Table 1 and the extant of area of salt affected soil in India as identified by CSSRI, Karnal are given in Table 2.

Table 1. Classification of salt affected soil (source: Handbook of Agriculture, ICAR, 2011)

Nature of Soil	USDA Classification			SSSA Classification	
	ECe (dS m^{-1})	pH	ESP	ECe (dS m^{-1})	SAR
Normal	< 4.0	< 8.5	< 15	< 2	< 13
Saline	> 4.0	< 8.5	< 15	> 2	< 13
Sodic	Variable	> 8.5	> 15	Variable	> 13
Saline-Sodic	> 4.0	> 8.5	Variable	< 2	> 13

Table 2. Extent and distribution of salt affected soil in India (source: CSSRI, Karnal)

State	Salt affected area (× 1000 Ha)			
	Canal Command	Outside Canal	Coastal	Total
Andhra Pradesh	139	391	283	813
Bihar	224	176	Nil	400
Gujarat	540	327	302	1169
Haryana	455	Nil	Nil	455
Karnataka	51	267	86	404
Kerala	NA	NA	26	26
Madhya Pradesh	220	22	Nil	242
Maharashtra & Goa	446	NA	88	534
Odisha	NA	NA	400	400
Punjab	393	127	Nil	520
Rajasthan	138	984	Nil	1122
Tamil Nadu	257	NA	84	341
Uttar Pradesh	606	689	Nil	1295
West Bengal	Nil	NA	800	800
Total	3469	2983	2069	8521

Soil salinity negatively affects the growth of many crop plants, and the continued salinization of arable land provides an increasing threat to global crop production, especially in irrigated systems (Munns and Tester, 2008). The soil degradation mainly salinity has resulted in 4.0-6.3% crop loss in India (Table 3). Thus, understanding the mechanism and increasing the salinity tolerance of crop plants will provide an important contribution to the maintenance of crop yields. The Na^+ toxicity of many crops is due to over accumulation of Na^+ in the shoot (Munns, 1993, 2002; Tester and Davenport, 2003; Møller and Tester, 2007). Na^+ is taken up from the soil by the root and transported to the shoot in the transpiration stream and shoot Na^+ accumulation is the net result of distinct Na^+ transport processes occurring in different organs and cell types (Tester and Davenport, 2003), and each of these processes contributes to the salinity tolerance of a plant.

Table 3. Impact of soil degradation on Indian Agriculture (Source: The cost of inaction: Valuing the economy-wide cost of environmental degradation in India, World Bank)

Crop	Percent loss
Paddy	2.7 – 4.7%
Wheat	3.9 – 6.4%
Barley	4.5 – 7.0%
Groundnut	2.8 – 4.4%
Gram	5.6 – 7.8%
Rapeseed & mustard	5.8 – 8.5%
Jowar	5.7 – 7.6%
Bajra	6.8 – 8.4%
Cotton	5.3 – 6.9%
Maize	3.2 – 4.9%
Sugarcane	4.5 – 7.9%
All other crops	4.0 – 6.3%
Total	4.0 – 6.3%

The knowledge of physiological and biochemical mechanism of salinity stress tolerance in crop plants is very important as understanding the basic responses of crop plants will help in identifying certain physiological, biochemical and molecular traits for screening of better genotypes tolerant against soil salinity which will help to identify genotypes suitable for cultivation in salt affected areas directly or through conventional breeding and biotechnological means. In this chapter an attempt was made to combine the existing knowledge of salinity tolerance in plants and the future prospect of research in this area.

2. BASIC RESPONSES OF PLANTS TO SALINITY STRESS

High salinity adversely affects germination, growth, physiology and productivity by causing ionic and osmotic stresses and oxidative damage (Iterbe-Ormaetxe *et al.*, 1998). Salt stress is also responsible for an increased respiration rate, ion toxicity (Sudhir and Murthy, 2004), changes in C and N metabolism (Kim *et al.*, 2004), mineral distribution, membrane instability (Marschner, 1986) and permeability (Gupta *et al.*,

2002), decreased biosynthesis of chlorophyll (Khan, 2003) and photosynthetic inefficiency (Munns, 2002), all of which ultimately leads to lowered economic productivity.

Accumulation of sugars and other compatible solutes that can serve as osmoprotectants, stabilizing biomolecules under stress conditions is a common phenomenon. Although use of ions for osmotic adjustment may be energetically more favorable, many plants accumulate organic osmolytes to tolerate osmotic stresses which include proline, betaine, polyols, sugar alcohols, and soluble sugars. Glycine betaine and trehalose act as osmoprotectants by stabilizing quaternary structures of proteins and highly ordered states of membranes. Proline serves as a sink for carbon and nitrogen and a free-radical scavenger, stabilizes sub cellular structures (membranes and proteins) and buffers cellular redox potential. A schematic diagram of the response of plant to salinity stress and possible changes and symptoms are given in Fig. 1.

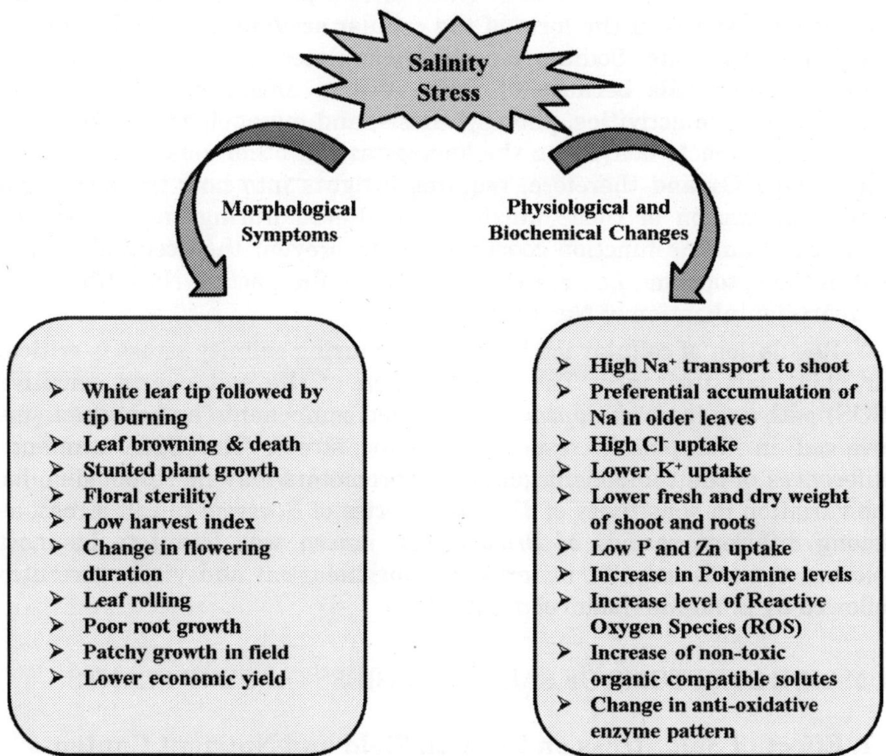

Fig. 1. Effect of soil salinity in crop plants

Salt-stress induces accumulation of reactive oxygen species (ROS), causing oxidative damage to membrane lipids, proteins, and nucleic acids. To combat oxidative stresses plants employ two strategies: i) antioxidants (e.g., ascorbate, glutathione, α-tocopherol, and carotenoids) and ii) detoxifying enzymes, viz., superoxide dismutase, catalase, and enzymes of ascorbate-glutathione cycle. The activity and expression levels of the genes encoding detoxifying enzymes are probably enhanced by ROS under abiotic stresses.

Salt stress has both osmotic (cell dehydration) and toxic (ion accumulation) effects on whole plant and leaf physiology (Flowers, 2004). Salinity reduces the supply of CO_2 to leaves, and further depresses the already low CO_2/O_2 in chloroplasts (Remorini *et al.*, 2009). The consequent accumulation of photoreducing power causes an excess of electrochemical energy in membranes (Zhu, 2001). This extra energy is canalized through the Mehler reaction, which generates ROS such as superoxide anion ($O_2^{\cdot-}$) and hydrogen peroxide (H_2O_2) (Herna´ndez *et al.*, 2001), ultimately provoking oxidative-stress syndrome (Herna´ndez *et al.*, 2000; Baltruschat *et al.*, 2008).

Exposure to higher levels of NaCl affects plant water relations and creates ionic stress in the form of the cellular accumulation of Cl^- and in particular, Na^+ ions. Sodium ions if accumulated in the cytoplasm are toxic to living cells because of their adverse effects on K^+ nutrition, cytosolic enzyme activities, photosynthesis, and metabolism. Besides, salt stress also impacts heavily on the homeostasis of other ions such as Ca^{2+}, Mg^{2+}, and NO_3^- and therefore, requires insights into how transport and compartmentation of these nutrients is altered during salinity stress. Three mechanisms function cooperatively to prevent the accumulation of Na^+ in the cytoplasm, i.e., restriction of Na^+ influx, active Na^+ efflux, and compartmentalization of Na^+ in the vacuole.

Regulation of cellular ion homeostasis during salinity stress is critical for plant salt tolerance. The identification of the salt overly sensitive (SOS) pathway in *Arabidopsis* has revealed components and mechanisms involved in the plant's response to ionic stress. The DNA sequence differences of the component genes and promoters may be responsible for the variation in sensitivity of different species of *Brassica* to salt stress, as among different species of *Brassica; B. juncea* was found to be most tolerant towards salinity in terms of morphological and yield attributes followed by *B. napus* (Islam *et al.*, 2001).

3. MAJOR SYMPTOMS OF SALINITY STRESS ON CROP PLANTS

3.1. Effect of Salt Stress on Growth, Yield and Nutrient Contents

Salinity affects crop production and agricultural sustainability in many regions of the world mainly by reducing the value and productivity of the affected land (Mohammed *et al.*, 1998). Water resources in arid areas are frequently brackish and constitute the only available alternative for crop production (Botella *et al.,* 1993). Soluble salts accumulate in irrigated soils because plants absorb only a small fraction of the minerals dissolved in irrigation water and only pure water evaporates from the soil surface, causing yield reductions (Dirksen, 1985). The species *Brassica napus* and *B. carinata* are moderately salt tolerant and have a good yield potential on the marginal lands, where as *B. juncea* and *B. compestris* are salt sensitive where reduction in stem diameter, number of siliqua per plant, 1000 seed weight and other yield attributing characters is common feature (Akhtar *et al.* 2002).

Gupta and Yadav (1986) reported that groundnut could be grown with water having EC up to 3.0 dS m^{-1}, but our recent study shows that groundnut plant starts facing salinity stress above 2.0 dS m^{-1} and EC above 4.5 dS m^{-1} kills the plants, however, as enough genotypic variation exists, the salinity level in between 3-4 dS m^{-1} during most of the cropping period was ideal for screening for salinity tolerance (Singh *et al.*, 2008). In groundnut the salinity caused accumulation of Na in leaves and to compensate that and maintain proper ratio of various nutrients there was accumulation of Ca and K content. Interestingly, the salinity tolerant cultivars showed comparatively less Na and K accumulation in their leaves than that of sensitive cultivars (Singh *et al.*, 2010a).

Some efforts have been made to study the performance of a few groundnut cultivars by recording germination and studying plant till vegetative phase in pots (Nautiyal *et al.*, 1989; Vadez *et al.*, 2005) and in field (Janila *et al.*, 1999) and very few till maturity in field (Mensah *et al.*, 2006; Nautiyal *et al.*, 2000). Also in a recent attempt screening protocoal using rate of survival under NaCl treatment in glass house (Vadez *et al.*, 2005) as well as in vitro regenerated shoots grown on media (Mungala *et al.*, 2008) has been made as a measure of their tolerance to salinity. Salinity tolerance is a relative term depending mainly upon its intensity and relative performance of cultivars. The groundnut cultivars with high field emergence followed by high plant stand and low mortality under saline conditions could be considered as tolerant of salinity stress. However, data on yielding ability is more vital as increasing salinity decreases pod yield (Singh *et al.*, 2008). Salinity caused severe reduction in plant stature and plant height with small immature pods followed by very high (56-100%) plant mortality at harvest (Singh *et al.*, 2010a).

Assessment of the mineral nutrient status of plants facing various abiotic stresses can be done using biochemical and enzymatic methods involving marker enzymes which are based on the fact that the activity of certain enzymes in deficient tissue is lower or higher (depending on the nutrient) than in normal tissue (Lavon and Goldschmidt, 1999; Vı́llora *et al.*, 2000; Singh and Chaudhari, 1993). Catalase and peroxidase are indicators of the nutritional status of Fe in plants and are also useful as indicators of Fe and Mn deficiencies (Moreno *et al.*, 2000; Valenzuela *et al.*, 1995). Carbonic anhydrase and ribonuclease activity is a good indicator for Zn (Dwivedi and Takkar, 1974; Sharma *et al.*, 1990) and Cu deficiency increased ribonuclease activity in leaves (Agarwala *et al.*, 1985,

1995). Under salinity stress decrease in micronutrient contents are correlated with decrease in activity of related enzymes.

Salinity stress results in ionic and nutritional imbalance due to competition of salt ions with nutrients. Saline condition influences the different steps of nitrogen (N) metabolism, its uptake, reduction and protein synthesis which are responsible for the reduction in plant growth (Frechill *et al.*, 2001) and decreased dry matter production. A negative correlation between concentrations of nitrate (NO_3) and Cl^- was seen in the shoots and roots (Abdelgadir *et al.*, 2005). Salinity reduced potassium (K^+), and calcium (Ca^{2+}) contents, and increased Na^+ and Cl^- content in leaves and stems (El-Hendawy *et al.*, 2005). Saline environment is generally deficient in nitrogen (Amonkar and Karmarkar, 1995) as a result there is reduction in NO_3^- uptake due to high Cl^- in saline condition. Addition of N to salinity, improved the growth and yield of plant, and increased salt tolerance (Dubey and Pessarakli, 1995).

The most common salinity effect is a general stunting of plant growth. As salt concentration increases above a threshold level both the growth rate and size of most plant species progressively decrease but all the plant parts are not affected equally. Shoot growth is often suppressed more that the root growth (Meiri *et al.*, 1970). Often vegetative growth response to salinity is not so reliable guide for predicting seed or grain yield. Yield of rice (Pearson, 1959) and corn (Kaddah and Gohowail, 1964) are greatly reduced without affecting straw yield under saline condition. Growth and yield of bell pepper at different B and salinity levels and the results from the experiments and from published data for wheat, tomato and chickpea indicated an antagonistic relationship for excess B and salinity (Yermiyahu *et al.*, 2008). Thus, toxic effects on growth and yield were less severe for combined B toxicity and salinity than what would be expected if effects of the individual factors were additive. Though the mechanism of relationships between B and salinity in plants is not clear, the possible explanations are reduced uptake of B in the presence of Cl and reduced uptake of Cl in the presence of B (Yermiyahu *et al.*, 2008).

The crop tolerance to high B has been attributed to reduced uptake of B as a result of B efflux from roots. Salinity interacts with B toxicity by a combined effect on B and water uptake and B partitioning within the plant (Wimmer *et al.,* 2001). Salinity usually reduces shoot B concentrations. The whole tissue B concentration is a poor indicator of B tolerance, but the tissue B distribution and subcellular ion compartmentation and B and salinity interactions is important criteria in wheat (Wimmer *et al.,* 2001). Under saline conditions, the total B concentration increased in leaf tips, decreased in roots and was not affected in basal leaf parts, however, soluble B concentrations in basal leaf parts increased in the combined salt/high B treatment compared to high B treatment alone in wheat (Wimmer *et al.,* 2001). Looking to the complex relations between salinity and B toxicity the interactions of soluble B with salinity was studied on physiological response of plants in tomato plants and several hypotheses are established by (Bastías *et al.*, 2010). The increase of aquaporin functionality due to the presence of B and Ca compared with NaCl-treated plants could be the most feasible, whereas there is currently no satisfactory explanation for the results for the cell wall amino acid composition. In addition, the elemental composition results revealed that, in addition the known interactions between B and Ca with respect to cell wall stability, Mg and Mn were also increased in NaCl+B and NaCl+Ca treatments, suggesting their possible involvement in the cell wall function necessary for plant growth (Bastías *et al.*, 2010).

3.2 Effect of Salinity on Pigment System of the Plant

Photosynthesis is the most important process affected in plants under saline conditions. Reduced photosynthesis under salinity is attributed to stomata closure leading to a reduction of intercellular CO_2 concentration and to non-stomatal factors also. There is strong evidence of salinity affecting photosynthetic enzymes, chlorophyll and carotenoid (Stepien and Klobus, 2006). Decrease in Chlorophyll and carotenoid contents of leaves in response to salt stress is a general phenomenon (Parida and Das, 2005). However, an increase in pigment content in *Amaranthus* sp has also been observed (Wang and Nil, 2000). The literature showed wide variations in pigment content depending on salt stress. In quite a few cases, the chlorophyll content was paralleled by changes in the Chl a/b ratio, which is an indicator of the antenna size of PS I and PS II. The core antenna contains only Chl a, whereas the outer antenna contains both Chl a and Chl b. A higher Chl a/b ratio therefore indicates a smaller antenna size and a lower ratio a larger antenna size. Chlorophyll a, b, total chlorophyll and carotenoid content decrease in response to salinity stress (Ahmad, 2009). Salt stress directly or indirectly affects the photosynthetic functions by changing the structural organization and physio-chemical properties of thylakoid membranes (Alia-Mohanti and Saradhi, 1992). Salinity stress reduces quantum yield and Fv/Fm ratio in naked oat (Zhao *et al.*, 2007).

3.3. Relative Water Content and Water Availability

Salinity and water stress have quite similar effects on the growth and cell viability (Lutts *et al.*, 2004). Salinity causes pronounced decrease in water uptake and plant growth in shoot and root (Misra and Dwivedi, 2004). High concentration of salt in the root zone (rhizosphere) reduces soil water potential and the availability of water (Llyod *et al.*, 1989), as a result reduction of the water content dehydration at cellular level and osmotic stress are obvious. The increased amount of Na^+ and Cl^- in the soil-water medium affects the uptake of many indispensable nutrients through competitive interactions and by affecting the ion selectivity of membranes.

Salt stress reduces both RWC and fresh weight in *Brassica* genotypes, but with varied degree of reduction depending upon genotypes

(Siddiqui *et al.*, 2008). The ability of plants to utilize water as well as changes in plant metabolic processes was reduced due to salinity in bean (Munns, 2002). Water potential (Ψ) decreased considerably in the 100 mM sodium chloride and sodium sulphate-treated plants due to salinity induced cellular water loss in *Phaseolus vugaris* (Kaymakanova and Steova, 2008). The decrease in the fresh weight of plants after both salinity and drought-induced water stress have been reported for many species as one of the physiological symptoms of stress (Passioura and Munns, 2000; Sucre and Sua´rez, 2011).

3.4 Electrolyte Leakage and Membrane Stability

Cell-membrane stability, an indicator of the structural integrity, is affected by dehydration and salt stresses (Thomas, 1997). Under saline conditions, plasma-membrane leakage (an indicator of cell plasma-membrane integrity) increases and there is a linear relationship between external salinity and membrane-leakage rate (Orcutt and Nielsen, 2000). Cell-membrane stability using leaf discs in maize, subjected to osmotic stress, showed correlation with the salinity and drought resistance of the plant (Simond and Orcutt, 1988).

Increase in electrolyte leakage in the leaves of *B. juncea* with increasing levels of salinity has been reported by Ahmad *et al.* (2009). High salt depositions in the soil generate a low water potential zone making it increasingly difficult for the plant to acquire both water as well as nutrients (Mahajan and Tuteja, 2005). Thus, salt stress essentially results in a water deficit condition in the plant and takes the form of a physiological drought. Salt stress causes disruption of ionic equilibrium, influx of Na^+, dissipates the membrane potential and facilitates the uptake of Cl^- down the chemical gradient, which is evident from the reduction of membrane stability of salt treated plants. High concentration of Na^+ causes osmotic imbalance, membrane disorganization, , inhibition of cell division and expansion leading to reduction in growth. High Na^+ levels also lead to reduction in photosynthesis and production of reactive oxygen species (Yeo *et al.*, 1998).

3.5 Polyamine Biosynthesis

Polyamines (PAs) are organic poly-cations of a specific group of cell growth and development regulators and are preferentially detected in actively growing tissues and under stress conditions. Plant PA metabolism is extremely sensitive to adverse environmental conditions and hence, PAs are considered as stress markers in plants (Liu *et al.*, 2007; Takahashi and Kakehi, 2010). The function of PAs is presumed to be protective, as they act as free radical scavengers, stabilize cellular membranes, and maintain cellular balance (Bors *et al.*, 1989; Besford *et al.*, 1993). However, the precise role and mode of action of PAs in plant stress have long been a matter of debate. Changes in the levels of PAs under stress cannot be predicted and may be affected by several factors, such as plant species or cultivar, duration of stress treatment, developmental stage of tissues, and stress intensity. In addition, cultivars of the same species but differing in stress sensitivity might also show different changes in the pattern of PAs under stress. In nine rice cultivars that differed in salt sensitivity, salt-tolerant cultivars accumulated high concentrations of Spd and Spm, while the salt-sensitive ones accumulated extensive Put and lower levels of Spd and Spm (Krishnamurthy and Bhagwat, 1989). Stress-tolerant plants generally have a large capacity to enhance PA biosynthesis in response to abiotic stresses, compared with intolerant plants (Kusano *et al.*, 2008; Gill and Tuteja, 2010). In *Lupinus luteus* (a drought tolerant species), Put and Spd were accumulated in leaves in osmotic and salt-stress conditions (Legocka and Kluk, 2005). Zapata *et al.* (2004) studied the effect of salinity on plant growth, ethylene production, and PA level in *Spinacia oleracea*, *Lactuca sativa*, *Cucumis melo*, *Capsicum annum*, *Brassica oleracea*, *Beta vulgaris*, and *Lycopersicon esculentum* (Legocka and Nowicka, 2012).

4. MECHANISM OF SALINITY STRESS TOLERANCE IN CROP PLANTS

4.1 Oxidative Stress and Antioxidant Enzyme Activities

Salt stress leads to stomatal closure, reducing CO_2 availability in the leaves and carbon fixation, exposing chloroplasts to excessive excitation energy, which in turn increases the generation of reactive oxygen species (ROSs) and induce oxidative stress (Parida and Das, 2005; Parvaiz and Satyawati 2008). These ROS have potential to interact with many cellular components, causing damage to membranes and other cellular structures. However, an elaborate and highly efficient network, composed of antioxidant enzymes and antioxidants, is responsible for maintaining the levels of ROS under tight control (Gao *et al.*, 2008).

Salt induced osmotic stress as well as sodium toxicity triggers the formation of reactive oxygen species (ROS) such as superoxide ($O_2^{\bullet -}$), hydrogen peroxide (H_2O_2), hydroxyl radical ($^{\cdot}OH$), and singlet oxygen ($^1O_2$), all of these disrups cellular structures by damaging mitochondria and chloroplasts (Mittler, 2002). Plants have developed a series of enzymatic and non-enzymatic detoxification systems to counteract ROS, and protect cells from oxidative damage (Sairam and Tyagi, 2004). The Antioxidant enzymes such as superoxide dismutase (SOD), catalase (CAT), peroxidase (POX) and glutathione reductase (GR) function in detoxification of super oxide and H_2O_2 (Mittler, 2002; Kholova *et al.*, 2009).

An antioxidant enzyme dependent salinity tolerance mechanism in plants is proposed in Figure 2. The SOD constitutes the primary step of cellular defence it dismutates $O_2^{\cdot -}$ to H_2O_2 and O_2. Further, the accumulation of H_2O_2 is restricted through the action of catalase or by the ascorbate glutathione cycle, where ascorbate peroxidase reduces it to H_2O.

Finally, glutathione reductase catalyzes the NADPH dependent reduction of oxidized glutathione (GSSG) to the reduced glutathione (GSH) (Noctor *et al.*, 2002). The tolerant and susceptible wheat genotypes show differences in antioxidant activity in response to salinity stress (Sairam *et al.*, 2005). Protective roles of the antioxidant enzymes in temperature and salt stress have been reported for a number of plants species (Jaleel *et al.*, 2007). The antioxidants ascorbate and glutathione are involved in scavenging H_2O_2 in conjunction with MDAR and GR, which regenerate ascorbate (Horemans *et al.*, 2000).

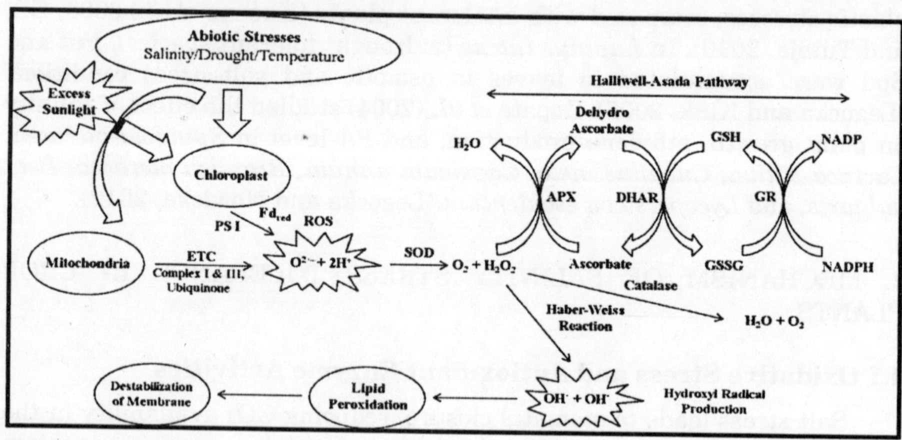

Fig. 2. Antioxidant enzyme dependent salinity tolerance mechanism in plants (modified from Mittler, 2002)

4.2 Osmotic adjustment and differential accumulation of various osmolytes

High salinity causes hyperosmotic stress and ion disequilibrium causing secondary effects (Hasegawa *et al.,* 2000; Zhu, 2001). Plants cope by either avoiding or tolerating salt stress and are either dormant during the salt episode or there is a cellular adjust to tolerate the saline environment. Tolerance mechanisms can be categorized as those that function to minimize osmotic stress or ion disequilibrium or alleviate the consequent secondary effects caused by these stresses. The chemical potential of the saline solution initially establishes a water potential imbalance between the apoplast and symplast that leads to turgor decrease, which, if severe enough, can cause growth reduction (Bohnert *et al.*, 1995). Growth cessation occurs when turgor is reduced below the yield threshold of the cell wall. Cellular dehydration begins when the water potential difference is greater than that can be compensated for by turgor loss (Taiz and Zeiger, 1998).

The cellular response to turgor reduction is osmotic adjustment which is achieved in sub-cellular compartments by accumulation of compatible osmolytes and osmoprotectants (Bohnert *et al.*, 1995; Bohnert and Jensen,

1996). However, Na^+ and Cl^- are energetically efficient osmolytes for osmotic adjustment and are compartmentalized into the vacuole to minimize cytotoxicity (Blumwald *et al.*, 2000; Niu *et al.*, 1995). Since plant cell growth occurs primarily because of directional expansion mediated by an increase in vacuolar volume, compartmentalization of Na^+ and Cl^- facilitates osmotic adjustment that is essential for cellular development. Filek *et al.* (2012) reported that Na^+ may directly or indirectly exert a positive influence on the accumulation of other compounds involved in osmotic adaptation. Such an assumption could explain the greater tolerance to osmotic stress applied in wheat seedlings grown on media containing NaCl.

Salt tolerance mechanism requires osmolytes/compatible solutes accumulation in the cytosol and organelles, where these function as osmotic adjustment and osmoprotection (Rhodes and Hanson, 1993). Some compatible osmolytes are essential ions, such as K^+, but the majority are organic solutes. Compatible solute accumulation as a response to osmotic stress is a ubiquitous process in organisms as diverse as bacteria to plants and animals. However, the solutes that accumulate vary with the organism and even between plant and genotypes. A major category of organic osmotic solutes consists of simple sugars (mainly fructose, glucose and sucrose), sugar alcohols (glycerol and methylated inositols) and complex sugars (trehalose, raffinose and fructans) (Bohnert and Jensen, 1996). Others are quaternary amino acid derivatives (proline, glycine betaine, β-alanine betaine, proline betaine), tertiary amines 1,4,5,6-tetrahydro-2-mehyl-4-carboxyl pyrimidine), and sulfonium compounds (choline sulphate, dimethyl sulfonium propironate) (Nuccio *et al.*, 1999). Many organic osmolytes presumed to be osmoprotectants, as their level of accumulation is insufficient to facilitate osmotic adjustment. Glycine betaine preserves thylakoid and plasma membrane integrity after exposure to saline solutions or to freezing or high temperatures (Rhodes and Hanson, 1993). Many of the osmoprotectants enhance stress tolerance of plants when expressed as transgene products (Bohnert and Jensen, 1996; Zhu, 2001). An adaptive biochemical function of osmo-protectants is the scavenging of ROS that are by products of hyperosmotic and ionic stresses, causing membrane dysfunction and cell death (Bohnert and Jensen, 1996).

The accumulation of proline has been observed due to salinity in several crop species in India. Salt stress stimulated the accumulation of glycine betaine (GB) and increase in levels of malondialdehyde (MDA) content in *Brassica juncea* (Ahmad, 2009). There was an abrupt rise in osmolytes accumulation in wheat cultivars under sailinity treatment (Sairam *et al.*, 2002). Tolerant genotypes of *Brassica* (CS 52 and CS 54) have inbuilt mechanism in the form of greater gene expression and activity of P5CS (pyrrolline-5-carboxylate synthetase), whose product proline provides osmotolerance in the form of retention of moisture (higher

RWC) and MSI, resulting in more yield stability (Chakraborty *et al.*, 2012b).

In plant kingdom the organisms ranging from bacteria to higher plants show a strong correlation between increased cellular proline levels and the capacity to survive both water deficit and salinity. The organic nitrogen may serve as the reserve that can be utilized during recovery from salinity. In *Lathyrus sativus*, a hardy grain legume, which can withstand drought, showed high proline accumulation in leaves and roots under water stress (Tyagi *et al.*, 1999). Though proline is synthesized either from glutamate or from ornithine, glutamate is the primary precursor in osmotically stressed cells. The biosynthetic pathway of proline accumulation consists of two important enzymes, viz. pyrroline carboxylic acid synthetase and pyrroline carboxylic acid reductase and transcripts corresponding to both cDNAs accumulate in response to NaCl treatment. Both these regulatory steps are keys in developing strategies for overproducing proline in selected plant species.

4.3 Expression of Osmolarity Related Genes

A common feature of compatible solutes is that these compounds can accumulate to high levels without disturbing intracellular biochemistry and have the capacity to persevere the activity of enzymes of plants growing in saline environment (Bohnert and Jensen, 1996). These compounds have minimal effect on pH or charge balance of the cytosol or luminal compartments of organelles. The synthesis of compatible osmolytes is often achieved by diversion of basic intermediary metabolites into unique biochemical reactions. Often, stress triggers this metabolic diversion. Higher plants synthesize glycine betaine from choline by two reactions that are catalyzed in sequence by choline monooxygenase (CMO) and betaine aldehyde dehydrogenase (BADH). Under salt stress expression of BADH increases (Rhodes and Hanson, 1993). Pinitol is synthesized from myo-inositol by the sequential catalysis of inositol-O-methyltransferase and ononitol epimerase (Bohnert and Jensen, 1996).

Glycine betaine accumulates in many species of the poaceae and chenopodiaceae (Flowers *et al.*, 2004) but is absent in many crop species such as rice and the model species in plant transformation, *Nicotiana tabacum*. Thus considerable work on engineering the production of glycine betaine in species that do not produce it naturally is in progress. The *E. coli* has a cluster of bet genes (betA: choline dehydrogenase, betB: betaine aldehyde dehydrogenase, bet1: a putative regulatory protein and betT: the choline transport system). Transformation of a cyanobacterium with the bet cluster resulted in glycine betaine synthesis (Nomura *et al.*, 1995).

In higher plants the pathway of glycine betaine synthesis is short and straight forward. In spinach choline monooxygenase (CMO) converts choline to betaine aldehyde and betaine aldehyde dehydrogenase (BADH) converts it to glycine betaine (Burnet *et al.,* 1995). The CMO is an iron-sulphur protein which in its native form is a homodimer (Burnet *et al.*, 1995). In sugarbeet the BADH activity increased 2 to 4-fold in leaves and roots with increase of NaCl from 0 to 500 mM and this increase in BADH activity is correlated with the level of translatable BADH mRNA. Two copies of BADH are reported in the haploid sugar beet genome and analysis of cDNA clones showed small nucleotide sequence differences consistent with the existence of two different BADH alleles (McCue and Hanson, 1992). Many species including tobacco lack both CMO and BADH. Spinach and sugarbeet cDNA sequences encoding BADH were expressed in tobacco, and even without a typical transit peptide BADH was still targeted to the chloroplast in the leaves of transgenic plants. Expressed levels and substrate affinity was comparable with the native enzyme and transgenic plants were able to synthesize glycine betaine from supplied betaine aldehyde showing a constitutive ability to transport betaine aldehyde into the chloroplast. The glycine betaine so synthesized was not metabolized further and accumulated to concentrations similar to those plants that accumulate it naturally. There is no information yet on its sub-cellular localization in transgenic plants. Betaine aldehyde is toxic when supplied exogenously in tobacco unless it was transgenic (Rathinasabapathi *et al.*, 1994). Accumulation of BADH transcripts was a common response to osmotic stress. There was 8-fold increase in BADH mRNA levels in leaves of barley under high-salt conditions, and these levels were maintained under sustained stress, but decreased when stress was removed (Ishitani *et al.*, 1995).

There are several other osmolytes than glycine betaine and one of them is trehalose, which play more than one role (Rhodes and Hanson, 1993; Yancey, 1994). TPS1 from *Saccharomyces cerevisiae* encodes trehalose–6-phosphate synthase, which is regarded as both a metabolic enzyme and a regulator (Serrano *et al.*, 1999). TPS1 has been reported to modulate the heat-shock response as well (Hazell *et al.*, 1995). Serrano *et al.* (1999) suggested that the products encoded by TPS1 may have a generalized role in activating stress-defence systems. Tobacco, transgenic for trehalose synthesis, was reported to have improved drought and salt tolerance (Romero *et al.*, 1997), but there was also a changed carbohydrate profile suggesting changes in basic biochemical pathways and the resultant average tissue concentration of trehalose (<0.5 mM) appeared too low to be performing a conventional osmoprotectant role. The transgenic plants for trehalose biosynthesis also showed linked severe morphological alterations including stunted growth.

Proline is another important metabolite which increases under salinity stress. The intermediates of proline biosynthesis and catabolism, such as glutamine and Δ-1-pyrroline-5-carboxylic acid could increase the expression of several osmotically regulated genes in rice (Iyer and Caplan, 1998). There is also evidence that degradation of proline in the mitochondria is directly coupled to respiratory electron transport system and ATP production. A pyrroline-5-carboxylate synthetase (P5CS) cDNA

from moth-bean was introduced into rice, where expression of this P5CS transgene under the control of a stress inducible promoter led to stress-induced over-production of the P5CS enzyme and proline accumulation in transgenic rice plants. Second generation (RI) transgenic plants showed an increase in biomass under salt and water stress conditions (Zhu, 1998).

4.4 Relative Distribution of K, Ca and Na in Different Plant Parts and Na Exclusion

The fundamental basis of the adaptation of plants to salinity stress is the control of transport of salt across membranes (Hasegawa *et al.*, 2000). In glycophytes, salt exclusion is the predominant strategy of adaptation to saline substrates, i.e., tolerance to salinity depends mainly on the ability of roots to limit transport of sodium (Na^+) to the leaves and shoot (Yahya, 1998). In sorghum (*Sorghum bicolor*), wheat (*Triticum aestivum*), barley (*Hordeum vulgare* L.) and corn (*Zea mays*) which are glycophytic in nature and where (Alberico and Cramer, 1993) Na^+ exclusion mechanism is present for salt tolerance.

Also, within the plant, salt are restricted from reaching sensitive organs (Watson *et al.*, 2001). Internal exclusion mechanisms can involve processes such as sequestering salt ions in specialized tissues by removing them from the transport stream (Blom-Zandstra *et al.*, 1998; Jeschke, 1981) and/or by effectively compartmentalizing Na within vacuoles in the stem and leaf (Leigh and Storey, 1993). In other word, if Na^+ ions are not strongly discriminated against at the root membrane, salinity of the xylem stream will increase under saline conditions. Some of the ions in the transpiration stream could be selectively accumulated by parenchyma cells in the xylem and then retranslocated back to the root via phloem (Orcutt and Nielsen, 2000). The K^+ content and K^+-Na^+ selectivity decreased and electrolyte leakage was increased in different cultivars of *Brassica napus* under salt stress. Sodium import, transport and deposition were increased by salinity stress but remobilization was decreased. The K^+ and Mg^{2+} import, deposition, and remobilization were also decreased (Rezaei *et al.*, 2006).

Under saline condition salt tolerance is usually related to ability to regulate Na^+ and Cl^- uptake by plant root and their subsequent translocation to the shoot. In contrast, the higher concentration of essential elements (particularly K^+ and Ca^{2+}) in leaf tissues may contribute to the salt tolerance ability of plants (Ashraf and McNeilly, 1990; He and Cramer, 1993). The mechanism of selectivity of ion transport appears to confer tolerance to salt stress in Indian mustard. Higher selectivity towards K^+ and restricted uptake of Na^+ was observed in tolerant genotypes of Indian mustard (Kumar, 1984; Garg *et al.*, 1997).

4.5 Expression of Salt Overly Sensitive (SOS) Pathway Genes

Zhu and his co-workers identified three genetically linked *Arabidopsis* loci (*SOS1, SOS2* and *SOS3*), which are components of a

stress signalling pathway that control ion homeostasis and salt tolerance (Hasegawa *et al*,. 2000; Sanders, 2000; Zhu, 2001). Genetic analysis of Na^+/Li^+ sensitivity established that *sos1* is epistatic to *sos2* and *sos3*. These *sos* mutants also exhibit a K^+ deficient phenotype in medium supplemented with µM $[K^+]_{ext}$ and $[Ca^{2+}]_{ext}$. Na^+ and K^+ deficiency of *sos2* and *sos3* is suppressed with mM $[Ca^{2+}]_{ext}$ (Zhu *et al.*, 1998). The *sos1* exhibits hyperosmotic sensitivity unlike *sos3* and *sos2*. Thus, the SOS signaling pathway regulates Na^+ and K^+ homeostasis and is Ca^{2+} activated. SOS3 encodes a Ca^{2+} binding protein with sequence similarity to the regulatory B subunit of calcineurin (protein phosphatase 2B) and neuronal Ca^{2+} sensors (Ishitani *et al.*, 2000; Liu and Zhu, 1998). Interaction of SOS3 with the SOS2 kinase (Liu *et al.*, 2000) and SOS2 activation is Ca^{2+} dependent (Halfter *et al.*, 2000). The *in planta* function of SOS3 as a salt tolerance determinant is dependent on Ca^{2+} binding and N-myristoylation (Ishitani *et al.*, 2000). After Zhu (*et al.*, 2001) we Propose a SOS Pathway for Na^+ homeostasis in plants as in Fig. 3.

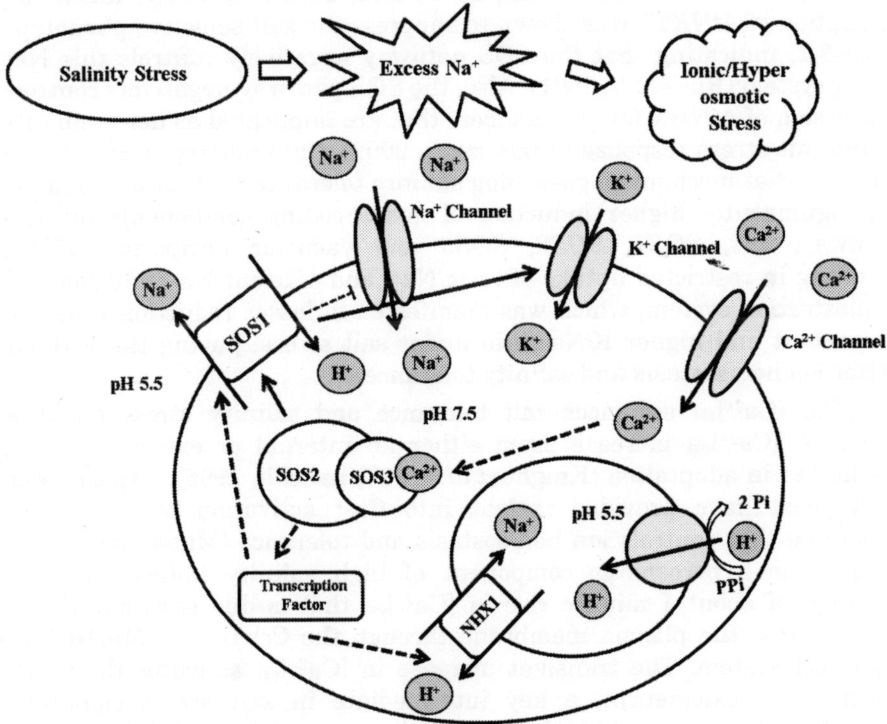

Fig. 3. Proposed Salt Overly Sensitive (SOS) Pathway for Na^+ homeostasis in plants (Modified from Zhu *et al.*, 2001)

The SOS2 serine/threonine kinase (446 amino acids) has a 267 amino acid N-terminal catalytic domain that is similar in sequence to yeast SNF1 (sucrose non-fermenting) kinase and the mammalian AMPK (AMP-activated protein kinase) (Liu *et al.*, 2000; Zhu, 2001). The kinase activity of SOS2 is essential for its salt tolerance determinant function (Zhu, 2001). The SOS2 C-terminal regulatory domain interacts with the kinase domain to cause autoinhibition. A 21 amino acid motif in the regulatory domain of SOS2 is the site where SOS3 interacts with the kinase and is the autoinhibitory domain of the kinase (Guo *et al.*, 2001). Binding of SOS3 to this motif blocks autoinhibition of SOS2 kinase activity. Deletion of the autoinhibitory domain results in constitutive SOS2 activation, independent of SOS3. Also, a Thr_{168} to Asp mutation in the activation loop of the kinase domain constitutively activates SOS2.

Genetic and biochemical evidences indicate that components of the SOS signal pathway function in the hierarchical sequence (Hasegawa *et al.*, 2000; Sanders, 2000; Zhu, 2001). The Ca^{2+} binds to SOS3 which leads to interaction with SOS2 and activation of the kinase. Among the SOS signal pathway outputs are transport systems that facilitate ion homeostasis. The plasma membrane sited Na^+/H^+ antiporter SOS1 is controlled by the SOS pathway at the transcriptional and post-transcriptional level (Guo *et al.*, 2001; Zhu, 2001). Recently, functional disruption of *AtHKT1* was shown to suppress the salt sensitive phenotype of *sos3-1*, indicating that the SOS pathway negatively controls this Na^+ influx system (Rus *et al.*, 2001). Also, the SOS pathway negatively controls expression of *AtNHX* family members that are implicated as determinants in the salt stress response (Yokoi *et al.*, 2002). Chakraborty *et al.* (2012a) reported that mechanism governing salinity tolerance in *Brassica* could be the significantly higher induction of genes coding components of SOS pathway, viz., *SOS1*, *SOS2*, *SOS3* and vacuolar antiporter *NHX1*, resulting in restricted uptake of toxic Na^+ ,and efficient Na-exclusion and sequestration system, which was manifested in lesser reduction in tissue K^+ content and higher K/Na ratio under salt stress, paving the way for better ion homeostasis and salinity tolerance.

The $[Ca^{2+}]_{ext}$ enhances salt tolerance and salinity stress elicits a transient $[Ca^{2+}]_{ext}$ increase, from either an internal or external source, implicated in adaptation (Knight *et al.*, 1997, Läuchli, 1996). Experiments with yeast have provided insight into Ca^{2+} activation of salt stress signalling that controls ion homeostasis and tolerance (Matsumoto *et al.*, 2001). The hyperosmotic component of high salinity induces a short duration of about 1 minute rise in $[Ca^{2+}]_{ext}$ that is due substantially to influx across the plasma membrane through the Cch1p and Mid1p Ca^{2+} transport system. The transient increase in $[Ca^{2+}]_{ext}$ activates the PP2B phosphatase calcineurin; a key intermediate in salt stress signalling controlling ion homeostasis, leading to the transcription of *ENA1*, which encodes the P-type ATPase primarily responsible for Na^+ efflux across the plasma membrane (Nakamura *et al.*, 1993; Mendoza *et al.*, 1994; Matsumoto *et al.*, 2001). The model proposes that the hyperosmotically-induced localized $[Ca^{2+}]_{ext}$ transient activates calmodulin that is tethered to Cch1p-Midp (Ehlers and Augustine, 1999; Sanders, 2000; Matsumoto *et al.*,2001). Calmodulin in turn activates signalling through the calcineurin pathway, which mediates ion homeostasis and salt tolerance (Matsumoto *et al.*, 2001).

Thus a paradigm for salt-induced Ca^{2+} signalling and the activation of the SOS pathway can be suggested. Components of the SOS pathway, either SOS3 or upstream elements, might be associated with an osmotically responsive channel through which Ca^{2+} influx could initiate signalling through the pathway. It is notable that a new elevated $[Ca^{2+}]_{ext}$ steady state is established in yeast cells, that are maintained in medium supplemented with NaCl, after the hyperosmotic induction of the short duration $[Ca^{2+}]_{ext}$ transient (Matsumoto *et al.*, 2001). The newly established $[Ca^{2+}]_{cyt}$ is likely to contribute to cellular capacity for growth in salinity. The vacuolar membrane H^+/Ca^{2+} antiporter Vcx1p and endomembrane localized Ca^{2+}-ATPases are pivotal effectors that regulate the amplitude and duration of the $[Ca^{2+}]_{ext}$ transient (Miseta *et al.*, 1999). The $[Ca^{2+}]_{ext}$ steady state established in salt containing medium presumably also involves coordination of channel activation that facilitates influx from external and internal sources and energy dependent transport systems that compartmentalize the divalent cation. It is reasonable to assume that the salt induced $[Ca^{2+}]_{ext}$ transient detected in plant cells (Knight, 1997) and, perhaps, a new $[Ca^{2+}]_{ext}$ steady-state are controlled by the ECA and ACA Ca^{2+}-ATPases and CAX1 and 2 transporters which are orthologs of Vcx1p (Sze *et al.*, 2000). Nevertheless, Ca^{2+} has at least two roles in salt tolerance, a pivotal signaling function in the salt stress response leading to adaptation and a direct inhibitory effect on a Na^+ entry system.

5. SALT TOLERANT CROP PLANTS

The restriction of uptake and transport and internal tolerance mechanisms are the two important criteria which plants employ to combat high external concentrations and hence tolerance could be attributed to the lower the content and lower uptake or accumulation of these in the root and shoot and high yield in toxic soils (Singh *et al.*, 2010b). Ameliorating high-mineral soils using soil amendments is expensive and extremely difficult. Use of tolerant crop genotypes, phytoremediation by tolerant crops, and inoculations of beneficial microorganisms are the solutions (Singh *et al.*, 2010b). The salt tolerance of a crop can best be described by plotting its relative yield as a continuous function of soil salinity (Fig 4). For most crops, this response function follows a sigmoidal relationship however, several crops may die before the seed or fruit yields decrease to zero, thus eliminating the bottom part of the sigmoidal curve (Maas and Hoffman, 1977). By plotting relative yield with increasing soil salinity level a threshold level of soil salinity is obtained for some of the crops which have its greatest value in providing general salt tolerance guidelines for crop management decisions. Farmers need to know the soil salinity levels that begin to reduce yield and how much yield will be reduced at levels above the threshold. However, more precise plant response functions would be advantageous for crop simulation modelling (van Genuchten and Hoffman, 1984). From the experimental data in general the plants were classified in four categories according to their response to salinity (Bernstein, 1962).

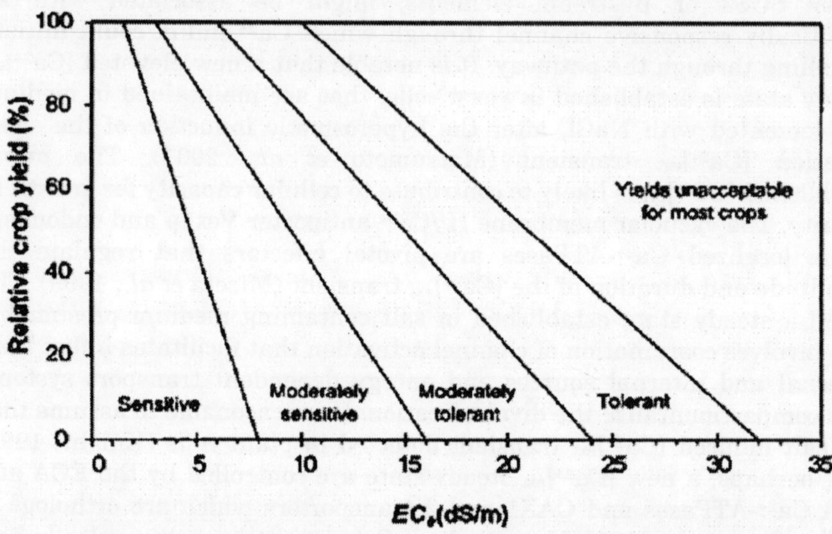

Fig. 4. Classification of the crop based on soil salinity level (source: FAO Report)

Accordingly, threshold salinity level which can cause damage, level of loss with further increase in salinity and some of the recommended crop cultivars suitable for growing in salt affected areas is listed for the important crops cultivated in India are listed in Table 4. The last column provides a qualitative salt tolerance rating that is useful in categorizing crops in general terms.

Table 4. List of field crops with their salinity tolerance level and suitable cultivars for cultivation in salt affected areas (source: FAO, ICAR, CSSRI reports)

Crop		Salinity tolerance level		Tolerant Cultivars	Over all nature of the crop*
Common Name	Botanical Name	Threshold level (dS m^{-1})	% loss per dSm^{-1} rise		
Rice	*Oryza sativa* L.	3.0	12.0	Lunishree, Sonamani, Pokkali, Nona Bokra, IR 8, Damodar, PVR 1, Basmati HBC19, Basmati 370, CSR 11, CSR 13, CSR 27, IR 28	S
Wheat	*Triticum aestivum* L.	6.0	7.1	KRL 1-4, KRL 19, Kharchia, KRL 210, KRL 213, KRL 99, KRL 35	MT

Indian Mustard	*Brassica juncea* L.	--	--	CS 52, CS 54, CS 56, Urvashi	MS
Barley	*Hordeum vulgare* L.	6.0	7.1	BH 924, RD 2508, Ratna, R 56	MT
Chick pea	*Cicer arietinum* L.	--	--	Pusa 256, Pusa 329, Pusa362, BGD 87, CSG 9651, BG 267, Karnal Chana 1, CSG 88101	MS
Soybean	*Glycine max* (L.) Merrill	5.0	20.0	CoSoy 2, DS 40, PalamSoy, Pusa 16	MT
Cotton	*Gossypium hirsutum* L.	7.7	5.2	Arya-Anubam, RAHS 14, Dhumad, Jayadhar, A 82-1	T
Sesame	*Sesamum indicum* L.	--	--	RT 54, RT 46, RT 127	S
Oat	*Avena sativa* L.	--	--	JHO 815, JHO 802, JHO 816, UPO 201	T
Groundnut	*Arachis hypogaea* L.	3.2	29.0	GG 4, MH 2, ICGV 86590, ICGS 44, Gangapuri, TG 37A, Kopergaon 3, VRI 4	MS
Sorghum	*Sorghum bicolor* (L.) Moench	6.8	16.0	JJ 1041, ICSB 707, CSV 15, S 35	MT
Sugarcane	*Saccharum officinarum* L.	1.7	5.9	Neeraj (Co 99006), Damodar (Co 99004), CoJ 20193	MS
Pigeon pea	*Cajanus cajan* (L.) Huth	--	--	H06-12, Pusa 991, JBP 110-B, ICPL 2037, ICPB 2039	S

* T: Tolerant, MT: Moderately Tolerant, S: Susceptible, MS: Moderately Susceptible

6. CONCLUSION AND FUTURE RESEARCH STRATEGIES

Soil salinity adversely affects plant growth and development accompanied by an increase in uptake of Na^+ and Cl^- ions and a decrease in uptake of K^+, Ca^{2+}, Mg^{2+} resulting in ionic imbalance, sodium ion injury and disturbed metabolic processes, changed concentration of biomolecules, photosynthetic activity and poor productivity. Although, the wild species are very much inferior in terms of yield and other agronomic characters but when it is to face the adverse environment, they show much more competence compared to cultivated agronomically superior genotypes. It is found that external application of potassium helps to maintain the K/Na ratio inside the plant tissue for smooth functioning of the cell. Sometimes, foliar application of certain plant growth regulators like salicylic acid, abscisic acid and polyamine viz. spermine, spermidine, putrescine, cadaverine etc. seems to ameliorate the adverse effect of salt stress in plants.

Tolerance capacity of the plants vastly depends up on its mechanism to overcome the salt stress through osmotic adjustment by accumulating compatible solutes inside the cell and thereby reducing the osmotic potential of the cell. This adaptation strategy helps the plants to still take up water from rhizosphere zone, when the water potential there goes well below normal level due to soil water salinity. Other most detrimental effect faced by the plants is sudden outburst of reactive oxygen species produced due to salinity stress, which disrupts the cellular structure and damages subcellular organelles, leading to cell death. The genotypes having better antioxidant defence capacity are more capable of combating the stress.

At molecular level, efficient operation of different signal proteins and various symporters and antiportes lying either in the plasma-membrane or tonoplast play important role in salinity tolerance. Activity of different Na^+/H^+ antiporters viz. SOS1 and NHX1 depends up on the activation of other signal proteins like SOS2, SOS3 and other calcium binding proteins. Apart from these, several transcription factors like *WRKY*, *DREB*, *CBF* also contribute towards salinity tolerance in plants. The physiological, biochemical and molecular biological knowledge on salinity tolerance in crop plants are important for understanding the responses and identification of the parameters for developing screening methodologies for all crop plants.

Incredible advances have been made over the last four decades to understand the response of plants towards soil and water salinity and their management, and most of the works are focused on revealing the inherent capacity of the plants to combat the ill effect of salt stress. However, only very few works has brought finished product either in the form of genetically improved varieties through biotechnological or conventional crop improvement means or in the form of some good management practices. This need consolidated effort to come out with answer to the problem of soil salinity.

Optimistically, discussing the avenues for future research strategies following approaches may be useful towards proving salinity tolerance in crop plants.

- Development of suitable methodologies for screening for salinity tolerance in all the crops and then its subsequent recommendations in saline areas.
- Exploration of all the available natural resources needs to be carried out to manage salinity in field.
- Use of known tolerant varieties/genotypes and their further improvement for yield and tolerance level.
- Mining of the differentially expressed genes from the wild relatives of the cultivated species, and subsequent transfer of those to cultivable species either through molecular or conventional plant breeding could be a possible way.

To characterise these wild genotypes for their tolerance behaviour to single or multiple stresses and search for the candidate genes responsible for the tolerance. The major setback faced by the breeders in this approach is that most of wild tolerant genotypes are often cross-incompatible with the cultivated species where there is a need to go for more precise method of gene transfer using biotechnological tools.

Search for novel salt tolerant genes or protein should be done not only in the related plant and crop species, rather sources of tolerance can also be explored in some known halophytic plants like mangroves and others which can thrive well under extremely saline environment, as nowadays it is possible to transfer target gene(s) from any organism to other. However, such type of work has been initiated by some of the reputed laboratories in India and abroad with a mission to successfully transfer the important genes imparting tolerance to soil salinity from mangrove gene pool to some of important crop plants like rice and others. Besides this, mining of the genes are also possible from wide range of microbial gene pool as well as that of *Archea*. Apart from the commonly known pathways that impart tolerance to eukaryotes including higher plants, there may be some other operating mechanism in these organisms which make them able to survive in the salty environment of sea or saline hot spring.

Few preliminary works suggest that by altering the promoter region of the genes it is possible to fetch tolerance towards various abiotic stresses. As these promoter elements work upstream to many of the genes involved in same or different pathways, so by up-regulating their function one can bring tolerance to susceptible genotypes. The proteins like dehydration-responsive element-binding proteins (DREBs), or C-repeat-binding proteins (CBFs), are among the first families of transcriptional regulators that are transcriptionally up-regulated by salinity, water deficit or low temperature stress. Also, the *WRKY* gene family has been suggested to play important roles in the regulation of transcriptional reprogramming associated with plant stress responses. Modification of the expression patterns of *WRKY* genes and/or changes in their activity contribute to the elaboration of various signalling pathways and regulatory networks.

The genotypes having superior antioxidant defence capacity in terms of either accumulation of antioxidants like ascorbic acid, glutathione, malonaldehyde etc. or higher activity of the enzymes viz. superoxide dismutase, ascorbate peroxidase, catalase, glutathione reductase, peroxidase etc. are more capable of withstanding salinity stress. Salinity stress cause osmotic and oxidative stress, hence genetic modifications in these areas could yield beneficial result in bringing salinity tolerance in crop plants. Incorporation of genes facilitating biosynthesis of compatible solutes whose accumulation will help in osmotic adjustment in the plant cell thereby maintaining better water balance inside the plant tissues when it is facing osmotic pressure from outside.

Though there is ample opportunity for research in this area, it needs multidisciplinary approaches to address all the component of the problem of salinity. Although, spectacular results leading to development of salt tolerant varieties capable enough to give comparable yield like agronomical superior varieties is unlikely. Integrated research programme soil scientists, plant breeders, agronomists, physiology and molecular biologists can yield some future product to solve the salinity problem.

REFERENCES

Abdelgadir, E; Ka, MA and Fujiyama, H 2005. Nitrogen nutrition of rice plants under salinity. *Biol. Plant.*, 49: 99–104.

Agarwala, SC; Chatterjee, C; Sharma, CP and Nautiyal, N 1985. Copper nutrition of sugarbeet. *J. Exp. Bot.*, 36: 881–888.

Ahmad, P 2009. Growth and antioxidant responses in mustard (Brassica juncea L.) plants subjected to combined effect of gibberellic acid and salinity. *Achives Agron. Soil Sc.*, (*iFirst* Article). 1-14.

Akhtar, J; Tanveer-ul-Haq; Saqib, M and Mahmood, K 2002. Effect of salinity on yield, growth and oil contents of four *Brassica* species. *Pak. J. Agri Sci.*, 39: 76-79.

Alberico, GL and Cramer, GR 1993. Is the salt tolerance of maize related to sodium exclusion? I. Preliminary screening of seven cultivars. *J. Plant Nutri.*, 16: 1289–1303.

Alia-Mohanty, P and Saradhi, PP 1992. Effect of sodium chloride on primary photochemical activities in cotyledonary leaves of *Brassica juncea*. *Biochem. Physiol.*, 188: 1-12.

Amonkar, DV and Karmarkar SM 1995. Nitrogen uptake and assimilation in halophytes. In: Nitrogen nutrition in higher plants, eds. H. S. Srivastava, and R. P. Singh. New Delhi, India: Associated Publishers Company. pp. 431–445.

Ashraf, M and McNeilly, T 1990. Responses of four *Brassica* species to sodium chloride. *Environ. Exp. Bot.*, 30: 475–487.

Baltruschat, H; Fodor, J; Harrach, BD; Niemczyk, E; Barna, B and Gullner, G 2008. Salt tolerance of barley induced by the root endophyte Piriformospora indica is associated with a strong increase in antioxidants. *New Phytol.*, 180: 501–10.

Bernstein, L 1962. Salt-affected soils and plants. In: Proc. of the 18th UNESCO Symposium on Problems of the Arid Zones. Paris. pp. 139-174.

Besford, RT; Richardson, CM; Campos, JL and Tiburcio, AF 1993. Effect of polyamines on stabilization of molecular complexes in thylakoid membranes of osmotically stressed oat leaves. *Planta,* 189: 201–206.

Blom-Zandstra, M; Vogelzang, SA and Veen, BW 1998. Sodium fluxes in sweet pepper exposed to varying sodium concentrations. *J. Exp. Bot.*, 49: 1863–1868.

Blumwald, E.; Aharon, GS and Apse, MP 2000. Sodium transport in plant cells. *Biochemica et Biophysica Acta,* 1465: 140-151.

Bohnert, HJ and Jensen, RG 1996. Strategies for engineering water stress tolerance in plants. *Trends Biotech.,* 14: 89-97.

Bohnert, HJ; Nelson, DE and Jensen, RG 1995. Adaptations to environmental stresses. *Plant Cell,* 7: 1099-1111.

Bors, W; Langebartels, C; Michel, C and Sandermann, JH 1989. Polyamines as radical scavengers and protectants against ozone damage. *Phytochemistry,* 28: 1589–1595.

Botella, MA; Cerda, AC and Lips, SH 1993. Dry matter production, yield, and allocation of carbon-14 assimilates by wheat as affected by Nitrogen sources and salinity. *Agron. J.*, 85: 1044-1049.

Burnet, M; LaFontaine, PJ and Hanson, AD 1995. Assay, purification, and partial characterisation of choline monooxygenase from spinach. *Plant Physiol.* 108: 581–588.

Chakraborty, K; Sairam RK and Bhattacharya, RC 2012b. Salinity induced expression of pyrrolline-5-carboxylate synthetase determine salinity tolerance in Brassica spp. *Acta Physiol. Plant.*, 34: 1935-41.

Chakraborty, K; Sairam, RK and Bhattacharya, RC 2012a. Differential expression of salt overly sensitive pathway genes determines salinity stress tolerance in Brassica genotypes. *Plant Physiol. Biochem.,* 51: 90-101.

Dirksen, C 1985. Relationship between root uptake weighted mean soil salinity and total leaf water potentials of alfalfa. *Irrig. Sci.*, 6: 39-50.

Dubey, RS and Pessarakli, M 1995. Physiological mechanisms of nitrogen absorption and assimilation in plants under stressful conditions. In: Handbook of Plant and Crop Physiology, eds. M. Pessarakli New York: Marcel Dekker, pp. 605–625.

Dwivedi, RS and Takkar, PN 1974. Ribonuclease activity as an index of hidden hunger of Zn in crops. *Plant Soil,* 40: 173–181.

Ehlers, MD and Augustine GJ 1999. Cell signaling. Calmodulin at the channel gate. *Nature,* 399: 105-107.

El-Hendawy, SE; Hu, Y and Schmidhalter, U 2005. Growth, ion content, gas exchange, and water relations of wheat genotypes differing in salt tolerances. *Aust. J. Agril. Res.,* 56: 123–134.

Filek, M; Walas, S; Mrowiec, H; Rudolphy-Sko´rska, E; Sieprawska, A and Biesaga-Kos cielniak J 2012. Membrane permeability and micro- and macro element accumulation in spring wheat cultivars during the short-term effect of salinity- and PEG-induced water stress. *Acta Physiol Plant.,* 34: 985–995.

Flowers, TJ 2004. Improving crop salt tolerance. *J.Exp.Bot.,* 55: 307–319.

Frechill, S; Lassa, B; Ibarretxe, L; Lamsfus, C and Aparicio Trejo, P 2001. Pea responses to saline stress in affected by the source of nitrogen nutrition (ammonium or nitrate). *Plant Growth Regul.,* 35: 171–179.

Gao, S; Ouyang, C; Wang, S; Xu, Y; Tang, L and Chen, F 2008. Effects of salt stress on growth, antioxidant enzyme and phenylalanine ammonia lyase activities in *Jatropha curcas* L. seedlings. *Plant Soil Environ.,* 54: 374–381.

Garg, BK; Kathju, S; Vyas, SP and Lahiri, AN 1997. Genotypic difference to salt stress in Indian mustard. In: Proc. Symposium on Recent Advances in Management of Arid Ecosystems, held at Central Arid Zone Research Institute, Jodhpur, 3–5 March. Rajasthan, India: Arid Zone Research Association of India.

Gill, SS and Tuteja, N 2010. Polyamines and abiotic stress tolerance in plants. *Plant Signal Behav.,* 5: 26–33.

Guo, Y; Halfter, U; Ishitani, M and Zhu, JK 2001. Molecular characterization of functional domains in the protein kinase SOS2 that is required for plant salt tolerance. *Plant Cell,* 13: 1383-400.

Gupta, IC and Yadav, JSP 1986. Crop tolerance to saline irrigation waters. *J. Ind. Soc. Soil Sci.,* 34: 379-386.

Gupta, NK; Meena, SK; Gupta, S. and Khandelwal, SK 2002. Gas exchange, membrane permeability, and ion uptake in two Species of indian jujube differing in salt tolerance. *Photosyntetica,* 40: 535-539.

Hasegawa, PM; Bressan, RA; Zhu, JK and Bohnert, HJ 2000. Plant cellular and molecular responses to high salinity. *Ann. Rev. Plant Physiol. Plant Mol. Biol.,* 51: 463–499.

Hazell, BW; Nevalainen, H and Attfield, PV 1995. Evidence that *Saccharomyces cerevisiae* CIF1 (GGS1/TPS1) gene modulates heat shock response positively. *FEBS Letters,* 377: 457–460.

He, T and Cramer, GR 1993. Growth and ion accumulation of the two rapid-cycling *Brassica* species differing in salt tolerance. *Plant Soil,* 153: 19–31.

Hernandez, JA; Ferrer, MA; Jimenez, A; Barcelo, AR and Sevilla F 2001. Antioxidant systems and O^{2-}/H_2O_2 production in the apoplast of pea leaves. Its relation with salt-induced necrotic lesions in minor veins. *Plant Physiol.,* 127: 817–831.

Hernandez, JA; Jimenez, A; Mullineaux, P and Sevilla, F 2000. Tolerance of pea (*Pisum sativum* L.) to long term salt stress is associated with induction of antioxidant defences. *Plant Cell Environ.,* 23: 853–862.

Horemans, N; Foyer CH; Potters, G and Asard, H 2000. Ascorbate function and associated transport systems in plants. *Plant Physiol. Biochem.,* 38: 531–540.

Ishitani, M; Liu, J; Halfter, U; Kim, CS; Shi, W and Zhu, JK 2000. SOS3 function in plant salt tolerance requires Nmyristoylation and calcium binding. *Plant Cell,* 12: 1667-1677.

Ishitani, M; Nakamura, T; Han, SY and Takabe, T 1995. Expression of the betaine aldehyde dehydrogenase gene in barley in response to osmotic stress and abscisic acid. *Plant Mol. Biol.,* 27: 307–315.

Islam, MR; Bhuiyan, AR; Prasad, B; Rashid, MH and Quddus, MA 2001. Salinity effect on yield and component characters in Rapeseed and Mustard varieties. *J. Biol. Sci.,* 1: 840-842.

Iterbe-Ormaetxe, I; Escuredo, PR; Arrese-Igor, C and Becana, M 1998. Oxidative damage in pea plants exposed to water deficit of paraquat. *Plant Physiol.,* 161: 173-181.

Iyer, S and Caplan, A 1998. Products of proline catabolism can induce osmotically regulated genes in rice. *Plant Physiol.,* 116: 203–211.

Jaleel, CA; Gopi, R; Manivannan, P and Panneerselvam, R 2007. Antioxidative potentials as a protective mechanism in *Catharanthus roseus* (L.) G.Don. plants under Salinity Stress. *Turk. J. Bot.,* 31: 245-251.

Janila, P; Rao, TN and Kumar, AA 1999. Germination and early seedling growth of groundnut (*Arachis hypogaea* L.) varieties under salt stress. *Annals Agric. Res.,* 20: 180–182.

Jeschke, WD and Nassery, H 1981. K^+-Na^+ selectivity in roots of Triticum, Helianthus and Allium. *Physiol. Plant.*, 52: 217-224.

Kaddah, MT and Ghowail, SI 1964. Salinity effects on the growth of corn at different stages of development. *Agron. J.*, 64: 214-217.

Kaymakanova, M and Steova, N 2008. Effect of salt stress on growth and photosynthesis rate of bean plants (*Phaseolus vulgaris* L.). *J. Cen. Europ. Agri.*, 9: 385-392.

Khan, NA 2003. NaCl inhibited chlorophyll synthesis and associated changes in ethylene evolution and antioxidative enzyme activities in wheat. *Biol. Plant.*, 47: 437-440.

Kholová J; Sairam, RK; Meena, RC and Srivastava GC 2009. Response of maize genotypes to salinity stress in relation to osmolytes and metal-ions contents, oxidative stress and antioxidant enzymes activity. *Biol. Plant.*, 53: 249-256.

Kim, Y; Arihara, J; Nakayama, T; Nakayama, N; Shimada, S and Usui, K 2004. Antioxidative responses and their relation to salt tolerance in *Echinochloa oryzicola* vasing and *Steraia virdis* (L.) Beauv. *Plant Growth Regul.*, 44: 87-92.

Knight, H; Trewavas, AJ and Knight, MR 1997. Calcium signaling in *Arabidopsis thaliana* responding to drought and salinity. *Plant J.*, 12: 1067-1078.

Krishnamurthy, R and Bhagwat, KA 1989. Polyamines as modulators of salt tolerance in rice cultivars. *Plant Physiol.*, 91: 500–504.

Kumar, D 1984. The value of certain plant parameters as an index for salt tolerance in Indian mustard *(Brassica juncea L.). Plant Soil,* 79: 261–272.

Kusano, T; Berberich T; Tateda C and Takanashi Y 2008. Polyamines: essentials factors for growth and survival. *Planta,* 228: 367–381.

Lauchli, A 1996. Calcium in Plant Growth and Development, eds. Leonald, R. T. and Hepler, P. K. The American Society of Plant Physiologists Symposium Series (American Society of Plant Physiologists. Rockvill, MD) **4**: 26-35.

Lavon, R and Goldschmidt, E 1999. Enzymatic methods for detection of mineral element deficiencies in Citrus leaves: a mini-review. *J. Plant Nutr.*, 22: 139–150.

Legocka, J and Kluk, A 2005. Effect of salt and osmotic stress on changes in polyamine content and arginine decarboxylase activity in *Lupinus luteus* seedlings. *J. Plant Physiol.*, 162: 662–668.

Legocka, J and Sobieszczuk-Nowicka, E 2012. Sorbitol and NaCl stresses affect free, microsome-associated and thylakoid-associated polyamine content in *Zea mays* and *Phaseolus vulgaris. Acta Physiol. Plant.,* 34: 1145–1151.

Leigh, RA and Storey, R 1993. Intercellular compartmentation of ions in barley leaves in relation to potassium nutrition and salinity. *J. Exp. Bot.,* 44: 755–762.

Liu, J and Zhu, JK 1998. A calcium sensor homolog required for plant salt tolerance. *Science,* 280: 1943–1945.

Liu, J; Ishitani, M; Halfter, U; Kim, CS and Zhu, JK 2000. The *Arabidopsis thaliana* SOS2 gene encodes a protein kinase that is required for salt tolerance. *Proc. Natl. Acad. Sci. USA,* 97: 3730–3734.

Liu, JH; Kitashiba, H; Wang, J; Ban, Y and Moriguchi, T 2007. Polyamines and their ability to prove environmental stress tolerance to plants. *Plant Biotechnol.*, 24: 117–126.

Maas, EV and Hoffman, GJ 1977. Crop salt tolerance - current assessment. *J. Irrig. Drainage Div.*, ASCE 103 (IR2): 115-134.

Mahajan, S and Tuteja, N 2005. Cold, salinity and drought stresses. *Archives Biochem. Biophy.*, 444: 139-158.

Marschner, H 1986. Mineral nutrition in higher plants. Academic Press, London, pp. 477-542.

Matsumoto, TK; Ellsmore, AJ; Cessna, SG; Low, P; Pardo, JM; Bressan, RA and Hasegawa, PM 2001. An osmotically induced cytosolic Ca^{2+} transient activates calcineurin signaling to mediate ion homeostasis and salt tolerance of *Saccharomyces cerevisiae*.

McCue, KF and Hanson, AD 1992. Salt-inducible betaine aldehyde dehydrogenase from sugar beet—cDNA cloning and expression. *Plant Mol. Biol.*, 18: 1–11.

Meiri, A and Poljakoff-Mayber, A 1970. Effect of various salinity regimes on growth, Leaf expansion and Transpiration rate of Bean plants. *Soil Sci.*, 109: 26-34.

Mendoza, I; Rubio, F; Rodriguez-Navarro, A and Pardo, JM 1994. The protein phosphatase calcineurin is essential for NaCl tolerance of *Saccharomyces cerevisiae*. *J. Biol. Chem.*, 269: 8792–8796.

Mensah, JK; Akomeah, PA; Ikhajiagbe, B and Ekpekurede, EO 2006. Effect of salinity on germination, growth and yield of five groundnut genotypes. *African J. Biotech.* 5: 1973–1979.

Miseta, A; Kellermayer, R; Aiello, DP; Fu, L and Bedwell, DM 1999. The vacuolar Ca^{2+}/H^{+} exchanger Vcx1p/Hum1p tightly controls cytosolic Ca^{2+} levels in *S. cerevisiae*. *FEBS Letters*, 451: 132-136.

Misra, N and Dwivedi, UN 2004. Genotypic difference in salinity tolerance of green gram cultivars. *Plant Sci.*, 166: 1135-1142.

Mittler, R 2002. Oxidative stress, antioxidants and stress tolerance. *Trends Plant Sci.*, 7: 405- 410.

Mohammed, M; Shibli, R; Ajlouni, M and Nimri, L 1998. Tomato root and shoot responses to salt stress under different levels of phosphorus nutrition. *J. Plant Nutr.*, 21: 1667- 1680.

Møller, IS and Tester, M 2007. Salinity tolerance of Arabidopsis: a good model for cereals? *Trends Plant Sci.*, 12: 534–540.

Moreno, DA; Vıllora, G; Pulgar, G and Romero, L 2000. Effect of nitrogen and potassium supply on concentration of iron and manganese and activities of catalase, peroxidase and aconitase in pepper plants. *J. Plant Nutr.*, 23: 1787–1795.

Mungala, AJ; Radhakrishnan, T and Dobaria, JR 2008. In vitro screening of 123 Indian peanut cultivars for sodium chloride induced salinity tolerance. *World J. Agric. Sci.*, 4: 574–582.

Munns, R 2002. Comparative physiology of salt and water stress. *Plant cell Environ.*, 25: 239-250.

Munns, R and Tester, M 2008. Mechanisms of salinity tolerance. *Ann. Rev. Plant Biol.* 59: 651-681.

Munns, RA 1993. Physiological processes limiting plant growth in saline soils: Some dogmas and hypothesis. *Plant Cell Environ.*, 16: 15–24.

Nakamura, T; Liu, Y; Hirata, D; Namba, H; Harada, S; Hirokawa, T and Miyakawa, T 1993. Protein phosphatise type 2B (calcineurin)-mediated, FK506-sensitive regulation of intracellular ions in yeast is an important determinant for adaptation to high salt stress conditions. *EMBO J.*, 12: 4063-4071.

Nautiyal, PC; Bandyopadhyay, A; Koradia, VG and Makad, M 2000. Performance of groundnut germplasm and cultivars under saline water irrigation in the soils of Mundra in Gujarat, India. *Intl. Arachis Newsletter,* 20: 80–82.

Nautiyal, PC; Ravindra, V and Joshi, YC 1989. Germination and early seedling growth of somegroundnut (*Arachis hypogaea* L.) cultivars under salt stress. *Ind. J. Plant Physiol.*, 32: 251–253.

Niu, X; Bressan, RA; Hasegawa, PM and Pardo, JM 1995. Ion Homeostasis in NaCl Stress Environments. *Plant Physiol.*, 109: 735–742.

Noctor, G; Gomez, L; Vanacker, H and Foyer, CH 2002. Interaction between biosynthesis, compartmentation and transport in the control of glutathione homeostasis and signalling. *J Exp Bot.*, 53: 1283–1308.

Nomura, M; Ishitani, M; Takabe, T; Rai, AK and Takabe, T 1995. *Synechococcus sp*. PCC7942 transformed with the *Escherichia coli* bet genes produces glycine betaine from choline and acquires resistance to salt stress. *Plant Physiol.*, 107: 703–708.

Nuccio, ML; Rhodes, D; McNeil, SD and Hanson, AD 1999. Metabolic engineering of plants for osmotic stress resistance. *Curr. Opin. Plant Biol.*, 2: 128-134.

Orcutt, DM and Nilsen, ET 2000. The Physiology of Plants Under Stress: Soil and Biotic Factors. John Wiley and Sons, New York, NY 683 p.

Parida, AK and Das, AB 2005. Salt tolerance and salinity effect on plants: a review. *Ecotoxicol. Environ. Saf.*, 60: 324–349.

Parvaiz, A and Satyawati, S 2008. Salt stress and phytobiochemical responses of plants – a review. *Plant Soil Environ.*, 54: 89–99.

Passioura, JB and Munns, R 2000. Rapid environmental changes that affect leaf water status induce transient surges or pauses in leaf expansion rate. *Aust. J. Plant Physiol.*, 27: 941–948.

Rathinasabapathi, B; McCue, KF; Gaga, DA and Hanson, AD 1994. Metabolic engineering of glycine betaine synthesis–plant betaine aldehyde dehydrogenases lacking typical transit peptides are targeted to tobacco chloroplasts where they confer betaine aldehyde resistance. *Planta.*, 193: 155–162.

Remorini, D; Melgar, JC; Guidi, L; Degl'Innocenti, E; Castelli, S and Traversi, ML 2009. Interaction effects of root zone salinity and solar irradiance on the physiology and biochemistry of *Olea europaea*. *Environ Exp.Bot.*, 65: 210–219.

Rezaei, H; Sima, KK; Malakouti, MJ and Pessarakli, M 2006. Salt Tolerance of Canola in Relation to Accumulation and Xylem Transportation of Cations. *J. Plant Nutr.*, 29: 1903–1917.

Rhodes, D and Hanson, AD 1993. Quaternary ammonium and tertiary sulfonium compounds in higher plants. *Ann. Rev. Plant Physiol. Plant Mol. Biol.*, 44: 357-384.

Romero, C; Bellees, JM; Vaya, JL; Serrano, R and Culianez-Macia, FA 1997. Expression of the yeast trehalose–6-phosphate synthase gene in transgenic tobacco plants: pleiotropic phenotypes include drought tolerance. *Planta,* 201: 293–297.

Rus, A; Yokoi, S; Sharkhuu, A; Reddy, M; Lee, BH; Matsumoto, TK; Koiwa, H; Zhu, JK; Bressan, RA and Hasegawa, PM 2001. *At*HKT1 is a salt tolerance determinant that controls Na+ entry into plant roots. *Proc. Natl. Acad. Sci. USA*, 98: 14150-14155.

Sairam, RK and Tyagi, A 2004. Physiology and molecular biology of salinity stress tolerance in plants. *Curr. Sci.,* 86: 407-421.

Sairam, RK; Rao, VB and Srivastava, GC 2002. Differential response of wheat genotypes to long term salinity stress in relation to oxidative stress, antioxidant activity, and osmolytes concentration. *Plant Sci.,* 163:1037-1046.

Sairam, RK; Srivastava, GC; Aggarwal, S and Meena, RC 2005. Differences in antioxidant activity in response to salinity stress in tolerant and susceptible wheat genotypes. *Biol. Plant.*, 49: 85-91.

Sanders, D 2000. Plant biology: The salty tale of Arabidopsis. *Curr. Biol.,* 10: 486-488.

Serrano, R; Culiañez-Maciá, FA and Moreno, V 1999. Genetic engineering of salt and drought tolerance with yeast regulatory genes. *Sci Hortic.,* 78: 261–269.

Sharma, PN; Chatterjee, C; Agarwala, SC and Sharma, CP 1990. Zinc deficiency and pollen fertility in maize (Zea mays). *Plant Soil,* 124: 221–225.

Siddiqui, ZS; Khan, MA; Kim, BJ; Huang, JS and Kwon, TR 2008. Physiological Responses of *Brassica* genotypes to Combined Drought and Salt Stress. *Plant Stress,* 2: 78-83.

Siddiqui, ZS; Khan, MA; Kim, BJ; Huang, JS and Kwon, TR 2008. Physiological Responses of *Brassica* genotypes to Combined Drought and Salt Stress. *Plant Stress,* 2: 78-83.

Simond, JM and Orcutt, DM 1988. Free and conjugated desmethylsterol composition of Zea mays hybrids method of determining cold injury. *Plant Physiol.,* 42: 423–1426.

Singh, AL and Chaudhari,V 1993. Screening of groundnut germplasm collection and selection of genotypes tolerant of lime-induced iron-chlorosis. *J. Agric. Sci. Camb.* 121: 205-211.

Singh, AL; Hariprasanna, K and Solanki, RM 2008. Screening of Groundnut Genotypes for Tolerance of Salinity Stress. *Aus. J. Crop Sci.* 1(3): 69-77.

Singh A.L., M.S. Basu and N.B. Singh, 2004. Mineral Disorders of Groundnut. National Research center for groundnut (ICAR), Junagadh India p 85.

Singh, AL; Hariprasanna, K; Chaudhari, V; Gor, HK and Chikani, BM 2010a. Identification of groundnut (*Arachis hypogaea* L.) cultivars tolerant of soil salinity. *J. Plant Nutr.,* 33: 1761-1776.

Singh, A.L., R.S. Jat, Vidya Chaudhari, Himansu Bariya, S.J. Sharma 2010b. Toxicities and Tolerance of Mineral Elements Boron, Cobalt, Molybdenum and Nickel in Crop Plants. In: Plant Nutrition and Abiotic Stress Tolerance II (Ed: Naser A. Anjum). Plant Stress 4 (Special Issue 2): 31-56, Global Science Books.

Stepien, P and Klobus, G 2006. Water relations and photosynthesis in *Cucumis sativus* L. leaves under salt stress. *Biologia Plantarum,* 50: 610-616.

Sucre, B and Sua´rez, N 2011. Effect of salinity and PEG-induced water stress on water status, gas exchange, solute accumulation, and leaf growth in Ipomoea pes-caprae. *Environ. Exp. Bot.,* 70: 192–203.

Sudhir, P and Murthy, SDS 2004. Effects of salt stress on basic processes of photosynthesis. *Photosynthetica,* 42: 481- 486.

Szabolcs, I 1994. Soils and salinization. In: Handbook of Plant and Crop Stress. Ed. M. Pessarakli, Marcel and Dekker, New York, pp 3-11.

Sze, H; Liang, F; Hwang, I; Curran, AC and Harper, JF 2000. Diversity and regulation of plant Ca^{2+} pumps: insights from expression in yeast. *Ann. Rev. Plant. Physiol. Plant Mol. Biol.,* 51: 433-462.

Taiz, L and Zeiger, E 1998. Plant Physiology. Sunderland, Massachusetts: Sinauer Associates, Inc. p. 792.

Takahashi, T and Kakehi, JI 2010. Polyamines: ubiquitous polycations with unique roles in growth and stress responses. *Ann. Bot.,* 105: 1–6.

Tester, M and Davenport, RJ 2003. Na^+ transport and Na^+ tolerance in higher plants. *Ann. Bot.,* 91: 503–527.

Thomas, H 1997. Drought resistance in plants. In: Mechanisms of environmental stress resistance in plants, eds. A.S. Basra and R.K. Basra, Amsterdam: Harwood Acad Publishers. pp 1–42.

Tyagi, A; Santha, IM and Mehta, SL 1999. Effect of water stress on proline content and transcript levels in *Lathyrus sativus. Ind. J. Biochem. Biophys.*, 36: 207–210.

Vadez, V; Srivastava, N; Krishnamurthy, L Aruna, R and Nigam, SN 2005. Standardization of a protocol to screen for salinity tolerance in groundnut. *Intl. Arachis Newsletter* 25: 42–47.

Valenzuela, JK; Alvarado, JJ; Sanchez, A and Romero, L 1995. Influence of N, P and K treatments on several physiological and biochemical iron indicators in melon plants irrigated with brackish water. In: Abadıa, J. (Ed.), Iron Nutrition in Soils and Plants. Kluwer Academic Publishers, The Netherlands, pp. 135–140.

van Genuchten, MT and Hoffman, GJ 1984. Analysis of crop salt tolerance data. In: I. Shainberg and J. Shalhevet, eds. Soil salinity under irrigation - process and management ecological studies 51. New York, Springer-Verlag. pp. 258-271.

Vıllora, G; Moreno, DA; Pulgar, G and Romero, L 2000. Yield improvement in zucchini under salt stress: determining micronutrient balance. *Sci. Hort.*, 86: 175–183.

Wang, Y and Nil, N 2000. Changes in chlorophyll, ribulose biphosphate carboxylase-oxygenase, glycine betaine content, photosynthesis and transpiration in Amaranthus tricolor leaves during salt stress. *J. Hortic. Sci. Biotechnol.*, 75: 623-627.

Watson, R; Pritchard, J and Malone, M 2001. Direct measurement of sodium and potassium in the transpiration stream of salt-excluding and non-excluding varieties of wheat. *J. Exp. Bot.*, 52:1873–1881.

Wimmer MA, Muehling KH, Lauchli A, Brown PH, Goldbach HE (2001) Interaction of salinity and boron toxicity in wheat (*Triticum aestivum* L.). In : Horst WJ, Schenk MK, Burkert A, Claassen N, Flessa H, Frommer WB, Goldbach H, Olfs HW, Romheld V(Eds) *Plant nutrition food security and sustainability of agro ecosystems through basic and applied research, 14th International Plant Nutrition Colloquium,* Hannover, Germany, pp 426-427.

Yahya, A 1998. Salinity effects on growth and on uptake and distribution of sodium and some essential mineral nutrients in sesame. *J. Plant Nutr.*, 21:1439–1451.

Yancey, PH 1994. Compatible and counteracting solutes. In: Strange K, ed. Cellular and molecular physiology of cell volume regulation. Boca Raton: CRC Press, pp. 82–109.

Yeo, AR 1998. Molecular biology of salt tolerance in the context of whole-plant physiology. *J. Exp. Bot.*, 49: 915-929.

Yermiyahu U, Ben GA, Keren R, Reid RJ (2008) Combined effect of salinity and excess boron on plant growth and yield. *Plant and Soil,* 304, 73-87.

Yokoi, S; Quintero, FJ; Bressan, RA; Pardo, JM and Hasegawa, PM 2002. Differential expression and function of *Arabidopsis thaliana* NHX Na^{+}/H^{+} antiporters in the salt stress response. *Plant J.*, 30: 529-39.

Zapata, PJ; Serrano, M; Pretel, MT; Amoros, A and Botella, MA 2004. Polyamines and ethylene changes during germination of different plants under salinity. *Plant Sci.*, 167: 781–788.

Zhao, GQ; Ma, BL and Ren, CZ 2007. Growth, gas exchange, chlorophyll fluorescence and ion content of naked oat in response to salinity. *Crop Sci.*, 47: 123-131.

Zhu, BC; Su, J; Chang, MC; Verma, DPS; Fan, YL and Wu, R 1998. Over expression of a delta-pyrroline-5-carboxylate synthetase gene and analysis of tolerance to water and salt stress in transgenic rice. *Plant Sci.*, 139: 41–48.

Zhu, JK 2001. Plant salt tolerance. *Trends Plant Sci.*, 6: 66-71.

Zhu, JK; Liu, J and Xiong, L 1998a. Genetic analysis of salt tolerance in arabidopsis. Evidence for a critical role of potassium nutrition. *Plant Cell*, 10: 1181–1191.

15

RESEARCH PARADIGM AND INFERENCE OF STUDIES ON HIGH TEMPERATURE STRESS IN RICE

Beena R.

Department of Plant Breeding and Genetics,
Regional Agricultutal Research Station, Kerala Agricultural University,
Pattambi, Palakkad, Kerala

INTRODUCTION

Rice is produced in at least 95 countries across the globe and provides a staple food for more than half of the world's current population (IRRI, 2002; Coats, 2003). There is a demand of 2000 million metric tons of rice by 2030 (FAO, 2002). World rice production must increase by ~1% annually to meet the growing demand for food that will result from population growth and economic development. In India, rice is grown on 43Mha under a diverse set of agro-environments varying in seasons, temperature, rainfall, soil types, hydrology, varieties and input management. This staple food contributes 43% of the total food. Rice is the staple food of most of the South Indian states. Meeting this ~35% increase in demand will require significant improvements in rice production. However, achieving these improvements will be challenge as the future climate changes and water scarcity increases (Bouman *et al.*, 2007).

Current Intergovernmental Panel on Climate Change (IPCC) projections indicate that atmospheric CO_2 concentration will increase over this century, reaching 730- 1020 ppm by 2100 (Meehl *et al.*, 2007). An increase in global temperature ranging from 1.1 to 6.4°C depending on global emissions scenarios, will accompany the rises in atmospheric CO_2 (Meehl *et al.*, 2007). Direct injuries due to high temperatures include protein denaturation and aggregation, and increased fluidity of membrane lipids. Indirect or slower heat injuries include inactivation of enzymes in chloroplast and mitochondria, inhibition of protein synthesis, protein degradation and loss of membrane integrity (Howarth, 2005). Heat stress also affects the organization of microtubules by splitting and/or elongation of spindles, formation of microtubule asters in mitotic cells, and elongation of phragmoplast microtubules (Smertenko *et al.,* 1997). These injuries eventually lead to starvation, inhibition of growth, reduced ion flux, production of toxic compounds and reactive oxygen species (ROS) (Howarth, 2005). CERES rice model projected that a 2°C increase in temperature would cancel out the positive effects of elevated CO_2 on rice yields, and that increasing water storage in combination with rising temperature will likely limit rice production even with rising CO_2 (Lal *et al.*, 1998). Therefore, it is important to determine how rice yield are influenced by elevated atmospheric CO_2 as it would have crucial implications on future food security of this region.

2. HIGH TEMPERATURE STRESS AND SPIKELET FERTILITY IN RICE

Rice is most sensitive to high temperatures at heading and next most sensitive at about 9 days before heading. One or two hours of high temperature at anthesis has a decisive effect on the incidence of sterility. High temperatures before or after anthesis have much less effect on sterility. Clear differences exist in the tolerance of rice varieties for high temperature induced sterility. At 35°C, N22, upland rice from India, has higher than 80% spikelet fertility, whereas BKN6624-46-2, a lowland selection from Thailand, has about 10% spikelet fertility. If defined as the temperature when spikelet fertility decreases below 80%, the critical temperature is estimated as 36.5°C for N22, 35°C for IR747B2-6, and 32°C for BKN6624-46-2. Thus, the difference in the critical high temperatures between heat-tolerant N22 and heat-susceptible BKN6624-46-2 is more than 4°C (Yoshida, 1981). Thus it is concluded that flowering (anthesis and fertilization), and to a lesser extent booting (microsporogenesis), are the most susceptible stages of development to temperature in rice (Satake and Yoshida, 1978; Farrell *et al.*, 2006). Sterility is caused by poor anther dehiscence and low pollen production, and hence low numbers of germinating pollen grains on the stigma (Matsui *et al.*, 2000, 2001; Prasad *et al.*, 2006). The anthers of heat tolerant cultivars dehisce more easily than those of susceptible cultivars under high temperature conditions (Satake and Yoshida 1978, Mackill *et al.* 1982, Matsui *et al.* 1997a,b, 2001). This is because of the tight closure of the locules by the cell layers, which delays locule opening and decreases spikelet fertility at high temperature (Matsui and Omasa 2002). It was suggested that spikelet fertility at high temperature can be used as a screening tool for heat tolerance during the flowering stage (Prasad *et al.* 2006). There is a lot of variation among the rice genotypes for heat tolerance. Tropical *indica* species are more tolerant than *japonica* species. Genotypes N22 (Yoshida *et al.*, 1981; Prasad *et al.*, 2006) and Akitakomachi (Matsui *et al.*, 2001) are the most tolerant genotypes found to date among indica and japonica spp, respectively. Jagadish *et al.* (2007) this study has shown that <1 h of exposure to high temperature is sufficient to induce sterility in rice (*O. sativa*). IR64 and Azucena exhibited different sensitivities and responses to temperature of spikelet fertility.

3. MECHANISMS OF HEAT TOLERANCE IN RICE

Heat tolerance is generally defined as the ability of the plant to grow and produce economic yield under high temperature (Wahid *et al.* 2007). As plants cannot move, the only option they have to defend themselves from various stresses is to make metabolic and structural adjustments (Yamanouchi *et al.* 2002). Relatively little is known about stress avoidance and true tolerance mechanisms in rice. For phenotyping and modelling, it is required that these two mechanisms, as well as the effects of temperature on the fertility of spikelets, are clearly understood and investigated further.

3.1 Heat Avoidance

The ability of a variety to decrease its spikelet temperature with reduced relative humidity can be considered as an avoidance mechanism. Other important mechanisms are changing leaf orientation, transpirational cooling, or alteration of membrane lipid compositions.

3.2 Heat Tolerance

True heat tolerance at sensitive stages might be conferred by protecting structural proteins, enzymes and membranes from heat damage. The role of heat shock proteins (HSPs) and other stabilizing factors is crucial in these processes (Maestri *et al.* 2002).

3.3 Heat Escape Mechanism

Heat escape is a strategy, whereby the heat sensitive process of anthesis-anther dehiscence, pollination, pollen germination, pollen tube growth and fertilization can be completed during early morning hours before the occurrence of high temperature episodes. This mechanism is shown by some of the tropical cultivars. In many crop plants, early maturation is closely correlated with smaller yield losses under high temperatures, which may be attributed to the engagement of an escape mechanism (Adams *et al.*, 2001). Rice cultivar CG14 (*O. glaberrima*) reached peak anthesis earlier in the morning (1.5 h after dawn) under both control (30°C) and high (38°C) temperature conditions than *O. sativa* genotypes (≥3 h after dawn). Early time of day of flowering should confer an advantage as temperatures are cooler in the early morning; this advantage can readily be exploited by inter-specific hybrids (so-called NERICAs — New Rice for Africa. Bala (which includes N22 in its background) and Co 39 exhibited moderate heat tolerance.

For drought, it appears that genotypes developed to date possess either better avoidance or better tolerance. A special research challenge to follow from identifying QTLs for avoidance and tolerance traits would be to combine these strategies in one genotype, ideally with appropriate phenology, so as to use the escape strategy as well.

3.4 Acquired or Induced Thermo-Tolerance

Under natural conditions, development of abiotic stress levels is a progressive phenomenon. Accordingly, plants initially experience a milder stress level before being exposed to severe stress. In response to this milder stress, several genes have been shown to be either up-regulated or down-regulated, which leads to the development of a considerable degree of tolerance in plants when stress becomes severe. Numerous studies have shown that acclimated plants survive exposure to a severe stress, which could otherwise be lethal. This phenomenon of acclimation response is termed as 'acquired thermotolerance' (Sun *et al.,* 2001). Acquired or induced thermo-tolerance is the ability induced by a sub-lethal heat stress to overcome subsequent exposure to lethal high temperatures (Burke *et al.*, 2000). This type of tolerance is a cell autonomous phenomenon that results from an exposure to a short period of sub-lethal temperature or other moderate stress treatment (Larkindale and Vierling, 2008). It depends mainly on the induction of specific pathways during the acclimation period and subsequent acquisition of thermo-tolerance (Klueva *et al.*, 2001).

The induction stress triggers several signalling pathways, resulting in the expression of an array of stress-responsive genes, whose gene products alter the physiological and biochemical processes relevant to stress tolerance. While several genes are suppressed, a large number of genes are up-regulated in response to stress. Based on the concept that seedlings acquire a certain degree of stress tolerance when exposed to milder stress levels, the novel and high-throughput 'temperature induction response' (TIR) technique has been developed and validated in many crop species. This involves treating very young seedlings to a mild induction temperature before exposing them to a potentially lethal temperature (Kumar *et al,* 1999; Senthil Kumar *et al.*, 2007). The induction temperature is known to trigger an array of cellular and molecular response mechanisms which would then lead to the survival of seedlings when subsequently exposed to the potentially lethal temperature. This novel TIR technique has been demonstrated to reveal genetic variability in intrinsic stress tolerance at the cellular level.

4. HEAT TOLERANCE OF PHOTOSYNTHESIS AND EXPRESSION OF RUBISCO UNDER HIGH TEMPERATURE STRESS

Photosynthesis is sensitive to heat stress. Inhibition of net photosynthesis (Pn) by moderate heat stress has been attributed to an inability of Rubisco activase to maintain Rubisco in an active form. Sharkey (2005) reported that there is no or little damage to photosystem II as a result of moderate heat stress even though moderate heat stress can reduce the photosynthetic rate to near zero. Moderate heat stress can stimulate dark reduction of platoquinone and cyclic electron flow in the light and increase thylakoid leakiness. Rubisco deactivates at moderately high temperature. There is an increase in the amount of H_2O_2 as a result of oxygenase side reactions during high temperature stress. The ability to withstand moderately high temperature can be improved by altering thylakoid lipid composition or by supplying isoprene.

5. PHYSIOLOGICAL TRAITS RELATED TO TOLERANCE OF HIGH TEMPERATURE STRESS

Different physiological mechanisms may contribute to heat tolerance in the field—for example, heat tolerant metabolism as indicated by higher photosynthetic rates, stay-green, and membrane thermo-stability, or heat avoidance as indicated by canopy temperature depression.

5.1. Canopy Temperature Depression

CTD shows high genetic correlation with yield and high values of proportion of direct response to selection (Reynolds *et al.*, 1998), indicating that the trait is heritable and therefore amenable to early generation selection. Plant canopy temperature is considered as an index of water stress in many crop species (Chaturvedi *et al.*, 1999). Many researchers have used canopy temperature as a means of determination of water stress due to inadequate soil moisture (Khera and Sandhu, 1986). Canopy temperature was related to yield and yield reduction in spring wheat (Diaz *et al.*, 1983), pearl millet (Singh and Kanemasu, 1983) and rice (Turner, 1986). Sensing the thermal radiation emitted by the canopy is one way of assessing water stress (Penuelas *et al.*, 1992; Araus *et al.*, 1993). Sanguineti *et al.* (1999) stated that maize plants with a deeper root system would maintain a cooler canopy temperature and ultimately higher yield under drought stress. Relatively lower canopy temperature in drought stressed rice indicates a relatively better capacity for taking up soil moisture or for maintaining a relatively better plant water status (Garrity and O'Toole, 1995).

5.2. Membrane Stability

Cell membrane stability (CMS) is a physiological index widely used for the evaluation of drought and temperature tolerance (Sullivan, 1972;

Martineau, 1979; Blum and Ebercon, 1981; Sadalla *et al.*, 1990; Reynolds *et al.* 1994; Fokar *et al.*, 1998). Premachandra *et al.* (1990) reported that cell membrane stability is an indicator of drought tolerance. Drought stress induces membrane deterioration leading to severe metabolic dysfunction (Nir *et al.*, 1969; Buttrose and Swift, 1975). The maintenance of membrane stability during desiccation is important for normal physiological metabolism to continue under low water potential. Tyagi *et al.* (1999) reported that the membrane stability index was higher in tolerant genotypes under water stress. Lower membrane stability or higher injury reflects the extent of membrane lipid peroxidation (Dhindsa *et al.*, 1981), which in turn is a consequence of higher susceptibility to oxidative stress due to various environmental stresses including drought (Leibler *et al.*, 1986). Although resistance to high temperatures involves several complex tolerance and avoidance mechanisms, the membrane is thought to be a site of primary physiological injury by heat (Blum, 1988), and measurement of solute leakage from tissue can be used to estimate damage to membranes. Since membrane thermostability is reasonably heritable (Fokar *et al.*, 1998) and shows high genetic correlation with yield.

6. INTERACTIVE EFFECT OF HIGH TEMPERATURE WITH HIGH CO_2 CONCENTRATION

Most of the CO_2 enrichment experiments conducted during the last decade have shown that rice yields increase in response to elevated atmospheric CO_2 (Baker *et al.*, 1990; Ziska and Teramura, 1992; Ziska *et al.*, 1997; Moya *et al.*, 1998; Kim *et al.*, 2001. 2003). The observed yield increase showed a wide variation (95-60%) which may be due to a variety of reasons such as the genotype and interactions between elevated atmospheric CO_2 and other environmental factors. In order to improve breeding efforts in future, it is increasingly important to quantify how rice physiological and yield parameters will respond to anticipated changes in the atmosphere.

Most of this increase must come from greater yield on existing crop land to avoid environmental degradation, destruction of natural ecosystems and loss of biodiversity. Achieving greater yields depends on increasing crop biomass, because there is little scope to future increase in the proportion that biomass allocated to grain. Total crop biomass is determined mainly by crop photosynthesis and respiration losses, both of which are sensitive to temperature. Future crop yields will be influenced by complex interactions between the effects of increases in atmospheric concentrations of CO_2 and trace gases such as ozone as well as the effects of temperature increases brought about by climate change. An increases in night temperature although small in magnitude, had a negative effect on the yield of irrigated rice in dry season. The results of the study highlight the need for greater fundamental understanding of the effects of night temperature on physiological processes governing crop growth and yield.

So it is important to develop heat tolerant varieties of rice. There is convincing evidence for the fact that stress responsive genes are predominantly expressed when plants experience sub lethal levels of stress. This would bring in certain specific alterations in metabolism, leading to increased tolerance when the plants subsequently experience a lethal level of stress. Genetic variability in stress tolerance is therefore, a result of the extent of stress gene expression when plants experience such lethal stresses. This phenomenon has extensively studied for high temperature stress tolerance. Based on the knowledge that a gradually increasing induction stress would trigger the expression of specific genes leading to greater level of tolerance to severe stress, a novel temperature induction response technique has been developed. Any genotype that shows superior expression of stress responsive genes would acquire higher tolerance to more severe stress levels and hence can be considered as having higher tolerance at the cellular level.

When soils are wet, Global change effects on production will largely mirror effects on photosynthesis. Rising CO_2 may increase yield substantially when plants are C limited or when photosynthates in excess of current requirements can be stored for later use (Allen *et al.*, 1991; Lawler and Mitchell, 1991). Yield increases at high CO_2 should occur most frequently in areas where temperatures approximate the optimum for crop growth. Future increases in temperature will reduce yields by reducing C gain and accelerating crop development. Effects of higher temperature and CO_2 concentration on plants often are not additive, meaning that the combined influence of these changes can't be predicted from knowledge of their individual effects (Idso *et al.*, 1987; Long. 1991). In addition, it appears that the magnitude and even direction of crop responses to CO_2 enrichment and temperature changes are species and even cultivar specific (Baker and Allen, 1993; Ziska *et al.*, 1997). Understanding interactive effects of rising CO_2 concentration and temperature for crop yields and water economy is among the major challenges confronting research.

7. STRATEGIES TO OVERCOME HIGH TEMPERATURE STRESS

7.1 Breeding for Temperature Tolerant Varieties

A mapping population of 98 lines (backcross inbred lines, BILs) derived from a backcross of Nipponbare/Kasalath// Nipponbare was planted at two experimental sites, Nanjing and Nanchang, and treated with high and optimal temperature during grain filling, respectively. The grain weight heat susceptibility index (GWHSI= (grain weight at optimum temperature - grain weight at high temperature) / grain weight at optimum temperature X 100) was employed to evaluate the tolerance of rice to heat stress. A genetic linkage map with 245 RFLP markers and a mixed linear-model approach was used to detect quantitative trait loci (QTLs) and their main effects, epistatic interactions and QTL X environment interactions (QXE). The threshold of LOD score=2.0 was used to detect the significance of association between marker and trait. A total of 3 QTLs controlling heat tolerance during grain filling were detected, on chromosomes 1, 4 and 7, with LOD scores of 8.16, 11.08 and 12.86, respectively, and they explained the phenotypic variance of 8.94, 17.25 and 13.50%, correspondingly. The QTL located in the C1100–R1783 region of chromosome 4 showed no QTL X environment interaction and epistatic effect, suggesting that it could be stably expressed in different environments and genetic backgrounds, and thus it would be valuable in rice breeding for heat tolerance improvement. This QTL allele, derived from Kasalath reduced 3.31% of the grain weight loss under heat stress. One located between R1613–C970 on chromosome 1 and the other between C1226–R1440 on chromosome 7, with additive effect 2.38 and 2.92%, respectively. The tolerance alleles of both these QTLs were derived from Nipponbare. Both of these QTLs had significant QTL X environment interactions, and the latter was involved in epistatic interaction also. Eight pairs of epistatic effect QTLs were detected, one pair each on chromosomes 1, 2, 3, 5, 7, 8, 10 and 12. The results could be useful for elucidating the genetic mechanism of heat-tolerance and the development of new rice varieties with heat tolerance during grain filling phase(Chang-lan *et al.*, 2005).

QTL mapping studies for heat tolerance have been conducted on various rice populations at booting (Zhao *et al.* 2006) and flowering stages (Cao *et al.* 2003, Chen *et al.* 2008, Zhang *et al.* 2008, 2009, Jagadish *et al.* 2010a, Xiao *et al.* 2011). QTL for heat tolerance at flowering stage have been mapped on all chromosomes except chromosome 6 and 7; however, confirmation and fine mapping of the identified QTL for heat tolerance have not been reported. Ye *et al.* (2011) conducted a study to identify quantitative trait loci (QTL) controlling heat tolerance in rice, the progeny of BC1F1 and F2 populations derived from an IR64 N22 cross were exposed to 38/24°C for 14 days at the flowering stage, and spikelet fertility was assessed. A custom 384-plex Illumina Golden Gate genotyping assay was used to genotype the F2 and selected BC1F1 plants. Four single nucleotide polymorphisms were associated with heat tolerance in the BC1F1 population using selective genotyping and single marker analysis, and four putative QTL were found to be associated with heat tolerance in the F2 population. Two major QTL were located on chromosome 1 (qHTSF1.1) and chromosome 4 (qHTSF4.1). These two major QTL could explain 12.6% (qHTSF1.1) and 17.6% (qHTSF4.1) of the variation in spikelet fertility under high temperature. Tolerant allele of qHTSF1.1 was from the susceptible parent IR64, and that of qHTSF4.1 was from tolerant parent N22. The effect of qHTSF4.1 on chromosome 4 was confirmed in selected BC2F2 progeny from the same IR64 · N22 cross, and the plants with qHTSF4.1 showed significantly higher spikelet fertility than other genotypes. Using identified genetic resources and QTL to improve heat tolerance in rice varieties has not been achieved. Further studies on QTL mapping using accurate phenotyping technology and validation of mapped major QTL in different populations are needed to identify candidates for map-based cloning of genes underlying heat tolerance in rice.

7.2 Genetic Engineering for Heat Tolerance

Heat stress has significant effects on protein metabolism, including degradation of proteins, inhibition of protein accumulation, and induction of certain protein synthesis, depending on the level and duration of heat stress (Monjardino *et al.*, 2005; He *et al.*, 2007). Moderate heat response involves down regulation of proteins functioning in lipid biogenesis, cytoskeleton structure, sulfate assimilation, amino acid biosynthesis, nuclear transport and antioxidant response (Ferreria *et al.*, 2006; Demirevska-Kepova *et al.*, 2005).While synthesis of most normal proteins and mRNAs is inhibited in heat stress conditions, the transcription and translation of a small set of proteins, called heat shock proteins (HSPs), may be induced or enhanced when plants are exposed to elevated temperatures (Vierling, 1991; Blumenthal *et al.*, 1990; Krishnan *et al.*, 1989). Singh *et al.*, 2008 reported that high temperature tolerance has been genetically engineered in plants mainly by over-expressing the heat shock protein genes or indirectly by altering levels of heat shock transcription factor proteins. Apart from heat shock proteins, thermo-tolerance has also been altered by elevating levels of osmolytes, increasing levels of cell detoxification enzymes and through altering membrane fluidity. It is suggested that Hsps may be directly implicated in thermo-tolerance as agents that minimize damage to cell proteins. The other three above approaches leading to thermo-tolerance in transgenic experiments though operate in their own specific ways but indirectly might be aiding in creation of more reductive and energy-rich cellular environment, thereby minimizing the accumulation of damaged proteins. Intervention in protein metabolism such that accumulation of damaged proteins is minimized thus appears to be the main target for genetically-engineering crops against high temperature stress. A few successful cases on incorporation of *HSP* genes to improve heat tolerance were reported in rice. For example, enhanced thermo-tolerance was achieved in transgenic rice over expressing an arabidopsis *HSP101* gene. Over expression of a rice chloroplast *sHSP* (*Oshsp26*) gene conferred better tolerance not only to heat stress but also to oxidative stress in *E. coli* (Lee *et al.*, 2000), and over expression of sHSP17.7 confers both heat tolerance and UV-B resistance to rice plants (Murakami *et al.*, 2004). The effectiveness of this strategy in breeding heat-tolerant perennial grasses needs to be further validated.

7.3 Management Strategies For Heat Tolerance

Under heat stress, Ca2+ requirement for growth is high to mitigate adverse effects ofthe stress. It has been shown that exogenous application of Ca^{2+} promotes plant's heat tolerance. Application of Ca^{2+} in the form of

$CaCl_2$ prior to the stress treatment elevated the content of lipid peroxidation product, MDA and stimulated the activities of guaiacol peroxidase, SOD and catalase, which could be the reasons for the induction of heat tolerance. Among the low molecular weight organic compounds, glycine betaine and polyamines have been successfully applied to induce heat tolerance in various plant species. For example, barley seeds pre-treated with glycine betaine led to plants with lower membrane damage, better photosynthetic rate, improved leaf water potential and greater shoot dry mass, compared to untreated seeds. Also, exogenous application of 4mM spermidine improved tomato heat resistance by improving chlorophyll fluorescence properties, hardening and higher resistance to thermal damage of the pigment-protein complexes structure, and the activity of PSII during linear increase in temperature. Thus, to improve plant heat tolerance, alternative approaches to genetic means would include pre-treatment of plants or seeds with heat stress or certain mineral or organic compounds. The success of such approach, however, depends on plant species and genotypes and must be studied on case basis. Similarly, a shift in the cropping pattern from flooded to aerobic rice is expected in the near future. Keeping the field wet but not flooded and with added organic matter reduces the global warming potential from rice fields without a decrease in yield (Yu *et al.*, 2004).

8. CONCLUSION

From theses studies it can be concluded that the main cause of floret sterility induced by high temperature at flowering is anther indehiscence. High temperature at flowering inhibits swelling of the pollen grains, which is the driving force behind anther dehiscence in rice. Anthers of high temperature tolerant cultivars dehisce more easily than those of susceptible cultivars and contribute to pollination under high temperature conditions. So strategies to overcome heat stress should be strengthened in the years to come with more emphasis to give for field studies. More focused study should be taken up to understand the fundamentals of the effects of night temperature on physiological processes governing crop growth and yield.

REFERENCES

Adamas, SR; Cocksbull, KE and Cave, CRJ. 2001. Effect of temperature on the growth and development of tomato fruits. *Annals of Bot.*, 88: 869-877.

Allen, LH Jr; Baker, JT; Albrecht, SL; Boote, KJ; Pan, D and Vu, JCV. 1995. Carbon dioxide and temperature effects on rice. In: Climate Change and Rice. Peng, S; Ingram, KT; Neue, HU and Ziska, LH (eds.). Springer-Verlag, Berlin, Heidelberg. pp. 256-277.

Araus, JL; Reynolds, MP and Acevedo, E. 1993. Leaf posture, grain yield, growth, leaf structure and carbon isotope discrimination in wheat. *Crop Sci.*, 33: 1273-1279.

Baker, JT and Allen, LH Jr. 1993a. Contrasting crop species responses to CO2 and temperature: Rice, soybean, and citrus. Vegetatio 104/105: 239-260. Also: pp. 239-260. In: CO2 and Biosphere. (*Advances in Vegetation Science* 14). Rozema, J; Lambers, H; Geijn SC van de and Cambridge, ML. (eds.). Kluwer Academic Publishers, Dordrecht.

Baker JT; Allen LH and Boote KJ. 1990. Growth and yield responses of rice to carbon-dioxide concentration. *J. Agric.Sci.*, 115, 313–320.

Baker, JT and Allen, LHJr. 1993b. Effects of CO_2 and temperature on rice: A summary of five growing seasons. *J. Agric. Meteorol., (Japan)* 48: 575-582.

Baker, JT; Allen Jr., LH and Boote, KJ. 1992. Temperature effects on rice at elevated CO_2 concentration. *J. Exp. Botany,* 43, 959–964.

Blum, A and Ebercon, A. 1981. Cell membrane stability as a measure of drought and heat tolerance in wheat. *Crop Sci.*, 21: 43-47.

Blum, A. 1988. Plant breeding for stress environments. CRC Press, Boca Daton, FL.

Blumenthal, C; Bekes, F; Wrigley, CW and Barlow, EW. 1990. The acquisition and maintenance of thermotolerance in Australian wheats. *Australian Journal of Plant Physiology*, 17:37–47.

Bouman, BAM; Humphreys, E; Tuong, TP and Barker, R. 2007. Rice and water. *Adv. in Agron.*, 92: 187–237.

Burke, J.J; O'Mahony, PJ and Oliver, MJ. 2000. Isolation of Arabidopsis mutants lacking components of acquired thermotolerance. *Plant Physiol.*, 123: 575-587.

Buttrose, MS and Swift, JG. 1975. Effects of killing by heat or desiccation membrane structure in pea roots. *Aust. J. Plant Physiol.*, 2: 225-233.

Cao, L; Zhao, J; Zhan, X ; Li, D; He, L and Cheng, S. 2003: Mapping QTLs for heat tolerance and correlation between heat tolerance and photosynthetic rate in rice. *Chin. J. Rice Sci.*, 17: 223-227.

Chang-lan Z; Ying-hui, X; Chung-ming, W; Ling, J; Hu-qu, Z and Jian-min,W. 2005. Mapping QTL for heat-tolerance at grain filling stage in rice. *Rice Sci.*, 12: 33-38.

Chaturvedi, GS; Tripathy, RP and Kumar, A. 1999. Comparative canopy temperature measurements in Indian mustard genotypes (*Brassica juncea* L.). *Indian J. Plant Physiol.*, 4:61-62.

Chen, Q; Yu, S; Li, C and Mou, T. 2008. Identification of QTLs for heat tolerance at flowering stage in rice. *Sci. Agric. Sin.*, 41: 315-32.

Coats, B. 2003. Global rice production. In: Rice Origin, History, Technology and Production (Eds. Smith, CW and Dilday, RH), pp. 247–470. Wiley, Hoboken, NJ, USA.

Demirevska-Kepova, K; H¨olzer, R; Simova-Stoilova, L and Feller, U. 2005.Heat stress effects on ribulose-1,5-bisphosphate carboxylase/oxygenase, Rubisco binding protein and Rubisco activase in wheat leaves. *Biologia Plantarum*, 49: 521–525.

Dhindsa, RS and Matowe, W. 1981. Drought tolerance in two mosses correlated with enzymatic defence against lipid peroxidation. *J. Expt. Bot.*, 32: 79-91.

Diaz, RA; Matthais, AD and Hanks, RJ. 1983. Evapotranspiration of spring wheat from canopy temperature. *Agron. J.*, 75: 805-810.

FAO. 2002. World Agriculture: Towards 2015/2030 Summary Report. FAO, Rome, Italy.

Farrell, TC; Fox, KM; Williams, RL and Fukai, S. 2006. Genotypic variation for cold tolerance during reproductive development in rice: screening with cold air and cold water. *Field Crops Res.*, 98: 178–194.

Ferreira, S; Hjernø, K and Larsen, M. 2006. Proteome profiling of *Populus euphratica* Oliv. upon heat stress. *Annals of Botany*, 98: 361–377.

Fokar, M; Nguyen, HT and Blum, A. 1998. Heat tolerance in spring wheat. I Estimating cellular thermotolerance and its heritability. *Euphytica*, 104: 1-8.

Garrity, DP and O'Toole, JC. 1995. Selection for reproductive stage drought avoidance in rice using infrared thermometry. *Agron. J.*, 87: 773-779.

He, Y and Huang, B. 2007. Protein changes during heat stress in three Kentucky bluegrass cultivars differing in heat tolerance. *Crop Sci.*, 47: 2513–2520.

Howarth, CJ. 2005. Genetic improvements of tolerance to high temperature. In: Abiotic stresses: plant resistance through breeding and molecular approaches. (Eds. Ashraf, M and Harris, PJC) 277–300, Haworth Press Inc., New York.

Idso, SB; Kimball, BA; Anderson, MG and Mauney, JR. 1987. Effects of atmospheric CO_2 enrichment on plant growth: The interactive role of air temperature. *Agric. Ecosystems Environ.*, 20: 1-10.

IRRI. 2002. Rice Almanac: Source Book for the Most Important Economic Activity on Earth, 3rd edn. CABI Publishing, Oxon, UK.

Jagadish, S; Cairns, J; Lafitte, R; Wheeler, T; Price, A and Craufurd, P. 2010a: Genetic analysis of heat tolerance at anthesis in rice. *Crop Sci.*, 50:1633-1641.

Jagadish, SVK; Craufurd, PQ and Wheeler, TR. 2007. High temperature stress and spikelet fertility in rice. *J. Exp. Bot.*, 58:1627–1635.

Katiyar-Agarwal, S; Agarwal, M and Grover, A. 2003. Heat tolerant basmati rice engineered by over-expression of hsp101. *Plant Molecular Biology*, 51:677–686.

Khera, KL and Sandhu, BS. 1986. Canopy temperature of sugarcane as influenced by irrigation régime. *Agric. Meterol.*, 37: 245-258.

Kim, HY; Lieffering, M; Kobayashi, K; Okada, M and Mitchell, MW. Gumpertz, M., 2003. Effects of free-air CO_2 enrichment and nitrogen supply on the yield of temperate paddy rice crops. *Field Crops Res.* 83: 261–270.

Klueva, NY; Maestri, E; Marmiroli, N and Nguyen, HT. 2001. Mechanisms of thermotolerance in crops. In: Crop Responses and Adaptations to Temperature Stress (Ed.)A.S. Basra, Food Products Press, Binghamton, NY, pp. 177–217.

Krishnan, M; Nguyen, HT and Burke, JJ. 1989. Heat shock protein synthesis and thermal tolerance in wheat. *Plant Physiology*, 90:140–145.

Kumar, G; Krishnaprasad, BT; Savitha, M; Gopalakrishna, R; Mukhopadhyay, K; Ramamohan, G and Udayakumar, M. 1999. Enhanced expression of heat shock proteins in thermo-tolerant lines of sunflower and their progenies selected on the basis of temperature induction response. (TIR). *Theor. and Appl. Gen.*, 99:359–367.

Lal, M; Singh, KK; Rathore, LS; Srinivasan, G and Saseendran, SA. 1998. Vulnerability of rice and wheat yields in NW India to future changes in climate. *Agricultural and Forest Meteorology*, **89:** 101–114.

Larkindale, J and Vierling, E. 2008. Core genome responses involved in acclimation to high temperature. *Plant Physiol.*, 146: 748-761.

Lawlor, DW and Mitchell, RAC. 1991. The effects of increasing CO2 on crop photosynthesis and productivity: a review of field studies. *Plant Cell Environ.*, 14:807-818.

Lee, BH; Won, SH and Lee, HS. 2000. Expression of the chloroplast-localized small heat shock protein by oxidative stress in rice. *Gene*, 245:283–290.

Liebler, DC; Kling, DD and Reed, DJ. 1986. Antioxidant protection of phosopholipid bilayers by tocopherol. Control of tocopherol status and lipid peroxidation by ascorbic acid and glutathione. *J. Biol. Chem.*, 261: 12114-12119.

Long, SP. 1991. Modification of the response of photosynthetic productivity to rising temperature by atmospheric CO2 concentrations: Has its importance been underestimated? *Plant Cell Environ.*, 14: 729-739.

Mackill, DJ; Coffman, WR and Rutger, JN. 1982. Pollen shedding and combining ability for high temperature tolerance in rice. *Crop Sci.*, 22: 730-733.

Maestri, E; Klueva, N; Perrota, C; Gulli, M; Nguyen, HT and Marmiroli, N. 2002. Molecular genetics of heat tolerance and heat shock proteins in cereals. *Plant Mol. Biol.*, 48: 667–681.

Martineau, JR; Williams, JH and Specht, JE. 1979. Temperature tolerance in soybeans. II Evaluation of segregating populations for membrane thermo stability. *Crop Sci.*, 19: 79-81.

Matsui, T and Omasa, K. 2002. Rice (*Oryza sativa* L.) cultivars tolerant to high temperature at flowering: anther characteristics. *Ann. Bot.*, 89: 683–687.

Matsui, T; Namuco, OS; Ziska, LH and Horie, T. 1997a. Effect of high temperature and CO2 concentration on spikelet sterility in indica rice. *Field Crops Res.*, 51:213–219.

Matsui, T; Omasa, K and Horie, T. 1997b. High temperature induced florets sterility of japonica rice at flowering in relation to air temperature, humidity and wind velocity condition. *Jpn. J. Crop. Sci.*, 66:449–455.

Matsui, T; Omasa, K and Horie, T. 2000. High temperature at flowering inhibit swelling of pollen grains, a driving force for thecae dehiscence in rice (*Oryza sativa* L.). *Plant Production Sci.*, 3: 430–434.

Matsui, T; Omasa, K and Horie, T. 2001. The difference in sterility due to high temperatures during the flowering period among japonica rice varieties. *Plant Production Sci.*, 4: 90–93.

Meehl, GA; Stocker, TF and Collins, WD. 2007. Global climate projections. In: Climate Change 2007: The Physical Science Basis. Contribution of Working Group I to the Fourth Assessment Report of the Inter-Governmental Panel on

Climate Change (Eds. Solomon, S; Qin, D and Manning, M) pp. 747–845. Cambridge University Press, Cambridge, UK and New York, NY, USA.

Monjardino, P; Smith, AG and Jones, RJ. 2005. Heat stress effects on protein accumulation of maize endosperm. *Crop Sci.*, 45: 1203–1210.

Moya, TB; Ziska, LH; Namuco, OS and Olszyk, D. 1998. Growth dynamics and genotypic variation in tropical, field-grown paddy rice (*Oryza sativa* L.) in response to increasing carbon dioxide and temperature. *Global Change Biol.*, 4:645–656.

Murakami, T; Matsuba, S and Funatsuki, H. 2004. Overexpression of a small heat shock protein, sHSP17.7, confers both heat tolerance and UV-B resistance to rice plants. *Molecular Breeding*, 13: 165–175.

Nir, I; Klein, S and Poijakoff-Mayber, A. 1969. Effects of moisture stress on submicroscopic structure of maize roots. *Aust. J .Biol. Sci.*, 22: 17-33.

Penuelas, J; Save, R; Marfa, O and Serrano, L. 1992. Remotely measured canopy temperature of greenhouse strawberries as indicator of water status and yield under mild and very mild water stress conditions. *Agric. Forest Meteorol.*, 58: 63-77.

Prasad, PVV; Boote, KJ; Allen, LH; Sheehy, JE and Thomas, JMG. 2006. Species, ecotype and cultivar differences in spikelet fertility and harvest index of rice in response to high temperature stress. *Field Crops Res.*, 95: 398–411.

Premchandra, GS; Sameoka, H and Ogata, S. 1990. Cell osmotic membrane – stability an indication of drought tolerance, as affected by applied nitrogen in soil. *J. Agric. Res.*, 115: 63-66.

Reynolds, MP; Balota, M; Delgado, MB; Amani, I and Fischer, RA. 1994. Physiological and morphological traits associated with spring wheat yield under hot, irrigated conditions. *Aust. J. Plant Physiol.*, 21: 717-730.

Sadalla, MM; Shanahau, JF and Quick, JS. 1990. Heat tolerance in winter wheat. I Hardening and genomic effects on membrane thermo stability. *Crop Sci.*, 30: 1249-1247.

Sanguineti, MC; Tuberosa, R; Landi, P; Salvi, S; Maccaferri, M; Casarini, E and Conti, S.1999. QTL analysis of drought - related traits and grain yiel d in relation to genetic variation for leaf abscisic acid concentration in field grown maize. *J. Exp. Bot.*, 50: 1289-1297.

Satake, T and Yoshida, S. 1978. High temperature induced sterility in indica rices at flowering. *Jpn. J. Crop Sci.*, 47: 6–17.

Senthil-Kumar, M; Kumar, G; Srikanthbabu, V and Udayakumar, M. 2007. Assessment of variability in acquired thermotolerance: A potential option to study genotypic response and the relevance of stress genes. *J. Plant Physiol.*, 164:111–125.

Sharkey, TD. 2005 Effect of moderate heat stress on photosynthesis: importance of thylakoid reactions, rubisco deactivation, reactive oxygen species and thermotolerance provided by isoprene. *Plant, Cell Environ.*, 28: 269-277.

Singh, P. and Kanemasu, ET. 1983. Leaf and canopy temperatures of pearl millet genotypes under irrigated and non irrigated conditions. *Agron. J.*, 75: 497-501.

Smertenko, A; Draber, P; Viklicky, V and Opatrny, Z. 1997. Heat stress affects the organization of microtubules and cell division in *Nicotiana tabacum* cells. *Plant Cell Environ.*, 20: 1534-1542.

Sullivan, CY. 1972. Mechanisms of heat and drought resistance in grain sorghum and method of measurements. In: Sorghum in the seventies. (Ed. Rao, NGP and House, LR).Oxford and IBH Publishing, New Delhi, pp 247-264.

Sun, W; Bernard, C; Van De Cotte, B; Van Montagu, M and Verbruggen, N. 2001. At- hsp17.6A, encoding a small heat shock protein in Arabidopsis, can enhance osmotolerance upon overexpression. *Plant J.*, 27: 407-415.

Turner, NC. 1986. Adaptation to water deficits: A challenging perspective. *Aust. J. Plant Physiol.*, 13: 175-190.

Tyagi, A; Kumar, N and Sairam, RK. 1999. Efficacy of RWC, membrane stability, osmotic potential, endogenous ABA and root biomass as indices for selection against water stress in rice. *Indian J. Plant Physiol.*, 4: 302-306.

Vierling, E. 1991. The roles of heat shock proteins in plants. *Annu. Rev. Plant Physiol. Plant Mol. Biol.*, 42: 579–620.

Wahid, A; Gelani, S; Ashraf, M and Foolad, M.R. 2007. Heat tolerance in plants: an overview. *Environt. Expt. Bot.*, 61: 199–223.

Xiao, Y; Pan, Y; Luo, L; Zhang, G; Deng, H; Dai, L; Liu, X; Tang, W; Chen, L and Wang, G. 2011. Quantitative trait loci associated with seed set under high temperature stress at the flowering stage in rice. *Euphytica*, 178: 331-338.

Yamanouchi, U; Yano, M; Lin, H; Ashikari, M and Yamada, K. 2002. A rice spotted leaf gene, Spl7, encodes a heat stress transcription factor protein. *Proc. Natl. Acad. Sci. U S A.*, 99: 7530-7535.

Yoshida, S; Satake, T and Mackill, DS. 1981. High temperature stress in rice. IRRI Research Paper Series, Vol. 67.

Yu, K; Chen, G and Patrick, WH Jr. 2004. Reduction of global warming potential contribution from a rice field by irrigation, organic matter, and fertilizer management. *Global Biogeochem. Cycles,* 18:1029-1038.

Zhang, G; Chen, L; Xiao, G; Xiao, Y; Chen, X and Zhang, S. 2009. Bulked segregant analysis to detect QTL related to heat tolerance in rice using SSR markers. *Agric. Sci. China,* 8: 482-487.

Zhang, T; Yang, L; Jiang, K; Huang, M; Sun, Q; Chen, W and Zheng, J. 2008. QTL mapping for heat tolerance of the tassel period of rice. *Mol. Plant Breed.*, 6: 867-873.

Zhao, Z; Zhang, L; Xiao, Y; Zhang, W; Zhai, H and Wan, J. 2006. Identification of QTLs for heat tolerance at the booting stage in rice. *Acta Agron. Sin.*, 32: 640-644.

Ziska, LH; Namuco, O; Moya, T and Quilang, J. 1997. Growth and yield response of field-grown tropical rice to increasing carbon dioxide and air temperature. *Agron. J.*, 89: 45-53.

Ziska LH and Teramura AH. 1992. CO_2 enhancement of growth and photosynthesis in rice (*Oryza sativa*) – modification by increased ultraviolet-B radiation. *Plant Physiol.*, 99, 473–481.

16

INFLUENCE OF HEAT STRESS ON PLANT RESPONSES : AN APPROACH TO PHYSICO-CHEMICAL, BIOTECHNOLOGICAL AND MOLECULAR ASPECTS

Anaytullah, Sananda Mondal*, Ashish Kumar Srivastava*** and Bandana Bose[1]***

**Department of Plant Physiology, Institute of Agricultural Sciences, B. H. U., Varanasi, (UP), India [[1] Corresponding author: bbosebhu@gmail.com]*
*** Nimbkar Agricultural Research Institute, P.O. Box-44, Lonand Road, Phaltan, Dist. Satara-415 523, Maharashtra, India*
****Area Office, National Seeds Corporation Limited, Court Compound, Kashipur, Samastipur, Bihar-848 101, India*

INTRODUCTION

On the planet earth plants due to their static nature are often subjected to more environment fluctuations among those salt stress, drought, heat stress, flooding, freezing, low temperature, UV light, nutrient deficiency, heavy metal and pathogenicity may display delayed growth and development, reduced productivity, and in severe cases, plant death. Plants display distinct changes in the morphological features in response to these abiotic stresses, including shortened life cycle (Porter, 2005). Sometimes morphological changes in lieu of abiotic stresses may occur in exposed plants to get rid off or to reduce the deleterious effects of stresses (Wang *et al.*, 2004). All of these in integrated form are known as stress effects. The prime cause of stress effects is regulated by water status. Water is a basic need from seed germination to plant maturation. Heat stress leads to increased evapo-transpiration of water from the plant surface and cause dehydration of aerial parts. High temperature hampers the cell water relations and limits growth in many plant species (Machado and Paulsen, 2001; Mazorra *et al.,* 2002; Wahid and Close, 2007). Elevated temperature is known to produce osmotic stress on the growing tissues due to reductions in root hydraulic conductance and tissue water content (Jiang and Huang, 2001; Morales *et al.*, 2003). Thus, stress, which occurs due to loss of water or unavailability of water to plants due to higher temperature comes under the perview of heat stress. Reddy *et al.* (1991) demonstrated that reduction of water damages the growth and development of cotton plants in sowing seasons. High temperature cause significant reduction in leaf growth of sorghum (*Sorghum bicolor*) and water status and water potential in wheat leaves. High temperature also hampers water, ions and organic solutes movement across the biological membranes. It affects photosynthetic and respiratory processes (Taiz and Zeiger, 2006), increases evapo-transpiration rate (Tsukaguchi *et al.*, 2003) and reduces the leaf osmotic potential and increases the leaf fluorescence. Heat stress induces the closure of stomata and negatively influences leaf water status. High temperature induced water stress is closely associated to reduction of soil water contents. In response to both biotic and abiotic stresses, plants synthesize a wide set of proteins including chitinases and peroxidases and up-regulate phenylalanine ammoniolyase, PAL (Dejaegher *et al.*, 1985). In fact, plants develop defensive reactions involving an activation of many genes (Sturm and Chrispeels, 1990). The study of Dell'Aquila and Spada (1994) has shown that hard wheat seeds have a good aptitude to germinate and seedlings establishment when seeds were exposed to optimal thermal conditions (25°C). On the other hand, low and high temperatures modify the capacity of germination (Bewley and Black, 1982). In India, heat stress is a major reason of yield decline in wheat due to delayed germination and plant establishment during its late sowing. Heat stress is a major challenge to wheat productivity in India. Enzyme activity, particularly, peroxidase considered as biochemical markers of stress (Castillo, 1992; Jbir *et al.*, 2001, 2002) varies according to temperature during germination of wheat seeds. In natural environment, plants are usually exposed to the combined effect of temperature and light. But excess of light could be harmful to the photosynthetic apparatus. Exposure of plants to temperature above the range of optimal temperature caused disturbance to the overall life cycle of the plant. High temperature provokes damage at the physiological, biochemical and molecular levels. Either light or heat stress or both together can generate oxidative stress by accumulating the Reactive Oxygen Species (ROS). Photosynthesis is sensitive to heat stress due to the damage in components of photosystem II which are located in thylakoid membranes of the chloroplasts (Suzuki and Mittler, 2006). During heat stress ROS production dramatically elevated; cellular injury and cell death occur with a very short span of time, representing a characteristic collapse of cellular organization (Schoffl *et al.,* 1999). At moderately high temperatures, injuries or death may occur only after long-term exposure. Direct injuries due to high temperatures include protein denaturation and aggregation, and increased fluidity of membrane lipids. Indirect or slower heat injuries include inactivation of enzymes in chloroplast and mitochondria, inhibition of protein synthesis, protein degradation and loss of membrane integrity (Howarth, 2005). Heat stress also affects the organization of microtubules by splitting and/or elongation of spindles, formation of microtubule asters in mitotic cells, and elongation of phragmoplast microtubules. These injuries eventually lead to starvation, inhibition of growth, reduced ion flux, production of toxic compounds and reactive oxygen species (ROS) (Schoffl *et al.,* 1999; Howarth, 2005). Immediately after exposure to high temperatures and perception of signals, changes occur at the molecular level altering the expression of genes and accumulation of transcripts, thereby leading to the synthesis of stress-related proteins as a stress tolerance strategy (Iba, 2002). Expression of heat shock proteins (HSPs) is known to be an important adaptive strategy in this regard (Feder and Hoffman, 1999). The HSPs, ranging in molecular mass from about 10 to 200 kDa, have chaperone like functions and are involved in signal transduction during heat stress (Schoffl *et al.,* 1999). The tolerance conferred by HSPs results in improved physiological phenomena such as photosynthesis, assimilate partitioning, water and nutrient use. Impact of heat stress is a series of metabolic dysfunctions and physical constraints that pile-up inside plants and become impossible for a plant to adjust, avoid or correct. The present review thus focuses on the stepwise consequences of heat stress on different metabolic pathways in plant system in respect to physiological, biochemical, biotechnological and molecular aspects.

2. STRESS

'Stress' or 'pressure' was introduced into the theory of elasticity as an amount of force for a given unit area (Cauchy, 1821). Since the 1930s, biologists have attempted to apply this terminology to biological systems, although the nature of the stresses will vary between living and non living organisms (Levitt, 1972). Every plant has some requirements of some physical factors viz. light, temperature, air, water etc. and for some chemical factors for example, inorganic, organic molecules, nutrients etc. in optimum amount and in available form at a right time as its nutritional requirements for proper growth and development. Any deviation from this optimal condition of any factor essential for its growth will lead to aberrant change in physiological processes and due to this, plant body will experience tension and this state can be referred as plant under stress. When this condition prevails for some time stress causes injury with or without producing visible symptoms which can be referred to as strain. The strain may be elastic or plastic. If the reaction is temporary and the plant returns to its optimum level when conditions return to its original, it is called elastic biological strain. However, if the reaction is permanent and plant functions do not return to normal, it is called plastic biological strain. The term plant stress has been defined by Lichtenthaler (1996) as 'any unfavorable condition or substance that affects or blocks a plant's metabolism, growth or development', by Strasser as 'a condition caused by factors that tend to alter an equilibrium', and by Larcher as 'changes in physiology that occur when species are exposed to extraordinary unfavorable conditions that need not represent a threat to life but will induce an alarm response' (Gaspar *et al.,* 2002).

Type of Plant stresses

There are a variety of stresses to which the plants are exposed. Some common stresses are as follows.

i. **Water stress:** It includes both the abundance (flood) and scarcity of water (drought).
ii. **Temperature stress**: includes high temperature stress (Heat stress) and low teimperature stress (freezing and chilling).
iii. **Chemical stress:** A large number of pollutants, toxic substances insecticides, pesticides, chemical fertilizers etc. present in some soils cause havoc to plant life. They act as a chemical stress. High concentration of salts present in some soils act as salt stress.
iv. **Light stress:** It includes stresses, caused by Ultraviolet (UV), Infrared (IR) and ionizing radiations.
v. **Physical stress:** Such as sound, electromagnetic forces etc. also influence the plant growth and development.

3. HEAT STRESS

Plants generally die when exposed to temperature of 50°C or above, but some of them tolerate high temperature above the 50°C. Prokaryotes, particularly bacteria have been found growing at boiling point of water at the bottom of the oil wells at the temperature above 100°C. Some algae growing in hot springs (50-60°C); cacti and some desert plants experience much higher temperature and yet survive. In general, the high temperature results in loss of water so that the plants experience drought stress. The photosynthesis stops above 70-72°C and the protein molecules denature. The plants which resist high temperature show (i) high protoplasmic viscosity, (ii) high levels of water of hydration, (iii) possess enzyme which are more stable to high temperature, (iv) decreased photosynthesis, (v) increased respiration, (vi) closed stomata, stops CO_2 capture, increased photo-respiration, (vii) major slow-down in transpiration (cooling process loss and internal temperature increase), (viii) cell membrane leakage (signals changes in protein synthesis), (ix) continued physical water loss, (x) plant starvation through rapid use of food reserves, inefficient food use, and inability to call on reserves when and where needed, (xi) toxins generated through cell membrane releases and respiration problems, (xii) membrane integrity loss and protein breakdown, and (xii) growth inhibition.

Heat is often defined as the rise in temperature beyond a threshold level for a period of time sufficient to cause irreversible damage to plant growth and development. Heat stress due to high ambient temperatures is a serious threat to crop production worldwide (Hall, 2001). While stress often defined as where, temperatures are hot enough for sufficient time that they cause irreversible damage to plant function or development. In addition, high temperature can increase the rate of reproductive development, which shortened the time for photosynthesis to contribute to fruit or seed production (Essemine *et al.,* 2010).

3.1 Heat Stress and Plant Responses

Many morpho-physiological and metabolic alterations are taken place in plant system under the influence of heat stress.

3.1.1 Morpho-physiological changes

In tropical climates, high temperatures and excess of radiation are often the most important factors affecting plant growth, development and final crop yield. High temperatures can cause considerable pre-and post-harvest damages including scorching of leaves and sunburns on leaves, branches and stems, leaf senescence and abscission, shoot and root growth inhibition, fruit discoloration and damage, and reduced yield (Guilioni *et al.,* 1997; Ismail and Hall, 1999; Vollenweider and Gunthardt-Goerg, 2005). Similarly, in temperate regions, heat stress has been reported as one of the most important causes of reduction in dry matter production and yield in many crops including maize (Giaveno and Ferrero, 2003). High temperature-induced modifications in plants may be direct or indirect in altering the pattern of development. For example, long-term effects of heat stress on developing seeds may include delayed germination or loss of vigor, ultimately leading to reduced emergence and seedling establishment. Under diurnally varying temperatures, coleoptile's growth in maize reduced at 40°C and ceased at 45°C (Weaich *et al.,* 1996). High temperatures caused significant decline in shoot dry mass, relative growth rate and net assimilation rate in maize, pearl millet and sugarcane, though leaf expansion was minimally affected (Ashraf and Hafeez, 2004; Wahid, 2007). Major impact of high temperatures on shoot growth is a severe reduction in the first internode length resulting in premature death of plants (Hall, 1992). For example, sugarcane plants grown under high temperatures exhibited smaller internodes, increased tillering, early senescence and reduced total biomass (Ebrahim *et al.,* 1998). Heat stress alone or in combination with drought, is a common constraint during anthesis and grain filling stages in many cereal crops of temperate regions. For example, heat stress reduced the duration of grain filling with reduction in kernel growth leading to losses in kernel density and weight by 7% in spring wheat (Guilioni *et al.,* 2003). Similar reductions occurred in starch, protein and oil contents of the maize kernel (Wilhelm *et al.,* 1999) and also grain quality in other cereal crops under heat stress (Maestri *et al.,* 2002). Grain weight and grain number both appeared to be sensitive to heat stress in wheat crop as the number of grains per ear at maturity declined with increasing temperature (Ferris *et al.,* 1998). In tomato, reproductive processes were adversely affected by high temperature, which included meiosis in both male and female organs, pollen germination and pollen tube growth, ovule viability, stigmatic and style positions, number of pollen grains retained by the stigma, fertilization and post-fertilization processes, growth of the endosperm, pre-embryo and fertilized embryo (Foolad, 2005). Also, the most noticeable effect of high temperatures on reproductive processes in tomato is the production of an excreted style (i.e., stigma is elongated beyond the anther cone), which may prevent self pollination. Poor fruit set at high temperature has also been associated with low levels of carbohydrates and growth regulators released in plant sink tissues (Kinet and Peet, 1997). Growth chamber and green house studies suggest that high temperature is most deleterious when flowers are first visible and sensitivity continues for 10–15 days. Reproductive phases most sensitive to high temperature are gametogenesis (8–9 days before anthesis) and fertilization (1–3 days after anthesis) in various plants (Foolad, 2005). Both male and female gametophytes are sensitive to high temperature and response varies with genotype; however, ovules are generally less heat sensitive than pollen (Peet and Willits, 1998). Overall, based on the available studies, it seems that plant responses to high temperature vary with plant species and phenological stages. Reproductive processes are markedly affected by high temperatures in most plants which ultimately affect fertilization and post-fertilization processes leading to reduced crop yield.

3.1.2 Changes in water status

Plant water status is the most important variable under changing ambient temperatures (Mazorra *et al.,* 2002). In general, plants tend to maintain stable tissue water status regardless of temperature when moisture is ample; however, high temperatures severely impair this tendency when water is limiting (Machado and Paulsen, 2001). Under field conditions, high temperature stress is frequently associated with reduced water availability (Simoes-Araujo *et al.,* 2003). In *Lotus creticus*, for example, elevated night temperatures caused a greater reduction in leaf water potential of water-stressed as compared to non-stressed plants (A˜non *et al.,* 2004). In sugarcane, leaf water potential and its components were changed upon exposure to heat stress even though the soil water supply and relative humidity conditions were optimal, implying an effect of heat stress on root hydraulic conductance (Wahid and Close, 2007).

Similarly, in tomato heat stress perturbed the leaf water relations and root hydraulic conductivity (Morales *et al.,* 2003). In general, day time enhanced transpiration induces water deficiency in plants, causing a decrease in water potential and leading to perturbation of many physiological processes (Tsukaguchi *et al.,* 2003). High temperatures seem to cause water loss in plants more during day time than night time.

3.1.3 Carbon metabolism

It deals with the photosynthetic activity of chloroplasts, considered to be most heat sensitive cell function (Berry and Bjo¨rkman 1980; Yordanov *et al.* 1986). Alterations in various photosynthetic attributes under heat stress are good indicators of thermo-tolerance of the plant as they show correlations with growth. There are three major stress-sensitive sites in the photosynthetic machinery (PM): the photosystems, mainly photosystem II (PSII) with its oxygen-evolving complex (OEC), the ATP generating, and the carbon assimilation processes (Aro *et al.* 1993; Bukhov and Mohanty 1999; Bukhov and Carpentier 2000; Carpentier 1999; Nishiyama *et al.* 2005, 2006; Murata *et al.* 2007; Mohanty *et al.* 2007). Thermal stress leads to an increase in the relative fraction of QB-restoring RC (Havaux *et al.* 1991; Kreslavski and Khristin 2003), loss of thylakoid membrane integrity, especially destacking of the thylakoid membranes (Gounaris *et al.* 1984; Semenova 2004) and dysfunction in the system of CO_2 assimilation (Sharkey 2005). Any constraint in photosynthesis can limit plant growth at high temperatures. Photochemical reactions in thylakoid lamellae and carbon metabolism in the stroma of chloroplast have been suggested as the primary sites of injury at high temperatures (Wise *et al.,* 2004). Chlorophyll fluorescence, the ratio of variable fluorescence to maximum fluorescence (Fv/Fm), and the base fluorescence (F_0) are physiological parameters that have been shown to correlate with heat tolerance (Yamada *et al.,* 1996). Schreiber *et al.* (1994) argued that one of the first responses of a plant to environmental stress is an increase in non radiative energy dissipation, which is reflected by the amount of chlorophyll fluorescence. Good correlations have been found between chlorophyll fluorescence and root growth potential, gas exchange, electrolyte leakage, visible leaf damage and leaf water potential (Larcher, 1994). Larcher (1994) suggested that for monitoring heat stress, chlorophyll fluorescence may be a more reliable measurement of photosynthesis than CO_2 exchange which can be influenced by stomatal closure not induced primarily by heat. Increasing leaf temperatures and photosynthetic photon flux density influence thermo-tolerance adjustments of PSII, indicating their potential to optimize photosynthesis under varying environmental conditions as long as the upper thermal limits do not exceed (Salvucci and Crafts- Brandner, 2004b; Marchand *et al.,* 2005). In tomato genotypes differing in their capacity for thermo tolerance as well as in sugarcane, an increased chlorophyll *a*:*b* ratio and a decreased chlorophyll : carotenoids ratio were observed in the tolerant genotypes under high temperatures, indicating that these changes were related to thermo tolerance of tomato (Camejo *et al.,* 2005; Wahid and Ghazanfar, 2006). Furthermore, under high temperatures, degradation of chlorophyll *a* and *b* was more pronounced in developed compared to developing leaves (Karim *et al.,* 1997, 1999). Such effects on chlorophyll or photosynthetic apparatus were suggested to be associated with the production of active oxygen species (Camejo *et al.,* 2006; Guo *et al.,* 2006). PSII is highly thermo labile and its activity is greatly reduced or even partially stopped under high temperatures (Bukhov *et al*., 1999; Camejo *et al.,* 2005) which may be due to the properties of thylakoid membranes where PSII is located (Mcdonald and Paulsen, 1997). Heat stress may lead to the dissociation of oxygen evolving complex (OEC) resulting in an imbalance between the electron flow from OEC toward the acceptor side of PSII in the direction of PSI reaction center (De Ronde *et al.,* 2004). Heat stress causes dissociation of manganese (Mn)-stabilizing 33-kDa protein at PSII reaction center complex followed by the release of Mn atoms (Yamane *et al.,* 1998). Heat stress may also impair other parts of the reaction center e.g. the D1 and/or the D2 proteins (De Las Rivas and Barber, 1997). In wheat, high temperatures and excessive light damaged different sites of PSII, this implied different pathways for the recovery of its functional activity (Sharkova, 2001). In barley, heat pulses abruptly damaged the PSII units and caused loss of their capacity of oxygen evolution leading to a restricted electron transport, which was totally abolished after four hours (Toth *et al.,* 2005). This implied that the degradation of the impaired PSII units occurred in the light during this period of time. Following this, *de novo* synthesis of PSII units in the light gave a gradual rise to the observed PSII activities. These effects can result from different events including inhibition of electron transport activity and limited generation of reducing powers for metabolic functions (Allakhverdieva *et al.,* 2001). In field-grown Pima cotton under high temperatures, leaf photosynthesis was functionally limited by photosynthetic electron transport and ribulose-1,5-bisphosphate (RuBP) regeneration capacity, but not rubisco activity (Wise *et al.,* 2004). On the other hand, under high temperatures, PSI stromal enzymes and chloroplast envelops are thermo stable and, in fact, PSI driven cyclic electron pathway capable of contributing to thylakoids proton gradient, is activated (Bukhov *et al.,* 1999). High temperature influences the photosynthetic capacity of C_3 plants more strongly than C_4 plants. It alters the energy distribution and changes the activities of carbon metabolism enzymes, particularly the rubisco, thereby altering the rate of RuBP regeneration by the disruption of electron transport and inactivation of the oxygen evolving enzymes of PSII (Salvucci and Crafts Brandner, 2004b). Heat shock reduces the amount of photosynthetic pigments (Todorov *et al.,* 2003), soluble proteins, rubisco binding proteins (RBP) and large- (LS) and small subunits (SS) of rubisco in darkness but increases them in light, indicating their roles as chaperones and HSPs (Kepova *et al.,* 2005). Moreover, under heat stress, starch or sucrose synthesis is greatly influenced as observed from reduced activities of sucrose phosphate synthase (Chaitanya *et al.,* 2001), ADP glucose pyrophosphorylase and invertase (Vu *et al.,* 2001). In any plant species, the ability to sustain leaf gas exchange under heat stress has a direct relationship with heat tolerance. During the vegetative stage, high day temperature can cause damage to compensated leaf photosynthesis, reducing CO_2 assimilation rates (Hall, 1992). Increased temperatures curtail photosynthesis and increase CO_2 transfer conductance between intercellular spaces and carboxylation sites. Stomatal conductance (gs) and net photosynthesis (Pn) are inhibited by moderate heat stress in many plant species due to decreases in the activation state of rubisco (Crafts-Brander and Salvucci, 2002; Morales *et al.,* 2003). Although with an increase in temperature rubisco catalytic activity increases, a low affinity of the enzyme for CO_2 and its dual nature as an oxygenase limits the possible increases in Pn. For example, in maize the Pn was inhibited at leaf temperatures above 38°C and inhibition was much more severe when temperature was increased abruptly rather than gradually. However, this inhibition was independent of stomatal response to high temperature (Crafts-Brander and Salvucci, 2002). Despite observed negative effects of high temperature, the optimum temperature for leaf photosynthesis is likely to increase with elevated levels of atmospheric CO_2. Several studies have concluded that CO_2-induced increases in crop yields are much more plausible in warm- than in cool-season crops. Thus, despite its other potential negative implications, global warming may not greatly affect the overall Pn. A well-known consequence of elevated temperature in plants is the damage caused by heat-induced imbalance in photosynthesis and respiration i.e., the whole carbon metabolism; in general, the rate of photosynthesis decreases while dark- and photo-respiration rates increase considerably under high temperatures. Also, rate of biochemical reactions decreases and enzyme inactivation and denaturation take place as the temperature increases leading to severely reduced photosynthesis (Nakamoto and Hiyama, 1999). However, the magnitude of such alterations in response to heat stress differs with species and genotypes (Mcdonald and Paulsen, 1997). Furthermore, it has been determined that the photosynthetic CO_2 assimilation rate is less affected by heat stress in developing leaves than in completely developed leaves. Heat stress normally decreases the duration of developmental phases leading to smaller organs, reduced light perception and carbon assimilation processes including transpiration, photosynthesis and respiration (Stone, 2001). Nonetheless, photosynthesis is considered as the physiological process most sensitive to high temperatures, and that rising atmospheric CO_2 content will drive temperature increases in many already stressful environments. In crux, the anabolism of carbon is the most important metabolic aspect in plants regulated by a number of enzymes hence any alteration in temperature (more than optimum level) may influence the process of photosynthesis.

3.1.4 Assimilate partitioning

Assimilate partitioning (AP) is influenced greatly under high temperature stress. A reduction in source and sink activities may lead a measurable decline in growth, economic yield and harvest index in response to low or moderate heat stress. AP taking place via apoplastic and symplastic pathways under high temperatures has significant effects on transport and transfer processes in plants (Taiz and Zeiger, 2006). However, Yang *et al.* (2002) reported a genotypic variation in AP in wheat. Gowronska *et al.* (1992) observed that high temperature reduced total dry matter production and altered dry matter distribution in favor of vines at the expense of tubers. High temperature reduces both total dry matter accumulation and the proportion of total dry matter partitioned to tubers (Burton, 1972; Ewing, 1981; Khedher and Ewing, 1985). Dinar and Rudich (1985) observed the effect of heat stress on AP in two cultivars of tomato. Basipetal transport to roots was inhibited more in Roma VF than heat tolerant variety, Saladette. AP within young truss was more affected in Roma VF in hest stress. To elucidate causal agents of reduced grain filling in wheat under high temperatures, Wardlaw (1974) examined three main components of the plant system including source (flag leaf blade), sink (ear) and transport pathway (peduncle). It was determined that photosynthesis had a broad temperature optimum from 20 to 30°C, however. it declined rapidly at temperatures above 30°C. The rate of ^{14}C assimilate movement out of the flag leaf (phloem loading) was optimum around 30°C, however, the rate of movement through the stem was independent of temperature from 1 to 50°C. It was concluded that at least in wheat temperature effects on translocation result indirectly from temperature effects on source and sink activities. From such results, increased mobilization efficiency of reserves from leaves, stem or other plant parts has been suggested as a potential strategy to improve grain filling and yield in wheat under heat stress. This suggestion, however, is based on present limited knowledge of physiological basis of assimilate partitioning under high temperature stress. Further investigation in this area may lead to improved crop production efficiency under high-temperature stress.

3.1.5 Cell membrane thermo stability

Membrane thermostability (MT) test is a widely used and acceptable method to evaluate heat tolerance. As the high temperature membrane injury was caused by free radicals, the higher activity of superoxide dismutase and catalase may be important in the higher membrane thermo stability in tolerant cultivars. Sustained function of cellular membranes under stress is pivotal for processes such as photosynthesis and respiration (Blum, 1988). Heat stress accelerates the kinetic energy and movement of molecules across membranes thereby loosening chemical bonds within molecules of biological membranes. This makes the lipid bilayer of biological membranes more fluid by either denaturation of proteins or an increase in unsaturated fatty acids (Savchenko *et al.,* 2002). The integrity and functions of biological membranes are sensitive to high temperature, as heat stress alters the tertiary and quaternary structures of membrane proteins. Such alterations enhance the permeability of membranes, as evident from increased loss of electrolytes. The increased solute leakage, as an indication of decreased cell membrane thermo stability (CMT), has long been used as an indirect measure of heat-stress tolerance in diverse plant species, including soybean (Martineau *et al.,* 1979), potato and tomato (Chen *et al.,* 1982), wheat (Blum *et al.,* 2001), cotton (Ashraf *et al.,* 1994), sorghum (Marcum, 1998), cowpea (Ismail and Hall, 1999) and barley (Wahid and Shabbir, 2005). Electrolyte leakage is influenced by plant/tissue age, sampling organ, developmental stage, growing season, degree of hardening and plant species. In maize, injuries to plasmalemma due to heat stress were much less severe in developing than in mature leaves (Karim *et al.,* 1997, 1999). CMT has been used to screen and evaluate different wheat genotypes for thermal tolerance (Blum and Ebercon, 1981, Saadalla *et al.*, 1990a). This method measures the increased electrolyte diffusion resulting from heat induced cell membrane permeability. Electrical conductivity has been used as an index of membrane stability to identify heat-tolerant genotypes in wheat (Blum and Ebercon, 1981) and for screening of heat-tolerant genotypes in different crops (Blum, 1988). When tissues are subjected to high temperature, electrical conductivity increases due to damage to the cell membrane and consequent solute leakage. It was determined that an increase in saturated fatty acids in mature leaves elevated melting temperature of plasma membranes thus reducing heat tolerance of the plant. In *Arabidopsis* plants grown under high temperature, total lipid content in membranes decreased to about one-half and the ratio of unsaturated to saturated fatty acids decreased to one-third of the levels at normal temperatures (Somerville and Browse, 1991). It should be noted, however, that in some species heat tolerance does not correlate with the degree of lipid saturation, suggesting that factors other than membrane stability might be limiting the growth at high temperatures. The relationship between CMT and crop yield under high temperatures may vary from plant to plant and invokes for study of individual crops before using it as an important physiological selection criterion. For example, whereas a significant relationship between CMT and yield was observed in a few plant species such as sorghum (Sullivan and Ross, 1979), no such relationship was observed in soybean (Martineau *et al.,* 1979) or wheat (Shanahan *et al.*, 1990). Thus, the major cause(s) of yield suppression under heat stress remain largely elusive and deserve further experimentation.

3.1.6 Osmotic regulation

A key adaptive mechanism in many plants grown under abiotic stresses including salinity, water deficit and extreme temperatures, is accumulation of certain organic compounds of low molecular mass, generally referred to as compatible osmolytes (Hare *et al.*, 1998; Sakamoto and Murata, 2002). Under stress, different plant species may accumulate a variety of osmolytes such as sugars and sugar alcohols (polyols), proline, tertiary and quaternary ammonium compounds, and tertiary sulphonium compounds (Sairam and Tyagi, 2004). Accumulation of such solutes may contribute to enhanced stress tolerance of plants, as briefly described below. Glycinebetaine (GB), an amphoteric quaternary amine, plays an important role as a compatible solute in plants under various stresses, such as salinity or high temperature (Sakamoto and Murata, 2002). Production of GB is plant specific. Some plants like maize and sugarcane showed a significant level of GB production (Quan *et al.*, 2004; Wahid and Close, 2007) during dessiccating condition of high temperature. In contrast, plant species such as rice (*Oryza sativa*), mustard (*Brassica* spp.), Arabidopsis (*Arabidopsis thaliana*) and tobacco (*Nicotiana tabacum*) naturally do not produce GB under stress conditions. However, genetic engineering has allowed the introduction of GB-biosynthetic pathways into GB-deficit species (Sakamoto and Murata, 2002; Quan *et al.*, 2004). Like GB, proline is also known to occur widely in higher plants and normally accumulates in large quantities in response to environmental stresses (Kavi Kishore *et al.*, 2005). In assessing the functional significance of accumulation of compatible solutes, it is suggested that proline or GB synthesis may buffer cellular redox potential under heat and other environmental stresses (Wahid and Close, 2007). Similarly, accumulation of soluble sugars under heat stress has been reported in sugarcane, which entails great implications for heat tolerance (Wahid and Close, 2007). Under high temperatures, fruit set in tomato plants failed due to the disruption of sugar metabolism and proline transport during the narrow window of male reproductive development (Sato *et al.*, 2006). Hexose sensing in transgenic plants engineered to produce trehalose, fructans or mannitol may be an important contributory factor to the stress-tolerant phenotypes (Hare *et al.*, 1998). Among other osmolytes, 4-aminobutyric acid (GABA), a non-protein amino acid, is widely distributed throughout the biological world to act as a compatible solute. GABA is synthesized from the glutamic acid by a single step reaction catalyzed by glutamate decarboxylase (GAD). An acidic pH activates GAD, a key enzyme in the biosynthesis of GABA. Episodes of high tempeartures increase the cytosolic level of Ca^{+2}, which leads to calmodulin-mediated activation of GAD (Taiz and Zeiger, 2006). Several other studies show that various environmental stresses increase GABA accumulation through metabolic or mechanical disruptions, thus leading to cytosolic acidification. Kinetics of GABA in plants shows a stress-specific pattern of accumulation, which is consistent with its physiological role in the mitigation of stress effects. Rapid accumulation of GABA in stressed tissues may provide a critical link in the chain of events stemming from perception of environmental stresses to timely physiological responses (Kinnersley and

Turano, 2000). In summary, because of significant roles of osmolytes in response to environmental stresses in plants, crop stress (e.g. heat) tolerance might be enhanced by increased accumulation of compatible solutes through traditional plant breeding, marker-assisted selection (MAS) or genetic engineering (GE) approaches (Ashraf and Foolad, 2007).

3.1.7 Antioxidant metabolism

In addition to tissue dehydration, heat stress may induce oxidative stress. For example, generation and reactions of activated oxygen species (AOS) including singlet oxygen ($1O_2$), superoxide radical (O_2^-), hydrogen peroxide (H_2O_2) and hydroxyl radical (OH^-) are symptoms of cellular injury due to high temperature (Liu and Huang, 2000). AOS causes the autocatalytic peroxidation of membrane lipids and pigments thus leading to the loss of membrane semi-permeability and modifying its functions (Xu *et al.,* 2006). Superoxide radical is regularly synthesized in the chloroplast and mitochondrion and some quantities are also produced in microbodies. The scavenging of O_2^- by superoxide dismutase (SOD) results in the production of H_2O_2, which is removed by APX or CAT. However, both O_2^- and H_2O_2 are not as toxic as the (OH^-), which is formed by the combination of O_2^- and H_2O_2 in the presence of trace amounts of Fe_2^+ and Fe_3^+ by the Haber–Weiss reaction. The OH^- can damage chlorophyll, protein, DNA, lipids and other important macromolecules, thus fatally affecting plant metabolism and limiting growth and yield (Sairam and Tyagi, 2004). Plants have developed a series of both enzymatic and non-enzymatic detoxification systems to counteract AOS, thereby protecting cells from oxidative damage (Sairam and Tyagi, 2004). For example, over expression of SOD in plants affect a number of physiological phenomena, which include the removal of H_2O_2, oxidation of toxic reductants, biosynthesis and degradation of lignin in cell walls, auxin catabolism, defensive responses to wounding, defense against pathogen or insect attack, and some respiratory processes (Scandalios, 1993). More specifically, expression and activation of APX is related to the appearance of physiological injuries caused in plants by thermal stress (Mazorra *et al.,* 2002). Decrease in antioxidant activity in stressed tissues results in higher levels of AOS that may contribute to injury (Fadzillah *et al.,* 1996). Protection against oxidative stress is an important component in determining the survival of a plant under heat stress. Studies on heat-acclimated versus non-acclimated cool season turf grass species suggested that the former had lower production of ROS as a result of enhanced synthesis of ascorbate and glutathione (Xu *et al.,* 2006). Available data suggest that some signaling molecules may cause an increase in the antioxidant capacity of cells (Gong *et al.,* 1997; Dat *et al.,* 1998). Certainly further research is necessary to identify the signaling molecules which enhanced the production of antioxidants in cells exposed to heat stress.

3.2 Heat Shock Protein (HsPs)

Plants vary greatly in the amount of expressed Hsps as well as their types (Hamilton *et al.*, 1996). The most studied species of plants is *Arabidopsis thaliana* where the response to heat shock treatment occurs through the participation of a number of different Hsps and those are - 13 (Hsp20), 8 (Hsp70), 7 (Hsp90) and 8 (Hsp100). Swindell *et al.* (2007) described about 21 transcription factors (Hsfs), among them in tomato at least 15 Hsfs are there (von Koskull-Do ring *et al.*, 2007). Higher plants are characterized by the presence of at least 20 types of sHsps, but one species could contain 40 types of these sHsps (Vierling, 1991). sHsps, which are usually undetectable in plant cells under physiological conditions, are induced upon stress and plant tolerance to stress, including drought, salinity, oxidized species and low temperatures (Low *et al.*, 2000; Hamilton and Heckathorn, 2001; Scharf *et al.*, 2001; Zhang *et al.*, 2008). It is believed that this diversification and abundance of the sHsps in a plant reflects an adaptation of the plant to heat stress (Waters *et al.*, 1996).

Table 1: Location of sHsps types in plant cell (after Waters *et al.*, 1996).

Chloroplast-localized proteins	
Arabidopsis thaliana:	HSP21
Glycine max:	HSP22
Petunia hybrid:	HSP21
Pisum sativium:	HSP21
Triticum aestivum:	HSP26A
Triticum aestivum:	HSP26B
Zea mays:	HSP26
Mitochondrial-localized protein	
Chenopodium rubrum:	HSP23
Endoplasmic reticulum-localized proteins	
Arabidopsis thaliana:	HSP22
Glycine max:	HSP22
Pisum sativum:	HSP22
Class I Cytosolically localized proteins	
Arabidopsis thaliana:	HSP17.6
Arabidopsis thaliana:	HSP17.4
Arabidopsis thaliana:	HSP18.2
Chenopodium rubrum:	HSP18.3
Daucus carota:	HSP18.0
Daucus carota:	HSP17.8
Glycine max:	HSP17.5
Glycine max:	HSP17.6
Glycine max:	HSP18.5
Helianthus annuus:	HSP17.6
Lycopersicon esculentum:	HSP17.8

Medicago sativa:	HSP18.1
Medicago sativa:	HSP18.2
Oryza sativa:	HSP16.9
Oryza sativa:	HSP17.4
Pisum sativum:	HSP18.1
Triticum aestivum:	HSP16.9A
Triticum aestivum:	HSP16.9B
Triticum aestivum:	HSP16.9C
Zea mays:	HSP17.2
Class I1 Cytosolically localized proteins	
Arabidopsis thaliana:	HSP17.6
Glycine max:	HSP17.9
Ipomea nil (Pharbatis nil):	HSP18.8
Ipomea nil (Pharbatis nil):	HSP17.2
Lilium longiflorum:	HSP17.6
Lilium longiflorum:	HSP16.5
Pisum sativum:	HSP17.7
Triticum aestivum:	HSP17.3
Zea mays:	HSP17.5
Zea mays:	HSP17.8

3.2.1 Role of heat-shock proteins

The function of any protein is determined by its formation and folding into three dimensional structures (Levitt *et al.*, 1997). Formation of three dimensional structures requires 50% of principle amino acids sequence (Dobson *et al.*, 1998). That is where the role of Hsps in the folding of other proteins is important. Morimoto and Santoro (1998) indicated that Hsps protect cells from injury and facilitate recovery and survival after a return to normal growth conditions. On the other hand, Timperio *et al.* (2008) specified that upon heat stress, the role of Hsp as molecular chaperones could be different: unfolding of proteins is not the main effect and protection from damage could occur in an alternative way apart from ensuring the maintenance of correct protein structure. It has been suggested that Hsps general role is to act as molecular chaperones (Fig. 1) regulating the folding and accumulation of proteins as well as localization and degradation in all plants and animal species (Feder and Hofmann, 1999; Schulze-Lefert, 2004; Panaretou and Zhai, 2008; Hu *et al.*, 2009; Gupta *et al.*, 2010). These proteins, as chaperones, prevent the irreversible aggregation of other proteins and participate in refolding proteins during heat stress conditions (Tripp *et al.*, 2009).

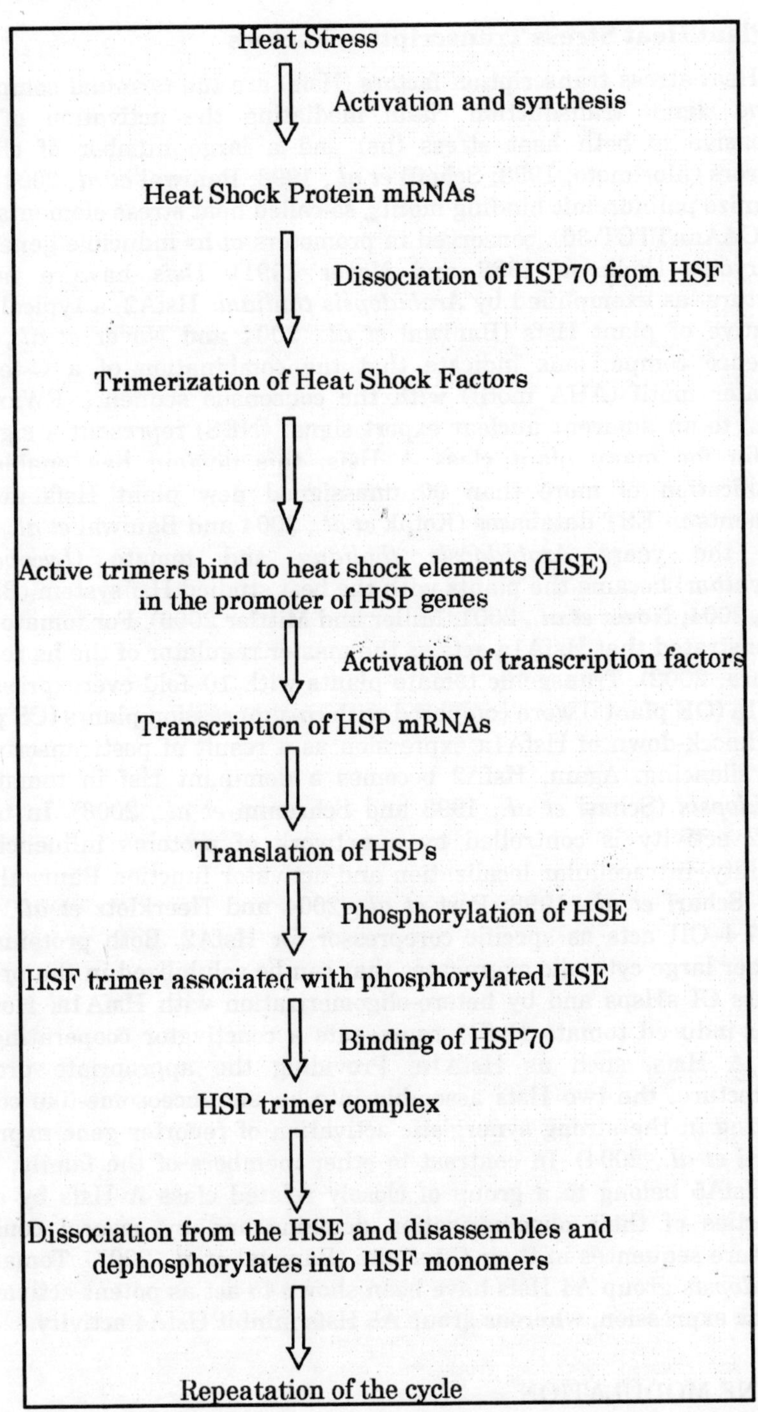

Fig. 1. Simple illustration of part of the chaperone machines that operate in the cytosol

3.3 Plant Heat Stress Transcription Factors

Heat stress transcription factors (Hsfs) are the terminal components of the signal transduction chain mediating the activation of genes responsive to both heat stress (hs) and a large number of chemical stressors (Morimoto, 1998; Schoffl *et al.*, 1998; Baniwal *et al.* 2004). They recognize palindromic binding motifs, so-called heat stress elements (HSE: 50-AGAAnnTTCT-30), conserved in promoters of hs inducible genes of all eukaryotes (Pelham, 1982 and Nover 1991). Hsfs have a modular structure, as exemplified by *Arabidopsis thaliana* HsfA2, a typical representative of plant Hsfs (Baniwal *et al.*, 2004 and Nover *et al.*, 2001). Sequence comparisons indicate that the combination of a C-terminal activator motif (AHA motif) with the consensus sequence FWxx (F/L) (F/I/L) to an adjacent nuclear export signal (NES) represent a signature domain for many plant class A Hsfs; this domain has enabled the identification of more than 90 unassigned new plant Hsfs and Hsf fragments in EST databases (Kotak *et al.*, 2004 and Baniwal *et al.*, 2007). Over the years *Arabidopsis thaliana* and tomato (*Lycopersicon esculentum*) became the plants with the best studied Hsf system (Baniwal *et al.*, 2004; Nover *et al.*, 2001; Miller and Mittler 2006). For tomato it was demonstrated that HsfA1a acts as the master regulator of the hs response (Mishra, 2002). Transgenic tomato plants with 10–fold overexpression of HsfA1a (OE plants) were compared with co-suppression plants (CS plants) with knock-down of HsfA1a expression as a result of posttranscriptional gene silencing. Again, HsfA2 becomes a dominant Hsf in tomato and *Arabidopsis* (Scharf *et al.*, 1998 and Schramm *et al.*, 2006). In tomato, HsfA2 activity is controlled by a network of proteins influencing its solubility, intracellular localization and activator function Baniwal *et al.*, 2004; Scharf *et al.*, 1998; Port *et al.*, 2004 and Heerklotz *et al.*, 2001). Hsp17.4–CII acts as specific corepressor for HsfA2. Both proteins form together large cytosolic aggregates that can be solubilized in the presence of class CI sHsps and by hetero-oligomerization with HsfA1a. However, the hs-induced tomato HsfB1 represents a coactivator cooperating with class A Hsfs, such as HsfA1a. Providing the appropriate promoter architecture, the two Hsfs assemble into an enhanceosome-like complex resulting in the strong synergistic activation of reporter gene expression (Bharti *et al.*, 2004). In contrast to other members of the family, HsfA4 and HsfA5 belong to a group of closely related class A Hsfs by special properties of their oligomerization domains and by several conserved signature sequences in their C-termini (Baniwal *et al.*, 2007). Tomato and *Arabidopsis* group A4 Hsfs have been shown to act as potent activators of hs gene expression, whereas group A5 Hsfs inhibit HsfA4 activity.

4. GENE MODULATION

Abiotic stresses including heat stress are regulated by various defense mechanisms representing the crucial role of transcriptional regulation. A number of genes involved in heat stress response and tolerance induced during this process have resulted from transcriptional analysis (Singh *et al.*, 2002; Chen and Zhu, 2004; Kotak *et al.*, 2007). Modification of gene expression improved stress tolerance in *Arabidopsis* and rice. The synthesis and accumulation of HSPs through HSFs network play an important role in plant heat tolerance (Wang *et al.*, 2004 and Kotak *et al.*, 2007). It is found that expression of specific gene is required when reproductive organ of plant is exposed to abiotic stresses (Sorensen *et al.*, 2003; Yamaguchi-Shinozaki and Shinozaki, 2000; Shinozaki *et al.*, 2003). Heat stress altered development and demarcation of pollen mother cell due to tissue specific alterations in gene expression (Sakata *et al.*, 2000; Abiko *et al.*, 2005). In maize reproductive parts like pollen development and growth of pollen tube are sensitive to heat stress. Heat shock elevated HSP70 and heat shock factor (hsf) genes are required at early pollen development. High temperature is required for hsfs gene expression. Studies showed that *hsp18* and *hsp70* genes are transcribed in response to heat shock (Gagliardi *et al.*, 1995; Mascarenhas and Crone, 1996; Nover *et al.*, 2001; Baniwal *et al.*, 2004). Further, Hopf *et al.* (1992) reported that the defective heat shock response of mature pollen of maize was due to reduced induction of heat shock gene transcription. *HSP70* assisted in translocation, proteolysis, protein translation, protein folding, aggregation and refolding of denatured proteins (Zhang *et al.*, 2005). Some reports showed that *HSP70* assists in ATP unfolding, folding, intracellular distribution, assembly, degradation of proteins and prevent the protein denaturation during high temperature stress (Iba, 2002; Gorantla *et al.*, 2007). Elevated temperature affects the gene expression in storage protein synthesis and starch metabolism during grain filling stage in rice (Yamakawa *et al.*, 2007). This effect down regulates several protein or starch synthesis related enzyme proteins such as GBSSI, BEIIb and prolamin (13-kD), and up regulates the enzymes such as α- amylase genes and HSPs (Lin *et al.*, 2005; Yamakawa *et al.*, 2007). Recently it is reported that α-amylase genes in seeds of rice reduced the seed weight and chalkiness during ripening under heat stress (Asatsuma *et al.*, 2006; Yamakawa *et al.*, 2007). In carrot, different genes such as tubulin gene, *hsp90* gene, oleosin gene, a LEA gene, DC8 and DC59 genes were observed in embryonic heat-shock cell lines. High temperature declined the expression of a LEA, an oleosin, DC8 and DC59 genes (Hatzopoulos *et al.*, 1990; Milioni *et al.*, 2001). Recently, transcriptome studies identified many stress-responsive genes and encoding transcriptional factors during environmental stresses. Changing the expression pattern of transcriptional factor may also greatly manipulate the stress tolerance in plant (Chen *et al.*, 2002). Plants practice oxidative stresses during primary adaptation stage to a stress. Definite responses require continued expression of many genes relating to processes definite to individual stresses (Sung *et al.*, 2003). Recent transgenic approaches may modify the heat tolerance in plants which is a multigenic characteristic. This multigenic phenomenon, modifying the expression pattern of transcription factors, motivates a series of genes (Dong *et al.,* 2003). In *Arabidopsis*, a new HSF3 depressed in response to heat-shock and over expression of it enhanced the thermo tolerance and increased the activity of APX in transgenic plants (Schöffl *et al.,* 1999; Panchuk *et al.,* 2002). Similarly, Hot1-Hot4 gene may function as heat shock protein and abolished acquired thermo tolerance phenotypically; *Hot1* is *Hsp101* in *A. thaliana* 16 (Hong and Vierling, 2000). HsfA1 acting as crucial regulator of thermo-tolerance in tomato that reduced the expression of heat shock genes in co-suppression lines (Mishra *et al.,* 2002). Heat shock induces many genes which are attributed to heat shock elements (HSE).These heat-shock elements (HSE) are situated in the promoter region of hsp genes (Hubel and Schöffl, 1994). Transgenic expression confirmed that Heat-shock transcription factor (HSF) binding to pentameric nucleotides (5`-nGAAn-3`) of HSE sequences (Perisic *et al.,* 1989; Sung *et al.,* 2003) and this HSF-HSE interaction and transcriptional activation is very conserved in nature. In HSP gene expression of maize, Ca^{2+} and calmodulins (CaM) are playing important role in the presence of the HSF. The primary characteristic of plant thermometer has not been recognized. It has been recommended that alteration in membrane fluidity can manipulate the gene expression. Membrane fluidity acts a key role in temperature sensing (Over *et al.,* 2000). In *Synechocystis*, *hsp17* gene act as a fluidity gene and membrane controls signaling mechanism in response to high temperature stress (Horvath *et al.,* 1998). A model for temperature sensing and regulation of heat stress response integrates the observed membrane alterations. It is recommended that thylakoid membrane acting as temperature sensor is physiological pertinent (Horvath *et al.,* 1998; Sung *et al.,* 2003). Many genes are concerned with signaling pathways and encode proteins particularly mitogen activated protein kinase (MAPK), histidine kinase, Ca^{2+} dependent protein kinase (CDPK), SOS3, Ca^{2+} sensor family and also various transcription factors are regulated under various biotic and abiotic stresses (Ichimura *et al.,* 2000). According to Larkindale and Knight (2002), acquired thermotolerance characteristically rises cytosolic Ca^{2+} during high temperature stress. It is related with the interplay of two enzymes i.e. MAPK and CDP. Kaur and Gupta (2005) demonstrated that high temperature transduced the signals to MAPK by cytosolic Ca^{2+} influxes, and eventually up regulated the calcium dependent (CDP) kinase. MAPKs may function as signaling conveys to transmit information within cell. It is reported that the induction of uptake of Ca^{2+} and some calmodulin related genes induced thermo tolerance in maize seedling (Gong *et al.,* 1997a, b).

5. BIOTECHNOLOGICAL CHANGES

Future research activities should be aimed at identification and breeding of heat tolerant germplasms exploiting the variation in both genotypic and morphological characters. Several approaches should be actively exploited to improve heat tolerance in current cultivars, including discovery and exploitation of new genes and alleles, improved breeding efficiency, marker-assisted selection and genetic modification.

Different approaches are made to produce stress resistant plants to tackle the global warming. The mechanism of molecular control of abiotic stress tolerance depends on activation and regulation of genes related to a particular stress. Abiotic stress tolerance in crop plants could practically be achieved by the combination of molecular techniques and traditional plant breeding in one program (Wang *et al.*, 2003; Vinocur and Altman, 2005). Temperature stress (high temperature) is considered as one of the major stresses on crop plants (Grover *et al.*, 2000).

Genetic engineering of heat shock factors (HSF) and antisense strategies are instrumental to the understanding of both the functional roles of HSPs and the regulation of HSFs. Manipulations of the HS-response in transgenic plants have the potential to improve common abiotic stress tolerance and this may have a significant impact on the exploitation of the inherent genetic potential of agronomically important plants (Table 2). Transgenic overexpression of certain HSFs and HSF-fusion proteins results in an expression of HSPs at normal temperature (Iba, 2002). The increased acquired thermotolerance of transgenic lines is attributed to the higher levels of HSP chaperones. It is also demonstrated that tomato *MT-sHSP* has a molecular chaperone function *in vitro* (Liu and Shono, 1999) and recently it has been demonstrated that *MT-sHSP* gene exhibits thermotolerance in transformed tobacco with the tomato *MT-sHSP* gene (Sanmiya *et al.*, 2004) at the plant level. Experimental data obtained from transgenic, reverse-genetics and mutation approaches in non-cereal species confirm causal involvement of HSPs in thermotolerance in plants (Queitsch *et al.*, 2000). However, the cellular targets of HSPs are still unknown.

The response of plants to heat shock resulted in changes in the level of enzymes, cellular membrane structure, photosynthesis activity and protein metabolism (Singla *et al.*, 1997). It has been reported that high temperature changed the properties of membranes of nucleus, endoplasmic reticulum, mitochondria, and chloroplasts of rice plant *O. sativa* (Pareek *et al.*, 1998). Increasing the degree of saturation could lead to heat tolerance (Grover *et al.*, 2000). In support of this hypothesis the possibility of using genetic engineering to have more tolerant plants to high temperatures is one option. This could be achieved by reducing the degree of saturation of fatty acids in membranes (Murakami *et al.*,2000) through silencing the enzyme φ3-fatty acid desaturase that is involved in the synthesis of triple bonds in fatty acids. There was some indications that changing the levels of expressing Hsps by changing the transcription factor of *Arabidopsis* (AtHsf) led to an increase in heat tolerance (Lee *et al.*, 1995). Consequently, the increase in the synthesis of osmolytes in the cell could participate in an increase in heat tolerance (Alia *et al.*, 1998).

Later, Sanmiya *et al.* (2004) reported that mitochondrial Hsps enhances thermotolerance in the transgenic plants of *N. tabacum* by the *MT-sHSP* gene from *L. esculentum*. Sung *et al.* (2003) take some attempts to develop transgenic plants which have the ability to resists temperature stress in tomato, *Arabidopsis*, maize, carrot, tobacco, barley.

Table 2: Transgenic attempts to enhance plant temperature stress tolerance

Phenotypes	Function	Plant	Gene
HS tolerant; HS sensitive	TF	Tomato	HsfA1
HS tolerant	TF	Arabidopsis	Hsf3
HT sensitive	HSP	Arabidopsis	Hsp70
Abolished acquired thermotolerance	HSP	Maize	Hsp101
HT tolerant	HSP	Carrot	Hsp17.7
HT tolerant	Fatty acid desaturation	Tobacco	Fad7
HS tolerant	AOS metabolism	Barley	Hvapx1

Abbreviations: AOS, active oxygen species; HS, heat shock; Hsf, heat-shock factor; HSP, heat-shock protein; HT, high temperature; TF, transcription factor; APX, ascorbate peroxidase; fad7, fatty acid desaturation.

7. FUTURE PROSPECTS

Agricultural crops suffer in presenting their optimum yield potential due to heat stress. Heat stress causes enormous alterations in various processes within the plant system; most of them are well understood from various physiological, biochemical and molecular studies. However, all the phases of plant's life show various modulations through changes in ontogeny, ROS production and its scavenging mechanism, morphophysiological characters, etc. in which the grain development is found to be affected maximum under heat stress. A plethora of these kinds of plant responses are finally regulated by the genes. Hence, more biotechnological and molecular studies including genomics, proteomics, metaboliomics and the cryptomics are necessary to elucidate the processes further.

REFERENCES

A˜non S, Fernandez JA, Franco JA, Torrecillas A, Alarc´on JJ, S´anchez-Blanco MJ (2004) Effects of water stress and night temperature preconditioning on water relations and morphological and anatomical changes of *Lotus creticus* plants. Sci Hortic 101: 333–342

Abiko M, Akibayashi K, Sakata T, Kimura M, Kihara M, Itoh k, Asamizu E, Sato S, Takahashi H, Higasitani A (2005) High temperature induction of male sterility during barley (*Hordeum vulgare* L.) anther development is mediated by transcriptional inhibition. Sexual Plant Production 18: 91-100

Alia HH, Sakamoto A, Murata M, (1998) Enhancement of the tolerance of *Arabidopsis* to high temperatures by genetic engineering of the synthesis of glycinebetaine. Plant J 16:155–161

Al-Khatib K, Paulsen GM (1999) High-temperature effects on photosynthetic processes in temperate and tropical cereals. Crop Sci 39:119-125

Allakhverdieva YM, Mamedov MD, Gasanov RA (2001) The effect of glycinebetaine on the heat stability of photosynthetic membranes. Turk J Bot 25: 11–17

Apel K, Hirt H (2000) Reactive oxygen species: Metabolism, oxidative stress and signal transduction. Annual Rev Plant Biol 55:373-399

Aro E-M, Virgin I and Andersson B (1993) Photoinhibition of photosystem II: inactivation, protein damage and turnover. Biochim Biophys Acta 1143:113–134

Asada K, Takahashi M (1987) Production and Scavenging of Active Oxygen in Chloroplasts. In: Photoinhibition: Topics in Photosynthesis, Kyle DJ, CB Osmond and CJ Arntzen (Eds.). Elsevier, Amsterdam, pp: 227-287

Asatsuma S, Sawada C, Kitajima A, Asakura T, Mitsui T (2006) α-Amylase affects starch accumulation in rice grains. J Appl Glycosci 53:187-192

Ashraf M, Foolad MR (2007) Roles of glycine betaine and proline in improving plant abiotic stress resistance. Environ Exp Bot 59: 206–216

Ashraf M, Hafeez M (2004) Thermotolerance of pearl millet and maize at early growth stages: growth and nutrient relations. Biol Plant 48:81–86

Ashraf M, Saeed MM, Qureshi MJ (1994) Tolerance to high temperature in cotton (*Gossypium hirsutum* L.) at initial growth stages. Environ Exp Bot 34:275–283

Baniwal SK *et al* (2004) Heat stress response in plants: a complex game with chaperones and more than twenty heat stress transcription factors. J Bio Sci 29, 471–487

Baniwal SK *et al* (2007) Role of heat stress transcription factor HsfA5 as specific repressor of HsfA4. J Biol Chem 282: 3605–3613

Baniwal SK, Bharti K, Chan KY, Fauth M, Ganguli A, Kotak S, Mishra SK, Nover L, Port M, Scharf KD, Tripp J, Weber C, Zielinski D, Von-Koss-Kull-Doring P (2004) Heat stress response in plant: A complex game withchaperon and more than twenty heat stress transcription factors. Journal of Biosciences 28:471-487

Ben Khedher M, Ewing EE (1985) Growth analysis of eleven potato cultivar grown in the greenhouse under long photoperiods with and without heat stress. Am Potato J 62: 537-554

Berry JA, Bjo¨rkman O (1980) Photosynthetic response and adaptation to temperature in higher plants. Annu Rev Plant Physiol 31:491–543

Bewley JD, Black M (1982) Viability and Longevity: Physiology and Biochemistry of Seeds. Vol. 2, Springer Verlag, Berlin-Heidelberg, New York, pp: 47

Bharti K *et al*. (2004) Tomato heat stress transcription factor HsfB1 represents a novel type of general transcription coactivator with a histone-like motif interacting with the plant CREB binding protein ortholog HAC1. Plant Cell 16: 1521–1535

Blum A (1988) Plant Breeding for Stress Environments. CRC Press Inc., Boca Raton, Florida, pp. 223

Blum A, Ebercon A (1981) Cell membrane stability as a measure of drought and heat tolerance in wheat. Crop Science 21: 43-47

Blum A, Klueva N, Nguyen HT (2001) Wheat cellular thermotolerance is related to yield under heat stress. Euphytica 117:117–123

Bukhov NG, Carpentier R (2000) Heterogeneity of photosystem II reaction centers as influenced by heat treatment of barley leaves. Physiol Plant 110:279–285

Bukhov NG, Mohanty P (1999) Elevated temperature stress effects on photosystems: characterization and evaluation of the nature of heat induced impairments. In: Singhal GS, Renger G, Sopory SK, Irrgang K-D, Govingjee (eds) Concepts in photobiology: photosynthesis and photomorphogenesis. Narosa Publishing House, New Delhi, pp 617–648

Bukhov NG, Wiese C, Neimanis S, Heber U (1999) Heat sensitivity of chloroplasts and leaves: leakage of protons from thylakoids and reversible activation of cyclic electron transport. Photosyn Res 59:81–93

Burton WG (1972) The response of potato plant and tuber to temperature. In: Crop Processes in Controlled Environments. Rees AE, Cockhull KL, Hand DW and Hurd RG (Eds). Academic Press, New York. pp 217-233

Camejo D, Jim´enez A, Alarc´on JJ, Torres W, G´omez JM, Sevilla F (2006) Changes in photosynthetic parameters and antioxidant activities following heat-shock treatment in tomato plants. Funct Plant Biol 33:177–187

Camejo D, Rodr´ıguez P, Morales MA, Dell'amico JM, Torrecillas A, Alarc´on JJ (2005) High temperature effects on photosynthetic activity of two tomato cultivars with different heat susceptibility. J Plant Physiol 162:281–289

Carpentier R (1999) Effect of high-temperature stress on the photosynthetic apparatus. In: Pessarakli M (ed) Handbook of plant and crop stress. Marcel Dekker Inc, New York, pp 337– 348

Castillo F (1992) Peroxidase and Stress. Plant peroxidases 1980-1990. In: Topics and Literature on Molecular, Biochemical, Physiological Aspects, Penel C, Gaspar T and Greppin H (Eds.). University of Geneva, Switzerland, pp: 187-203

Cauchy AL (1821) Cours d'analyse de l'Ecole Royale Polytechnique. Paris, France: Imprimerie Royale

Chaitanya KV, Sundar D, Reddy AR (2001) Mulberry leaf metabolism under high temperature stress. *Biol Plant* 44: 379–384

Chen THH, Shen ZY, Lee PH (1982) Adaptability of crop plants to high temperature stress. *Crop Sci* 22:719–725

Chen WJ, Zhu T (2004) Networks of transcription factors with roles in environmental stress response. Trends Plant Sci 9:591–596

Chen WQ, Provart J, Glazebrook F, Katagiri HS (2002) Expression profile matrix of Arabidopsis transcription factor genes suggests their putative functions in response to environmental stresses. Plant Cell 14: 559–574

Crafts-Brander C, Salvucci ME (2002) Sensitivity to photosynthesis in the C4 plant, maize to heat stress. Plant Cell 12:54–68

Dat JF, Foyer CH, Scott IM (1998) Changes in salicylic acid and antioxidants during induction of thermo tolerance in mustard seedlings. Plant Physiol 118: 1455–1461

De Las Rivas J, Barber J (1997) Structure and thermal stability of photosystem II reaction centers studied by infrared spectroscopy. Biochemistry 36: 8897–8903

De Ronde JAD, Cress WA, Kruger GHJ, Strasser RJ, Staden JV (2004) Photosynthetic response of transgenic soybean plants containing an *Arabidopsis P5CR* gene, during heat and drought stress. J Plant Physiol 61: 1211–1244

Dejaegher G, Boyer N, Gaspar T (1985) Thigmomorphogenesis in Bryonia dioica; change in soluble and wall peroxidase, phenylalanine ammonialyase activity, cellulose, lignin content and monomeric constituents. Plant Growth Regul 3: 133-148

Dell' Aquila A, Spada P (1994) Effects of high and low temperatures on protein synthesis patterns of germinating wheat embryos. Plant physiol Biochem 32: 65-73

Dinar M, Rudich J (1985) Effect of heat stress on assimilate partitioning in tomato. Ann Bot 56: 239-248

Dobson CM, Sali A, Karplus M (1998) Protein folding: a perspective from theory and experiment. Angew Chem. Int. Ed. 37: 868–893

Dong A, Zhu Y, Yu K, Cao Y, Sun C, Shen WH (2003) Regulation of biosynthesis and intracellular localization of rice and tobacco homologues of nucleosome assembly protein 1. Planta 216: 561-570

Ebrahim MK, Zingsheim O, El-Shourbagy MN, Moore PH, Komor E (1998) Growth and sugar storage in sugarcane grown at temperature below and above optimum. J Plant Physiol 153: 593–602

Ewing EE (1981) Heat stress and the tuberization stimulus. Am Potato J 58: 31-49

Fadzillah NM, Gill V, Finch RP, Burdon RH (1996) Chilling, oxidative stress and antioxidant responses in shoot cultures of rice. Planta 199: 552–556

Feder ME, Hofmann GE (1999) Heat-shock proteins, molecular chaperones, and stress response: evolutionary and ecological physiology. Annu Rev Physiol 61: 243–282

Ferris R, Ellis RH, Wheeeler TR, Hadley P (1998) Effect of high temperature stress at anthesis on grain yield and biomass of field grown crops of wheat. Plant Cell Environ 34: 67–78

Foolad MR (2005) Breeding for abiotic stress tolerances in tomato. In: Ashraf, M., Harris, P.J.C. (Eds.), Abiotic Stresses: Plant Resistance Through Breeding and Molecular Approaches. The Haworth Press Inc., New York, USA, pp. 613–684

Gagliardi D, Breton C, Chaboud A, Vergne P, Dumas C (1995) Expression of heat shock factor and heat shock protein 70 genes during maize pollen development. Plant Mol Biol 29: 841-856

Gaspar T, Franck T, Bisbis B, Kevers C, Jouve L, Hausman JF, Dommes J (2002) Concepts in plant stress physiology. Application to plant tissue cultures. Plant Growth Regulation 37: 263–285

Gawronska H, Michael K, Thornton, Dwelle RB (1992) Influence of heat stress on dry matter production and photoassimilate partitioning by four potato clones. 653-665

Giaveno C, Ferrero J (2003) Introduction of tropical maize genotypes to increase silage production in the central area of Santa Fe, Argentina. Crop Breed Appl Biotechnol 3: 89–94

Gong M, Chen SN, Song YQ, Li ZG (1997a) Effect of calcium and calmodulin on intrinsic heat tolerance in relation to antioxidant systems in maize seedlings. Aust J Plant Physiol 24: 371-379

Gong M, Li YJ, Dai Y, Tian M, Li ZG (1997b) Involvement of calcium and calmodulin in the acquisition of heat-shock induced thermotolerance in maize seedlings. J Plant Physiol 150: 615-621

Gorantla M, Babu PR, ReddyLachagari VB, Reddy AMM, Wusirika A, Bennetzen JL (2007) Identification of stress-responsive genes in an indica rice (*Oryza sativa* L.) using ESTs generated from drought-stressed seedlings. J Exp Bot 58: 253-265

Gounaris K, Brain ARR, Quinn PJ, Williams WP (1984) Structural reorganization of chloroplast thylakoid membranes in response to heat stress. Biochim Biophys Acta 766:198–208

Grover A, Agarwal M, Katiyar-Agarwal S, Sahi C, Agarwal S (2000). Production of high temperature tolerant transgenic plants through manipulation of membrane lipids. Curr Sci 79: 557–559

Guilioni L, W´ery J, Lecoeur J (2003) High temperature and water deficit may reduce seed number in field pea purely by decreasing plant growth rate. Funct. Plant Biol 30:1151–1164

Guilioni L, Wery J, Tardieu F (1997) Heat stress-induced abortion of buds and flowers in pea: is sensitivity linked to organ age or to relations between reproductive organs. Ann Bot 80: 159–168

Guo Y-P, Zhou H-F, Zhang L-C (2006) Photosynthetic characteristics and protective mechanisms against photooxidation during high temperature stress in two citrus species. Sci Hort 108: 260–267

Gupta SC, Sharma A, Mishra M, Mishra R, Chowdhuri DK, (2010) Heat shock proteins in toxicology: how close and how far? Life Sci 86: 377–384

Hall AE (1992) Breeding for heat tolerance. Plant Breed Rev 10: 129–168

Hall AE (2001) Crop Responses to Environment. CRC Press LLC, Boca Raton, Florida.

Hamilton EW, Heckathorn SA (2001) Mitochondrial adaptations to NaCl. Complex I is protected by anti-oxidants and small heat shock proteins, whereas complex II is protected by proline and betaine. Plant Physiol 126: 1266–1274

Hamilton EW, Heckathorn SA, Downs CA, Schwarz TE, Coleman JS, Hallberg RL (1996) Heat shock proteins are produced by field-grown naturally occurring plants in the summer in the temperate northeast US Bulletin of the Ecologic. Soc. Am. 77, Suppl. Part 2: 180 (Abstr.)

Hare PD, Cress WA, Staden JV (1998) Dissecting the roles of osmolyte accumulation during stress. Plant Cell Environ 21: 535–553

Hatzopoulos P, Franz G, Choy L, Sung R Z (1990) Interaction of nuclear factors with upstream sequences of a lipid-body membrane protein gene from carrot. Plant Cell 2: 457-467

Havaux M (1998) Carotenoids as membrane stabilizers in chloroplasts. Trends Plant Sci 3: 147–151

Havaux M, Greppin H, Strasser RJ (1991) Functioning of photosystems I and II in pea leaves exposed to heat stress in the presence or absence of light. Planta 186:88–98

Heerklotz D *et al.* (2001) The balance of nuclear import and export determines the intracellular distribution of tomato heat stress transcription factor HsfA2. Mol Cell Biol 21: 1759–1768

Hong SW, Vierling E (2000) Mutant of *Arabidopsis thaliana* defective in the acquisition of tolerance to high temperature stress. Proceeding of the National Academy Science, USA 97:7392-4397

Hopf N, Plesofsky-Vig N, Brambl R. (1992). The heat shock response of pollen and other tissues of maize. Plant Mol Biol 19: 623–630

Horvath I, Glatz A, Varvasovszki V, Torok Z, Pali T, Balogh G, Kovacs E, Nadasdi L, Benko S, Joo F, Vigh L (1998) Membrane physical state controls the signaling mechanism of the heat shock response in Synechocystis PCC 6803: identification of hsp17 as a fluidity gene. Proc Natl Acad Sci 95: 3513-3518

Howarth CJ (2005) Genetic improvements of tolerance to high temperature. In: Ashraf, M., Harris, P.J.C. (Eds.), Abiotic Stresses: Plant Resistance Through Breeding and Molecular Approaches. Howarth Press Inc., New York.

Hu W, Hu G, Han B (2009) Genome-wide survey and expression profiling of heat shock proteins and heat shock factors revealed overlapped and stress specific response under abiotic stresses in rice. Plant Sci 176: 583–590

Hubel A, Schöffl F (1994) Arabidopsis heat-shock factor: isolation and characterization of the gene and the recombinant protein. Plant Mol Biol 26: 353–362

Iba K (2002) Acclimative response to temperature stress in higher plants: Approaches of gene engineering for temperature tolerance. Annu Rev Plant Biol 53: 225-245

Ichimura KT, Mizoguchi R, Yoshida RT, Yuasa, Shinozaki K (2000) Various abiotic stresses rapidly activate Arabidopsis MAP kinases ATMPK4 and ATMPK6. Plant J 24: 655-665

Ismail AM, Hall AE (1999) Reproductive-stage heat tolerance, leaf membrane thermostability and plant morphology in cowpea. Crop Sci 39: 1762–1768

Jbir N, Chaibi W, Ammar S, Jemmali A, Ayadi A (2001) Root growth and lignification of two wheat species differing in their sensivity to NaCl, in response to salt stress.. Comptes Rendus Acad. Sci Ser. III Sci. de la Vie, 324: 863-868

Jemma Essemine, Saida Ammar, Sadok Bouzid (2010) Impact of heat stress on germination and growth in higher plants: physiological, biochemical and molecular repercussions and mechanism of defense. Journal of Biological Sciences 10(6): 565-572

Jiang Y, Haung B (2001) Plants and the environment. Effects of calcium on antioxidant activities and water relations associated with heat tolerance in two cool-season grasses. J Exp Bot 52: 341–349

Karim MA, Fracheboud Y, Stamp P (1997) Heat tolerance of maize with reference of some physiological characteristics. Ann Bangladesh Agri **7:** 27–33

Karim MA, Fracheboud Y, Stamp P (1999) Photosynthetic activity of developing leaves in *Zea mays* is less affected by heat stress than that of developed leaves. Physiol Plant 105: 685–693

Kaur N, Gupta AK (2005) Signal transduction pathways under abiotic stresses in plants. Curr Sci 88: 1771-1780

Kavi Kishore PB, Sangam S, Amrutha RN, Laxmi PS, Naidu KR, Rao KRSS, Rao S, Reddy KJ, Theriappan P, Sreenivasulu N (2005) Regulation of proline biosynthesis, degradation, uptake and transport in higher plants: its implications in plant growth and abiotic stress tolerance. Curr Sci 88: 424–438

Kepova KD, Holzer R, Stoilova LS, Feller U (2005) Heat stress effects on ribulose-1,5-bisphosphate carboxylase/oxygenase, Rubisco binding protein and Rubisco activase in wheat leaves. Biol Plant 49: 521–525

Kinet JM, Peet MM (1997) Tomato In: Wien HC (Ed.), The Physiology of Vegetable Crops. CAB International, Wallingford, UK, pp. 207–258

Kinnersley AM, Turano FJ (2000) Gamma aminobutyric acid (GABA) and plant responses to stress. Crit Rev Plant Sci 19: 479–509

Kotak S *et al* (2004) Characterization of C-terminal domains of Arabidopsis heat stress transcription factors (Hsfs) and identification of a new signature combination of plant class A Hsfs with AHA and NES motifs essential for activator function and intracellular localization. Plant J 39: 98–112

Kotak S, Larkindale J, Lee U, von Koskull-Doring P, Vierling E, Scharf KD (2007) Complexity of the heat stress response in plants. Current Opinion in Plant Biology 10: 310–316.

Kreslavski VD, Khristin MS (2003) After effect of heat shock on fluorescence induction and low-temperature fluorescence spectra of wheat leaves. Russ J Biophys 48:865–872

Larcher W (1994) Photosynthesis as a tool for indicating temperature stress events, p. 261–277. In: E.-D. Schulze and M.M. Caldwell (eds.). Ecophysiology of photosynthesis. Ecological studies 100. Springer-Verlag, Berlin

Larkindale J, Knight MR (2002) Protection against heat stress-induced oxidative damage in Arabidopsis involves calcium, abscisic acid, ethylene and salicylic acid. Plant Physiol 128: 682-695

Lee JH, Hubel A, Schoffl F (1995) Derepression of the activity of genetically engineered heat shock factor causes constitutive synthesis of heat shock proteins and increased thermotolerance in transgenic *Arabidopsis*. Plant J 8: 603–612

Levitt J (1972) Responses of plants to environmental stresses. London, UK: Academic Press, 2078–2089

Levitt M, Gerstein M, Huang E, Subbiah S, Tsai J (1997) Protein folding: the endgame. Annu Rev Biochem 66: 549–579

Lichtenthaler HK (1996) Vegetation stress: an introduction to the stress concept in plants. J Plant Physiology 148: 4-14

Lin SK, Chang MC, Tsai YG, Lur HS (2005) Proteomic analysis of the expression of proteins related to rice quality during caryopsis development and the effect of high temperature on expression. Proteomics 5: 214-215

Liu J, Shono M (1999) Characterization of mitochondria-located small heat shock protein from tomato (*Lycopersicon esculentum*). Plant Cell Physiol 40: 1297–1304

Liu X, Huang B (2000) Heat stress injury in relation to membrane lipid peroxidation in creeping bent grass. Crop Sci 40: 503–510

Low D, Brandle K, Nover L, Forreiter C (2000) Cytosolic heatstress proteins Hsp17.7 class I and Hsp17.3 class II of tomato act as molecular chaperones *in vivo*. Planta 211: 575–582

Machado S, Paulsen GM (2001) Combined effects of drought and high temperature on water relations of wheat and sorghum. Plant Soil 233

Marchand FL, Mertens S, Kockelbergh F, Beyens L, Nijs I (2005) Performance of high arctic tundra plants improved during but deteriorated after exposure to a simulated extreme temperature event. Global Change Biol 11:

Marcum KB (1998) Cell membrane thermo stability and whole plant heat tolerance of Kentucky bluegrass. Crop Sci 38: 1214–1218

Martineau JR, Specht JE, Williams JH, Sullivan CY (1979) Temperature tolerance in soybean. I. Evaluation of technique for assessing cellular membrane thermostability. Crop Sci 19: 75–78

Mascarenhas JP, Crone DE (1996) Pollen and the heat shock response. Sex Plant Rep 9: 370-374

Mazorra LM, Nunez M, Echerarria E, Coll F, S´anchez-Blanco MJ (2002) Influence of brassinosteriods and antioxidant enzymes activity in tomato under different temperatures. Plant Biol 45: 593–596

Mcdonald GK, Paulsen GM (1997) High temperature effects on photosynthesis and water relations of grain legumes. Plant Soil 196: 47–58

Milioni D, Franz G, Sung R, Hatzopoulos P (2001) Gene expression during heat-shock in embryogenic carrot cell lines. Plant Cell Tiss Organ Cult 65: 221-228

Miller G, Mittler R (2006) Could heat shock transcription factors function as hydrogen peroxide sensors in plants? Ann Bot (Lond.) 98: 279–288

Mishra SK (2002) In the complex family of heat stress transcription factors HsfA1 has a unique role as master regulator of thermotolerance in tomato. Genes Dev 16: 1555–1565

Mishra SK, Tripp J, Winkelhaus S, Tschiersch B, Theres K, Nover L, Scharf K-D (2002) In the complex family of heat stress transcription factors, HsfA1 has a unique role as master regulator of thermotolerance in tomato. Genes Dev 16: 1555-1567

Mohanty P, Allakhverdiev SI, Murata N (2007) Application of low temperature during photoinhibition allows characterization of individual steps in photodamage and repair of photosystem II. Photosynth Res 94:217–234

Morales D, Rodr´ıguez P, Dell'amico J, Nicol´as E, Torrecillas A, S´anchez- Blanco MJ (2003) High-temperature preconditioning and thermal shock imposition affects water relations, gas exchange and root hydraulic conductivity in tomato. Biol Plant 47: 203–208

Morimoto RI (1998) Regulation of the heat shock transcriptional response: cross talk between a family of heat shock factors, molecular chaperones and negative regulators. Genes Dev 12: 3788–3796

Morimoto RI, Santoro MG (1998) Stress-inducible responses and heat shock proteins: new pharmacologic targets for cytoprotection. Nat Biotechnol 16: 833–838

Murakami Y, Tsuyama M, Kobayashi Y, Kodama H, Iba K (2000) Trienoic fatty acids and plant tolerance of high temperature. Science 287: 476–479

Murata N, Takahashi S, Nishiyama Y, Allakhverdiev SI (2007) Photoinhibition of photosystem II under environmental stress. Biochim Biophys Acta 1767:414–421

Nakamoto H, Hiyama T (1999) Heat-shock proteins and temperature stress. In: Pessarakli, M. (Ed.), Handbook of Plant and Crop Stress. Marcel Dekker, New York, pp. 399–416

Nishiyama Y, Allakhverdiev SI, Murata N (2005) Inhibition of the repair of photosystem II by oxidative stress in cyanobacteria. Photosynth Res 84:1–7

Nishiyama Y, Allakhverdiev SI, Murata N (2006) A new paradigm for the action of reactive oxygen species in the photoinhibition of photosystem II. Biochim Biophys Acta 1757:742–749

Nover L (ed.) (1991) Heat Shock Response CRC Press

Nover L *et al* (2001) Arabidopsis and the heat stress transcription factor world: how many heat stress transcription factors do we need? Cell Stress Chaperones 6: 177–189

Nover L, Bharti K, Doring P, Mishra SK, Ganguli A, Scharf N-K (2001) Arabidopsis and the Hsf world: How many heat stress transcription factors do we need? Cell Stress Chaperones 6: 177-189

Over BL, Veena S, Franz O, Rajinder D (2000) Early steps in cold sensing by plant cells: the role of actin cytoskeleton and membrane fluidity. Plant J 23: 785-794

Panaretou B, Zhai C, (2008) The heat shock proteins: their roles as multi-component machines for protein folding. Fungal Biol Rev 22: 110–119

Panchuk II, Volkov RA, Schoffl F (2002) Heat stress and heat-shock transcription factor-dependent expression and activity of ascorbate peroxidase in *Arabidopsis*. Plant Physiology 129: 838-853

Pareek A, Singla SL, Grover A (1998) Plant Hsp90 family with special reference to rice. J Biosci 23:361–367

Peet MM, Willits DH (1998) The effect of night temperature on greenhouse grown tomato yields in warm climate. Agric Forest Meteorol 92: 191–202

Pelham HRB (1982) A regulatory upstream promoter element in the Drosophila hsp70 heat-shock gene. Cell 30, 517–528

Perisic O, Xiao H, Lis LS (1989) Stable binding of Drosophila heat shock factor to head-to-head and tail-to tail repeats of a conserved 5 bp recognition unit. Cell 59: 797-806

Port M *et al*. (2004) Role of Hsp17.4–CII as coregulator and cytoplasmic retention factor of tomato heat stress transcription factor HsfA2. Plant Physiol 135: 1457–1470

Porter JR (2005) Rising temperatures are likely to reduce crop yields. Nature 436: 174

Quan R, Shang M, Zhang H, Zhao Y, Zhang J (2004) Engineering of enhanced glycine betaine synthesis improves drought tolerance in maize. Plant Biotech J 2: 477–486

Queitsch C, Hong SW, Vierling E, Lindquest S (2000) Heat shock protein 101 plays a crucial role in thermotolerance in *Arabidopsis*. Plant Cell 12: 479–492

Rainey K, Griffiths P (2005) Evaluation of *Phaseolus acutifolius* A. Gray plant introductions under high temperatures in a controlled environment. Genet. Resour. Crop Evol 52: 117–120

Saadalla MM, Quick JS, Shanahan JF (1990a) Heat tolerance in winter wheat: II. Membrane thermostability and field performance. Crop Science 30:1248-1251

Sairam RK, Tyagi A (2004) Physiology and molecular biology of salinity stress tolerance in plants. Curr Sci 86: 407–421

Sakamoto A, Murata N (2002) The role of glycine betaine in the protection of plants from stress: clues from transgenic plants. Plant Cell Environ 25: 163–171

Sakata T, Takahashi H, Nishiyama I, Higashitani A (2000) Effects of high temperature on the development of pollen mother cells and microspores in barley (*Hordeum vulgare* L.). J Plant Res 113: 395-402

Salvucci ME, Crafts-Brandner SJ (2004b) Inhibition of photosynthesis by heat stress: the activation state of Rubisco as a limiting factor in photosynthesis. Physiol Plant 120: 179–186

Sanmiya K, Suzuki K, Egawa Y, Shono M, (2004) Mitochondrial small heat-shock protein enhances thermotolerance in tobacco plants. Febs Lett 557: 265–268

Sato S, Kamiyama M, Iwata T, Makita N, Furukawa H, Ikeda H (2006) Moderate increase of mean daily temperature adversely affects fruit set of *Lycopersicon esculentum* by disrupting specific physiological processes in male reproductive development. Ann Bot 97: 731–738

Savchenko GE, Klyuchareva EA, Abrabchik LM, Serdyuchenko EV (2002) Effect of periodic heat shock on the membrane system of etioplasts. Russ J Plant Physiol 49: 349–359

Scandalios JG (1993) Oxygen stress and superoxide dismutases. Plant Physiol 101: 7–12

Scharf KD, Siddique M, Vierling E (2001) The expanding family of *Arabidopsis* thaliana small heat stress proteins and a new family of proteins containing alpha-crystalline domains (Acd proteins). Cell Stress Chaperon 6: 225–237

Scharf K-D *et al.* (1998) The tomato Hsf system: HsfA2 needs interaction with HsfA1 for efficient nuclear import and may be localized in cytoplasmic heat stress granules. Mol Cell Biol 18: 2240–2251

Schoffl F *et al.* (1998) Regulation of the heat-shock response. Plant Physiol 117: 1135–1141

Schoffl F, Prandl R, Reindl A (1999) Molecular Responses to Heat Stress. In: Molecular Responses to Cold, Drought, Heat and Salt Stress in Higher Plants, Shinozaki, K. and K. Yamaguchi-Shinozaki (Eds.). R.G. Landes Co., Austin, Texas, pp: 81-98

Schramm F *et al.* (2006) The heat stress transcription factor HsfA2 serves as a regulatory amplifier of a subset of genes in the heat stress response in Arabidopsis. Plant Mol Biol 60: 759–772

Schreiber U, Bilger W, Neubauer C (1994) Chlorophyll fluorescence as a non-destructive indicator for rapid assess assessment of *in vivo* photosynthesis, p. 49–70. In: E-D Schulze and MM Caldwell (eds.). Ecophysiology of photosynthesis. Ecological studies 100. Springer-Verlag, Berlin

Schulze-Lefert P (2004) Plant immunity: the origami of receptor activation. Curr Biol 14: R22–R24

Semenova GA (2004) Structural reorganization of thylakoid systems in response to heat treatment. Photosynthetica 42:521–527

Shanahan JF, Edwards IB, Quick JS, Fenwick JR (1990) Membrane thermo stability and heat tolerance of spring wheat. Crop Sci 30: 247–251

Sharkey TD (2005) Effects of moderate heat stress on photosynthesis: importance of thylakoid reactions, rubisco deactivation, reactive oxygen species, and thermotolerance provided by isoprene. Plant Cell Environ 28:269–277

Sharkova VE (2001) The effect of heat shock on the capacity of wheat plants to restore their photosynthetic electron transport after photoinhibition or repeated heating. Russ J Plant Physiol 48: 793–797

Shinozaki K, Yamaguchi-Shinozaki K, Seki M (2003) Regulatory network of gene expression in the drought and cold sress responses. Curr Opin Plant Biol 6: 410-417

Simoes-Araujo JL, Rumjanek NG, Margis-Pinheiro M (2003) Small heat shock proteins genes are differentially expressed in distinct varieties of common bean. Braz J Plant Physiol 15: 33–41

Singh KB, Foley RC, Onate-Sanchez L (2002) Transcription factors in plant defense and stress responses. Curr Opin Plant Biol 5: 430–436

Singla SL, Preek A, Grover A (1997) High temperature. In: Prasad, M.N.V. (Ed.), Plant Ecophysiology. John Wiley, New York., pp. 101–127

Somerville C, Browse J (1991) Plant lipids, metabolism and membranes. Science 252: 80–87

Sørensen JG, Kristensen TN, Loeschcke V (2003) The evolutionary and ecological role of heat shock proteins. Ecol Lett 6: 1025–1037

Stone P (2001) The effects of heat stress on cereal yield and quality. In: Basra, A.S. (Ed.), Crop Responses and Adaptation to Temperature Stress. Food Products Press, Binghamton, NY, pp. 243–291

Sullivan CY, Ross WM (1979) Selecting for drought and heat resistance in grain sorghum. In: Mussell, H., Staple, R. (Eds.), Stress Physiology in Crop Plants. Wiley, New York, pp. 263–281

Sung D-Y, Kaplan F, Lee KJ, Guy CL (2003) Acquired tolerance to temperature extremes. *Trends Plant Sci.,* **8:** 179-187.

Suzuki N, Mittler R (2006) Reactive oxygen species and temperature stresses: A delicate balance between signaling and destruction. Physiol Plant 126: 45-51

Swindell WR, Huebner M, Weber AP (2007) Transcriptional profiling of *Arabidopsis* heat shock proteins and transcription factors reveals extensive overlap between heat and non-heat stress response pathways. BMC Genomics 8: 125

Taiz L, Zeiger E (2006) Plant Physiology. Sinauer Associates Inc. Publishers, Massachusetts

Timperio AM, Egidi MG, Zolla L (2008) Proteomics applied on plant abiotic stresses: role of heat shock proteins (HSP). J Proteomics 71: 391–411

Todorov DT, Karanov EN, Smith AR, Hall MA (2003) Chlorophyllase activity and chlorophyll content in wild type and *eti 5* mutant of *Arabidopsis thaliana* subjected to low and high temperatures. Biol Plant 46: 633–636

Toth SZ, Schansker G, Kissimon J, Kovacs L, Garab G, Strasser RJ (2005) Biophysical studies of photosystem II-related recovery processes after a heat pulse in barley seedlings (*Hordeum vulgare* L.). J Plant Physiol 162: 181–194

Tripp J, Mishra SK, Scharf K-D (2009) Functional dissection of the cytosolic chaperone network in tomato mesophyll protoplasts. Plant Cell Environ 32: 123–133

Tsukaguchi T, Kawamitsu Y, Takeda H, Suzuki K, Egawa Y (2003) Water status of flower buds and leaves as affected by high temperature in heat tolerant and heat-sensitive cultivars of snap bean (*Phaseolus vulgaris* L.). Plant Prod Sci 6: 4–27

Vara Prasad PV, Craufurd PQ, Summerfield RJ (1999) Fruit number in relation to pollen production and viability in groundnut exposed to short episodes of heat stress. Ann Bot 84: 381–386

Vierling E (1991) The role of heat shock proteins in plants. Annu Rev Plant Phys 42: 579–620

Vinocur B, Altman A (2005) Recent advances in engineering plant tolerance to abiotic stress: achievements and limitations. Curr Opin Biotechnol 16: 123–132

Vollenweider P, Gunthardt-Goerg MS (2005) Diagnosis of abiotic and biotic stress factors using the visible symptoms in foliage. Environ Pollut 137: 455–465

von Koskull-Do¨ring P, Scharf KD, Nover L (2007) The diversity of plant heat stress transcription factors. Trends Plant Sci 12: 452– 457

Vu JCV, Gesch RW, Pennanen AH, Allen LHJ, Boote KJ, Bowes G (2001) Soybean photosynthesis, Rubisco and carbohydrate enzymes function at supra-optimal temperatures in elevated CO_2. J Plant Physiol 158: 295–307

Wahid A, Close TJ (2007) Expression of dehydrins under heat stress and their relationship with water relations of sugarcane leaves. Biol Plant 51: 104–109

Wahid A, Ghazanfar A (2006) Possible involvement of some secondary metabolites in salt tolerance of sugarcane. J Plant Physiol 163: 723–730

Wahid A, Shabbir A (2005) Induction of heat stress tolerance in barley seedlings by pre-sowing seed treatment with glycinebetaine. Plant Growth Reg 46: 133–141

Wang H-D, Kazemi-Esfarjani P, Benzer P (2004) Multiplestress analysis for isolation of Drosophila longevity genes. Proc Natl Acad Sci USA 101: 12610–12615

Wang W, Vinocur B, Altman A (2003) Plant responses to drought, salinity and extreme temperatures: towards genetic engineering for stress tolerance. Planta 218: 1–14

Wardlaw IF (1974) Temperature control of translocation. In: Bielske, R.L., Ferguson, A.R., Cresswell, M.M. (Eds.), Mechanism of Regulation of Plant Growth. Bull. Royal Soc., New Zealand, Wellington, pp. 533–538

Waters ER, Lee GJ, Vierling E (1996) Evolution, structure and function of the small heat shock proteins in plants. J Exp Bot 47: 325–338

Weaich K, Briston KL, Cass A (1996) Modeling preemergent maize shoot growth. II. High temperature stress conditions. Agric J 88: 398–403

Wilhelm EP, Mullen RE, Keeling PL, Singletary GW (1999) Heat stress during grain filling in maize: effects of kernel growth and metabolism. Crop Sci 39: 1733–1741

Wise RR, Olson AJ, Schrader SM, Sharkey TD (2004) Electron transport is the functional limitation of photosynthesis in field-grown Pima cotton plants at high temperature. Plant Cell Environ 27: 717–724

Xu S, Li J, Zhang X, Wei H, Cui L (2006) Effects of heat acclimation pretreatment on changes of membrane lipid peroxidation, antioxidant metabolites, and ultrastructure of chloroplasts in two cool-season turf grass species under heat stress. Environ Exp Bot 56: 274–285

Yamada M, Hidaka T, Fukamachi H (1996) Heat tolerance in leaves of tropical fruit crops as measured by chlorophyll fluorescence. Sci Hortic 67: 39–48

Yamaguchi-Shinozaki UT, Shinozaki K (2000) Two component systems in plant signal transduction. Trends Plant Sci 5: 67-74

Yamakawa H, Hirose T, Kuroda M, Yamaguchi T (2007) Comprehensive expression profiling of rice grain filling-related genes under high temperature using DNA microarray. Plant Physiol 144: 258-277

Yamane Y, Kashino Y, Koike H, Satoh K (1998) Effects of high temperatures on the photosynthetic systems in spinach: oxygen-evolving activities, fluorescence characteristics and the denaturation process. Photosynth Res 57: 51–59

Yang J, Sears RG, Gill BS, Paulsen GM (2002) Genotypic differences in utilization of assimilate sources during maturation of wheat under chronic heat and heat shock stresses. Euphytica 125: 179–188

Yordanov IS, Dilova R, Petkova T, Pangelova V, Goltsev V and Su¨ss K-H (1986) Mechanisms of the temperature damage and acclimation of the photosynthetic apparatus. Photobiochem Photobiophys 12: 147–155

Zhang J-H, Huang WD, Liu Y-P, Pan Q-H (2005) Effects of temperature acclimation pretreatment on the ultrastructure of mesophylls in young grape plants (*Vitis vinifera* L. cv. Jingxiu) under cross-temperature stresses. J Integr Plant Biol 47: 959-970

Zhang J-H, Wang L-J, Pan Q-H, Wang Y-Z, Zhan J-C, Huang W-D (2008) Accumulation and subcellular localization of heat shock proteins in young grape leaves during cross-adaptation to temperature stresses. Scientia Horticulturae 117: 231–240.

17

IMPACT OF AIR POLLUTION ON PLANT METABOLISM

A.K. Trivedi and Jyoti Sati

National Bureau of Plant Genetic Resources, Regional Station
Bhowali – 263 132 Nainital, Uttarakhand (INDIA)

INTRODUCTION

Plants provide an enormous leaf area for impingement, absorption and accumulation of air pollutants to reduce the pollutant level in the air environment, which varies in extent for different species (Liu and Ding, 2008). The use of plants as monitors of air pollution has long been established as plants are the initial acceptor of air pollution. They act as the scavengers for many air born particulates in the atmosphere (Joshi and Swami, 2009). Air pollution affects plant processes in different ways. Impact of air pollutants on plant metabolism may be studied in the following heads:

2. DEPOSITION AND UPTAKE OF AIR POLLUTANTS

The major difference among laboratory and field condition is that in laboratory experiments are conducted under control condition but in the field pollutant play an important role, the concentration of pollutant vary time to time (Mccormick and Holzworth, 1975). A second aspect of pollution climate is the presence in the atmosphere of dynamic mixture of gases and dissolved pollutants. While it is convenient to undertake controlled environment with one or even few variables (e.g. SO_2 + NO_2 or O_3 + acid mist) but the real situation is much more complex. This is not important for atmospheric chemists only; there is evidence that plant

responses to pollution mixture may be additive, antagonistic or synergistic i.e., different than response to a single pollutant or more pollutants separately (Ormord, 1982).

The removal of air pollutants from the atmosphere is occurring by wet deposition, which is influenced by gravity. Gases and particles may enter solution by a number of different routes, but the most important condition is when particles act as condensation nuclei for water droplet (Fowler, 1980). The relative importance of wet deposition as a vector for pollution depends up to large extent on the amount of rain; consequently there are large input of acidity and other ions in mountainous areas which usually receive high rainfall. Some of the factors which lead to enhancement of the amount of rain in such areas may also lead to deposition. The concentration of pollutant in the cloud may be much greater than usually found in the rain (Choularton *et al.*, 1987). The compositions of rainfall vary between and within events, but depend on physical climate and weather type (Raynor and Hayes, 1981).

Dry deposition involves the transfer of gases and particles from the atmosphere to the ground by atmospheric turbulence and diffusion. As compared to wet deposition, dry deposition is more complex, as it may include specific properties of the ground or vegetation surface as well as atmospheric factors such as wind speed and temperature gradients (Lehmhaus *et al.*, 1985). Dry deposition gases over large areas and long time cannot generally be measured directly, although some attempts have been made to estimate dry deposition from the elemental budget in forest ecosystem (Lindberg *et al.*, 1986). The overall deposition of pollutant as wet deposition, dry deposition or occult deposition is a process which removes pollutants from the atmosphere. The effect of such deposition on vegetation can be measured by determining how much of the deposited pollutant is directly taken up by the plant and how much moves through the ecosystem.

Plant responses do not actively control the uptake of pollutants from rain or snow. Deposition of pollutant on vegetation does not depend on the physiological character of plants but it solely depends on morphological character of plants. The water droplet interact to leaf surface depends on the nature of cuticle that form the wax layer on the outer surface of leaf (Holloway, 1969) and also depend on the environmental condition in which the plant survive. The contact of water to leaf surface may act as a source or sink like relation in favour of many inorganic and organic compounds. Due to this many compounds leach out from the leaf and many get absorbed from the solution by leaf probably by ion exchange processes (Tukey, 1970). Many air pollutants present in the rain water that come in contact with leaf surface cause increase or decrease in acidity of leaf. As rain water passes through canopy nitrate and ammonia may be taken up and sulphate may be leached out from the plants. The exchange of substances between a plant and wet deposition is to some extant depend on the physiological and nutritional state of the plant. The deposition and uptake of pollutants on the plant surface are two different phenomenons. The substance deposited on the surface of the plants may never be absorbed. The uptake of pollutant gas by a plant occurs as a result of three steps. First, molecules of the gas must be brought to the leaf surface, a process normally affected by atmospheric turbulence. Second, the gas molecules must diffuse through the non-turbulent boundary layer of air close to the leaf surface before finally (i.e., third step) reaching the leaf surface or being absorbed by the stomata (Unsworth, 1981). The rate of dry deposition of many pollutant gases are strongly controlled by stomatal response and net fluxes of gases such as SO_2 and O_3 to the foliage have been shown to follow a strong diurnal cycle as stomata open and close (Fowler and Unsworth, 1979). This means that uptakes of any substances depend on environmental factors which influence stomatal function, such as drought stress, photoperiod, temperature and humidity. In drought stress condition plant protect itself from ozone injury by closing its stomata or by limiting stomatal uptake (Mansfield and Freer Smith, 1984). The uptake of pollutant by individual leaf depends on the phenology of crop such as number of leaves, physiological maturity. Other growing condition such as light, water, nutrition may also influence uptake by modifying stomatal aperture. The major pathway of uptake is likely to be through stomata, the major part of dry deposition may be on external plant surface (Fowler and Unsworth, 1979).

3. MECHANISM OF PENETRATION THROUGH CUTICLE

The plant cuticle prevents water loss from the plant leaf and protects the plant from the effect of air pollution. It acts as a barrier to the movement of air pollutants, in gaseous and dissolved form but at the same time sorption of gases, water and dissolved solutes as well as the deposition of particulate matter occurs. Therefore, harmful effect of air pollutant on plants depends on the permeability of cuticle.

Plant cuticle consists of four layers, epicuticular waxes, the cuticle proper, the outer and inner cuticular layers. Electron micrographs of cuticle proper shows lamellate structure of this layer (Linn, 1980; Willison, 1980) and this is important for water permeability of the cuticle (Schmidt and Schoherr, 1982).

Ion penetration through the cuticle is strongly depending upon the nature of the acceptor solution. For example, when the cuticular membrane is regarded as a ion exchange membrane and deionised water is applied as a receiver then the permeability to cation is very slow (Chamel *et al.*, 1984; Heiska and Huttumen, 1986). The heavy metal ions, viz., copper, zinc, cadmium and manganese, all do exhibit high cuticular retention and competition for the same site of sorption (Chamel *et al.*, 1984).

3.1 Factor Influencing the Permeability of the Cuticle to the Air Pollutants

Environmental factor such as temperature, humidity, or light intensity can influence the rate of solute uptake from the leaf surface (Price, 1982; Reed and Tukey, 1982). For solute penetration the most important factor are cuticutar thickness and tortuosity of the pathway, the wettability, the chemical composition and arrangement of membrane constituents (Merida *et al.,* 1982) other factor which need to be considered are the molecular radius of the solute, its affinity to the cuticular membrane and the concentration inside and outside of the cell (Price, 1982). The water permeability of cuticle is determined by the waxes (extractable lipids). These waxes along with those in the epicuticle protect the plant from excess loss of the water (Schonherr, 1982).

3.2 Influence of Air Pollution on Cuticular Permeability

Leaves from heavily polluted areas exhibit higher level of transpiration and enhance water loss (Eveling and Bataille, 1984). At the same time changes in wax structure have been observed (Heiska and Huttunen, 1986; Godzik and Halbwacks, 1986). The change in wax composition also changes the water permeability. Sulphur dioxide, ozone, and acid rain are known to increase the rate of leaching of mineral ions. But it is not known to what extent this is a result of change in cuticular properties (Ayazloo, 1982 and Prinz, 1984).

4. EFFECT ON STOMATAL PHYSIOLOGY

Stomatal responses involve managing water loss and carbon gain by the plants. Plant continuously sense the environment and alter the stomatal conductance to maximize the water use efficiency (Farquhar *et al.*, 1980). A change in stomatal conductance in plants exposed to chronic elevated O_3 has been attributed to a direct effect of O_3 on photosynthesis, which results in increased internal CO_2 and in turn lower stomatal conductance (Reich and Amundson, 1985). Exposure of acute O_3 results in a rapid transient decrease in stomatal conductance (within 3–6 min of exposure) accompanied by a burst of ROS in the guard cells, followed by a slower recovery to initial rates of stomatal conductance (Kollist *et al.*, 2007; Vahisalu *et al.*, 2010).

4.1 Effect of SO_2 and O_3 on Stomatal Conductance

Black and Unsworth (1980) conducted the experiment to study the effect of different concentration of SO_2 on *Vicia faba* leaves and observed that if the concentration of SO_2 increases the conductance of the leave stomata is increases, and show that SO_2 causes the closure of stomata. Increase in conductance cause increase in photosynthesis (Jones, 1983). Measuring photosynthetic response to increased CO_2 following the stress treatment, such as air pollution can be used to identify the stomatal component of photosynthetic inhibition. This is important because gaseous pollutant inhibit photosynthesis by two ways, the pollutant may cause the stomatal closures which result to reduced photosynthesis and the pollutant may be absorbed into the leaves mesophyll where the biochem-ical process associated with the CO_2 fixation are impaired (Wong *et al.*, 1979).

5. MECHANISM OF MEMBRANE DAMAGE

Ozone alters the permeability by increasing the leakiness of the membrane (Heath, 1980) and by inhibiting pump and transporters (Dominy and Heath, 1985). Ozone may also alter the control elements, which consists of membrane potential by depolarization (Heath, 1987), diffusion by altering fluidity (Pauls and Thompson, 1981) and energy resources by depleting ATP levels (Pell and Brenner, 1973). The interaction of air pollutant, photochemical oxidant in particulars, with the cell surface in there periplasmic space of plant leaves can result in the generation of superoxide anions and peroxides, as has been shown in aqueous solution during ozone degradation (Heath, 1987). These oxidants may not be reactive enough to be toxic in them; they certainly are potential intermediate in the generation of hydroxyl radical, which will attack on membrane lipids and proteins (Wolf *et al.*, 1986). The high reactivity of hydroxyl radical with proteins suggests that a limited radical attack on critical transport proteins in the plasma membrane might cause a greatly amplified effect on the ionic homeostasis of the cells. If plants are exposed to ozone which causes change in membrane permeability, which result in the ionic imbalance of K^+ (Heath, 1987) most probably other ions, such as Ca^{++}, could also be affected. Calcium is known to pass through membrane via a discrete channel (Reuter, 1983) and activities of these channels would be functionally impaired by ozone and its products. The free calcium ion is act as a major intercellular regulator of many different chemical processes, acting as a second messenger and regulating many interrelated cellular function of the plants such as cell elongation, cell division, protoplasmic streaming, enzyme secretion and in regulating the stomatal response (Schwartz, 1985). Calcium can act alone or by binding to the receptor proteins which seem to accentuate its properties.

By changing Ca^{++} fluxes through the membrane surrounding the cytoplasm, air pollutant or secondary chemicals e.g., Aluminium (Zhao *et al.*, 1987) would be also able to alter the cytosolic concentration of free Ca^{++} and also able to change the biochemical and physiological reaction which depend on the Ca^{++} ion concentration (Heath, 1980).

6. EFFECT ON FREE RADICAL REACTION

Oxidative stresses such as ozone and sulphur dioxide gives rise to free radicals within the plants. Cellular maintenance and/or repair processes occur as a consequence of this free radical production. Oxidising molecular species such as ozone or sulphur dioxide can form potentially toxic free radical product in chain reaction in the environment of photosynthetic ally active plants. In case of sulphur dioxide, a primary pathway for the oxidation of sulphite to sulphate in the chloroplast (Asada, 1980) can give rise to the formation of superoxide anions ($O_2^{\cdot -}$). The oxidation is initiated by light and mediated by the photosynthetic electron chain. Oxygen is consumed in the process and superoxide anion is formed. Other free radicals which are more powerful oxidant such as hydroxyl radical are formed in a course of the chain reaction. Superoxide related active oxygen species also are major product formed from ozone in aqueous solution (Heath, 1987). Since the superoxide anion is an initiator of a chain reaction which generate highly reactive free radical species. In organic solvent or within biological membranes superoxide is a powerful nucleophile and base. It can react with proton to form the more toxic hydroperoxyl radical (HOO·) or act as a reducing agent with thylakoid component such as plastocyanin (Halliwell and Gutteridge, 1985). In the environment of membranes it can also react nonenzymatically with hydrogen peroxide forming the hydroxyl radical, a species which can destroy phospholipids. The nonenzymatic reaction between superoxide and hydrogen peroxide can only takes place in the presence of transition metal ions (Elstner, 1982). Copper and iron both present in the thylakoid membrane can serve this function.

High exposure of air pollutant forces chloroplast into an excessive excitation energy level, which in turn increases the generation of ROS and induces oxidative stress (Woo *et al.*, 2007). The deleterious effect of pollutant are caused by the production of reactive oxygen species in plants, which cause peroxidative destruction of cellular constituents (Tiwari *et al.*, 2006). It has been reported that proline act as a scavenging reaction of ROS with other amino acid, such as typtophane, tyrosine, histidine etc are more effective as compared to proline. Proline is of special interest because of its extensive accumulation in plants during environmental stress (Wang *et al.*, 2009).

7. BIOCHEMISTRY OF FREE RADICAL SCAVENGING

Cells are protected, at least in part, from reactive oxygen species by enzymatic and non- enzymatic defence mechanisms (Bowler *et al.*, 1992 ; Elstner and Osswald 1994). Defence against reactive oxygen species is provided by the scavenging properties of molecules, such as ascorbic acid, α-tocopherol, phenolic compounds, and glutathione. Superoxide dismutases (SODs) catalyze the reduction of superoxide to hydrogen peroxide. Hydrogen peroxide is then converted to H_2O by the action of catalases and peroxidases. Of particular importance is the ascorbate-specific peroxidase localized in the chloroplast. Acting in concert, ascorbate peroxidase, dehydroascorbate reductase, and glutathione reductase remove H_2O_2 in a series of reactions called the Halliwell–Asada pathway (Aono *et al.*, 1993).

Environmental extremes may either accelerate the production of reactive oxygen species or impair the normal defense mechanisms that protect cells from reactive oxygen species. In water-deficient leaves, for example, greater oxygen photoreduction by photosystems I and II increases superoxide production, and the pool of glutathione, as well as the activity of glutathione reductase, increases presumably as part of the cell defense mechanism. In contrast, levels of ascorbate, another antioxidant, generally decline with mild water stress. Transgenic plants overexpressing mitochondrion superoxide dismutase (Mn-SOD), the isozyme localized in the mitochondrial matrix, show less water-deficit damage and, significantly, improved survival and yield under field conditions (McKersie *et al.*, 1996).

The morpho-physiological responses of plants may vary according to the nature and the dose of the pollutant and the species. However, at biochemical and molecular levels, there appears to be a similarity among different pollutants and also among most of the environmental stresses. Several lines of evidences indicate that most air pollutants enter the plant tissues and act primarily through the production of reactive oxygen species (ROS) also called oxidative-free radicals, so is the case with most other abiotic stresses. Three important ROS, superoxide anion (O_2^-), hydroxy-free radical (OH^-) and H_2O_2 are highly toxic and cause changes in DNA, proteins and lipids and in membrane organizations. The generation of superoxide anion in chloroplasts is believed to take place in normal unstressed photosynthetic conditions also, by photoreduction of O_2 at PS I and PS II, when the energy from the triplet excited state of chlorophyll is transferred to O_2. The increase in enzymes and metabolites involved in the scavenging of ROS in response to some air pollutants and other oxidative stresses. The first reaction in detoxification of superoxide anion is its conversion (a dismutation reaction) to H_2O_2 by the enzyme superoxide dismutase. Initially this enzyme was considered to be an intracellular enzyme. At least three types of SODs with several isoforms are present in plants. These are: (i) chloroplastic or cytosolic Cu–Zn SOD; the cytosolic Cu–Zn SOD is referred to as Cu–Zn SOD and while chloroplastic one is referred to as Cu–Zn SOD II. (ii) mitochondrial Mn SOD and (iii) chloroplastic Fe SOD. The amino acid sequence of Fe–SOD and Mn–SOD apoproteins are similar, whereas Cu–Zn SOD is different. The genes for different types of SODs are also identified. Hydrogen peroxide can also be formed non-enzymatically from O_2^-, especially in the peroxisomes and mitochondria. Increase in one or the other type of SOD activities in response to air pollutants has been demonstrated in some investigations at both acute as well as chronic dosage of the pollutant (Asada *et al.*, 1987).

Ascorbate peroxidise is one of the important peroxidases, of ubiquitous occurrence in plants. It is regarded as a universal housekeeping protein in the cytosol and chloroplasts of plant cells. In the cytosol of nitrogen fixing root nodules, where ROS are produced under unstressed conditions also, it may constitute up to 1% of the total protein. Ascorbate peroxidases use ascorbate as a substrate and are believed to scavenge excess of H_2O_2 formed in plant cells under both normal and stress conditions. The product of ascorbate oxidation by ascorbate peroxidase is an ascorbate-free radical which is reduced back to dehydroascorbate by the enzyme monohydroascorbate reductase with NAD(P)H as the electron donor. Activity of ascorbate peroxidise increases in response to air pollutants specially with O_3. Concentration ascorbate in the chloroplasts is quite high (10–20 mM) and apart from its obvious role as a substrate for ascorbate peroxidase, it can react chemically with almost all forms of ROS. Rapid changes in the concentration or redox status of ascorbate have also been reported in response to O_3 (Barnes and Can, 1972; Luwe *et al.*, 1993) and to NO_2 (Ramge *et al.*,1993).

Polyamines are present in plant cells in millimolar concentrations, and they can act as free radical scavengers either directly or after interacting with other molecules such as free ferulic and caffeic acids. Increase in polyamine level in response to air pollutants has been observed in pea in response to SO_2 (Preibe *et al.*, 1978).

8. EFFECTS ON THYLAKOID MEMBRANE REACTION

Air pollutant has harmful effect on plants and perturb plant metabolism, including photosynthesis, nitrogen metabolism, phloem loading, stomatal movement and reduce plant growth and crop production. SO_2 has been known to inhibit photosynthetic CO_2 fixation in plants (Malhotra and Khan, 1984). NO_2 is known to inhibit photosynthetic CO_2 fixation in intact leaves. The mechanism of this inhibition of photosynthetic CO_2 fixation by NO_2 fumigation not been clarified. One possibility is that NO_2 compete with NADP in accepting electron from reducing side of the photosystem I another possibility is that acidification in the stroma inhibit CO_2 fixation at the enzyme step of fructose and sedoheptulose bisphosphate because HNO_2 penetrate easily in the chloroplast and release H^+ (Purczeld *et al.,* 1978). O_3 is known to inhibit photosynthetic activity in plants (Heath 1980). The effect of O_3 on photosynthesis assayed by chlorophyll fluorescence induction showed that the inhibition of water splitting enzyme system was the first sign of O_3 injury in bean (Schreiber *et al.*, 1978).

9. EFFECTS ON MITOCHONDRIA

Mitochondria are organelle which provides energy for cellular functioning. This process involved the oxidation of organic acids through the enzymes of krebs cycle, located in the matrix space. Reduced coenzyme provided by the dehydrogenases is reoxidized in the membrane through a respiration chain whose last step is the reduction of oxygen to water by cytochrome oxidase. Malate dehydrogenase activity is reduced in jack pine after exposure to SO_2. Ozone changes the pathway of carbohydrate metabolism (Tingey, 1974). Glycolysis is reduced thus allowing a major role to the pentose phosphate pathway. Decrease oxidation activity in mitochondria is observed when plant exposed to ozone before visible signs of injury appeared (Macdowall, 1965). The rate of succinate oxidation, initially unchanged, decrease in parallel with the extent of inhibition of oxidative phosphorylation as the duration of O_3 exposure increased. When O_3 was directly applied to a suspension of isolated mitochondria oxidizing succinate, the inhibition of both O_2 uptake and oxidative phosphorylation was clearly visible. The same result was observed with other krebs cycle substrate malate, citrate, α ketoglutrate (Lee, 1967). Freebairn (1957) showed the inhibitory effect of O_3 on the O_2 uptake of mitochondria. When symptoms of ozone injury develop, the swelling of mitochondria occur (Lee, 1968; Swanson *et al.*, 1973). The efficiency of oxidative phosphorylation was also decreased when mitochondria are treated with $CdCl_2$, electron transport and oxidative phosphorylation was inhibited (Miller *et al.*, 1973). In addition to fixing less CO_2, plants growing in elevated O_3 commonly have higher rates of mitochondrial respiration (Hayes *et al.*, 2007).

10. EFFECTS ON CARBOHYDRATE METABOLISM

Exposure of plants to gaseous air pollutants generally increases stomatal resistances and decreases net photosynthesis. The effects of combinations of pollutants on photosynthesis are usually greater than that exposure to single pollutant. Pollutants have different target sites to disrupting the plant metabolism. Within the chloroplast, such disruption can occur in the stromal reactions responsible for assimilating CO_2, as well as in the photo system that generate the ATP and NADPH needed in stromal reaction. SO_2 usually inhibits Photo System II and O_3 and NO_2 inhibit both Photo System I and Photo System II (Sugahara 1984). SO_2, NO_2 and O_3 combination was found to first increase the photosynthesis but after some time it inhibit the total photosynthesis activity (Sugehara, 1984). Concentration of total and soluble sugars decreased significantly in plants sensitive to the air pollution. The decrease in total sugar content of damaged leaves probably corresponds with the photosynthesis inhibition or stimulation of respiration (Tzvetkova and Kolarov, 1996).

The concentration of soluble sugars is indicative of the physiological activity of a plant and it determines the sensitivity of plants to air pollution. Reduction in soluble sugar content in polluted sites can be attributed to increased respiration and decreased CO_2 fixation because of chlorophyll deterioration. It has been mentioned that pollutant like SO_2, NO_2 and H_2S under hardening condition can cause more depletion of soluble sugars in the leaves of plants grown in polluted area. The reaction of sulphide with aldehyde and ketone of carbohydrates can also cause reduction in carbohydrate content (Tripathi and Gautam, 2007). Non-structural carbohydrates essential for growth, including sucrose and starch, also decreased on exposure to higher concentration of O_3. O_3-induced decrease in primary metabolism is well correlated with the capacity at the cellular level for CO_2 fixation, based on studies of Rubisco transcript level, protein level and enzyme activity. Similar changes have been detected in other components of the photosynthetic machinery and Calvin Benson-cycle enzymes, including RuBisCO activase, ATP synthase, the oxygen-evolving subunit of photosystem II, aldolase, phosphoglycerate kinase and NADP-glyceraldehyde-3-phosphate dehydrogenase (Agrawal *et al.*, 2002; Sarkar *et al.*, 2010).

11. EFFECTS ON PROTEINS AND AMINO ACID METABOLISM

In polluted environment, SO_2 may be used as a source of sulphur and exposing plants to low concentration of gas can result in increase in concentration of sulphate, glutathione, cysteine and methionine (Maldotra and Sarkar, 1979). SO_2 readily diffuses into leaves and dissolve in the moisture available in the mesophyll cells, forming sulphite and bisulphite both of which are highly toxic species of sulphur (Puckett *et al.*, 1973). Many proteins have enzymatic activity and are vital for the rapid rate of biochemical reactions in the cells. Two other important functions of the proteins are as major H^+ ion buffer and structural component of the cells (Larsen, 1980). The amino acids can provide secondary structure to protein molecule by molecular bonding. Pollutant may attack these bonds and so expose molecule for further attack (Igntenko and Cherenkevich, 1985). Damage to the membrane protein, perhaps by oxidation of susceptible amino and sulfhydryl groups on the surface of the membrane protein, may be the alterations in the permeability of the plasma lemma (Dominy and health, 1985). Pollution effects have been linked to the acceleration in senescence (Kohut *et al.,* 1986). Leaf injury symptoms may cause increase in proteolysis in polluted leaves which may be due to promotion of proteases activity in the leaf (Watanabe and Konodo, 1984).

Proteins are also found in conjugated form with the prosthetic groups e.g. glycoproteins which are bound with carbohydrates. An important class of conjugated proteins are the chromo proteins, which include the chlorophyll proteins, where pigment form the prosthetic group. The chlorophyll may be stabilised by this protein moiety and the latter can be affected by sulphite from SO_2 and O_3, so reducing the chlorophyll complex stability (Sugahara *et al.,* 1980). Conformational changes in enzymatic protein molecules due to air pollution may cause deleterious effect on the functioning of the cell. Alterations in the active site configuration could affect the metabolic function of the enzyme (Rhodes *et al.,* 1979). In absence of proper enzyme function protein synthesis and its accumulation in the grains will be adversely affected. Moreover, protein content is very important in grain quality. Grain quality change has been observed in response to pollution (Kohut *et al.*, 1987).

Proline accumulation might play an important role in inhibiting air pollution induced peroxidation. Proline accumulation often occurs in variety of plants in the presence of different stresses. For example proline accumulation in leaves of plants exposed to SO_2 fumigation (Tankha and Gupta, 1992), heavy metals (Wang *et al.*, 2009) and stress condition (Woodward and Bennett, 2005 and Wang *et al.*, 2009) has been reported.

12. EFFECT ON SECONDARY METABOLITES

Acetaldehyde, ethylene, ethane, ethanol and the polyamine puterscine all are used as a indicators of plant response to air pollutants. Ethylene is a plant hormone present in low level in younger plants but its level increases as the plant senesce (Abeles, 1973). Ethylene level increases immediately following exposure to the air pollution and decline prior to cell death (Bucher, 1984). Ethane is produced by stressed plants and is related to lipid peroxidation (Peiser and Yang, 1979). Ethane emission follows that of ethylene and is associated with visual injury. Detection of ethane has been associated with many of the air pollutants such as ozone, sulphur dioxide and acid rain (Bucher, 1984; Chia *et al.*, 1984; Peiser and Yang, 1979).

Secondary metabolites compose broad classes of plant biochemicals with wide roles within the plant (Bell, 1980). Acetaldehyde and ethanol may serve as important early indicators of stress response. Plants exposed to sulphur dioxide emit detectable level of ethanol and acetaldehyde prior the emission of ethylene or ethane (Kimmerer and Kozlowski, 1982). Polyamines play important role in air pollution tolerance. Puterscine, a diamine has been associated with several pollutant stresses including sulphur dioxide, cadmium, low external pH and magnesium deficiency (Weinstein *et al.*, 1986). The production of some secondary metabolites by vegetation can alter quality of plants. Change in certain secondary metabolites could lead to alter crop quality, changes in plant susceptibility to disease or attractiveness to insect (Skarby, 1984).

13. EFFECTS ON GENE EXPRESSION AND POST TRANSLATIONAL MODIFICATION

Cell metabolism is an interactive and well co-ordinated biochemical network. Metabolic pathways are functionally compartmented but continuously interconnected. These are optimized by feedback loops of regulatory information and can thus perform the physiological functions at cellular level. In response to alterations in plant environment, balances between these pathways are continuously controlled and adequately adjusted. Metabolic efficiency in plant system is achieved by two complementary properties: conservative homeostasis and adaptive flexibility. Both proceed from the network type of structure in cell metabolism, not from a particular reaction and ensure the dynamic adaptiveness of cell functions.

Free radicals like superoxide anion arising from ozone interaction with the unsaturated lipids in the membrane can cause breaks in DNA strands (Lesko *et al.*, 1980). The level of superoxide dismutase increases after exposure to ozone indicating free radical scavenging activity. This plays important role to protect the plant from ozone effect (Bennett *et al.*, 1985). As a part of metabolic modifications in response to pollution, synthesis of different forms of an enzyme takes place. Low level SO_2 fumigation of crop plants alters the isozyme pattern of glutamate dehydrogenase in the absence of visible injury (Varshney and Varshney, 1984).

Post – translational covalent modifications of proteins by phosphorylation/ dephosphorylation is a frequent, well known process of enzyme regulation by metabolic interactions. Exposure of plants to reactive oxygen species stimulates the transcription and translation of genes that encode enzymes involved in protection mechanisms. In Arabidopsis, exposure to low levels of O_3 induces the expression of several genes that encode enzymes associated with protection from reactive oxygen species, including SOD, glutathione S-transferase and phenylalanine ammonia lyase. In transgenic tobacco transformed with a gene from Escherichia coli to give additional glutathione reductase activity in the chloroplast, short-term exposure to high levels of SO_2 is much less damaging than for wild-type tobacco (Aono *et al.*, 1993).

14. EFFECTS ON CARBON GAINS, ALLOCATION AND GROWTH OF PLANTS

The amount of carbon gained by plants cannot be translated directly into biomass increase since carbon losses through maintenance and construction respiration must be taken into account. The incremental gain in dry matter through time is generally dependent on the amount of carbon that is allocated to the leaf versus non photosynthetic tissue. Allocation of carbon to leaf does result in compounding of dry matter increase whereas this is not the case for non-photosynthetic tissues such as stems and roots (Monsi, 1968).

Air pollution can substantially reduce plant growth under controlled condition, and even at ambient level occurring over large regions (Heck *et al.*, 1983). Air pollution affects allocation and that it does so in a predictable manner. Roots have proportionately greater growth reduction when exposed to air pollutants than shoot (Godzik *et al.*, 1985). Any environmental factor that cause a reduction in leaf photosynthetic capacity alter allocation patter in a manner that favour the production of new leaf over new root (Winner *et al.*, 1985). In addition to decreased carbon availability from O_3-mediated changes in primary metabolism, plant carbon balance is further impacted by indirect costs associated with the detoxification needed to counter the ROS increase generated by O_3. Although the dissolution chemistry of O_3 in the apoplast is not completely understood, the ability of the apoplast to quench ROS generated from O_3 depends upon the concentration of radical-scavenging metabolites and enzymes in the apoplast, the rate of their reactions with O_3 and the rate of regeneration of the reduced compounds (Luwe *et al.*, 1993).

Leaf longevity studies have also shown that senescence is induced by elevated O_3 and represents lost opportunity for carbon gain (Pell *et al.*, 1999). These metabolic changes can alter source-sink relations, with reduced root biomass commonly reported following chronic O_3 exposure (Andersen, 2003; Fiscus *et al.*, 2005). All of these responses to O_3 have an energetic cost to the plant that contributes to the overall decrease in growth and biomass (Ashmore, 2005).

The decrease in total sugar content of damaged leaves probably corresponded with the photosynthetic inhibition or stimulation of respiration rate. Higher starch accumulation in damaged leaves may result both in the higher resistance of their photosynthetic apparatus (Prokopiev, 1978) and low starch export from the mesophyll. The negative effects of heavy metals on carbon metabolism result from their possible interaction with the reactive centre of ribulose bisphosphate carboxylase (Stiborova *et al.*, 1987).

References

Abeles, F.B. (1973). Ethylene in plant biology. Academic Press. New York.

Agrawal, G., Rakwal, R., Yonekura, M., Kubo, A. and Saji, H. (2002). Proteome analysis of differentially displayed proteins as a tool for investigating ozone stress in rice (*Oryza sativa* L.) seedlings. *Proteomics* **2**:947–59

Andersen, C.P. (2003). Source-sink balance and carbon allocation below ground in plants exposed to ozone. *New Phytol.* **157**:213–28

Aono, M., Kubo, A., Saji, H., Tanaka, K. and Kondo, N. (1993) Enhanced tolerance to photo oxidative stress of transgenic *Nicotiana tabacum* with high chloroplastic glutathione reductase activity. *Plant Cell Physiol.* **34**: 129–135

Asada, K. (1980). Formation and scavenging of superoxide in chloroplast with relation to injury by sulphur dioxide studies on the effect of air pollutants on plants and mechanism of phytotoxicity. Natl. Inst. Environ. Stud. **11**: 165-179.

Asada, K. and Kiso, K. (1973). Initiation of aerobic oxidation of sulphite by illuminated spinach chloroplast. *Eur. J. Biochem.* **33**: 253-257.

Asada, K. and Takahashi, M. (1987) In Photo inhibition (eds. Kyle, D., Osmond, C. B. and Arntzen, C.), Elsevier Science Publishers, Amsterdam, pp. 227–287.

Ashmore, M.R. (2005). Assessing the future global impacts of ozone on vegetation. *Plant Cell Environ.* **28**: 949–64

Ayazloo, M., Garsed, S.G. and Bell, J.N.B. (1982). Studies on the tolerance to sulphur dioxide of grass populations in polluted areas. II. Morphological and physiological investigations. *New Phytol.* **90**: 109-126.

Barnes, R. L. (1972). Effect of chronic exposure to ozone on soluble sugars and ascorbic acid contents of pine seedlings. *Can. J. Bot.* **50**: 215–219.

Bell, E.A. (1980). The possible significance of secondary compound in plant. In secondary Plant Product, Eds. E.A. Bell and B.V. Charlwood. Springer Verlag New York 11-21.

Bennett, J.H., Lee, E.H. and Heggstad, H.E. (1985). Biochemical aspect of plant tolrrance to ozone and oxyradicals, superoxide dismutase. In gaseous air pollutant and plant metabolism. Butterworths, London.

Black, V.J. and Unsworth, M.H. (1980). Stomatal response to sulphure dioxide and vapour presser deficit. *Exp. Bot.* **31**: 667-677.

Bowler, C., Van Montagu, M. and Inze, D. (1992), *Annu. Rev. Plant Physiol. Plant Mol. Biol.* **43**: 83–116.

Bucher, J. B. (1984). Emission of volatiles from plant under air pollution stress.In gaseous air pollution and plant metabolism. Butterworth, London. 399-412.

Chamel, A.R., Gambonnet, B., Genova, C. and Jourdain, A. (1984).Cuticular behaviour of cadmium studied using isolated plant cuticles. *Environ, Quality*. **13**: 483-487.

Chia, L.S., Mayfield, C.I. and Thompson, J.E. (1984). Stimulated acid rain induce lipid peroxidation and membrane damage in foliage. *Plant Cell Environ.* **7**: 333-338.

Choularton, T.W., Gay, M.J., Jones, A., Fowler, D., Cape, J.N. and Leith, I.D. (1987). The influence of attitude on wet deposition. Atmospheric environment.

Dominy, P.J. and Heath, R,L. (1985). Inhibition of the stimulated ATPase of the plasma lemma of Pinto Beans leaves by ozone. *Plant physiol.* **77**: 43-45.

Doncheva, S., Nikolov, B. and Pajeva, J. (1993). Influence of copper excess upon the synthetical activity of root meristem cells. Comp. rend. Acad. bulg. Sci., 46 (11), 105–108.

Elstner, E.F. (1982). Oxygen activation and oxygen toxicity.Ann. Rev. Plant Physiol. **33**: 73-96.

Eveling, D. and Bataille, A. (1984).The effect of deposition of small particle on the resistance of leaves and petals to water loss. *Environ. Pollut.* **36**: 229-238.

Farquhar, G.D., Schulze, E.D. and Kupers, M. (1980). Response to humidity by stomata of *Nicotiana glauca* L. and *Corylisa vellana* L. are consistent with the optimisation of carbon dioxide uptake with respect to water loss. *Plant Physiol.* **7**: 315-327.

Fiscus, E.L., Booker, F.L. and Burkey, K.O. (2005). Crop responses to ozone: uptake, modes of action, carbon assimilation and partitioning. *Plant Cell Environ.* **28**: 997–1011.

Fowler, D. (1980). Removal of sulphur and nitrogen compound from the atmosphere in rain and dry deposition. In ecological impact of Acid precipitation, D. Drablos and A. Tollan (eds). 22-32. SNSF, Oslo-As. Choularton, T.W., Gay,

M.J., Jones, A., Fowler, D., Cape, J.N. and Leith, I.D. (1987). The influence of altitude on wet deposition. Atmospheric environment.

Freebairn, H.T. (1957). Reversal of inhibitory effect of ozone on oxygen uptake of mitochondria.*Science.* **126**: 303.

Godzik, S. and Halbwachs, G. (1986). Structural alteration of *Aesculus hippocastannum* L. leaf surface by air pollutants. *Z. Pflanzenkrank, Pflanzensch.* **931**: 590-594.

Godzik, S., Ashmore, M.R. and Bell, J.N.B. (1985). Reponses of radish cultivars to long term and short term exposure to sulphur dioxide and their mixture. *New Phytol.* **100**: 191-197.

Halliwell, B. and Gutteridge, J.M.C. (1985). Free radical in biology and medicine, Clarendon, Oxford, 340pp

Hayes, F., Jones, M.L.M., Mills, G. and Ashmore, M. (2007). Meta-analysis of the relative sensitivity of semi-natural vegetation species to ozone. *Environ. Pollut.* **146**:754–62.

Heath, R.L. (1980). Initial events in injury to plants by air pollution. *Plant Physiol.* **31**: 395-431.

Heath, R.L. (1987).The biochemistry of ozone attack on the plasma membrane of plant cells. *Adv. In Phytochem.* 21.

Heck, W.W., Adams, R.M., Cure, W.W., Heagle, A.S., Heggestad, H.E., Kohut, R.J., Kress, L.W., Rawling, J.O. and Taylor, O.C. (1983). A reassessment of crop loss from ozone. *Environ. Sci. Technol.* **17**: 573-581.

Heiska, E. and Huttumen, S. (1986). Preliminary measurement of ion penetration through isolated cuticular membrane of pine needles. In press.

Hill, A.C. and Bennett, J.H. (1970). Inhibition of apparent photosynthesis by nitrogen oxides. *Atmos. Environ.* **4**: 341-348.

Holloway, P.J. (1969). Effect of superficial wax on leaf wettability. *Annals Applied Biology.* **63**:145-153.

Igntenko, A.V. and Cherenkevich, S.N. (1985). Reactivity of amino acid and proteins in reaction with ozone. *Kinetics and catalysis* **26 (6)**: 1145-1148.

Jones, H.G. (1983). Plant and microclimate. Cambrige Univ Press, London. P. 104-105.

Joshi, P.C. and Swami, A. (2009). Air pollution induced changes in the photosynthetic pigments of selected plant species. *J. Environ. Boil.* **30**: 295-298.

Khout, R. J., Amundson, R. G. and Laurence, J. A. (1986). Evolution of growth and yield of soyabean exposed to ozone in the field. *Environmental Pollution Sereies A* **41**: 219-234.

Khout, R. J., Amundson, R. G., Laurence, J. A., Colavito, L., Van Leuken, P. and King, P. (1987). Effect of ozone and sulphure dioxide on yield of winter wheat. *Phyto. Path.* **77**: 71-74.

Kollist, T.,Moldau, H., Rasulov, B., Oja, V. and Ramma, H. (2007). A novel device detects a rapid ozoneinduced transient stomatal closure in intact Arabidopsis and its absence in abi2 mutant. *Physiol. Plant.* ***129***:796–803

Larsen, P.O. (1980). Physical and chemical properties of amino acids. *Biochem. Plants* **5**: 226-269.

Lee, T.T. (1967). Inhibition of oxidative phosphorylation and respiration by ozone in tobacco mitochondria. *Plant Physiol*. **42**: 691.

Lee, T.T. (1968). Effect of ozone on swelling of tobacco mitochondria. *Plant Physiol.* **43**: 133.

Lesko, S.A., Lorentzen, R.J. and Tso POP. (1980). Role of superoxide in deoxyribonucleic acid strand scission. *Biochemistry* **19**: 3023-3028.

Lindberg, S.E., Lovett, G.M., Richter, D.D. and Johnson, D.W. (1986). Atmospheric deposition and canopy interaction of major ions in a forest. *Science* **231**: 141-145.

Linn, P. (1980). Microscopic structure of cuticle of Cliviaminiata. (Manuscript).

Liu, Y.J. and Ding, H. (2008). Variation in air pollution tolerance index of plants near a steel factory: Implication for landscape-plant species selection for industrial areas. *Wseas Trans. Environ*. Dev. **4**: 24-32

Luwe, M.W.F., Takahama, U. and Heber, U. (1993). Role of ascorbate in detoxifying ozone in the apoplast of spinach (*Spinacia oleracea* L.) leaves. *Plant Physiol*. **101**:969–76

Macdowall, F.D.H. (1965). Stages of ozone damage to respiration of tobacco leaves. *Can. J. Bot*. **43**: 419

Malhotra, S.S. and Khan, A.A. (1984). Biochemical and Physiological impact of major pollutants. In Treshow, M. (ed.) Air Pollution and Plant Life: 113-157. Wiley, Chichester

Malhotra, S.S. and Sarker, S.K. (1979).Effect of sulphur dioxide on sulphur and free amino acid content on pine seedling. *Physiologia Plantarum*. **47**: 223-228.

Mansfield, T.A. and Freer Smith, P.H. (1984).The role of stomata in resistance mechanisms. In Gaseous Air Pollutant and Plant Metabolism, M.J. Koziolanf F.R. Whatley (eds.). 131-146. Butterworths, London.

Mccormick, R.A. and Holzworth, G.C. (1975). Air pollution climatology, in air pollution, A. Sterm, (ed.). 643-700. Academic Press, New York.

McKersie, B., Bowley, S. R., Harjanto, E. and Leprince, O. (1996) Water-deficit tolerance and field performance of transgenic alfalfa over expressing superoxide dismutase. *Plant Physiol*. ***111***: 1177–1181.

Merida, T., Schonherr, J. and Schmidt, H.W. (1981).Fine structure of plants cuticle in relation to water permeability. The fine structure of (*Clivia miniata*) leaves. *Planta*.***152***: 259-267.

Miller, R.J., Bittell, J.E. and Koeppe, D.E. (1973).The effect of cadmium on electron transfer reaction in corn mitochondria. *Physiol. Plant*. **28**:166.

Monsi, M. (1968). Mathematical models of plants communities. 131-150.

Ormord, D.P. (1982). Air pollutant intractions in mixtures. In effect of air pollution in agriculture and horticulture, M.H. Unsworth and D.P. Ormord (eds.). 307-332. Butterworths, London.

Pauls, K.P. and Thompson, J.E. (1981).Effect of invitro treatment with ozone on physical and chemical properties of membrane. *Physiol. Plant*. **53**: 255-262.

Pell, E.J. and Brennan, E. (1973). Change in Respiration, Photosynthesis, ATP and total Adenylate content of ozonated Pinto Beans Foliage as they relate to symptom expression. *Plant Physiol.* **53**: 255-262.

Pell, E.J., Sinn, J.P., Brendley, B.W., Samuelson, L. and Vinten-Johansen, C. (1999). Differential response of four tree species to ozone-induced acceleration of foliar senescence. *Plant Cell Environ.* **22**:779–90

Pieser, G. D. and Yang, S. F. (1979). Ethylene and ethane production from sulphur dioxide injured plants. *Plant Physiol.* **63**: 142-145

Preibe, A., Klein, H. and Jager, H-J. (1978). *J. Exp. Bot.* **26**: 1045–1050.

Price, C.E. (1982). A review of the factor influencing the penetration of pesticides through plant leaves. Cutler *et al.* Academic Press, London, 237-252.

Prinz, B. (1985). Symptomatic and mogliche Ursachen der Waldschaden. LIS-Berichte, **57**: 7-25.

Prokopiev, E. (1978). Afforestation of Industrial Areas. Sofia, Zemizdat, 208 p. (In Bulg.).

Puckett, K.J., Nieboer, E., Flora, W.P. and Richardson, D. H. S. (1973). Sulphur dioxide. Its effect on photosynthesis C14fixation in lichens and suggested mechanism of phytotoxicity. *New Phytologist.***72**: 141-154.

Purczeld, P., Chon, C.J., Protis, A.R., Heldt, H.W. and Heber, U. (1978). The mechanism of carbon fixation by the pH of chloroplast stroma. Studied with nitrite mediated proton transfer across the envelop. *Biochem.Biophys. Acta.***501**: 488-498.

Ramge, P., Badeck, F-W., Plochl, M. and Kohlmaier, G. H. (1993). *New Phytol.***125**: 771–785.

Raynor, G.S. and Hayes, J.V. (1981). Acidity and conductivity of precipitation on long island, New York in relation to meteorological variables.Water, Air and Soil pollution.**15**: 119-245.

Reed, D. and Tukey, J.R.H.B. (1982). Permeability of Brussels sprouts and carnation cuticles from leaves development in different temperature and light intensities. In: Cutler *et al.* Academic Press, London, 267-278.

Reich, P.B. and Amundson, R.G. (1985). Ambient levels of ozone reduce net photosynthesis in tree and crop species. *Science* **230**:566–70

Rhodes, D., Sims, A.P. and Stewart, G.P. (1979). Glutamine synthetase and control of nitrogen assimilation in *Lemna minor* L. in Nitrogen assimilation in plants. 501-520.

Sarkar, A., Rakwal, R., Agrawal, S., Shibato, J. and Ogawa, Y. (2010). Investigating the impact of elevated levels of ozone on tropical wheat using integrated phenotypical, physiological, biochemical, and proteomics approaches. *J. Proteome Res.* **9**: 4565–84.

Schmidt, H.W. and Schonherr, J. (1982). Development of plant cuticles: occurrences and role of non ester bonds in cutin of Cliviaminiata leaves. *Planta.* **156**: 380-384.

Schonherr, J. (1982). Resistance of plant surface to water loss: transport properties of cutinsuberin and associated lipids. *In: Encyclopedia of Plant Physiology,* N.S. **12B**: 153-179.

Schreiber, U,.Vidaver, W., Runeckles, V.C. and Rosen, P. (1978). Chlorophyll fluorescence assay for ozone injury in intact plants. *Plant Physiology*.**61**: 81-84.

Schwartz, A. (1985). Role of Ca^{++} and EDTA on stomatal movement in *Commelina communis* L. *Plant Physiol*. **79**: 1003-1005.

Skarby, L. (1984). Changes in the nutritional quality of crops. Butterworths., London. 351-359.

Stiborova, M., Ditrichova, M. and Brezinova, A. (1987). Effect of heavy metal ions on growth and biochemical characteristics of photosynthesis of barley and maize seedlings. *Biol. Plant*. **29**: 453–467.

Sugehara, K. (1984). Effect of air pollution on light reaction in chloroplast.In gases air pollution and plant metabolism, M. J. Kozio and F.R. Whatley. Butterworth, London 169-180.

Sugehara, K., Uchida, S. and Takimoto, M. (1980). Effect of sulphite ion on water soluble chlorophyllprotiens. *Inst. Environ. Stud*. **11**: 103-112

Swanson, E.S., Thomson, W.W. and Mudd J. B. (1973). The effect of ozone on leaf cell membrane.*Can. J. Bot*. **51**: 1213.

Tankha, K. and R.K. Gupta, (1992). Effect of Water deficit and SO_2 on total soluble protein, nitrate reductase activity and free proline content in sun flower leaf. *Biol. Planta*. **34**: 305-310.

Tingey, D.T. (1974).ozone induced alteration in metabolite pools and enzyme activities of plants. In Air pollution effect on plant growth, M. Dugger eds., ACS Symp. Ser. 3, Am. Chem. Soc., Washington, 40-57.

Tiwari, S., Agrawal, M. and Marshall, F. M. (2006). Evaluation of ambient air pollution impact on carrot plants at a sub urban site using open top chambers. *Environ. Monit. Assess*.119: 15-30.

Tripathi, A.K. and Gautam, M. (2007). Biochemical parameters of plants as indicators of air pollution. *J. Environ. Biol*.**28**: 127-132.

Tukey, H.B. (1970). The leaching of substances from plants. *Ann. Review Plant Physiology*. **21**: 305-324.

Tzvetkova, N. and Kolarov, D. (1996). Effect of air pollution on carbohydrate and nutrient concentrations in some deciduous tree spicies. *Bulg. J. Plant Physiol*. **22**: 53-63.

Unsworth, M.H. (1981). The exchange of carbon dioxide and air pollutants between vegetation and the atmosphere. In plants and their Atmospheric Environment, J. Grace, E.D. Ford and P.G. Jarvis (eds) 111-138. Blackwell Scientific, Oxford.

Vahisalu, T., Puzorjoa, I., Brosche, M., Valk, E. and Lepiku, M. (2010). Ozone-triggered rapid stomatalresponse involves the production of reactive oxygen species and is controlled by SLAC1 and OST1.*Plant J*. **62**:442–53

Varshney, S.R.K. and Varshney, C.K. (1984). Effect of SO_2 on ascorbic acid in crop plants. *Environ. Pollut*, (Ser. A) **35**: 285-290.

Wang, F., Zeng, B., Sun, Z. and Zhu, C. (2009). Relationship between proline and Hg2+-induced oxidative stress in a tolerant rice mutant. *Arch. Environ. Contam.Toxicol*. **56**: 723-731

Watanabe, T. and Kondo, N. (1984).The change in leaf proteinase and proteinase inhibition activities by air pollution. *Inst. Environ. Stud. Jpn.* **65**: 241.

Weinstein, L.H., Kaur Sawhney, R., Rangam, M.V., Wettlaufer, S.H. and Galston, A.W. (1986). Cadmium induced accumulation of putrescine in oat and bean leaves. *Plant Physiol.* **82**: 641-645

Willison, J.M.M. (1980). Freeze fractured cuticle of plantd. *Plant Sci.* **18**: 121-126.

Winner, W.E., Mooney, H.A., Willliam, K. and Caemmerer, S. (1985). Measuring and assessing SO_2 effect on photosynthesis and plant growth. 118-139

Wolf, S.P., Garner, A. and Dean, R.T. (1986). Free radicals, lipid and protein degradation. Biochemical

Woo, S.Y., Lee, D.K. and. Lee, Y.K. (2007). Net photosynthetic rate, ascorbate peroxidase and glutathione reductase activities of Erythrinaorientalis in polluted and non-polluted areas. *Photosynthetica* **45**: 293-295.

Woodward, A.J. and Bennett, I.J. (2005). The effect of salt stress and abscisic acid on proline production, chlorophyll content and growth of in vitro propagated shoots of Eucalyptus camaldulensis. *Plant Cell Tissue Organ Cult.* **82**: 189-200

Zhao, X.J., Sucoff, E. and Stadelmann, E.J. (1987). Al^{3+} and Ca^{++} alteration of membrane permeability of Quercusrubra root cortex cell. *Plant. Physiol.* **83**:159-162.

Section III:

TECHNOLOGICAL INNOVATIONS IN AGRICULTURE

18

CAN NANOTECHNOLOGY BE HARBINGER FOR SECOND GREEN REVOLUTION?

A.K. Trivedi

National Bureau of Plant Genetic Resources, Regional Station Bhowali (Niglat) – 263132, Nainital Uttarakhand (INDIA)

INTRODUCTION

Nanotechnology is the manipulation, control and restructuring of matter at dimensions of roughly 1 to 100 nanometers, where new phenomena enable new applications. At this size, matter can take on different chemical and physical properties, giving the products characteristics useful to industry, medicine and technology. Nanotechnology is a new technology and emerging branch of science. Term Nanotechnology was first time used by K. Eric Drexler in 1986 in his book entitled "Engines of Creation: The Coming Era of Nanotechnology". Drexler visualized the possibility that human- designed molecular robots could replicate themselves just about the same way cells build copies of themselves in order to reproduce. Nanotechnology is a new field that combines fundamental science, materials science and engineering at nanometer scales (billionth of a meter) to the design and manufacture of structures and devices with dimensions about the size of a molecule. Nanotechnology techniques provide new tools for biomolecular studies. Developments in nanotechnology have resulted from the concerted efforts of scientists and engineers working in separate fields such as biotechnology, chemistry, microelectronics and photonics e.g., nanofabrication integrates material science, engineering and biology.

Nanometer will likely to play key role in the 21st century similar to the one played by the micrometer in the 20th century (Vettiger *et al*. 2002). Nanotechnology has been engorged among academia and researchers as a revolutionary technology for the next green revolution. Nanotechnology has several attributes in practice that have impacts on future agricultural production.

Potential of nanotechnology in diverse industries such as agriculture, electronics and health is considerable (Opara, 2002a). The prospects for nanoprobes, atom-moving devices, nanoresistors, nanowires, and nanorobots for industrial applications have been demonstrated by numerous researchers (Scientific American, 2001). Greater fusion between nanotechnology with other emerging high technologies offers greater potentials for realizing these potential technological and economic benefits (Lievonen, 2000). Experimentation in nanotechnology begins with the presumption that the resulting structures may possess superior properties including chemical, electrical, mechanical and optical. Although biomedical research and defense for fighting cancer and building weapons have led investments in nanotechnology, the development of tools and techniques for characterization and building nanostructures has influential applicability across many industries including agriculture. Government funding and private investors provide billions of research dollars for the development of new materials and applications. The potential utility of these technologies is such that they are expected to be a trillion-dollar industry within the next 10 years (Curtis *et al*., 2006).

By interventions of technological innovations agriculture and other land-based human endeavors have been transformed into industries that produce variety of products as well as services. Beginning from food gatherers, hunters use of improved technology for crop, livestock and fisheries management have enabled humans to exploit natural resources for fulfilling his needs. The limitations and toil of human muscle as power source for agriculture were later overcome by the domestication of animals which provided more efficient and reliable power sources for various agricultural operations, transportation and agro-processing. However, it was not until the beginning of the use of engine-powered mechanical technology that humans reached a momentous turning point in the application of engineering and technological innovations in agriculture and rural development.

Recent advancements in science and technology have had both positive and negative consequences on humankind and the environment. The impacts of technological innovations have transformed the way we practice agriculture. Advances in the use of science and technology in agriculture have helped to avoid the human catastrophe predicted during the Malthusian era. In spite of this the proportion of humankind affected by food insecurity and malnutrition continues to rise in many regions of the developing world and has been projected to continue in the foreseeable future. In view of persistent food insecurity and related poverty which rural agricultural population facing in the developing countries, the future of agriculture on a global scale remains a subject of intense debate. Moreover, its trade is declining in comparison with other industries and its negative impacts on the environment are a subject of scrutiny and debate. Apparently, modern innovative technologies are essential in deal with these problems. Old technologies that are now part of our socio-economic fabric have had remarkable impacts on agriculture. For example, new irrigation techniques for water delivery and conservation have enabled us to grow crops where it was almost impossible few decades back, particularly in arid climates.

2. IMPACT OF TECHNOLOGICAL INNOVATIONS IN AGRICULTURE

Technological innovation has become the strategic weapon for competitive advantage and multi pronged progress. Through application of modern technological developments, agriculture has undergone extensive changes during the past century. Recent agricultural production has been characterized by intensive application of new production inputs such as engine-powered devices, agro-chemicals use of new plant and animal materials such as high-yielding varieties. As a consequence, new agricultural practices, handling and processing techniques have been developed to meet growing consumer demand for consistently high quality, safe, diverse and nutritious food products.

Priorities of consumer demand for food and other agricultural commodities/products are changing time to time. In addition to the demand for steady supply of top quality products and services, consumers are placing preferences for safety, functionality, and sustainability of agricultural practices. Many consumers of agricultural products now demand for information about their source as well as the method of production in terms of sustainability and impact on the environment. Many consumers increasingly demand evidence of organic plant food products. These market-pull factors have placed new demands for assurance of traceable agricultural supply chains. The preceding development assures the future for innovative technologies to adequately address the numerous economic and socio-cultural problems facing agriculture and farming in the 21st century and beyond.

During the past century agricultural productivity has increased and methods of agricultural production changed extensively. Changes in agricultural technology have been a major factor shaping agriculture during this period (Sunding and Zilberman, 2000). On a global scale, remarkable changes in agricultural production patterns have taken place. Today, farming community which is a small fraction of total population of the country produces enough food for the entire population as well as for export also. This has been possible through the application of modern labor saving technologies for intensive agriculture, on-farm mechanization, irrigation, post harvest handling and processing and the use of improved crop varieties and animal genetics.

From a technological and engineering viewpoint, several distinct eras in agricultural development can be identified and characterized. Progress in agricultural engineering and technology has resulted in new machinery and process techniques for the production, post harvest handling and utilization of food and agricultural raw materials. With the evolution of Agricultural Engineering as a profession and academic discipline, the activities have grown tremendously to include information systems technology for production, handling, and processing and improved methods for efficient and sustainable utilization of land, water and other natural resources.

In the view of rapidly increasing human population and low agricultural productivity per labor force, the mechanization of agriculture helped to transform vast unproductive land into highly productive estates and thereby removed the hard work of manual labor. It was therefore, befitting that agricultural engineering shared the limelight at the end of the 20th century when agricultural mechanization was named among the top 20 engineering achievements of the past millennium (Anon, 2000; Shirley, 2000). Agricultural mechanization stood alongside such other engineering feats as electricity, automobiles, computers, safe and abundant water supply. Several agricultural machines such as tractor, reaper and combine harvester tremendously increased farm efficiency and productivity and thereby freed agricultural labor to industrialize the world. Some of the existing and future opportunities for nano-agriculture based on recent technological advances and future prospects are as following:

2.1 Nano-Emulsions in Agriculture

Nano-emulsions are systems that are metastable. The challenge is to stabilize them against crystallisation, agglomeration and sedimentation by means of sufficient amounts of suitable surfactants and additionally protective colloids. They can consist of lipid or polymeric vesicles or particles, in the size range of 20-200 nm. They can have multiple phases, the simplest being oil-in-water. Nano-emulsions of solid particles can be dried through evaporation of the outer phase, to leave particulates, which can then be suspended in water for application. Nano-emulsions can be produced by both high energy (mechanical process using rotators, ultrasound, or pressure homogenizers) (Meleson *et al.*, 2004) and low energy means (either spontaneous emulsification, due to solvent diffusion as a result of mixing or dilution or by phase inversion temperature, a process which is controlled by specific surfactants, such as polyethoxylated surfactants, in response to temperature change (Anton *et al.*, 2008; Wang *et al.*, 2007). In terms of agricultural applications, nano-emulsions could be used for hydrophilic and hydrophobic pesticides. Examples are pyrethroids such as γ-cyhalothrin and β-cypermethrin (Wang *et al.*, 2007) and others like *Artemisia arborescens* L. essential oil (Lai *et al.*, 2006a). Potential advantages are the solubilisation of hydrophobic pesticides (hence no need for toxic organic solvents). Nanoparticluate active ingredient itself is not sufficient for perfect performance; it still needs special adjuvants as door openers for entering the leaf, in case of systemic actives but their basic role remains valuable.

2.2 Nanoparticles in Agriculture

Silica is known to be biocompatible and it has been applied in biological systems for last several years. It can also be engineered as hollow nanoparticles with different pore diameters and shells of different thickness. A number of research groups are working on its potential as a delivery vehicle for pesticides, where it has been shown to afford protection against UV degradation and controlled release dependent on pore diameter and shell thickness (Liu *et al.*, 2006; Li *et al.*, 2006b). In research experiments nano silica has been reported to provide insecticide activity on its own, through desiccation of insects cuticles. It has also been applied successfully as a thin film to improve cereal germination and decrease fungal growth. But due to its toxicity upon inhalation nano silica is not a preferred auxiliary in crop protection products. Nanoparticles have been used in the controlled release of bioactive substances in wood (Liu *et al.*, 2002 a, b, c) and the use of micro and nanoparticles as a tool for plant science (Taylor and Fauquet, 2002). It has been observed that silica nanoparticles are to be effective in the controlled release of substances into protoplasts within plant cells (Torney *et al.*, 2007). Corredor *et al.* (2009) has clearly described the mechanisms of interaction between plant cells and nanoparticles.

2.3 Nanodispersions in Agriculture

Nanodispersions of insoluble compounds can be created which have properties similar to that of solutions because micron size suspensions show poor activity. By combining processing technique of modified emulsion templating and freeze drying; the resulting powder composites remain stable, highly porous and form nanodispersions when added to water. Zhang *et al.* (2008) has demonstrated antimicrobial activity of Triclosan using this technique.

2.4 Nanoclays in Agriculture

Nano clays and layered double hydroxides can be developed and used for biological applications (Choy *et al.*, 2007). Both these materials show good biocompatibility, low toxicity along with the potential for controlled release. Chemicals can be loaded between layers of both the materials (an arrangement that can be influenced by buffer conditions in particular pH. In the case of hydrophobic chemicals, this arrangement prevents recrystallization, increases solubility and therefore bioavailability. Layered double hydroxides have been used in different ways such as pesticides, growth regulators, plant nutrients and slow-release fertilizer (Olanrewaju *et al.*, 2000; Lakraimi *et al.* 2000). α- naphthalene acetate, a plant growth regulator has also been intercalated with nanoclays to explore storage and controlled release of the plant growth regulator (bin Hussein *et al.* 2002). For nano clays controlled release can be engineered through coating with different polymers. These polymers manipulate electrostatic interactions between the chemical load and the clay particles (Lee and Fu 2003). In addition, nano clays can protect against UV-degradation of pesticides (El-Nahhal *et al.*, 1999). Layered double hydroxides have high affinity for anionic species. These are dissolved in acidic conditions.

2.5 Plant Transformations

Mesoporous silica nanoparticles deliver DNA and chemicals into plants (Torney, 2007). With the mesoporous nanoparticles two biogenic species can be delivered at a time. Gene can be brought and induce it in a controlled manner at the same time and at the same location. The controlled release will improve the ability to study gene function in plants. It is expected that, in future scientists will be able to use this technology to deliver imaging agents or chemicals inside cell walls. This would provide plant biologists with a window into intracellular events.

Plant transformations occur mostly with the use of a gene gun, not through endocytosis. In order to use the gene gun to introduce the nanoparticles to walled plant cells, the chemists synthesized even smaller gold particles to cap the nanoparticles. These 'golden gates' prevent chemical leakage and add weight to the nanoparticles, enabling their delivery into the plant cell with the standard gene gun. The most remarkable advantage of this technology is that one can deliver several things into a plant cell at the same time and release them whenever want.

2.6 Nanocomposite Material as Photoanode

Artificial photosynthesis, using solar energy to split water generating hydrogen and oxygen, is often considered a 'Holy Grail' of chemistry which can offer a clean and portable source of energy supply as durable as sunlight. It takes about 2.5 volts to break a single molecule of water down into oxygen along with negatively charged electrons and positively charged protons. The extraction and separation of these oppositely charged electrons and protons from water molecules provides the electric power. The photocatalytic splitting of water into hydrogen and oxygen using solar light is a potentially clean and renewable source for hydrogen fuel. So many efforts have been made in this area; many researchers have faced different types of challenges ranging from fundamental sciences to engineering. Titanium dioxide has been considered one of the most promising photocatalytic materials due to its relatively low cost, chemical stability, and photo stability. However, the catalytic property of titanium dioxide is limited with ultraviolet (UV) regions which accounts for only a small fraction of the incoming solar energy and thus renders the overall process impractical. Tungsten trioxide has been focused on as a new photo anode material, or as a mixture material with titanium dioxide for water splitting. Tungsten trioxide can offer comparatively small band gap (2.5 eV) and corrosion stability in aqueous solution. Although tungsten trioxide has shown great potential such as photo oxidation of water with visible light and high photocurrent with nano crystals, the quantum yield is still low. Titanium oxide nanotubes coated with tungsten oxide have been prepared to harvest more solar light. The tungsten trioxide coatings significantly enhanced the visible spectrum absorption of the titanium dioxide nanotube array, as well as their solar-spectrum induced photocurrents. This nano composite material can be used as efficient and stable solar-driven photoanode for oxidation of water.

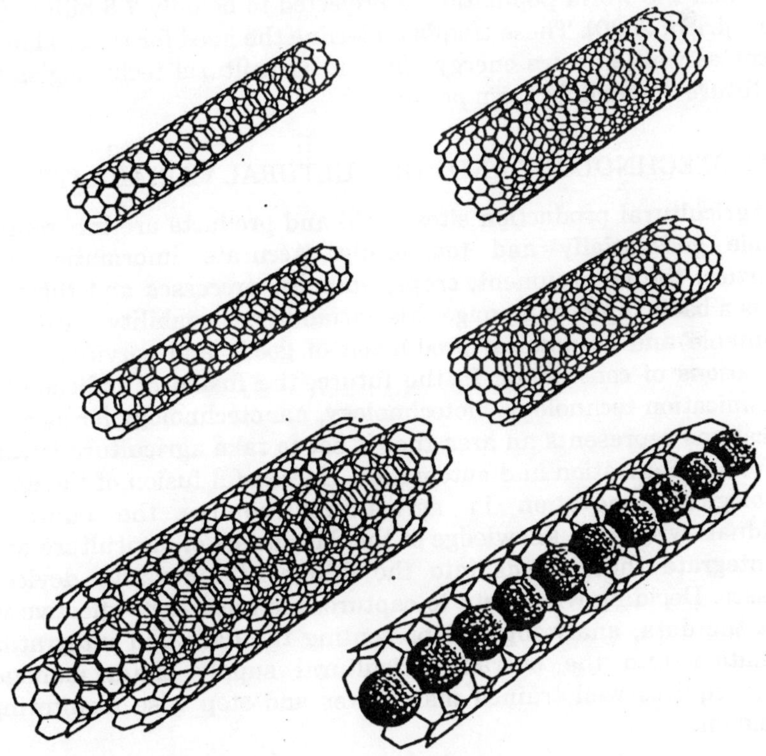

Fig. 1. Multiwalled and metal-atomfilled carbon nanotube structures.

3. ENERGY UTILIZATION IN AGRICULTURE

Energy input is a major constraint in agricultural modernization, especially with respect to the use of electromechanical systems. Increase in energy use in agriculture reduces hard work and allows individuals to engage in higher productivity roles and other off-farm productive activities (Opara, 1989). The impact of improved farm power in agricultural modernization and productivity improvement is well illustrated by Clarke and Friedrich (2000): "In very broad terms a farmer using only his own labor can feed himself and 3 other persons, using draught animal power he can feed 6 people, and using a tractor he can feed up to 50 and more. It is not a realistic approach to think that the burgeoning world population can be nurtured by peasant farmers, 80% of whom are using family labor only".

Furthermore, if the major part of the energy supply used is from non-renewable, fossil fuel sources, the world food system becomes no longer sustainable. Based on future projections of current food production growth rates and using currently available technologies it has been estimated that food production for 12.8 billion people in 2025 would be feasible at a time when the world population is projected to be only 7.8 billion (Clarke and Fridrich, 2000). These trends underline the need for renewable energy sources as well as more energy efficient agricultural technologies to cope with future demand for farm power.

4. NANOTECHNOLOGY AND AGRICULTURAL VARIABILITY

Agricultural production sites, yield and products are inherently very variable geospatially and temporally. Accurate information on soil properties, the environment, crops, animals, processes and their interaction is a basic input to manage this variability. Variability is essential for sustainable and economical production of goods and services that meet expectations of consumers. In the future, the fusion of information and communication technology, biotechnology, nanotechnology for agricultural applications represents an area that is set to take agriculture beyond the feats of mechanization and automation. Successful fusion of the emerging technological innovation in agriculture requires the education of individuals with good knowledge of the data needs of agriculture and who can integrate these needs into the design of innovative devices and processes. Deciding what data to capture, the most cost- effective way to collect the data, analyzing and presenting the data and integrating the information into the overall agricultural supply chain management system requires well trained man power and step wise monitoring and assessment.

4.1 Agrinfortronics

Agriculture is intrinsically diverse in space and time with respect to fields, crops, animals, weather etc. It is crucial to manage this variability to optimize both inputs and outputs and reduce adverse impacts on the environment. The future of agriculture in the knowledge economy (Opara, 2002a) relies on the sensing, acquisition, analysis, storage and transmission of reliable and accurate data about the plant/ animal and production/ handling environment to meet the demand for high yield and good quality products. This might be possible by successful fusion of information and communication technology and mechatronics for agricultural applications, which is referred as agrinfortronics (Opara, 2002a). Development of a new scanning-probe-based data-storage concept called "millipede" has been done which combines ultrahigh density, terrabit capacity, small form factor, and high data rate (Vittiger *et al.*, 2002). Such nanotechnology has considerable applications in successful development and application of agrinfortronics.

4.2 Nanotechnology in Spinning Cotton

Recent advances in materials science and chemistry have produced mastery in nanoparticle technology, with wide implications in the field of agriculture. In the cotton industry, where current techniques of spinning cotton are quite wasteful. From harvesting the cotton to finalizing the fabric it is made into over 25% of the cotton fiber is lost to scrap or waste. However, new nanotechnology based technique called electrospinning makes good use of the scrap material that would otherwise be used to make low-value products like cotton balls, yarn and cotton batting. The technique relies on electrical rather than mechanical forces to form fibers. Thus special properties are required of polymer solutions for electro-spinning including the ability to carry electrical charges. The fiber that is produced is less than 100 nanometers in diameter, which is 1,000 times smaller than what is produced in conventional spinning.

Fig. 2: Close-up of nanofibers from electrospun cotton.

4.3 Food Packaging

A major problem in food processing is determining and developing an effective packaging material. Using nanoparticle technology, Bayer has developed an even more airtight plastic packaging that will keep food fresher and longer than any available packaging material. Researchers at Bayer Polymers refer to this new plastic as a "hybrid system" as it is enriched with an enormous number of silicate nanoparticles. When this plastic is processed into a thin film and wrapped over food, it does a better job than previous plastics of preventing food from going bad on the shelf and it helps prevent odors from one food mixing with another. Nanotechnology has the potential to revolutionize agriculture and food production. Potential applications of the technology include controlled nutraceutical delivery systems for food; on farm applications to deliver drugs or pesticides to livestock or crops; and smart-sensing devices for agriculture-environment interactions. Currently, most prominent agrifood nanoproducts are nanocomposites for food packaging. These nanomaterials provide barriers to oxygen and carbon dioxide, thus protecting food quality for longer time.

5. SUSTAINABILITY OF AGRICULTURE

Consumer demand for sustainable agricultural practices and waste minimization has become imperative in modern agribusiness. This is reflected in rising global trade on eco-foods and demand for traceability from field to plate (Opara and Mazaud, 2001). However, in regions where poverty and food insecurity are major problems, there is concern that new innovations are needed to enhance yields and to control the incidence of pests and diseases under sustainable farming practices. New measurement technologies are needed to detect and segregate products from different production systems such as organic versus conventional agriculture. As people become more aware about sustainability concern (including environmental, water resources, land resources, air and biodiversity) and agriculture is implicated as a major cause of extensive resource depletion, biotechnologists and nanotechnologists will come under more public scrutiny to focus and study the sustainability issues on priority basis and adopt sustainability as a professional ethic. Nanotechnology can be useful to address the problem as the nanotechnological devices might be helpful in differentiation of foods in the market grown under different production systems as it is difficult based on sensory attributes.

5.1 Bio-Security

Safety of the food products is as important as production and preservation of food in the agricultural field. It has been observed that often food contains bacteria and viruses, which frequently ends in illness and sometimes fatality. The most likely area in which nanotechnology can assist bio-security is the world of analysis and detection, such as biosensors to detect the quality and health of agricultural products as well as livestock. Without nano scale detection techniques it will not be possible to maintain bio-security standards in agriculture and food industry.

5.2 Integration of Agricultural Biotechnology, Bio-Engineering and Nanobiology

Agriculture is a part of the biological science. Most of the measurements in biology are at the scale of microns and below where the sphere of nanotechnology resides. Convergence of biotechnology, bioengineering and nanobiology to solve practical problems of agriculture is rational. Today agricultural biotechnology has created incredible possibilities in crop and animal production through genetic manipulation of species. It can be presumed that development in nanobiology will certainly influence future agricultural technologies. Successful arrangement of control proteins on DNA within a cell has been already reported by Austin *et al.* (2002). These advances open up remarkable scope for nanofabrication in modern molecular biology. Furthermore, nanofabricated devices offer the scope for their injection into plants and animals to detect tissue parts affected by rare phenomena such as diseases, nutrient deficiencies and developmental abnormalities.

Nanotechnology holds major implications in agriculture and food production. It offers the tools to understand and transform biosystems, sub-cellular dynamics; regeneration mechanisms; genome description and food characterization. So far, the use of nanotechnology in agriculture has been mostly theoretical, but it has begun and will continue to have a significant impact on the main areas of the food production such as development of new functional materials, product development, design of methods, instrumentation for food safety as well as bio-security. The effects on society as a whole will be dramatic.

5.3 Agricultural Diagnostics and Drug Delivery with Nanotubes

Recent researches in nanomaterial sciences and technology has resulted in the development of several devices which have potential applications in agriculture and related biological fields. Nanotube is such a device which can be integrated with other chemical, mechanical or biological systems and can be an excellent candidate for electrical sensing of individual biomolecules. Nanotube electronic devices function accurately under certain extreme biological conditions such as saline (salty) water and have dimensions comparable to typical biomolecules such as DNA, whose width is approx 2nm (McEuen *et al.*, 2002; Bohr, 2002). In spite of practical problems in achieving reliable, rapid and reproducible nanofabrication of complex arrays of nanotubes, such devices have the potential for use in site-specific control of pests and diseases. Such devices will be great use in agriculture and other fields in days to come.

5.4 Nano Agricultural Mechanization

Developments in electronics in combination with mechanical engineering paved the way for automation and control of various agricultural operations. Such electro-mechanical systems have enhanced equipment maneuverability and control environmental quantification such as greenhouses. In this regard, silicon-based integrated circuits have played a crucial role. Today technical progress of semiconductor industry has exemplified by leading edge products such as microprocessors and memory chips (Bohr, 2002).

Prospects of transistors at the molecular and atomic scale further enlarge the opportunities in the evolution and application of nanotechnology in the development of the next phase of miniaturized micro-electro machinery for agricultural mechanization. Nanoagricultural mechanization may revolutionaries the agriculture that extends the horizon where the current farm mechanization, symbolized by the mechanical tractor ends. In view of scaling transistors all the way down to the size of individual molecules (Bohr, 2002; Reed, 1999) and incorporating this into future agricultural machinery is a stimulating application of nanotechnology which is set to have wide applications in agricultural automation and control.

5.5 Profitability of Small-Scale Farm Mechanization

The majority of the human population still relies on small-scale subsistence agriculture. Despite the resounding success achieved in medium to large scale agriculture, profitable mechanization of small scale agriculture has evaded mankind. Agricultural biotechnologies have the potential to improve the situation through the availability of high-yielding plant genetic resources that have resistance properties to pests and diseases. Agricultural engineering has the experience to continue to lead this charge with the help of other emerging branches of science and technology like information and communication technology, biotechnology and nanotechnology.

5.6 Nanostructural Basis of Food Production and Monitoring

In the face of increasing awareness about health and nutrition, consumers are consistently demanding for supply of quality products having good organoleptic attributes to meet their health and nutritional needs. Measuring and predicting quality reliably is therefore a tough task and a challenge in postharvest processing of agri-foods. Recent developments in nanotechnology applied to food science provide new tools and insights into structure and function in materials. In spite of being in infancy, the future of nanotechnology in agriculture appears to be dazzling. Success in nanotechnology may result in nanobots, nanostructures, nanowires and nanocircuits (Opara, 2002a). Nanobots (miniature/micro robots of the size of human blood cells or even smallar) can explore every capillary and even be guided in for close-up inspections of neural details in animals during breeding and on- farm diagnostics. Using wireless connections, the nanobots would communicate with one another and with other computers that are compiling the scan database. Such data base will be very precise and of great help to diagnostic and research. Nanostructures, such as smart nano-cards that collect and store data about products and process history, can be implanted into plants and animals during growth and development to collect and transmit vital real-time data such as growth rates and physiological activities that provide clues on performance, productivity and exposure to chemical, environmental and physical hazards. Such smart nano-cards will further facilitate screening of plant genetic resources for abiotic and biotic stresses as well as climate change study. Agriculture and allied disciplines generate bulk volume of data on the environment, crops, animals, inputs and processes. The emergence of bioinformatics facilitates easy tools to handle and manage bulk amount of data. As most farms and other agricultural enterprises are located in rural areas distant from main communication centers, the development of high-capacity information networks will facilitate the collection, analysis and transmission of vital information, which is the backbone of precision agriculture. Enhancing the carrying capacity of existing optical networks using nanowires and nanocircuits could make it possible to achieve mining and transmission of huge data for multiple applications including agriculture. There is ample possibility of using nanostructures or miniaturized electromechatronics that have the capacity to work under extreme conditions unsuitable for humans such as contaminated sites and lethal atmospheres. Such tiny micro-machinery could also permit greater accuracy and high efficiency in various operations in the field or parts of a plant, animal or their food components.

5.7 Efficient Delivery of Agri-Inputs:

Rise in energy prices in the future will be a driving force for increasing efficiency in both production and application of agrochemicals. Highly efficient nano-based agrochemicals can significantly contribute to use of less fossil fuel intensive agricultural inputs. Moreover, agriculture places considerable demand on fresh water and in turn, contributes significantly to pollution of groundwater through the use of pesticides and fertilizers (Kruger *et al.*, 1993). Therefore, it is a driver for more effective, targeted and controlled use of agrochemicals.

6. SOCIO-ECONOMIC CHALLENGES FACING GLOBAL AGRICULTURE

Despite the tremendous impacts of technology applications in increasing agricultural productivity, modern agriculture faces somewhat unprecedented challenges in the years ahead, ranging from economic and social dilemma to environmental and ethical issues as well as technological challenges. Some of the socio-economic challenges are as following:

6.1 Nanotechnology to Improve Declining Agricultural Trade

One of the major challenges facing agriculture is its decline in economic importance relative to other industries such as the manufacturing and service sector. This problem is most acute in the agricultural production sector (farm-gate), which now accounts for less than 20% of the economic value of finished agricultural products at the consumer level in developed countries and over 50% in developing countries. This increasing importance of the post harvest sector (innovative handling systems and value- adding) underscores the need for a new approach to agricultural farm commodity trading to integrated supply chain agribusiness. The problem of declining terms of trade facing agriculture can be exemplified by the change in prices of agricultural commodities in last two – three decades. It has been observed that cost of agriculture commodities doubled during the two – three decades period but prices received by the farmer remained unchanged or slightly changed. A synergy of factors including the continuing decline in profitability of agriculture, the huge capital investment required in modern farming, poor image and low appeal of farming among young men and women have continued to diminish the competitiveness of agriculture in comparison to other sectors. This can be interpreted that current agricultural progress has become the victim of its own success by avoiding major catastrophe of massive global food insecurity that was faced in the 20th century. Nanotechnology based farm mechanization may be one of the remedial measure to enhance sustainable production and attract young educated people in agriculture.

6.2 Food insecurity, poverty and Malnutrition in Less Developed Countries

Twenty percent of the world's over 6 billion population (1.2 billion) lives on less than US$1 per day; 70% of these are rural and 90% live in Asia and sub-Saharan Africa (Thirtle *et al.*, 2003). The World Bank (2001) poverty survey shows that almost half of the world's population lives on less than US$2 per day. Both governments and development agencies have therefore recognized the need to focus on the rural resource-poor farmers as a strategy to alleviate the problem of food insecurity and poverty facing these regions.

Today, agriculture in most parts of the developing regions remains largely in the hands of resource-poor rural dwellers who barely make a living due to poor yield and very low returns. The socio-economic quandary facing these farmers remains a challenge for them and the entire global community. The problem of food insecurity is further compounded by the rising world population, which is expected to nearly double from its present 6 billion people by the year 2035, with most of the growth occurring in the developing countries. As these developing and transitional economies emerge; life expectancies are also on the rise with consumers having additional disposable income. Under this scenario, the demand for food supply is expected to increase. Concurrently, as population continues to rise, land available for agriculture is decreasing, more land is being utilized for roads, railway tracks, industries and townships. This further exacerbates the need for increased agricultural productivity necessary to meet the dietary and nutritional needs of the rapidly expanding population. It has been presumed that like applications of biotechnology in enhancing agricultural production in 20th century, nanotechnology will help to produce more food grains from less land in future.

6.3 Globalization and World Trade

Globalization of the world food system is progressing at unprecedented pace, particularly with the emergence of multinational and transnational food chains that dominate the majority of global agrifood market. The companies increasingly source their raw materials from distant locations and from variety of sources. Using modern transportation facilities the consumption of food products grown in distant locations is increasing. Globalization of the world agrifood system through trade raises crucial challenges in terms of fair trade and competition between farmers in different parts of the world. Farmers in developing countries have very limited access to improved technologies as compared to farmers in developed countries who commonly have access to both modern technology and government subsidy. It is expected that Nano technological devices will help to bridge the gap through dissemination of agricultural information and farm mechanization.

6.4 Environmental and Ecological Sustainability and Welfare Issues

Modern agricultural production methods have contributed to the depletion of natural resources and environmental degradation. Consumers are increasingly aware that modern intensive agriculture and food production practices have considerable negative impacts on the environment and ecology. Farm mechanization has significantly contributed to soil erosion, environmental pollution (Clark and Friedrich, 2000) and disruption of vital eco-systems. Now environmental awareness has prompted consumers to demand and support the development, promotion and introduction of sustainable agriculture. Consumers expect food to be produced and processed in line with good agricultural practices, which embodies greater care of the environment and welfare of animals (Opara, 2000).

6.5 Emergence of Food-Related Safety Hazards and Scares

Nymand-Christensen (1999) in a survey found that for 68% of consumers most important issue was the safety of the food they eat, this response was admittedly after the Bovine Spongiform Encephalopathy (BSE) crises, but before the dioxin crises in Europe. Numerous microbial, physical, and chemical hazards occur in the human food chain, which contribute to major safety issue. Concerns over pesticides and other chemical residues have led to the ban of many agro-chemicals used in food production and preservation. Physical contaminants such as stone, wood, soil, plastic, broken glasses and metal parts like nuts, bolts and nails can cause illness or injury to the consumer if they become embedded inside food.

Several major food-related health hazards, have raised public concern on the way we practice agriculture as well as the way we handle and process food. Food poisoning incidents and death due to Salmonella in poultry and eggs, *Escherichia coli* 0157 (*E. coli*) contamination of meat and meat products, fresh, minimally processed and processed fruit and vegetables and the occurrence of other emerging food pathogens have reduced consumer confidence in the safety of food systems. The increase in food regulations following recent food scares have challenged agribusiness and other life science industries to examine their production and processing practices. Developing countries are also making safety legislations more stringent.

6.6 Impacts of Genetically Modified Organisms (GMOs) and Other Food Materials

The ongoing crisis of confidence in the food chain has been exacerbated by the introduction of food products derived from genetically modified organisms (GMOs) or materials. Food safety concern surrounding biotechnology is controversial in both the public and scientific sphere; but consumers are generally worried of the uncertainty of introducing new gene sequences which might impel or provoke other factors that are hazardous to humans and the environment.

6.7 Technological Challenges Facing Global Agriculture

There are many technological challenges and opportunities facing agriculture in both developed and developing countries which must be addressed to counter some of the socio-economic problems outlined above.

6.8 Farming of Marginal Lands and Fragile Ecosystems

As world population mainly in developing countries continues to increase day by day and poverty limits their ability to access food and other resources, there is a need to increase the productivity of existing farms. Unfortunately, most of these farmlands have marginal fertility due to over-cropping and unfavorable agro-climatology. New scientific and technological inputs are needed to improve the productivity of these farms in order to raise food production in the affected areas.

Globalization of the world agricultural and food supply system demands greater vertical integration within and between firms in order to adequately manage variability and remain competitive. With ever increasing and changing consumer demand and consumption patterns for food products, the assurance of consistent supply of top quality, safe and traceable food products remains high on the trade agenda. Developing cost-effective supply chain technologies for precision agriculture, including site-specific farming, will provide farmers and postharvest operators the necessary tools to exercise better control over yields, product quality and safety. Recent food scares and public concern over GM foods and the effects of these concerns in reducing public confidence in the safety of the human food chain have made traceability an important index of trade in agribusiness (Opara, 2002b).

6.9 Reforms in Educational Curriculum and Research Agenda

To realize the potential opportunities in agriculture that the emerging knowledge-based technological innovations offer, there is a need to integrate these innovations such as nanotechnology into the curriculum and research agenda of engineering and technology education. The importance of the knowledge-innovation nexus and the role of education and research in generating innovative ideas are well documented (Opara, 2002a).

Curriculum reforms that provide future engineers and technologists with new knowledge and skills which broaden their employment prospects would make such academic programs and professional disciplines more attractive to young school leavers (Opara, 2003b,c).

6.10 Research and Development Needs for Nanoagriculture

During the past two decades, precision agricultural technologies have evolved that match farm input to variation over space and time and thereby reduce residues. The use of large-scale precision agricultural technologies is developing rapidly in advanced countries. New developments in computer, geospatial technology and remote sensing suggest vast potential but so far it has had limited use, presumably due to its high costs. One challenge in improving nanoagriculture will be to integrate the hardware (nanomachines), software and management tools that will take advantage of new information about agriculture. Another important issue associated with nanoagriculture is the vision for improving environmental quality. It is expected that widespread adoption of nanoagriculture in the future may be encouraged by environmental regulation to promote sustainable agriculture.

CONCLUSIONS

Technological innovations have had profound impact on the evolution of the agricultural sector. Nanotechnology has become one of the most exceedingly exciting disciplines of science and technology. The strong interest in nanotechnology has been generally driven by visions of a flow of new commercial applications of nanotechnology that will lead to a new industrial revolution that is destined to pervade almost every industry, including the oldest one (agriculture). Academia, governments and the private sector have shown considerable interest in the development of nanotechnology as well as its applications in diverse industries including agriculture.

Increasing globalization, multi-functionality of agriculture, changing consumer and public perceptions and expectations about agriculture, recent food safety panic, rising demand for sustainable agricultural practices and declining terms of trade have brought considerable pressure on agriculture. Nanotechnology represents a new technological frontier to assist in addressing these socio-economic and geo-political challenges facing agriculture and similar land-based industries.

At present, agriculture is undergoing a technological revolution as evidenced by the introduction of biotechnology, information and communication technology. Developments in the information and communication technology revolution have already spearheaded the emergence of precision agriculture, which is finding applications in various aspects of modern agriculture, ranging from precision farming, bioresource management and supply chain optimization. Several significant changes have been observed in agriculture from the emergence of biotechnology (Zilberman *et al.*, 1998), and it is envisioned that the application of nanotechnology in agriculture is set to bring about more changes.

The evolution of nanotechnology suggests that the universities and private enterprise research and development companies are key players in its industrialization. The majority of the value of nanotechnology is currently embodied in specific knowledge. Therefore, some of the questions, which need to be addressed as we move ahead, include the role of the farmers (particularly resource-poor farmers) as stakeholders in research related to nanotechnology applications in agriculture. This is particularly relevant given that the process of product innovation is the key to nanotechnology.

Ongoing research and developments in nanotechnology represent technology-push factors that have numerous potential applications in agriculture, both on-farm and postharvest handling and processing. The convergence of a technology triad (information and communication technology, biotechnology and nanotechnology) that is currently driving innovation across industry sectors will provide greater opportunity for success in advancing and realizing these potentials. Recent experiences gained by researchers, industry owners; government organizations and the wider society from the development of other cutting-edge technologies (such as agricultural biotechnology) should provide guidance in policy formulation and implementation in tackling the socio-cultural and legal issues which could pause the evolution of nano-agriculture.

Modern innovative technologies have had mixed impacts in addressing the productivity problems facing agriculture in many parts of the developing world. The debate on the benefits and demerits of agricultural biotechnology still rages on, while the problems of declining agricultural productivity and increasing food insecurity continue in many regions. The prospects for miniaturization of electromechanical systems in agriculture offer yet another opportunity for the successful mechanization of small-scale farms, particularly in developing countries. There is a need to bring about collaborative and farmer-oriented adaptive research and development on nano-agriculture for the transformation of traditional subsistence agriculture into modern market-oriented agribusiness. Effective communication of the goals, benefits and potential risks of nano agriculture to users and the general public will ultimately determine the future of nanotechnology in agriculture.

Nanotechnology is still in its early stage compared to for example biotechnology. However, the lower cost of some production equipment, research and development based on nanotechnology is sought to allow for its more diversified applications in agriculture. There are some records of the first nano agrochemicals in development, which are nano-reformulations of existing pesticides, fungicides, plant, soil and seed treatments. Combined fertilizers and pesticide in terms of nanoclay capsule containing growth stimulants and biocontrol agents are being developed. Such product will have the property of slow release of active ingredients and therefore, the crop treatment will require only one application over its life cycle. Agrochemical companies are reducing the existing chemical emulsions to the nanoscale and substituting active ingredients with their encapsulated nano sized equivalents in attempt to bring a number of benefits into potential applications of nanotechnology to pesticides, fertilizers and plant growth regulators. Employing nano materials in this industrial and research area could reduce use of certain agrochemicals such as pesticides, and further provide a better ability to control the application and dosage of active substance to the target.

References

Anon (2000). Agricultural mechanisation named among top 20 engineering achievements. *Resource* 7(4): 17.

Anton, N., Benoit, J. P. and Saulnier, P. (2008). Design and production of nanoparticles formulated from nano-emulsion templates - A review. *Journal of Controlled Release* 128, 185-199.

Austin, R.H., J.O. Tegenfeldt, H. Cao, S.Y. Chou, and E.C. Cox (2002). Scanning the Controls: Genomics and nanotechnology. *IEEE Transactions on Nanotechnology* 1(1): 12-18.

Bin Hussein, M.Z., Zainal, Z., Yahaya, A.H., Foo, D.W.V. (2002). Controlled release of a plant growth regulator, alpha-naphthaleneacetate from the lamella of Zn–Al-layered double hydroxide nanocomposite. *J. Control Release* 82, 417–427.

Bohr, M.T. (2002). Nanotechnology goals and challenges for electronic applications. IEEE *Transactions on Nanotechnology* 1(1):56-62.

Choy, J. H., Choi, S. J., Oh, J. M. and Park, T. (2007). Clay minerals and layered double hydroxides for novel biological applications. *Applied Clay Science* 36, 122-132.

Clarke, L.J. and T. Friedrich. (2000). Increasing food production and protecting resources. The role of agricultural engineers. Agricultural Engineering Branch, Food and Agriculture Organization (FAO) of the United Nations, Rome, 8 pages.

Corredor E, Testillano PS, Coronado MJ, Gonzalez-Melendi P, Fernandez-Pacheco R, Marquina C, Ibarra MR, de la Fuente JM, Rubiales D, Perez-de-Luque A, Risueno MC (2009). Nanoparticle penetration and transport in living pumpkin plants: in situ subcellular identification. *BMC Plant Biology* 9:45

Curtis, J., Greenberg, M., Kester, J., Phillipos, S., Krieger, G. (2006). Nanotechnology and Nanotoxicology: A Primer for Clinicians. *Toxicol. Rev.* 25 (4): 245 – 260.

El-Nahhal, Y., Nir, S., Margulies, L. and Rubin, B. (1999). Reduction of photodegradation and volatilization of herbicides in organo-clay formulations. *Applied Clay Science* 14, 105-119.

Kruger, E. L., Somasundaram, L., Kanwar, R. S. and Coats, J. R. (1993). Persistence and degradation of [C-14] Arazine and [C-14] Deisopropylatrazine as affected by soil depth and moisture conditions. *Environmental Toxicology and Chemistry* 12, 1959-1967.

Lai, F., Wissing, S. A., Muller, R. H. and Fadda, A. M. (2006a) *Artemisia arborescens* L essential oilloaded solid lipid nanoparticles for potential agricultural application: Preparation and characterization. *Aaps Pharmscitech* 7.

Lakraimi, M., Legrouri, A., Barroug, A., De Roy, A., Besse, J.P. (2000). Preparation of a new stable hybrid material by chloride-2,4- dichlorophenoxyacetate ion exchange into the zinc–aluminium– chloride layered double hydroxide. *J. Mater. Chem.* 10, 1007–1011.

LEE, W.F. & FU, Y.T. (2003). Effect of montmorillonite on the swelling behavior and drug-release behavior of nanocomposite hydrogels. Journal of Applied Polymer *Science* 89, 3652–3660.

Li, Z. Z., Xu, S. A., Wen, L. X., Liu, F., Liu, A. Q., Wang, Q., Sun, H. Y., Yu, W. and Chen, J. F. (2006b). Controlled release of avermectin from porous hollow silica nanoparticles: Influence of shell thickness on loading efficiency, UV-shielding property and release. *Journal of Controlled Release* 111: 81-88.

Lievonen, J.T. (2000). Innovation opportunities form fusing high technologies. Tech Monitor (Sept-Oct), pp. 22-28.

Liu Y, Laks P, Heiden P (2002a). Controlled release of biocides in solid wood. I. Efficacy against brown rot wood decay fungus (*Gloeophyllum trabeum*). *J Appl Polym Sci.* 86: 596-607.

Liu Y, Laks P, Heiden P (2002b). Controlled release of biocides in solid wood. II. Efficacy against Trametes versicolor and *Gloeophyllum trabeum* wood decay fungi. *J Appl Polym Sci.* 86: 608-614.

Liu Y, Laks P, Heiden P (2002c). Controlled release of biocides in solid wood. III. Preparation and characterization of surfactantfree nanoparticles. *J Appl Polym Sci.* 86: 615-621.

Liu, F., Wen, L. X., Li, Z. Z., Yu, W., Sun, H. Y. and Chen, J. F. (2006). Porous hollow silica nanoparticles as controlled delivery system for water-soluble pesticide. *Materials Research Bulletin* 41, 2268-2275.

McEuen, P.L., M.S. Fuhrer, and H. Park. (2002). Single-walled carbon nanotube Electronics. *IEEE Transactions on Nanotechnology* 1(1): 78-85.

Meleson, K., Graves, S. and Mason, T. G. (2004). Formation of concentrated nanoemulsions by extreme shear. *Soft Mater* 2, 109.

Nymand-Christensen, J. (1999). How should consumer concerns be addressed? Teagasc, Irish Agriculture and Food Development Authority. Accessed 18 Dec. 2001. (http://www.teagasc.org/publications/agrifood1999/paper11.htm.)

Olanrewaju, J., Newalkar, B.L., Mancino, C., Komarneni, S., (2000). Simplified synthesis of nitrate form of layered double hydroxide. *Mater. Lett.* 45, 307–310.

Opara, L.U. (1989). Computer-based model for selective mechanisation of Nigerian Agriculture. Unpublished MEng thesis, University of Nigeria, Nuskka.

Opara, L.U. (2002a). Agricultural Engineering education and research in knowledge-based Economy. In: Kosutic, S (Ed.). Proc. of the 30th International Symposium on Agricultural Engineering, Zagreb, Croatia, pp. 33-46.

Opara, L.U. (2002b). Engineering and technological outlook on traceability of agricultural Production and products. Agricultural Engineering International: the CIGR Journal of Scientific Research and Development, Vol. IV: 1-12.

Opara, L.U. (2003b). 'From Farmers to Blue-Colar Professionals': Agricultural Engineering students' perspectives on their career prospects, critical skills, and discipline name change. *World Trans. Engg. Techn. Edu.* 2(3): 399-402.

Opara, L.U. (2003c). Image is everything: Undergraduate students' contributions towards better understanding and improvement of public attitudes and perceptions of Agricultural Engineering. *World Trans. Engg. Techn. Edu.* 2(3): 457- 460.

Opara, L.U. and F. Mazaud. (2001). Food traceability from field to plate. *Outlook on Agriculture* 30(4): 239-247.

Reed, M. (1999). Molecular-scale electronics. *Proc. IEEE* 87, 625-658.

Scientific American. (2001). Nanotech: the science of small gets down to business. *Scientific American* 285, Number 3.

Shirley, M. (2000). Agricultural mechanisation among century's top engineering achievements. *Institutions of Agricultural Engineers – Newsletter* (Spring), 3-5.

Sunding, D. and D. Zilberman. (2000). The Agricultural Innovation Process: Research and Technology Adoption in a Changing Agricultural Sector. In: Handbook of Agricultural Economics.

Taylor NJ and Fauquet CM (2002). Microparticle bombardment as a tool in plant science and agricultural biotechnology. *DNA Cell Biol*. 21:963-977.

Thirtle, C., L. Lin and J. Piesse (2003). The impact of research-led agricultural productivity growth on poverty reduction in Africa, Asia and Latin America. *World Development* 31(12): 1959-1975.

Torney F, Trewyn BG, Lin VS-Y, Wang K (2007). Mesoporous silica nanoparticles deliver DNA and chemicals into plants. *Nature Nanotech* 2:295-300.

Vettiger, P., G. Cross, M. Despont, U. Dreschler, U. Durig, B. Gotsmann, W. Haberle, M.A. Lantz, H.E. Rothuizen, R. Stutz, and G.K. Binnig. (2002). The "Millipede" – Nanotechnology entering data storage. *IEEE Transactions on Nanotechnology* 1(1): 39- 55.

Wang, L., Li, Z., Zhang, G., Dong, J. and Eastoe, J. (2007). Oil-in-water nanoemulsions for pesticide formulations. *J. Colloid Interface Sci.* 314 (2007) 230–235.

Zhang H, Wang D, Butler R, Campbell NL, Long J, Tan B, Duncalf DJ, Foster AJ, Hopkinson A, Taylor D, Angus D, Cooper AI, Rannard SP (2008). Formation and enhanced biocidal activity of water-dispersable organic nanoparticles. *Nature Nanotechnol.* 3, 506–511.

Zilberman, D., C. Yarkin and A. Heiman (1998) Agricultural biotechnology: Economic and international implications, Mimeograph, Department of Agricultural and Resource Economics, University of California, Berkeley.

Subject Index